AF365428

Failure Mechanisms in Alloys

Failure Mechanisms in Alloys

Special Issue Editor

George A. Pantazopoulos

MDPI • Basel • Beijing • Wuhan • Barcelona • Belgrade • Manchester • Tokyo • Cluj • Tianjin

Special Issue Editor
George A. Pantazopoulos
ELKEME Hellenic Research
Centre for Metals S.A.
Greece

Editorial Office
MDPI
St. Alban-Anlage 66
4052 Basel, Switzerland

This is a reprint of articles from the Special Issue published online in the open access journal *Metals* (ISSN 2075-4701) (available at: https://www.mdpi.com/journal/metals/special_issues/fail_mech_alloy).

For citation purposes, cite each article independently as indicated on the article page online and as indicated below:

LastName, A.A.; LastName, B.B.; LastName, C.C. Article Title. *Journal Name* **Year**, *Article Number*, Page Range.

ISBN 978-3-03928-276-0 (Pbk)
ISBN 978-3-03928-277-7 (PDF)

Cover image courtesy of George A. Pantazopoulos.
SEM micrograph showing ductile fracture surface of a low carbon structural steel grade (C15), after tensile overload. See page 156.

© 2020 by the authors. Articles in this book are Open Access and distributed under the Creative Commons Attribution (CC BY) license, which allows users to download, copy and build upon published articles, as long as the author and publisher are properly credited, which ensures maximum dissemination and a wider impact of our publications.
The book as a whole is distributed by MDPI under the terms and conditions of the Creative Commons license CC BY-NC-ND.

Contents

About the Special Issue Editor

George A. Pantazopoulos graduated from the Chemical Engineering Department of the National Technical University of Athens (NTUA) in 1992 (order of graduation considered as equivalent to "summa cum laude") with a specialization in Materials Science. He received his Ph.D. from the Mechanical Engineering Department of NTUA, in the field of Manufacturing Technology and Processing of Advanced Materials in 1999. He served so far in the metals industry for more than twenty (20) years as a Quality Control Engineer, as a senior R&D project leader and lately as Quality and Technology Manager. His principal scientific research interests are focused on the field of failure and fracture analysis, materials processing, mechanical testing, and tribology. He is the author and co-author of more than 150 papers in the above areas, in International Journals and Conferences, as well as 2 books and various monographs/book chapters while he was an invited speaker in Conferences and Academic Institutions. He is Subject Editor/Editorial Board member and reviewer in Journals in the field of materials science and failure analysis. He is the recipient of various awards and distinctions, including a best paper and a dissertation thesis prize.

Preface to "Failure Mechanisms in Alloys"

Άνευ αιτίου ουδέν εστίν

("Nothing is without a cause")

Aristotle

According to various textbooks, the definition of "failure" is the lack of satisfaction of a purpose (lack of fitness for purpose) or goals or the inability of the material/structure or process to serve certain requirements (e.g., to fulfill standard specifications). The process aiming at the determination of the root cause(s) of the failure in order to establish certain prevention measures is usually denoted as "failure analysis". The failure investigation process is a fascinating logic procedure; encompassing an intensive thinking action, starting from the final event and going backwards, explores the subtle relationships of a chain of events, linked by a "cause-and-effect" joint. The evidence provided by the extracted findings builds the necessary documentation to support the failure hypothesis. The stronger and more complete the documented objective evidence is, the more solid and realistic the failure hypothesis is.

Although, at a first glance, the term "failure" provokes negative feelings, regarding the consequences of a failure event, a latent positive character exists from a different perspective. Failure analysis is a problem solving and improvement technique which constitutes a common approach for a broad spectrum of industries, improving the integrity of safety and accident prevention, environmental performance, plant maintenance, and quality assurance. The utilization of failure prevention tools and techniques meets many applications of risk assessment in engineering applications, from nuclear reactors to automotive/aerospace industries and from financial/banking systems to service sectors. In addition to the academic research, industrial expertise offers a profuse knowledge resource and, through the participation to complex and multifaceted failure analysis projects, a solid learning system is constructed, ensuring sustainability and excellence with a strongly positive impact to the society.

The interdisciplinary nature of failure analysis makes this topic very challenging, encouraging team working and knowledge sharing from various professional and scientific areas. In the specific volume of Metals, emphasis is placed in metallic materials (metals and alloys) and the various disciplines assisting in the analysis and interpretation of failures, such as materials science, applied mechanics, corrosion, tribology, manufacturing technology, modeling, and simulation. The Special Issue "Failure Mechanisms in Alloys" is a rich collection of high-quality research papers from esteemed authors, including twenty five articles (25), two reviews (2) and one editorial (1) which "opens" the volume, introducing the contributing papers and providing a classification of the topics provided. The contents, which cover a broad spectrum of failure mechanisms (e.g., erosion corrosion, fatigue, creep fatigue, creep, corrosion, and environmentally-assisted cracking, wear), investigation techniques (both experimental and numerical/stochastic), and industrial manufacturing processes (such as metal forming and machining), make this laborious treatise an attractive "Memoir" for researchers and professionals, from both the academic and industrial world.

The rewarding results of the creation of this Special Issue are shared among the main contributors: the authors, the reviewers and the MDPI Metals Editorial team, whose valuable efforts are reflected in this amazing knowledge journey. As a final note, the continued support of the colleagues and management team of the ELKEME—Hellenic Research Centre for Metals S.A. is gratefully acknowledged.

George A. Pantazopoulos
Special Issue Editor

Editorial

Failure Mechanisms in Alloys

George A. Pantazopoulos

ELKEME Hellenic Research Centre for Metals S.A., 61st km Athens—Lamia National Road, 32011 Oinofyta, Greece; gpantaz@elkeme.vionet.gr; Tel.: +30-2262-60-4463

Received: 6 January 2020; Accepted: 10 January 2020; Published: 13 January 2020

1. Introduction and Scope

The era of lean production and excellence in manufacturing, while advancing with sustainable development, demands the rational utilization of raw materials and energy resources, adopting cleaner and environmentally friendly industrial processes. In view of the new industrial revolution (through digital transformation), the exploitation of smart and sophisticated materials systems, the need for minimizing scrap and increasing efficiency, reliability, and lifetime, on the one hand, and the pursuit of fuel economy and the limitation of the carbon footprint, on the other hand, are absolute necessary conditions for the imminent growth in a globalized and highly competitive economy. These parameters require the development and fabrication of high-resistance metals and alloys and the explicit knowledge of their potential damage and degradation processes, in order to ensure long-lasting service, avoiding undesired, costly, and catastrophic failures.

The occurrence of unexpected failures in critical industrial and transportation sectors could lead to serious accidents or even to catastrophes, having a crucial impact on infrastructure and society in general. Profound knowledge about failure mechanisms in metals and alloys is an absolute prerequisite, which leads to the understanding and determination of the root source of the failure(s) and to their successful prevention. Together with failure mechanisms, the recognition of service conditions (temperature, type or nature of the environmental parameters, loading conditions, assembly parameters, and interactions with neighboring components) and the information pertaining to the industrial production processes involved in the fabrication of the metal component are of critical significance in order to put all the failure "puzzle pieces" in a meaningful and reasonable order. The "reconstruction" of a failure event is mainly focused on the interpretation of its natural complexity and the unfolding of the logical sequence of the events involved. These failure-linking stages, which occurred concurrently, intermittently, or successively, gave rise to the final ultimate failure, in a "backward-thinking" procedure. The organization of a failure investigation is a multi-step, structured, and disciplined process that must be limited in time and budget, according to the requirements and the project objectives. The collection of the "pieces" and "clues" during a failure investigation is a diligent and multi-tasking process. In the real world, missing pieces disrupt the continuity and clarity of the "cause-and-effect" chain relationships, converting them from deterministic to fuzzy or stochastic.

Failure analysis is an interdisciplinary scientific topic, reflecting the opinions and interpretations coming from a systematic evidence-gathering procedure, embracing various important sectors, imparting knowledge, and substantiating improvement practices. The deep understanding of a material or component's role and properties is of central importance for "fitness for purpose" in certain industrial processes and applications. The "scheme" presenting the interaction loci of the failure analysis area can be simply presented as the "knowledge triangle", illustrated in Figure 1.

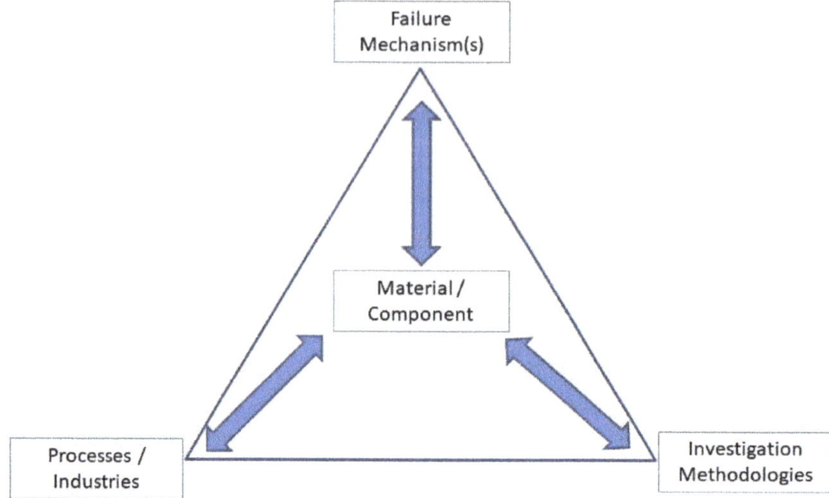

Figure 1. The failure analysis "knowledge triangle", presenting the interaction areas and the central role of the understanding of a material or component's role and properties.

2. Contributions

The Special Issue "Failure Mechanisms in Alloys" contains in total twenty-seven (27) research articles [1–27], with two review papers among them (see [9] and [24]). The contents of this collection cover a wide spectrum of the cross-disciplinary fields of the entire domain of failure analysis, providing valuable contributions in diverse and challenging topics exhibiting interests in the investigation of failure mechanism, industrial processes, and approach methods (Figures 2–4). Moreover, different groups of materials are involved in the presented studies (Table 1).

Almost the complete range of the general types of failure mechanisms has been addressed in the published works gathered in this Special Issue. Instant overload as well as progressive failure modes are included (Figure 2a). More specifically, the broad category of static overload includes more generic subjects also from the field of manufacturing-related topics, where the effect of deformation and fracture was studied as an important and undetached ingredient of the fabrication process per se (e.g., hot and cold working, machining). Therefore, the mostly "intense area" (Overload/Static) comprises studies concerning general deformation and fracture phenomena, as the result of instant loading/testing conditions [4,9,10,14,15,19] and studies related to manufacturing and production processes [6,7,12,13,16,20,22,25,26]. Testing and modeling procedures addressing the evolution of deformation and fracture during forming [7,13,26], the impact toughness, [4] and certain production process characteristics [25] are also included. Nevertheless, shear fracture processes that emerged in machining and chip formation are also part of this broader group of studies, relevant to manufacturing topics (see [12,16,20]).

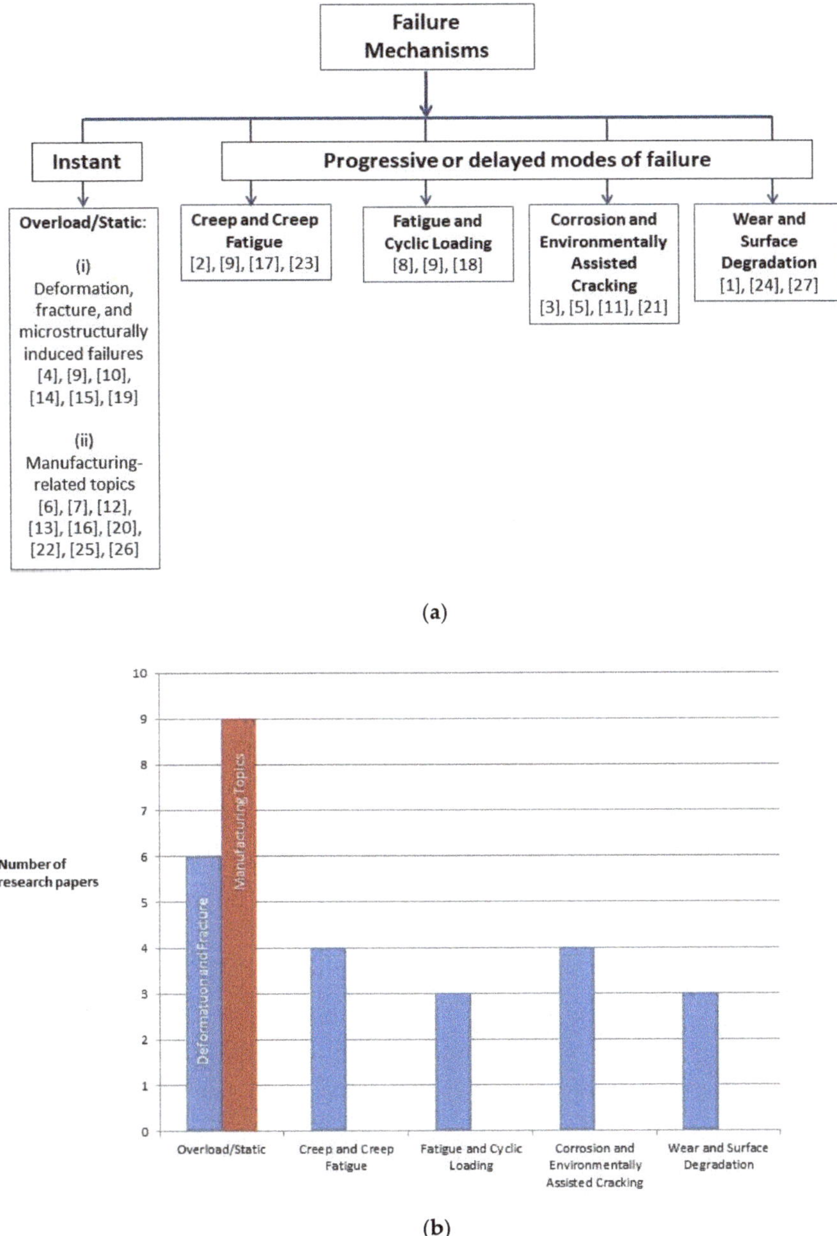

Figure 2. (a) Categorization of the subject "Failure Mechanism", as it was addressed by the studies contained in this Special Issue; (b) distribution graph presenting the number of papers focused on the various failure mechanisms, which were ad hoc considered herein for simplicity purposes.

"Progressive" or "delayed modes of failure" possess exceptional interest in various industrial sectors, such as the chemical, mechanical, and manufacturing industries and in plant machinery. Four basic categories can be distinguished (Figure 2a):

a.   Creep and Creep Fatigue failure modes [2,9,17,23];

b.   Fatigue and Cyclic Loading [8,9,18];

c.   Corrosion and Environmentally Assisted Cracking [3,5,11,21];

d.   Wear and Surface Degradation [1,24,27].

The relevant number of studies' distribution graph is presented in Figure 2b.

It can be seen from Figure 2b that manufacturing-related topics occupy a significant percentage of the content, reflecting the importance of failure assessment and the emergence of failure and damage prognosis and prevention, as the basic component of the knowledge and learning processes required for quality improvement in industry.

Considering the "Process/Industry" as a classification criterion, the published studies are relevant with certain industries or processes involved as far as the observed failure mechanism is concerned (Figure 3). More specifically, the main industrial processes involved are presented as follows:

(i)   Casting and Metal Forming [6,7,11,13,15,19,24–26]

(ii)   Machining [12,16,20]

(iii)   Chemical/Petrochemical [3,17,21,23]

(iv)   Heat Treatment [22]

(v)   General Plant Machinery [1,9,18,24]

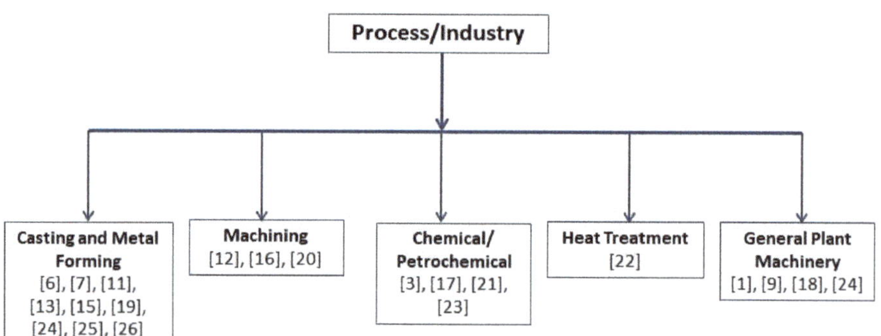

Figure 3. Classification according to the "Process/Industry" relevant criterion, addressed in the studies of this Special Issue.

The metal component manufacturing industry (casting, metal forming, and machining) seems to be highly involved, as one of the primary industrial component production sectors, in highlighting the efforts in understanding material behavior and taking actions for failure prediction and prevention.

Based on the "Investigation Method/Approach", several broad categories can be distinguished (Figure 4):

1.   Phenomenological and Experimental [1–9,11–27]

2.   Numerical Modeling [6,7,10,12,13,15,25,26]

3.   Statistical and Stochastic [8,16,20,22,25]

4.   Systems or Quality [9]

5.   Combined (experimental, analytical, numerical model, etc.) [3,6–8,12–16,18–20,22,25,26]

As can be readily observed, the experimental and empirical approach is the dominant methodology of failure investigation. In addition, the emergence of numerical simulation, using finite element modeling (FEM), tends to be very popular in the prediction of material behavior and potential failure prevention. The contribution of quality and organization systems is very promising in the case of complex processes, where teamwork in process planning, risk assessment, resource allocation, and implementation of improvement actions is a key concept in modern quality assurance and management.

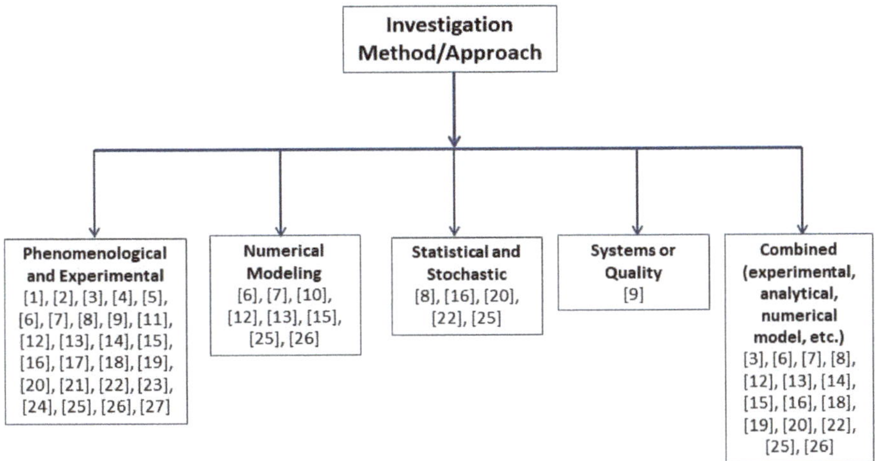

Figure 4. Classification based on the relevant "Investigation Method/Approach" followed in the studies of this Special Issue.

A broad range of materials was covered in the present studies (see Table 1). Ferrous metals (structural steels, stainless steels, and special resistance steels) are more frequently investigated, indicating the importance of this material type in engineering constructions and in severe process environments.

Table 1. Representative types of engineering material types of investigation.

Material Type	Reference
Structural and pipeline steels	[1,5,9,13,16,20,22,27]
Cast iron	[14,19]
Special resistance and stainless steels	[3,4,9,11,17,18,21,23,24]
Wear resistant coatings and surface layers	[24,27]
Ti, Ti alloys	[7,12]
Al alloys	[13,25,26]
Mg alloy	[13]
Cu, Cu alloys	[6,9,10]
Nanomaterials	[10]
W–Cu composite	[15]

3. Conclusions

The present collection of studies reflects the profound interest in the specific field of research, covering a wide range of failure mechanisms, processes, and investigation methodologies. The majority of the studies followed a phenomenological and experimental approach, while it seems that there was a general contribution tendency toward numerical simulation, to further enrich the value of the research results and broaden the application perspectives, especially in the case of larger-scale metal-forming and manufacturing processes. Ferrous alloys (structural, special purpose or resistance, stainless steels, cast irons) constitute the majority of the studied materials, since they are considered one of the principal sources of construction components and are used in various industrial sectors.

On a final note, it is hoped and strongly believed that the accumulation of additional knowledge in the field of failure mechanisms and the adoption of the principles, philosophy, and deep understanding of the failure analysis process approach will strongly promote the learning concept as a continuously evolving process, leading to personal and social progress and prosperity.

Funding: This research received no external funding.

Acknowledgments: The Guest Editor expresses his deep gratitude to the contributing authors, who shared their precious research works in this collection of papers, embracing and expanding the "field of view" of the fascinating area of the research in the field of failure mechanisms, as well as shedding light on significant industrial applications and challenges. Moreover, the assistance and support provided plentifully by the MDPI *Metals* Editorial Team and, especially, by the Managing Editor, Harley Wang, made this journey a joyful knowledge endeavor, offering a rich and a valuable account.

Conflicts of Interest: The author declares no conflict of interest.

Dedication: This Editorial is dedicated to the memory of an exceptional scientist, researcher, academic teacher, and mentor, Dimitrios I. Pantelis, Professor of the School of Naval Architecture and Marine Engineering of the National Technical University of Athens (NTUA), who passed away in October 2019. His legacy of academic, scientific, and human values will stand as a model of inspiration.

References

1. UlHaq Toor, I.; Muzammil Irshad, H.; Mohamed Badr, H.; Abdul Samad, M. The effect of impingement velocity and angle variation on the erosion corrosion performance of API 5L-X65 carbon steel in a flow loop. *Metals* **2018**, *8*, 402. [CrossRef]
2. Liu, D.; John Pons, D. Crack propagation mechanisms for creep fatigue: A consolidated explanation of fundamental behaviours from initiation to failure. *Metals* **2018**, *8*, 623. [CrossRef]
3. Haidemenopoulos, G.N.; Kamoutsi, H.; Polychronopoulou, K.; Papageorgiou, P.; Altanis, I.; Dimitriadis, P.; Stiakakis, M. Investigation of stress-oriented hydrogen-induced cracking (SOHIC) in an amine absorber column of an oil refinery. *Metals* **2018**, *8*, 663. [CrossRef]
4. Kim, K.; Park, M.; Jang, J.; Chan Kim, H.; Moon, H.-S.; Lim, D.-H.; Bae Jeon, J.; Kwon, S.-H.; Kim, H.; Jun Kim, B. Improvement of strength and impact toughness for cold-worked austenitic stainless steels using a surface-cracking technique. *Metals* **2018**, *8*, 932. [CrossRef]
5. Wu, W.; Yin, H.; Zhang, H.; Kang, J.; Li, Y.; Dan, Y. Electrochemical investigation of corrosion of X80 steel under elastic and plastic tensile stress in CO_2 environment. *Metals* **2018**, *8*, 949. [CrossRef]
6. Pashos, G.; Pantazopoulos, G.A.; Contopoulos, I. A comprehensive CFD model for dual-phase brass indirect extrusion based on constitutive laws: Assessment of hot-zone formation and failure prognosis. *Metals* **2018**, *8*, 1043. [CrossRef]
7. Mei, H.; Lang, L.; Liu, K.; Yang, X. Evaluation study on iterative inverse modeling procedure for determining post-necking hardening behavior of sheet metal at elevated temperature. *Metals* **2018**, *8*, 1044. [CrossRef]
8. Woo, S.; O'Neal, D.L. Improving the reliability of mechanical components that have failed in the field due to repetitive stress. *Metals* **2019**, *9*, 38. [CrossRef]
9. Pantazopoulos, G.A. A short review on fracture mechanisms of mechanical components operated under industrial process conditions: Fractographic analysis and selected prevention strategies. *Metals* **2019**, *9*, 148. [CrossRef]
10. Bazios, P.; Tserpes, K.; Pantelakis, S.G. Numerical computation of material properties of nanocrystalline materials utilizing three dimensional Voronoi models. *Metals* **2019**, *9*, 202. [CrossRef]
11. Li, S.; Wang, Y.; Wang, X. In situ observation of the deformation and fracture behaviors of long-term thermally aged cast duplex stainless steels. *Metals* **2019**, *9*, 258. [CrossRef]
12. Lampropoulos, A.D.; Markopoulos, A.P.; Manolakos, D.E. Modeling of Ti6Al4V alloy orthogonal cutting with smooth particle hydrodynamics: A parametric analysis on formulation and particle density. *Metals* **2019**, *9*, 388. [CrossRef]
13. Xiao, Y.; Hu, Y. An extended iterative identification method for the GISSMO model. *Metals* **2019**, *9*, 568. [CrossRef]
14. Gu, J.; Chen, P.; Li, K.; Su, L. A macroscopic strength criterion for isotropic metals based on the concept of fracture plane. *Metals* **2019**, *9*, 634. [CrossRef]
15. Hao, Z.; Liu, J.; Cao, J.; Li, S.; Liu, X.; He, C.; Xue, X. Enhanced ductility of a W-30Cu composite by improving microstructure homogeneity. *Metals* **2019**, *9*, 646. [CrossRef]

16. Monkova, K.; Monka, P.P.; Sekerakova, A.; Hruzik, L.; Burecek, A.; Urban, M. Comparative study of chip formation in orthogonal and oblique slow-rate machining of EN 16MnCr5 steel. *Metals* **2019**, *9*, 698. [CrossRef]

17. Haidemenopoulos, G.N.; Polychronopoulou, K.; Zervaki, A.D.; Kamoutsi, H.; Alkhoori, S.I.; Jaffar, S.; Cho, P.; Mavros, H. Aging phenomena during in-service creep exposure of heat-resistant steels. *Metals* **2019**, *9*, 800. [CrossRef]

18. González-Ciordia, B.; Fernández, B.; Artola, G.; Muro, M.; Sanz, Á.; de Lacalle, L.N.L. Failure-analysis based redesign of furnace conveyor system components: A case study. *Metals* **2019**, *9*, 816. [CrossRef]

19. Angella, G.; Zanardi, F. Validation of a new quality assessment procedure for ductile irons production based on strain hardening analysis. *Metals* **2019**, *9*, 837. [CrossRef]

20. Monkova, K.; Monka, P.P.; Sekerakova, A.; Tkac, J.; Bednarik, M.; Kovac, J.; Jahnatek, A. Research on chip shear angle and built-up edge of slow-rate machining EN C45 and EN 16MnCr5 steels. *Metals* **2019**, *9*, 956. [CrossRef]

21. Bahrami, A.; Taheri, P. A Study on the failure of AISI 304 stainless steel tubes in a gas heater unit. *Metals* **2019**, *9*, 969. [CrossRef]

22. Mihaliková, M.; Zgodavová, K.; Bober, P.; Sütoová, A. Prediction of bake hardening behavior of selected advanced high strength automotive steels and hailstone failure discussion. *Metals* **2019**, *9*, 1016. [CrossRef]

23. Bahrami, A.; Taheri, P. Creep failure of reformer tubes in a petrochemical plant. *Metals* **2019**, *9*, 1026. [CrossRef]

24. Psyllaki, P.P. An introduction to wear degradation mechanisms of surface-protected metallic components. *Metals* **2019**, *9*, 1057. [CrossRef]

25. Qu, F.; Xu, J.; Jiang, Z. Finite element analysis of forward slip in micro flexible rolling of thin aluminium strips. *Metals* **2019**, *9*, 1062. [CrossRef]

26. Vazdirvanidis, A.; Pressas, I.; Papadopoulou, S.; Toulfatzis, A.; Rikos, A.; Katsivarda, M.; Symeonidis, G.; Pantazopoulos, G. Examination of formability properties of 6063 alloy extruded profiles for the automotive industry. *Metals* **2019**, *9*, 1080. [CrossRef]

27. Urban, M.; Monkova, K. Research of tribological properties of 34CrNiMo6 steel in the production of a newly designed self-equalizing thrust bearing. *Metals* **2020**, *10*, 84. [CrossRef]

© 2020 by the author. Licensee MDPI, Basel, Switzerland. This article is an open access article distributed under the terms and conditions of the Creative Commons Attribution (CC BY) license (http://creativecommons.org/licenses/by/4.0/).

Article

The Effect of Impingement Velocity and Angle Variation on the Erosion Corrosion Performance of API 5L-X65 Carbon Steel in a Flow Loop

Ihsan UlHaq Toor [1,2,*], Hafiz Muzammil Irshad [1], Hassan Mohamed Badr [1] and Mohammed Abdul Samad [1]

[1]   Department of Mechanical Engineering, King Fahd University of Petroleum & Minerals (KFUPM), Dhahran 31261, Saudi Arabia; muzammil.duke@gmail.com (H.M.I.); badrhm@kfupm.edu.sa (H.M.B.); samad@kfupm.edu.sa (M.A.S.)
[2]   Center of Research Excellence in Nanotechnology, King Fahd University of Petroleum & Minerals (KFUPM), Dhahran 31261, Saudi Arabia
*   Correspondence: ihsan@kfupm.edu.sa; Tel.: +966-13-860-7493; Fax: +966-13-860-2949

Received: 13 May 2018; Accepted: 29 May 2018; Published: 31 May 2018

Abstract: Erosion corrosion performance of API 5L-X65 carbon steel was investigated at three different impingement velocities (3, 6 & 12 m/s), five different angles (15, 30, 45, 60, & 90°), and with/without solid particles (average particle size of 314 μm). The experiments were conducted in 0.2 M NaCl solution at room temperature for a duration of 24 h and the results showed that the maximum erosion corrosion rate was observed at 45° irrespective of the velocity. The highest erosion corrosion rate at 45° was due to the balance between the shear and normal impact stress at this angle. Ploughing, deep craters, and micro-forging/plastic deformation were found to be the main erosion corrosion mechanisms. The maximum wear scar depth measured using optical profilometery was found to be 51 μm (average) at an impingement angle of 45°.

Keywords: impingement; erosion corrosion; API 5L-X65; flow loop; wear scar

1. Introduction

Erosion-corrosion is a combined material degradation mechanism in which material is removed by the mechanical process of erosion coupled with the electrochemical process of corrosion. This is one of the serious problems being faced by many industries and is responsible for multibillion-dollar losses due to premature equipment failures [1]. Erosion corrosion is ranked as the 5th most common degradation issue in fluid handling systems (pumps, compressors, piping systems in offshore oil/gas facilities, desalination plants, etc.) and has direct consequences in terms of equipment safety [2–6]. This problem gets more aggravated by the presence of high amounts of solid/sand particles of different morphologies in the flowing fluid [7,8]. These high velocity solid particles strike the stationary and rotating equipment in the presence of fluid mixed with water and/or oxygen and thus cause unplanned plant shutdowns, resulting in multibillion-dollar loss [9]. The seawater in desalination plants, cooling systems, fire-fighting systems, and power generation industries is very corrosive and if present together with solid particles, can reduce the equipment design life significantly by the combined effect of erosion and corrosion.

So, therefore, in order to overcome this problem of erosion corrosion in the industries, a detailed erosion corrosion database is required which can assist in developing erosion corrosion models for better prediction. The development of efficient prediction models depends on realistic and reliable experimental data obtained by simulating the real-time industrial conditions. The erosion corrosion behavior of carbon steel being used a lot as pipeline material can be investigated by conducting the experiments in flow loops. The design of such flow loops enables the researchers to investigate the erosion corrosion problem by controlling different parameters such as angle, velocity, fluid type, temperature and amount/type of solid particles etc. However, due to its high cost of maintenance, construction challenges and space limitations within the research laboratories, such types of experimental facilities are limited.

According to ASTM G73-98, corrosion is the progressive loss of original material from a solid surface due to continued exposure to impacts by liquid jets. The problems usually arise when a stream of water impinges on, or flows over components especially whenever there is a change of flow direction and/or a change in the cross-sectional area of the flow passage [10,11]. In erosion, solid particles entrained in high-velocity jets are repeatedly impacted on the metal at oblique angles, resulting in material removal from the surface, whereas corrosion is a material loss because of chemical or electrochemical reactions [12]. Erosion-corrosion caused by flowing fluid with/without solid particles is a form of tribo-corrosion material removal mechanism as it damages both the surface layers, (passive film or corrosion products) as well as the base metal. The degradation mechanism resulting from the combined effect of electrochemical and mechanical processes seems very complex [12–15]. The material removal during erosion corrosion process can be either by chemical dissolution or by erosion simply due to fluid flow alone or due to the impingement of slurry mixed with solid particles. There can be more intricate situations in which electrochemical corrosion can have synergistic effects [13,14] and the synergistic effects of erosion-corrosion can be considerably higher than the individual effect of erosion or corrosion.

So therefore, the main objective of this study was to evaluate the erosion corrosion behavior and underlying erosion corrosion mechanism in one of the most commonly used pipeline grade carbon steel material API 5L-X65. The experiments were conducted to examine the role of various parameters such as impingement velocity (3~12 m/s) and impingement angle (15°~90°) on the erosion corrosion behavior of this steel in 0.2 Mol NaCl solution at room temperature with/without solid particles for a duration of 24 h. The eroded surfaces were examined using Field Emission Scanning Electron Microscopy (FE-SEM) and optical profilometer to understand the erosion corrosion mechanism.

2. Experimental Procedure

2.1. Erosion-Corrosion Test Apparatus

In order to carry out the detailed experimental investigations, a state of the art mini-flow loop was designed and manufactured locally as shown in Figure 1. A centrifugal pump was used to circulate the fluid in the flow loop. To investigate the effect of solid particles, the discharge pipe was equipped with a venture-tube with a controlled stream of solid particles. The two-phase mixture flows into the erosion-corrosion test cell and then to a cyclone separator for separating the solid particles before returning back to the main reservoir. The test cell is equipped with a fixed convergent nozzle (3 mm diameter nozzle made of 316 L stainless steel) to produce the required liquid jet velocity. The cell also has a sample holder, which can be used to fix the specimens at five different impingement angles. All of the components of the flow loop such as pipes, storage tank, cyclone separator, valves, pump etc., were made of corrosion resistant 316 L stainless steel. Fluid jet velocity was calculated using ultrasonic water flow meter (Model: Wprime-280W) and pump speed was controlled with the help of 650 V SSD drive control panel. Taylor 9940 N temperature gauge was used to monitor the temperature, whereas 15 kW pump (Lowara Company: Vicenza, Italy) was used for fluid pumping.

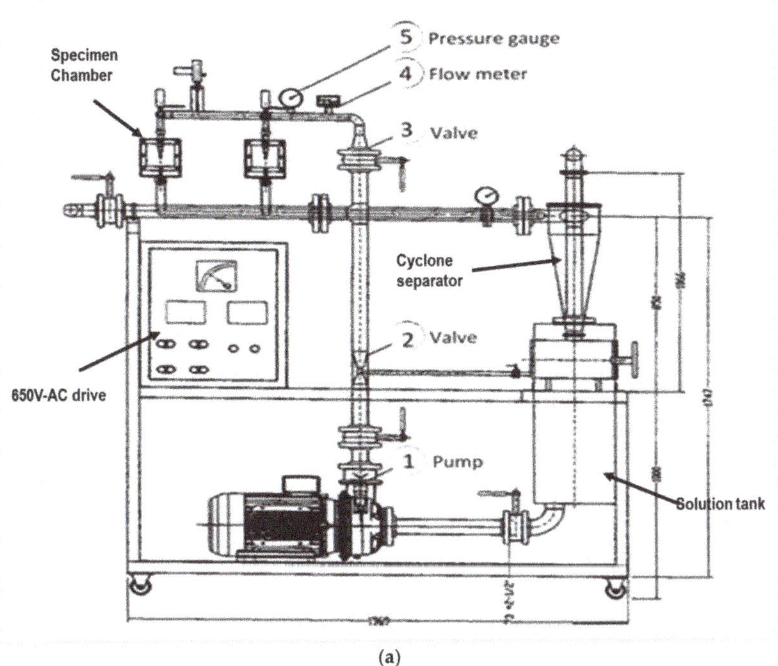

(a)

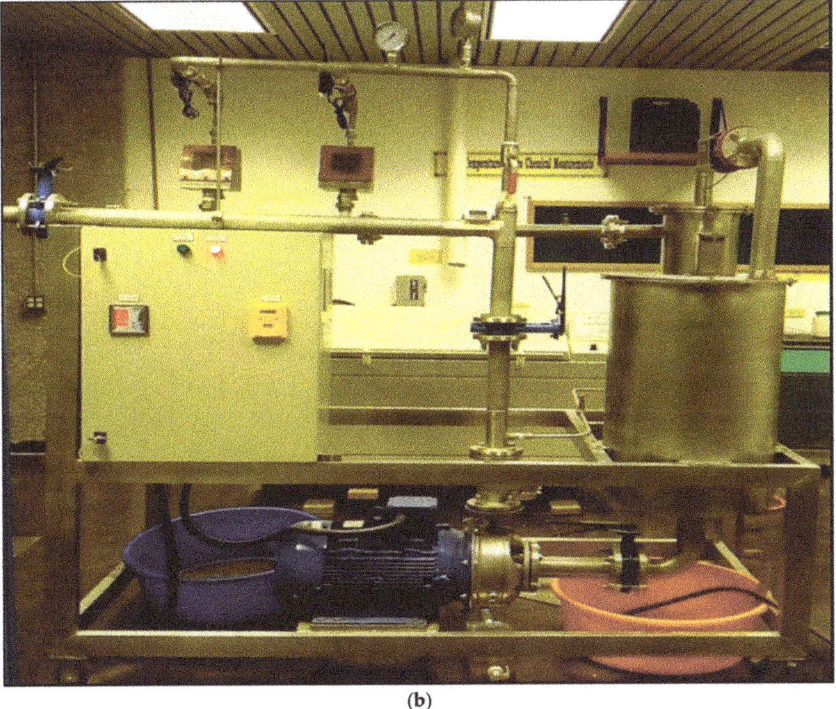

(b)

Figure 1. (**a**) Schematic of the flow loop showing the details of its components; (**b**) State of the art in-house designed mini flow loop.

2.2. Test Specimens and Sand Particle Characterization

Table 1 shows the composition of pipeline grade API 5L-X65 carbon steel used in this study. This material is one of the most commonly used piping material for transportation of fluids in petroleum, desalination and many other industrial applications due to its distinct physical properties. It has appreciable ductility, strength, weld-ability, and can be heat treated to achieve desired properties. Its average Vickers hardness (HV) was measured using a CSM Micro Combi hardness tester (Diamond indenter) under 2N load (P) over an indentation time of 10 s and was found to be 298 HV. The specimen preparation involved cutting, mounting and grinding. The specimens were cut into 20 mm × 20 mm × 5 mm dimension, hot mounted in Lucite material to expose only (20×20) mm^2 surface area to the impinging fluid. The hot mounted specimens were ground up to 600 grit size paper. The samples were cleaned with acetone, rinsed with distilled water and subsequently dried with a drier before conducting each experiment. The specimens were weighed to an accuracy of 0.01 mg. Figure 2 gives an overview of the whole process of specimen preparation.

Table 1. Detailed chemical composition of API 5L-X65 carbon steel used in the experiments.

Elements	C	Mn	Fe	Cu	Mo	Ni	Cr	S	V	Co	P
Wt.% Compositions	0.165	1.29	98.1	0.0548	0.0092	0.0183	0.0275	0.0062	0.0076	0.0063	0.0015

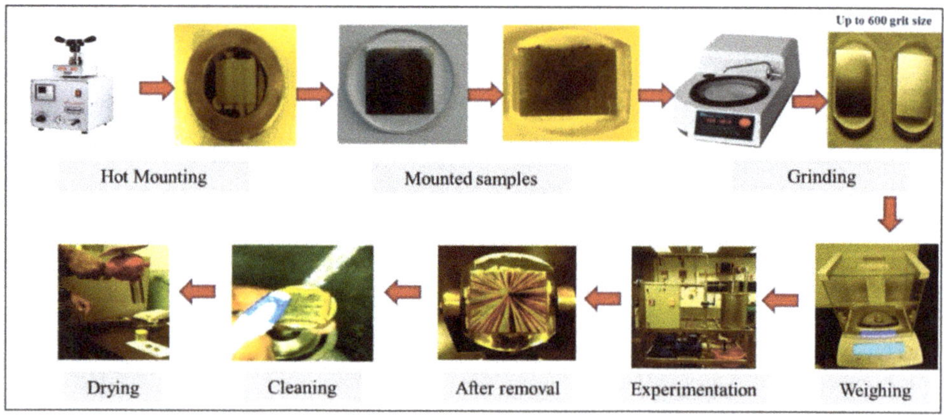

Figure 2. Specimen preparation process showing all the steps involved from the start till end.

Sand particles with an average particle size of 314 μm were used for erosion corrosion investigations. These silica sand particles were from Riyadh region of Saudi Arabia and supplied by a local company, based in Jeddah.

2.3. Test Procedure

Erosion corrosion tests were performed according to ASTM-G-73-98 and tap water from sweat water line was used to prepare the solution for experiments. The solution tank (Figure 1) was filled with an approximately 69 L of 0.2 Mol NaCl solution. The specimens were hot mounted, ground to 600 grit size SiC paper and subsequently weighed before fixing in the test cell. The specimens were fixed in an-inhouse designed specimen holder (Figure 1) inside the test chamber under a 3 mm circular nozzle for fluid impingement. Erosion corroison experiments were carried out at three different velocities (3 m/s, 6 m/s, and 12 m/s) and at five different impingement angle (15°, 30°, 45°, 60° and 90°) for a duration of 24 h. After each experiment, the specimens were cleaned using a soft toothbrush to remove any loosely attached corrosion products, subsequently cleaned using acetone, rinsed in distilled water and dried with a drier (Figure 2). The weight before and after the experiments was measured using a digital balance up to 0.01 mg and erosion corrosion rate calculation was done as per equation 1 based on ASTM G1-03 standard. The K is a contact to have the corrosion rate in desired units and a value of 3.45×10^6 was used in these calculations. Each experiment was performed thrice under each experimental condition for data repeatability.

$$\text{Corrosion Rate} = \frac{\text{Weight loss (mg/g)} * \text{K}}{\text{Density of sample (g/cm}^3) * \text{Exposed Area (cm}^2) * \text{Exposure time (hr)}} \tag{1}$$

3. Results and Discussion

3.1. Effect of Impact Angle and Velocity on Corrosion Rate

Figure 3 shows the corrosion rate (without solid particles) of API 5L-X65 carbon steel as a function of impingement angle at three different velocities. The maximum and minimum corrosion rates were observed at 45° and 90° angles respectively. It is reported in the published literature that both the shear and normal stress play a vital role during fluid impingement and shear stress is dominant at lower angles and vice versa [8]. The highest corrosion rate observed at 45° is due to the balance between shear and impact force [8]. Corrosion rates were increased with an increase in velocity, though overall trend remains the same at all velocities. There was no significant difference in the corrosion rate between 3 m/s and 6 m/s, however, corrosion rate was significantly increased at 12 m/s, which is due to an increased wall shear stress values at this velocity. The shear stress is reported to be one of the important factors, which increases the corrosion rate at high fluid jet velocities. This finding, which is reported extensively in research, states that mass transport of oxygen increases with an increase in fluid velocity and that will increase the corrosion rate of the material [16].

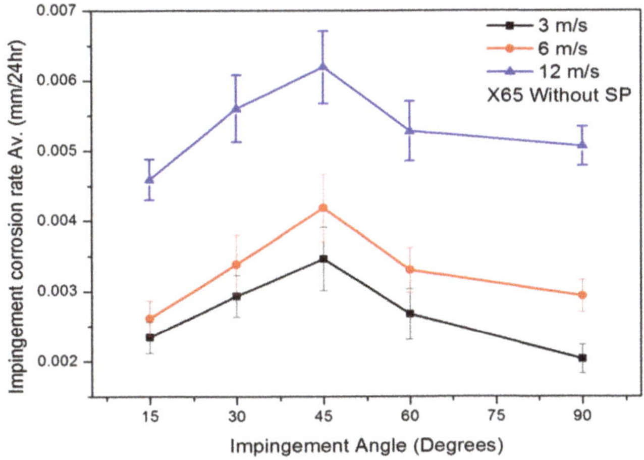

Figure 3. Effect of impingement velocity and angle on the corrosion rate (without solid particles) of API 5L-X65 steel.

Surface Morphology of Corroded Surfaces

Surface morphologies of corroded specimens were examined using FE-SEM as shown in Figure 4a–e. It is interesting to note that corrosion pattern and wear scar morphology was changed with impingement angle indicating that different stresses are acting at different angles. Impingement points (circled in Figure 4a–e) changed from elliptical at lower angles to more round/circular at higher angles. Figure 4a shows the presence of plate-like circular holes in specimens eroded at 15°. This was because of the fact that the corrosive slurry attacked the target material surface locally at this angle. On the other hand, the fluid has its maximum velocity and forms a wide stagnant layer which by covering the entire surface of the specimen promotes susceptibility towards localized material loss/corrosion [17]. The wear scars were found to be neither continuous nor deeper and observed to have gaps among each other in the direction of the fluid movement. Figure 4a,b show an elliptical to circular morphology at the impingement point which can be attributed to the maximum tangential force/or shearing stress at lower angles. With an increase in impingement angle, normal impact/extrusion force was increased as well. The wear scars observed at 30° were found to be wider, deeper and continuous as compared to those observed at 15° as shown in Figure 4b. The electrolyte stagnation may have occurred which ultimately resulted in pitting [4]. However with continued fluid flow of high shear stress, elongated corrosive wear scars were observed. At 45° impingement angle, number of wear scars and the total affected area was much more when compared both with lower than and higher than 45° (Figure 4c) angle. The wear scars were found to be continuous. As the impingement angle was increased (>45°), normal impact component became dominant. At 60° angle, the wear scars were less in number and shallower in depth (Figure 4d). On the other hand, at 90° angle the normal stress component was maximum and after the impact, the fluid does not have enough energy to erode the surface deeper (Figure 4e).

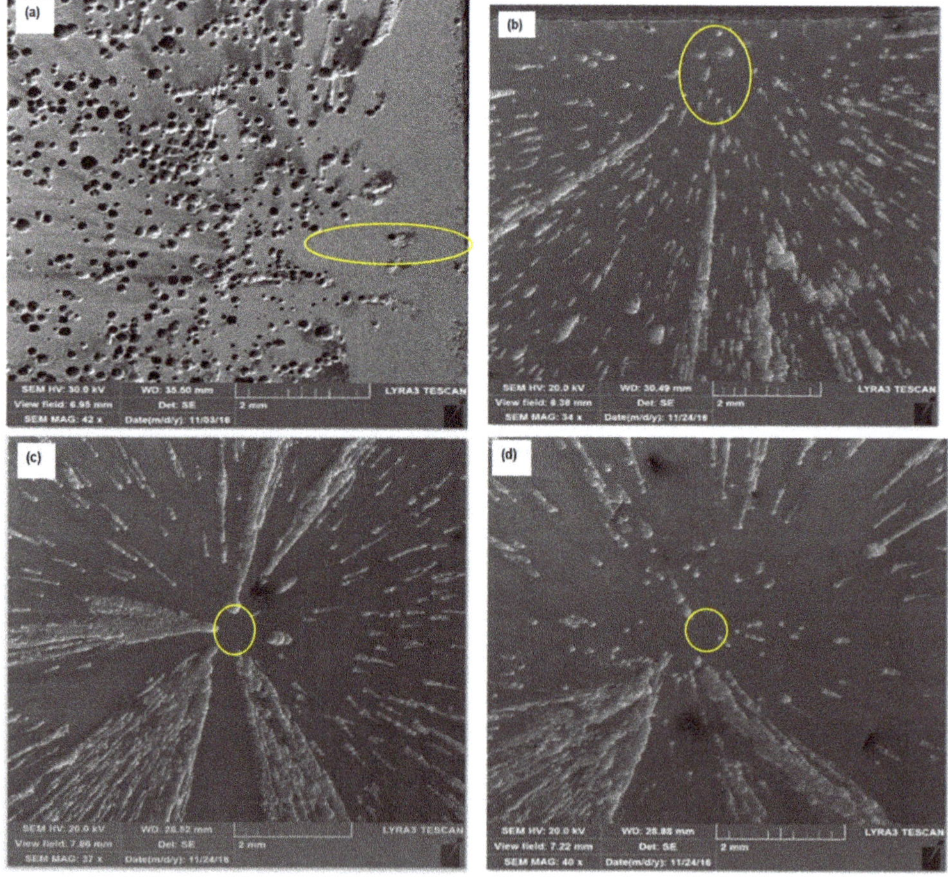

Figure 4. *Cont.*

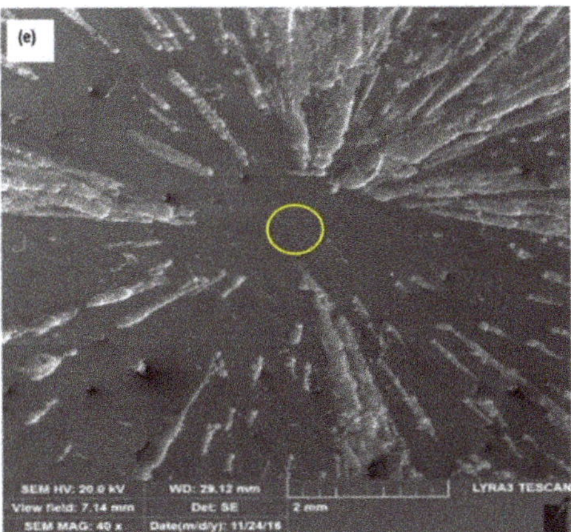

Figure 4. Field Emission Scanning Electron Microscopy (FE-SEM) micrographs (**a–e**) of wear scars after corrosion experiments (without solid particles) at 15–90° angles (circled points show high-velocity regions).

3.2. Erosion Corrosion and Mechanism

Figure 5 shows the erosion corrosion rate (with solid particles) of API 5L-X65 carbon steel, as a function of impingement angle and velocity. The trend observed was similar to the one observed in Figure 3, however the values of erosion corrosion rates were higher than the ones without the presence of solid particles. The fluid jet velocity is an important parameter which can increase the erosion-corrosion rate and both are related through a relationship $E = K(V)^n$. In this equation, E is the erosion rate, K is the material constant (K depends on particle size and impact angle), V is the fluid velocity, and n is the velocity exponent. The value of n in the case of slurry erosion corrosion was reported to be around two [18]. Both the horizontal and vertical components of kinetic energy play a vital role during erosion corrosion of the materials. It is reported that vertical component of kinetic energy (KE) is responsible for the solid particle's penetration into the material, while horizontal component controls the ploughing during erosion corrosion experiments [18].

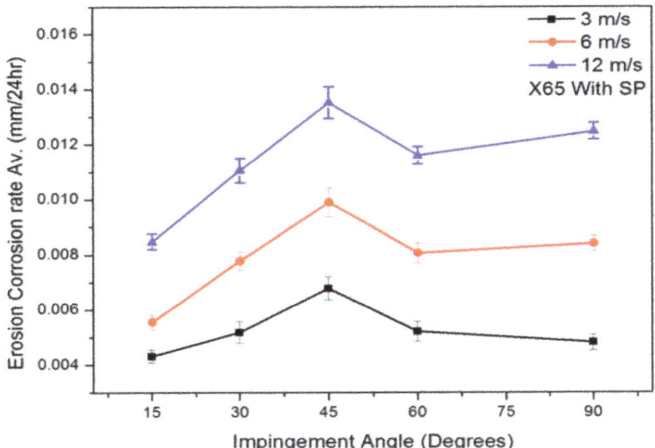

Figure 5. Effect of impingement velocity and angle on erosion corrosion rate (with solid particles) of API 5L-X65 steel.

Figure 6a,b shows that longer erosion scars of ploughing were observed at oblique angles of 15° and 30°, suggesting that horizontal component of KE was dominant at these angles. Another observation is that larger surface area is affected by abrasive particles at lower angles than at higher angles. Severe plastic deformation together with deep ploughing and increased surface activation was observed at 45° angle, which was mainly due to the balance between shear and normal impact stress. The effect of microcutting was not that significant at 60° as confirmed by the presence of shallow and shorter erosion scars. Some wider erosion scars were also observed due to an increased friction force of sliding abrasive particles and this resulted in high normal impact stress. At normal impingement angle, the particles hit the surface with maximum kinetic energy and imparted highest normal impact force which resulted in the formation of flakes and platelets. These flakes were subsequently removed due to multiple strikes of the incoming particles (Figure 6e). Furthermore, the erosion scars became wider and some micro cracks also found at higher angles because of increased surface hardness due to work hardening. After multiple strikes of the sand particles, average hardness was found to be 303 HV and slight increase in weight loss was observed after 60° angle.

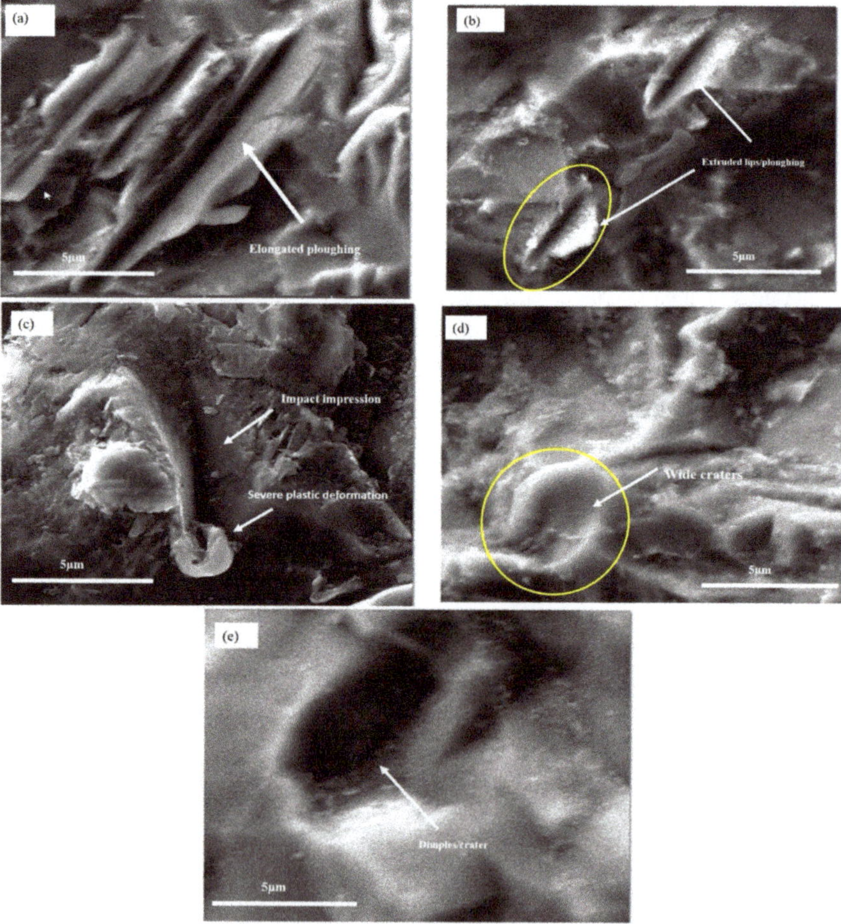

Figure 6. FE-SEM micrographs (**a**–**e**) showing different features of material removal during erosion corrosion experiments.

3.3. Correlation with Erosion Corrosion Model

Correlations are important to develop reliable erosion corrosion prediction models. The prediction models presented as a function of impingement velocity, angle, particle size/shape and substrate material properties [19–22]. One of the early erosion prediction models was the one developed by Finnie et al. [18]. The model shows the following relationship;

$$ER = AF_s V^n f(\theta)$$

(2)

where ER is the erosion rate (mg/mg), A is an empirical constant, V is the particle impingement velocity, and n is an empirical coefficient. It was reported that a value of 1.73 can be used safely for the empirical coefficient for various oilfield materials [23]. F_s is a particle shape coefficient and it is equal to 1 for angular sand particles. While $f(\theta)$ is a function of the impact angle as below;

$$f(\theta) = x\cos^2\theta\sin(w\theta) + y\sin^2\theta + z$$

(3)

Whereas w, x, y, and z are empirical constants which depend on the material being eroded. The suitable values of the model constants for carbon steel, assuming V has units of ft/s are discussed elsewhere [23]. In this study, the impingement erosion data were correlated using above-mentioned model. The values of A and z were determined by performing a regression analysis using MATLAB software. Lower and upper bounds for A were found to be 0.012 and 0.0396 whereas for z, they were 0.428 and 0.572 respectively. Values of the correlation coefficients were found to be $R^2 > 99\%$, showing a good fit using the given parameters [23]. Furthermore, using known parameters as mentioned above, erosion rates were calculated at v = 12 m/s at 15° to 90° angles respectively. Figure 7 shows the comparison between experimentally obtained erosion rate versus the one calculated using Finnie's model. The calculated erosion rate showed a good agreement with the present experimental data of API 5L-X65 carbon steel. It was deduced that Finnie's model can be safely used to determine the erosion rate with a reasonable accuracy. However, the parameters for this model given and calculated are specific to current test materials and test conditions.

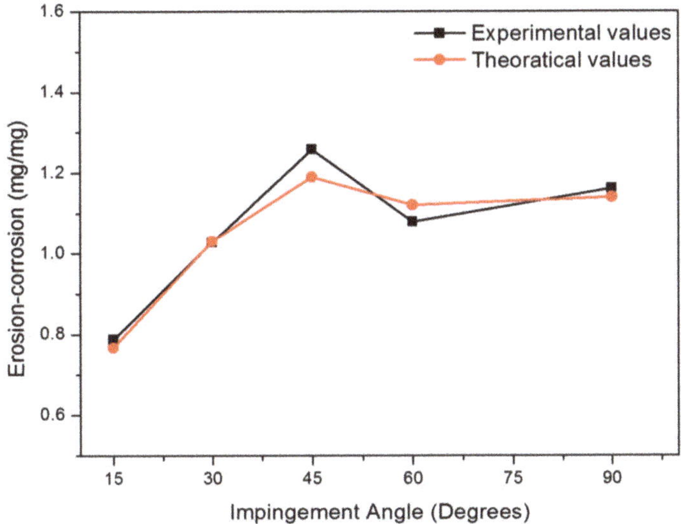

Figure 7. A comparison between the experimental erosion corrosion rate and theoretical data calculated using Finnie model at 12 m/s fluid impact velocity.

3.4. Effect of Erosion on Corrosion

Figure 8 shows the individual as well the combined effects of erosion-corrosion. It is clear that erosion has a considerable contribution to the overall material loss in terms of erosion corrosion. It was observed that the introduction of solid particles increased the erosion-corrosion rate of API 5L-X65 steel at all impingement angles. However, the synergistic effect at 45° and 90° angle was more significant as compared to other angles.

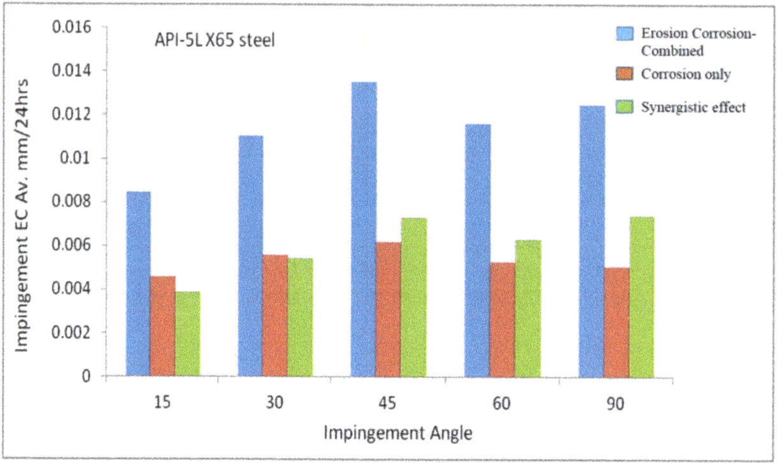

Figure 8. The synergistic effect of erosion on corrosion with the introduction of solid particles on material loss.

At 45° angle, particles penetrated much deeper into the specimens due to the combined effect of shear and impact stresses, which increased the erosion corrosion (Figure 8). It was found that at 90°, due to extensive extrusion and fracture of platelets, erosion enhanced the corrosion significantly. Corrosion rate was considered here as a pure corrosion damage (without solid particles) and solid particle effect was considered as erosion corrosion (the combined effect of erosion and corrosion). So the effect of erosion on corrosion was obtained by subtracting the corrosion values from total material loss observed during erosion corrosion experiments. With the introduction of solid particles, surface roughness was increased and that accelerated the localized attack. There was a close relationship between contact force exerted by the particles on the surface and predicted material degradation rate as reported elsewhere [24,25].

3.5. Wear Scar Penetration Depths Using Optical Profilometer

Figure 9 shows the average penetration depth of wear scars at the impingement point (high-velocity region close to impingement points) when the experiments were conducted at 12 m/s at five different angles for a duration of 24 h. These regions (impingement points) were selected expecting that the effect of shear and/or normal impact stress will be the highest at these points. Typical 3D image analysis of wear scars and their penetration depths were recorded using A GTK-A 3D optical profilometer from Bruker Co. (Billerica, MA, USA) as shown in Figure 10.

It is clear from Figure 9, that with an increase in impingement angle, corrosion (without solid particles) rate was increased and that was quite obvious with an observed increase in penetration depth as well (Figure 10). The maximum depth was observed at 45° and it was found to be 36 μm ± 5. As explained earlier, the maximum depth was expected at 45° due to the balance between shear stress and normal stress as is also seen in Figure 10A. However, in the case of erosion corrosion (with solid particles), the particle ploughed deeper and the maximum depth of 51 μm ± 5 was observed at 45°. At 90° angle particles strike with their maximum kinetic energy and activate the surface by extrusion which results in platelets. These platelets later fractured and removed by the subsequent impact of the particles as shown in Figure 10B. The 3D image of Figure 10B shows that the affected area exactly at impingement point for angle 90° was also eroded and recessed scars were wider with maximum depth of 35 μm ± 3.

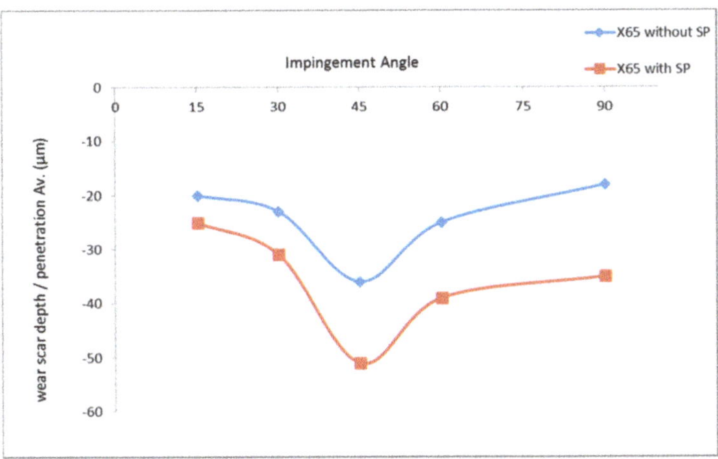

Figure 9. Variation of wear scar penetration depth (μm) during corrosion and erosion corrosion experiments at different angles.

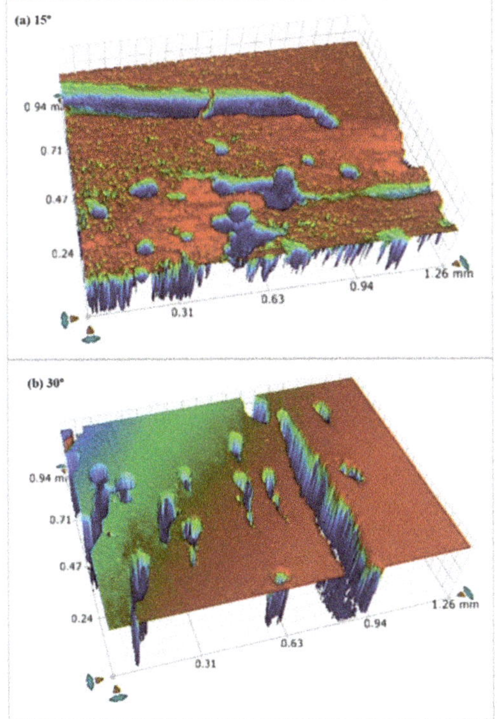

Figure 10. *Cont.*

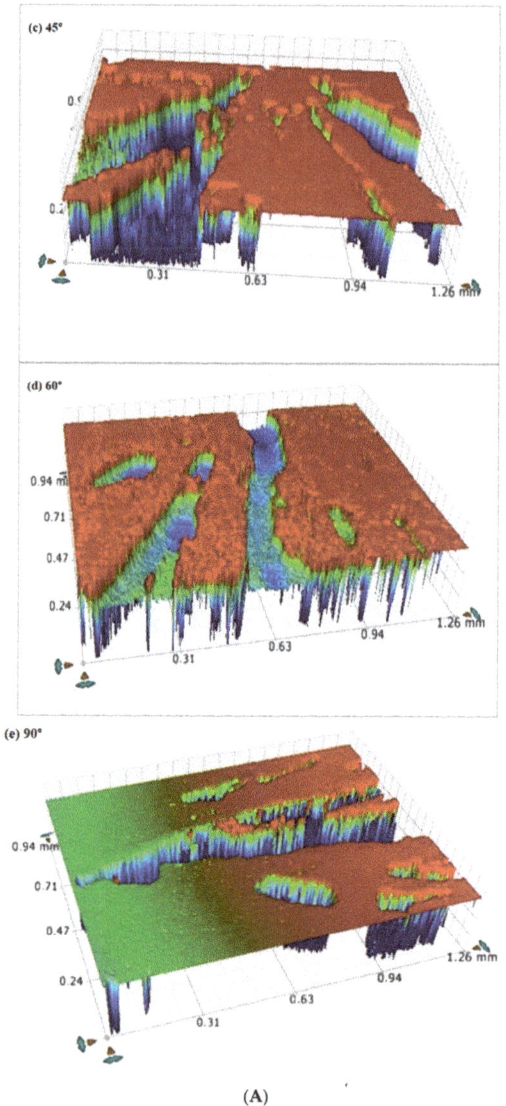

Figure 10. *Cont.*

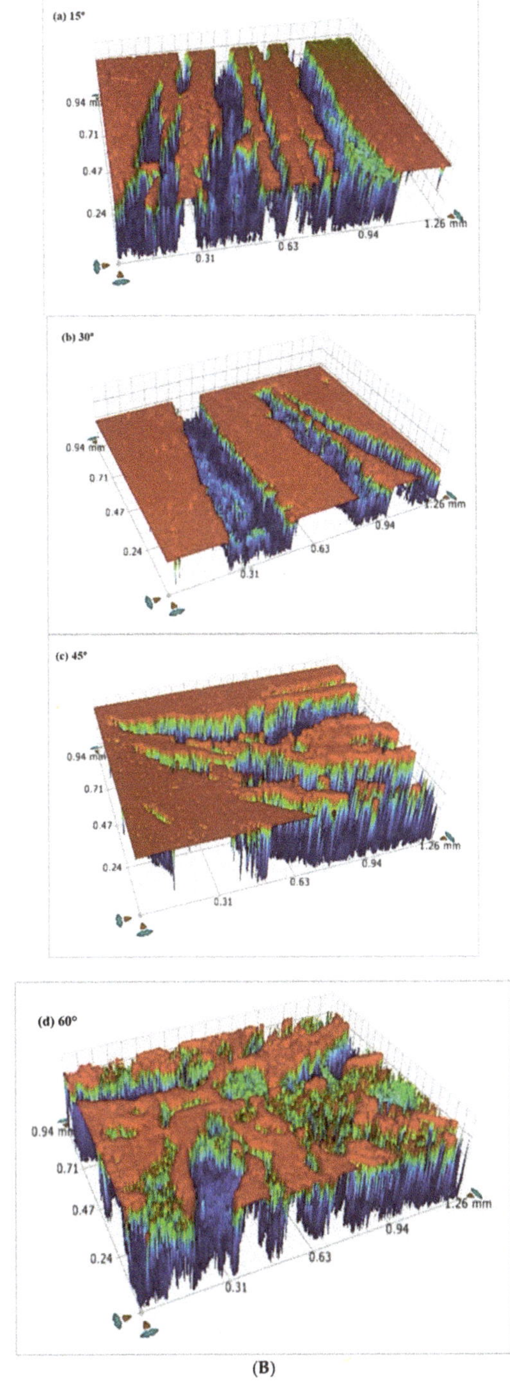

Figure 10. (**A**) 3D images obtained using an optical profilometer at high velocity regions near the impingement points in the absence of solid particles; (**B**) 3D images obtained using an optical profilometer at high velocity regions near the impingement points in presence of solid particles.

4. Conclusions

An experimental study was performed using a Flow loop to investigate the corrosion and erosion-corrosion behavior of API 5L-X65 at five different impingement angles (15°, 30°, 45°, 60° and 90°) and three different jet velocities, i.e., 3 m/s, 6 m/s and 12 m/s at each angle. The eroded surfaces and wear scars were examined to determine surface morphology/penetration depths. The following conclusions can be drawn from this work;

1. The maximum corrosion and erosion corrosion rates were observed at 45° angle due to a balance between shearing force and normal impact force. However the synergistic effect at 45° and 90° angle was more significant as compared to other angles.
2. The effect of erosion on corrosion is quite significant as striking solid particles cut the surface and activate the localised sites and thus accelerate the corrosion damage.
3. Corrosion and erosion corrosion rate was increased with an increase in the impingement velocity. This was mainly due to high shear and normal impact stresses in the absence of solid particles and due to the high kinetic energy in the presence of solid particles causing more mass loss.
4. Ploughing, elongated erosion scars, and metal cutting were the dominant erosion corrosion mechanisms at oblique angles, whereas extrusion, flattening of ridges and fracture were dominant at high angles.
5. The corrosion product or oxide layer was not stable under these severe conditions of high velocity fluid impingement which resulted in more material loss.

Author Contributions: I.U.T., H.M.B. and H.M.I. designed the detailed experimental plan for this research. H.M.I. carried out the experiments and data analysis under the supervision of I.U.T. and H.M.B. M.A.S. and H.M.I. designed the 3D optical profilometric studies and subsequent data analysis. I.U.T. and H.M.I. prepared the manuscript with help of other coauthors.

Acknowledgments: The authors gratefully acknowledge the financial support provided by King Fahd University of petroleum & Minerals (KFUPM) Saudi Arabia, under the research grant# IN141022 in conducting this research.

Conflicts of Interest: The authors declare no conflicts of interest.

References

1. Fontana, M.G. *Corrosion Engineering*; Tata McGraw-Hill Education: New York, NY, USA, 2005.
2. Levy, A.V. *Solid Particle Erosion and Erosion-Corrosion of Materials*; ASM International: Geauga County, OH, USA, 1995.
3. Heidersbach, R. *Metallurgy and Corrosion Control in Oil and Gas Production*; John Wiley & Sons: Hoboken, NJ, USA, 2010.
4. Pasha, A.; Ghasemi, H.; Neshati, J. Synergistic Erosion–Corrosion Behavior of X-65 Carbon Steel at Various Impingement Angles. *J. Tribol.* **2017**, *139*, 011105. [CrossRef]
5. Neville, A.; Reyes, M.; Xu, H. Examining corrosion effects and corrosion/erosion interactions on metallic materials in aqueous slurries. *Tribol. Int.* **2002**, *35*, 643–650. [CrossRef]
6. Javaherdashti, R.; Nwaoha, C.; Tan, H. *Corrosion and Materials in the Oil and Gas Industries*; CRC Press: Boca Raton, FL, USA, 2013.
7. Levy, A. Erosion and erosion-corrosion of metals. *Corrosion* **1995**, *51*, 872–883. [CrossRef]
8. Liang, G. Erosion-corrosion of carbon steel pipes in oil sands slurry studied by weight-loss testing and CFD simulation. *J. Mater. Eng. Perform.* **2013**, *22*, 3043–3048. [CrossRef]
9. Mazumder, Q.H. Prediction of Erosion due to solid particle Impact in Single-Phase and Multiphase Flows. *J. Press. Vessel Technol.* **2007**, *129*, 576–582. [CrossRef]
10. Giourntas, L.; Hodgkiess, T.; Galloway, A. Enhanced approach of assessing the corrosive wear of engineering materials under impingement. *Wear* **2015**, *338*, 155–163. [CrossRef]
11. Sasaki, K.; Burstein, G. Erosion-corrosion of stainless steel under impingement by a fluid jet. *Corros. Sci.* **2007**, *49*, 92–102. [CrossRef]

12. Khan, M.I.; Yasmin, T. Erosion-Corrosion of Low Carbon (AISI 1008 Steel) Ring Gasket Under Dynamic High Pressure CO2 Environment. *J. Fail. Anal. Prev.* **2014**, *14*, 537–548. [CrossRef]

13. Barker, R.; Hu, X.; Neville, A.; Cushnaghan, S. Flow-induced corrosion and erosion-corrosion assessment of carbon steel pipework in oil and gas production. *Corrosion* **2013**, *69*, 193–203. [CrossRef]

14. Neville, A.; Hodgkiess, T.; Xu, H. An electrochemical and microstructural assessment of erosion—Corrosion of cast iron. *Wear* **1999**, *233*, 523–534. [CrossRef]

15. Neville, A.; Hodgkiess, T.; Dallas, J. A study of the erosion-corrosion behaviour of engineering steels for marine pumping applications. *Wear* **1995**, *186*, 497–507. [CrossRef]

16. Jingjun, L.; Yuzhen, L.; Xiaoyu, L. Numerical simulation for carbon steel flow-induced corrosion in high-velocity flow seawater. *Anti-Corros. Methods Mater.* **2008**, *55*, 66–72. [CrossRef]

17. Guanghong, Z.; Hongyan, D.; Yue, Z.; Nianlian, L. Corrosion-erosion wear behaviors of 13Cr24Mn0.44N stainless steel in saline-sand slurry. *Tribol. Int.* **2010**, *43*, 891–896. [CrossRef]

18. Lin, F.; Shao, H. Effect of impact velocity on slurry erosion and a new design of a slurry erosion tester. *Wear* **1991**, *143*, 231–240. [CrossRef]

19. Finnie, I.; Kabil, Y. On the formation of surface ripples during erosion. *Wear* **1965**, *8*, 60–69. [CrossRef]

20. Wong, C.Y. Predicting the material loss around a hole due to sand erosion. *Wear* **2012**, *276*, 1–15. [CrossRef]

21. Sundararajan, G.; Shewmon, P. A new model for the erosion of metals at normal incidence. *Wear* **1983**, *84*, 237–258. [CrossRef]

22. Hutchings, I. A model for the erosion of metals by spherical at normal incidence. *Wear* **1981**, *70*, 269–281. [CrossRef]

23. Chen, X.; McLaury, B.S.; Shirazi, S.A. Application and experimental validation of a computational fluid dynamics (CFD) based erosion prediction model in elbows and plugged tees. *Comput. Fluids* **2004**, *33*, 1251–1272. [CrossRef]

24. Nguyen, B.Q.; Nguyen, V.B.; Lim, C.Y.; Gupta, L.M. Effect of impact angle and testing time on erosion of stainless steel at higher velocities. *Wear* **2014**, *321*, 87–93. [CrossRef]

25. Ige, O.; Umoru, L. Effects of shear stress on the erosion-corrosion behaviour of X-65 carbon steel: A combined mass-loss and profilometry study. *Tribol. Int.* **2016**, *94*, 155–164. [CrossRef]

© 2018 by the authors. Licensee MDPI, Basel, Switzerland. This article is an open access article distributed under the terms and conditions of the Creative Commons Attribution (CC BY) license (http://creativecommons.org/licenses/by/4.0/).

 *metals* 

Article

Crack Propagation Mechanisms for Creep Fatigue: A Consolidated Explanation of Fundamental Behaviours from Initiation to Failure

Dan Liu and Dirk John Pons *

Department of Mechanical Engineering, University of Canterbury, Christchurch 8041, New Zealand;
dan.liu@pg.canterbury.ac.nz
* Correspondence: dirk.pons@canterbury.ac.nz; Tel.: +64-3369-5826

Received: 10 July 2018; Accepted: 2 August 2018; Published: 8 August 2018

Abstract: *Background*—Creep-fatigue damage is generally identified as the combined effect of fatigue and creep. This behaviour is macroscopically described by crack growth, wherein fatigue and creep follow different principles. *Need*—Although the literature contains many studies that explore the crack-growth path, there is a lack of clear models to link these disparate findings and to explain the possible mechanisms at a grain-based level for crack growth from crack initiation, through the steady stage (this is particularly challenging), ending in structural failure. *Method*—Finite element (FE) methods were used to provide a quantitative validation of the grain-size effect and the failure principles for fatigue and creep. Thereafter, a microstructural conceptual framework for the three stages of crack growth was developed by integrating existing crack-growth microstructural observations for fatigue and creep. Specifically, the crack propagation is based on existing mechanisms of plastic blunting and diffusion creep. *Results*—Fatigue and creep effects are treated separately due to their different damage principles. The possible grain-boundary behaviours, such as the mismatch behaviour at grain boundary due to creep deformation, are included. The framework illustrates the possible situations for crack propagation at a grain-based level, particularly the situation in which the crack encounters the grain boundary. *Originality*—The framework is consistent with the various creep and fatigue microstructure observations in the literature, but goes further by integrating these together into a logically consistent framework that describes the overall failure process at the microstructural level.

Keywords: creep fatigue; crack growth; grain boundary

---

1. Introduction

The process of fatigue failure is physically and macroscopically described by crack-growth behaviour. The situation is complicated when creep is also present, because this provides an additional mechanism for plastic deformation and crack growth.

In the case of creep-fatigue, an engineering structure fails when the crack length achieves a critical value. The total damage is the accumulated effect of cycle count, temperature, and period (frequency).

Some of the ways this has been modelled is to prove a mathematical representation of the various components of facture mechanics [1–3] or to apply curve fitting [4–6]. However, this is not the present topic. Here we are more interested in the progression of crack growth at the microstructural level [7–10], specifically the way the crack navigates through the grains and along grain boundaries under the combined effects of cyclic loading and creep. Existing works do not provide an integrative treatment of this process.

This paper develops a microstructural theory for crack growth. It does so by adapting existing mechanisms of plastic blunting and diffusion creep into an integrative conceptual framework.

This provides a logically consistent framework that describes the creep-fatigue failure in terms of microstructural damage.

2. Background Literature

2.1. Phases of Crack Growth

It is generally accepted that the process of crack growth under cyclic loading is divided into three phases (Figure 1) [8]: crack initiation, crack propagation, and structural fracture. Specifically, the first stage presents a threshold, below which there is no crack growth. The second stage shows a relatively steady state, in which the crack growth rate increases steadily with the increasing number of cycles (this is always described by Paris' Law [11]). The final stage represents an unstable situation, where the engineering structure fails within a small number of cycles.

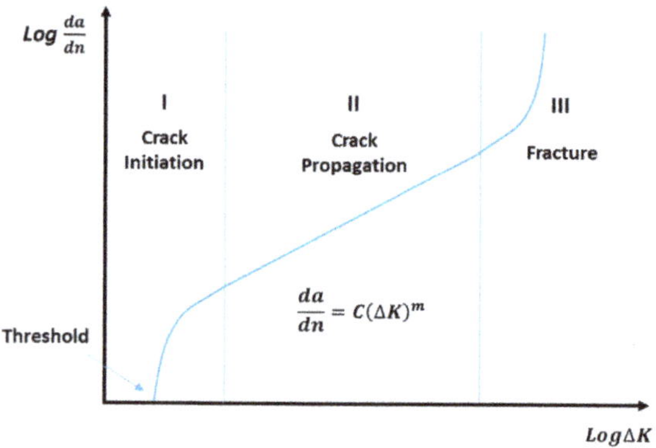

Figure 1. Three stages of crack growth.

2.2. Established Principles in the Literature

Crack growth caused by fatigue effect is attributed to plastic deformation at a crack-tip plastic zone, wherein the plastic deformation occurs via dislocations [7]. Ordinarily there is a preferred plane, and in that plane there are specific directions along which dislocation motion occurs; the slip plane. The axial (tension or compression) loading is decomposed into a normal stress, which is perpendicular to the slip plane, and a shear stress, which is parallel to the slip plane. This shear stress then results in the movements along the slip plane. This implies that shear stress contributes to plastic dislocation, and then results in crack growth. In this case, the axial loading is transformed into the shear stress to produce fatigue propagation.

The simplest form of dislocation is edge dislocation, see Figure 2, which represents a simplistic grain unit. Specifically, the dislocation starts from the extra half-plane of atoms (plane I), which is identified as a defect. When a shear stress is applied, and its magnitude is sufficient, the bond between atom A and atom B is cut. Then, the bottom of all planes move to the right, and atom B is bonded with its neighbouring atom (atom C). In this case, a new half-plane of atoms (plane II) is generated. Under the shear stress, this irreversible process is successively and repeatedly presented, with the final result being the creation of a step at the edge of this unit. The process is irreversible because the other atoms in the grain also move into new positions (not shown in the figure) due to other disturbances, hence the atoms do not spontaneously revert to their original positions (plastic deformation). Hence, the dislocation is an enduring feature of the grain. It can, however, be sent in a new direction should the orientation of the shear stress change, as occurs in reversed loading. By this means, the plastic deformation is produced due to the motion of numbers of dislocations. In addition, there may be atomic displacement perpendicular to the plane, hence causing a screw dislocation. In practical situations, the behaviours of edge and screw dislocations are mixed, and both contribute to the overall plastic deformation of the grain and ultimately of the macroscopic deformation of the part.

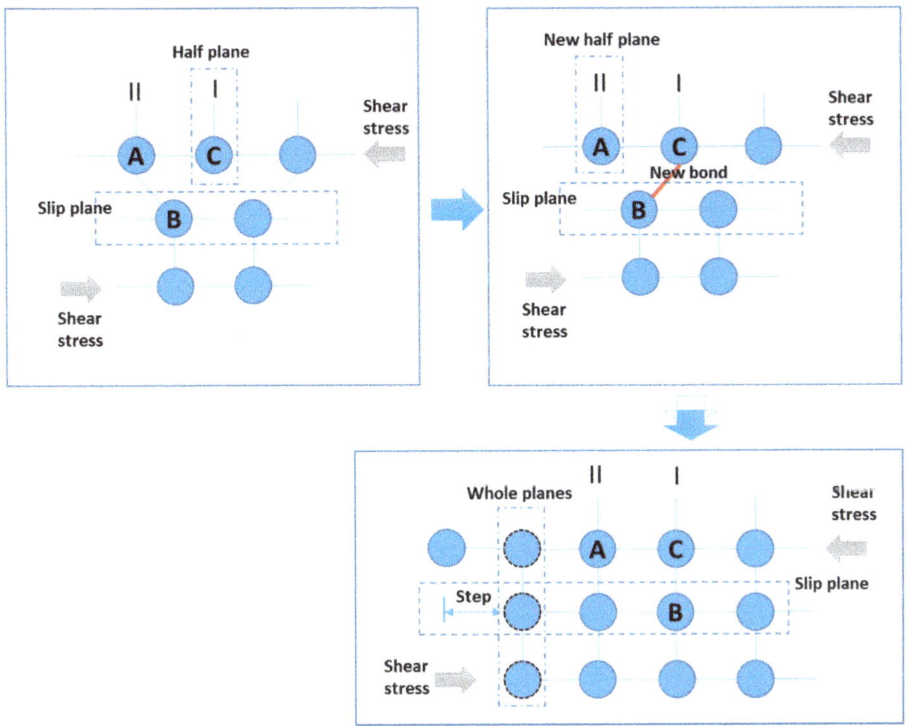

Figure 2. Edge dislocation process.

As mentioned above, the half-plane of atoms is regarded as the source of dislocation. The edge of this half plane is terminated within a grain; hence, a large number of dislocations are piled up, and then plastic deformation arises for the grain as a whole. This implies that the cracks caused by fatigue are initiated within the grains and then are propagated through the grains, causing the observed transgranular fracture phenomenon.

However, the creep mechanism gives a different crack-growth behaviour; crack growth along the grain boundary. In general, crack growth caused by creep is attributed to diffusion behaviour, which is presented as the movements of vacancies [12,13]. Under the creep situation, a constant applied loading leads to stress concentration at the triple points which are formed by the three adjacent grains (Figure 3). Normally, diffusion is a behaviour of atomic shifts from an area with high stress concentration to an area with low stress concentration. In this case, the atomic movements due to the diffusion mechanism results in the convergence of vacancies at the triple points, which then segregate the triple grain boundaries.

By this means, the micro-crack is initiated at the grain boundary. After the process of micro-crack initiation, the localised stress is highly concentrated at the crack tip, which then recruits more vacancies from the bulk of the grain, and these are diffused to the crack tip along the grain boundary. This leads to further propagation of the creep crack along the grain boundary. High temperature provides more favourable conditions for diffusion behaviour, hence advances the creep crack along grain boundaries. Thus, creep failure is strongly temperature dependent.

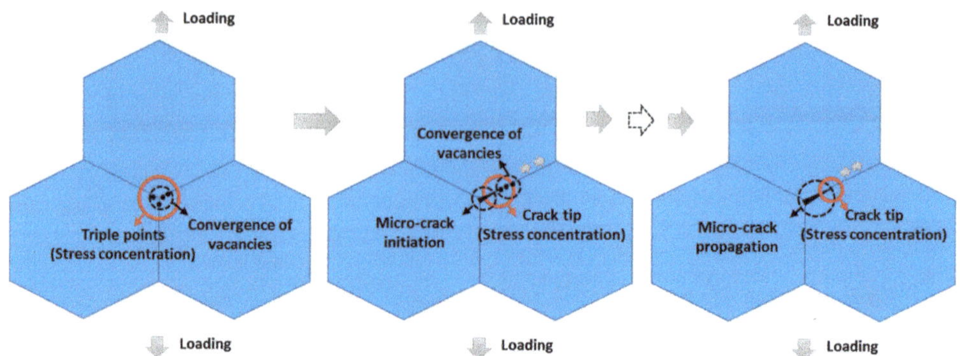

Figure 3. Crack growth by diffusion mechanism.

Overall, crack-growth behaviours of fatigue and creep follow different principles; specifically, fatigue effect occurs via cracks through the grains, while the creep effect involves fracture along the grain boundary.

Significant research effort has been exerted to explore the crack-growth path, through performing fatigue tests and observing the fracture surfaces (fractography). These research efforts mainly focus on the specific features of crack growth for fatigue or creep, such as the creep damage caused by triple points [14–16] and grain-boundary effects for fatigue-crack growth [17,18].

In summary, the literature proposes multiple mechanisms at the microstructural level to describe the crack-growth process.

Stage I: Crack initiation

(a) Dislocation behaviour [19] is a key mechanism at crack initiation. This effect is widely accepted [19]. This effect causes the slip planes to extrude or intrude, which results in stress concentration at the surface and then leads to crack nucleation.

(b) The strain energy release rate [20,21] determines the crack-growth rate. This mechanism underpins the unstable stage after the threshold of crack initiation. In this stage, increased energy accelerates the increase of crack-growth rate at the beginning phase, which is then retarded due to reduced energy and finally achieves a stable state.

(c) The shear effects [22,23] provide opportunistic directions for crack growth. This mechanism involves the direction oriented at 45° providing a more favourable condition for crack growth at the initial phase of stage I.

Stage II: Crack propagation

(d) The plastic blunting process [24–27] has asymmetrical effects in the tensile and compressive loading cycles. This mechanism describes crack-growth behaviour in one loading cycle. The tensile loading blunts the crack tip, and the new crack surface is created due to a shearing effect. Then, when the loading is reversed, the surface created under tensile loading remains (crack extension) because of a crushing effect.

(e) The crack tip plastic zone [28–30] enhances crack growth. This mechanism involves a highly localised stress around the crack tip. This, on the one hand, results in plastic deformation (crack opening) at the crack tip; on the other hand, it gives a more favourable situation for diffusion.

(f) Diffusion creep [12,13] provides a mechanism that leads to grain elongation and then further crack opening. In addition, since atoms diffuse from a high- to a low-concentration region, more vacancies are generated and converged at the crack tip, where highly localised stress is presented, which provides a more favourable situation for creep damage.

(g)   Existing precipitates [31] cause fatigue resistance. The mechanism is mainly that precipitates are the obstacle to dislocation, and hence restrain the process of crack propagation.

(h)   Crack deflection (change of direction) [32] preferentially occurs at the grain boundary. This is determined by the twist angle and the tile angle.

(i)   Triple-points [14–16] provide a mechanism for significant stress concentration, which provides a more favourable stress field for creep damage.

(j)   A region with a high density of micro-cracks is particularly weak, and supports crack-branching activities [33].

(k)   The slip bands within grains [34] may cause the crack to re-direct within grains.

Stage III: Structural failure

(l)   The plastic energy [23] available exceeds the need for producing the new crack surface. The excess energy is applied to form voids, and thus further worsens creep-fatigue resistance.

2.3. Gaps in the Body of Knowledge

Although the existing literature [22,23] indicates three stages of crack growth from initiation to failure, the implications of this for the behaviour of the crack at the grain-boundary effect is unclear. The grain boundary is known to play an important role in both fatigue-crack and creep-crack growth in different ways; hence, a grain-boundary effect cannot be ignored. Despite the apparent comprehensiveness of the above list of mechanisms, the literature contains many disparate findings without an obvious integration into a holistic theory for the crack propagation process.

Hence, there is a need to develop a framework to link these details together, to propose a coherent set of mechanisms for the steady stage of crack growth at the grain level, and to involve grain-boundary effects. This ideally needs to consider all the loading effects (fatigue, creep, and creep fatigue) at the microstructural level.

3. Approach

The approach in the present work was to start with existing principles of fatigue-crack and creep-crack growth (Section 2.2), and then conceptually develop a microstructural model of the creep-fatigue crack-growth process. This was achieved by imagining the process from the perspective of atoms in orderly grains and at the grain boundaries (where the orderly arrangement between atoms breaks down). The bonds between atoms inside a grain are strong, relative to the bonds between atoms at grain boundaries. We sought to identify a coherent set of mechanisms that would affect the bonds between atoms, for all the loading regimes. We integrated multiple existing concepts at the microstructural level, such as the plastic blunting process, the mechanisms of dislocation, and the diffusion creep process. These processes are extant in the literature, and our contribution is to integrate and apply them to the fine scale to give a new understanding of the interaction between the underlying mechanisms. In addition, a number of new mechanisms for crack growth were also proposed to construct a comprehensive conceptual framework.

Since the principles of crack growth of fatigue and creep are different (the fatigue effect occurs via cracks through the grains, while the creep effect involves the grain boundary cracking), the fatigue and creep crack-growth behaviours were discussed and illustrated separately. In this microstructural conceptual framework, three stages of crack growth from crack initiation to structural fracture were included. The second stage of steady crack growth is the primary focus. This is where the majority of the lifetime is consumed. It also presents multiple situations regarding the position of the crack relative to the grain boundary. The result is a proposed theory of the failure mechanisms, and this is represented in the form of a schematic representation, see Section 4.

4. Results

The microstructural conceptual framework mainly focuses on the second stage of crack growth. On the one hand, this is because this stage presents a stable process of crack growth and is numerically described by the well-known Paris' Law [11]. On the other hand, this is because the majority of the lifetime is consumed at the second stage of crack growth for low cycle fatigue, and this fatigue regime is an important consideration when the creep effect is activated during the design process [8]. In this framework, the crack-growth behaviours under the tensile partition and compressional partition are discussed separately, and crack-growth behaviours caused by the fatigue effect and creep effect are described separately.

4.1. New Propositions

The challenge is showing how these multiple premises are integrated together at the microstructural level. There are no models and conceptual frameworks in the literature that provide such integration. To solve this problem, we propose several new effects. Our propositions of mechanisms that operate at the microstructural level during crack growth are:

A   That **energy dynamics** explain the crack initiation and the following unstable phase, and is based on a liberation of energy due to coalescence of dislocations. In general, when the internal energy stored in the structure due to cyclic loading arrives at a critical value, the barrier to initiate the crack is overwhelmed. After this, the released energy is gradually consumed to produce new crack surfaces at an accelerated crack growth rate.

B   That **grain-mismatch** occurs due to grain elongation under the tensile loading, resulting in relative movement between two neighbouring grains. Then, the shear stress along the grain boundary is increased and a weaker region along the grain boundary is created. This results in a mismatch band at the grain boundary. This widens the crack body and enhances its growth.

C   That **bonding crushing effects** exist at the finer scale. During the process of compression, the atomic bonds, which are distorted and rearranged in the tensile phase, may be further damaged and become potential failure sites for next loading cycle. This process is irreversible, since the atoms cannot return to their original position under the compressional loading.

D   That a **crack net** is caused by the aggregation of micro-cracks. This crack net probes a larger volume of material for weaknesses, and then promotes the main crack.

E   That **irregular configuration of the grain boundary** causes stress/strain pile up at the grain boundary, and then results in a large driving force for extending the crack tip into the neighbouring grain.

Furthermore, it is proposed that these mechanisms are integrated by the following assumptions:

F   That the **primary mechanism** for failure in the creep-fatigue loading regime is crack growth by mechanisms of crack blunting (including shearing and crushing); hence, fatigue effects.

G   That the **supporting mechanisms** that augment the extent of damage are grain elongation and diffusion; hence, creep effects.

We formulated these propositions as part of an iterative process of seeking a consolidated framework. Hence, the above are not so much lemmas that were asserted at the outset, but rather principles that were acquired during the framework development.

The resulting conceptual framework, shown below, is a fusion of the principles (Section 2.2) and propositions (Section 4.1) into a more extensive conceptual framework.

4.2. Conceptual Framework of Temporal Development of Crack-Growth for Creep Fatigue

4.2.1. Stage I: Crack Initiation

The first stage (stage I) is that of crack initiation. We summarise the crack initiation mechanics as follows. Fundamentally, crack initiation is caused by stress concentration, where two situations are presented. A typical situation is that the crack starts at a surface defect such as a machining mark. This provides a highly localised stress concentration, hence a pre-existing micro-crack [8]. By this means, a crack is initiated at this specific point, and then the localised plastic strain caused by the ductile micro-tearing events provides opportunity for further propagation.

Second, for a situation with perfect surfaces (defect-free), dislocations play an important role [19] for the initiation of a fatigue crack [35] (Figure 1.9 therein), [36] (Figure 1 therein). During this process, loading cycles cause dislocations to pile up at the microstructural level. These dislocations are in the crystal lattice at the nanoscale. Under loading, they align and coalesce to produce thicker slip planes through the crystal and eventually slip bands at the macrostructural level [37]. Under cyclic loading, there is more relative movement between the bands, so the effect becomes more pronounced—the bands are further displaced. We suggest that this is due to irreversibility caused by the localised hardening effects of the dislocations.

In the area of persistent slip bands, the slip planes extrude or intrude to the surface of the object, see Figure 4. This results in tiny steps in the surface, where the stress concentration then results in crack nucleation.

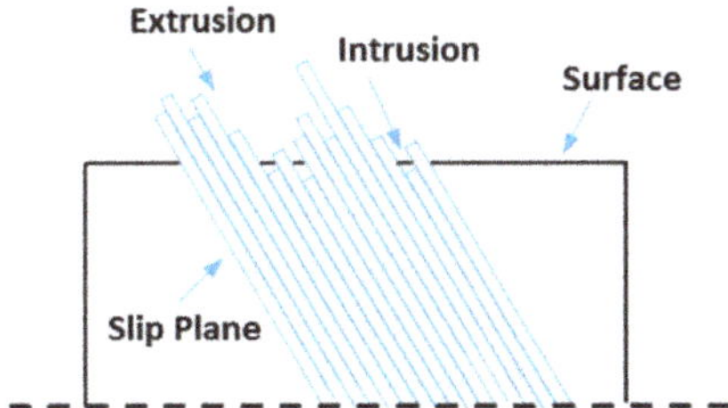

Figure 4. Crack initiation due to dislocation.

In stage I (the crack-initiation stage), a crack-growth threshold exists. In general, there are two types of thresholds [38]: microstructural and mechanical. The microstructural threshold describes the initiation of micro-cracks at different microstructural features, such as slip bounds and cavities. The mechanical threshold is related to macroscopic geometry for long crack growth, and is discussed in the present work. Below this latter threshold, there is ordinarily no crack growth [8]. We attribute this to the cracks being too short to result in a meaningful stress concentration effect (which is primarily an effect of proportional geometry). Also, if there are many superficial defects at the surface—which is usually the case—then the material does not have a geometrically exact boundary at the nanoscale, and hence the stress avoids this region: the many small defects effectively decrease the stiffness of the superficial layer, and hence this layer is to some extent unloaded.

After the threshold stress intensity is exceeded, there is a regime of high acceleration of crack-growth rate. Then, following this high-acceleration regime, a transition to a slower acceleration of crack-growth rate is presented.

On the one hand, this process is attributed to the concept of strain energy release rate, which describes the energy consumed per unit of newly created crack surface. This implies that the more strain energy is released, the more new crack surfaces are produced. Normally, the strain energy release rate could be numerically related to the M-integral [20,21]. The relationship between the number of cycles and M-integral was investigated by Margaritis [39] (Figure 7 therein), where the

M-integral increases with the increasing number of cycles at the early phase, then it decreases over the peak, and finally it reaches a constant situation. This trend is consistent with observation of the initiation stage shown in Figure 1. Specifically, after the threshold, increased energy (released strain energy) accelerates the increase of crack-growth rate at the beginning phase, then reduced energy retards the acceleration of the crack-growth rate, and finally a steadily increasing crack-growth rate is achieved under a situation with constant energy (this is viewed as the crack-propagation stage).

We suggest that the underlying energy dynamics are based on a rapid liberation of energy due to coalescence of strain. Specifically, the internal energy, which is stored in the structure in the form of the irreversible formation of dislocations that increase in number due to cyclic loading. Then, when this energy arrives at a critical value, the vacancies in the dislocations coalesce and form a crack—hence breaking through the barrier to initiate the crack. This qualitative leap from perfection to imperfection requires internal energy. This process is accomplished during a short time period by releasing large amounts of energy. After this, the released energy is consumed to produce new crack surfaces in a high acceleration of crack-growth rate, and the whole system is forced into an unsteady condition. Naturally, the need to re-balance this system gradually reduces the amount of released energy (caused by applied loading) until a balanced condition is reached. During this process, the high acceleration of crack-growth rate is gradually retarded, and then this acceleration reaches stabilisation (the whole system is re-balanced). The growth of the crack causes further large-scale deformation of the grain as a whole, which causes more dislocations to occur within the grain, and repeats the cycle.

An alternative explanation is that the unstable behaviour of crack growth in stage I is due to shear mechanisms [22,23]. Generally, a crack is propagated through two different methods: the plastic deformation around the crack tip and the shear-stress effect at the planes oriented at 45° to the loading direction. During the initial phase of crack growth, a small plastic zone and a small stress field are presented around the crack tip because of the small magnitude of stress intensity. In this case, the mechanism of plastic deformation around the crack tip may not be significant enough to become the driving force for crack growth. Instead, the shear stress at the planes oriented at 45° provides more favourable conditions for crack growth. This is because the shear stress at these planes and the relative movements between these planes under cyclic loading provide more vulnerable areas for crack growth. Prior damage and lattice imperfections may exist on these planes from previous loading cycles. In this case, a crack grows along the planes oriented at 45° with a minimum of effort, hence resulting in a high acceleration of the crack-growth rate.

As the stress intensifies, the shear-stress effect is gradually suppressed by the plastic-deformation mechanism; hence, the acceleration of crack-growth rate is reduced. This is because, during this process, the direction of the crack growth gradually deviates from the surface of the planes oriented at 45°, and then the behaviour of penetrating the grain boundary results in the deceleration of crack-growth rate. Finally, crack-growth behaviour is stably caused by the crack-tip plastic zone, and a steady situation is achieved (stage II is initiated).

We suggest that both mechanisms apply, and complement each other. The key elements (mechanisms) in the crack initial stage are presented by Figure 5.

We summarise the crack initiation mechanics as follows, see Figure 6. Crack nucleation starts at the surface due to highly localised stress concentration. In the case of smooth surfaces, it is attributed to the presence of persistent slip bands (extrusion and intrusion). Stage I (the stage of crack initiation) has a threshold value, below which no crack growth occurs. Once a crack is initiated, it grows quickly at the early phase, then the acceleration of crack-growth rate is reduced, and the crack-growth behaviour finally becomes steady (see next section). This process is proposed to be underpinned by the above mechanisms of strain energy release and shear stress.

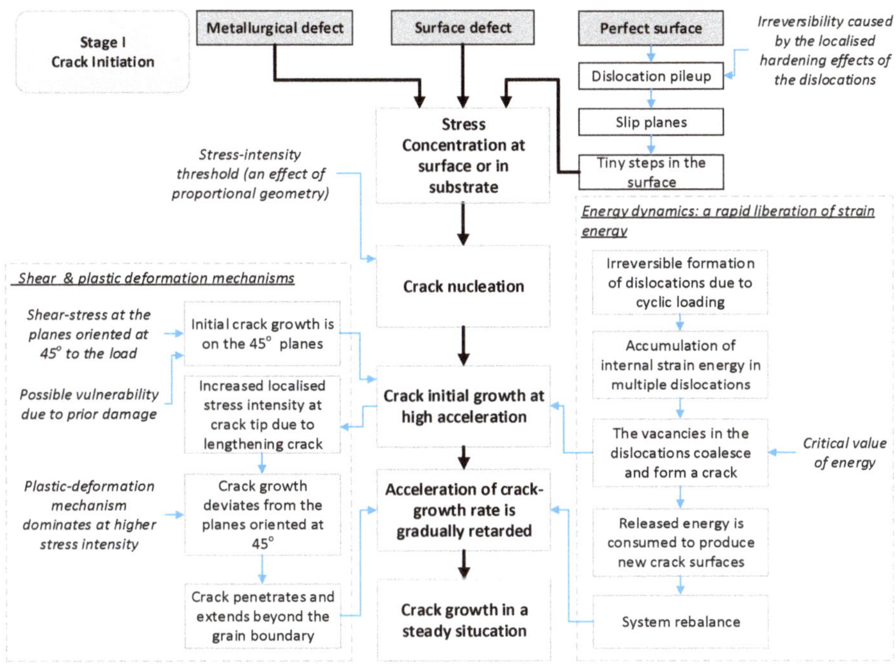

Figure 5. Key elements (mechanisms) in the crack initial stage.

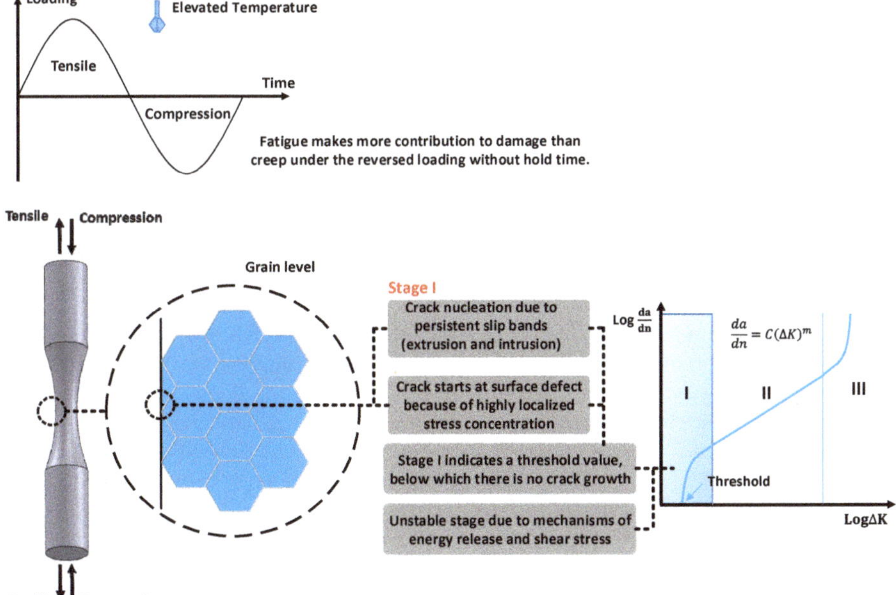

Figure 6. Stage I: Crack initiation.

4.2.2. Stage II: Crack Propagation

The second stage (stage II) shown in Figure 1 presents the behaviour of crack propagation, wherein the crack growth undergoes a relatively steady process. This stage is numerically presented by Paris' Law (Equation (1)) [11], which shows a power-law relationship between the crack growth rate and the range of the stress intensity factor during the fatigue cycle:

$$\frac{da}{dn} = C(\Delta K)^m \tag{1}$$

where $\frac{da}{dn}$ is the crack growth rate; ΔK is the effective stress intensify factor, which is identified as the difference between maximum and minimum stress intensify factors for one cycle; K is the stress intensify factor; a is the crack growth; n is the number of cycles; and C and m are constants.

The crack-growth process of creep fatigue is mainly discussed under the second stage of crack growth, and this discussion is based on the idea of the plastic blunting process [24–27].

At the beginning of the loading cycle, the crack tip is sharp (Figure 7a). Then, the applied tensile stress leads to crack opening, and the stress concentration at crack tip causes the slip along planes at 45° (Figure 7b). This leads to the gradual increase of crack-tip area, and finally the crack tip widens to its maximum plastic deformation (Figure 7c) due to the plastic shear effect. At the same time, the crack tip becomes blunt and grows longer because a new crack surface is created.

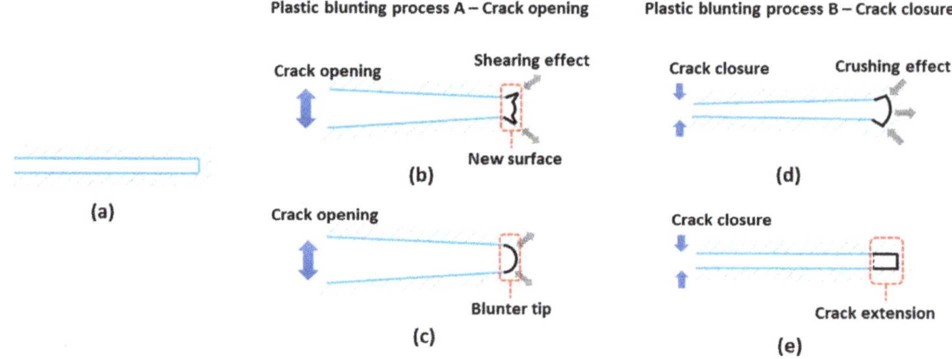

Figure 7. Plastic blunting process.

When the loading is changed to a compressional situation, the crack is closed, and the slip at the crack tip is reversed (Figure 7d). Finally, the crack surfaces are crushed together, and the new crack surface created under tensile loading remains (Figure 7e) (crack extension). Consequently, the sub-cracks promoted by each loading cycle are accumulated, and the structure fails as the total damage achieves a critical value.

The idea of plastic blunting process was integrated with a general understanding of creep mechanism to show crack-growth behaviour of creep fatigue. The process of crack propagation (Figures 8–19) is described (at a grain-based level) by three typical situations: (1) crack tip within a grain (Section 4.2.2.1); (2) crack tip at the grain boundary (Section 4.2.2.2 (1)) or at the triple point (Section 4.2.2.2 (2)); and (3) crack tip through grain boundary (Section 4.2.2.3).

In the present work, fatigue is identified as crack growth by *overloaded local atomic bonds*, and creep is regarded as the *flow of atoms relative to each other* with transfer of bonding to new partners.

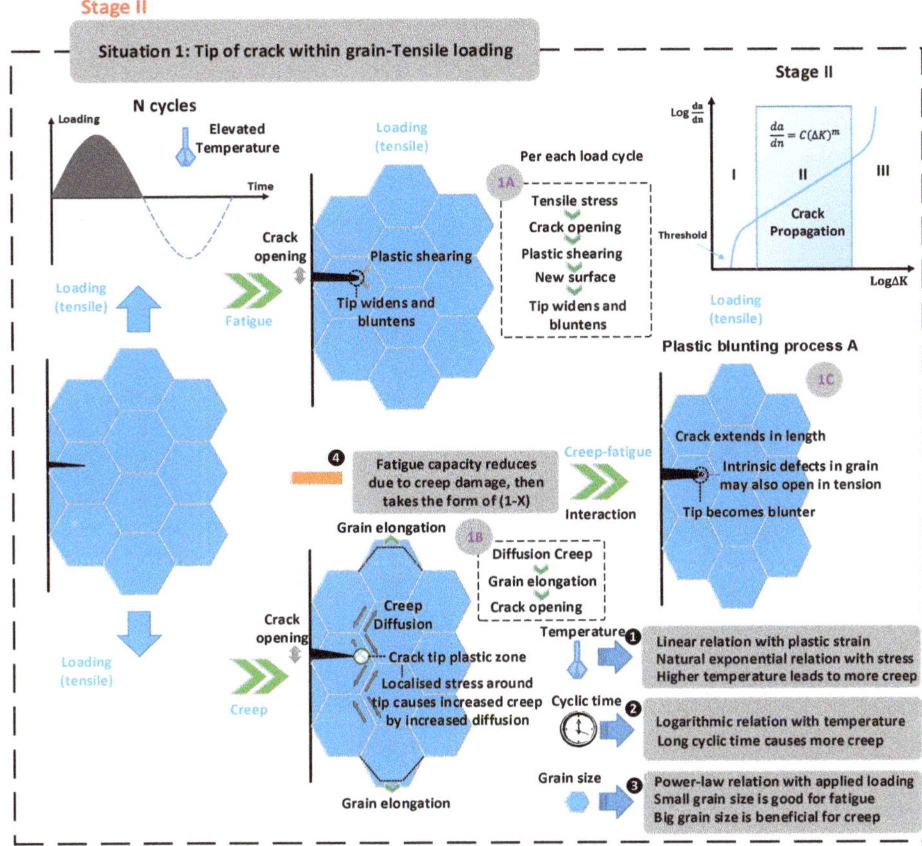

Figure 8. Crack tip within a grain—Tensile loading.

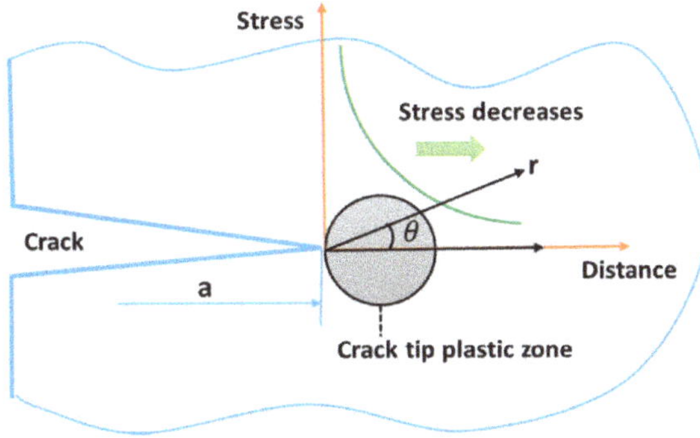

Figure 9. Stress around crack tip.

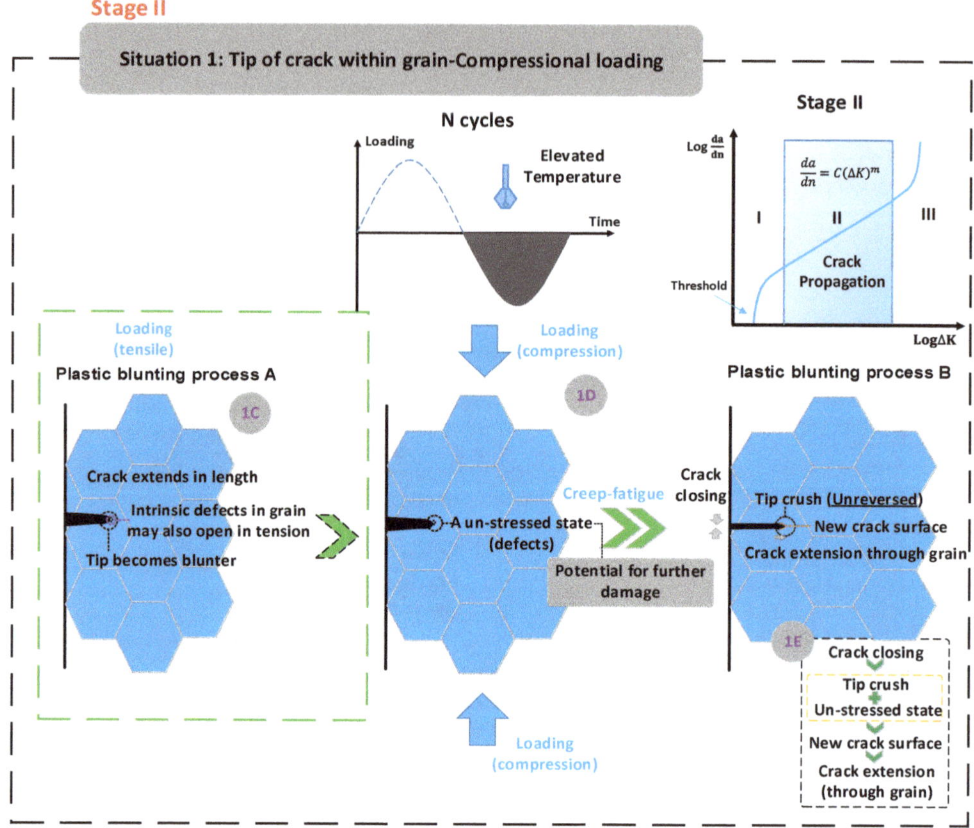

Figure 10. Crack tip within a grain—Compressional loading.

4.2.2.1. Crack Tip within a Grain

- Conceptual framework for crack growth in the tensile regime

Under tensile loading (Figure 8), the fatigue component follows the idea of the plastic blunting process. In this process (1A in Figure 8), tensile stress leads to crack opening, and shear behaviour is presented at the crack tip. Then, by the plastic shear effect, the new surface is created; meanwhile, the crack tip widens and becomes blunt (the blunting behaviour was illustrated in Figure 12 in Ref. [40] and Figure 32 in Ref. [41], where a stretch region is presented). Diffusion creep is identified as the main creep mechanism for creep-fatigue damage in the present work. It is generally believed that diffusion creep leads to grain elongation, which then results in a further opportunity for crack opening (1B in Figure 8).

In addition, diffusion creep has a dependency on applied loading, and larger loading produces more creep damage. Specifically, the idea of a crack tip plastic zone [28–30] indicates that the localised stress is highly significant around the crack tip, and gradually decreases in the direction of crack growth (Figure 9). In particular, under the theory of linear elasticity, stress distribution around the crack tip can be numerically presented as (Equation (2)):

$$\sigma_{ij} = \sigma\sqrt{\frac{a}{2r}}f(\theta) = \frac{K}{\sqrt{2\pi r}}f(\theta) \tag{2}$$

where σ_{ij} is the stress distribution, a is the crack length, K is the stress intensity factor, and r and θ are polar coordinates. This equation encapsulates a stress-concentration phenomenon around the crack tip. This relation is commonly accepted as being applicable to the creep-fatigue condition on the assumption that the crack-tip plastic zone is small. Higher localised stress around the crack tip causes more creep damage due to increased diffusion, and thus contributes to crack opening. Specifically, atoms normally diffuse from a region of high concentration to a region of low concentration. In this case, more vacancies are generated and converged at the crack tip, and then a more favourable situation for creep damage is presented.

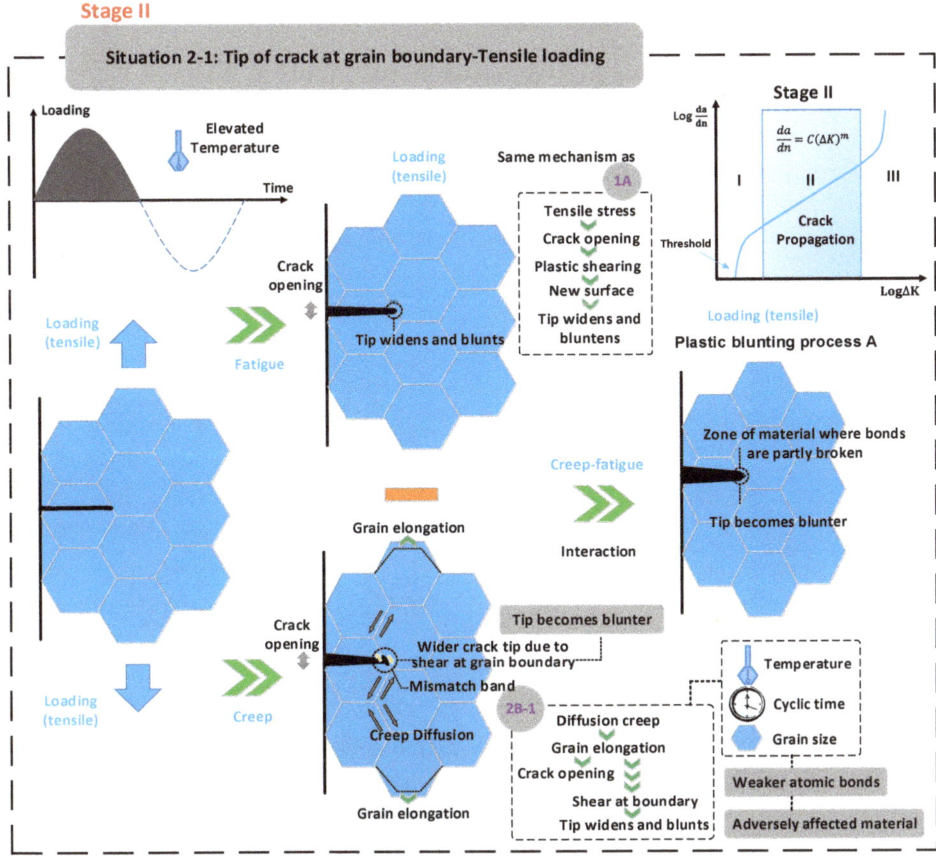

Figure 11. Crack tip at grain boundary—Tensile loading.

Consequently, the crack tip surface caused by the fatigue effect is further extended and becomes blunter due to the creep effect (1C in Figure 8), which implies the combined effects of fatigue and creep. A defected-related state, such as intrinsic defects, around the crack tip may also promote crack growth (negative to fatigue resistance), where the weakest zone of atomic bonds is represented. An example of presenting a defected-related effect is shown in Figure 32 from [41], where the crack is promoted due to void coalescence and the bridges between voids are sheared off. In addition, precipitates may also be considered to be defected-related effects, but ones which are normally beneficial to fatigue resistance. On the one hand, this benefit results from the obstacle to further dislocation [31], whereby the precipitates restrain the process of crack propagation. On the other hand, this benefit could also be explained by the fatigue-softening behaviour [42]. Generally, softening behaviour is caused by disordering the stacking sequence of precipitates due to reversed loading, whereby the crack encounters a tougher situation to penetrate this area. In addition, the fatigue-softening behaviour can also be attributed to the reduced area intercepted by the dislocation of the slip planes due to the gradual consumption of precipitates under the reversed loading (the precipitates are cut into smaller sizes under reversed loading, and thus are dissolved).

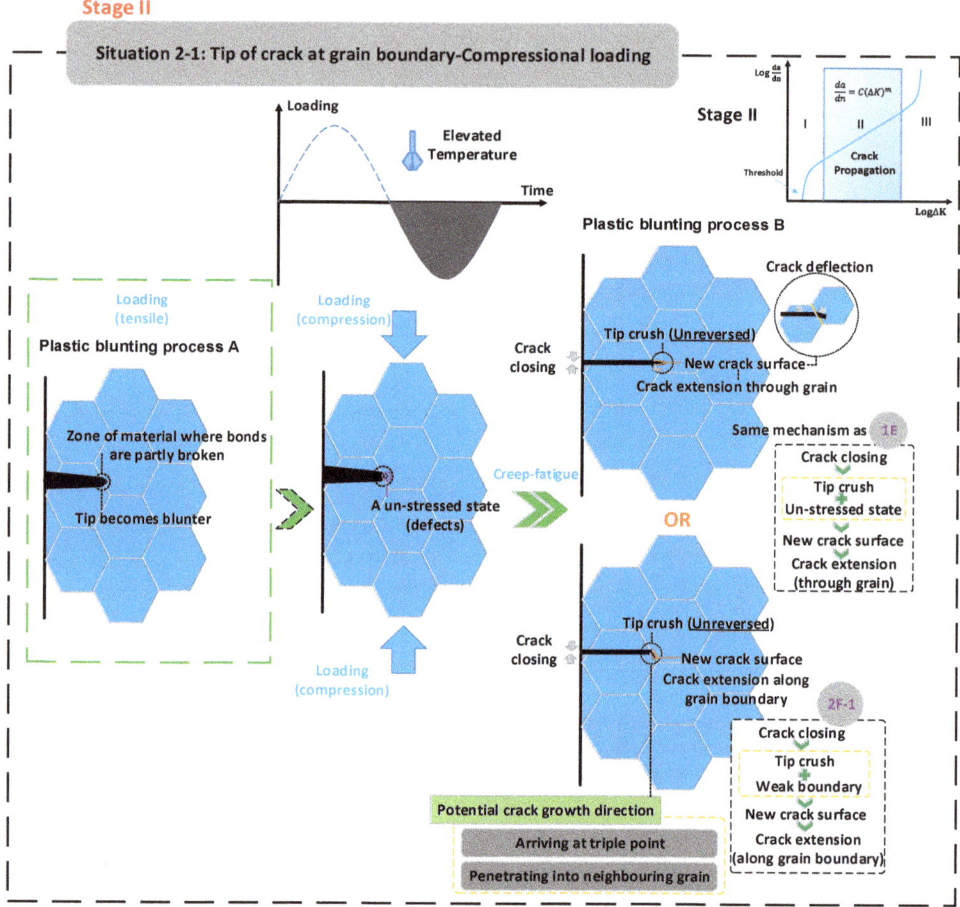

Figure 12. Crack tip at grain boundary—Compressional loading.

Creep effect has strong dependencies on temperature, cyclic time and grain size [43]. Generally, a higher temperature leads to more creep damage, and a linear relation between temperature and applied loading is presented (❶ in Figure 8). In addition, a longer cycle time gives more creep damage, and a logarithmic relation between cyclic time and temperature is presented (❷ in Figure 8). Furthermore, a small grain size is positive for fatigue while big grain size is beneficial for creep resistance, and a power-law relationship between grain size and applied loading is presented ❸ in Figure 8). Consequently, the negative creep effect caused by elevated temperature, or prolonged cyclic time, or smaller grain size gives weaker atomic bonds; then, this results in adversely affected material.

The concept of 'fatigue capacity' indicates that the total fatigue capacity is gradually consumed by creep effect. This process is numerically presented by taking the form of "1-X" (❹ in Figure 8) [43,44].

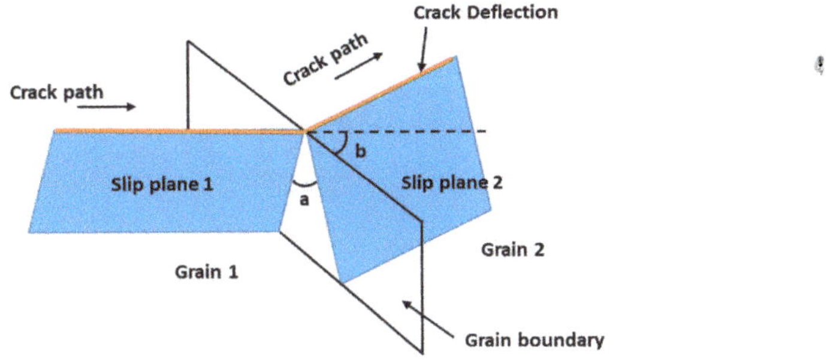

Figure 13. Crack deflection.

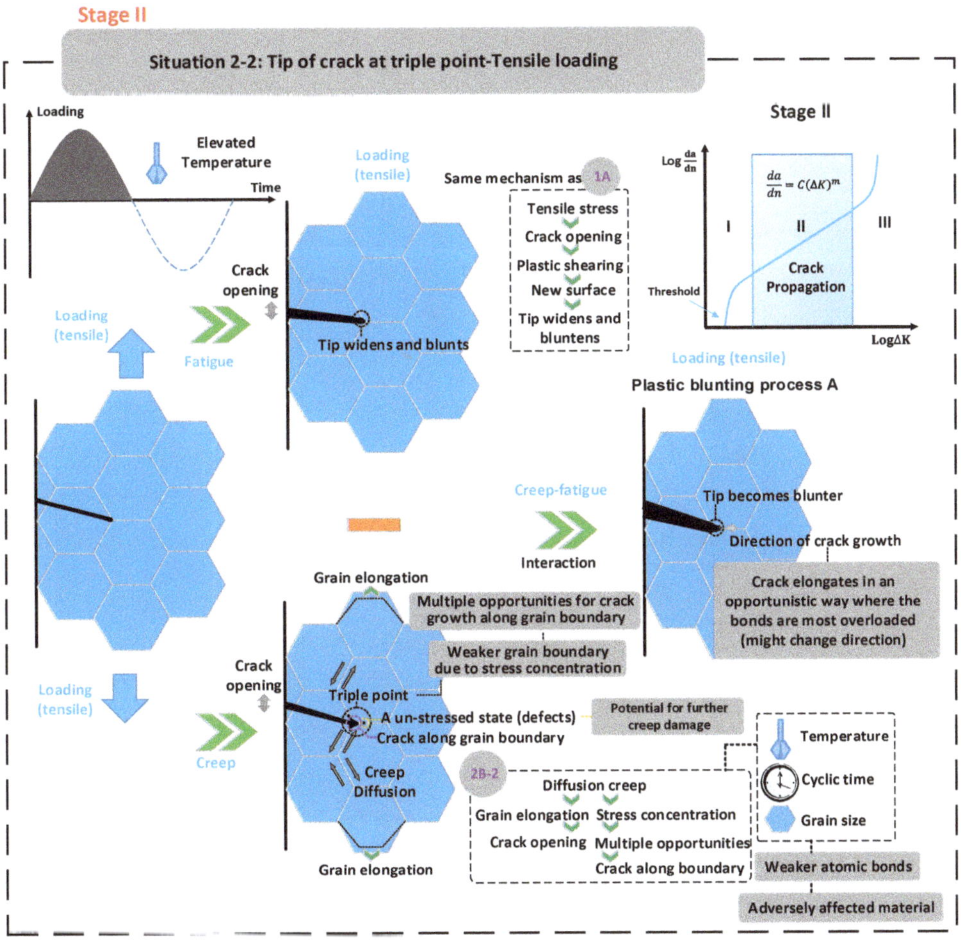

Figure 14. Crack tip at triple point—Tensile loading.

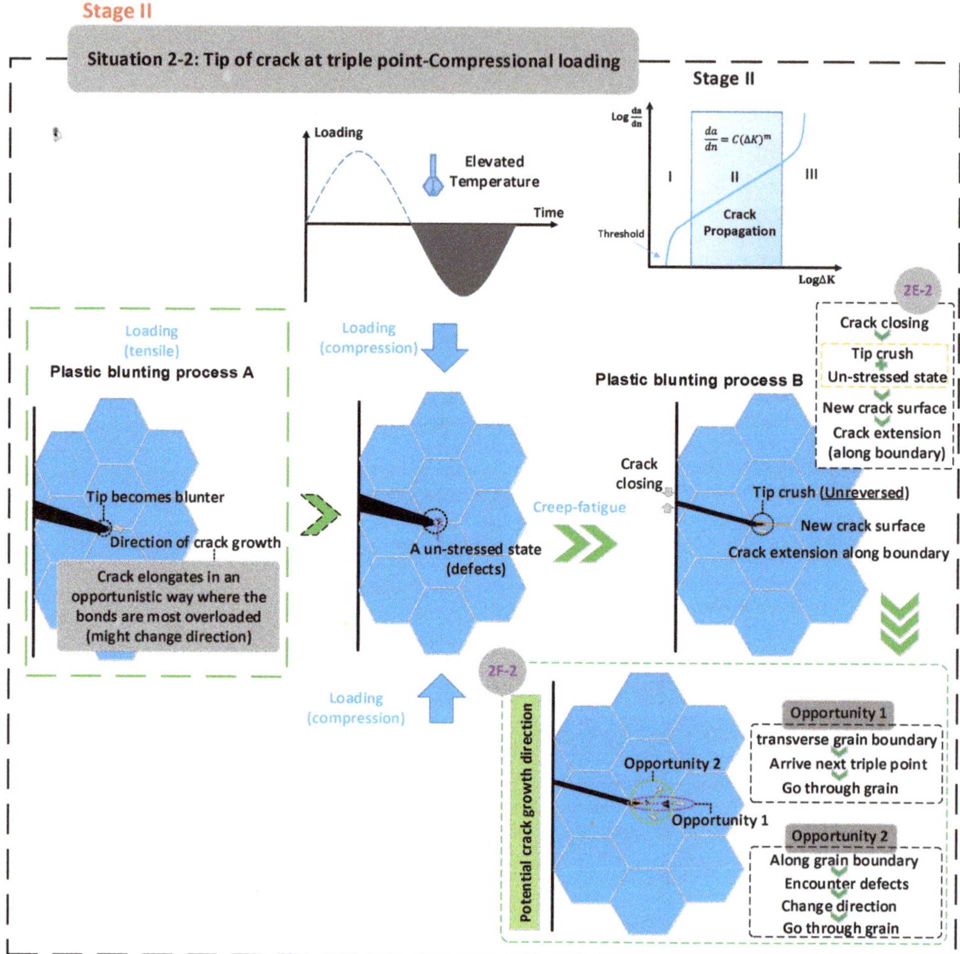

Figure 15. Crack tip at triple point—Compressional loading.

- Conceptual framework for crack growth in the compressional regime

Under compressional loading, crack closure is presented based on the idea of the plastic blunting process, see Figure 10. In this process, the crack surface, especially the tip zone, is crushed together. Since the crushing process is inelastic and irreversible, the new crack surface created under tensile loading remains (1E in Figure 10). This implies that compressional loading also contributes to crack growth.

This phenomenon may be explained by the observation of the crack tip opening displacements [45,46] under compressional loading, where the crack is not entirely closed in this loading regime due to the plastic effect (this is illustrated in Figure 3 of Ref. [45]).

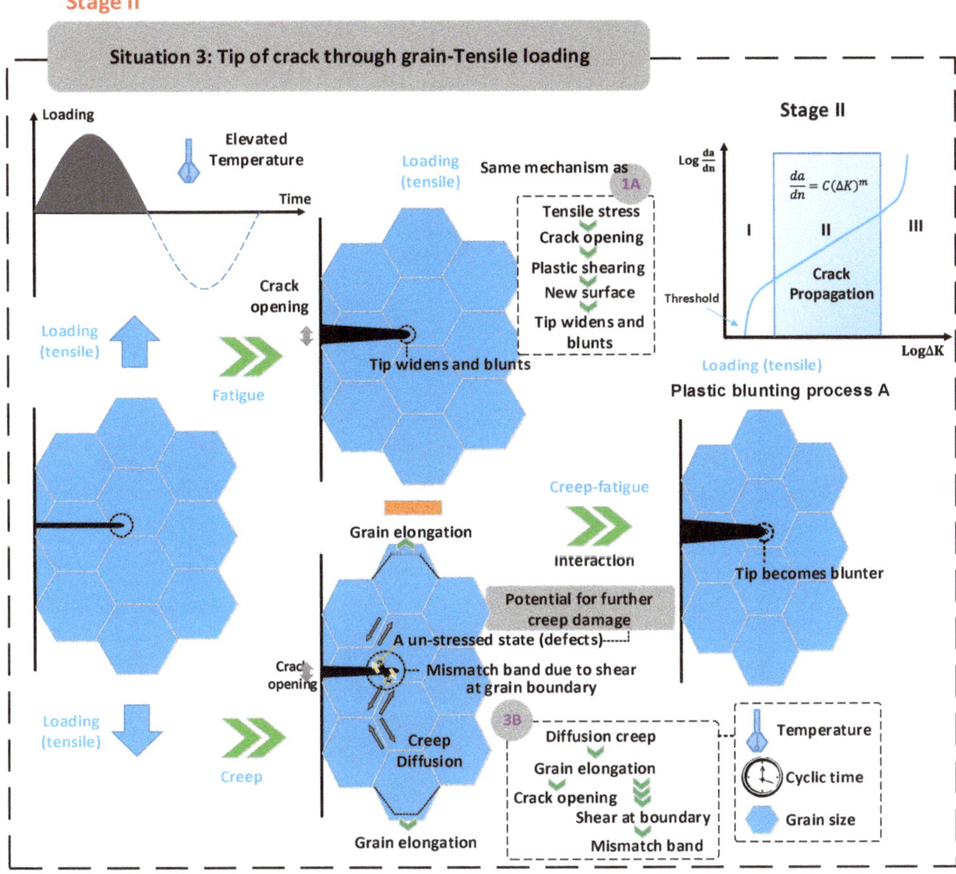

Figure 16. Crack tip through grain boundary—Tensile loading.

In addition, the contribution of compressional loading may also be explained by the behaviour of atomic bonds. Specifically, creep fatigue does its damage in the tensile phase by changing the landscape of atom arrangements and hence the bond vulnerability (which implies the bonds are extended and become weaker). During the process of compression, because of the crushing effect, some of these partly broken bonds may be further damaged and become potential failure sites for the next loading cycle, while others may be totally damaged and then lead to crack extension. This is significantly not a reversible process, since the atoms cannot return to their original positions after one tension-compression cycle. Furthermore, the defect-related state around the crack tip may further propagate cracks due to the crushing effect. Specifically, the crushing effect causes the crack tip to be squeezed, and then the tip is possibly extruded towards the zone with the weak atomic bonds. In this case, the crack is further extended (1D in Figure 10) by cutting these vulnerable bonds with the assistance of the crushing effect. Consequently, the sub-crack under one loading cycle is formed, which contributes to the final fracture.

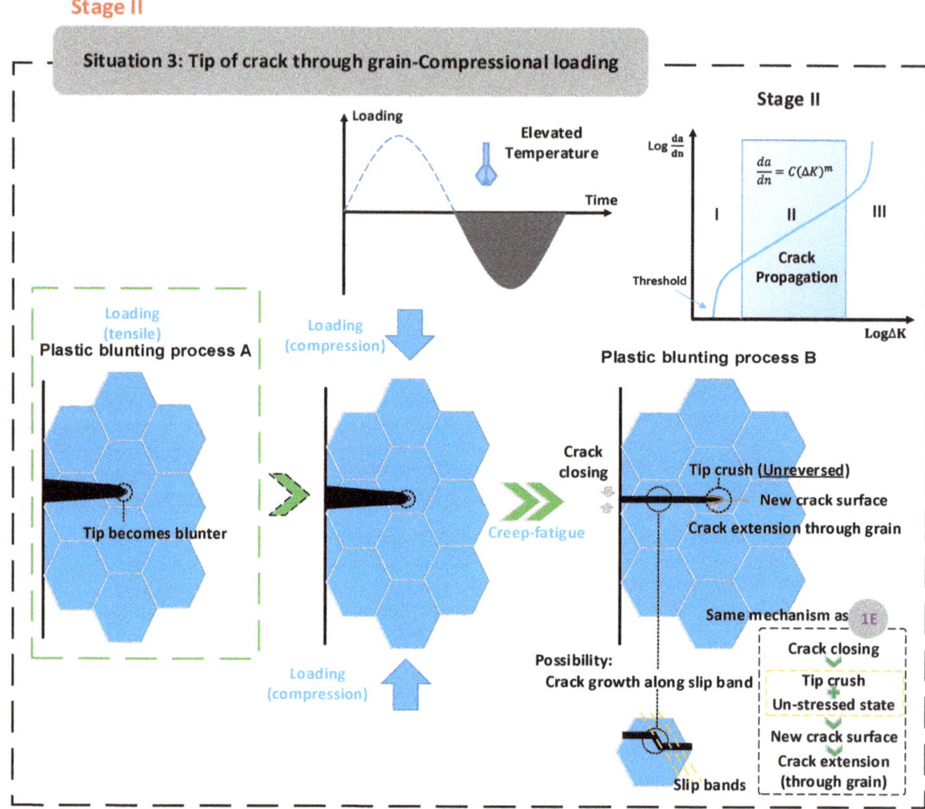

Figure 17. Crack tip through grain boundary—Compressional loading.

After N cycles, the accumulated crack arrives at the grain boundary or it encounters the triple point. This results in the following situation.

4.2.2.2. Crack Tip at Grain Boundary or Triple Point

(1)   Crack tip at grain boundary

- Conceptual framework for crack growth in the tensile regime

Under tensile stress (Figure 11), the fatigue component experiences the same process shown in 'situation 1-1A', wherein the new crack surface is created due to the plastic shearing effect. Then, this causes the crack tip to widen and become blunt. For the creep component (2B-1 in Figure 11), diffusion creep results in grain elongation along the stress direction. On the one hand, this results in crack opening, which was discussed in Section 4.2.2.1; on the other hand, shear stress, which results from grain elongation, leads to elongation of the grain boundary, which then causes the crack tip to widen; meanwhile, a mismatch band appears [47,48]. Specifically, the grain boundary presents a zone of irregular atomic bonds, and thus the bonds at the grain boundary present anisotropic behaviour. Therefore, the relative movement of points caused by grain elongation at the grain boundary may present different results, the extension or breakage of atomic bonds. This results in a zone of material where bonds are partly weakened, and provides potential for further damage.

Consequently, creep intensifies the plastic blunting process caused by the fatigue effect, and a blunter crack tip arises.

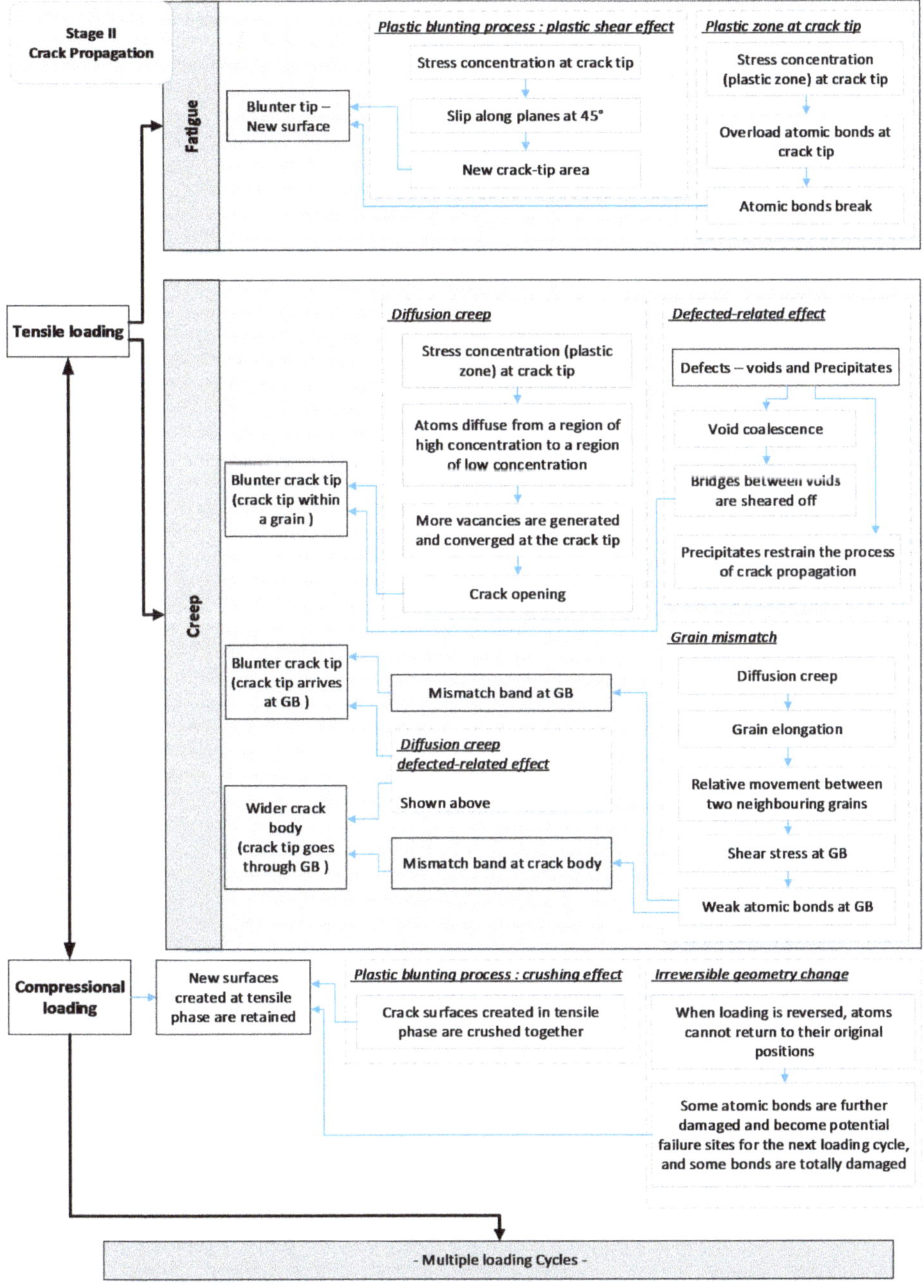

Figure 18. Key elements for one loading cycle in the stage of crack for tension and compression.

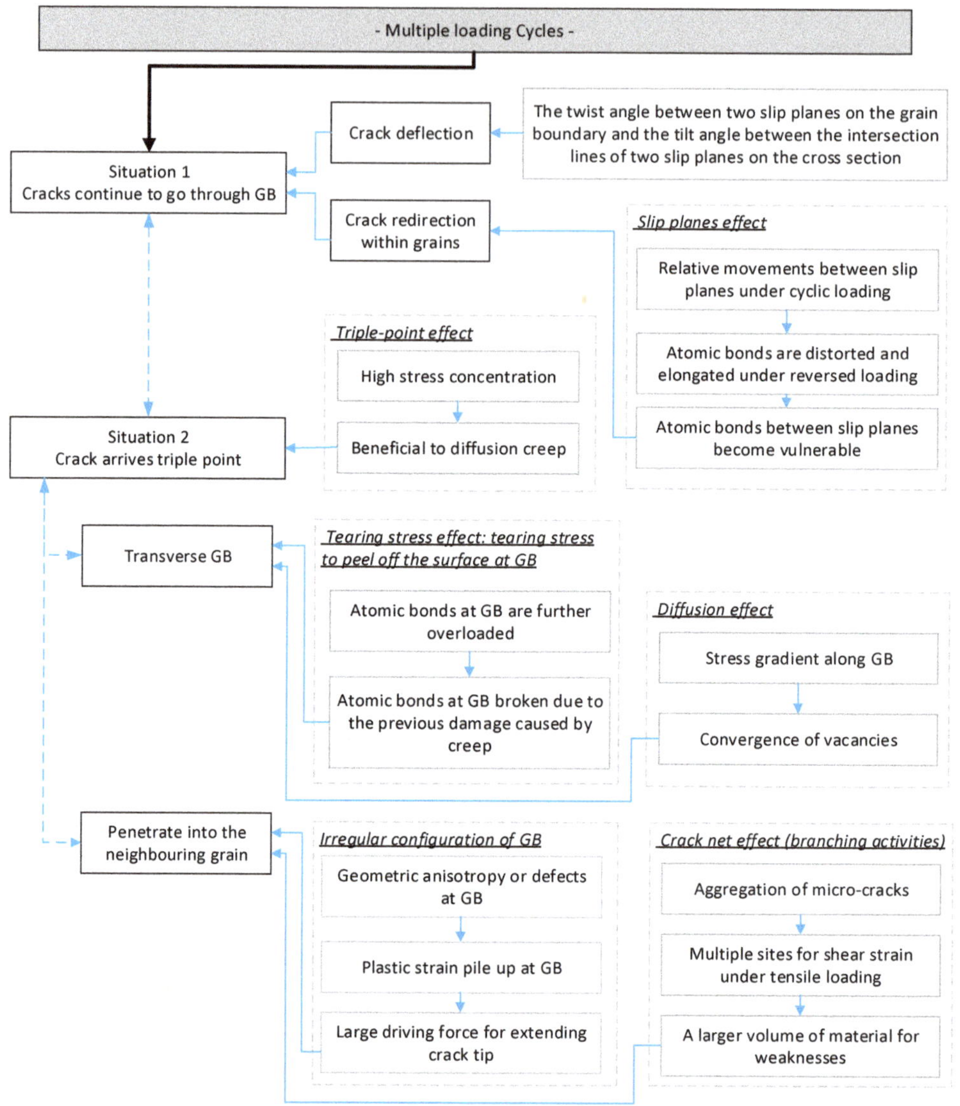

Figure 19. Key elements for multiple load cycles in the stage of crack propagation.

- Conceptual framework for crack growth in the compressional regime

In the situation with compressional loading (Figure 12), the same creep-fatigue process shown in 'situation 1-1E' is presented, where reversed loading crushes the material ahead of the crack, even across the grain boundary. As a result, the new crack surface caused by the tensile loading remains, due to the tip-crush effect, and the crack is further extended across the grain boundary.

In particular, the crack path may be deflected (this behaviour is illustrated in Figure 12 of Ref. [49]) at the boundary. Specifically, crack deflection at the grain boundary is generally determined by the twist angle (a) between two slip planes on the grain boundary and the tilt angle (b) between the intersection lines of two slip planes on the cross section [32]. This behaviour is illustrated in Figure 13, where the crack path is redirected at the grain boundary of the sample surface.

Alternatively, the tip-crush effect during the process of crack closure may lead to crack growth along the grain boundary. Specifically, grain band sliding, which occurs during crack opening, may result in a weaker region along the grain boundary. This then provides a more likely direction for crack propagation than the direction passing through the grain boundary. After this, the crack may arrive at a triple point or may penetrate into the neighbouring grain (deviated progression along the grain boundary). These two situations will be discussed in the next section.

(2) Crack tip at triple point

- Conceptual framework for crack growth in the tensile regime

The triple point, which is defined as the connection point among three adjacent grains, provides a different crack-growth mechanism for fatigue and creep. Specifically, the fatigue effect under tensile stress (Figure 14) presents the same process shown in 'situation 1-1A', where a blunt tip is presented at the triple point according to the idea of plastic blunting process. Compared with the situation shown in Section 4.2.2.2 (1), creep presents a different behaviour at the triple point. On the one hand, diffusion creep results in grain elongation along the direction of tensile stress, and then this leads to crack opening (2B-2 in Figure 14); on the other hand, the triple point presents a significant stress concentration [14–16], which provides an effective stress field for crack growth along the grain boundary (2B-2 in Figure 14). In this situation, multiple opportunities arise for crack propagation along the grain boundary. Generally, the weakest microstructural zone will finally determine the direction of crack growth. Specifically, the weakest zone shows an intensified state, which is normally caused by defects. These include vacancies, which are beneficial to diffusion creep [10,50], and micro-cracks, which aggravate the concentration of stress and hence provide favourable conditions for diffusion. These defect-related effects may also change the direction of crack growth. In addition, according to the idea of the crack tip plastic zone (shown in Figure 9), the stress concentration is presented at the crack tip, where a strong high-stress field is provided hence more creep damage is produced. This implies that this area/direction may have more potential for crack growth. As a result, with the combined effect of fatigue and creep, a blunter crack tip is presented. The crack grows opportunistically in the direction where the grain boundaries are most vulnerable (for propagation into the grain see below).

- Conceptual framework for crack growth in the compressional regime

Under compressional loading (Figure 15), according to the idea of the plastic blunting process, the crushing effect maintains the new crack surface created under the tensile situation, and thus the crack is extended along the grain boundary (2E-2 in Figure 15).

Crack growth in the next cycle or the next few cycles may have two possible paths (2F-2 in Figure 15). Specifically, on the one hand, a crack may traverse (or grow along) the grain boundary, where the weakest atomic bonds are present. This may be attributed to stress concentration at the grain boundary, since the existence of the stress gradient results in the convergence of vacancies [12,13], and in particular, the vacancies at the grain boundary provide favourable conditions for diffusion creep [10,50]. In this case, the crack along the grain boundary gradually accumulates, and then reaches the next triple point, where the opportunity for re-direction is provided.

On the other hand, the crack may be deflected away from the grain boundary. This behaviour is illustrated in Figure 3 of Ref. [47] and Figure 7 of Ref. [48]. This phenomenon may result from defects, such as vacancies in the grain or on its boundary and micro-cracks in the vicinity of the crack. These provide locations for re-direction of crack growth. Specifically, vacancies provide better conditions for diffusion creep, as discussed in the previous sections. In addition, the aggregation of micro-cracks, which are attached to the main crack, also contributes to crack deflection at the grain boundary. This phenomenon is identified as branching activities, and is illustrated in Figure 10 of Ref. [33] and Figure 13 of Ref. [51]. In this situation, a crack-based tree is constructed at the microstructural level, where the main crack is the trunk of this tree and micro-cracks form the branches. The weakest area normally appears at the region with the highest density of micro-cracks, and it plays an important role in crack propagation. This is because the aggregation of micro-cracks weaves a crack net which offers multiple sites for shear strain under tensile loading. The shear stress at the inter-atomic level for any one crack is not decreased by having more micro-cracks. Consequently, the crack net probes a larger volume of material for weaknesses than a single crack could do on its own. This promotes the main crack to automatically grow in a direction most favourable to increasing the total strain (hence most injurious to the integrity of the part).

In addition, the deviating progression may also be caused by the stress concentration at the grain boundary due to irregular configuration thereof. Specifically, this grain-boundary condition results in stress/strain pileup at the boundary (weak boundary condition is presented) and then leads to larger driving force for extending the crack tip into the neighbouring grain, whereby the barrier of the grain boundary is overcome. This is consistent with the general understanding that cracks always propagate towards the direction which requires the minimum energy (stress).

These two situations reflect the fact that fatigue makes a greater contribution than creep in the present case, where the cyclic loading is without hold time. Specifically, the first option implies that the tensile stress, which is perpendicular to the grain boundary, provides a tearing stress to peel off the surface at the grain boundary. This is a significant fatigue behaviour, where the bonds at the grain boundary are further overloaded and then broken due to the previous damage caused by creep. In addition, the surface of the grain boundary is normally not ideally smooth; the geometric anisotropy or defects at the grain boundary may cause that plastic strain to pile up at one specific point, which means that less stress is needed at this point to break the barrier and penetrate the grain boundary, thereby providing for re-direction for crack growth [52]. This transgranular behaviour (which is consistent with the principle of fatigue-crack growth) is illustrated in Figure 4 of Ref. [53]. This combined intergranular and transgranular process represents the mixture of fatigue and creep effects [54]. The intergranular behaviour may become more significant when the dwell time is applied and prolonged [54,55].

After this process, the crack goes through the grain boundary, and the crack tip is extended to the next grain, which is the third situation and is examined below.

4.2.2.3. Crack Tip through Grain Boundary

- Conceptual framework for crack growth in the tensile regime

Under tensile loading (Figure 16), the fatigue component experiences the same process shown in situation 1-1A (Figure 8), where the new crack surface is produced due to plastic shearing effect, and then the crack tip widens and becomes blunt. For the creep partition (3B in Figure 16), on the one hand, diffusion creep results in grain elongation, which leads to crack opening due to the configurational effect of the grain. On the other hand, the extension of the grain boundary caused by grain elongation gives a relative movement between two neighbouring grains, which generates shear stress along the grain boundary. In this case, a mismatch band at the grain boundary is generated due to the distortion of the existing crack under shearing conditions. This band may be further intensified due to pre-existing damage caused by the grain-boundary effect (shown in Section 4.2.2.2 (1), situation 2-1). As a result, this mismatch band widens the crack body and further promotes crack opening. Finally, the combination of fatigue and creep results in a blunter crack tip.

- Conceptual framework for crack growth in the compressional regime

Under compressional loading (Figure 17), the crack-closure process maintains the new crack surface created under tensile loading. Meanwhile, the fatigue crack is extended during the process of crushing, and this is enhanced by any nearby defects in the material.

It is notable that the slip band may result in a re-direction of the crack within grains, following the slip bands inside the grain. We propose that this is caused by the relative movements between slip planes under cyclic loading, which provides a better condition for crack growth along the slip planes. Microstructurally, at the surfaces of the slip bands, the atomic bonds are distorted and elongated under the reversed loading. Thus, these bonds become vulnerable, and then the weakest region (the surfaces of the slip bands) is formed. In this case, at these surfaces, the atomic bonds require the least effort to cut, which subsequently results in crack propagation along the slip bands. The slip band activities are illustrated in Figure 11 of Ref. [34], wherein the crack behaviour of re-direction within a grain is presented.

The above mechanisms for one loading cycle in the stage of crack propagation for tension and compression are summarised in Figure 18, and for multiple load cycles in Figure 19.

4.2.3. Stage III: Structural Failure

The final stage shown in Figure 20 illustrates the fracture of an engineering structure:

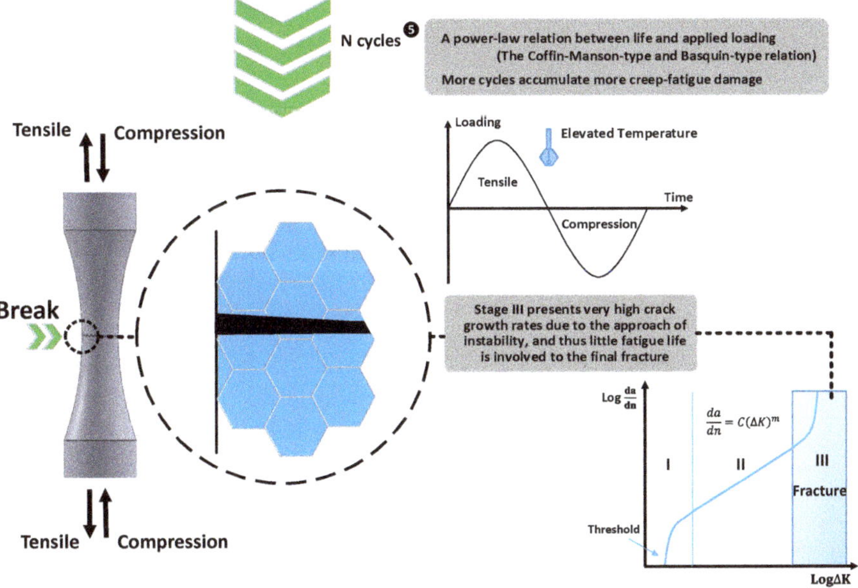

Figure 20. Structural failure.

In this stage, crack growth rate rapidly increases, and part-separation commences when the total crack length reaches a critical value (final facture). This stage represents an unstable state, where the equilibrium shown in the second stage of crack growth is broken due to intensified accumulation of damage and reduction of remaining cross-sectional area. This is attributed to a high stress field at the crack tip due to high stress intensity [23]; hence, the plastic zone becomes large compared to the crack size. Under this stress state, the plastic energy available exceeds that needed for producing the new crack surface. In this case, on the one hand, a part of the plastic energy is applied to create new crack surface; one the other hand, the rest of plastic energy is applied to form voids [23] (see Figure 4.11 therein). Then, under the applied stress, these voids are further expanded and then coalesced by internal necking [56]. This phenomenon worsens the creep-fatigue resistance, and then leads to rapid growth of the crack. Therefore, in this stage, a small contribution of loading cycles leads to big creep-fatigue damage, which then results in failure.

The key elements (mechanisms) in the stage of structural failure are presented in Figure 21.

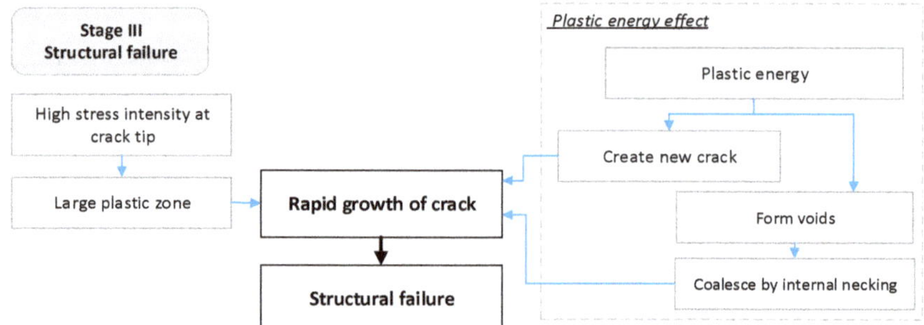

Figure 21. Key elements in the stage of structural failure.

During the process of crack growth, the creep-fatigue damage gradually accumulates with the increasing number of cycles, and this process is presented in a power-law relation (❺ in Figure 20). As discussed in Section 4.2.2.1, the idea of the crack tip plastic zone implies that the stress concentration around the crack tip plays an important role in crack growth. Specifically, the more the stress is concentrated, the larger the plastic zone, and then the more the crack is promoted. Therefore, we assume that crack growth is a geometry-related behaviour, and the crack growth rate (crack growth in one cycle) may be related to the size of the plastic zone. Normally, the area of a zone can be presented as a second-order power relation with a certain dimension, such as radius for circle and the length of a side for square. In this case, since the stress amplitude is directly related to size of the plastic zone, we believe that the applied loading could be related to the crack growth rate in a power-law form. In addition, this relation is also consistent with the idea of damage accumulation shown in Figure 22. In the first stage, the original steady state (a perfect body) is suddenly broken due to the large amount of accumulated energy, which implies at this stage a small amount of damage is produced with a large number of loading cycles (stage I in Figure 22). Then, after the process of re-balancing, the damage accumulation gives a relatively steady state, where the rate of accumulation stably increases (stage II in Figure 22). Finally, when the total damage reaches or is close to a critical value, the load-bearing capacity collapses in a short time. This stage implies that much damage is produced within a small number of loading cycles (stage III in Figure 22).

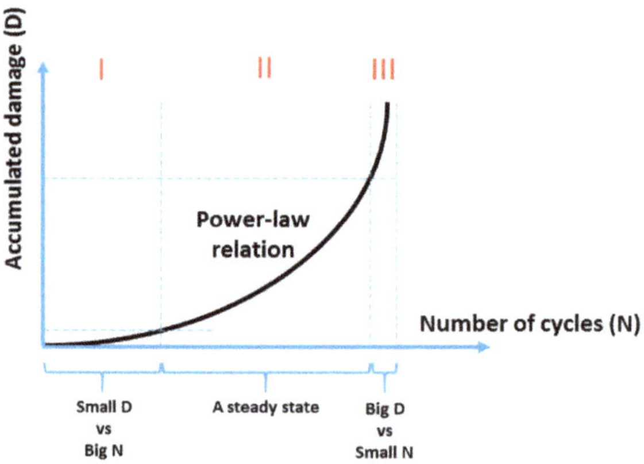

Figure 22. Damage accumulation

5. Discussion

5.1. Summary of the Crack Growth Process and Mechanisms

In the present work, a microstructural graphical-based model for creep fatigue is proposed. In this microstructural conceptual framework, we propose that different crack growth mechanisms are active for different stages in the temporal evolution of the crack (stages I to III), different parts of the tension-compression cycle, and for different regions within the microstructure (grain boundary, triple point, inside the grain).

In general, the temporal evolution of the crack is mainly based on the mechanisms of the plastic blunting process, stress-concentration state and diffusion-creep behaviour. The illustration of the crack-growth process is presented in three different stages:

Stage I: Crack initiation

In the first stage (the crack initial stage), a threshold is presented, below which no significant crack is detected. The existence of this threshold can be attributed to the effect of stress concentration. For a surface with defects (such as a machining mark), the stress concentration is caused by the pre-existing micro-cracks. For a defect-free surface, the behaviours of extrusion and intrusion between slip plans give some tiny steps in the surface, which then results in the stress concentration.

Stage II: Crack propagation

In the second stage (the steady stage of crack growth), the crack-growth behaviours under tensile loading and compressional loading were discussed separately. The primary differentiating factor between the tension-compression cycles can be explained by the plastic blunting process. Specifically, tensile loading creates a new crack surface due to the shearing effect around the crack-tip area, and then the crushing effect results in crack extension (by maintaining the new surface created by the tensile loading) under the compressional loading. Regarding the microstructure, the plastic blunting process implies that the damage produced within one tension-compression cycle is not a reversed process since the atoms cannot return to their original positions after one cycle.

In addition, the crack-growth behaviours of fatigue and creep were also illustrated separately. Specifically, the fatigue component was generally presented by the plastic blunting process shown above, and the creep component was explained by the mechanism of diffusion creep. On the one hand, stress concentration at the crack tip provides a favourable condition for diffusion to form voids due to the stress-gradient effect, and then the void coalescence by shearing of the bridges between voids results in crack growth. On the other hand, diffusion creep gives grain elongation under tensile loading, which then further opens the crack created by fatigue effect. In addition, grain elongation also gives the shear stress at the grain boundaries between two adjacent grains, which probably results in the mismatch behaviour of the crack due to weak atomic bonds at the boundaries; then, the crack widens.

Furthermore, the crack-growth behaviour with respect to the grain-boundary effect was discussed, and this effect plays a more important role in creep-crack growth than fatigue behaviour. In this case, due to the shear stress between two adjacent grains caused by grain elongation, the grain-boundary effect results in a blunter crack tip or wider crack body (mismatch). In particular, this effect can be presented by a special situation, the triple-point effect. In the triple points, the high stress concentration provides better conditions for diffusion.

Stage III: Structural failure

In the final stage (structural failure), the fracture occurs within small loading cycles. This can be attributed to the high stress field at the crack tip. Specifically, under this stress state, the plastic energy exceeds the needs for producing the new crack surface. In this case, a part of energy is applied to create the new surface, and the rest of the energy is applied to form voids, which further promote crack growth through internal necking.

5.2. Original Contributions

The original contribution of the present work is the proposal of a microstructural conceptual framework for the multiple crack propagation mechanisms at a grain-based level. In general, the plastic blunting process was introduced to describe fatigue-crack growth, the mechanism of diffusion was included to present creep-crack growth, and the grain-boundary effect is proposed to describe the phenomena taking place when a crack encounters a grain boundary. While the ideas of crack growth are well known, the literature does not provide a comprehensive explanation of the process from initiation to failure. The novel contribution of the above conceptual framework is therefore the integration of multiple mechanisms at the microstructural level, inclusion of the grain boundary effect, and coverage of the stages from initiation to failure. To achieve this, some new propositions (such as the grain mismatch and the bonding crush effect) were proposed, and some existing concepts (such as the energy dynamics and the micro-cracks aggregation) were extended to explain the crack-growth behaviours in the present work. These principles connect the existing but separate crack-growth principles into a consolidated explanation of the whole failure process.

We have proposed a number of new mechanisms (propositions A-G in Section 4.1) and phenomena for crack growth:

(1) We proposed the grain-mismatch mechanism which occurs at the grain boundary under the tensile loading. At the microstructural level, tensile loading results in relative movement between two neighbouring grains, which then generates shear stress along the grain boundary. This shear stress may be caused by the stretch of atomic bonds. In this case, a weak region arises at the grain boundary. On the one hand, when a crack arrives at the grain boundary, the grain-mismatch behaviour results in a wider crack tip, which creates more new surface under tensile loading, and then intensifies the crack-extension process due to the crushing effect under the compressional loading. On the other hand, when a crack goes through the grain boundary, the grain-mismatch behaviour widens the crack body, and then a mismatch band is formed. This further promotes the crack-opening behaviour under the tensile loading, then a blunter crack tip is created. The new surface generated in this stage is then retained under compressional loading. The grain-mismatch mechanism gives a wide-ranging explanation of crack-growth behaviour at the grain boundary. This effect is transient and is not expected to be easy to detect.

(2) We proposed the idea of the bonding crushing effect. Specifically, during the compressional phase, the distorted and rearranged atomic bonds caused by tensile loading are further damaged. Then, the atomic bonds may break during the compressional process or become potential failure sites for the next loading cycle. The compressional phase is also presented by the existing principle of plastic blunting process, which indicates that the crack is extended because the new surface created in tensile phase is crushed together at the compressional phase. This existing principle gives a general understanding of crack extension, but not to the atomic level. Our crushing effect offers a deeper explanation of the process of crack closure at the atomic level. This key element of the bonding crushing effect is that the damage (distorted and rearranged atomic bonds) under the tensile loading cannot be recovered when the loading is reversed. Then, the damage is retained by this irreversible process, since the atoms cannot return to their original positions under compressional loading.

(3) We proposed that the crack growth is a geometry-related behaviour. In general, crack-growth rate is determined by the size of the plastic zone, and a larger plastic zone gives longer crack growth. This is combined with our other propositions, namely that fatigue-crack growth is caused by overloaded local atomic bonds, and creep-crack growth is defined as the flow of atoms relative to each other with transfer of bonding to new partners. These two damage behaviours are stress-motivated processes. On the one hand (for the fatigue process), larger plastic zone results in more overloaded atomic bonds. Thus, more bonds become vulnerable. They may totally break under the compressional process, or may become potential failure sites during the next loading cycles. This implies that more completed or potential crack surfaces are formed. On the other hand (for the creep process), a larger plastic zone leads to more space and enhanced stress conditions for diffusion. Thus, at a

given volume, more atoms are transferred to new positions and then form new bonds with other atoms. This phenomenon is more significant in the crack-tip area. During the process of crack growth, the increased crack length leads to the expansion of the plastic zone, and higher localised stress around the crack tip. Hence, more vacancies are generated and converged in this area, and a more favourable situation arises for creep. Hence, the plastic-zone-based explanation for crack growth presents an explanation of the loading contribution for fatigue and creep behaviours at the atomic level.

(4) We proposed a mechanism for crack behaviour at the triple point. This situation may not happen during the whole process from crack initiation to component failure. However, it is a possible situation; thus, it is valuable to discuss it. At a triple point, the stress concentration (which arises from the geometry of the grains) results in weaker atomic bonding between the grains, and thus a more favourable situation for both fatigue and creep damage. In particular, we proposed two possible paths of crack growth over one or more cycles. On the one hand, the crack may be deflected away from the grain boundary. This phenomenon may be explained by two existing mechanisms—the crack-net behaviour and the irregular configuration of the grain boundary—which have been extended in the present work. Specifically, the crack net around the grain boundary gives a larger volume of material for weaknesses, and then promotes and deflects the main crack. In this case, a crack penetrates into its neighbouring grain, where the crack net is highly concentrated. The irregular configuration of a grain boundary causes stress/strain pile up at the grain boundary, and then leads to a larger driving force for extending the crack tip into the neighbouring grain. On the other hand, a crack may grow along the grain boundary. This phenomenon may result from the larger contribution of creep than fatigue. In this case, convergence of vacancies caused by diffusion along the grain boundary gives a favourable condition for crack growth in this area. The situation of the crack tip at a triple point introduces further opportunities for changes in the direction and extent of crack growth.

(5) We applied the mechanism of energy dynamics to qualitatively describe the whole process of crack growth from initiation to failure. Although this mechanism is extant, it was then further extended in the present work. Generally, in the initial stage of crack growth, internal energy is continually stored in the structure under the cyclic loading in the form of irreversible dislocations. Dislocations are vacancies in the atomic lattice; hence, they comprise localised vacancies and strain between atoms. When this energy arrives at a critical energy level, the vacancies in the dislocations coalesce and form a crack. This results in a cascade effect of a rapid liberation of energy, which is consumed to form new crack surfaces. During this process, the high crack-growth rate is gradually reduced, and then achieves a steady situation due to the need of the system to rebuild new dislocations. We propose that this cyclic energisation process underpins the steady second stage of crack growth. When the accumulated energy/damage is large enough, the crack is re-energised even as it progresses; hence, the growth rate accelerates. This process is consistent with the representation of the final stage of crack growth. In this case, the completed process from crack initiation to component failure is described in the perspective of energy accumulation and release.

Using these, it has been possible to construct a comprehensive conceptual framework of the crack growth process at the microstructural level. This conceptual framework is also consistent with the existing concepts in the literature (premises a-l in Section 2.3). In itself, this does not *prove* that these propositions are true, but it does show that these concepts are useful in better understanding the creep-fatigue mechanisms.

5.3. Implications for Practitioners

The conceptual framework presents a theory for the multiscale mechanisms of crack development and growth. This theory has implications for various practitioners. For those who are studying microstructure, the implications are that an explanatory framework has been provided. This integrates multiple existing microstructural phenomena, and hence it is expected that existing microstructural observations would be able to be reconciled to the theory. For those who are designing engineering components, e.g., design engineers, the implications are more abstract, in that it emphasizes the need to avoid the reverse loading condition, where possible. Practical materials used by engineers are invariably full of crack initiators; hence, the default state of the material is Stage I, with many of the cracks propagating into Stage II. The aim of engineering design is to avoid Stage III failure, as this terminates the life of the part/machine and can have catastrophic consequences for human safety. Hence, from the engineering perspective, there is a need to keep the part in the Stage II regime for as long as possible. Designers affect this by controlling the geometry of the part, the external loading, the material selection and treatment. The theory shows, per Figures 19 and 20, that the combination of tension and compression is key to crack growth. This is already well known, but the theory makes the relationships more explicit. In particular, this theory makes explicit the mechanisms whereby the compressional process contributes crack growth. This is because the distorted and rearranged atomic bonds caused by the tensile loading are further damaged, such that some atomic bonds break during the compressional process, and others become weak areas for the next loading cycle.

5.4. Limitations

The analysis presented here does not completely include all possible loading situations, though it is believed to be broadly representative.

This schematic representation of creep-fatigue crack growth is based on the assumption that fatigue damage makes a greater contribution to failure than creep damage. In this case, the crack mainly grows through the grains. The creep behaviour is regarded as an assistant effect. This may not be true of all situations, especially not at higher temperatures, longer cyclic times, or cyclic loading with hold time, where creep begins to dominate. In these situations, the creep effect may contribute to more damage, and crack growth may take a different direction. This is because the creep damage gradually increases with the crack growing due to increasing number of creep voids, but the fatigue damage gradually decreases during this process [57,58].

For a creep-dominated structure, we believe that the crack growth along the grain boundary is more significant. In particular, at the situation of the cyclic loading with hold time, a component of pure creep effect is introduced. This period may result in a more complex situation, where the stress relaxation occurs under strain-controlled loading. In this case, the stress-related mechanisms, such as the stress condition around the crack tip, may need further examination. In addition, at the situation of the cyclic loading with hold time, the effect of oxidation may become more significant. This effect normally results in brittle crack surfaces, and then reduces the failure life. The research [55] shows that this reduction in life is more severe at the creep-fatigue-oxidation condition than the creep-oxidation condition. In this case, the combination of creep, fatigue and oxidation affects crack growth in a more complex manner. Therefore, other possibilities for crack growth, such as the influences of stress relaxation and oxidation on the fatigue component, may need to be considered. These may be an opportunity for future research.

5.5. Implications for Further Research

The graphical-based crack-growth model proposed in this paper is a microstructural conceptual framework. Some of the fatigue and creep behaviours have been well demonstrated and explained in previous research, such as stress distribution at the crack tip for fatigue and the stress concentration at the triple point for creep. However, there are still some behaviours which are difficult to validate through experiments, such as the situation with the crack tip at the grain boundary. Our microstructural conceptual framework has offered an explanation of these behaviours, but this work is conjectural. It would be valuable to test the validity of this at the microstructural level. To achieve this, it would be necessary to combine fractography and fatigue tests.

Overall, the present work presents some unanswered questions that require experiments. This opens multiple opportunities for future research:

(1)   The crack-growth behaviours at the grain boundary when crack tip at/through the grain boundary. The question is that it is difficult and complex to empirically examine and investigate

the phenomena at the grain boundary since it will be difficult to dynamically capture those moments when the crack tip arrives at the grain boundary. To catch this microstructural behaviour, a new experimental method may also need to be developed.

(2)   The grain-mismatch behaviour due to grain elongation under the tensile loading. It is valuable to explore the shape change of grains under tensile loading, and observe the relative movements and investigate the interfacial phenomena between two neighbouring grains. This research may also be extended to the situation of compressional loading.

(3)   The bonding crushing effects during the process of compression ahead of the crack. This is a call for an atomic-level investigation, e.g., using atomic force microscopy. Existing approaches have been focussed on statically describing the force-distance relation and force field between two neighbouring bonds. This provides valuable information on the local material properties [59–61]. However, the dynamical process (such as the time-depended force/distance variances and the trace of atomic movements) under a cyclic or constant loading have not been addressed. In addition, it may also be valuable to explore both the situations of tensile and compressional loadings.

(4)   The stress/strain pile-up at the grain boundary because of the irregular configuration of the grain boundary. This is the question of investigating the dislocation pile-up [62] and measuring the stress field at the grain boundary [47]. The future research may start from the existing theories and methods, and then go further to evaluate the configuration-based effects at the grain boundary. In addition, finite element analysis may also be involved to investigate the stress/strain distributions and the shape distortion at the simulative grain boundaries.

(5)   Numerical modelling and its experimental validation. This could be done at several levels. At a high level of abstraction, there are already formulations for describing the failure process. For example, Paris's law [11] gives the crack growth in one cycle based on the stress intensity factor, the Arrhenius equation [63,64] presents the creep behaviour at the steady stage based on diffusion creep, and the conventional loading-life equations (the Basquin [65,66] equation and Coffin-Manson [67,68] equation) model life based on empirical data. At the level of detail providing in this paper, as represented in Figures 19 and 20, the crack-growth process is graphically described in the microstructural level.

An ideal model would be one that represented the physics of the interactions between atoms within the finer level of the grain. Of interest are the interactions between the imperfections in the crystal lattice (e.g., dislocations), the grain boundaries, and the way the cracks propagate. However, this is likely to be a challenging proposition for some time. It would be ideal to observe the stages of crack growth using methods such as atomic force microscopy (AFM). Some work in this area includes the investigation of the force-distance relation and force field between two neighbouring bonds [59–61]. However, the difficulties are the extremely small area of examination afforded by AFM, and the practical difficulties of arranging a crack to grow in the sample. Furthermore, AFM measures free surfaces, whereas cracks have a three-dimensional interaction with the grains.

Possibly a more immediately achievable model could be to use finite element methods to represent the grains and grain boundary regions. It would be necessary to incorporate the anisotropy of the crystal lattice. This may be able to explore the interaction of the crack with the microstructure. Potentially this could be validated by microscopy. Neither observed fracture surfaces nor cross-sections show the full development of the crack architecture in the 3D volume. Also, fracture surfaces only show the Stage III development and termination of the cracks. A FEA approach might help better understand the evolution of the crack architecture across the Stages.

Our framework has proposed a number of variables, and the general direction of causality. If it were possible—which does not currently appear to be the case—to measure all these variables, then a statistical method such as factor analysis or structural path modelling might be used to determine the relative importance of the variables and the strength of the relationships.

6. Conclusions

Based on the general understanding of crack-growth behaviour, a microstructural conceptual framework has been proposed, and represented graphically, wherein fatigue and creep effects are treated separately due to the different damage principles. This microstructural model illustrates the process of damage accumulation from crack initiation to crack propagation then to structural failure. In particular, the analysis of crack propagation is mainly based on existing mechanisms of plastic blunting and diffusion creep. The possible grain-boundary behaviours, such as the mismatch behaviour at grain boundary due to creep deformation, are included. This conceptual framework is consistent with the various creep and fatigue microstructure observations in the literature, but goes further by integrating these premises and our proposed propositions together into a logically consistent framework that describes the overall failure process.

Author Contributions: D.L. and D.J.P. conceptualised the overall framework, D.L. worked out the details and produced the graphical model, D.L. and D.J.P. contributed to writing the paper.

Funding: This research received no external funding.

Conflicts of Interest: The authors declare no conflict of interest. The research was conducted without personal financial benefit from any funding body, and no such body influenced the execution of the work.

References

1. June, W. A continuum damage mechanics model for low-cycle fatigue failure of metals. *Eng. Fract. Mech.* **1992**, *41*, 437–441. [CrossRef]
2. Metzger, M.; Nieweg, B.; Schweizer, C.; Seifert, T. Lifetime prediction of cast iron materials under combined thermomechanical fatigue and high cycle fatigue loading using a mechanism-based model. *Int. J. Fatigue* **2013**, *53*, 58–66. [CrossRef]
3. Seifert, T.; Riedel, H. Mechanism-based thermomechanical fatigue life prediction of cast iron. Part I: Models. *Int. J. Fatigue* **2010**, *32*, 1358–1367. [CrossRef]
4. Jing, H.; Zhang, Y.; Xu, L.; Zhang, G.; Han, Y.; Wei, J. Low cycle fatigue behavior of a eutectic 80Au/20Sn solder alloy. *Int. J. Fatigue* **2015**, *75*, 100–107. [CrossRef]
5. Shi, X.; Pang, H.; Zhou, W.; Wang, Z. Low cycle fatigue analysis of temperature and frequency effects in eutectic solder alloy. *Int. J. Fatigue* **2000**, *22*, 217–228. [CrossRef]
6. Solomon, H. Fatigue of 60/40 solder. *IEEE Trans. Compon. Hybrids Manuf. Technol.* **1986**, *9*, 423–432. [CrossRef]
7. Callister, W.D.; Rethwisch, D.G. *Materials Science and Engineering*; John Wiley & Sons: Hoboken, NJ, USA, 2011; Volume 5.
8. Dowling, N.E. *Mechanical Behavior of Materials: Engineering Methods for Deformation, Fracture, and Fatigue*; Prentice: London, UK, 2012; p. 954.
9. Evans, R.W.; Wilshire, B. *Introduction to Creep*; The Institute of Materials, University of Michiganb: Ann Arbor, MI, USA, 1993; p. 115.
10. Kassner, M.E. *Fundamentals of Creep in Metals and Alloys*; Butterworth-Heinemann: Oxford, UK, 2015; p. 337.
11. Paris, P.; Erdogan, F. A critical analysis of crack propagation laws. *J. Basic Eng.* **1963**, *85*, 528–533. [CrossRef]
12. Liu, C.; Han, Y.; Yan, M.; Chaturvedi, M. Creep crack growth behaviour of alloy 718. *Superalloys* **1991**, *178*, 537–548.
13. Sadananda, K. A theoretical model for creep crack growth. *Metall. Mater. Trans. A* **1978**, *9*, 635–641. [CrossRef]
14. Cocks, A.; Pontern, A. *Mechanics of Creep Brittle Materials 1*; Springer Science & Business Media: Berlin/Heidelberg, Germany, 1989; p. 310.
15. Ejaz, N.; Qureshi, I.; Rizvi, S. Creep failure of low pressure turbine blade of an aircraft engine. *Eng. Fail. Anal.* **2011**, *18*, 1407–1414. [CrossRef]
16. Král, P.; Dvořák, J.; Kvapilová, M.; Svoboda, M.; Sklenička, V. Creep damage of Al and Al-Sc alloy processed by ecap. *Acta Metall. Slov. Conf.* **2013**, *3*, 136–144. [CrossRef]
17. Dai, C.; Zhang, B.; Xu, J.; Zhang, G. On size effects on fatigue properties of metal foils at micrometer scales. *Mater. Sci. Eng. A* **2013**, *575*, 217–222. [CrossRef]

18. Hanlon, T.; Kwon, Y.-N.; Suresh, S. Grain size effects on the fatigue response of nanocrystalline metals. *Scr. Mater.* **2003**, *49*, 675–680. [CrossRef]

19. Laird, C. Mechanisms and theories of fatigue. *Fatigue Microstruct.* **1979**, 149–203.

20. Banks-Sills, L.; Motola, Y.; Shemesh, L. The m-integral for calculating intensity factors of an impermeable crack in a piezoelectric material. *Eng. Fract. Mech.* **2008**, *75*, 901–925. [CrossRef]

21. Warzynek, P.; Carter, B.; Banks-Sills, L. *The M-Integral for Computing Stress Intensity Factors in Generally Anisotropic Materials*; National Aeronautics and Space Administration: Washington, DC, USA, 2005.

22. Miller, K. Materials science perspective of metal fatigue resistance. *Mater. Sci. Technol.* **1993**, *9*, 453–462. [CrossRef]

23. Rodopoulos, C.A. Fatigue damage map as a virtual tool for fatigue damage tolerance. In *Virtual Testing and Predictive Modeling*; Springer: Berlin/Heidelberg, Germany, 2009; pp. 73–104.

24. Chowdhury, P.; Sehitoglu, H. Mechanisms of fatigue crack growth—A critical digest of theoretical developments. *Fatigue Fract. Eng. Mater. Struct.* **2016**, *39*, 652–674. [CrossRef]

25. Laird, C. The influence of metallurgical structure on the mechanisms of fatigue crack propagation. In *Fatigue Crack Propagation*; ASTM International: West Conshohocken, PA, USA, 1967.

26. Laird, C.; de La Veaux, R. Additional evidence for the plastic blunting process of fatigue crack propagation. *Metall. Mater. Trans. A* **1977**, *8*, 657–664. [CrossRef]

27. Peralta, P.; Choi, S.-H.; Gee, J. Experimental quantification of the plastic blunting process for stage ii fatigue crack growth in one-phase metallic materials. *Int. J. Plast.* **2007**, *23*, 1763–1795. [CrossRef]

28. Shi, K.; Cai, L.; Qi, S.; Bao, C. A prediction model for fatigue crack growth using effective cyclic plastic zone and low cycle fatigue properties. *Eng. Fract. Mech.* **2016**, *158*, 209–219. [CrossRef]

29. Wang, G. The plasticity aspect of fatigue crack growth. *Eng. Fract. Mech.* **1993**, *46*, 909–930. [CrossRef]

30. Weertman, J. Fatigue crack propagation theories. In *Fatigue and Microstructure*; American Society for Metals: Geauga County, OH, USA, 1979; Volume 279.

31. Ham, R.; Broom, T. The mechanism of fatigue softening. *Philos. Mag.* **1962**, *7*, 95–103. [CrossRef]

32. Zhai, T.; Jiang, X.; Li, J.; Garratt, M.; Bray, G. The grain boundary geometry for optimum resistance to growth of short fatigue cracks in high strength al-alloys. *Int. J. Fatigue* **2005**, *27*, 1202–1209. [CrossRef]

33. Kinloch, A.; Wang, C.; Wu, S.; Ladani, R.; Zhang, J.; Bafekrpour, E.; Ghorbani, K.; Mouritz, A. Aligning Graphene Nanoplatelets with an External Electric Field to Improve Multifunctional Properties of Epoxy Nanocomposites. *Carbon* **2015**, *94*, 607–618. [CrossRef]

34. Villechaise, P.; Cormier, J.; Billot, T.; Mendez, J. Mechanical behaviour and damage processes of udimet 720li: Influence of localized plasticity at grain boundaries. In Proceedings of the 12th International Symposium on Superalloys, Pennsylvania, PA, USA, 9–13 September 2012.

35. Claude Bathias, A.P. *Fatigue of Materials and Structures: Fundamentals*; John Wiley & Sons: Hoboken, NJ, USA, 2013; p. 512.

36. Sangid, M.D. The physics of fatigue crack initiation. *Int. J. Fatigue* **2013**, *57*, 58–72. [CrossRef]

37. Wang, Z.; Beyerlein, I.; LeSar, R. Slip band formation and mobile dislocation density generation in high rate deformation of single fcc crystals. *Philos. Mag.* **2008**, *88*, 1321–1343. [CrossRef]

38. Zerbst, U.; Vormwald, M.; Pippan, R.; Gänser, H.-P.; Sarrazin-Baudoux, C.; Madia, M. About the fatigue crack propagation threshold of metals as a design criterion—A review. *Eng. Fract. Mech.* **2016**, *153*, 190–243. [CrossRef]

39. Margaritis, G.; Botsis, J. Energy evaluations during crack initiation. *Eng. Fract. Mech.* **1991**, *40*, 1123–1134. [CrossRef]

40. Lach, R.; Adhikari, R.; Weidisch, R.; Huy, T.; Michler, G.; Grellmann, W.; Knoll, K. Crack toughness behavior of binary poly (styrene-butadiene) block copolymer blends. *J. Mater. Sci.* **2004**, *39*, 1283–1295. [CrossRef]

41. Zerbst, U.; Klinger, C.; Clegg, R. Fracture mechanics as a tool in failure analysis—Prospects and limitations. *Eng. Fail. Anal.* **2015**, *55*, 376–410. [CrossRef]

42. Fournier, D.; Pineau, A. Low cycle fatigue behavior of inconel 718 at 298 k and 823 k. *Metall. Mater. Trans. A* **1977**, *8*, 1095–1105. [CrossRef]

43. Liu, D.; Pons, D.J. Physical-mechanism exploration of the low-cycle unified creep-fatigue formulation. *Metals* **2017**, *7*, 379.

44. Liu, D.; Pons, D.; Wong, E.-h. The unified creep-fatigue equation for stainless steel 316. *Metals* **2016**, *6*, 219. [CrossRef]

45. Pippan, R.; Grosinger, W. Fatigue crack closure: From lcf to small scale yielding. *Int. J. Fatigue* **2013**, *46*, 41–48. [CrossRef]

46. Prasad, K.; Kumar, V.; Rao, K.B.S.; Sundararaman, M. A comparative assessment of crack closure mechanisms in timetal 834 near α titanium alloy under isothermal and thermomechanical fatigue loading. *J. Alloys Compd.* **2016**, *688*, 8–11. [CrossRef]

47. Guo, Y.; Collins, D.; Tarleton, E.; Hofmann, F.; Tischler, J.; Liu, W.; Xu, R.; Wilkinson, A.; Britton, T. Measurements of stress fields near a grain boundary: Exploring blocked arrays of dislocations in 3d. *Acta Mater.* **2015**, *96*, 229–236. [CrossRef]

48. McMurtrey, M.; Was, G.; Cui, B.; Robertson, I.; Smith, L.; Farkas, D. Strain localization at dislocation channel–grain boundary intersections in irradiated stainless steel. *Int. J. Plast.* **2014**, *56*, 219–231. [CrossRef]

49. Pineau, A. Crossing grain boundaries in metals by slip bands, cleavage and fatigue cracks. *Philos. Trans. R. Soc. A* **2015**, *373*, 20140131. [CrossRef] [PubMed]

50. Kumar, Y.; Venugopal, S.; Sasikala, G.; Albert, S.K.; Bhaduri, A. Study of creep crack growth in a modified 9cr–1mo steel weld metal and heat affected zone. *Mater. Sci. Eng. A* **2016**, *655*, 300–309. [CrossRef]

51. Pretty, C.J.; Whitaker, M.T.; Williams, S.J. Thermo-mechanical fatigue crack growth of rr1000. *Materials* **2017**, *10*, 34. [CrossRef] [PubMed]

52. Kacher, J.; Eftink, B.; Cui, B.; Robertson, I. Dislocation interactions with grain boundaries. *Curr. Opin. Solid State Mater. Sci.* **2014**, *18*, 227–243. [CrossRef]

53. Benz, J.K.; Wright, R.N. *Fatigue and Creep Crack Propagation Behaviour of Alloy 617 in the Annealed and Aged Conditions*; Idaho National Laboratory (INL): Idaho Falls, ID, USA, 2013.

54. Tang, Z.; Jing, H.; Xu, L.; Zhao, L.; Han, Y.; Xiao, B.; Zhang, Y.; Li, H. Creep-fatigue crack growth behavior of g115 steel under different hold time conditions. *Int. J. Fatigue* **2018**, *116*, 572–583. [CrossRef]

55. Zhao, L.; Nikbin, K. Characterizing high temperature crack growth behaviour under mixed environmental, creep and fatigue conditions. *Mater. Sci. Eng. A* **2018**, *728*, 102–114. [CrossRef]

56. Scheyvaerts, F.; Pardoen, T.; Onck, P. A new model for void coalescence by internal necking. *Int. J. Damage Mech.* **2010**, *19*, 95–126. [CrossRef]

57. Xu, L.; Rong, J.; Zhao, L.; Jing, H.; Han, Y. Creep-fatigue crack growth behavior of g115 steel at 650 °C. *Mater. Sci. Eng. A* **2018**, *726*, 179–186. [CrossRef]

58. Xu, L.; Zhao, L.; Han, Y.; Jing, H.; Gao, Z. Characterizing crack growth behavior and damage evolution in p92 steel under creep-fatigue conditions. *Int. J. Mech. Sci.* **2017**, *134*, 63–74. [CrossRef]

59. Butt, H.-J.; Cappella, B.; Kappl, M. Force measurements with the atomic force microscope: Technique, interpretation and applications. *Surf. Sci. Rep.* **2005**, *59*, 1–152. [CrossRef]

60. Neuman, K.C.; Nagy, A. Single-molecule force spectroscopy: Optical tweezers, magnetic tweezers and atomic force microscopy. *Nat. Methods* **2008**, *5*, 491–505. [CrossRef] [PubMed]

61. Williams, J.M.; Han, T.; Beebe, T.P. Determination of single-bond forces from contact force variances in atomic force microscopy. *Langmuir* **1996**, *12*, 1291–1295. [CrossRef]

62. Kato, M. Hall–petch relationship and dislocation model for deformation of ultrafine-grained and nanocrystalline metals. *Mater. Trans.* **2014**, *55*, 19–24. [CrossRef]

63. Arrhenius, S. Über die dissociationswärme und den einfluss der temperatur auf den dissociationsgrad der elektrolyte. *Zeitschrift Physikalische Chemie* **1889**, *4*, 96–116. [CrossRef]

64. Arrhenius, S. Über die reaktionsgeschwindigkeit bei der inversion von rohrzucker durch säuren. *Zeitschrift Physikalische Chemie* **1889**, *4*, 226–248. [CrossRef]

65. Basquin, O. The exponential law of endurance tests. *Am. Soc. Test. Mater.* **1910**, *10*, 625–630.

66. Wöhler, A. Theorie rechteckiger eiserner brückenbalken mit gitterwänden und mit blechwänden. *Zeitschrift Bauwesen* **1855**, *5*, 121–166.

67. Coffin, L.F., Jr. *A Study of the Effects of Cyclic Thermal Stresses on a Ductile Metal*; Knolls Atomic Power Lab.: New York, NY, USA, 1953.

68. Manson, S. Behavior of Materials Under Conditions of Thermal Stress. Available online: https://ntrs.nasa.gov/archive/nasa/casi.ntrs.nasa.gov/19930092197.pdf (accessed on 8 March 2018).

© 2018 by the authors. Licensee MDPI, Basel, Switzerland. This article is an open access article distributed under the terms and conditions of the Creative Commons Attribution (CC BY) license (http://creativecommons.org/licenses/by/4.0/).

Article

Investigation of Stress-Oriented Hydrogen-Induced Cracking (SOHIC) in an Amine Absorber Column of an Oil Refinery

Gregory N. Haidemenopoulos [1,2,*], Helen Kamoutsi [2], Kyriaki Polychronopoulou [1], Panagiotis Papageorgiou [2], Ioannis Altanis [3], Panagiotis Dimitriadis [3] and Michael Stiakakis [3]

[1] Department of Mechanical Engineering, Khalifa University of Science and Technology, Abu Dhabi 127788, UAE; Kyriaki.polychrono@ku.ac.ae
[2] Department of Mechanical Engineering, University of Thessaly, 38334 Volos, Greece; ekamoutsi@mie.uth.gr (H.K.); panos.papa.konos@gmail.com (P.P.)
[3] Motor Oil Hellas, Korinth Refinery, 20100 Korinth, Greece; altaniio@moh.gr (I.A.); dimitrpa@moh.gr (P.D.); stiakami@moh.gr (M.S.)
* Correspondence: hgreg@mie.uth.gr; Tel.: +30-6947523717

Received: 4 August 2018; Accepted: 22 August 2018; Published: 24 August 2018

Abstract: Stress-oriented hydrogen-induced cracking (SOHIC) of an amine absorber column made of a Hydrogen Induced Cracking (HIC) resistant steel and operating under wet H_2S service was investigated. SOHIC was not related to welds in the column and evolved in two steps: initiation of HIC cracks in the rolling plane and through-thickness linking of the HIC cracks. Both the original HIC cracks as well as the linking cracks propagated with a cleavage mechanism. The key factors identified were periods with high hydrogen charging conditions, manifested by high H_2S/amine ratio, and stress triaxiality, imposed by the relatively large thickness of the plate. In addition, the mechanical properties of the steel away from cracked regions were unaffected, indicating the localized nature of SOHIC.

Keywords: hydrogen-assisted cracking; corrosion; SOHIC; cleavage fracture

---

1. Introduction

Amine absorber columns in oil refineries are columns where a downflowing amine solution absorbs H_2S from the upflowing sour gas stream. The H_2S-rich amine solution, when exiting the absorber, is routed to a regenerator to produce an H_2S-lean amine that is recycled for use in the absorber. Under specific operation conditions, where the ratio of H_2S/amine is high, H_2S corrosion of the steel shell can take place, producing hydrogen, which can then enter the steel and cause hydrogen blistering and hydrogen-induced cracking. According to American Petroleum Institute standard API 571 [1], it is atomic hydrogen that enters the material and diffuses through the lattice. The hydrogen is then concentrated to various microstructural sites, such as interfaces between the matrix and inclusions or interfaces between the matrix and other phases. The local pressure increases and decohesion takes place, generating internal blistering or so-called hydrogen-induced cracking (HIC). In most cases, HIC cracks are oriented parallel to the rolling plane. Under the action of applied or residual stresses, the HIC cracks arrange in a vertically (through thickness) stacked array. Subsequently, the array is joined by cracks between the individual HIC cracks, which run perpendicular to the main applied stress. The interconnected HIC cracks, which run through the thickness of the plate, form a significant through-thickness crack. This cracking is classified as stress-oriented hydrogen-induced cracking (SOHIC). As mentioned above, hydrogen is produced by a corrosion reaction, such as when steel is exposed to wet H_2S service.

Hydrogen damage in wet H_2S environments is classified as blistering/HIC/SOHIC/Sulfide Stress Cracking (SSC) in API 571 recommended practice [1], while the assessment of hydrogen damage is performed in accordance with API 579 [2], following the work of Buchheim et al. [3] on the development of fitness-for-service rules for the assessment of HIC and SOHIC damage. While hydrogen blistering and HIC is a frequent problem when steel operates in wet H_2S service, SOHIC is a rather rare phenomenon and when occurring is mostly associated with residual stresses at welds. SOHIC in pipelines and pressure vessels has been thoroughly reviewed by Pargeter [4]. Most of the case studies reported associated with SOHIC are related with SOHIC at welds and concern mostly pipelines. There is only one case of SOHIC in an amine absorber column, reported by McHenry et al. [5] regarding the Chicago refinery incident in 1984. Even in this case, SOHIC originated from the Heat Affected Zone (HAZ) of a repair weld, which did not receive a stress-relieving post-weld heat treatment. Following the Chicago refinery incident, the Occupational Safety and Health Administration (OSHA) of the Department of Labor issued in 1986 a memorandum [6] stating, among others, that in a survey of similar refinery vessels and associated equipment conducted by the National Association of Corrosion Engineers (NACE), approximately 60% of 24 amine absorbers evaluated exhibited cracking. In addition, 12 of 14 monoethanolamine (MEA) units and three of five diethanolamine (DEA) units exhibited cracking. Sixteen instances of cracking were reported in associated equipment (i.e., regeneration units and piping) exposed to a chemically similar environment. Additionally, a similar survey by the Japan Petroleum Institute indicated that cracking had occurred in 72% of the amine gas treatment facilities which had responded to the survey.

As mentioned above, most failure cases involving SOHIC refer to welded piping, such as the work by Anezi et al. [7] in spiral welded pipes. The effects of loading and microstructure have been discussed by Kobayashi et al. [8] as well as by Koh et al. [9], while the effects of heat treatment on SOHIC of pressure vessel steels have been discussed by Tsuchida et al. [10]. Most reported results refer to HIC, as does the work of Findley et al. [11] on the mechanism of HIC in pipeline steels and the work of Gan et al. [12] on hydrogen trapping in H_2S environments. In situ observation of HIC propagation is reported by Fujishiro et al. [13], while corrosion-induced microcracking in H_2S environments has been discussed by Okonkwo et al. [14]. A recent review of HIC in pipelines and pressure vessel steels is presented by Ghosh et al. [15]. The reported cases of SOHIC failures in pressure vessels are limited and mostly associated with welds. The present case refers to SOHIC in the base plate of a pressure vessel, away from welds. This is a rare case and it is, therefore, very important to investigate the conditions and contributing factors of this type of damage.

A decommissioned amine absorber column, which exhibited hydrogen blistering in the internal wall, was made available for study. Blistering is shown in Figure 1. The aim of the present work is to investigate the underlying SOHIC damage and to identify the contributing factors that led to cracking.

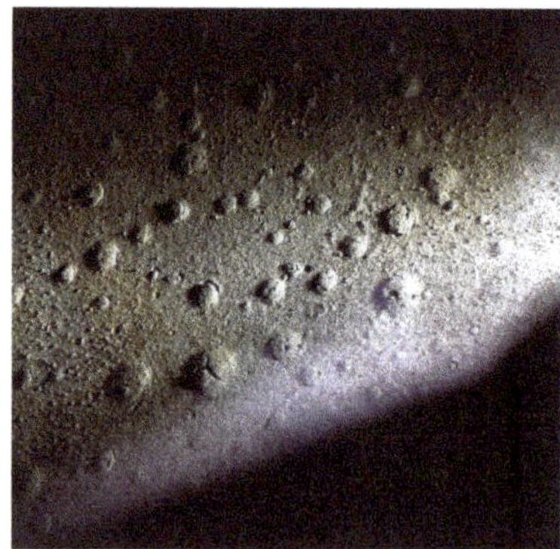

Figure 1. Blistering in internal surface.

2. Materials and Methods

2.1. Operating Conditions and Material of Construction

The column length, without the end cups, is 19 m. The internal diameter is 2.2 m and the shell thickness 91 mm. The service is characterized as wet H_2S, H, and amine service. The amine was methyldiethanolamine (MDEA). The operating pressure was 80 kg/cm^2 and the operating temperature was 70 °C. The column had been field-hydrotested before operation. In addition, all the welds had been 100% radiographically inspected, while a stress-relieving Post Weld Heat Treatment (PWHT) was performed in all welds. The column operated for 11 years. Following the detection of blistering in the inside diameter (ID), shown in Figure 1, and based on a relevant API level-3 fitness-for-service analysis, a decision was made to replace the vessel. In the 6-month period before replacement, hydrogen permeation measurements were conducted at the outside diameter (OD), indicating hydrogen flux values ranging from 10 to 120 pl/cm^{2}s. It is to be noted that hydrogen flux is related to corrosion activity, by H_2S, at the ID. During the 11 years, the vessel operated occasionally at high H_2S/MDEA molar ratio, above the normal ratio of 0.3 and reaching values up to 1. It is anticipated that this led to higher corrosion activity and higher hydrogen generation and entry in the material. The system of axes used for reference is shown in Figure 2. A cylindrical coordinate system is selected. The axis of the column is the z axis, while the θ axis is the circumferential direction and the r axis is the thickness direction.

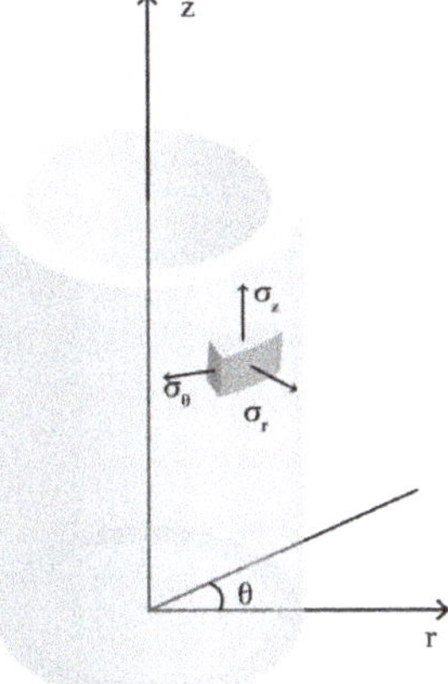

Figure 2. System of axes used for the positioning of specimens.

The material of construction was SA516-60 HIC-resistant steel. Typical mechanical properties of the material are: yield strength 283 MPa, ultimate tensile strength 452 MPa, and elongation 39%. The steel was designated as a HIC-resistant steel since special melting practices, such as vacuum induction melting (VIM) and vacuum arc remelting (VAR), were employed during fabrication. Gases dissolved in the liquid during melting, such as oxygen, nitrogen, and hydrogen, escape from the liquid to the vacuum chamber. This reduces porosity and additionally provides for: (a) minimization of inclusion content and (b) inclusion shape control. Normalizing has been applied as the final heat treatment process. In addition, the steel was subjected to a HIC test as per NACE TM-0284 [16].

2.2. Sampling Positions

A section of the column shell was cut for investigation. The position of the section was 5 m from the bottom of the column. In the inside diameter, blistering was evident (Figure 1). The shell section is shown in Figure 3. The section was inspected by ultrasonic testing (UT). Locations with UT signals are indicated by thick black horizontal lines (UT) in Figure 3. The positions of the metallographic sections are indicated by perpendicular white lines. Three positions were considered in this work: positions 500, 577, and 604, which pass through the UT signals. The numbers indicate distances, in mm, from a reference point on the vessel shell. It should be noted that the UT signal from position 577 was scattered and not concentrated in a line, as the signals from positions 500 and 604. As will be shown below, the UT signals from positions 500 and 604 correspond to SOHIC cracking, while the signal from position 577 corresponds to HIC, i.e., isolated HIC cracks without linking. An explanation is provided in Figure 3b, correlating the UT signal with the respective metallographic section.

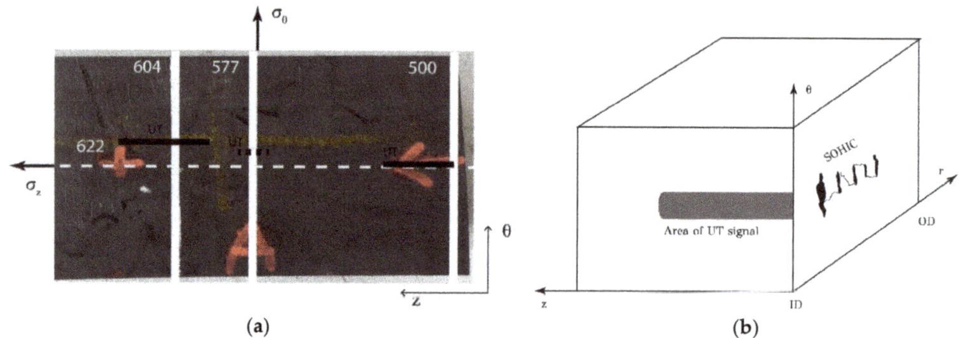

Figure 3. (**a**) Shell section for analysis (top view) indicating locations of metallographic sections and positions of UT signals; (**b**) correlation of SOHIC with UT signal (SOHIC: Stress Oriented Hydrogen Induced Cracking, ID: Inside Diameter, OD: Outside Diameter, UT: Ultrasonic Testing).

2.3. Experimental Procedures

Chemical analysis of the steel was performed by optical emission spectroscopy. Metallographic analysis was carried out on transverse cross sections at locations 500, 577, and 604. Specimen preparation included cutting with Struers "Accutom 2" (Struers, Ballerup, Denmark), grinding with SiC papers 120, 320, 500, 800, and 1000 grit, and polishing with diamond paste of 3 and 1 μm diameter. Etching was performed with 4% Nital reagent. Examination of the metallographic specimen was performed on an optical metallographic microscope, Leitz "Aristomet" (Leica Camera AG, Germany) at magnifications 50×–500×. Mechanical testing involved tensile testing according to ASTM EN 10002-1 specification and Charpy-V-notch (CVN) impact testing in several locations throughout the thickness, according to ASTM E23 specification. Specimens for mechanical testing were extracted from a region with no UT signals, i.e., a region free from cracking (HIC or SOHIC). The exact locations of the tensile specimens relative to the inside diameter (ID) and outside diameter (OD) are shown in Figure 4. Two tensile specimens were tested for each location. For the CVN testing, three specimen orientations were used, depicted in Figures 4 and 5. In the CVN-A configuration, the specimen length is along the transverse direction (θ) and the notch is in the thickness (r) direction. In the CVN-B configuration, the specimen length is along the longitudinal axis (z) of the vessel shell and the notch in the transverse (θ) direction. In the CVN-C configuration, the specimen length is along the thickness direction of the shell (r) and the notch in the transverse direction (θ). Two impact test specimens were tested for each location.

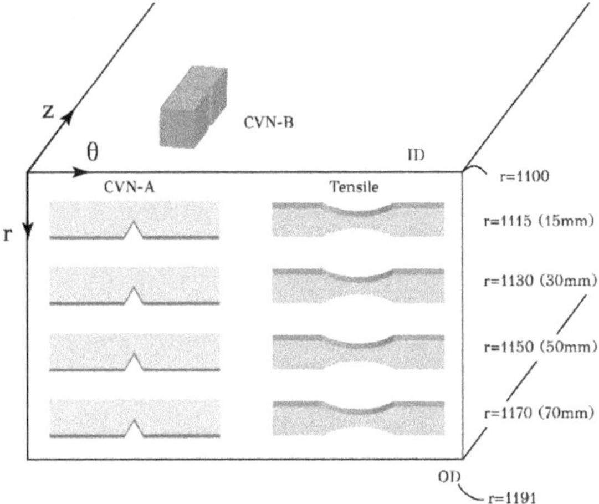

Figure 4. Positions of tensile specimens as well as Charpy-V-Notch, CVN-A and CVN-B specimens relative to ID and OD.

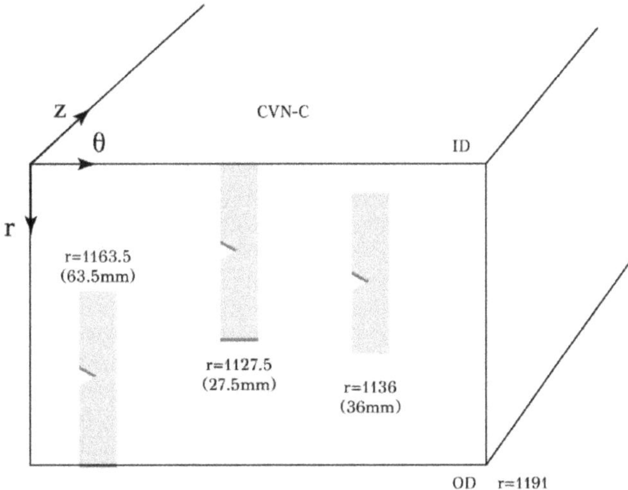

Figure 5. Positions of the CVN-C specimens relative to ID and OD.

The crack at location 604 was opened for fractographic analysis. In order to open the crack, a section containing the crack was cut and cooled to $-30\ ^\circ$C for 24 h. Then, the section containing the crack was put in an anvil. The crack was opened by impact on one side of the section with a hammer. A small section of "fresh" fracture area was generated. Investigation of the fracture surface was performed in a JEOL JSM-7610F SEM (JEOL, Tokyo, Japan) equipped with a field emission gun. The operating voltage was 20 kV.

3. Results

3.1. Material Characterization

The chemical composition of the material is, in wt %: 0.16 C/0.20 Si/1.13 Mn/0.01 P/(<0.01 S). The tensile properties are shown in Table 1. In addition to yield strength, ultimate tensile strength, and elongation, the elastic modulus was determined to be 210 GPa. In general, the measured properties agree with the manufacturer's data. Although there is a slight decrease in the ultimate tensile strength relative to the manufacturer's data, the elongation of the material is high and indicates a ductile material. This argument is also in agreement with the impact test results, which are shown in Tables 2–4 for CVN-A, CVN-B, and CVN-C configurations, respectively. All notch configurations exhibited high impact values that do not indicate any deterioration of the notch ductility of the material in the regions away from cracking. In other words, the observed cracking was highly localized.

Table 1. Tensile test results at different locations throughout the thickness of the plate.

Location (mm from ID)	Yield Strength (MPa)	UTS (MPa)	Elongation (%)
15	290	432	46.4/45.06
30	287	430	42.26/37.2
50	291	439	38.0/39.4
70	286	438	40.4/38.5
Manufacturer Data	283–285	445–452	38–39

Table 2. Results from CVN-A impact test configuration.

Notch Position (mm from ID)	Specimen ID	CVN Energy (J)	CVN Average (J)
15	A1/A5	186/194	190
30	A2/A6	184/180	182
50	A3/A7	170/186	178
70	A4/A8	168/168	168

Table 3. Results from CVN-B impact test configuration.

Notch Position (mm from ID)	Specimen ID	CVN Energy (J)	CVN Average (J)
15	B1/B5	186/178	182
30	B2/B6	216/188	202
50	B3/B7	190/192	191
70	B4/B8	184/182	181

Table 4. Results from CVN-C impact test configuration.

Notch Position (mm from ID)	Specimen ID	CVN Energy (J)	CVN Average (J)
27.5	C3/C6	170/186	178
36	C2/C5	154/160	156
63.5	C1/C4	186/190	188

3.2. Metallographic Characterization of SOHIC

The microstructure of the steel is depicted in the micrograph of Figure 6, indicating that the microstructure consists of ferrite and pearlite. The metallographic section at location 604 is shown in Figure 7, which is an assembly of micrographs starting from the inside diameter (ID) of the shell. This is a typical form of stress-oriented hydrogen-induced cracking (SOHIC). Although, microscopically, the crack direction changes several times during propagation, macroscopically, it is a through-thickness crack with a direction perpendicular to the applied stress. The crack appears to interconnect blisters/HIC cracks, which have formed at various levels across the thickness of the plate. The HIC cracks lie on the rolling plane and are stacked normal to the rolling plane. The HIC cracks are linked with cracks which run in a direction normal to the rolling plane and perpendicular to the applied hoop stress. The initiation is at 5 mm from the ID, while the overall crack length is 23 mm.

The metallographic section at location 577 is depicted in Figure 8a. Blisters/HIC cracks have formed, but they are not interconnected, at least at the plane of observation. This can explain the scattered mode of the UT signal at position 577. The propagation of HIC cracks is through the ferrite phase, between the pearlite colonies at ferrite/pearlite interfaces, as depicted in Figure 8b. A similar crack propagation mechanism has been also observed by Gingel and Garat [17] in an API 5L grade X60 pipeline steel.

The metallographic section at location 500 is depicted in Figure 9. The crack has initiated at 5 mm from the ID and has an overall length of 30 mm. This type of cracking is also classified as SOHIC.

Both cracks at locations 500 and 604 (Figures 7 and 9) are, therefore, classified as SOHIC. They interconnect blisters/HIC cracks and they propagate from the ID towards the OD under the action of the applied stress. According to API RP 571 [1], SOHIC consists of arrays of small HIC cracks, initiated at internal blisters in the rolling plane. These HIC cracks are stacked in a direction normal to the applied hoop stress and are linked by cracks normal to the stress. SOHIC should not be confused with stepwise hydrogen-induced cracking (SWC), which exhibits a not-aligned stepped morphology in the absence of stress. The limited cases of SOHIC failures that are reported in Pargeter's review [4] indicate that SOHIC requires a combination of severe hydrogen cracking conditions and stress. In the present case, the amine absorber column was operated under high H_2S/MDEA ratios for certain time periods, promoting high hydrogen charging conditions. In addition, it has been suggested [4] that triaxial loading can encourage the formation of small HIC cracks by increasing the hydrogen solubility in the steel. Stress triaxiality can be induced by the large plate thickness (91 mm). It is expressed by the ratio $\sigma_h/\overline{\sigma}$, where σ_h is the hydrostatic stress and $\overline{\sigma}$ is the von Mises equivalent stress. The triaxiality ratio was calculated (see Appendix A) for the operating conditions and was found as 0.5, which is a number indicating that the material was subjected to a moderate triaxial stress state. In addition, it is interesting to note that the HIC cracks form even though the radial stress is compressive. However, these cracks do not propagate in the rolling plane, but instead they are linked in the radial direction under the action of the hoop stress.

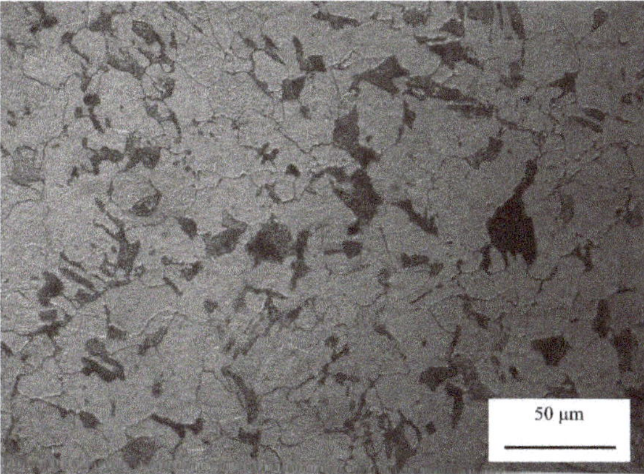

Figure 6. Ferrite–pearlite microstructure of the investigated steel.

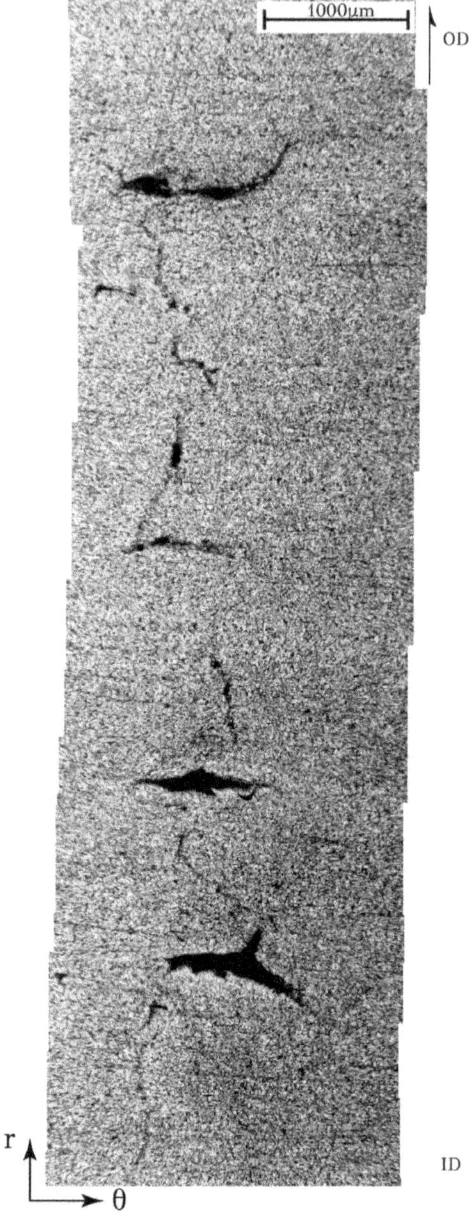

Figure 7. SOHIC cracking at location 604.

Figure 8. (a) HIC cracks at location 577; (b) tip of HIC crack indicating propagation in ferrite between
pearlite colonies.

Figure 9. SOHIC at location 500.

3.3. Fractographic Analysis of Opened SOHIC Crack

As discussed above, the SOHIC crack at location 500 was opened for fractographic analysis.
A low-magnification assembly of SEM micrographs is shown in Figure 10. The region of "fresh" fracture
resulting by the opening procedure and the region of SOHIC fracture are indicated. The boundary
between the fresh fracture and SOHIC is depicted in Figure 11. It is clear that there are two fracture
modes. The fresh fracture area is characterized by dimple fracture, also indicated in Figure 12, while the
SOHIC area is characterized by cleavage, indicated in Figure 13. In addition, several blisters/HIC
cracks can be seen on the fracture surface. The entire SOHIC region, including the areas between
blisters, is characterized by cleavage as indicated in Figure 14. No black deposit, related to FeS,
was found on the fracture surface. In addition, Energy Dispersive X-ray (EDX) analysis performed
on the cleavage areas did not detect the presence of sulfur. The absence of black deposits and sulfur
indicate clearly that cracking is not related to sulfide stress cracking (SSC). A large blister/HIC crack is
depicted in Figure 15. It was possible to approach the interior of this blister in order to investigate the
morphology of the blister wall as it resulted by the HIC crack opening due to the buildup of hydrogen

pressure. The internal wall is depicted in Figure 16, indicating a cleavage fracture surface. Thus, blistering and HIC cracking proceeded with cleavage of the material.

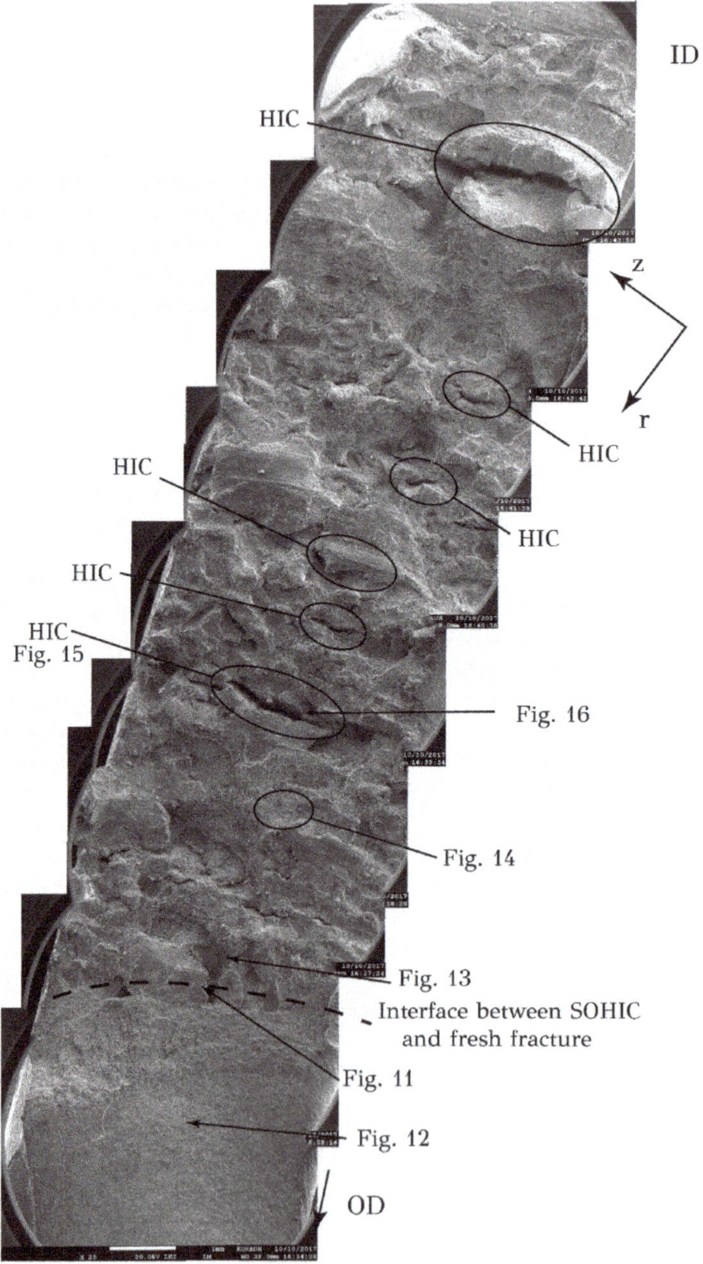

Figure 10. Assembly of SEM micrographs depicting the fracture surface of the opened SOHIC crack at position 500. The location of HIC cracks and individual SEM micrographs is indicated.

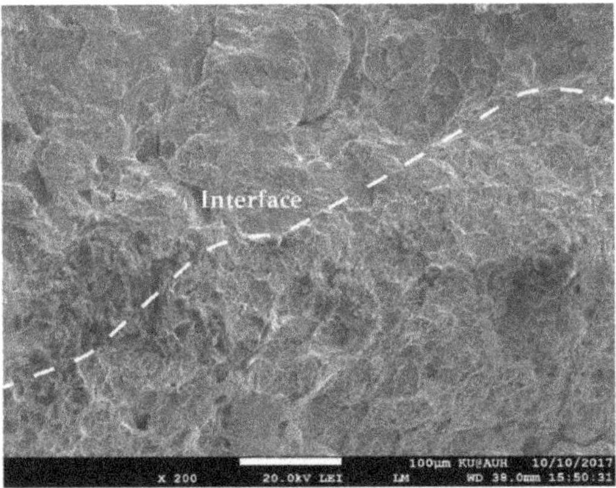

Figure 11. Boundary between SOHIC and "fresh" fracture.

Figure 12. Dimple fracture in the "fresh" fracture region.

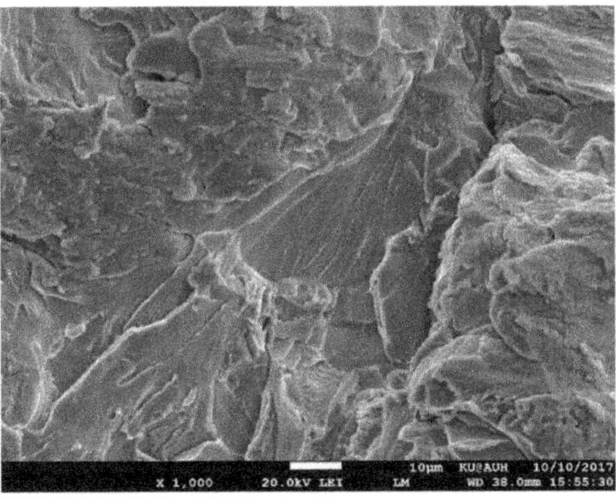

Figure 13. Cleavage fracture at SOHIC region.

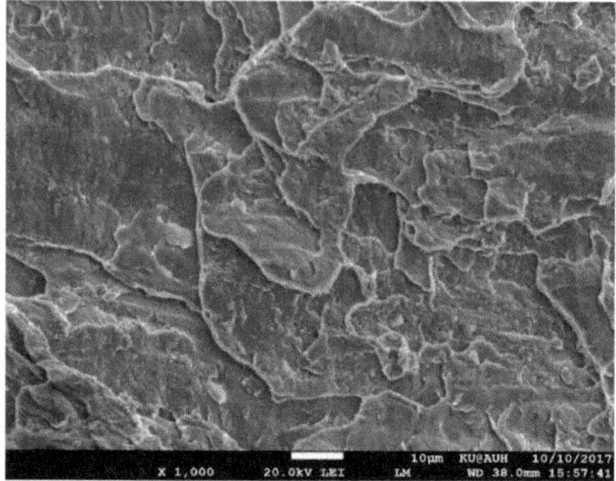

Figure 14. Cleavage fracture in SOHIC region between HIC cracks.

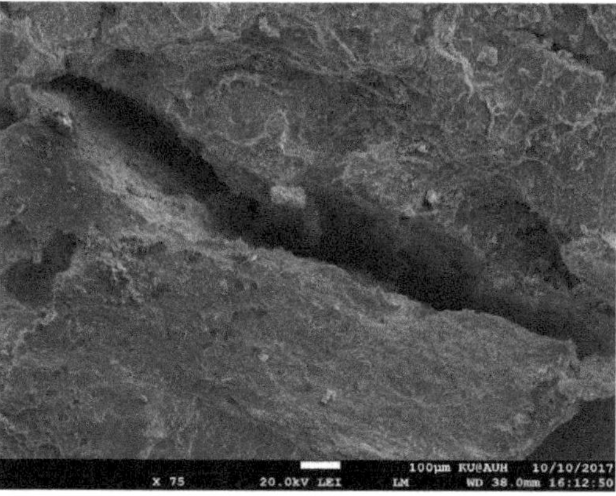

Figure 15. A large HIC crack at the SOHIC fracture region.

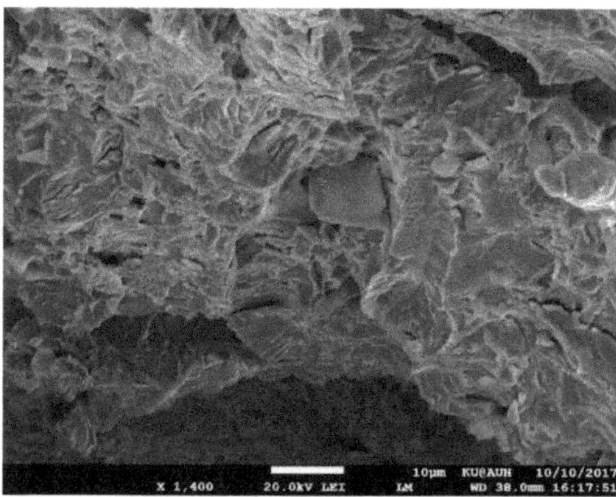

Figure 16. Cleavage fracture morphology of the internal blister/HIC crack wall.

It is important to note that location 577 exhibited only stacked HIC cracks with no evidence of crack linking. This indicates that the SOHIC sequence involves the formation of HIC cracks first, followed by through-thickness linking of the HIC cracks, in agreement with observations by Crolet and Adam [18] and Ohki et al. [19]. Both HIC formation and linking involve transgranular cleavage fracture mechanisms as indicated by the fractographic analysis of the opened crack at location 500. The transgranular propagation mode is in agreement with the findings of Bruckhoff et al. [20]; however, the cleavage mode of linking is not in agreement with the suggestion of Pargeter [4] that linking is through a slip mechanism, or that of Azevedo [21], who observed ductile fracture mode in linking hydrogen blisters in an API 5L X46 steel. Slip may be involved in the transport of hydrogen, as shown by several studies [22,23], but the propagation of the linking cracks is clearly through a cleavage mechanism. The cleavage mode of the linking cracks confirms the important influence of stress triaxiality in SOHIC, since triaxiality introduces plastic constraint and promotes cleavage fracture. Triaxiality effects have been discussed by other investigators, such as the effect on hydrogen concentration by Toribio et al. [24], the effect on ductility by Mirza et al. [25], and the effect on fracture behavior by Borvik et al. [26].

4. Discussion: Implications for H_2S/Amine Service

The results presented above indicated that the failure mechanism of the absorber column, which operated under wet H_2S conditions, was SOHIC. The main characteristic of SOHIC is the presence of HIC cracks lying on the rolling plane, stacked one on top of the other, and linked with cracks running perpendicular to the hoop stress. The SOHIC sequence involves the formation of HIC cracks first, followed by through-thickness linking of the HIC cracks. The major contributing factors have been: (a) high hydrogen charging conditions, since for some time periods, the vessel operated under high H_2S/MDEA ratios; and (b) stress triaxiality, imposed by the relatively large thickness of the plate. The role of these two factors have been discussed thoroughly above. An additional factor that may have to be considered is the type of steel used for the construction of the vessel. As discussed in the review by Ossai et al. [27], there is a need for high-quality steel, free from inclusions and other microstructural defects that could act as sites for hydrogen-related crack initiation. For wet H_2S service, it is typical to use a HIC-resistant steel. However, as indicated by the resulting cracking, a HIC-resistant steel might not be SOHIC-resistant, in particular when subjected to high hydrogen charging conditions. This argument has also been raised by Pargeter [4]. As mentioned above, inclusions could act as potential sites for HIC. It appears that clean steels, such as the HIC-resistant steels, contain fewer stringer-type inclusions than conventional steels. In these cases, HIC is rather initiated on other interfaces, such as ferrite–pearlite interfaces. The metallographic analysis at the 577 position revealed that HIC cracks propagate in the ferrite matrix between pearlite colonies, as also observed in [28]. It is apparent that the use of a HIC-resistant steel should not be the only measure for mitigating SOHIC. Current practice in the fabrication of pressure vessels operating in similar environments calls for the use of austenitic stainless-steel lining in order to reduce hydrogen diffusion and hydrogen entry to the steel plate. In addition, operating conditions should be carefully monitored to minimize service under high H_2S/amine ratios.

Finally, it is important to note that the results of mechanical testing indicated that the ductility of the steel in regions away from the observed cracking has not deteriorated. Tensile elongation and notch ductility were maintained at high levels. However, even in a highly localized form, SOHIC did take place, indicating the type of hydrogen damage that can take place in an otherwise ductile material.

5. Conclusions

Taking into account the results presented above, the following conclusions can be drawn:

(1) The cracking in the amine absorber column shell is classified as stress-oriented hydrogen-induced cracking (SOHIC).

(2) SOHIC proceeds in two steps: (a) initiation of small HIC cracks lying in the rolling plane and stacked in a direction normal to the applied stress; (b) through-thickness linking of HIC cracks.

(3) The propagation of the HIC cracks as well as the through-thickness link cracks is associated to cleavage fracture mechanisms.

(4) The key factors identified in this failure were: (a) short periods of high hydrogen charging conditions as manifested by high $H_2S/MDEA$ ratios and (b) stress triaxiality imposed by the relatively large thickness of the plate.

(5) The results indicate that a HIC-resistant steel might not be immune from SOHIC. Under high hydrogen charging conditions, HIC cracks can initiate at interfaces other than stringer-type inclusions, such as ferrite–pearlite interfaces in the microstructure of the steel.

Author Contributions: G.N.H. supervised the work and wrote the manuscript; H.K., P.P. and K.P. performed various aspects of the experimental work. I.A., D.P. and M.S. provided operational data and specimens for the investigation. All authors contributed to the analysis of the results.

Funding: This research received no external funding.

Acknowledgments: The provision of operational data and specimens for this investigation by Motor Oil Hellas is greatly appreciated.

Conflicts of Interest: The authors declare no conflict of interest.

Appendix A

The stresses in the shell of a thick-wall cylindrical vessel subjected to an internal pressure are

$$\sigma_z = (p_i r_i^2 - p_o r_o^2)/(r_o^2 - r_i^2)$$

$$\sigma_\theta = [(p_i r_i^2 - p_o r_o^2)/(r_o^2 - r_i^2)] - [r_i^2 r_o^2(p_o - p_i)/(r^2(r_o^2 - r_i^2))]$$

$$\sigma_r = [(p_i r_i^2 - p_o r_o^2)/(r_o^2 - r_i^2)] + [r_i^2 r_o^2(p_o - p_i)/(r^2(r_o^2 - r_i^2))]$$

where σ_z, σ_θ, and σ_r are the axial, circumferential (hoop), and radial stresses, respectively; p_i and p_o are the internal and external pressures, respectively; and r_i and r_o are the internal and external radii of the cylinder, respectively. Inserting p_i = 8 MPa, p_o = 0.1 MPa, r_i = 1100 mm, and r_o = 1191 mm, then at position r = 1110 mm, i.e., 10 mm from the internal surface, the stresses are σ_z = 45.8 MPa, σ_θ = 98.5 MPa, and σ_r = −7 MPa. Stress triaxiality is defined by the ratio $\sigma_h/\overline{\sigma}$, where σ_h is the hydrostatic stress and $\overline{\sigma}$ is the von Mises equivalent stress given by:

$$\sigma_h = \frac{1}{3}(\sigma_z + \sigma_\theta + \sigma_r)$$

$$\overline{\sigma} = \sqrt{\frac{1}{2}[(\sigma_z - \sigma_\theta)^2 + (\sigma_\theta - \sigma_r)^2 + (\sigma_r - \sigma_z)^2]}$$

Inserting the above values for stresses, the triaxiality ratio is derived as $\sigma_h/\overline{\sigma}$ = 0.5.

References

1. American Petroleum Institute. *API RP 571, Damage Mechanisms Affecting Fixed Equipment in the Refining and Petrochemical Industries*, 2nd ed.; American Petroleum Institute: Washington, DC, USA, 2011.

2. American Petroleum Institute. *API RP 579, Fittness-for-Service*; American Petroleum Institute: Washington, DC, USA, 2007.

3. Buchheim, G.M.; Osage, D.A.; Staats, J.C. Development of fitness-for-service rules for the assessment of hic and SOHIC damage in API 579-1/ASME FFS-1. In Proceedings of the ASME 2008 Pressure Vessels and Piping Conference, Chicago, IL, USA, 27–31 July 2008; pp. 761–775.

4. Pargeter, R.J. Susceptibility to SOHIC for linepipe and pressure vessel steels—review of current knowledge. In Proceedings of the International Corrosion Conference, Nashville, TN, USA, 11–15 March 2007.

5. McHenry, H.I.; Shives, T.R.; Read, D.T.; McColskey, J.D.; Brady, C.H.; Purtscher, P.T. *Examination of a Pressure Vessel that Ruptured at the Chicago Refinery of the Union Oil Company on July 23 1984*; Occupational Safety & Health Administration, US Department of Labor: Washington, DC, USA, 1986.

6. Occupational Safety and Health Administration (OSHA). *Potentially Hazardous Amine Absorber Pressure Vessels Used in Refinery Processing, Memorandum to OSHA Regional Administrators*; US Department of Labor: Washington, DC, USA, 1986. Available online: https://www.osha.gov/laws-regs/standardinterpretations/1986-04-11 (accessed on 28 October 2017).

7. Al-Anezi, M.A.; Rao, S. Failures by SOHIC in sour hydrocarbon service. *J. Fail. Analysis Prev.* **2011**, *11*, 363–371. [CrossRef]

8. Kobayashi, K.; Dent, P.; Fowler, C.M. Effects of stress conditions and microstructure on SOHIC susceptibility. In Proceedings of the International Corrosion Conference, San Antonio, TX, USA, 9–13 March 2014.

9. Koh, S.U.; Jung, H.G.; Kang, K.B.; Park, G.T.; Kim, K.Y. Effect of microstructure on hydrogen-induced cracking of linepipe steels. *Corrosion* **2008**, *64*, 574–585. [CrossRef]

10. Tsuchida, Y.; Naruoka, Y.; Tokunaga, Y. Effects of heat treatment conditions on SOHIC in normalized steel plates for pressure vessel use. *J. High Press. Inst. Jpn.* **1996**, *34*, 9–15. [CrossRef]

11. Findley, K.O.; O'Brien, M.K.; Nako, H. Critical assessment 17: Mechanisms of hydrogen induced cracking in pipeline steels. *Mater. Sci. Technol.* **2015**, *31*, 1673–1680. [CrossRef]

12. Gan, L.; Huang, F.; Zhao, X.; Liu, J.; Cheng, Y.F. Hydrogen trapping and hydrogen induced cracking of welded X100 pipeline steel in H_2S environments. *Int. J. Hydrogen Energy* **2018**, *43*, 2293–2306. [CrossRef]

13. Fujishiro, M.T.; Hara, D.T. In-situ observation of hydrogen induced cracking propagation behavior. *Corrosion* **2018**. Available online: http://corrosionjournal.org/doi/abs/10.5006/2757 (accessed on 16 June 2018). [CrossRef]

14. Okonkwo, P.; Shakoor, R.; Benamor, A.; Amer Mohamed, A.; Al-Marri, M. Corrosion behavior of API X100 steel material in a hydrogen sulfide environment. *Metals* **2017**, *7*, 109. [CrossRef]

15. Ghosh, G.; Rostron, P.; Garg, R.; Panday, A. Hydrogen induced cracking of pipeline and pressure vessel steels: A review. *Eng. Fract. Mech.* **2018**, *199*, 609–618. [CrossRef]

16. NACE International. *Evaluation of Pipeline and Pressure Vessel Steels for Resistance to Hydrogen-Induced Cracking*; NACE International: Houston, TX, USA, 2003.

17. Gingell, A.; Garat, X. Observations of damage modes as a function of microstructure during NACE TM-01-77/96 tensile testing of API 5L grade X60 linepipe steels. In Proceedings of the International Corrosion Conference, San Antonio, TX, USA, 25–30 April 1999.

18. Crolet, J.L.; Adam, C. SOHIC without H_2S. *Mater. Perform.* **2000**, *39*, 86–90.

19. Ohki, T.; Tanimura, M.; Kinoshita, K.; Tenmyo, G. Effect of inclusions on sulphide stress cracking. In Proceedings of the Symposium on Stress Corrosion—New Approaches, Montreal, QC, Canada, 22–27 June 1975; pp. 399–419.

20. Bruckhoff, W.; Geier, O.; Hofbauer, K.; Schmitt, G.; Steinmetz, D. Rupture of a sour gas line due to stress orientated hydrogen induced cracking-Failure analyses, experimental results and corrosion prevention. In Proceedings of the NACE CORROSION/85 Conference, Boston, MA, USA, 25–29 March 1985; p. 389.

21. Azevedo, C.R.F. Failure analysis of a crude oil pipeline. *Eng. Fail. Anal.* **2007**, *14*, 978–994. [CrossRef]

22. Miyoshi, E.; Tanaka, T.; Terasaki, F.; Ikeda, A. Hydrogen-induced cracking of steels under wet hydrogen-sulfide environment. *J. Eng. Ind.* **1976**, *98*, 1221–1230. [CrossRef]

23. Robertson, I.M.; Sofronis, P.; Nagao, A.; Martin, M.L.; Wang, S.; Gross, D.W.; Nygren, K.E. Hydrogen embrittlement understood. *Metall. Mater. Trans. B* **2015**, *46*, 1085–1103. [CrossRef]

24. Toribio, J.; Kharin, V.; Vergara, D.; Lorenzo, M. Hydrogen diffusion in metals assisted by stress: 2D numerical modelling and analysis of directionality. *Solid State Phenom.* **2015**, *225*, 33–38. [CrossRef]

25. Mirza, M.S.; Barton, D.C.; Church, P. The effect of stress triaxiality and strain-rate on the fracture characteristics of ductile metals. *J. Mater. Sci.* **1996**, *31*, 453–461. [CrossRef]

26. Børvik, T.; Hopperstad, O.S.; Berstad, T. On the influence of stress triaxiality and strain rate on the behaviour of a structural steel. Part ii. Numerical study. *Eur. J. Mech.* **2003**, *22*, 15–32. [CrossRef]

27. Ossai, C.I.; Boswell, B.; Davies, I.J. Pipeline failures in corrosive environments—A conceptual analysis of trends and effects. *Eng. Fail. Anal.* **2015**, *53*, 36–58. [CrossRef]
28. Cayard, M.S.; Kane, R.D.; Cooke, D.L. An exploratory examination of the effect of sohic damage on the fracture resistance of carbon steels. In Proceedings of the International Corrosion Conference, New Orleans, LA, USA, 9–14 March 1997.

© 2018 by the authors. Licensee MDPI, Basel, Switzerland. This article is an open access article distributed under the terms and conditions of the Creative Commons Attribution (CC BY) license (http://creativecommons.org/licenses/by/4.0/).

Article

Improvement of Strength and Impact Toughness for Cold-Worked Austenitic Stainless Steels Using a Surface-Cracking Technique

Kwangyoon Kim [1], Minha Park [1], Jaeho Jang [1], Hyoung Chan Kim [1], Hyoung-Seok Moon [1], Dong-Ha Lim [1], Jong Bae Jeon [1], Se-Hun Kwon [2], Hyunmyung Kim [3] and Byung Jun Kim [1,*]

[1] Energy Plant R&D Group, Korea Institute of Industrial Technology, Busan 46938, Korea; kky0858@kitech.re.kr (K.K.); pmh0812@kitech.re.kr (M.P.); jngjho@kitech.re.kr (J.J.); chancpu@kitech.re.kr (H.C.K.); hyoungseok.moon@kitech.re.kr (H.-S.M.); dongha4u@kitech.re.kr (D.-H.L.); jbjeon@kitech.re.kr (J.B.J.)

[2] School of Materials Science and Engineering, Pusan National University, Busan 46241, Korea; sehun@pusan.ac.kr

[3] Department of Nuclear & Quantum Engineering, Korea Advanced Institute of Science and Technology, Daejeon 34141, Korea; h46kim@kaist.ac.kr

* Correspondence: jun7741@kitech.re.kr; Tel.: +82-10-5136-7743

Received: 25 October 2018; Accepted: 9 November 2018; Published: 12 November 2018

Abstract: For cryogenic applications, materials must be cautiously selected because of a drastic degradation in the mechanical properties of materials when they are exposed to very low temperatures. We have developed a new technique using a cold-working and surface-cracking process to overcome such degradation of mechanical properties at low temperatures. This technique intentionally induced surface-cracks in cold-worked austenitic stainless steels and resulted in a significant increase in both strength and fracture at low temperatures. According to the microstructure observations, dissipation of the crack propagation energy with surface-cracks enhanced the impact toughness, showing a ductile fracture mode in even the cryogenic temperature region. In particular, we obtained the high strength and toughness materials by a surface-cracking technique at 5% cold-worked specimen with surface-cracks.

Keywords: cold-working process; surface-cracking process; impact toughness; strength; low temperatures; austenitic stainless steels

---

1. Introduction

The study of very low temperature environments in terms of cryogenics is one field in which the materials play a major part in desirable performances in severe conditions. Cryogenic technologies have various applications in many different fields such as in the power industry, chemistry, electronics, manufacturing, transportation, and food processing by refrigeration [1–7]. For cryogenic applications, materials must be cautiously selected because of a drastic degradation in mechanical properties of the materials when they are exposed to very low temperatures [8]. Generally, austenitic steels with face centered cubic (FCC) structures such as stainless steels [9], Al alloys [10], Ni alloys [11], and Ti alloys [12] are used as low temperature materials because they have high impact toughness at low temperatures [13–15]. Although these alloys with FCC structure have excellent mechanical properties at low temperatures, brittle fracture may occur in welds and rectangular structures due to the stress concentration [16]. A brittle fracture is usually very dangerous because it occurs abruptly with little or no warning, resulting in serious economic losses and potentially a loss of many lives. Therefore, it is in high demand to develop new technologies in order to improve the mechanical properties of FCC alloys for use in severe environments such as in the cryogenics field.

In particular, materials with good mechanical properties are required for use in severe environments such as the cryogenics field because material characteristics are significantly degraded and changed at low temperatures. Work hardening, also known as strain hardening or cold-working, is a method to strengthen a metal by plastic deformation [17–19]. In detail, the strengthening by cold-working occurs because of a decrease in the mobility of dislocations during plastic deformation of metals. The strain hardening is caused by an increase in the dislocation density within the austenitic structure [20]. Furthermore, the strengthening effect by phase transformation in austenitic stainless steels where the strained-induced martensitic transformation from austenitic phase shows a substantial strengthening effect [21–25]. It has also been found that the yield and tensile strength of austenite stainless steels is gradually improved by increasing the cold-working level. Therefore, a cold-working process is typically considered to be an important technique for increasing the strength of steels [26,27]. However, negative effects such as the unfavorable material embrittlement can be caused by reductions in ductility and impact toughness after a cold-working process [28].

Generally, impact toughness is reduced with an increase in work hardening and precipitates, whereas it typically shows a positive correlation with increases in ductility [29]. In recent research, the grain size refinement was an important technique for improving the impact toughness for high-strength steels [30–34]. Thermal heat treatments affected enhanced impact toughness by an ultra-fine grained structure [29]. Another approach to enhancing impact toughness is to provide multiple pathways for crack propagation, for example, by adding ultra-fine particles within the fine granular structure. Delamination, or splitting, resulting from anisotropic microstructures such as crystalline grains and secondary phases is well known to improve the toughness of metals at low temperatures [23,28,29]. Toughening mechanisms are for distributing the stress near the crack tips by allowing the delamination fracture of the grains from one another, instead of brittle fractures in bulk materials [35–37]. Most engineering designs require materials with high strength and impact toughness to avoid dangerous failure due to brittle fracture at low temperatures [38]. However, it is difficult to obtain tougher and stronger steels because materials typically show opposing characteristics for ductility and brittleness. Therefore, the achievement of a simultaneous enhancement of strength and toughness is a challenge [38,39].

Higher strength and toughness are key requirements for steels used in various structural applications, such as for aircraft, buildings, and heavy machinery including cryogenics applications, in order to satisfy the increasing demands for reliability, durability, and safety. Toughness and strength do not always have opposite characteristics [40,41]. Increasing the toughness of metals without sacrificing other properties is critical for their economic competitiveness [31,38,42,43]. It is true that for intrinsically ductile materials, such as metals, the improvement in strength usually comes at the expense of toughness [44,45]. However, the present results show the possibility and potential of improving toughness and strength at the same time. In this study, we have developed a new technique using a cold-working and surface-cracking process to increase the strength and toughness at the same time. Our proposed method is called a surface-cracking technique which has two processes with the cold-working and surface-cracking process. Cold-working (CW) is one of the typical methods which imparts higher strength to steel, but decreases in ductility accompanied by toughness degradation are generally unavoidable. However, our work shows that enhanced toughness can be achieved using the surface-cracking technique without sacrificing strength. We applied this technique to austenitic stainless steels including base metal and weld metal.

2. Materials and Methods

The chemical compositions of stainless steels used in this study are shown in Table 1. A thick plate of STS304 was welded with a STS308L welding electrode (ESAB SEAH, Gyeongnam, Korea) by the submerged-arc welding (SAW) method. The dimension of the welded plate was in the width of 100 mm, the length of 1000 mm, and the thickness of 25 mm. For the post-weld heat treatment (PWHT), the austenitic stainless steels were processed at 105 °C for 2 h, and subsequently quenched in water.

In order to observe the microstructures, the test samples were mechanically polished using a 2000 grit SiC paper and then micro-polished using a 3 μm and 1 μm diamond paste. Then, sample surfaces were etched in a solution containing 40% distilled water, 30% HNO_3, and 30% NO_3.

Tensile testing was carried out at room temperature using a 5 kN full-automatic INSTRON 4204 tensile machine (INSTRON, Norwood, MA, USA) at a cross head speed of 0.03 mm/s. The dimensions of the miniaturized tensile specimens were 5 mm in gauge length, 1.2 mm in width, and 0.5 mm in thickness.

Table 1. The chemical compositions of the specimens used in this study (wt %).

Alloys	C	Mn	Si	Cr	Ni	Mo	Al	Cu	Fe
Base Metal (STS304)	0.046	1.19	0.42	18.23	8.02	0.149	0.003	0.223	Bal.
Weld Metal (STS308L)	0.02	1.98	0.41	19.71	10.79	0.03	-	0.13	Bal.

Vickers hardness tests were performed at room temperature with a load of 4.95 N with 10 s of dwell time. At least 10 measurements were made to calculate the average micro-hardness values by excluding the minimum value and maximum value. We prepared Charpy V-notch impact specimens (10.0 mm × 10.0 mm × 55.0 mm) without or with surface-cracks to evaluate the improvement of impact toughness by surface-cracks (Figure 1). Charpy V-notch impact tests were performed to measure the absorbed energy at temperatures from 20 to −180 °C using a full-automatic Zwick Charpy impact test machine according (Zwick/Roell, Ulm, Germany) to ASTM E23 [46].

Our proposed method is called a surface-cracking technique which involves two processes including a cold-working and a surface-cracking process. The first step was a typical method for improving the strength of stainless steel by a cold-working process. From the as-received (AS) STS304, a plate of steel with 20 mm thickness was cold worked into different thicknesses of the plates CW05, CW10, and CW30, with reductions of plate thickness by 5, 10 and 30%, respectively. In order to minimize the effect of directionality by cold-working, a cold-working process was performed several times at the same thickness reduction rate in both the longitudinal and transverse directions. The configurations and process conditions of the test specimens used in this study are summarized in Table 2.

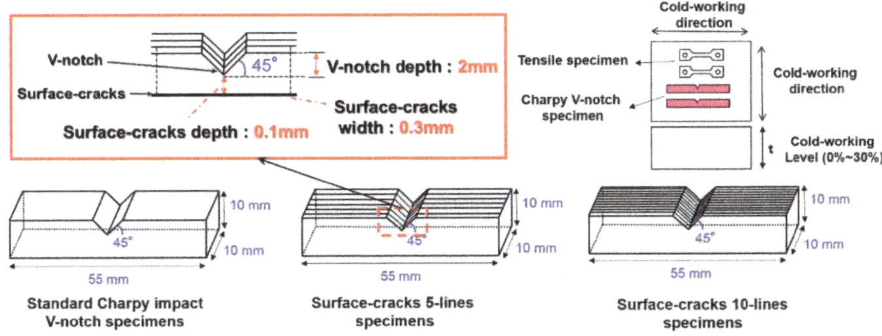

Figure 1. Configurations of the specimens used in this study.

The second step was a surface-cracking process for improving the impact toughness of stainless steel by inducing the surface-cracks on the specimen surface. This method was applied to Charpy V-notch impact test specimens to evaluate the enhancement of impact toughness for stainless steels. In this method, surface-cracks were formed in the direction perpendicular to the V-notch direction by wire cut machining V850G (EXCETEK, Taichung, Taiwan). This work was machined to a depth of 0.1 mm to disperse the tri-axial stress at the V-notch tip using a wire with a width of 0.3 mm (as shown in Figure 1). In order to observe the effect of surface-cracks, surface-cracks were fabricated with

5 and 10 lines and compared with standard specimens without surface-cracks. The overall process of a surface-cracking technique is shown in Figure 2.

Table 2. Configurations of the specimens used in this study.

Specimen ID	Region	Cold-Working Level	Number of Added Surface-Cracks
BM-AS	Base metal	-	-
BM-AS-L5	Base metal	-	5 lines
BM-AS-L10	Base metal	-	10 lines
BM-CW05	Base metal	5%	-
BM-CW05-L10	Base metal	5%	10 lines
BM-CW10	Base metal	10%	-
BM-CW10-L10	Base metal	10%	10 lines
BM-CW30	Base metal	30%	-
BM-CW30-L10	Base metal	30%	10 lines
WM	Weld metal	-	-
WM-L5	Weld metal	-	5 lines
WM-L10	Weld metal	-	10 lines

To analyze the phase transformation of austenitic stainless steels after a cold-working process, the volume fraction of martensite was measured by a FERITSCOPE FMP30 (Fischer, Windsor, CT, USA). The volume fraction was measured 8 times per specimen in total. The fracture surface of the Charpy impact test specimen was observed with a FE-SEM HITACHI S-4800 (HITACHI, Tokyo, Japan).

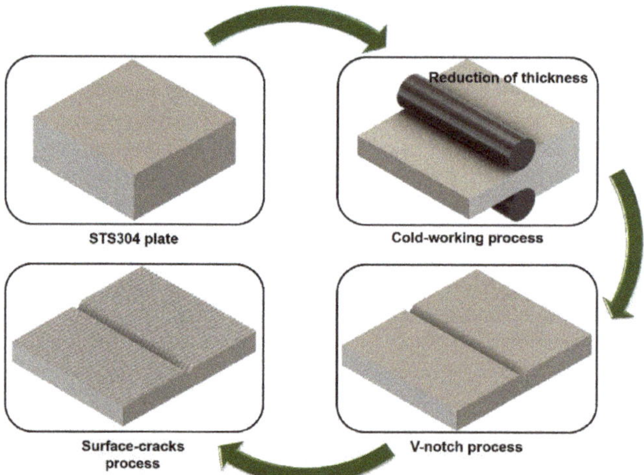

Figure 2. Schematic diagram of a surface-cracking technique.

3. Results and Discussion

3.1. Strengthening by a Cold-Working Process

Figure 3 shows the microstructure of base metal for austenitic stainless steel STS304 before and after a cold-working process. The microstructure of the base metal before cold-working was typically an austenitic structure with 160 μm grain size, as shown in Figure 3a. After the cold-working process, the microstructure was changed from austenite single phase to dual phases which consisted of austenite and martensite (Figure 3b–d). The fraction of martensitic phase increased with an increase in the cold-working level from 5% to 30%.

Figure 3. The microstructure of the base metal for austenitic stainless steel STS304 before and after a cold-working process: (**a**) AS-BM, (**b**) BM-CW05, (**c**) BM-CW10, and (**d**) BM-CW30.

Figure 4 shows the results of tensile properties for STS304 before and after a cold-working process. The ultimate tensile strength (UTS) of highly cold-worked steel (BM-CW30) with a reduction of plate thickness by 30% was about two times higher than that of the as-received steels (BM-AS). The tensile strength of cold-worked steels increased with an increase of the cold-working level from 5% to 30%. However, the elongation of cold-worked steels decreased with reductions in ductility.

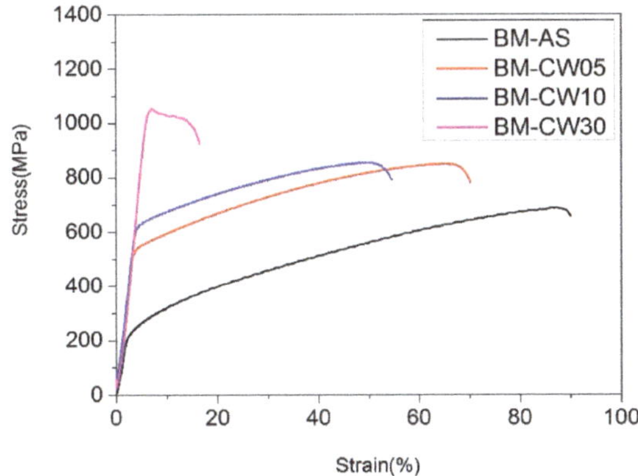

Figure 4. The stress-strain curves of STS304 before and after a cold-working process.

3.2. Mechanism of Strengthening of a Cold-Working Process

Cold-working generally leads to an increase in the yield strength of stainless steels because of the increased number of dislocations [43]. Figure 5 shows the results of the quantitative analysis for the martensite volume fraction measured by the ferrite scope. As-received steels (BM-AS) with a fully austenite structure were close to a 0% volume fraction of martensite. The volume fraction of martensite increased with the cold-working level from 5% to 30% (Figure 5). In the case of 30% cold-worked specimens (BM-CW30), the volume fraction of martensite increased to about 8%. Therefore, strength was increased by the phase transformation from austenite to martensite and the increase in dislocations by plastic deformation [20].

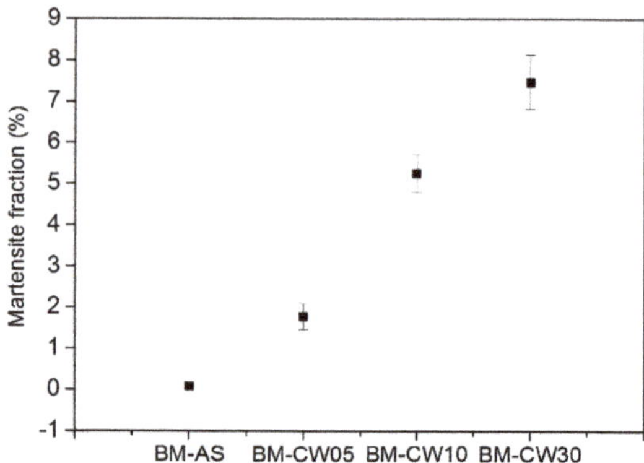

Figure 5. Martensite volume fractions (in %) of STS304 before and after a cold-working process.

Figure 6 shows the result of the Vickers hardness test before and after a cold-working process. The hardness value of as-received steel (BM-AS) was about 161 HV. Similar to the tensile test results, hardness values increased with an increase in the level of cold-working. The increase of the hardness value after a cold-working process was owed to the same mechanism as that in the tensile results which have the phase transformation of martensite from austenite and the increase in dislocations by plastic deformation. Therefore, results from tensile and hardness tests were in good agreement with quantitative analysis of the volume fraction of martensite as measured by the ferrite scope. Table 3 summarizes the test results of the tensile properties and Vickers hardness values obtained in this work.

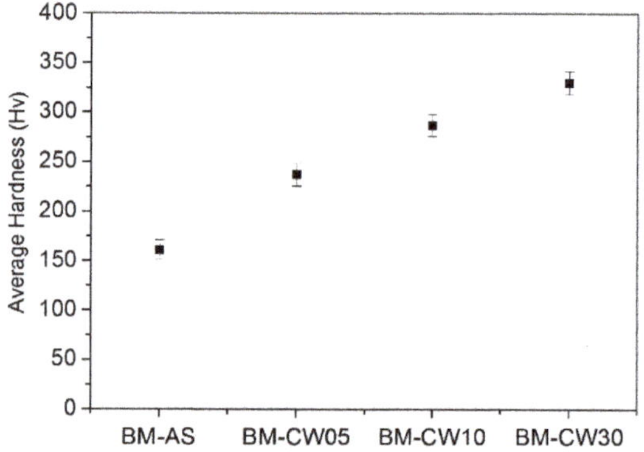

Figure 6. Hardness of STS304 before and after a cold-working process.

Table 3. Tensile properties and Vickers hardness values of the specimens used in this study.

Specimen ID	Yield Strength, σy (MPa)	Tensile Strength, σt (MPa)	Elongation, (%)	Average Vickers Hardness (HV)
BM-AS	212	689	91	161
BM-CW05	548	848	70	237
BM-CW10	611	854	56	286
BM-CW30	1002	1053	17	330

3.3. Toughening by a Surface-Cracking Process

Generally, ordinary cracks in materials weaken their mechanical characteristics due to the stress concentration at crack tips. Although the presence of flaws or cracks causes fractures due to the concentration of stress, a number of aligned cracks in a matrix can distribute the fracture energy in reverse. The absorbed energy vs. test temperature curves for specimens with a surface-cracking process is shown in Figure 7. At room temperature, the absorbed energy of as-received specimens was about 400 J. Although austenite stainless steel with face centered cubic (FCC) structure does not show the ductile-to-brittle temperature (DBTT) behavior, the absorbed energy tends to decrease with decreasing temperature. In the case of specimens with surface-cracks, the absorbed energy was almost the same comparing the as-received specimen at room temperature. However, the specimen with surface-cracks had a higher absorbed energy compared to the as-received specimen without surface-cracks at the low temperature region (below $-60\,°C$) As the number of surface-cracks increased from 0 lines (as-received) to 10 lines, the absorbed energy significantly increased at below $-60\,°C$. At a very low temperature ($-180\,°C$), the absorbed energy of the specimen with 10 surface-cracks (BM-AS-L10) was two times higher than that of the as-received specimen without surface-cracks. Therefore, this effect of enhanced toughness by a surface-cracking process occurred at very low temperatures when the material changed to having brittle properties.

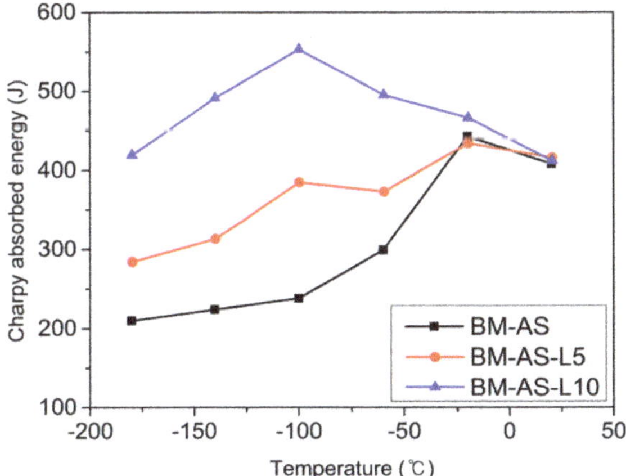

Figure 7. The absorbed energy vs. test temperature curves for specimens with surface-cracks.

3.4. Toughening by a Surface-Cracking Process

Figure 8 shows the fracture shape of as-received specimens and specimens with surface-cracks after the Charpy impact test. The facture shape of all specimens with or without surface-cracks observed large plastic deformation at room temperature (Figure 8a–c). There was no significant difference between the as-received specimens and specimens with surface-cracks near the V-notch. We also could not observe any differences in plastic deformation by the number of micro-cracks. However, the effect of surface-cracks was clearly observed at the cryogenic temperature ($-180\,°C$) (Figure 8d–f). Compared to the results at room temperature, as-received specimens were completely fractured and accompanied with slight plastic deformation at the cryogenic temperature (Figure 8d). However, the specimens with surface-cracks significantly increased plastic deformation around the V-notch and surface-cracks. The plastic deformation also increased with the increasing number of surface-cracks. From the results of the Charpy impact test in Figure 7, the effect of surface-cracks could be confirmed through the fractured formations. Therefore, the fracture shape of the fractured Charpy specimens was in good agreement with the results of the absorbed energy from the Charpy impact test at room temperature and cryogenic temperature.

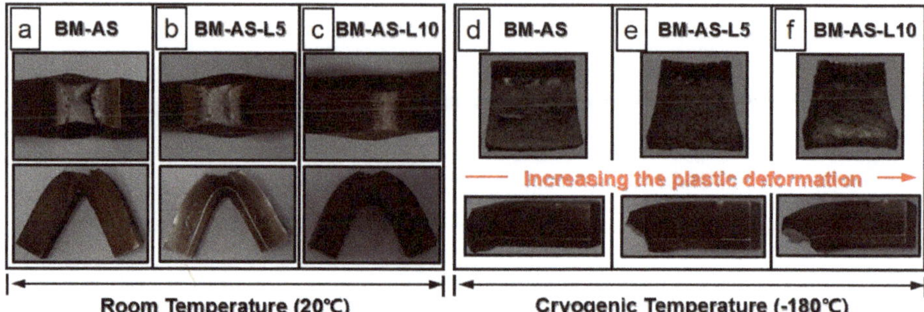

Figure 8. The fracture shape of the as-received specimen and the specimen with surface-cracks after Charpy impact test at room temperature (20 °C) and cryogenic temperature (−180 °C): (**a**) BM-AS (20 °C), (**b**) BM-AS-L5 (20 °C), (**c**) BM-AS-L10 (20 °C), (**d**) BM-AS (−180 °C), (**e**) BM-AS-L5 (−180 °C), and (**f**) BM-AS-L10 (−180 °C).

Figure 9 shows the fracture surface of as-received specimens and specimens with surface-cracks at 20 °C. The facture surface of all specimens with or without surface-cracks was typically observed with ductile fracture by large plastic deformation at room temperature. There was no significant difference in the overall fracture surface due to the large deformation around the V-notch in all specimens. The effect of enhanced toughness by a surface-cracking process did not happen in room temperature regions with high ductility properties. Since plastic deformation completely occurred around V-notch tip at room temperature, the additional effect of stress dissipation from the surface-cracks was not significant. Figure 10 shows the fracture surface of as-received specimens and specimens with surface-cracks at −180 °C. The fracture surface of the as-received specimen showed the mixed fracture modes, both ductile and brittle, at −180 °C (Figure 10a) because it became very brittle in the low temperature region despite the ductile behavior at room temperature. However, the fracture mode with surface-cracks was predominantly the ductile mode at very low temperatures (as shown in Figure 10b–c). The rate of ductile fracture increased with the increasing number of surface-cracks. The fracture surface of the as-received specimen observed short stretch zones by blunting at the V-notch tip (Figure 10a, left). However, the fracture surface of specimens with surface-cracks showed the largest extension of stretch zones near the V-notch and surface-cracks, and ductile tearing occurred at very low temperatures (Figure 10c, left). This indicates that vertical surface-cracks in the V-notch direction serve to distribute the stress by increasing the blunting effect at the V-notch tip.

Here, toughening by a surface-cracking process could be explained by the effects of the stress dissipation induced by surface-cracks near the V-notch tip. For example, a number of surface-cracks induced by a surface-cracking process led to the enhancement of impact toughness around the V-notch where the surface-cracks in the vertical direction to the V-notch acted as barriers to propagate the main crack and disperse the principal stresses of the V-notch. As the number of surface-cracks increased, the absorbed energy increased with an increase in the dissipation of the crack propagation energy (Figures 7–10). In other words, impact toughness was enhanced by the change in the stress condition from a plain strain to a plain stress condition due to the reduction of effective thickness by surface-cracks at the V-notch tip. Therefore, the dissipation of the crack propagation energy in the materials with surface-cracks enhanced the impact toughness, showing a ductile fracture mode in the cryogenic temperature region.

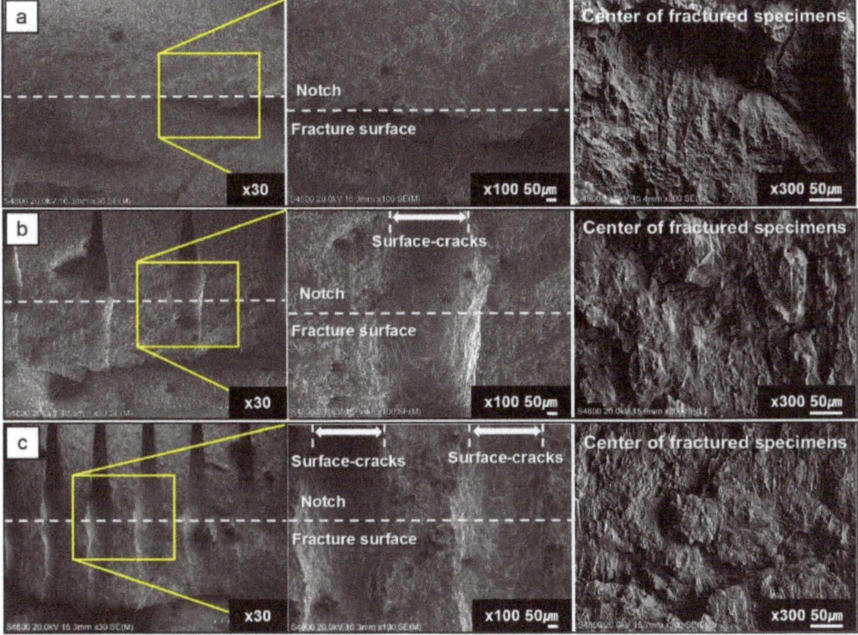

Figure 9. The fracture surface of the as-received specimen and the specimen with surface-cracks at room temperature (20 °C): (**a**) BM-AS, (**b**) BM-AS-L5, and (**c**) BM-AS-L10.

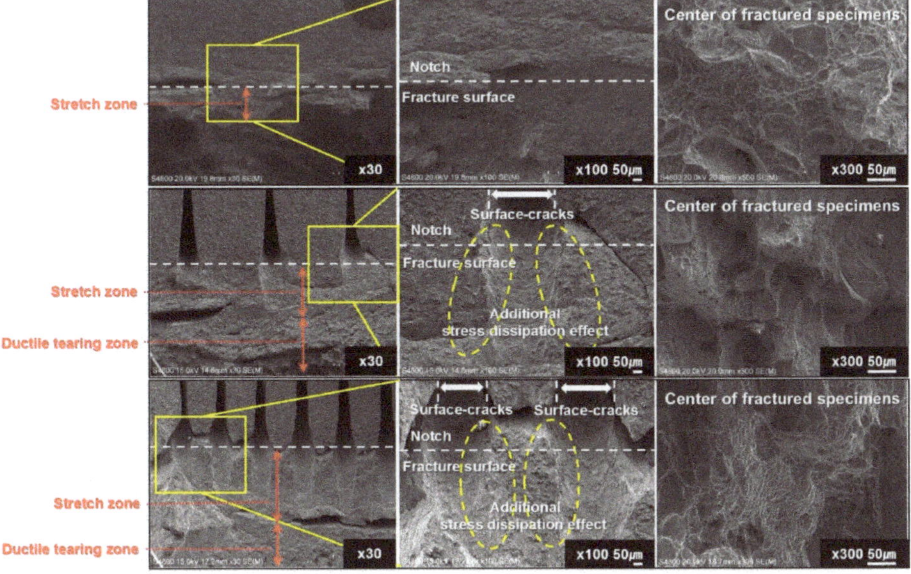

Figure 10. The fracture surface of the as-received specimen and the specimen with surface-cracks at cryogenic temperature (−180 °C): (**a**) BM-AS, (**b**) BM-AS-L5, and (**c**) BM-AS-L10.

3.5. High Strength and Toughness Steels by a Surface-Cracking Technique

The absorbed energy vs. test temperature curves for cold-worked steels are shown in Figure 11. At room temperature, the absorbed energy of cold-worked steels decreased with a cold-working process. The absorbed energy also decreased with the cold-working level from 5% to 30% (solid lines in Figure 11). As we have already mentioned in the Section 3.2, austenite stainless steels after a cold-working process generally resulted in a higher yield strength and lower ductility owing to the increase in dislocation density and martensitic transformation from the austenite phase. Therefore, the result of the Charpy impact test provided the decrease in absorbed energy due to the reduction of ductility by a cold-working process. When comparing the surface-cracking effect in the same cold-working level, the absorbed energy of the specimen with surface-cracks was about two times higher than that of the specimen without surface-cracks at low temperatures (below −60 °C). In particular, strength and toughness were simultaneously increased in the 5% cold-worked specimen with surface-cracks. The interesting thing was that the effect of enhanced toughness by a surface-cracking technique happened at all temperature regions by excluding the as-received results (BM-AS).

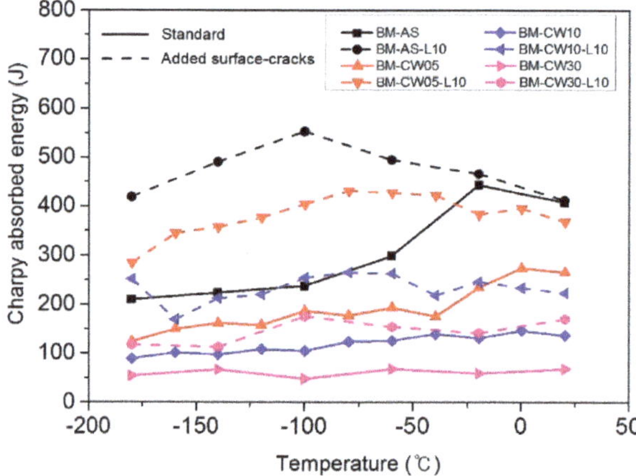

Figure 11. The absorbed energy vs. test temperature curves for cold-worked steels with a surface-cracking technique.

Figure 12 shows the fracture surface of cold-worked specimens with surface-cracks at 20 °C. The as-received specimen was fractured in ductile modes by large plastic deformation (Figure 9), while cold-worked steel appeared to be in a mixture of brittle and ductile modes, showing both cleavage facets and dimples (Figure 12a,c,e). The fraction of brittle fracture increased with the cold-working level from 5% to 30% at room temperature. However, in case of the cold-worked steels with surface-cracks, the fraction of the ductile fracture increased due to the effect of surface-cracks after the surface-cracking technique (Figure 12b,d,f). Figure 13 shows the fracture surface of cold-worked specimens with surface-cracks at −180 °C. Mostly brittle fracture and a small proportion of dimples were observed in the 5% and 10% cold-worked specimens without surface-cracks (Figure 13a,c), and 30% cold-worked specimens also showed almost brittle fractures as shown in Figure 13e. However, in case of the cold-worked steels with surface-cracks, the fraction of ductile fracture increased due to the effect of surface-cracks after a surface-cracking technique even at very low temperatures (Figure 13b,d,f). Therefore, the results of the Charpy impact test and fracture surface analysis were in good agreement with the effect of enhanced impact toughness by a surface-cracking technique.

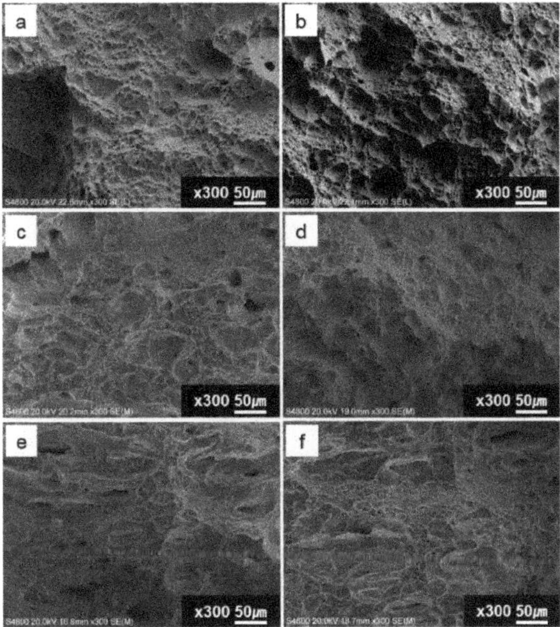

Figure 12. Fracture surface of cold-worked specimens without or with surface-cracks at room temperature (20 °C): (**a**) BM-CW05, (**b**) BM-CW05-L10, (**c**) BM-CW10, (**d**) BM-CW10-L10, (**e**) BM-CW30, and (**f**) BM-CW30-L10.

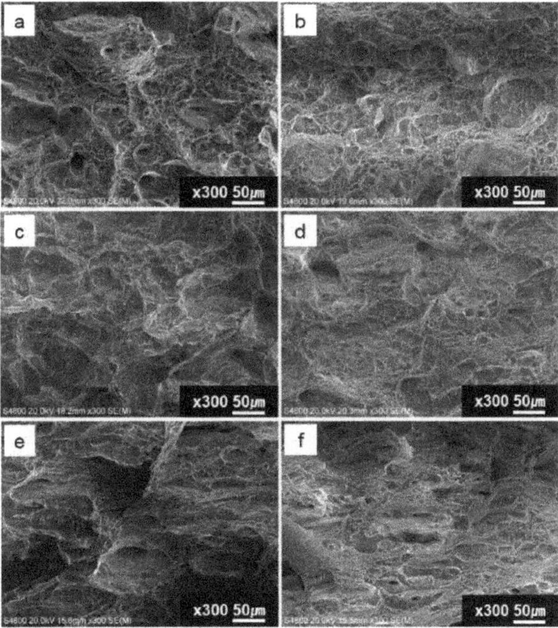

Figure 13. Fracture surface of cold-worked specimens without or with surface-cracks at cryogenic temperature (−180 °C): (**a**) BM-CW05, (**b**) BM-CW05-L10, (**c**) BM-CW10, (**d**) BM-CW10-L10, (**e**) BM-CW30, and (**f**) BM-CW30-L10.

3.6. Application of a Surface-Cracking Technique in Weld Metal

From the results of the experiments, the surface-cracking technique seems to be effective on some brittle materials rather than the materials with fully ductile properties. In particular, it is very important to improve the mechanical properties of the weld metal because the weld metal is more brittle than the base metal. Therefore, we applied the surface-cracking technique to improve the impact toughness of weld metal. The absorbed energy vs. test temperature curves for welded specimens with surface-cracks is shown in Figure 14. The absorbed energy of welded specimens had a relatively lower value compared to the as-received specimen. After a surface-cracking process, the absorbed energy increased through the effect of surface-cracks at all temperature regions. As the number of surface-cracks increased from 0 lines to 10 lines, the absorbed energy significantly increased. Similar to the results of the cold-worked specimen, the effect of enhanced toughness in welded specimens with surface-cracks happened at all temperature regions. It was clear that this effect occurred in brittle conditions such as low temperatures, cold-worked steel, and weld metal.

This surface-cracking technique has enormous potential for improving the mechanical properties of structure materials. For example, this technique is also applicable to corner areas of rectangular structures. The rectangular structure has a high risk of fracture due to the concentration of stress on the corner areas. It is possible to improve the toughness by inducing surface-cracks at the corner of rectangular structures. We can also control the mechanical properties with this technique because they are dependent on a cold-working level and surface-cracks. The application of this technique to industrial products is designed to change their physical properties and thus improving the products' toughness, resistance, performance, and durability. This may offer many benefits to the optimized design of high-strength ductile metallic materials in the future.

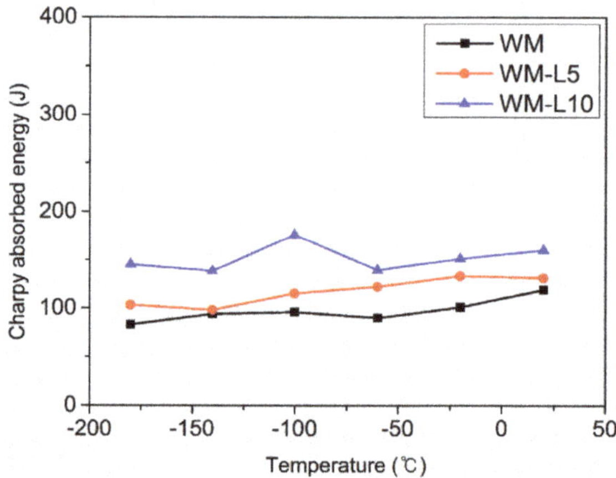

Figure 14. The absorbed energy vs. test temperature curves for welded specimens with surface-cracks.

4. Conclusions

In this study, we have developed a new technique using a cold-working process and surface-cracking process to increase the strength and toughness of stainless steels at the same time. This surface-cracking technique has an enormous potential for improving the mechanical properties of structure materials. We can also control the mechanical properties with this technique because they are dependent on cold-working levels and surface-cracks. The application of this technology in industrial products is designed to change their mechanical properties and thus improving the products' toughness, strength, and integrity. This may offer many benefits in the optimized design of high-strength ductile metallic materials in the future. The main results obtained in this study are as follows:

1. After a cold-working process, the microstructure was changed from austenite single phase to dual phases which consisted of austenite and martensite. The tensile strength and hardness increased with an increase in the cold-working level. The phase transformation from austenite to martensite and the increase in dislocations by plastic deformation led to the strengthening of cold-worked STS304 steels.

2. The specimen with surface-cracks had a higher absorbed energy compared to the as-received specimen without surface-cracks at low temperature regions (below $-60\ °C$). The dissipation of the crack propagation energy with surface-cracks enhanced the impact toughness, showing a ductile fracture mode even in the cryogenic temperature region.

3. The absorbed energy of cold-worked steels decreased with a cold-working level. Comparing the surface-cracking effect of specimens with the same cold-working level, the absorbed energy of specimens with surface-cracks was two times higher than that of the specimen without surface-cracks. In particular, the strength and toughness were simultaneously increased at 5% cold-worked specimens with surface-cracks. The absorbed energy of the welded specimen had a relatively lower value compared to the as-received specimen. After a surface-cracking process, the absorbed energy was increased by effect of a surface-cracking process. Similar to the result of the cold-worked specimen, toughening by a surface-cracking process in the welded specimen happened in all temperature regions.

Author Contributions: K.K. and B.J.K. conceived and designed the experiments; K.K. and M.P. performed the experiments; J.J., H.C.K., H.-S.M., D.-H.L., J.B.J., H.K., S.-H.K., and B.J.K. analyzed and discussed the data; K.K. and B.J.K. wrote the paper.

Funding: This study was supported by the R&D Program of the Korea Institute of Industrial Technology (KITECH) as "Fundamental research and development on three-dimensional nano-porous catalysts and gas sensor for hydrogen production and detection" (KITECH EO-18-0022).

Conflicts of Interest: The authors declare no conflict of interest.

References

1. Pacio, J.C.; Dorao, C.A. A review on heat exchanger thermal hydraulic models for cryogenic applications. *Cryogenics* **2011**, *51*, 366–379. [CrossRef]

2. Raab, J.; Tward, E. Northrop Grumman aerospace systems cryocooler overview. *Cryogenics* **2010**, *50*, 572–581. [CrossRef]

3. Haberbusch, M.S.; Nguyen, C.T.; Stochl, R.J.; Hui, T.Y. Development of No-Vent™ liquid hydrogen storage system for space applications. *Cryogenics* **2010**, *50*, 541–548. [CrossRef]

4. Chen, Q.-S.; Wegrzyn, J.; Prasad, V. Analysis of temperature and pressure changes in liquefied natural gas (LNG) cryogenic tanks. *Cryogenics* **2004**, *44*, 701–709. [CrossRef]

5. Honda, A.; Okano, F.; Ooshima, K.; Akino, N.; Kikuchi, K.; Tanai, Y.; Takenouchi, T.; Numazawa, S.; Ikeda, Y. Application of PLC to dynamic control system for liquid He cryogenic pumping facility on JT-60U NBI system. *Fusion Eng. Des.* **2008**, *83*, 276–279. [CrossRef]

6. Kim, S.B.; Takano, R.; Nakano, T.; Imai, M.; Hahn, S.Y. Characteristics of the magnetic field distribution on compact NMR magnets using cryocooled HTS bulks. *Phys. C Supercond.* **2009**, *469*, 1811–1815. [CrossRef]

7. Qiu, L.L.; Zhuang, M.; Hu, L.B.; Xu, G.F.; Yuan, C.Y. Operational performance of EAST cryogenic system and analysis of it's upgrading. *Fusion Eng. Des.* **2011**, *86*, 2821–2826. [CrossRef]

8. Lee, K.J.; Chun, M.S.; Kim, M.H.; Lee, J.M. A new constitutive model of austenitic stainless steel for cryogenic applications. *Comput. Mater. Sci.* **2009**, *46*, 1152–1162. [CrossRef]

9. Marshall, P. *Austenitic Stainless Steels: Microstructure and Mechanical Properties*; Springer Science & Business Media: Dordrecht, The Netherlands, 1984.

10. Kim, Y.G.; Han, J.M.; Lee, J.S. Composition and temperature dependence of tensile properties of austenitic Fe–Mn–Al–C alloys. *Mater. Sci. Eng. A* **1989**, *114*, 51–59. [CrossRef]

11. Nakanishi, D.; Kawabata, T.; Aihara, S. Effect of dispersed retained γ-Fe on brittle crack arrest toughness in 9% Ni steel in cryogenic temperatures. *Mater. Sci. Eng. A* **2018**, *723*, 238–246. [CrossRef]

12. Yumak, N.; Aslantas, K.; Pekbey, Y. Effect of cryogenic and aging treatments on low-energy impact behaviour of Ti–6Al–4V alloy. *Trans. Nonferrous Met. Soc. China* **2017**, *27*, 514–526. [CrossRef]
13. Barucci, M.; Ligi, C.; Lolli, L.; Marini, A.; Martelli, V.; Risegari, L.; Ventura, G. Very low temperature specific heat of Al 5056. *Phys. B Condens. Matter* **2010**, *405*, 1452–1454. [CrossRef]
14. Dai, Q.; Yuan, Z.; Chen, X.; Chen, K. High-cycle fatigue behavior of high-nitrogen austenitic stainless steel. *Mater. Sci. Eng. A* **2009**, *517*, 257–260. [CrossRef]
15. Abe, A.; Nakamura, M.; Sato, I.; Uetani, H.; Fujitani, T. Studies of the large-scale sea transportation of liquid hydrogen. *Int. J. Hydrogen Energy* **1998**, *23*, 115–121. [CrossRef]
16. Nam, T.-H.; An, E.; Kim, B.J.; Shin, S.; Ko, W.-S.; Park, N.; Kang, N.; Jeon, J.B. Effect of post weld heat treatment on the microstructure and mechanical properties of a submerged-arc-welded 304 stainless steel. *Metals* **2018**, *8*, 26. [CrossRef]
17. Markandeya, R.; Nagarjuna, S.; Sarma, D.S. Effect of prior cold work on age hardening of Cu–3Ti–1Cr alloy. *Mater. Charact.* **2006**, *57*, 348–357. [CrossRef]
18. Karlsen, W.; Van Dyck, S. The effect of prior cold-work on the deformation behaviour of neutron irradiated AISI 304 austenitic stainless steel. *J. Nuclear Mater.* **2010**, *406*, 127–137. [CrossRef]
19. Karjalainen, L.P.; Taulavuori, T.; Sellman, M.; Kyröläinen, A. Some strengthening methods for austenitic stainless steels. *Steel Res. Int.* **2008**, *79*, 404–412. [CrossRef]
20. Shintani, T.; Murata, Y. Evaluation of the dislocation density and dislocation character in cold rolled Type 304 steel determined by profile analysis of X-ray diffraction. *Acta Mater.* **2011**, *59*, 4314–4322. [CrossRef]
21. Milad, M.; Zreiba, N.; Elhalouani, F.; Baradai, C. The effect of cold work on structure and properties of AISI 304 stainless steel. *J. Mater. Process. Technol.* **2008**, *203*, 80–85. [CrossRef]
22. Shen, Y.; Li, X.; Sun, X.; Wang, Y.; Zuo, L. Twinning and martensite in a 304 austenitic stainless steel. *Mater. Sci. Eng. A.* **2012**, *552*, 514–522. [CrossRef]
23. Xu, D.; Wan, X.; Yu, J.; Xu, G.; Li, G. Effect of cold deformation on microstructures and mechanical properties of austenitic stainless steel. *Metals* **2018**, *8*, 522. [CrossRef]
24. Järvenpää, A.; Jaskari, M.; Karjalainen, L.P. Reversed microstructures and tensile properties after various cold rolling reductions in AISI 301LN steel. *Metals* **2018**, *8*, 109. [CrossRef]
25. Wang, Y.; Wu, X.; Wu, W. Effect of α' martensite content induced by tensile plastic prestrain on hydrogen transport and hydrogen embrittlement of 304L austenitic stainless steel. *Metals* **2018**, *8*, 660. [CrossRef]
26. Kim, T.K.; Kim, S.H. Study on the cold working process for FM steel cladding tubes. *J. Nuclear Mater.* **2011**, *411*, 208–212. [CrossRef]
27. Sivasankaran, S.; Sivaprasad, K.; Narayanasamy, R. Microstructure, cold workability and strain hardening behavior of trimodaled AA 6061–TiO_2 nanocomposite prepared by mechanical alloying. *Mater. Sci. Eng. A.* **2011**, *528*, 6776–6787. [CrossRef]
28. Mallick, P.; Tewary, N.K.; Ghosh, S.K.; Chattopadhyay, P.P. Microstructure-tensile property correlation in 304 stainless steel after cold deformation and austenite reversion. *Mater. Sci. Eng. A* **2017**, *707*, 488–500. [CrossRef]
29. Panwar, S.; Goel, D.B.; Pandey, O.P. Effect of cold work and aging on mechanical properties of a copper bearing microalloyed HSLA-100 (GPT) steel. *Bull. Mater. Sci.* **2007**, *30*, 73–79. [CrossRef]
30. Tsuji, N.; Okuno, S.; Koizumi, Y.; Minamino, Y. Toughness of ultrafine grained ferritic steels fabricated by ARB and annealing process. *Mater. Trans.* **2004**, *45*, 2272–2281. [CrossRef]
31. Kimura, Y.; Inoue, T.; Yin, F.; Tsuzaki, K. Inverse temperature dependence of toughness in an ultrafine grain-structure steel. *Science* **2008**, *320*, 1057–1060. [CrossRef] [PubMed]
32. Lu, K.; Lu, L.; Suresh, S. Strengthening materials by engineering coherent internal boundaries at the nanoscale. *Science* **2009**, *324*, 349–352. [CrossRef] [PubMed]
33. Takaki, S.; Kawasaki, K.; Kimura, Y. Mechanical properties of ultra fine grained steels. *J. Mater. Process. Technol.* **2001**, *117*, 359–363. [CrossRef]
34. Stolyarov, V.V.; Valiev, R.Z.; Zhu, Y.T. Enhanced low-temperature impact toughness of nanostructured Ti. *Appl. Phys. Lett.* **2006**, *88*, 041905. [CrossRef]
35. Rao, K.T.V.; Ritchie, R.O. Fatigue of aluminium—Lithium alloys. *Int. Mater. Rev.* **1992**, *37*, 153–186. [CrossRef]
36. Benzeggagh, M.L.; Kenane, M. Measurement of mixed-mode delamination fracture toughness of unidirectional glass/epoxy composites with mixed-mode bending apparatus. *Comp. Sci. Technol.* **1996**, *56*, 439–449. [CrossRef]

37. Guo, W.; Dong, H.; Lu, M.; Zhao, X. The coupled effects of thickness and delamination on cracking resistance of X70 pipeline steel. *Int. J. Pres. Vessel. Pip.* **2002**, *79*, 403–412. [CrossRef]

38. Morris, J.W. Stronger, tougher steels. *Science* **2008**, *320*, 1022–1023. [CrossRef] [PubMed]

39. Jafari, M.; Kimura, Y.; Tsuzaki, K. Enhanced upper shelf energy by ultrafine elongated grain structures in 1100 MPa high strength steel. *Mater. Sci. Eng. A* **2012**, *532*, 420–429. [CrossRef]

40. Lu, K. The future of metals. *Science* **2010**, *328*, 319–320. [CrossRef] [PubMed]

41. Launey, M.E.; Ritchie, R.O. On the fracture toughness of advanced materials. *Adv. Mater.* **2009**, *21*, 2103–2110. [CrossRef]

42. Ritchie, R.O. The quest for stronger, tougher materials. *Science* **2008**, *320*, 448. [CrossRef] [PubMed]

43. Rajasekhara, S.; Ferreira, P.J.; Karjalainen, L.P.; Kyröläinen, A. Hall–Petch behavior in ultra-fine-grained AISI 301LN stainless steel. *Metall. Mater. Trans. A* **2007**, *38*, 1202–1210. [CrossRef]

44. Dumont, D.; Deschamps, A.; Brechet, Y. On the relationship between microstructure, strength and toughness in AA7050 aluminum alloy. *Mater. Sci. Eng. A* **2003**, *356*, 326–336. [CrossRef]

45. Yuan, S.; Liu, G.; Wang, R.; Zhang, G.-J.; Pu, X.; Sun, J.; Chen, K.-H. Effect of precipitate morphology evolution on the strength-toughness relationship in Al–Mg–Si alloys. *Scr. Mater.* **2009**, *60*, 1109–1112. [CrossRef]

46. American Society for Testing and Materials (Filadelfia). *Standard Test Methods for Notched Bar Impact Testing of Metallic Materials*; ASTM: West Conshohocken, PA, USA, 2002.

© 2018 by the authors. Licensee MDPI, Basel, Switzerland. This article is an open access article distributed under the terms and conditions of the Creative Commons Attribution (CC BY) license (http://creativecommons.org/licenses/by/4.0/).

Article

Electrochemical Investigation of Corrosion of X80 Steel under Elastic and Plastic Tensile Stress in CO$_2$ Environment

Wei Wu [1],*, Hailong Yin [2,3], Hao Zhang [2], Jia Kang [2], Yun Li [2] and Yong Dan [1]

[1] School of Chemical Engineering, Northwest University, Xi'an 710069, China; danyong@nwu.edu.cn
[2] School of Chemical Engineering and Technology, Xi'an Jiaotong University, Xi'an 710049, China; yinhl@shccig.com (H.Y.); upczhang@stu.xjtu.edu.cn (H.Z.); kangjia1009@stu.xjtu.edu.cn (J.K.); yunli@mail.xjtu.edu.cn (Y.L.)
[3] General Department, Shaanxi Coal and Chemical Industry Group Co., Ltd., Xi'an 710065, China
* Correspondence: wuwei@nwu.edu.cn; Tel.: +86-29-88302632

Received: 23 October 2018; Accepted: 10 November 2018; Published: 14 November 2018

Abstract: An investigation into the electrochemical corrosion behavior of X80 pipeline steel under different elastic and plastic tensile stress in a CO$_2$-saturated NaCl solution has been carried out by using open-circuit potential, potentiodynamic polarization, electrochemical impedance spectroscopy, and surface analysis techniques. The results show that the corrosion rate of X80 steel first increases and then slightly decreases with the increase of elastic tensile stress, whereas the corrosion rate sharply increases with the increase of plastic tensile stress. Both elastic and plastic tensile stress can enhance steel corrosion by improving the electrochemical activity of both anodic and cathodic reactions. Moreover, compared with elastic tensile stress, plastic tensile stress has a more significant effect. Furthermore, electrochemical reactions for CO$_2$ corrosion and mechanoelectrochemical effect are used to reasonably explain the corrosion behavior of stressed X80 steel in CO$_2$ environment.

Keywords: pipeline steel; tensile stress; corrosion; potentiodynamic polarization; EIS

---

1. Introduction

Pipeline has been acknowledged as a very efficient way for long-distance transportation of oil and natural gas due to its advantages of large transport capacity, low cost, and high reliability. With the increasing demand for oil and natural gas around the world, pipeline materials that are able to satisfy the requirements of higher operating pressure, larger diameter and thinner wall thickness have been paid more attention to. Therefore, the low-alloy high-strength pipeline steels like X70 and X80 become the best selection for new pipeline projects in China and some other countries due to their excellent mechanical properties (strength, deformability, and toughness) and the superior weldability of these steels [1]. However, in practice, pipelines have to suffer from severe environment conditions. More specifically, during the oil or natural gas extraction and transmission process, the aggressive impurities, especially carbon dioxide and chlorides, exist in produced oil and gas, and are unable to be completely removed. Such an environment would easily result in severe CO$_2$ corrosion, which may reduce the service life of the pipeline and lead to the leakage of the pipeline [2]. Many studies so far have focused on corrosion of high-strength pipeline steels in oil and natural gas transportation environments [2–11].

Besides the aggressive medium, pipelines would also be subject to various stress conditions resulting from internal pressure, soil movement, and so on [12], which play an important role in corrosion behaviors. Some research efforts have been devoted to the effect of applied stress on the corrosion of pipeline steel. Xu and Cheng [12,13] investigated the effect of uniaxial elastic stress and plastic strain on the corrosion of X100 pipeline steel in a near-neutral pH NS4 solution (a simulated soil solution). They discovered that the static elastic stress has little effect on electrochemical corrosion of steel, whereas plastic strain would enhance the steel corrosion by increasing corrosion activity of the steel during plastic deformation. Boven et al. [14] reported that residual stress and applied stress can both accelerate the corrosion process of X65 steel immersed in near-neutral pH NS4 solution. Bao et al. [15] studied the electrochemical behavior of tensile stressed oil tube steel P110 exposed to a brine solution with dissolved CO_2. It was found that tensile stress can improve the surface thermodynamic activity of P110 steels and accelerate the anodic and cathodic reactions. Li and Cheng [16] investigated the corrosion behavior of X70 steel in carbonate-bicarbonate solution, and found that tensile stress enhance the steel dissolution more significantly than compressive stress. Li et al. [17] studied the corrosion of elastic-stressed X70 steel exposed to CO_2- and H_2S-contained brine solutions and the results show that elastic tensile stress has less effect on the corrosion rate of X70 steel. Wang et al. [18] investigated the effect strain on the corrosion behavior of X80 steel in NaCl aqueous solution. The above researches provided valuable results of the synergism of stress and corrosive medium on pipeline corrosion. However, systematic research on corrosion behavior of stressed high-strength pipeline steel in the typical CO_2 corrosion environment is still scarce, especially on the effect of different levels of plastic stress on the corrosion of pipeline steel.

This study aims to investigate the effect of different levels of elastic and plastic tensile stress on the corrosion behavior of the typical high-strength pipeline steel X80 in a CO_2-saturated NaCl solution. The electrochemical characteristics and corrosion rates of X80 steel under different tensile stress were obtained, and the corrosion morphology was observed. The corrosion mechanism, taking account into tensile stress, were analyzed to better understand the corrosion changes of stressed high-strength pipeline steel in a CO_2 environment.

2. Materials and Methods

2.1. Material and Solution

The experimental specimens were cut from a commercial LSAW (longitudinal submerged arc welding) pipe made of X80 steel. Its chemical composition is listed in Table 1. The stress-strain curve of the steel was measured on a rod specimen through an electronic tensile testing machine (HH-YFML-EDC, Sinotest Equipment Co., Ltd., Changchun, China) with a drawing rate of 3 mm/min, as shown in Figure 1. The mechanical parameters of the X80 steel are as follows: Yield strength (σ_{ys}): 640 MPa; tensile strength (σ_b): 743 MPa; and yield-tensile ratio (λ): 0.861. The geometry and size of the sample used in this work is shown in Figure 2. The surface of the specimen was coated by insulating anticorrosive rubber except only 10 mm long surface at the middle part of the sample as the working electrode. Before the electrochemical experiments, the working electrode surface would be polished up to 1200-grit with SiC paper, washed with distilled water and alcohol, and then dried in a desiccator until use.

Table 1. Chemical composition of X80 steel used in this work (wt%).

Steel	C	Si	Mn	P	Nb	Ti	Al	Mo	Ni	Cu	Cr	Fe
X80	0.05	0.22	1.78	0.01	0.055	0.015	0.044	0.26	0.256	0.143	0.03	Bal.

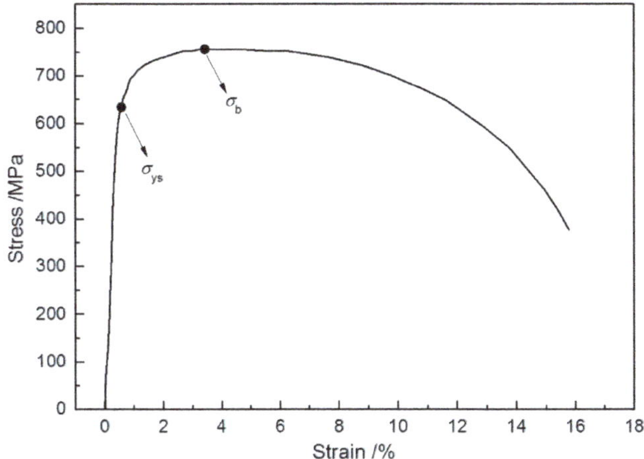

Figure 1. Stress-strain curve of X80 steel at a drawing rate of 3 mm/min in air.

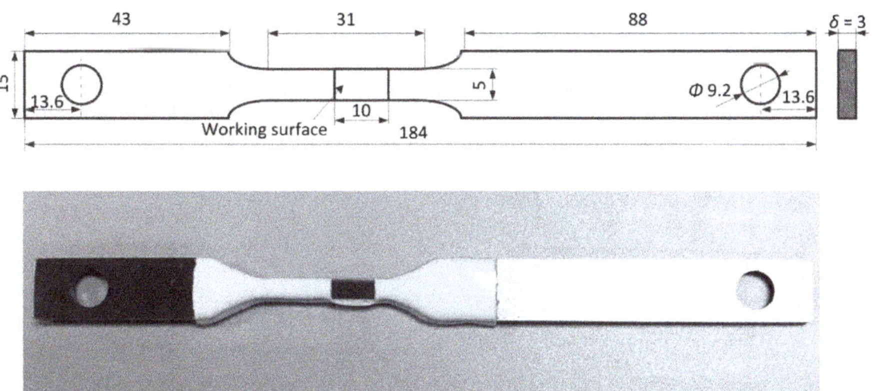

Figure 2. The geometry and the size of the X80 steel specimen (unit: mm).

The specimens were applied with a constant stress at 0%, 30%, 60%, 90%, 103%, 105%, and 108% of its yield strength, respectively. At least three samples were used in each test to avoid accidental error. The electrolyte solution was a 3.5 wt% NaCl aqueous solution (Tianjin Hengxing Chemical Reagent Co., Ltd., Tianjin, China) dissolved with CO_2. To eliminate the influence of oxygen, N_2 was bubbled through the electrolyte to deaerate the aqueous solution for at least one hour. Then, CO_2 was injected at a constant gas flow until the experiment ended. The experimental temperature was controlled to be $30 \pm 1\ ^\circ$C during the measurement.

2.2. Experimental Set-Up

The experimental set-up used in this work is shown in Figure 3. It contains an electronic tensile testing machine (HH-YFML-EDC, Sinotest Equipment Co., Ltd., Changchun, China) and a plexiglas-made electrochemical cell. The electrochemical measurements were performed by using a CorrTest CS350 electrochemical workstation (Wuhan Corrtest Instruments Corp., Ltd., Wuhan, China) with a conventional three-electrode system, where the working electrode (WE) is the X80 steel specimen, the reference electrode is a saturated calomel electrode (SCE) and the counter electrode (CE) is a platinum plate.

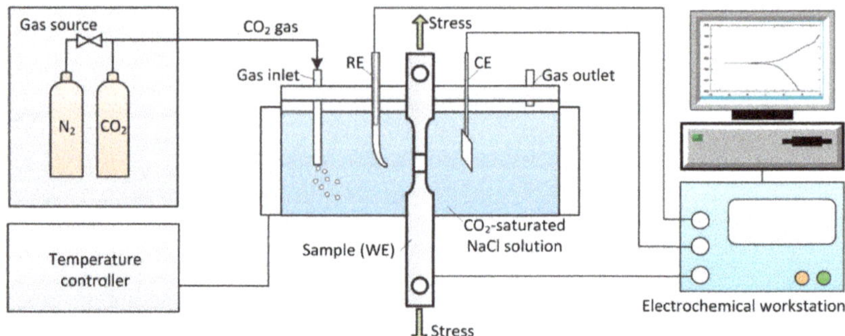

Figure 3. Schematic of the experimental set-up for the electrochemical measurement on the tensile specimen.

2.3. Electrochemical Measurements

The electrochemical methods used in this study include open-circuit potential test (OCP), potentiodynamic polarization scan (PDS), and electrochemical impedance spectroscopy measurement (EIS). The details of each method are described as follows.

OCP refers to the potential in the working electrode comparative to the reference electrode when there is no polarization current or potential existing in the cell. It reflects the electrochemical thermodynamic stability of the steel. The OCP of each specimen was measured from initial immersion to stable value. The measurement time lasted about one hour.

After the stability of OCP, PDS was performed with a scan rate of 0.5 mV/s. The scanning potential range is from -200 mV to 200 mV (vs. OCP). From the potentiodynamic polarization curve, the information of electrochemical kinetics of the steel can be obtained by data fitting.

EIS is one of the widely used electrochemical techniques for studying electrode kinetics of electrode materials due to its non-destructive nature [19,20], from which the electrolyte-electrode interfacial information such as double-layer capacitance, charge-transfer resistance and diffusion impedance can be obtained. In order to investigate the characteristics of electrode reaction of anode and cathode, the EIS would be performed in the weak anodic and cathodic polarization regions, respectively. The optima herefore, in this work, the EIS of each specimen was measured at -60 mV and $+60$ mV (vs. OCP) with the sinusoidal perturbation amplitude of 10 mV and a frequency in the range from 100 kHz to 0.01 Hz.

2.4. Surface Observation

The samples with 90 min of immersion under different tensile stress were examined on a JSM-6390 (JEOL, Akishima, Tokyo, Japan) Scanning Electron Microscope (SEM). The corroded surface morphology of the samples was observed.

3. Results

3.1. Open-Circuit Potential with Time

Figure 4 shows the open-circuit potentials of the X80 steel immersed in 3.5 wt% NaCl aqueous solution with saturated CO_2 under different elastic and plastic tensile stress. It can be seen that the OCP of each sample gradually tends to be a steady-state value with time, though the obvious fluctuation was observed in the beginning of immersion time, which indicates that the X80 steel under different tensile stress reached the thermodynamic stable state. Additionally, in the elastic stress range, the steady-state value of the OCP first decreases and then increases with the increasing tensile stress, while in the plastic stress range, the OCP monotonously decreases with the increase of tensile stress. Overall, the OCPs of stressed X80 steel are more negative than that without tensile stress.

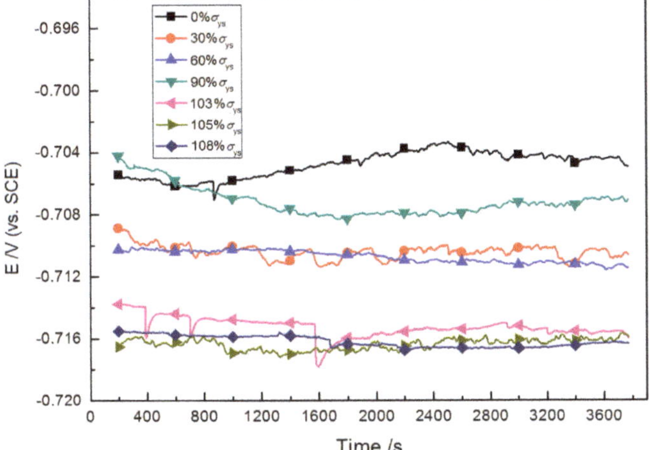

Figure 4. Open-circuit potential of X80 steel in 3.5 wt% NaCl aqueous solution with saturated CO_2 at 30 °C under different elastic and plastic tensile stress.

3.2. Potentiodynamic Polarization

Figure 5 gives the potentiodynamic polarization curves of the tensile stressed X80 steel in 3.5 wt% aqueous solution with saturated CO_2 at 30 °C. The results reveal that all the polarization curves have similar shapes; moreover, there is not an obvious cathodic Tafel region. The anodic Tafel slope of each sample was obtained by fitting the polarization curves, as listed in Table 2.

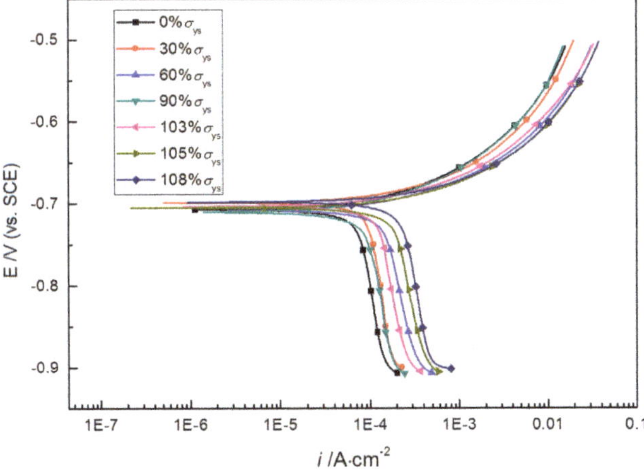

Figure 5. Potentiodynamic polarization curves of X80 steel exposed to 3.5 wt% NaCl aqueous solution with saturated CO_2 at 30 °C under different elastic and plastic tensile stress.

Table 2. Anodic Tafel slope obtained by Tafel fitting of polarization curves for X80 in 3.5 wt% NaCl aqueous solution with saturated CO_2 at 30 °C under different elastic and plastic tensile stress.

Tensile Stress	0%σ_{ys}	30%σ_{ys}	60%σ_{ys}	90%σ_{ys}	103%σ_{ys}	105%σ_{ys}	108%σ_{ys}
b_a/mV·dec^{-1}	40.29	34.54	42.02	35.28	38.52	42.06	39.69

In addition, the corrosion rate of each X80 steel sample in 3.5 wt% NaCl aqueous solution with saturated CO_2 at 30 °C under different tensile stress was calculated based on the corrosion current density through Faraday's law:

$$v_d = 8.76 \times 10^3 \times \frac{i_{corr}M}{zF\rho} \ (\text{mm·a}^{-1}), \tag{1}$$

where M is the molar mass of metal (g·mol^{-1}), ρ is the density of metal (g·cm^{-3}), F is the Faraday constant (96,485 C·mol^{-1}), and i_{corr} is the corrosion current density (A·cm^{-2}) and it is obtained by fitting the Tafel curves shown in Figure 5. The calculated results are shown in Figure 6. It is shown clearly that, in the elastic stress range, the corrosion rate increases from 0.845 mm·a^{-1} at 0%σ_{ys} to 1.442 mm·a^{-1} at 60%σ_{ys} with the increasing tensile stress, and then slightly decreases to 1.110 mm·a^{-1} at 90%σ_{ys}. In the plastic stress range, the corrosion rate increases sharply up to 2.631 mm·a^{-1} at 108%σ_{ys}, which is about three times higher than that without tensile stress. The change tendency of the corrosion rate is consistent with the observation from OCP. The above results indicate that both elastic and plastic tensile stress can accelerate corrosion reaction rate of X80 steel in the experimental solution, furthermore, plastic tensile stress plays a more important role in the corrosion process, as compared to elastic tensile stress.

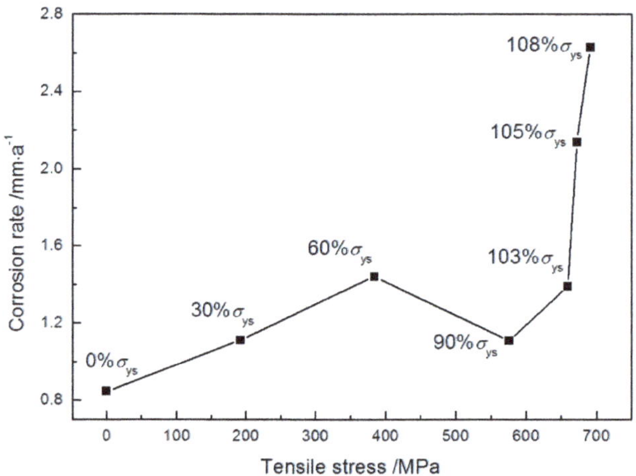

Figure 6. Corrosion rates of X80 steel exposed to 3.5 wt% NaCl aqueous solution with saturated CO_2 under different elastic and plastic tensile stress at 30 °C by fitting potentiodynamic polarization curves.

3.3. Electrochemical Impedance Spectroscopy

3.3.1. Anodic EIS

Figure 7a,b presents the Nyquist and Bode plots of X80 steel measured at +60 mV (vs. OCP) in 3.5 wt% NaCl aqueous solution with saturated CO_2 at 30 °C under elastic and plastic tensile stress, respectively. Each Nyquist plot in elastic stress range is composed of a capacitive loop at high frequency, an inductive loop, and a capacitive loop at low frequency. However, under plastic stress conditions, the Nyquist plots do not have the capacitive loop in the low frequency range. In general, the capacitive loop at high frequency is associated with the electron transfer process on the surface as well as the charge-discharge process of electrical double layer between the electrode surface and electrolyte solution, while the data in the low frequency may be related to the adsorption of electrochemical reaction intermediates.

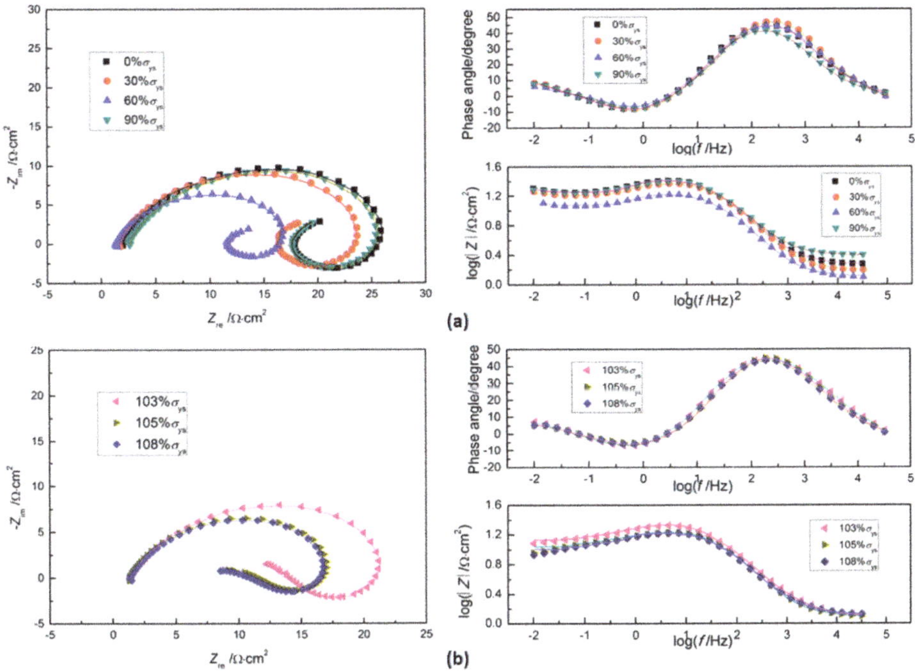

Figure 7. Electrochemical impedance spectroscopy (EIS) curves of X80 steel measured at +60 mV (vs. open-circuit potential, OCP) in 3.5 wt% NaCl aqueous solution with saturated CO_2 under different (**a**) elastic and (**b**) plastic tensile stress at 30 °C. Symbol: Experimental data; and line: Simulation data.

All the obtained EIS data for the X80 steel in the solution under elastic and plastic tensile stress were fitted to the equivalent circuit models as indicated in Figure 8a,b, respectively. In Figure 8, the $R_{s,a}$ is the aqueous solution resistance, $R_{t,a}$ the charge transfer resistance that occurs on the electrode surface, $R_{L,a}$ the resistance which may result from intermediate reactions on electrode surface. Generally, the electrical double layer is expressed as a capacitance component $C_{dl,a}$, but taking into account of the dispersion effect, a constant phase angle element (CPE$_{dl,a}$) $Q_{dl,a}$ is used to replace the $C_{dl,a}$ in the fitting procedure. The impedance of CPE$_{dl,a}$ is expressed as [19]

$$Z_{\text{CPE}_{dl,a}} = \frac{1}{Y_{dl,a}(j\omega)^{n_{dl,a}}}, \tag{2}$$

where $Y_{dl,a}$ is the constant in $\Omega^{-1}\cdot cm^{-2}\cdot s^{n}$, $n_{dl,a}$ is the number constant between 0 and 1, and ω is the angular frequency. Thus, the total impedance of Figure 7a can be expressed by Equation (3), and that of Figure 7b by Equation (4):

$$Z = R_{s,a} + \frac{1}{Y_{dl,a}(j\omega)^{n_{dl,a}} + \dfrac{1}{R_{t,a} + \frac{1}{1/j\omega L_a + 1/R_{L,a}} + \frac{1}{j\omega C_a + 1/R_{C,a}}}}, \tag{3}$$

$$Z = R_{s,a} + \cfrac{1}{Y_{dl,a}(j\omega)^{n_{dl,a}} + \cfrac{1}{R_{t,a} + \cfrac{1}{1/j\omega L_a + 1/R_{L,a}}}},$$

(4)

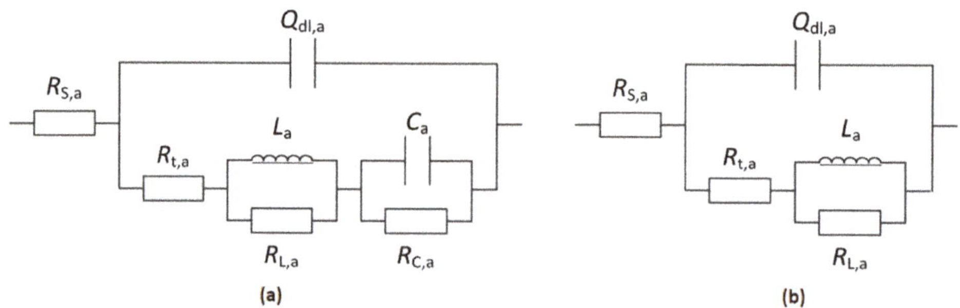

Figure 8. Equivalent circuit model for fitting the EIS measured at +60 mV (vs. OCP) under different (**a**) elastic and (**b**) plastic tensile stress at 30 °C.

According to the equivalent circuit models shown in Figure 8, the impedance spectra for the anodic dissolution of X80 steel have been fitted and some parameters are listed in Tables 3 and 4.

Table 3. Electrochemical impedance spectroscopy (EIS) parameters of X80 steel obtained by fitting the EIS curves of X80 steel measured at +60 mV (vs. OCP) in 3.5 wt% NaCl aqueous solution with saturated CO_2 under elastic tensile stress at 30 °C.

Tensile Stress	$R_{s,a}/\Omega \cdot cm^2$	$Y_{dl,a}/\Omega^{-1} \cdot cm^{-2} \cdot s^n$	$n_{dl,a}$	$R_{t,a}/\Omega \cdot cm^2$	$L_a/H \cdot cm^2$	$R_{L,a}/\Omega \cdot cm^2$	$C_a/F \cdot cm^{-2}$	$R_{C,a}/\Omega \cdot cm^2$
0	1.858	0.0005508	0.7866	16.30	1.5200	9.809	3.870	6.825
30%σ_{ys}	1.575	0.0004375	0.8186	14.95	1.3050	8.360	4.046	7.034
60%σ_{ys}	1.273	0.0007162	0.7938	10.65	0.8289	5.958	6.738	2.879
90%σ_{ys}	1.682	0.0005311	0.7905	15.51	1.3490	9.387	3.751	97.718

Table 4. EIS parameters of X80 steel obtained by fitting the EIS curves of X80 steel measured at +60 mV (vs. OCP) in 3.5 wt% NaCl aqueous solution with saturated CO_2 under plastic tensile stress at 30 °C.

Tensile Stress	$R_{s,a}/\Omega \cdot cm^2$	$Y_{dl,a}/\Omega^{-1} \cdot cm^{-1} \cdot s^n$	$n_{dl,a}$	$R_{t,a}/\Omega \cdot cm^2$	$L_a/H \cdot cm^2$	$R_{L,a}/\Omega \cdot cm^2$
103%σ_{ys}	1.308	0.0007835	0.7555	12.94	1.59	8.573
105%σ_{ys}	1.268	0.0006818	0.7937	10.25	1.274	6.282
108%σ_{ys}	1.355	0.0007329	0.7872	9.52	1.477	6.588

The charge transfer resistance $R_{t,a}$ listed in Tables 3 and 4 reflect the anodic reaction rate caused by tensile stress, that is, the larger the $R_{t,a}$, the lower the reaction rate. From the Tables, it can be seen that the value of $R_{t,a}$ first decreases and then increases with the increase of elastic tensile stress, and decreases in plastic stress state, which is consistent with the change tendency observed from the polarization curves.

3.3.2. Cathodic EIS

Figure 9 shows the Nyquist and Bode plots of the X80 steel measured at −60 mV (vs. OCP) in 3.5 wt% NaCl aqueous solution with saturated CO_2 at 30 °C under different tensile stress. Similar shapes of the Nyquist and Bode plots are observed from 0%σ_{ys} to 108%σ_{ys}, suggesting that the cathodic reactions of all samples follow the same mechanism, which is also supported by the polarization curves. As shown in Figure 9, each Nyquist plot consists of a high-frequency capacitance and a low-frequency line (Warburg impedance). The former is related to electron transfer and charge-discharge of double layer capacitance, and the latter is produced by the diffusion of cathode active species. According to the electrochemical features of the cathodic polarization region and cathodic EIS of samples, it can be obtained that cathodic reaction process is mixed controlled by charge transfer and active species diffusion.

The impedance spectra were fitted by an equivalent circuit of Figure 10 where $R_{S,c}$ represents the resistance of the aqueous solution, $R_{t,c}$ the charge transfer resistance, $Q_{dl,c}$ the constant phase angle element used to replace $C_{dl,c}$, and $Z_{W,c}$ the Warburg impedance suggesting a diffusion-controlled reaction. The Warburg impedance $Z_{W,c}$ is expressed by Equation (5)

$$Z_{W,c} = A_{W,c}(j\omega)^{-0.5}, \tag{5}$$

where $A_{W,c}$ is the coefficient of $Z_{W,c}$. The total impedance of Figure 9 can be expressed by Equation (6).

$$Z = R_{s,c} + \frac{1}{Y_{dl,c}(j\omega)^{n_{dl,c}} + \frac{1}{R_{t,c}+Z_{W,c}}}, \tag{6}$$

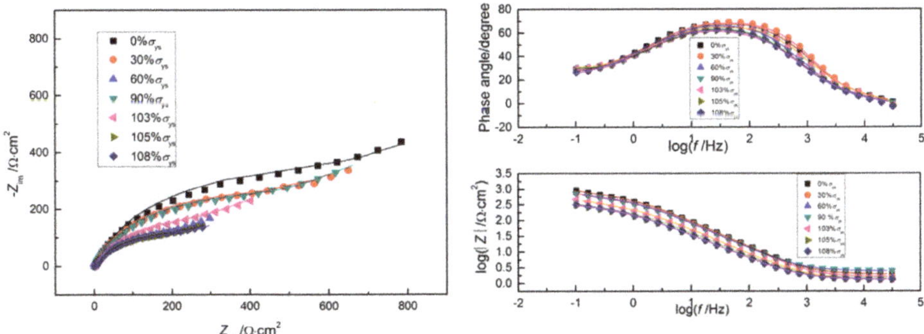

Figure 9. EIS curves of X80 steel measured at −60 mV (vs. OCP) in 3.5 wt% NaCl aqueous solution with saturated CO_2 under different elastic and plastic tensile stress at 30 °C. Symbol: Experimental data; Line: simulation data.

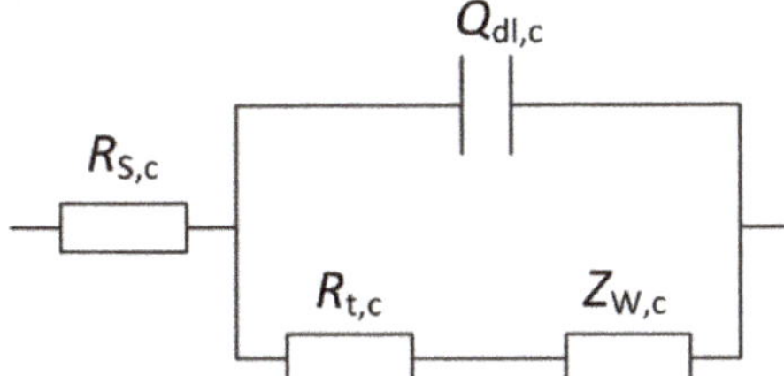

Figure 10. Equivalent circuit model for fitting the EIS measured at −60 mV (vs. OCP) under different elastic and plastic tensile stress at 30 °C.

Table 5 lists the fitted results of the impedance spectra in Figure 9. Here, $R_{t,c}$ reflects the cathodic reaction rate caused by tensile stress. Table 5 shows that the value of $R_{t,c}$ decreases from 573.3 $\Omega\cdot cm^2$ at 0%σ_{ys} to 228.0 $\Omega\cdot cm^2$ at 60%σ_{ys}, then goes up to 519.2 $\Omega\cdot cm^2$ at 90%σ_{ys}, while under plastic stress conditions it decreases from 260.7 $\Omega\cdot cm^2$ at 103%σ_{ys} to 196.9 $\Omega\cdot cm^2$ at 108%σ_{ys}. While $R_{t,c}$ is not monotonously decreased with the increasing tensile stress, the $R_{t,c}$s of all stressed samples are smaller than that without tensile stress, which means both the elastic and plastic tensile stress are able to enhance the cathodic reactions.

Table 5. EIS parameters of X80 steel obtained by fitting the EIS curves of X80 steel measured at −60 mV (vs. OCP) in 3.5 wt% NaCl aqueous solution with saturated CO_2 under different elastic and plastic tensile stress at 30 °C.

Tensile Stress	$R_{S,c}$ /$\Omega \cdot cm^2$	$Y_{dl,c}$ /$\Omega^{-1} \cdot cm^{-2} \cdot s^n$	$n_{dl,c}$	$R_{t,c}$ /$\Omega \cdot cm^2$	$A_{W,c}$ /$\Omega \cdot cm^2 \cdot s^{-0.5}$
0	1.914	0.0003185	0.8375	571.3	0.002394
30%σ_{ys}	1.599	0.0003293	0.8463	465.5	0.003021
60%σ_{ys}	1.449	0.0008707	0.8009	228.0	0.006291
90%σ_{ys}	1.446	0.0003825	0.8295	474.6	0.003360
103%σ_{ys}	1.319	0.0006090	0.8200	260.7	0.004155
105%σ_{ys}	1.278	0.0001056	0.7876	212.0	0.007714
108%σ_{ys}	1.314	0.0009652	0.8065	196.9	0.006539

3.4. Characterizations of Corroded Samples

Figure 11 presents the SEM photographs of corroded surface of X80 steel 3.5 wt% NaCl aqueous solution with saturated CO_2 under different elastic and plastic tensile stress. It is shown that without tensile stress (Figure 11a), several corrosion products are on the steel surface. Under elastic tensile stress range, a little more corrosion product is observed (Figure 11b–d). However, under plastic tensile stress range, not only corrosion products but also pits appear on the surface of the steel (Figure 11e–g), indicating that the surface activity of the samples under plastic tensile stress is higher than that under elastic tensile stress.

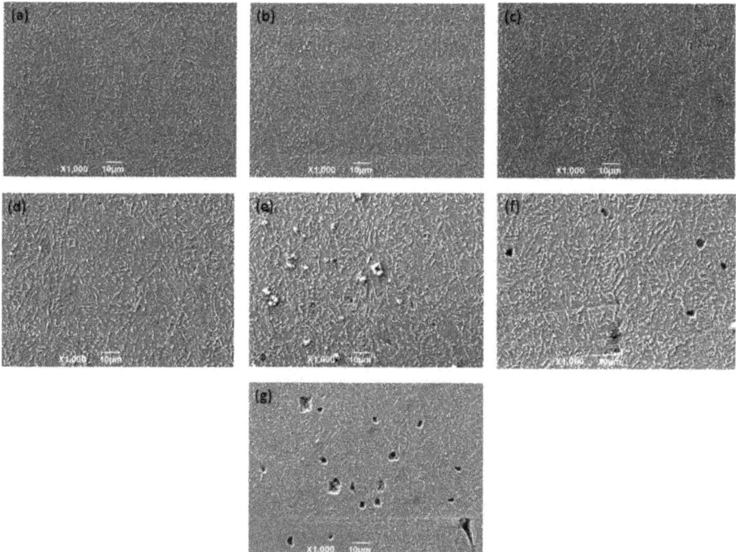

Figure 11. Scanning electron microscope (SEM) photographs of corroded surface of X80 steel in 3.5 wt% NaCl aqueous solution with saturated CO_2 at 30 °C under different tensile stress: (**a**) 0%σ_{ys}; (**b**) 30%σ_{ys}; (**c**) 60%σ_{ys}; (**d**) 90%σ_{ys}; (**e**) 103%σ_{ys}; (**f**) 105%σ_{ys}; and (**g**) 108%σ_{ys}.

4. Discussion

4.1. Electrochemical Reactions for CO₂ Corrosion

The experimental solution in this study is a typical CO_2 corrosion environment. As far as CO_2 corrosion goes, it is a complex process involving a number of chemical, electrochemical, and mass transport processes [21]. When CO_2 is injected into an aqueous solution, it is first hydrated to generate a weak acid (H_2CO_3):

$$CO_2 + H_2O \leftrightarrow H_2CO_3, \tag{7}$$

Then, the H_2CO_3 is dissociated in two steps:

$$H_2CO_3 \leftrightarrow H^+ + HCO_3^-, \tag{8}$$

$$HCO_3^- \leftrightarrow H^+ + CO_3^{2-}, \tag{9}$$

which results in an acidic and corrosive solution. Subsequently, electrochemical corrosion reactions take place on the electrode surface, where the anodic reaction refers to the oxidation of iron:

$$Fe \rightarrow Fe^{2+} + 2e, \tag{10}$$

There are several possible mechanisms proposed for explaining iron oxidation reaction in acidic media [2,22], where the consecutive mechanism proposed by Bockris et al. [23] has been commonly used to describe the anodic current in CO_2 corrosion of mild steel. In this study, the anodic polarization curves shown in Figure 5 have a Tafel slope (b_a) of around 40 mV (see Table 2), which is in accordance with the consecutive mechanism expressed by the reactions (11)–(13).

$$Fe + H_2O \rightarrow FeOH_{ads} + H^+ + e, \tag{11}$$

$$FeOH_{ads} \rightarrow FeOH_{ads}^+ + e, \tag{12}$$

$$FeOH_{ads}^+ \rightarrow Fe^{2+} + OH^-, \tag{13}$$

The rate-determining step for the consecutive mechanism is the reaction (12) in which the adsorption of the reaction intermediate $FeOH_{ads}$ results in the low-frequency capacitance C_a and low-frequency inductance L_a of the anodic Nyquist plots, as shown in Figure 7.

For the CO_2 corrosion system, the cathodic reaction refers to the hydrogen evolution reaction which is a family of cathodic reactions that all have molecular hydrogen as their product, which contains the reduction of H^+, H_2CO_3, HCO_3^-, and H_2O:

$$2H^+ + 2e \rightarrow H_2, \tag{14}$$

$$2H_2CO_3 + 2e \rightarrow H_2 + HCO_3^-, \tag{15}$$

$$2HCO_3^- + 2e \rightarrow H_2 + 2CO_3^{2-}, \tag{16}$$

$$2H_2O + 2e \rightarrow H_2 + 2OH^-, \tag{17}$$

Reaction (14) is the most important cathodic reaction in an acidic solution. However, in the CO_2 system, the cathodic reaction is related to the CO_2 partial pressure and the pH value of the experimental solution [21]. At low pH (<4), reaction (14) is dominant owing to the high concentration of H^+. When pH lies between 4 and 6, besides the H^+ reduction, reaction (15) becomes important, and the total cathodic current is contributed by reactions (14) and (15) [24]. In this study, the pH value of the experimental solution was measured to be 4.0, which indicates that the reduction of H^+ and H_2CO_3 are the main cathodic reactions. Therefore, the capacitance loop in the high-frequency zone in the cathodic Nyquist plots, as shown in Figure 9, reflects the electron transfer of H^+ reduction and H_2CO_3 reduction as well as the charge-discharge of double layer capacitance. What is more, besides electron charge transfer, the mass transfer of H^+ and H_2CO_3 has an important influence on cathodic current of reactions (14) and (15), which gives rise to a deviation from cathodic Tafel behavior, as shown in Figure 9, and brings the Warburg impedance in the low-frequency zone in the cathodic Nyquist plots (see Figure 9).

4.2. Mechanoelectrochemical Effect

According to the mechanoelectrochemical theory proposed by Gutman [25], the applied load can affect the electrochemical properties of the electrode. The change of equilibrium electrochemical potential by applied load is expressed as follows [25]

$$\Delta\varphi_e^0 = -\frac{\Delta P V_M}{zF},$$

(18)

where P is the excess pressure (Pa), V_M is the molar volume of metal ($m^3 \cdot mol^{-1}$), and z is the number of transferred electron. This equation indicates that an applied load would decrease the potential of the steel and increase its electrochemical thermodynamic activity.

When the tensile stress reaches the plastic zone, the plastic stress leads to the appearance of the dislocation pile-up group at the barrier through the movement and proliferation of dislocation, which would affect the electrode potential by the following equation [25,26]:

$$\Delta\varphi_p^0 = -\frac{n\Delta\tau R}{\bar{\alpha}k N_{max}zF},$$

(19)

where $\Delta\varphi_p^0$ is the change in equilibrium potential owing to plastic deformation of the electrode, $\Delta\tau$ reflects the extent of work hardening, numerically equaling the residual stress caused by dislocations, n is the number of dislocation in a dislocation pile-up, R is the ideal gas constant, k is the Boltzmann constant, $\bar{\alpha}$ is the average dislocation density, and N_{max} is the maximal number of dislocation in unit volume. It is a big challenge to calculate the $\Delta\varphi_p^0$ by Equation (19), since the parameters involved in Equation (19) are difficult to be measured accurately in experiment. However, based on this equation, it is undoubted that with the increase of the plastic stress, the number of dislocations increases, and as a result, the electrode potential decreases and the steel corrosion is enhanced.

In this study, a negative shift of corrosion potential is observed in both elastic and plastic regions, as shown in Figure 4; moreover, the potential in the plastic stress region is more negative than that in the elastic stress region, which indicates the electrochemical corrosion activity of the steel was improved by the tensile stress, and plastic stress is more significant for corrosion activity compared with elastic stress.

4.3. Corrosion of Stressed Pipeline Steels in CO_2 Environment

According to the above analysis, we attempted to describe, qualitatively, the corrosion behavior of X80 steel under tensile stress in CO_2 environment. Figure 12 shows the schematic diagram of the possible corrosion process of stressed X80 steel in CO_2 environment. Without applied stress, the corrosion process of the steel is just controlled by the combination of cathodic charge transfer and active species diffusion (Figure 12a). Under elastic tensile stress, stress would take effect according to mechano-electrochemical theory. At relatively low tensile stress, only a few dislocation in the steel start to move by tensile stress, which does not affect the corrosion process, but with the increase of tensile stress, the edge dislocation would appear on the steel surface, which results in the increases of the active sites on the surface (Figure 12b). At very high elastic stress (i.e., 90%σ_{ys} in this study), the pre-existing dislocation pile-up in the steel would be relaxed and no new ones formed [18], therefore, the corrosion activity of the steel slightly decreases (Figure12c) and hence its corrosion morphology is not obviously different from that at other elastic stress states (Figure 11c,d). In general, elastic tensile stress has a limited effect on the corrosion of X80 steel. However, under plastic tensile stress conditions, slipping begins to take place and several slip planes emerge at the surface of steel, which largely increases the corrosion activity of the steel surface. The active dissolution would preferentially occur from local areas of exit of slip planes, as a result, the corrosion is significantly accelerated (Figure 12d).

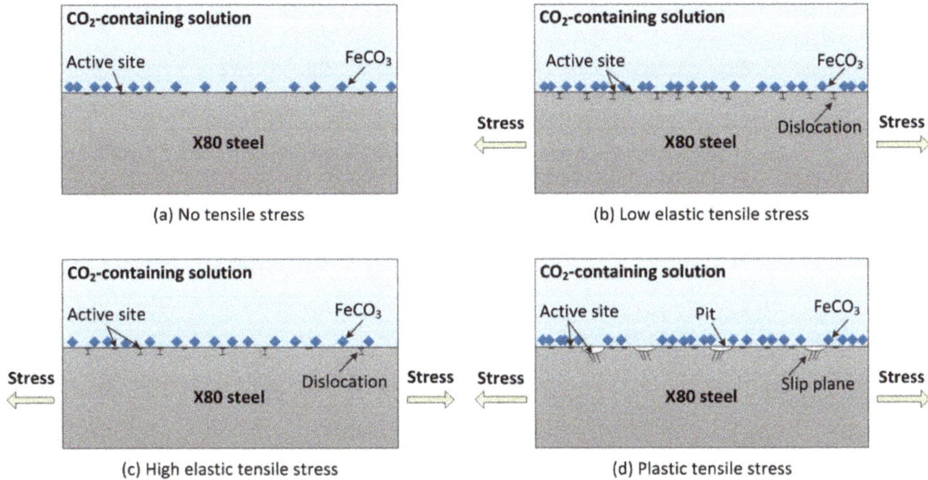

Figure 12. Schematic of corrosion of stressed X80 steel in CO_2 environment.

5. Conclusions

(1) Corrosion rate of stressed X80 steel in 3.5 wt% NaCl aqueous solution with saturated CO_2 at 30 °C increases from 0.845 mm·a^{-1} at 0%σ_{ys} to 1.442 mm·a^{-1} at 60%σ_{ys}, and then slightly decreases to 1.110 mm·a^{-1} at 90%σ_{ys} due to partial stress relaxation and dislocation annihilation at very high elastic stress. In the plastic stress range, corrosion rate increases sharply up to 2.631 mm·a^{-1} at 108%σ_{ys}, which is about three times higher than that without tensile stress.

(2) Both elastic tensile stress and plastic tensile stress can decrease the electrode potential and improve the electrochemical corrosion activity of X80 steel in CO_2-saturated NaCl aqueous solution. Moreover, plastic tensile stress has a more significant effect than elastic tensile stress.

(3) According to potentiodynamic polarization and EIS analysis, both anodic and cathodic reactions are accelerated by tensile stress.

(4) According to the mechano-electrochemical theory, the corrosion of stressed X80 steel is significantly affected by the surface activity, which has much to do with the microscopic defect (like dislocation and slip plane) on the surface caused by tensile stress.

Author Contributions: W.W. proposed this research idea, design the experimental scheme and drafted the article. H.Y., H.Z. and J.K. conducted electrochemical tests. Y.L. and Y.D. conducted the analysis and interpretation of experimental data. All authors read and approved the final manuscript.

Funding: This research was funded by the National Natural Science Foundation of China, grant numbers 51605368 and 21576224.

Conflicts of Interest: The authors declare no conflict of interest.

References

1. Liang, P.; Li, X.; Du, C.; Chen, X. Stress corrosion cracking of X80 pipeline steel in simulated alkaline soil solution. *Mater. Des.* **2009**, *30*, 1712–1717. [CrossRef]

2. Nešić, S. Key issues related to modelling of internal corrosion of oil and gas pipelines—A review. *Corros. Sci.* **2007**, *49*, 4308–4338. [CrossRef]

3. Song, F.M. A comprehensive model for predicting CO_2, corrosion rate in oil and gas production and transportation systems. *Electrochim. Acta* **2010**, *55*, 689–700. [CrossRef]

4. Eliyan, F.F.; Mahdi, E.S.; Alfantazi, A. Electrochemical evaluation of the corrosion behaviour of API-X100 pipeline steel in aerated bicarbonate solutions. *Corros. Sci.* **2012**, *58*, 181–191. [CrossRef]

5. Xue, H.B.; Cheng, Y.F. Passivity and pitting corrosion of X80 pipeline steel in carbonate/bicarbonate solution studied by electrochemical measurements. *J. Mater. Eng. Perform.* **2010**, *19*, 1311–1317. [CrossRef]

6. Mohammadi, F.; Eliyan, F.F.; Alfantazi, A. Corrosion of simulated weld HAZ of API X-80 pipeline steel. *Corros. Sci.* **2012**, *63*, 323–333. [CrossRef]

7. Wang, Y.; Cheng, G.X.; Wu, W.; Qiao, Q.; Li, Y.; Li, X.F. Effect of pH and chloride on the micro-mechanism of pitting corrosion for high strength pipeline steel in aerated NaCl solutions. *Appl. Surf. Sci.* **2015**, *349*, 746–756. [CrossRef]

8. Kermani, M.B.; Morshed, A. Carbon dioxide corrosion in oil and gas Production—A compendium. *Corrosion* **2013**, *59*, 659–683. [CrossRef]

9. Eduok, U.; Ohaeri, E. Electrochemical and surface analyses of X70 steel corrosion in simulated acid pickling medium: Effect of poly (N-vinyl imidazole) grafted carboxymethyl chitosan additive. *Electrochim. Acta* **2018**, *278*, 302–312. [CrossRef]

10. Eduok, U.; Jossou, E.; Szpunar, J. Enhanced surface protective performance of chitosanic hydrogel via nano-CeO_2, dispersion for API 5L X70 alloy: Experimental and theoretical investigations of the role of CeO_2. *J. Mol. Liq.* **2017**, *241*, 684–693. [CrossRef]

11. Ohaeri, E.; Omale, J.; Eduok, U.; Szpunar, J. Effect of Thermomechanical Processing and Crystallographic Orientation on the Corrosion Behavior of API 5L X70 Pipeline Steel. *Metall. Mater. Trans. A* **2018**, *49*, 2269–2280. [CrossRef]

12. Xu, L.Y.; Cheng, Y.F. An experimental investigation of corrosion of X100 pipeline steel under uniaxial elastic stress in a near-neutral pH solution. *Corros. Sci.* **2012**, *59*, 103–109. [CrossRef]

13. Xu, L.Y.; Cheng, Y.F. Corrosion of X100 pipeline steel under plastic strain in a neutral pH bicarbonate solution. *Corros. Sci.* **2012**, *64*, 145–152. [CrossRef]

14. Boven, G.V.; Chen, W.; Rogge, R. The role of residual stress in neutral pH stress corrosion cracking of pipeline steels. Part I: Pitting and cracking occurrence. *Acta Mater.* **2007**, *55*, 29–42. [CrossRef]

15. Bao, M.; Ren, C.; Lei, M.; Wang, X.; Singh, A.; Guo, X. Electrochemical behavior of tensile stressed P110 steel in CO_2 environment. *Corros. Sci.* **2016**, *112*, 585–595. [CrossRef]

16. Li, M.C.; Cheng, Y.F. Corrosion of the stressed pipe steel in carbonate–bicarbonate solution studied by scanning localized electrochemical impedance spectroscopy. *Electrochim. Acta* **2008**, *53*, 2831–2836. [CrossRef]

17. Li, K.; Wu, W.; Cheng, G.X.; Li, Y.; Hu, H.J.; Zhang, H. Corrosion behavior of X70 pipeline steel and corrosion rate prediction under the combination of corrosive medium and applied pressure. In Proceedings of the ASME 2017 Pressure Vessels and Piping Conference, Waikoloa, HI, USA, 16–20 July 2017.

18. Wang, Y.; Zhao, W.; Ai, H.; Zhou, X.; Zhang, T. Effects of strain on the corrosion behaviour of X80 steel. *Corros. Sci.* **2011**, *53*, 2761–2766. [CrossRef]

19. Lasia, A. *Electrochemical Impedance Spectroscopy and Its Applications*; Springer: New York, UK, 2014; ISBN 978-1461489320.

20. Bard, A.J.; Faulkner, L.R. *Electrochemical Methods: Fundamentals and Applications*, 2nd ed.; John Wiley & Sons: New York, UK, 2001; ISBN 978-0471043720.

21. Nordsveen, M.; Nešić, S.; Nyborg, R.; Stangeland, A. A mechanistic model for carbon dioxide corrosion of mild steel in the presence of protective iron carbonate films—Part 1: Theory and verification. *Corrosion* **2012**, *59*, 616–628. [CrossRef]

22. Liu, Q.Y.; Mao, L.J.; Zhou, S.W. Effects of chloride content on CO_2 corrosion of carbon steel in simulated oil and gas well environments. *Corros. Sci.* **2014**, *84*, 165–171. [CrossRef]

23. Bockris, J.O.; Drazic, D.; Despic, A.R. The electrode kinetics of the deposition and dissolution of iron. *Electrochim. Acta* **1961**, *4*, 325–361. [CrossRef]

24. Nešić, S.; Postlethwaite, J.; Olsen, S. An electrochemical model for prediction of corrosion of mild steel in aqueous carbon dioxide solutions. *Corrosion* **1996**, *52*, 280–294. [CrossRef]

25. Gutman, E.M. *Mechanochemistry of Materials*; Cambridge International Science Publishing: Cambridge, UK, 1998; ISBN 1-898326-32-0.

26. Xu, L.Y.; Cheng, Y.F. Development of a finite element model for simulation and prediction of mechanoelectrochemical effect of pipeline corrosion. *Corros. Sci.* **2013**, *73*, 150–160. [CrossRef]

© 2018 by the authors. Licensee MDPI, Basel, Switzerland. This article is an open access article distributed under the terms and conditions of the Creative Commons Attribution (CC BY) license (http://creativecommons.org/licenses/by/4.0/).

 *metals*

Article

A Comprehensive CFD Model for Dual-Phase Brass Indirect Extrusion Based on Constitutive Laws: Assessment of Hot-Zone Formation and Failure Prognosis

George Pashos, George A. Pantazopoulos * and Ioannis Contopoulos

ELKEME Hellenic Research Centre for Metals S.A., 61st km Athens—Lamia National Road,
32011 Oinofyta, Viotias, Greece; gpashos@elkeme.vionet.gr (G.P.); jkontopoulos@elkeme.vionet.gr (I.C.)
* Correspondence: gpantaz@elkeme.vionet.gr; Tel.: +30-2262-60-4463

Received: 13 November 2018; Accepted: 6 December 2018; Published: 8 December 2018

Abstract: A numerical method for the precise calculation of temperature, velocity and pressure profiles of the α-β brass indirect hot extrusion process is presented. The method solves the Navier–Stokes equations for non-Newtonian liquids with strain-rate and temperature-dependent viscosity that is formulated using established constitutive laws based on the Zener–Hollomon type equation for plastic flow stress. The method can be implemented with standard computational fluid dynamics (CFD) software, has relatively low computational cost, and avoids the numerical artifacts associated with other methods commonly used for such processes. A response surface technique is also implemented, and it is thus possible to build a reduced order model that approximately maps the process with respect to all combinations of its parameters, including the extrusion speed and brass phase constitution. The reduced order model can be a very useful tool for production, because it instantaneously provides important quantities, such as the average pressure or the temperature of hot-spots that are formed due to the combined effect of die/billet friction and the generation of heat from plastic deformation (adiabatic shear deformation heating). This approach can assist in the preliminary evaluation of the metal flow pattern, and in the prediction and prevention of critical extrusion failures, thus leading to subsequent process and product quality improvements.

Keywords: brass extrusion; CFD simulation; extrusion failures; plastic deformation processing

---

1. Introduction

Brasses are Cu–Zn alloys, characterized by high corrosion resistance, electrical and thermal conductivity, superior mechanical properties, and formability. The main industrial applications of solid or hollow brass rods are based on their superior performance in machining operations, required for high precision and productivity manufacturing processes, especially for alloys with lead [1–4].

The emerging requirements for the replacement of leaded brasses by new lead-free or low-lead brasses (foreseen by the relevant regulations, especially for drinking water installations), as well as the prevention of critical failures and defects (e.g., during hot working), become strictly specified and necessary for the establishment of sustainable and economically viable processing routes with superior productivity and exceptional product quality. The following actions may be considered mandatory in achieving such goals:

(a) Rigorous review and re-engineering (revision) of brass rod production process conditions and product performance evaluation. The study of the evolution of microstructure and mechanical properties that control service performance, such as fracture resistance of brass rods used in mechanical, electrotechnical and hydraulic applications was systematically studied in Ref. [5–8].

(b) Assessment and deployment of optimum final machining process conditions, in order to achieve superior machinability properties directly applicable to final product manufacturing, as it is dictated by the widespread applications of solid and hollow brass rods. Such studies were mainly focused in the optimization of various machining criteria under certain machining process parameters variation, using parametric and non-parametric studies, e.g., design of experiments (DOE) and analysis of variance (ANOVA) approaches (see relevant Ref. [9–11]).

The production sequence of brass rods embraces the following metallurgical processes/stages [2]: (a) casting, (b) hot extrusion, (c) pickling and (d) cold drawing. Extra heat treatment stages (final or intermediate) may be added depending on the requirements of the final application. An informative flow chart of a typical brass rod production process is illustrated in Ref. [2]. Defects and failures that may be inherited from casting and/or extrusion processes cannot be easily eliminated. On the contrary, critical production defects and failures remain in the final product. These constitute significant sources of quality rejections (during production, "in-process" failures), and are considered as insidious since they could skip final inspection undetected, and thus "pass freely" in the field of application, where the risk of failure and the relevant implications (material, economic and other losses) may be enormous. A review of such brass rods and component defects and failures is presented systematically in Ref. [12]. In this study, the various types of common failures, originating from casting, extrusion, drawing, as well as those originating in the field of service are analytically presented in a "cause-and-effect" approach. More specifically, two main extrusion defects/failures are accounted: the "back-end" defect and the "hot-shortness/overheating" damage, which are both related to metal flow abnormalities during hot extrusion. The latter is a combination of frictional heat dissipation and deformation heat generation, due to intense adiabatic shear, which gives rise to the creation of hazardous "hot-spots", resulting subsequently in the deterioration of microstructure, softening, strain localization and final rupture. Very often, the local temperature spikes reaches and surpasses solidus temperature, leading to inevitable localized melting and severe cracking.

Highlighting the explicit demand of the product quality optimization, via the defect/failure minimization during hot extrusion, a precise knowledge of the metal-flow pattern and the relationships with the principal process parameters is of paramount importance.

To this end, a simulation of hot extrusion plastic flow was attempted here, treated as a fluid dynamics process, coupled with the theory of plastic deformation of solid materials. This was based on the plausible assumption that the hot metal is past its elastic behavior regime and flows like a viscous fluid, while obeying a certain law of hot creep [13]. Moreover, in order for the computational fluid dynamics (CFD) software to be applicable, it is prerequisite that plastic stress does not depend on the material strain but, instead, it depends only on the strain rate and temperature [14]. In the next section, a model is presented, based on fundamental conservation equations and the Zener–Hollomon parameter, which is of major significance for the mathematical formulation of stress as a function of strain rate. The deployment of fluid dynamics models and governing equations could be utilized for the treatment of plastic flow of metals [15–17].

The extrusion of brass for the production of solid and hollow rods covers numerous applications in the field of the manufacturing of mechanical, hydraulic and electrical components. The presence of critical production defects, associated with extrusion abnormalities, could affect acutely the life expectancy of critical parts and compromise the reliability and safety of installations and machine operation during their service time. The brass extrusion process, using material constitutive models, has been already studied [18–20]. However, the comprehensive investigation of the indirect extrusion process of dual-phase brass alloys, combining the utilization of a CFD numerical model and constitutive equations (Zener–Hollomon type), in the frame of featuring and predicting potential defects or failures during production, through visualization of flow pattern and temperature distribution is a novel and original approach which has not been published elsewhere according to the best of our knowledge.

The present CFD-based approach of extrusion with constitutive laws is rigorous and original. Its role is to retrieve valuable information and provide preliminary indications of high failure risks by analyzing thoroughly the metal flow pattern. Furthermore, the present study demonstrates a significant industrial value, since the construction of a reduced order model provides sufficient simplicity and functionality, especially in a plant-based environment, expanding the perspectives of product and process quality improvements. Therefore, models based on the above methodology ("metal-flow-mapping") could result in preliminary predictions of quality and could evaluate failure propensity in the plant environment, thus resulting effectively in the achievement of process consistency and stability leading to the minimization and prevention of final product defects.

2. Numerical Model and Experimental Method

This section presents the fundamentals and technical details of the numerical method as well as some information of the experimental aspects, e.g., in-situ temperature measurements, metallography and fractography. Section 2.1.1 presents the fundamental mathematical formulation, which is essentially a boundary value problem of fluid flow and heat transfer, adjusted to accommodate the rheology of the hot extrusion of brass. This section is complemented by the Appendix A, where the adjustment of the momentum transfer equations is presented in detail. In Sections 2.1.2–2.1.3 we present the details of the boundary value problem and specifically the computational domain and the boundary conditions. In Section 2.2 we present the utilized model of plastic stress for dual-phase brasses, based on a simple mixture law, together with Zener–Hollomon type equations. Sections 2.1.1–2.2 fully describe the numerical model and Section 2.3 lists its assumptions. Section 2.4 describes briefly the technique used to yield a reduced order model from the original numerical model and Section 2.5 lists the input and output parameters of the reduced order model. Section 2.6 contains a table with the utilized thermo-physical properties and Section 2.7 provides details of the computational mesh and also some preliminary validation of the model. Finally, Section 2.8 briefly explains the experimental methods used in this work.

2.1. Model Desciption—Mathematical Formulation

2.1.1. Governing Equations

The material is considered incompressible and, therefore, the continuity equation is applied in the following form:

$$\nabla \cdot \underline{v} = 0,\tag{1}$$

where $\underline{v}$ is the velocity field. Equation (1) is coupled with the momentum transfer equation:

$$\rho\frac{\partial \underline{v}}{\partial t} + \rho\underline{v}\cdot\nabla\underline{v} = -\nabla p + 2\mu(\dot{\varepsilon},\,T)\nabla\cdot\underline{\dot{\varepsilon}},\tag{2}$$

where ρ is the density, p is the pressure, $\underline{\dot{\varepsilon}}$ is the rate-of-strain tensor, defined as

$\underline{\dot{\varepsilon}} = 1/2\left(\nabla\underline{v} + (\nabla\underline{v})^{\mathrm{T}}\right)$, μ is the dynamic viscosity that is a function of temperature (denoted as T) and the equivalent strain rate (denoted as $\dot{\varepsilon}$); the equivalent strain rate is a magnitude of the tensor, $\underline{\dot{\varepsilon}}$, which is consistent with the von Mises yield criterion and is computed using the second invariant of $\underline{\dot{\varepsilon}}$ (details in the Appendix A). Equation (2) is similar to the momentum transfer equation for non-Newtonian liquids [15,16]. In fact, we treat the hot plastically deforming metal as a non-Newtonian liquid with variable viscosity that is derived from semi-empirical laws of creep [14,17]. The expression for μ is:

$$\mu(\dot{\varepsilon},\,T) = \frac{\sigma_p(\dot{\varepsilon},\,T)}{3\dot{\varepsilon}},\tag{3}$$

where σ_p is the plastic stress computed from the Garofalo equation; its formula is presented is Section 2.2 (for more details on the association between the dynamic viscosity and the plastic stress through (3) refer to the Appendix A). The simultaneous solution of (1)–(2) with the heat transfer equation is required to obtain the temperature field, T:

$$\rho c_p \frac{\partial T}{\partial t} + \rho c_p \underline{v} \cdot \nabla T = k \nabla^2 T + 2 C \mu(\dot{\gamma}, T) \underline{\dot{\varepsilon}} : \nabla \underline{v}, \tag{4}$$

where c_p is the specific heat and k is the thermal conductivity. The last term on the right-hand side (RHS) of (4) is the heating rate due to viscous dissipation [16]; its contribution is significant in cases of very viscous liquids with high shear rates. The constant, C, is the conversion ratio of viscous shear stresses to heat (a value of $C = 0.9$ is used [17], i.e., 90% conversion to heat).

2.1.2. Solution Domain

Equations (1)–(4) are solved in the axisymmetric domain Ω_1, depicted in Figure 1 that represents an indirect extrusion process. The process is simulated over a quasi-steady-state window of operation that starts shortly after the press reaches a constant speed and terminates before the ram gets too close to the die. During the steady-state operation all the relevant process quantities (temperature, pressure, etc.) remain constant and therefore, the transient terms on the left hand side (LHS) of (2) and (4) may be ignored.

To simulate the steady-state, a small domain, Ω_1, is only required, which captures the flow field near the die. Far from the die, the flow field is uniform. This implies that the billet moves at a constant velocity and does not deform, i.e., at a certain distance from the die, the material is not affected by the process. This is a good approximation for indirect extrusion where the container moves together with the billet and there is virtually no friction between them. If there is significant friction with the container walls, then the entire billet is deformed and the solution domain should contain the entire billet. This is the case for direct extrusion.

The domains Ω_2 and Ω_3, correspond to the extrusion container and die, respectively. The model includes only the innermost steel layer of the container surface area (liner). The die is modeled as a single uniform item, instead of two distinct steel parts (die & die holder). In Ω_3 only the heat transfer Equation (4) is solved, by eliminating all the terms except for the heat conduction term (first term in the RHS), whereas in Ω_2 the convective term is retained (second term in the LHS) with uniform horizontal velocity equal to the extrusion speed S, because of the equality between the container and billet speeds.

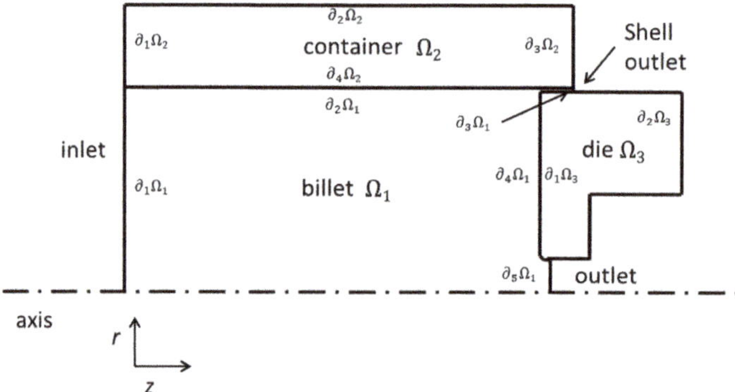

Figure 1. Solution domain for the steady-state indirect extrusion problem. Only a small segment of the billet near the die is considered, where the effects of the process are significant; far from the die the billet is relatively unaffected by the process. The domain consists of three subdomains, Ω_1, Ω_2 and Ω_3 (billet, container and die respectively), with their corresponding boundaries denoted as $\partial_i \Omega_1$, $\partial_i \Omega_2$ and $\partial_i \Omega_3$.

2.1.3. Boundary Conditions

The conditions imposed on the boundaries of Ω_1—denoted $\partial_i\Omega_1$—are:
Uniform inlet velocity on the left boundary of Ω_1,

$$v_r\big|_{\partial_1\Omega_1} = 0 \tag{5}$$

$$v_z\big|_{\partial_1\Omega_1} = S, \tag{6}$$

where v_r and v_z are the velocity components (radial and axial, respectively) and S is the extrusion speed.
Zero pressure on the outlet:

$$p\big|_{\partial_5\Omega_1} = 0. \tag{7}$$

The extruded shell sticks to the container walls and therefore, its relative velocity with respect to the container is zero. Since the relative velocity of the container with respect to the die is 0 (radial component) and S (axial component), thereby the velocity of the shell (prescribed at the boundary denoted "shell outlet") is:

$$v_r\big|_{\partial_3\Omega_1} = 0 \tag{8}$$

$$v_z\big|_{\partial_3\Omega_1} = S. \tag{9}$$

It should be noted that by using the boundary conditions (8) and (9), we explicitly restrict any back-extrusion through the gap between the ram and the container. However, we do not observe any back-extrusion in the current process and therefore, Equations (8) and (9) are valid.
Prescribed tangential viscous stress (skin friction) on the die walls is,

$$\tau_{nt}\big|_{\partial_4\Omega_1} = f\underline{v} \cdot \underline{t}, \tag{10}$$

where $\underline{t}$ is the unit tangent vector to the die walls, f is the friction factor derived from a simple model of hydrodynamic lubrication and τ_{nt} is the tangential component of the viscous stress vector ($\tau_{nt} = \underline{\tau}_n \cdot \underline{t} = \underline{n} \cdot \underline{\tau} \cdot \underline{t}$; $\underline{n}$ is the unit normal vector to the die walls). The lubrication model considers laminar flow between two parallel plates, one of them stationary and the other sliding. Thus, the friction factor is equal to the dynamic viscosity (μ) of the lubricant over the size of the gap (L) between the parallel plates.

$$f = \frac{\mu}{L}. \tag{11}$$

For a viscous lubricant such as a graphite paste that is used in the process, a viscosity in the order of 10^2 Pa s is expected, with a layer thickness in the order of microns. Therefore, the friction factor is expected to be in the order of 10^8 Ns/m^3. It should be noted, however, that f cannot be determined from (11), since the lubrication layer thickness is generally unknown. Instead, we can adjust f to match experimental measurements of the process quantity that is mostly sensitive to it (see Section 3.5), which is the maximum temperature of the extruded product.
Zero tangential viscous stress (slip condition) on the container walls:

$$\tau_{nt}\big|_{\partial_2\Omega_1} = 0. \tag{12}$$

Uniform inlet temperature:

$$T\big|_{\partial_1\Omega_1} = T_{\text{in}}. \tag{13}$$

Temperature continuity on the container and die walls that are in contact with the billet:

$$T\big|_{\partial_2\Omega_1} = T\big|_{\partial_4\Omega_2} \tag{14}$$

$$T|_{\partial_4 \Omega_1} = T|_{\partial_1 \Omega_3}. \tag{15}$$

Zero conductive heat flux on the outer die walls and on the left/right container walls:

$$\underline{n} \cdot \nabla T|_{\partial_1 \Omega_2} = \underline{n} \cdot \nabla T|_{\partial_3 \Omega_2} = \underline{n} \cdot \nabla T|_{\partial_2 \Omega_3} = 0. \tag{16}$$

Heat dissipation on the outer container wall via a simple cooling law:

$$\underline{n} \cdot k \nabla T|_{\partial_2 \Omega_2} = h(T - T_\infty), \tag{17}$$

where h is the heat transfer coefficient and T_∞ is the ambient temperature; T_∞ is prescribed at 27 °C for the present work. It should be noted that the cooling law has little physical meaning here, since $\partial_2 \Omega_1$ is not an outer wall that is exposed to the ambient air, but an inner wall between layers of steel. However, this equation can be used in lieu of a more accurate formulation, with the goal of approximating the average measured temperature of the inner steel liner. In other words, the heat transfer coefficient h is adjusted till the average temperature of Ω_2 matches the measured temperature.

2.2. Plastic Stress

The plastic stress in (3) is calculated through the Garofalo equation [18,19]:

$$A \left[\sin h \left(a \sigma_p \right) \right]^n = Z, \tag{18}$$

where A (s^{-1}), a (MPa^{-1}) and n (dimensionless) are parameters that depend on the composition of the material and Z is the Zener–Hollomon parameter [20]:

$$Z = \dot{\varepsilon} \exp\left(\frac{Q}{RT} \right), \tag{19}$$

where $\dot{\varepsilon}$ is the equivalent strain rate, Q (J/mol) is the activation energy, R (J/(molK)) is the universal gas constant and T (K) is the absolute temperature. The parameters of Equations (18) and (19) are usually derived from hot torsion experiments with variable temperature and torsion speed [18–22]. By substituting (19) in (18) and solving for σ_p, we acquire the plastic stress that is used in (3):

$$\sigma_p(\dot{\varepsilon}, T) = \frac{\mathrm{arcsinh}\left(\left(\frac{\dot{\varepsilon} \exp\left(\frac{Q}{RT} \right)}{A} \right)^{\frac{1}{n}} \right)}{a} \tag{20}$$

Equation (20) is valid for a single phase alloy. Otherwise, if there are multiple phases with considerably different mechanical properties, e.g., α and β phases of brass, then (20) loses its validity. This happens with the current case of the CW626N alloy, which could be approximated with a CuZn33 brass that also includes other alloying elements. The pure CuZn33 may be an α-brass, but in our case and based on the analysis of the brass billets in production, the equivalent Zn [23] (by virtue of the minor alloying elements and especially Al) is approximately 40%, which corresponds to a binary CuZn40. The equivalent Zn shifts the alloy composition to the regime of ($\alpha + \beta$)-brass and as such Equation (20) cannot be used.

Spigarelli et al. [18] suggested that two Garofalo equations could be used instead, one for each brass phase, with a different set of parameters each. The individual plastic stresses, coming from the two equations, contribute to the true plastic stress through a volumetric mixing law:

$$\sigma_p = (1 - f_\beta)\sigma_{p,\alpha} + f_\beta \sigma_{p,\beta} \tag{21}$$

where f_β is the β-phase volume fraction, $\sigma_{p,\alpha}$ is the α-phase plastic stress and $\sigma_{p,\beta}$ is the β-phase plastic stress. The single-phase brass plastic stress ($\sigma_{p,\alpha}$ and $\sigma_{p,\beta}$) is computed using (20), with parameters obtained from hot torsion experiments of single-phase brasses. There is no definite selection of the composition of those brasses, but a reasonable option is to choose single-phase brasses that are close to the transition to dual phase ($\alpha + \beta$)-brass. For $\sigma_{p,\beta}$ a good selection of pure β-brass is CuZn44 that is marginally close to the transition to dual ($\alpha + \beta$)-brass. Similarly for $\sigma_{p,\alpha}$ a good selection is CuZn36.

We use Equation (21) in the simulations, with $\sigma_{p,\alpha}$ and $\sigma_{p,\beta}$ that correspond to the plastic stresses of α-brass CuZn36 and β-brass CuZn44, respectively. Each individual plastic stress is computed by (20) using the parameters in Table 1 (reported in [19]). The use of a mixing law is essential, because at the process temperature (715 °C) and according to (21), the α-phase is three to four times harder (depending to the strain rate) than the β-phase. This large discrepancy of the mechanical properties between the two phases, reflects on the process parameters as well, especially on the extrusion pressure and temperature. For example, a pure α-phase brass develops an average pressure and maximum temperature during the process, approximately 380 MPa and 820 °C, whereas a pure β-phase develops only 140 MPa and 760 °C ($S = 0.015$ m/s, $T_{\text{in}} = 715$ °C).

Table 1. Parameters of Equation (20) for the single-phase brasses, CuZn36 (α-brass) and CuZn44 (β-brass).

Brass	Q (J/mol)	A (s^{-1})	n (Dimensionless)	a (MPa^{-1})
CuZn36	157,000	6.60×10^9	4.5	6.58×10^{-3}
CuZn44	92,000	1.44×10^7	3.0	9.21×10^{-3}

2.3. Model Assumptions

The following model assumptions were used:

- Brass phases in the billet are distributed evenly (no spatial variation).
- There is no phase transition during the extrusion process, i.e., α-brass to β-brass or vice-versa.
- The temperature of the billet, far from the die (inlet temperature), is uniform.
- The process is in steady-state regime.
- Heat generation due to skin friction between the billet and the die is neglected. Heat is only produced due to the deformation of the billet.
- The alloying elements contribute only to the equivalent Zn.
- There is no friction between container and billet.
- The friction between billet and die is based on a simple hydrodynamic lubrication model.

2.4. Reduced Order Model

The CFD model is reduced, using rigorous mathematical techniques (not described here). The reduced order model operates as a replacement of the CFD model and its purpose is to be used by production engineers, without any prior training, neither extensive knowledge of the extrusion process (physics, mathematics, etc.).

The reduced order model is based on the response surface method, initially introduced by Box and Wilson [24] and currently implemented with the DesignXplorer tool (version 18.2, Ansys Inc., Canonsburg, PA, USA) [25]. In essence, it relates output process parameters such as average outlet temperature, extrusion pressure, etc. (in the method nomenclature, these are the response parameters), with input parameters such as the extrusion speed, friction factor, etc. Unlike the CFD model which is computationally intensive and may require execution times from hours to days, the reduced order model, once established, produces results instantaneously and can be used to extensively investigate the process characteristics with respect to several input parameters (see Section 3.6). It should be noted, however, that the establishment of the reduced order model requires design of (computational) experiments (DoE) [25,26] and as such, it is by itself a computationally intensive task.

2.5. Process Parameters

The process parameters are separated in input parameters (imposed on the process, e.g., extrusion speed), and output parameters (dictated by the process and dependent on the input parameters, e.g., average temperature of the extruded product).

2.5.1. Input Parameters

The considered input parameters are shown in Table 2. The extrusion speed, S, is directly set by the control system of the hydraulic press. The inlet temperature, T_{in}, is equal to the temperature of the billet (considered uniform) right before the start of the extrusion process. It should be noted that currently there are no means to determine whether T_{in} is in fact equal to the temperature of the billet before it enters the container, e.g., there may be a significant decrease and non-uniformity of the billet temperature if it stays idle in the container.

The β-phase volume fraction, f_β, is determined by the equivalent Zn of the billet. The simplest approach to calculate f_β, would be to assume phase equilibrium at T_{in} and consult the binary Cu–Zn phase diagram. Given the equivalent Zn of the brass used in the process (~40%), the volume fraction of β-phase at the extrusion temperature is approximately 0.5.

In the current model, there is no method to determine h and therefore, it was approached in a different manner, described in Section 2.1.3: h was adjusted until the average temperature of the container liner reaches the measured temperature (measurements performed with an optical pyrometer); for this case study the average temperature is ~550 °C and h was determined (by trial-and-error) to about 1.7 W (m² K).

The friction factor, f, is yet another unknown input parameter that depends on the lubricant type and may be determined using an approach similar to that used for the determination of h. f has a strong impact on the temperature of the extruded product (see Section 3.5) and, thus, it may be adjusted until the calculated average temperature at the outlet of the container, reaches the measured temperature.

Table 2. Input parameters.

Input Parameter	Notation	Alternate Notation (Reduced Order Model)	Units
Extrusion speed	S	P3	m/s
Friction factor	f	P4	Ns/m³
Heat transfer coefficient on the liner outer walls	h	P7	W/(m² K)
β-phase volume fraction	f_β	P13	dimensionless
Inlet (initial) temperature of the billet	T_{in}	P14	K

2.5.2. Output Parameters

The considered output parameters are shown in Table 3. In order to convert the pressure of the hydraulic press (bar) to the pressure inside the container (extrusion pressure in Pa), it has to be multiplied by 1.68×10^6; this factor is calculated from the specifications of the hydraulic press.

Table 3. Output parameters.

Output Parameter	Notation	Alternate Notation (Reduced Order Model)	Units
Extrusion pressure	extrpress	P1	Pa
Maximum temperature (hot spots—see Section 3.2)	maxTemp	P2	°C
Average temperature of the extruded product	aveTemp	P9	°C
Average temperature of the container liner	contemp	P10	°C
Average temperature of the die	dietemp	P11	°C
Average temperature of the brass shell	sleevetemp	P12	°C

2.6. Thermo-Physical Properties

The thermo-physical properties of the model are shown in Table 4 [27].

Table 4. Thermo-physical properties. The specific heat and thermal conductivity of CW626N is approximated with that of CuZn30. Typical thermo-physical properties of tool steels were used.

Property	Brass—CW626N	Steel (Die & Container)
Density (kg/m^3)	8500	8030 *
Specific heat (J/(kg·K))	355 + 0.136T [27]	502.48 *
Thermal conductivity (W/(m·K))	140.62 + 0.011214T [27]	25 **
Viscosity (Pa·s)	Equation (3)	N/A

* ANSYS Fluent Database; ** Typical value for tool steel grades.

2.7. Mesh Sensitivity—Model Validity

The utilized mesh is shown in Figure 2. It consists of ~100,000 elements with boundary layers of smaller elements at the container and die walls, in order to resolve the steep solution profiles (temperature, velocity, etc.), especially near the abrupt geometrical features of the die. The current mesh is excessively refined, because the computational cost for the solution of a 2D problem is not considerable, even for very refined meshes. In fact, by coarsening the mesh more than half its original size (~40,000 elements), the relative difference of the calculated average pressure was less than 5% and the absolute difference of the calculated maximum and average temperatures of the billet was less than 1 °C. Therefore, we deem the solution independent of the mesh for the purposes of this work.

The accuracy of the model is examined with the aid of ongoing production trials. At the current stage we are able to successfully reproduce the actual hydraulic pressure during the steady-state operation. To elaborate, experiments were performed (Table 5), during the normal production process that involved the measurement of the container liner temperature and the billet temperature, before and after the extrusion. With the establishment of these temperatures that provide the average temperature of important components (billet: 715 °C, liner: 550 °C), along with the steady extrusion speed (1.5 cm/s) and the brass composition, we calculate the hydraulic pressure, which turns out to be identical to the actual one (approximately 160 bar).

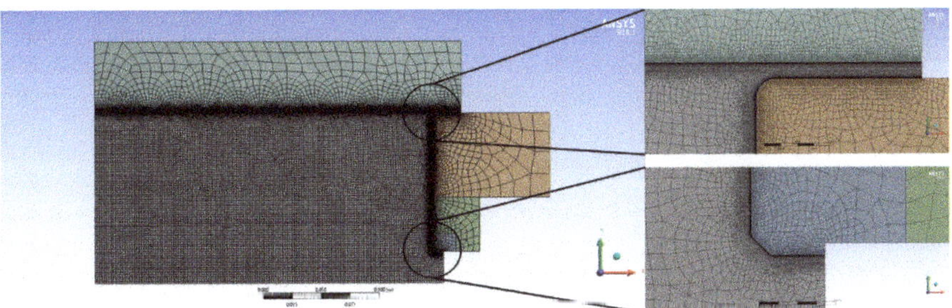

Figure 2. Computational mesh consisting of ~100,000 elements. The mesh is refined near the container and die walls, with 20 layers that span ~1.5 mm; the wall adjacent layer has a thickness of 10 μm.

Table 5. Validation experiments in the production process. The measurements of the hydraulic pressure are juxtaposed with the model results. The measured hydraulic pressure is obtained via a pressure gage with an estimated precision of 2.5 bar. The inlet temperature is the average of three measurements at the front, middle and back of the billet, using an optical pyrometer with estimated precision of 0.5 °C.

Billet Number	Inlet Temperature (°C)	Extrusion Speed (cm/s)	Hydraulic Pressure (bar)	Calculated Hydraulic Pressure (bar)
1	713.5	1.45	155	154.2
2	716.0	1.45	150	152.9
3	717.0	1.50	155	157.6
4	721.5	1.45	150	150.1

The validation of the model will continue using measurements from production, with variations of extrusion speed and/or brass composition.

2.8. Experimental Characterization—Metallography and Fractography

Metallographic characterization of a typical dual-phase ($\alpha + \beta$) brass was conducted on transverse cross-sections, after wet grinding up to 1200 grit SiC paper, followed by fine polishing using diamond and silica suspensions and finishing with ethanol and compressed-air drying. Subsequently, immersion chemical micro-etching, for approximately 5 s at room temperature, was performed using $FeCl_3$-based solution (Merck, Darmstadt, Germany) according to the ASTM E407 standard. Light optical metallography was performed using a Nikon Epiphot 300 (Nikon, Tokyo, Japan) inverted optical microscope using image analysis software, Image Pro Plus Version 7.0 (Rockville, MD, USA) for phase (area) fraction measurements. High-magnification fracture surface observations, utilizing a FEI XL40 SFEG scanning electron microscope (FEI, Eindhoven, The Netherlands), were realized using secondary electron (SE) imaging under 20 kV accelerating voltage.

3. Results and Discussion

3.1. Microstructure

A typical dual phase structure of a common dual phase brass alloy is shown in Figure 3. The phase structure consisted of approximately 65% α-phase and 35% β-phase. The α-phase represents the Cu solid solution with Zn (exhibiting a face-centered-cubic crystal lattice), while β-phase is the ordered non-stoichiometric intermetallic compound CuZn possessing a body-centered-cubic (bcc) lattice. Lead particles (seen as black-dots of size of 1 μm approximately) are well distributed mainly in the α/β interface areas, which act as high surface energy positions. The phase structure for these alloys depends on alloy chemical composition as well as on casting and hot working conditions, i.e., casting temperature, casting speed, extrusion temperature and cooling rate. Pb-particle size and distribution is predominantly influenced by casting and solidification conditions, while the phase structure, such as volume fraction of β-phase and grain size, is influenced mainly by hot extrusion conditions. The presence of β-phase significantly affects the mechanical behavior and machinability of the extruded brass [5,28].

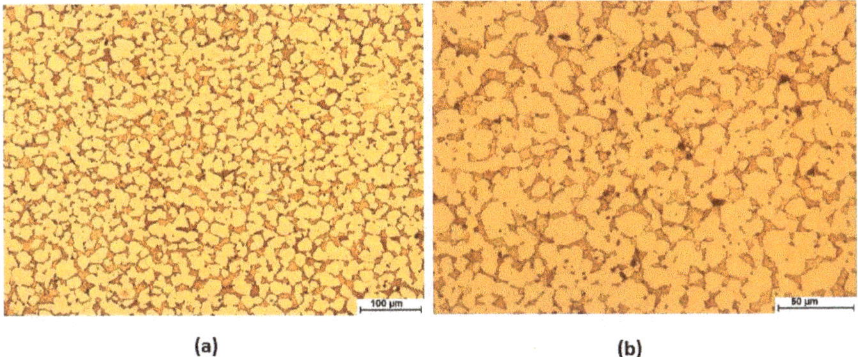

(a)   **(b)**

Figure 3. (**a**) Optical micrograph of the microstructure of a typical dual-phase brass on transverse section. (**b**) Detail of (**a**). Bright areas represent α-phase and dark areas represent β-phase. In (**a**) there is an appreciable amount of Pb particles, which appeared as black dots. Immersion etching in FeCl$_3$-based solution.

3.2. Equivalent Strain Rate

An important calculated quantity that affects the process characteristics and especially the extrusion pressure and the generation of heat due to deformation, is the equivalent strain rate (appears in Equations (2)–(4)). The equivalent strain rate has a typical profile shown in Figure 4a. The strain rate is large near the die and especially at locations, where there is an abrupt change of the die geometry (edges and slopes). Also, in cases where there is a high friction factor on the die walls, we observe large strain rates on the horizontal die walls near the outlet (Figure 4b). This occurs because the extruded metal sticks to the die, thus straining the metal layers adjacent to the die walls. The locally high strain rates near the die are also responsible for elevating the temperature considerably at the outer layers of the extruded product (see Section 3.3). The phase structure of the reference dual phase alloy consists of 65% α and 35% β under room temperature conditions. The β-phase fraction (f_β) is taken to be approximately equal to 0.5, which corresponds to an equi-volume phase structure, consisting of 50% α and 50% β at the extrusion temperature (T_{in} = 715 °C).

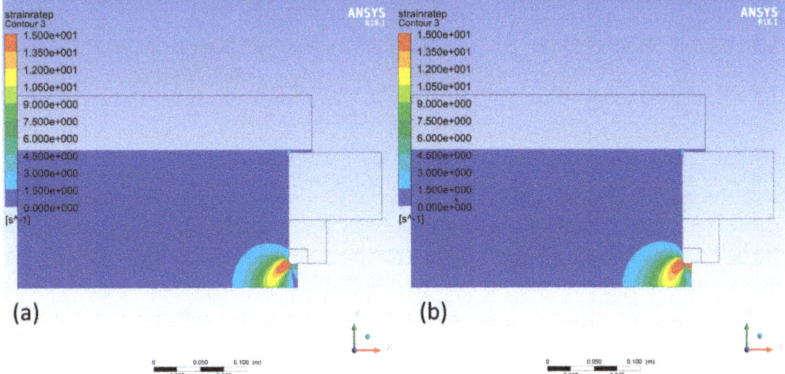

(a)   **(b)**

Figure 4. Contours of the equivalent strain rate (values larger than 15 s^{-1} are clipped to the red color contour). (**a**) Solution with no friction $f = 0$ Ns/m^3, $S = 1.5$ cm/s, $f_\beta = 0.5$, $T_{in} = 715$ °C, $h = 1.7$ W/(m^2 K). (**b**) Solution with friction $f = 5 \times 10^8$ Ns/m^3, $S = 1.5$ cm/s, $f_\beta = 0.5$, $T_{in} = 715$ °C, $h = 1.7$ W/(m^2 K).

3.3. Temperature

The temperature of the billet decreases near the container walls, due to the conduction of heat through the steel liner, but also increases near the die, due to the plastic deformation energy that is converted to heat (see heat transfer Equation (4)). The model shows a significant increase of temperature at the outlet—compared to the inlet temperature—in the neighborhood of 50 °C (average temperature of the extruded product).

The temperature increases towards the outlet (Figure 5a), where the equivalent strain rate is large, regardless of the friction of the billet with the die walls. If however, the friction factor is large, then the local strain rate is large near the horizontal die walls as well (Figure 4b), which consequently translates to the formation of hot spots in the outer layers of the extruded product (Figure 5b). The temperature of the hot spots may surpass the inlet temperature by more than 100 °C, depending on the process parameters (extrusion speed, friction factor, etc.).

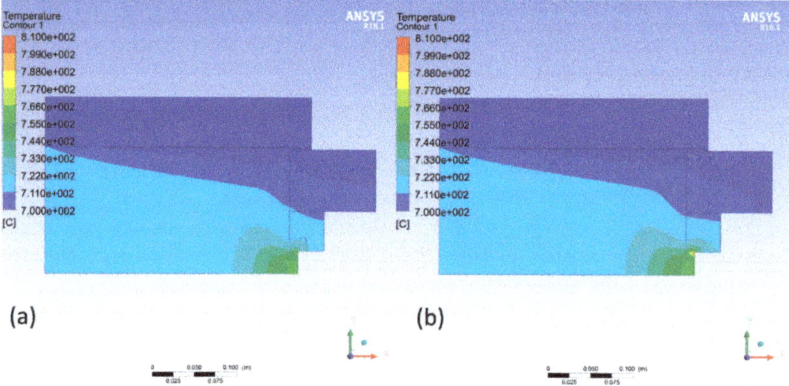

Figure 5. Contours of temperature (values smaller than 700 °C are clipped to the blue color contour). (a) Solution with no friction $f = 0$ Ns/m^3, $S = 1.5$ cm/s, $f_\beta = 0.5$, $T_{in} = 715$ °C, $h = 1.7$ W/(m^2 K). (b) Solution with friction $f = 5 \times 10^8$ Ns/m^3, $S = 1.5$ cm/s, $f_\beta = 0.5$, $T_{in} = 715$ °C, $h = 1.7$ W/(m^2 K). Notice the formation of a small hot spot—due to high friction—near the outlet and on the outer layer of the extruded product.

This may be detrimental to the process, because a very high temperature at the surface of the extruded product may cause severe defects, such as surface cracks (hot shortness); see [29] and Section 3.7.

3.4. Pressure/Flow Profile

A typical flow profile is shown in Figure 6a. The streamlines begin to converge towards the outlet at about ~15 cm from die. This suggests that the billet moves as an unyielding solid body, until it reaches a threshold at a certain distance before the die. This justifies our decision to simulate the process only in a small computational domain containing only the region beyond the threshold.

Moreover, streamlines diverge near the shell outlet, forming a flow stagnation point on the surface of the die holder, which translates in a localized pressure increase (Figure 6b). Far from the die, the pressure is virtually uniform and equal to the extrusion pressure, which as we discussed above is related to the measured pressure of the hydraulic press (see Section 2.5.2).

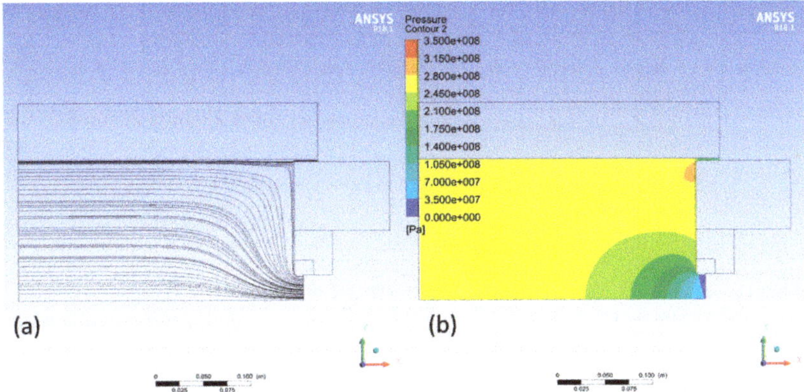

Figure 6. (**a**) Streamlines. (**b**) Pressure contours. Parameter values: $f = 5 \times 10^8$ Ns/m^3, $S = 1.5$ cm/s, $f_\beta = 0.5$, $T_{in} = 715$ °C, $h = 1.7$ W/(m^2 K).

3.5. Sensitivity Analysis

The sensitivity of the output parameters to changes of the input parameters is presented in Figure 7. Each bar represents the change of a single output parameter with respect to the change of a single input parameter for a given interval, e.g., we allow that the extrusion speed goes from 0.01 to 0.02 m/s. For that change of extrusion speed, the extrusion pressure increases by an amount represented by the first red bar. On the other hand, we allow that f_β changes from 0 to 1, which has the effect of lowering the extrusion pressure by an amount represented by the first inverted green bar (negative effect). The green bar is longer than the red bar and therefore, by going from pure α-phase to pure β-phase, the impact on the extrusion pressure is larger than by increasing the speed from 0.01 to 0.02 m/s. The intervals of the input parameters that are used for the construction of the chart in Figure 7, are shown in Table 6.

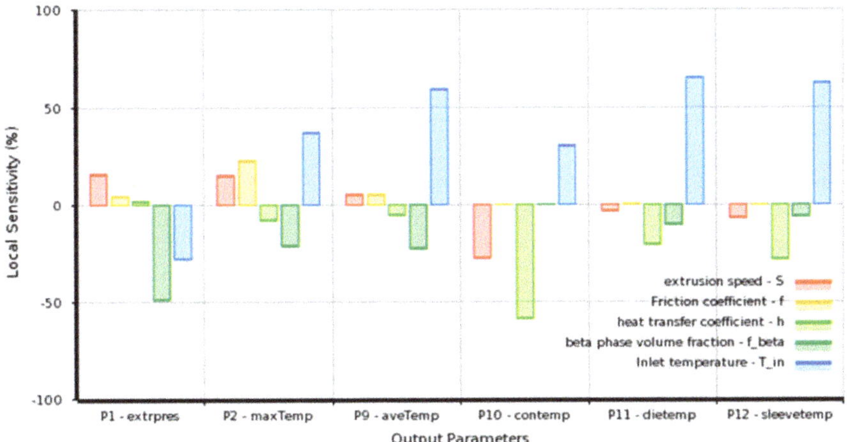

Figure 7. Sensitivity chart of the output parameters (designated by column—horizontal axis) with respect to the input parameters (designated by color bars). The analysis is performed in the neighborhood of a reference point with input parameter values: $f = 5 \times 10^8$ Ns/m^3; $S = 1.5$ cm/s, $f_\beta = 0.5$, $T_{in} = 715$ °C, $h = 1.7$ W/(m^2 K).

Table 6. Input parameter intervals.

Input parameter	Interval	Units
S	[0.01, 0.02]	m/s
f	[0, 10^9]	Ns/m^3
h	[0, 3]	W/(m^2 K)
f_β	[0, 1]	dimensionless
T_{in}	[900, 1050]	K

3.6. Parametric Study

The reduced order model enables a very extensive parametric study. In Figure 8 selected charts of the process output parameters vs. input parameters were presented, produced by the reduced order model. These charts encapsulate the importance of the reduced order model, since we can acquire results on-demand without using a complex numerical model. For example, let us consider an increase of extrusion speed for the sake of increasing the production capacity. The increase is only modest (~13%), going from 1.5 cm/s to 1.7 cm/s. From Figure 8a,c, using the default parameter values, we gather that the speed increase results in ~8% increase of the extrusion pressure and the increase of the maximum temperature by ~10 °C. Another example would be the reduction of the preheating furnace temperature down to 664 °C from 715 °C. By consulting Figure 8b, we find that this alteration results in the increase of the extrusion pressure up to 310 MPa and therefore 184 bar hydraulic pressure.

3.7. Phenomenological Aspects of Flow-Induced Extrusion Failures

High and/or localized impurity levels may result in cracking and discontinuities after extrusion. Bismuth (Bi) in the alloy can lead to the formation of brittle inter-granular films that aggravate hot shortness and adversely affect the hot workability of the alloy. On the other hand, excess antimony (Sb) impairs cold workability [30,31].

A back-defect that generally appears as a hole at the end of the extruded rod results from a combination of the metal flow-pattern and the introduction of surface oxides into the interior of the rod [29]. The flow pattern established by the differences in flow rates draws surface oxides into the center of the rod and the successive metal layers cannot weld together. The lack of "welding" between the mating surfaces results in the formation of such discontinuities and central rod defects.

Removing a certain length from the end of the extruded rod restricts the inheritance of this defect to the downstream processes and eventually to the final product. The back-defect is not common for indirect extrusion processes.

Hot-shortness failures appear in the form of surface cracks along the length of the extrusion. Hot-shortness results from overheating due to the high localized temperature (hot-spots), as a result of billet/die friction. This phenomenon raises the metal temperature to the point where localized regions of segregation may melt or become hot-short. The frictional heat dissipated during extrusion may also cause hot-tears on the metal surface. The surfaces of such tears oxidize and do not re-join during the metal processing operations. The non-bonded regions lead to further cracking or delaminations.

Other studies highlight the importance of the average shear strain rate criterion during the extrusion of brass (e.g., CuZn30), see Ref. [31]. Exceeding a certain threshold (e.g., 0.9 s^{-1} for CuZn30), the risk of transverse cracking is significant, as it can be illustrated by the relevant extrusion ratio-speed process curves that show a characteristic safe/failure boundary. A decrease in the extrusion temperature and in the extrusion speed will generally eliminate defects introduced due to hot shortness phenomena that result in intergranular fractures [12,32,33]. Typical intergranular fracture morphology of brass rod, induced due to high speed extrusion which resulted in overheating is illustrated in Figure 9. The presence of Pb might exaggerate the hot-shortness phenomena that are a result of localized melting and subsequent tearing due to "internal" liquid metal embrittlement. The presence of Pb particles could lead to the generation of voids, followed by growth, realized by an adsorption-induced decohesion, or dislocation emission process that will finally result in coalescence and necking (as is diligently outlined in Ref. [34]).

The utilization of the CFD-based extrusion model is expected to become a plant-based practice aiming to predict and prevent extrusion temperature rises that lead to the aggravation of hot-shortness/overheating-related defects formation. This is a preliminary step that will find promising applications only following systematic validation in actual production.

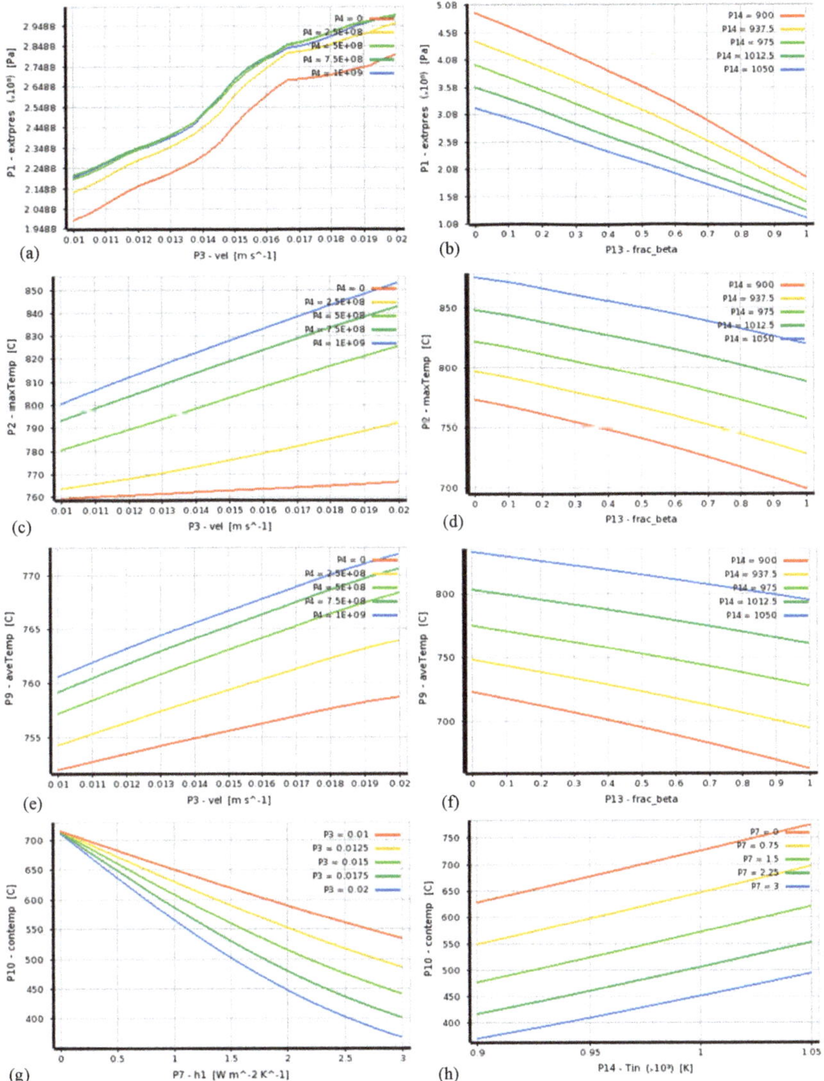

Figure 8. Process charts showing the dependence of the process output parameters on the input parameters. (**a**) Extrusion pressure (*P*1) vs. extrusion speed (*S*). (**b**) Extrusion pressure (*P*1) vs. β-phase fraction (f_β). (**c**) Max. extrusion temperature (*P*2) vs. extrusion speed (*S*). (**d**) Max. extrusion temperature (*P*2) vs. β-phase fraction (f_β). (**e**) Average extrusion temperature (*P*9) vs. extrusion speed (*S*). (**f**) Average extrusion temperature (*P*9) vs. β-phase fraction (f_β). (**g**) Container temperature (*P*10) vs. heat transfer coefficient (*h*). (**h**) Container temperature (*P*10) vs. inlet temperature (T_{in}). The input parameters have the default values: $f = 5 \times 10^8$ Ns/m^3, $S = 1.5$ cm/s, $f_\beta = 0.5$, $T_{in} = 715$ °C, $h = 1.7$ W/(m^2 K), unless otherwise stated by each chart.

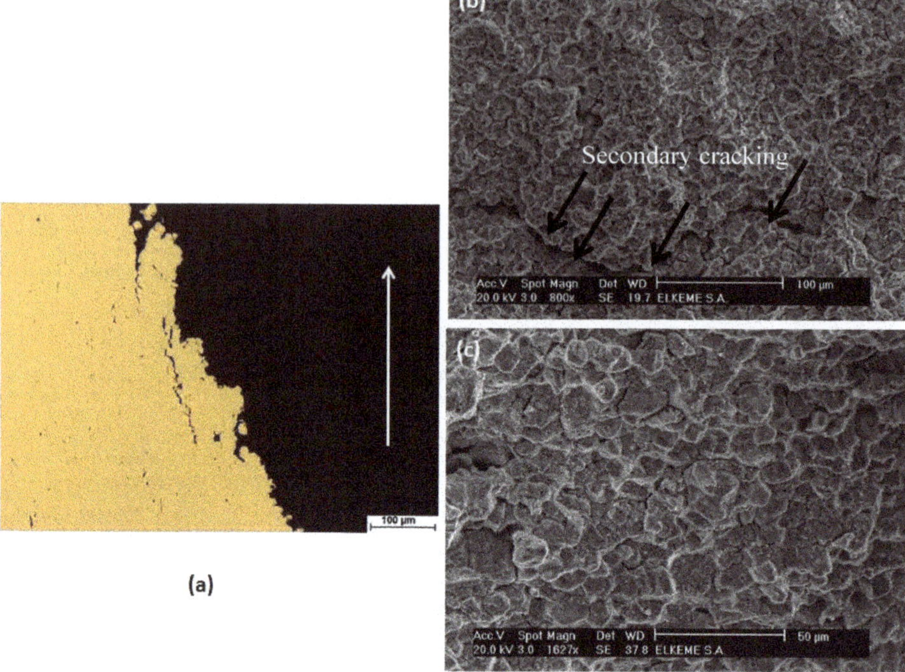

Figure 9. (**a**) Optical micrograph (as-polished/bright field illumination) showing extensive surface cracking occurred in extruded brass CW614N-CuZn39Pb3 (hexagonal rod of 6 mm across flats), (**b**) and (**c**) scanning electron microscope (SEM) fractographs (secondary electron imaging) showing a typical intergranular fracture mode and secondary, transverse cracking due to overheating. The white arrow shows the extrusion direction. This failure case history was presented in detail in Ref. [32].

4. Conclusions

A CFD-based model of dual phase brass extrusion was presented that is based on constitutive laws and the Zener–Hollomon parameter that connects plastic stress and strain rate. The main scope of the present research work is the investigation of the metal-flow pattern during indirect extrusion, and the calculation of the main output parameters such as extrusion pressure and temperature. The main results of the present study can be summarized as follows:

(1) A reduced order model was attempted, which is an approximation of the CFD model that relates the process input and output parameters, such as extrusion pressure vs speed (velocity). The reduced order model can be readily used by plant production personnel, since it can be embedded in a simple spreadsheet.

(2) The CFD model accommodates the rigorous theory of plastic deformation, and takes also into account the different plastic responses during extrusion of the two phases in equilibrium (α and β).

(3) The CFD model predicts the formation of hot-zones near the die (due to the local high strain rates), which may induce severe defects in the final product.

Knowledge of the extrusion metal flow pattern, as this is derived solving the constitutive laws together with the ad hoc assumption of dual phase plastic stress behavior ("rule of mixtures") and its relation to viscosity, could become a promising first step for the development of an integrated defect/failure prognostic tool that will effectively assist in the improvement of the product and process quality. Further adjustment/validation of the CFD model is required, based on crucial plant information that will associate the model output parameters and the formation of critical defects.

Author Contributions: Conceptualization and Simulation, G.P.; Supervision, I.C.; Project Consulting, Review and Paper Editing, G.A.P.

Funding: This research received no external funding.

Acknowledgments: The authors wish to express special thanks to the Management and Quality Division of FITCO Metal Works S.A. for the continuous support assisting in the project's realization and completion and for granting permission for the current publication.

Conflicts of Interest: The authors declare no conflict of interest.

Appendix A Relationship between Plastic Stress and Viscosity

The foundation of the approach that converts the plastic stress of a yielded solid material to an equivalent viscosity—with the purpose of using fluid dynamics to simulate the extrusion of a solid material—is based on the von Mises yield criterion. According to this, a solid material yields (i.e., flows plastically), when a function of the second invariant of the deviatoric stress tensor, referred to as equivalent von Mises stress, reaches a critical value, referred to as yield stress. Therefore, when the solid material yields, the equivalent von Mises stress is equal to the yield stress, or according to our notation, it is equal to the plastic stress:

$$\sigma_v = \sigma_y \equiv \sigma_p, \tag{A1}$$

where σ_v is the equivalent von Mises stress and σ_y is the yield stress, which is identical to the plastic flow stress, σ_p.

The von Mises theory has two fundamental quantities that describe the stresses and strain rates in solids under multiaxial loading conditions. Those are the equivalent stress, σ_v (described above), and the equivalent strain rate, $\dot{\varepsilon}$. They are formulated as [35]:

$$\sigma_v = \sqrt{\frac{3}{2}\underline{\underline{\tau}} : \underline{\underline{\tau}}} \tag{A2}$$

and

$$\dot{\varepsilon} = \sqrt{\frac{2}{3}\underline{\underline{\dot{\varepsilon}}} : \underline{\underline{\dot{\varepsilon}}}}, \tag{A3}$$

where $\underline{\underline{\tau}}$ is the deviatoric stress tensor, also called viscous stress tensor in the field of fluid dynamics and $\underline{\underline{\dot{\varepsilon}}}$ is the rate-of-strain tensor. It should be noted that σ_v and $\dot{\varepsilon}$ come from more general relations, involving the second invariant of their corresponding tensors, e.g., (A2) comes from $\sigma_v = \sqrt{3J_2}$, where J_2 is the second invariant of $\underline{\underline{\tau}}$. This yields $\sigma_v = \sqrt{3\left[\frac{1}{2}\mathrm{tr}\left(\underline{\underline{\tau}}^2\right)\right]}$, given that $\underline{\underline{\tau}}$ is traceless ($\mathrm{tr}(\underline{\underline{\tau}}) = 0$), and hence $\sigma_v = \sqrt{\frac{3}{2}\underline{\underline{\tau}} : \underline{\underline{\tau}}}$. A similar procedure is applied for (A3) also, however, $\underline{\underline{\dot{\varepsilon}}}$ is not generally traceless and as such (A3) would not hold. However, under the constraint of incompressibility (Equation (1)), $\mathrm{tr}(\underline{\underline{\dot{\varepsilon}}}) = 0$ and (A3) is valid.

We use a general equation from fluid dynamics, connecting $\underline{\underline{\tau}}$ with $\underline{\underline{\dot{\varepsilon}}}$ [16] that reads:

$$\underline{\underline{\tau}} = 2\mu\underline{\underline{\dot{\varepsilon}}}, \tag{A4}$$

where μ is the dynamic viscosity. The substitution of (A4) in (A2), yields:

$$\sigma_v = \sqrt{6\mu^2\underline{\underline{\dot{\varepsilon}}} : \underline{\underline{\dot{\varepsilon}}}}. \tag{A5}$$

The combination of (A5) and (A3) yields:

$$\sigma_v = 3\mu\dot{\varepsilon}. \tag{A6}$$

And finally the introduction of (A1) into (A6), yields the expression for the viscosity,

$$\mu = \frac{\sigma_p}{3\dot{\varepsilon}} \tag{A7}$$

References

1. Kuyucak, S.; Sahoo, M. A review of the machinability of copper-base alloys. *Can. Met. Q.* **1996**, *35*, 1–15. [CrossRef]
2. Pantazopoulos, G. Leaded brass rods C38500 for automatic machining operations. *J. Mater. Eng. Perform.* **2002**, *11*, 402–407. [CrossRef]
3. Gane, N. The effect of lead on the friction and machining of brass. *Philos. Mag.* **1981**, *43*, 545–566. [CrossRef]
4. Garcia, P.; Rivera, S.; Palacios, M.; Belzunce, J. Comparative study of the parameters influencing the machinability of leaded brasses. *Eng. Fail. Anal.* **2010**, *17*, 771–776. [CrossRef]
5. Pantazopoulos, G.; Toulfatzis, A. Fracture modes and mechanical characteristics of machinable brass rods. *Met. Microstruct. Anal.* **2012**, *1*, 106–114. [CrossRef]
6. Toulfatzis, A.; Pantazopoulos, G.; Paipetis, A. Fracture behavior and characterization of lead-free brass alloys for machining applications. *J. Mater. Eng. Perform.* **2014**, *23*, 3193–3206. [CrossRef]
7. Toulfatzis, A.; Pantazopoulos, G.; Paipetis, A. Microstructure and properties of lead-free brasses using post-processing heat treatment cycles. *Mater. Sci. Technol.* **2016**, *32*, 1771–1781. [CrossRef]
8. Toulfatzis, A.; Pantazopoulos, G.; Paipetis, A. Fracture mechanics properties and failure mechanisms of environmental-friendly brass alloys under impact, cyclic and monotonic loading conditions. *Eng. Fail. Anal.* **2018**, *90*, 497–517. [CrossRef]
9. Toulfatzis, A.; Pantazopoulos, G.; David, C.; Sagris, D.; Paipetis, A. Final heat treatment as a possible solution for the improvement of machinability of lead-free brass alloys. *Metals* **2018**, *8*, 575. [CrossRef]
10. Toulfatzis, A.I.; Pantazopoulos, G.A.; Besseris, G.J.; Paipetis, A.S. Machinability evaluation and screening of leaded and lead-free brasses using a non-linear robust multifactorial profiler. *Int. J. Adv. Manuf. Technol.* **2016**, *86*, 3241–3254. [CrossRef]
11. Toulfatzis, A.; Pantazopoulos, G.; David, C.; Sagris, D.; Paipetis, A. Machinability of eco-friendly lead-free brass alloys: Cutting-force and surface-roughness optimization. *Metals* **2018**, *8*, 250. [CrossRef]
12. Pantazopoulos, G. A review of defects and failures in brass rods and related components. *Pract. Fail. Anal.* **2003**, *3*, 14–22. [CrossRef]
13. Valberg, H.S. *Applied Metal Forming—Including FEM Analysis*; Cambridge University Press: Cambridge, UK, 2010.
14. Bozzi, S.; Vedani, M.; Lotti, D.; Passoni, G. Extrusion of aluminium hollow pipes: Seam weld quality assessment via numerical simulation. *Met. Sci. Technol.* **2009**, *27*, 20–29.
15. Mitsoulis, E. Flows of viscoplastic materials: Models and computations. *Rheol. Rev.* **2007**, *2007*, 135–178.
16. Bird, R.B.; Stewart, W.E.; Lightfoot, E.N. *Transport Phenomena*, 2nd ed.; Wiley: Hoboken, NJ, USA, 2002.
17. Schmitter, E.D. Modelling massive forming processes with thermally coupled fluid dynamics. In Proceedings of the COMSOL Multiphysics User's Conference, Bostion, MA, USA, 23–25 October 2005.
18. Spigarelli, S.; El Mehtedi, M.; Cabibbo, M.; Gabrielli, F.; Ciccarelli, D. High temperature processing of brass: Constitutive analysis of hot working of Cu–Zn alloys. *Mater. Sci. Eng. A* **2014**, *615*, 331–339. [CrossRef]
19. Spigarelli, S.; El Mehtedi, M.; Cabibbo, M. Characterization of Hot Deformation of CW602N Brass. *Acta Phys. Pol.* **2015**, *128*, 726–729. [CrossRef]
20. Pernis, R.; Kasala, J.; Boruta, J. High temperature plastic deformation of CuZn30 brass—Calculation of the activation energy. *Kov. Mater.* **2010**, *48*, 41–46. [CrossRef]
21. Yang, L.-C.; Pan, Y.-T.; Chen, I.-G.; Lin, D.-Y. Constitutive relationship modeling and characterization of flow behavior under hot working for Fe–Cr–Ni–W–Cu–Co super-austenitic stainless steel. *Metals* **2015**, *5*, 1717–1731. [CrossRef]
22. Orman, L. Tensile Behaviour of Alpha Brasses at Elevated Temperature. Master's Thesis, Department Metallurgical Engineering, University of British Columbia, Salt Lake City, UT, USA, 1982.
23. Gulyaev, A.P. *Physical Metallurgy*; MIR Publishers: Moscow, Russia, 1980; Volume 2.
24. Box, G.E.P.; Wilson, K.B. On the experimental attainment of optimum conditions (with discussion). *J. R. Stat. Soc. Ser. B* **1951**, *13*, 1–45.
25. Ansys Inc. *DesignXplorer User's Guide*; Ansys Inc.: Canonsburg, PA, USA, 2017.
26. Fluent, A. *Theory Guide*; Ansys Inc.: Canonsburg, PA, USA, 2017.

27. Valencia, J.J.; Quested, P.N. Thermophysical properties. In *ASM Handbook*; ASM International: Novelty, OH, USA, 2008; Volume 15, pp. 468–481.

28. Pantazopoulos, G.; Vazdirvanidis, A. Characterization of microstructural aspects of machinable α-β phase brass. *Microsc. Anal.* **2008**, *22*, 13–16.

29. Laue, K.; Stenger, H. *Extrusion—Processes Machinery, Tooling*; ASM International: Novelty, OH, USA, 1981.

30. Higgins, R. *Engineering Metallurgy-Applied Physical Metallurgy*, 6th ed.; Edward Arnold: London, UK, 1998.

31. Pernis, R.; Kasala, J.; Pernis, I. Surface defects of brass bars. In Proceedings of the International Conference "Metal 2011", Brno, Czech Republic, 18–20 May 2011.

32. Pantazopoulos, G.; Vazdirvanidis, A. Failure analysis of a fractured leaded-brass (CuZn39Pb3) extruded hexagonal rod. *J. Fail. Anal. Prev.* **2008**, *8*, 218–222. [CrossRef]

33. Pantazopoulos, G.; Vazdirvanidis, A. Fracture analysis and embrittlement phenomena of machined brass components. *Procedia Struct. Integr.* **2017**, *5*, 476–483. [CrossRef]

34. Lynch, S.P. Failures of structures and components by metal-induced embrittlement. *J. Fail. Anal. Prev.* **2008**, *8*, 259–274. [CrossRef]

35. Sugimoto, M.; Igaki, H.; Saito, K. Equivalent stress (rate) and equivalent strain rate for work-hardening materials: Theory of anisotropic plasticity based on maximum shear stress hypothesis. *Trans. Jpn. Soc. Mech. Eng.* **1973**, *39*, 1164–1174. [CrossRef]

© 2018 by the authors. Licensee MDPI, Basel, Switzerland. This article is an open access article distributed under the terms and conditions of the Creative Commons Attribution (CC BY) license (http://creativecommons.org/licenses/by/4.0/).

Article

Evaluation Study on Iterative Inverse Modeling Procedure for Determining Post-Necking Hardening Behavior of Sheet Metal at Elevated Temperature

Han Mei [1,2], Lihui Lang [2,3,*], Kangning Liu [3] and Xiaoguang Yang [1,2]

[1] School of Energy and Power Engineering, Beihang University, Beijing 100191, China; mh7196@126.com (H.M.); yxg@buaa.edu.cn (X.Y.)
[2] Collaborative Innovation Center of Advanced Aero-Engine, Beijing 100191, China
[3] School of Mechanical Engineering and Automation, Beihang University, Beijing 100191, China; lkn-514@163.com
[*] Correspondence: lang@buaa.edu.cn; Tel.: +86-010-823-16821

Received: 13 November 2018; Accepted: 5 December 2018; Published: 10 December 2018

Abstract: The identification of the post-necking strain hardening behavior of metal sheet is important for finite element analysis procedures of sheet metal forming process. The inverse modeling method is a practical way to determine the hardening curve to large strains. This study is thus focused on the evaluation of the inverse modeling method using a novel material performance test. In this article, hot uniaxial tensile test of a commercially pure titanium sheet with rectangular section was first conducted. Utilizing the raw data from the tensile test, the post-necking hardening behavior of the material is determined by a FE-based inverse modeling procedure. Then the inverse method is compared with some classical hardening models. In order to further evaluate the applicability of the inverse method, biaxial tensile test at elevated temperatures was performed using a special designed cruciform specimen. The cruciform specimen could guarantee that the maximum equi-biaxial deformation occurs in the center section. By using the inverse modeling procedure, the hardening curves under biaxial stress state are able to be extracted. Finally the stress-strain curves obtained from the two experiments are compared and analysis studies are provided.

Keywords: finite element analysis; inverse modeling; post-necking hardening; biaxial tensile test; elevated temperature

---

1. Introduction

Nowadays the Finite Element Analysis (FEA) is used widely in the design stage of sheet metal forming operations with the aim of reducing the number of trial steps, controlling dimension accuracy and finally obtaining high-quality sheet forming products [1] and is almost indispensable in the manufacturing industries. The simulation of metal forming process is a highly non-linear problem involving large material deformation and complex boundary conditions [2] which requires accurate hardening behavior of the investigated sheet metal over a wide range of plastic strains. Till now the most popular method to obtain the stress-strain relationship of metal sheet, represented either in the form of a set of discrete data points or analytical constitutive models, is typically by conducting standard uniaxial tensile test with rectangular cross-section [3,4]. The stress and strain values of the uniaxial tensile test can be obtained by using an extensometer that measures the elongation of a certain gauge length but the calculated stress value is accurate on the assumption that the test specimen is subjected to homogeneous state of uniaxial loading, which means that the stress-strain relationship identified by standard uniaxial tensile test is valid only before the so-called diffuse necking point. However, many sheet forming processes usually involve large deformation and can generate strains far beyond the necking point, so the stress-strain curve up to the necking point is not sufficient for numerical simulation procedures. Generally, the available pre-necking stress-strain curves are extrapolated to large strains using different hardening models [5]. It is obvious that the predicted post-necking hardening curve greatly depends on the choice of phenomenological constitutive models and one model that is best fitted to a certain material may not suits for another [6].

Recently, numerous investigations have been made to acquire the hardening curve beyond diffuse necking by many researchers [7–11]. The initial attempt was made by Bridgman [7], who derived a set of analytical models for the distribution of stress and strain across the diffuse necking region for a round bar. However, the correction and compensation for stress and stress values of Bridgman's method is based on the measurement of evolving geometrical parameters of the necking area, which requires much experimental effort. Ling [8] and Zhang et al. [9] extended the research of Bridgman by proposing new models for determining the post-necking hardening behavior of strip specimen with rectangular section. Other researchers [10,11] further studied the necking problem experimentally based on Bridgman's work in more sophisticated ways. Currently, a new optical-numerical measurement techniques, i.e., the Digital Image Correlation (DIC) method, is used widely for obtaining full-field information for both in-plane displacements and strains of the test specimen and has been applied by many researchers for determining the post-necking hardening curve of uniaxial test [12,13], Scheider et al. [14] investigated the necking phenomenon based on finite element analysis with the help of DIC technique and proposed a new post-necking model which has a higher accuracy than Zhang's [9] model. Another effort to identify the post-necking stress-strain curve of uniaxial tensile test is concentrated on the inverse modeling procedures. In most cases, the stress-strain points or the unknown parameters of constitutive model are determined iteratively with the help of FEA or the Virtual Fields Method (VFM). The finite element based inverse method is perhaps the most popular way in the iterative modeling procedures for determining stress-strain curve after necking and substantial studies are focused on it. Kajberg et al. [2] combined inverse modeling with in-plane displacement fields measured by DIC method and determined the parameters of parabolic hardening model beyond necking point. Kamaya et al. [15] determined the stress–strain curve including post-necking strain using hourglass type specimens of different notch radius. Faurholdt [16] and Koc [17] respectively designed a least-square cost function which represents the discrepancy between the experimental result and the FEM computed response. Other researchers [3,18,19] focus on the non-linear VFM which could replace finite element method in the inverse modeling process. The key point of VFM is to determine the hardening behavior by minimizing the discrepancy between the internal and the external work in the region where the diffuse neck develops [3]. The VFM-based inverse method was first proposed by Coppieters et al. [6] who identified the parameters of the Voce and the Swift hardening laws for a mild deep drawing steel using tensile test data from the pre- and post-necking region. Rossi et al. [18] presented a VFM-based procedure to extract the constitutive parameters of a plasticity model at large plastic strains using three dimensional displacement field. The proposed procedure method was then validated by conducting finite element simulation of uniaxial tensile test. Due to the insufficient accuracy of the common hardening laws when implementing VFM-based inverse modeling process, Coppieters et al. [3] proposed a p-model which can reflect the accurate stress-strain relationship both in the pre- and post-necking region of a cold rolled interstitial-free steel sheet. Kim et al. [19] applied the diffuse approximation (DA) method to decrease the noise effect from the measured full-field displacements when characterizing the post-necking strain hardening behavior using VFM.

The FE-based and VFM-based method have been experimentally validated in the pre-necking regime of uniaxial tensile test [20,21]. However, the evaluation of the inverse modeling methods by an independent material test is much more difficult and is currently lacking [3], as in a tensile test, the true strain at the smallest cross-section in the necked region may exceed 1.5 just before ductile fracture happens [20]. So it is important to select a material performance test capable of achieving very large strains under uniform deformation. Coppieters et al. [3,22] used the multi-axial tube expansion test designed by Kuwabara and Sugawara [23] to acquire uniaxial strain hardening behavior beyond necking point for the validation study of the VFM-based method and good agreement was obtained. However, very few investigation concentrates on the evaluation test at relatively high temperatures, which is a main target of this research.

This study is focused on the evaluation of the inverse modeling method using a novel material performance test at elevated temperature. In this article, hot uniaxial tensile test of a commercially pure titanium sheet with rectangular section was first conducted. Utilizing the raw data from the tensile test, the post-necking hardening behavior of the material is determined by an FE-based inverse modeling procedure. In order to further evaluate the applicability of the inverse method, biaxial tensile test at elevated temperature was performed using a special designed cruciform specimen. The cruciform specimen could guarantee that the maximum equi-biaxial deformation occurs in the center section. By using the inverse modeling procedure, the hardening curves under biaxial stress state are able to be extracted. Finally the stress-strain curves obtained from the two experiments are compared and analysis studies are provided.

2. Inverse Modeling Method

2.1. Experimental Details

The material used in this study is a commercial pure titanium (Grade 2). The titanium and Ti-alloys are well known for their high strength-to-weight ratio, good heat and corrosion resistance [24], and are used widely for high quality thin-walled components [25]. Specimens used in the hot tensile test were machined by wire-electrode cutting with the geometry of the specimen illustrated in Figure 1. The isothermal tensile tests were conducted on a ZWICK universal testing machine (constant-speed loading, the maximum load capacity is 20 kN) at 500 °C and three different strain rates, 0.1 s^{-1}, 0.02 s^{-1}, 0.005 s^{-1}. For a uniform temperature distribution, all specimens were held for 5 min before deformation. The raw data from the hot tensile test of traction force versus elongation is plotted in Figure 2. It is obvious from the figure that the tensile test specimen has a larger elongation and a smaller maximum traction force when the deformation speed is lower, which is a typical character for the titanium at elevated temperatures.

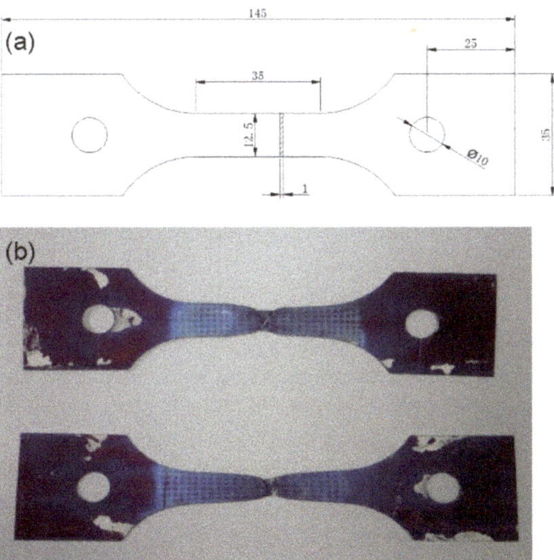

Figure 1. (a) Geometry of tensile specimen (unit: mm), (b) specimen after fracture.

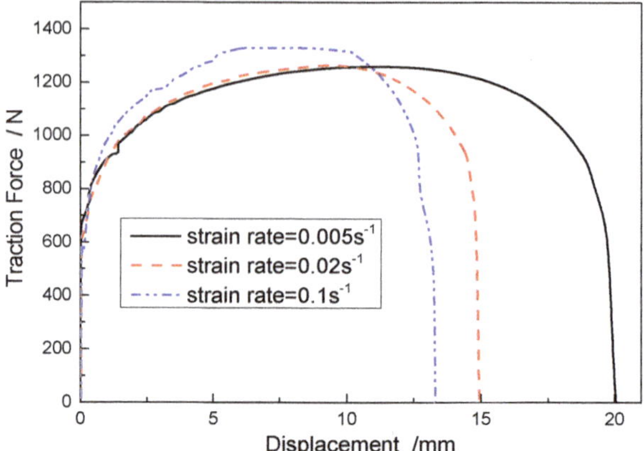

Figure 2. Traction force versus displacement in tensile test.

2.2. FE-Based Inverse Modeling Procedure

An Epsilon high temperature extensometer was used in the tensile test to measure the uniform deformation of the specimen, but the stress-strain curve of the material after necking is no longer reliable and is thus calculated by an FE-based inverse modeling procedure in this research. Many researchers [2,4,20] have proposed several FE-based inverse methods respectively and we used a modified inverse method based on the early work of Joun [20].

By utilizing the commercial finite element software ABAQUS, the uniaxial tensile test can be simulated and the force and elongation of the specimen can be calculated, so the main idea of the inverse method is to iteratively improve the input stress-strain data of the numerical model to better conform to the measured force and elongation $\left\{P_i^{ex}, U_i^{ex}\right\}$ from the experiment. Supposing at the Nth iterative step, the modified input stress-plastic strain data is $(\sigma^N, \varepsilon^N)$, where $\sigma^N = \{\sigma_i\}^N$ and $\varepsilon^N = \left\{\varepsilon_i^{pl}\right\}^N$, i stands for each point of the hardening curve. When the largest plastic strain in the specimen reaches a certain value of ε_i, the calculated reaction force and elongation by FEM simulation are represented here as P_i^{FE} and U_i^{FE} respectively. In order to reduce the difference between the calculated and measured force P_i^{FE}, P_i^{ex} at the same elongation $U_i^{ex} = U_i^{FE}$, the stress update algorithm as expressed in Equation (1) is adopted [20] to calculate the new iterative stress value corresponding to the plastic strain ε_i.

$$\sigma_i^{new} = \sigma_i^{old} \cdot P_i^{ex} / P_i^{FE} \tag{1}$$

A schematic diagram of the inverse method is shown in Figure 3, and the iterative procedure can be summarized in the following:

(1) The initial guess of the stress-plastic strain data $\left\{\sigma_i^N, \varepsilon_i^N\right\}$ ($N = 1$) pre-necking point is obtained from the experiment using an extensometer. Although the measured hardening curve by using extensometer has been proved accurate by some researchers [19], the iterative optimization method in this research still covers both pre- and post-necking region of the hardening curve to further exam the reliability of the inverse method.

(2) By conducting FEM simulation, the traction force and elongation of the specimen at the Nth step can be calculated as $\left\{P_{i,N}^{FE}, U_{i,N}^{FE}\right\}$. In order to compare the simulation result with the experiment, the experimental force value P_i^{ex} at the same $U_{i,N}^{FE} = U_i^{ex}$ is determined by interpolation. By comparing the measured force P_i^{ex} and predicted force $P_{i,N}^{FE}$, the improved stress at plastic strain ε_i^N is corrected as $\sigma_i^{N+1} = \sigma_i^N \cdot P_i^{ex} / P_{i,N}^{FE}$.

(3) Calculate the new simulation model using the updated input hardening curve $\left\{\sigma_i^{N+1}, \varepsilon_i^{N+1}\right\}$ and extract the $(N+1)$th force and elongation data of the specimen $\left\{P_{i,N+1}^{FE}, U_{i,N+1}^{FE}\right\}$.

(4) After several iterative steps, the predicted force and elongation data will get close to the experimental result. In order to evaluate the discrepancy between the calculated and the measured data, the relative mean square error Φ is calculated in this research, which has the expression as described in Equation (2).

$$\Phi = \sqrt{\sum (P_{i,N}^{FE} - P_i^{ex})^2 / M} / \left(\sum_i P_i^{ex} / M\right) \tag{2}$$

where M is the total number of the data points.

(5) If the relative mean square error Φ is lower than 0.3%, the last stress-plastic strain data $\left\{\sigma_i^N, \varepsilon_i^N\right\}$ is then regarded as the final effective hardening curve determined by the inverse modeling method, else go back to step (2).

Throughout the inverse modeling procedure, the elongation and traction force of the specimen are the only two necessary measured data sets from uniaxial tensile test, which are easy to be obtained. However, there is a problem arise from this method: When the elongation of the specimen gets to the value of U_i^{ex} in the tensile test, the actual maximum plastic strain in the specimen is denoted as ε_i^{ex} here, but without the application of the DIC method, the ε_i^{ex} value cannot be determined from experiment, then the stress value σ_i^N in the numerical simulation is corrected using Equation (1) at the elongation $U_i^{FE} = U_i^{ex}$ corresponding to plastic strain ε_i^N which may be different from the ε_i^{ex} value. But it will be discovered later that this discrepancy between the two plastic strain values from experiment and simulation is not important for the final convergence and accuracy of the determined hardening curve, which will be introduced in Section 3.

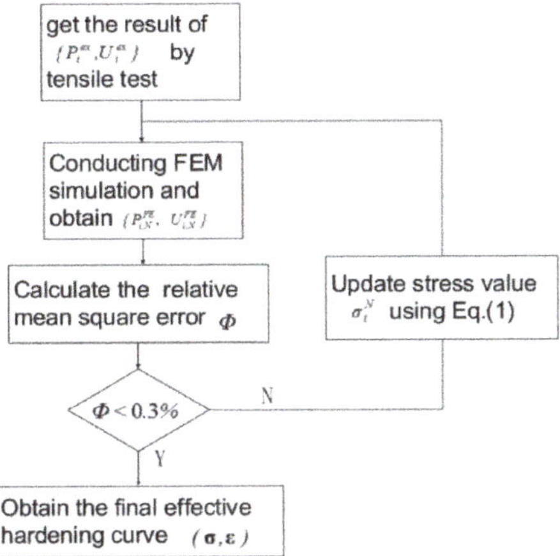

Figure 3. Schematic diagram of the inverse method.

3. Analysis of the Method

3.1. Convergence Analysis

The improved stress-plastic strain curves of some iterative steps by the inverse modeling method at different strain rates are illustrated in Figure 4. It is obvious from the figure that the stress-strain data measured by the extensometer is identical with the pre-necking curve obtained by the inverse method, which means that the inverse method is reliable and accurate in the pre-necking region. The flow stress demonstrates a relatively stable hardening rate after diffuse necking point and no work-softening is observed in each condition. At strain rate of 0.005 s^{-1}, the proposed algorithm converges to the final optimized curve after twelfth iteration. The eleventh improved and the twelfth improved curve are very close with each other, implying that the discrepancy of the traction force between values predicted by the FEA and by experimental measured data is very small in the last numerical simulations. From Figure 4b,c, we can see that the final effective stress-strain curve is obtained after the ninth and seventh iterative step respectively. It can be found from the figure that the approximation process is not monotonous, for example, at strain rate of 0.005 s^{-1}, the updated hardening curve of each step does not get close to the final curve from one side, on the contrary, the improved hardening curves appear on both side of the final curve, this is due to the uncertainty of the relationship between the elongation and maximum plastic strain of the test specimen. At strain rate of 0.005 s^{-1} and 0.1 s^{-1}, diffuse necking will occur at plastic strain of around 0.23 and 0.12 respectively, which indicates that a lower strain rate will guarantee a more uniform deformation of the specimen.

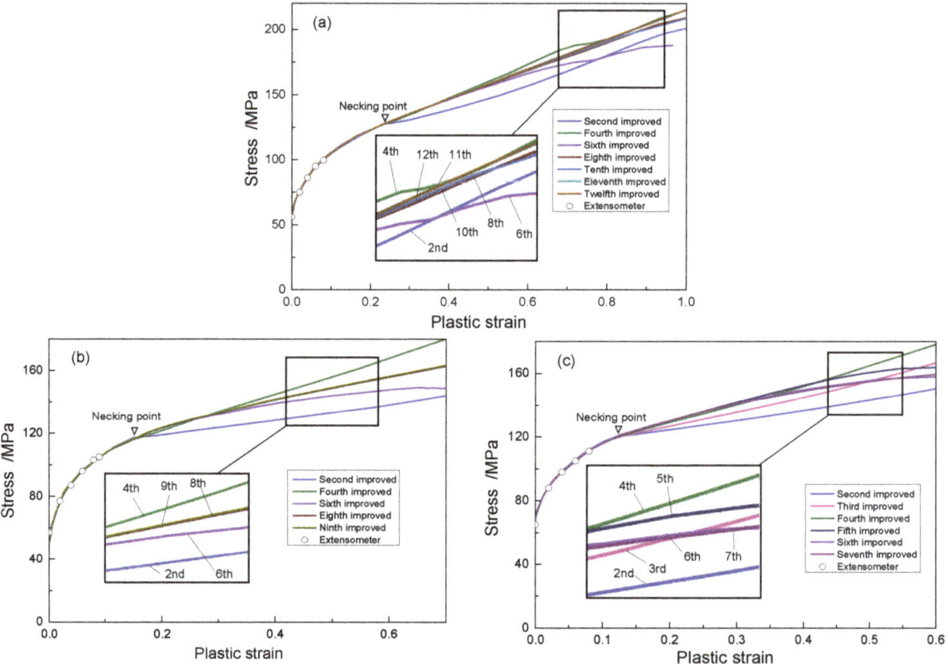

Figure 4. Improved stress-plastic strain curves by inverse method at different strain rates (**a**) 0.005 s^{-1} (**b**) 0.02 s^{-1} (**c**) 0.1 s^{-1}.

Figure 5 shows the calculated plastic strain corresponding to a certain elongation in each conditions. As has mentioned earlier, the proposed inverse modeling process does not take into consideration of the relationship between the maximum strain and the elongation of the specimen. However, after several iteration steps, the predicted plastic strain corresponding to different elongation in each condition all converges to a certain value, which means that the internal relation between the two variables could be automatically obtained by the proposed iterative procedure without the application of any non-contact optical measurement. Furthermore, it can be found from Figure 5a that the algorithm will take more iteration steps to get close to the final strain value if the elongation of the specimen is larger. Comparing Figure 5a,b it can be discovered that when the specimen is stretched by 10 mm, the corresponding maximum plastic strain is 0.2 and 0.37 at strain rate of 0.005 s^{-1} and 0.1 s^{-1} respectively, which again demonstrates that the specimen with a lower strain rate will deform more uniformly than the one with a higher strain rate. The curves of the relative mean square error Φ versus iterative step at different conditions are plotted in Figure 6, the Φ values of each inverse modeling procedure all drop to less than 0.3% after a certain number of iteration steps.

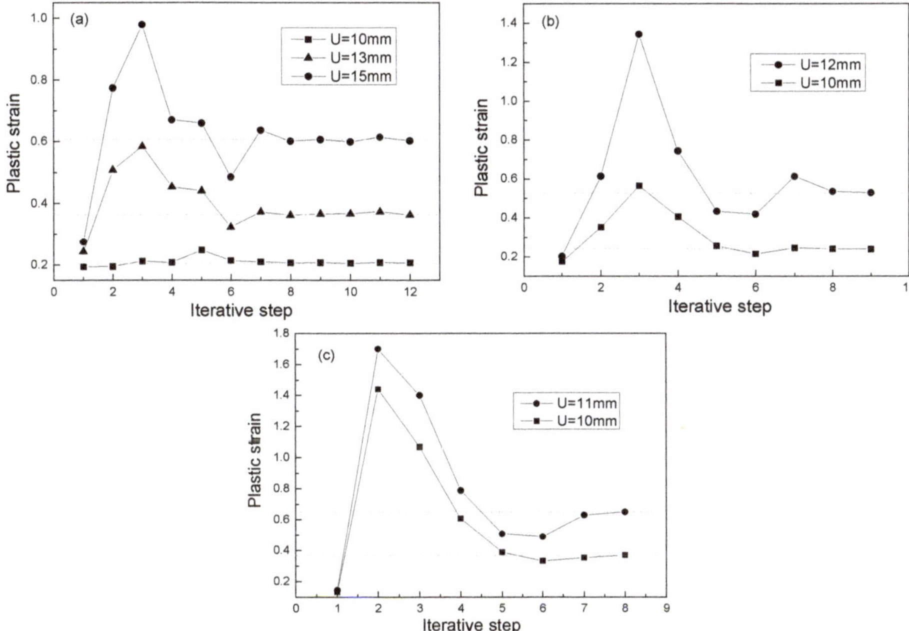

Figure 5. Relationship between elongation and plastic strain at different strain rates (**a**) 0.005 s^{-1} (**b**) 0.02 s^{-1} (**c**) 0.1 s^{-1}.

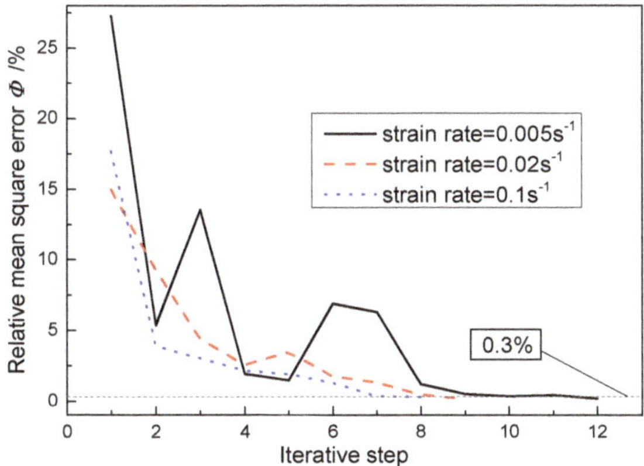

Figure 6. Relative mean square error Φ versus iterative step at different strain rates.

3.2. Comparison with Classical Hardening Laws

In this research, the flow stress of the material in the uniaxial tensile test is represented by a set of discrete points. During the iterative steps, every stress-strain point in the hardening curve is corrected based on Equation (1) by comparing the discrepancy of the FEM result with the experimental data. When performing the inverse modeling procedure, other researchers [6,19] tend to assume that the hardening curve after necking point could be described by some kind of hardening laws. In their approaches, the inverse modeling procedure is aimed at obtaining the best parameters for each hardening mode which can best fit to the experimental data, so the accuracy of these approaches greatly depend on the property of the selected hardening models. In order to evaluate the predicting ability of some classical hardening laws for describing the flow behavior of material under uniaxial tensile test. The stress-plastic strain curve obtained by the proposed inverse modeling process was fitted using three classical constitutive models, i.e., the Swift, Voce and Ludwik model with the expressions listed in Table 1.

Table 1. Other phenomenological constitutive models.

Model	Expression
Swift	$\sigma = K(p + \varepsilon_0)^n$
Voce	$\sigma = \sigma_s - (\sigma_s - \sigma_1)e^{np}$
Ludwik	$\sigma = \sigma_0 + Kp^n$

where σ is the Cauchy stress, p is the plastic strain and the others are the fitting parameters.

Figure 7 shows the comparison of the stress-plastic strain curve at strain rate of 0.005 s^{-1} obtained by the proposed inverse modeling method with the hardening curves fitted by the Swift, Voce and Ludwik model. The parameters of each hardening law are calculated using the Levenberg-Marquardt (L-M) algorithm with the parameter values for each model listed in Table 2. It is obvious that none of the selected three model can best describe the hardening behavior of the material. The flow stress curve by the inverse method presents a relatively stable hardening rate beyond the diffuse necking point, however, all the three fitted curves demonstrate a decreasing hardening rate and the accuracy of the pre-necking hardening curve is not very high except the Ludwik model as illustrate in the local view of Figure 7. In other words, if the stress-strain curve in the proposed inverse modeling method is represented by the selected hardening laws and the parameter of the model is then updated based on some type of optimization algorithm in each iterative step, the convergence condition cannot be finally satisfied.

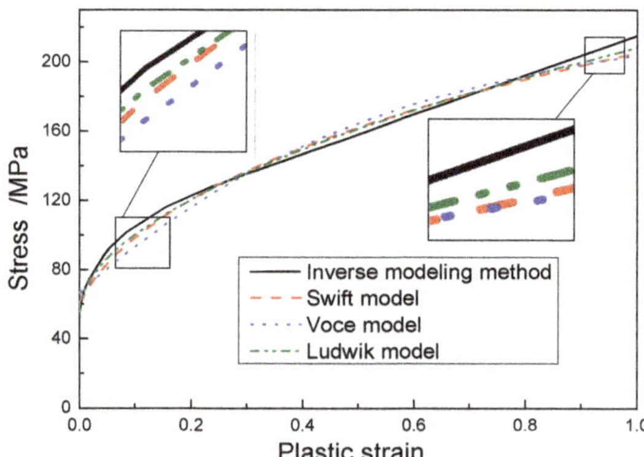

Figure 7. The entire stress-plastic strain curve fitted by classical hardening laws.

Table 2. Parameter values for each constitutive models.

Model	Swift	Voce	Ludwik
Parameter	$K = 202.93$ $\varepsilon_0 = 0.03435$ $n = 0.3586$	$\sigma_s = 230.84$ $\sigma_1 = 66.31$ $n = -1.812$	$\sigma_0 = 55.620$ $K = 152.73$ $n = 0.5318$

As has mentioned earlier, the hardening curve to large strain range can be obtained by extrapolating the available pre-necking stress-strain curves using different hardening models, this method is once very popular in the past when few available experiment could be applied to determine the post-necking hardening curve of sheet metal. However, the reliability of the extrapolation method is greatly dependent on the phenomenological constitutive model that is selected. In order to demonstrate this point, we further investigate the extrapolation properties of the selected three phenomenological models with the parameter values listed in Table 3. Using the pre-necking stress-plastic strain data, the three fitted curves are plotted in Figure 8. It can be seen that the three hardening models can describe the pre-necking hardening behavior of the material with very good accuracy, but after diffuse necking occurs, the constitutive models show very different hardening behavior compared with the optimized curve, the Swift and Voce model seem to underestimate the flow stress curve a lot whereas the fitted curve by the Ludwik model falls on the upper side of the flow stress curve. Actually, none of the selected hardening laws could describe the flow stress curve with relatively stable hardening rate. Thus it can be concluded that if the extrapolated hardening curve is used in the FEA procedure to represent the mechanical behavior of the material, unreliable numerical simulation results may be obtained.

Table 3. Parameter values for each constitutive models.

Model	Swift	Voce	Ludwik
Parameter	$K = 178.75$ $\varepsilon_0 = 0.00731$ $n = 0.2375$	$\sigma_s = 121.25$ $\sigma_l = 57.37$ $n = -15.000$	$\sigma_0 = 53.437$ $K = 169.21$ $n = 0.5186$

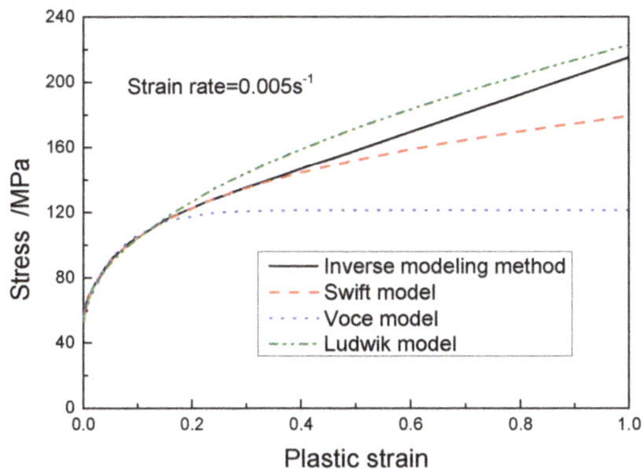

Figure 8. Extrapolation of three classical hardening laws.

4. Evaluation by Biaxial Tensile Test

4.1. Experimental Details of the Biaxial Tensile Test

The biaxial tensile test of metal sheet is becoming increasing popular recently in sheet metal forming industries [26,27]. As in sheet metal forming operations, the material is often subjected to deformation in more than one plane or axis. So the mechanical properties obtained by uniaxial tensile test are inadequate for predicting the material's deformation under states of biaxial stress [28]. Hannon et al. [28] reviewed several types of biaxial tensile test machines developed by Makinde [29], Kuwabara [30], etc. However, in the current literatures, few biaxial tensile test equipment could handle the testing procedure of metal sheet at relative high temperatures (e.g., above 400 °C). Lang et al. [31] developed a servo-hydraulic 100-kN hot biaxial tensile test machine in 2015 (as shown in Figure 9) on which the biaxial test was performed. The maximum displacement of each axis is 100 mm with the displacement control accuracy of each axis higher than 0.02 mm. The heating furnace is placed at the center of the pedestal, it could generate and maintain a constant temperature from room temperature to 800 °C.

Figure 9. Biaxial tensile test machine.

The design of biaxial test specimen is a challenging task, many cruciform specimens have been proposed to obtain the biaxial stress state of the test material at the center section of the specimen and to avoid stress concentrations outside the gauge area [28,32], but up to now, there is no standard specimen geometry in the literature [33], which makes it difficult to compare test results from different research institutes. In this study, a special designed thickness-tapered cruciform specimen is adopted in the test as illustrated in Figure 10. To concentrate strains in the central zone, the central region of the specimen is machined to make the upper surface of the test piece be part of a sphere with the radius of around 250.4 mm. The minimum thickness in the center is 0.2 mm, which could ensure that the final failure occurs in this section. In order to avoid strain localization at the junction of two arms, the rounding radius at the intersection of two arms is made 12 mm. The hot biaxial tensile test was performed at the same condition with the hot uniaxial tensile test, i.e., at 500 °C and at different strain rates (0.1 s^{-1}, 0.02 s^{-1} and 0.005 s^{-1}). It should be noted that the accurate strain value of the test specimen is difficult to be acquired, so the relatively consistent strain rate is achieved with the application of finite element analysis which will be discussed in Section 4.2. The result of the test is plotted in Figure 11. We can observe that the material presents a rate-dependent property as well at equi-biaxial tensile condition. The fractured specimen after the test is shown in Figure 12. It can be discovered that the special designed specimen with a reduced center section could guarantee that the maximum deformation occurs in the center.

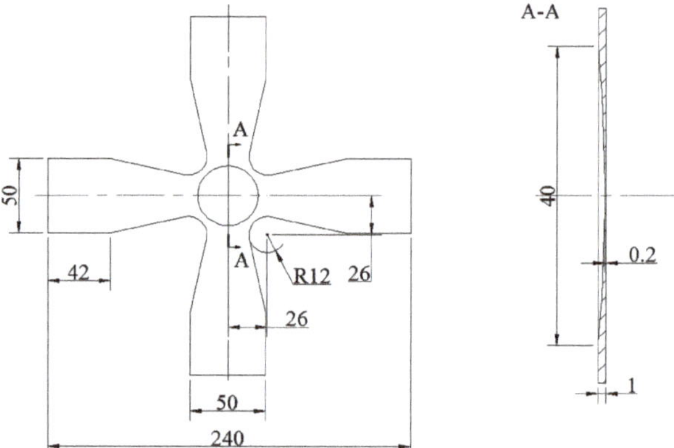

Figure 10. Specimen geometry of the biaxial tensile (unit: mm), where A-A is the cross section.

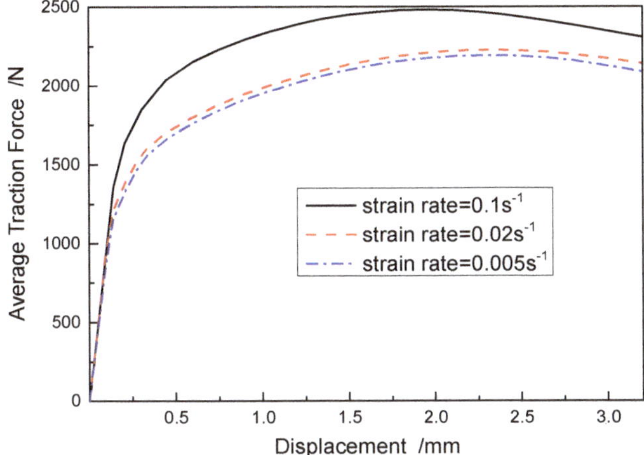

Figure 11. Specimen geometry of the biaxial tensile (unit: mm).

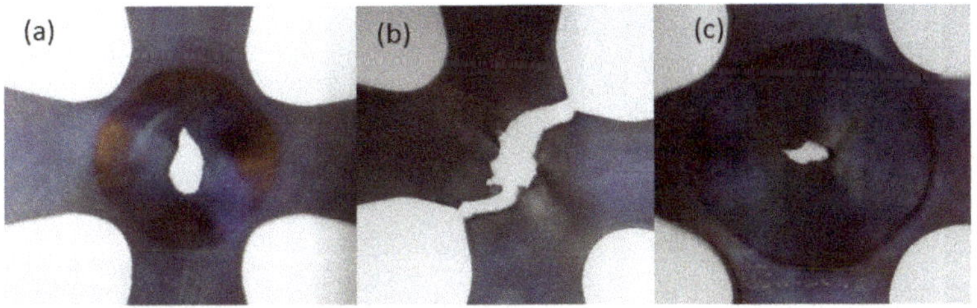

Figure 12. Fractured specimen after biaxial test (**a**) 0.1 s^{-1} (**b**) 0.02 s^{-1} (**c**) 0.005 s^{-1}.

4.2. FE Analysis of Biaxial Tensile Test

The accurate stress and strain of the special designed cruciform specimen is difficult to be calculated. So we adopted FE analysis to evaluate the hardening curves at biaxial stress state. The isotropic elastic-plastic numerical model is established by using the commercial FEA program ABAQUS as shown in Figure 13. The cruciform specimen is modeled with 8-node linear reduced integration element (C3D8R). Due to the symmetrical properties of the specimen, only one-quarter of the model is modeled (Figure 13). Figure 14 shows the von Mises equivalent stress distribution of the specimen when the displacement of each axis reach 3.0 mm. The simulation result also demonstrates that the central section of the specimen experiences the maximum deformation. In addition, with the help of FEA, the relationship between the grip displacement and maximum strain could be established. Using Equation (3), the relatively accurate strain rate could be achieved by controlling the grip speed.

$$\dot{\bar{\varepsilon}} = d\varepsilon/dt = d\varepsilon/dU \cdot dU/dt = d\varepsilon/dU \cdot v \tag{3}$$

where U is the grip displacement and v is the grip speed.

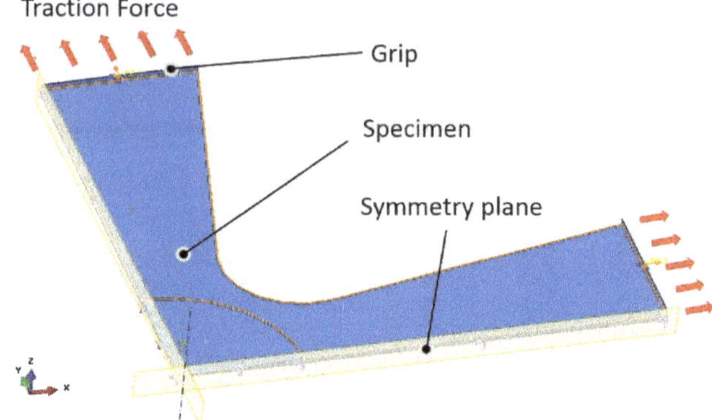

Figure 13. FE analysis of cruciform specimen.

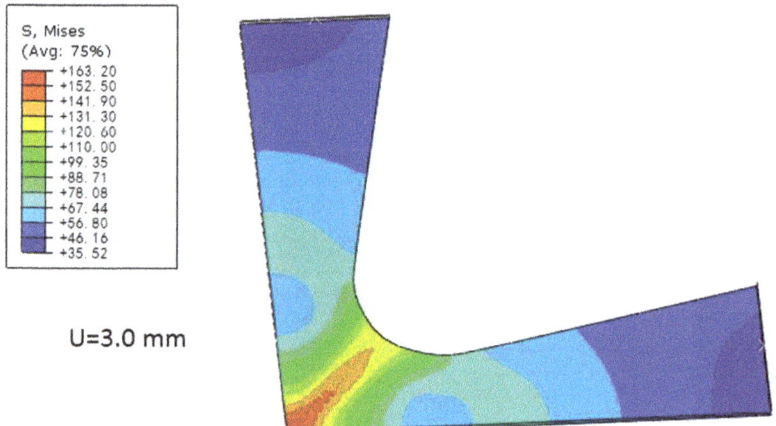

Figure 14. FE simulation result of cruciform specimen.

4.3. Validation of the Inverse Model

Figure 15 shows the improved stress-plastic strain curves of the investigated material in the biaxial tensile test by the inverse modeling method as described in Section 2.2. The hardening curves of the uniaxial tensile test obtained by the inverse method at corresponding conditions are also plotted in the figure. In order to reduce the iteration steps, the post-necking hardening curve calculated in Section 3 is regarded as the initial guess in the iterative step. We can see from the figure that the iterative process converges to a certain value after two or three steps. At strain rate of $0.1\ \mathrm{s}^{-1}$ and $0.02\ \mathrm{s}^{-1}$, the flow stress of the biaxial test obtained by the inverse method is close to the uniaxial test result. It should be noted that the achieved minimum relative mean square errors Φ are 0.97%, 1.01% and 2.94% for strain rate of $0.1\ \mathrm{s}^{-1}$, $0.02\ \mathrm{s}^{-1}$ and $0.005\ \mathrm{s}^{-1}$ respectively, which is much larger than the optimized values calculated in Section 3. In other words, the proposed inverse modeling method is able to obtain the hardening curve using complex specimens such as the special designed cruciform test piece, but the precision is not very high.

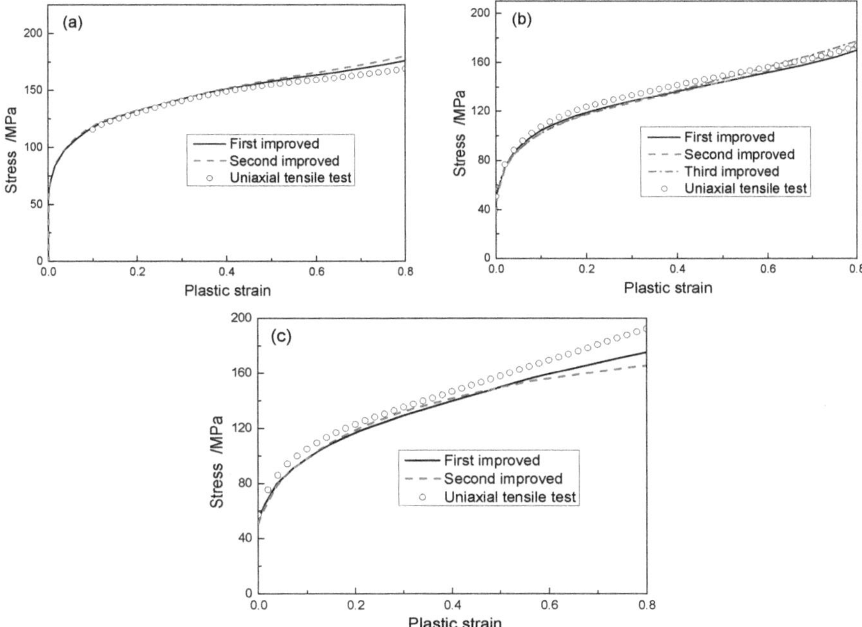

Figure 15. Improved stress-plastic strain curves and uniaxial test data at different strain rates. (**a**) 0.1 s^{-1} (**b**) 0.02 s^{-1}(**c**) 0.005 s^{-1}.

5. Conclusions

The post-necking hardening behaviors of uniaxial tensile test and biaxial tensile test are determined by using the inverse modeling method. The following conclusions can be derived from this study:

(1). By using the proposed inverse method, the internal relation between the plastic strain and elongation of the uniaxial tensile test could be automatically obtained without the application of any non-contact optical measurement.

(2). If the stress-strain curve in the inverse modeling method is represented by simple classical hardening laws, the convergence condition of the inverse iteration procedure cannot be finally satisfied.

(3). The extrapolated hardening curve using phenomenological constitutive models greatly depends on the model that is selected and it will be in poor accuracy for predicting the post-necking hardening curve of metal sheet.

(4). The proposed inverse modeling method could be used to obtain the hardening curve of metal sheet using complex specimens such as the special designed cruciform test piece, but the calculation accuracy is not very high.

Author Contributions: H.M. and L.L. conceived and designed the experiments; H.M. performed the experiments; H.M. and K.L. analyzed the data; H.M. wrote the paper; L.L. and X.Y. contributed to the revision of the paper.

Funding: This research received no external funding.

Conflicts of Interest: The authors declare no conflict of interest.

References

1.	Huang, C.Q.; Liu, L.L. Application of the Constitutive Model in Finite Element Simulation: Predicting the Flow Behavior for 5754 Aluminum Alloy during Hot Working. *Metals* **2017**, *7*, 1–12.

2.   Kajberg, J.; Lindkvist, G. Characterisation of materials subjected to large strains by inverse modelling based on in-plane displacement fields. *Int. J. Solids Struct.* **2004**, *41*, 3439–3459. [CrossRef]

3.   Coppieters, S.; Kuwabara, T. Identification of Post-Necking Hardening Phenomena in Ductile Sheet Metal. *Exp. Mech.* **2014**, *54*, 1355–1371. [CrossRef]

4.   Grajcar, A.; Kozlowska, A.; Grzegorczyk, B. Strain Hardening Behavior and Microstructure Evolution of High-Manganese Steel Subjected to Interrupted Tensile Tests. *Metals* **2018**, *8*, 1–12. [CrossRef]

5.   Enami, K. The effects of compressive and tensile prestrain on ductile fracture initiation in steels. *Eng. Fract. Mech.* **2005**, *72*, 1089–1105. [CrossRef]

6.   Coppieters, S.; Cooreman, S.; Sol, H.; Houtte, P.V.; Debruyne, D. Identification of the post-necking hardening behaviour of sheet metal by comparison of the internal and external work in the necking zone. *J. Mater. Process. Technol.* **2011**, *211*, 545–552. [CrossRef]

7.   Bridgman, P.W. *Studies in Large Plastic Flow and Fracture*; McGraw Hill: New York, NY, USA, 1952; pp. 15–20.

8.   Ling, Y. Uniaxial true stress-strain after necking. *AMP. J. Tech.* **1996**, *5*, 37–48.

9.   Zhang, Z.L.; Hauge, M.; Ødegård, J.; Thaulow, C. Determining material true stress–strain curve from tensile specimens with rectangular cross-section. *Int. J. Solids Struct.* **1999**, *36*, 3497–3516. [CrossRef]

10.   Mirone, G. A new model for the elastoplastic characterization and the stress–strain determination on the necking section of a tensile specimen. *Int. J. Solids Struct.* **2004**, *41*, 3545–3564. [CrossRef]

11.   Cabezas, E.E.; Celentano, D.J. Experimental and numerical analysis of the tensile test using sheet specimens. *Finite Elem. Anal. Des.* **2004**, *40*, 555–575. [CrossRef]

12.   Tong, W.; Tao, H.; Zhang, N.; Jiang, X.Q.; Manuel, P.M.; Louis, G.H.; Xiaohong, Q.G. Deformation and Fracture of Miniature Tensile Bars with Resistance-Spot-Weld Microstructures. *Metall. Mater. Trans. A* **2005**, *36*, 2651–2669. [CrossRef]

13.   Yang, S.Y.; Tong, W.A. Finite Element Analysis of a Tapered Flat Sheet Tensile Specimen. *Exp. Mech.* **2009**, *49*, 317–330. [CrossRef]

14.   Scheider, I.; Cornec, A.; Brocks, W. Procedure for the Determination of True Stress-Strain Curves from Tensile Tests with Rectangular Cross-Section Specimens. *J. Eng. Mater. Technol.* **2004**, *126*, 70–76. [CrossRef]

15.   Kamaya, M.; Kawakubo, M. A procedure for determining the true stress–strain curve over a large range of strains using digital image correlation and finite element analysis. *Mech. Mater.* **2011**, *43*, 243–253. [CrossRef]

16.   Faurholdt, T.G. Inverse modelling of constitutive parameters for elastoplastic problems. *J. Strain Anal. Eng. Des.* **2000**, *35*, 471–478. [CrossRef]

17.   Koc, P.; Štok, B. Computer-aided identification of the yield curve of a sheet metal after onset of necking. *Comput. Mater. Sci.* **2004**, *31*, 155–168. [CrossRef]

18.   Rossi, M.; Pierron, F. Identification of plastic constitutive parameters at large deformations from three dimensional displacement fields. *Comput. Mech.* **2012**, *49*, 53–71. [CrossRef]

19.   Kim, J.H.; Serpantié, A.; Barlat, F.; Pierron, F.; Lee, M.G. Characterization of the post-necking strain hardening behavior using the virtual fields method. *Int. J. Solids Struct.* **2013**, *50*, 3829–3842. [CrossRef]

20.   Joun, M.S.; Eom, J.G.; Min, C.L. A new method for acquiring true stress–strain curves over a large range of strains using a tensile test and finite element method. *Mech. Mater.* **2008**, *40*, 586–593. [CrossRef]

21.   Joun, M.; Choi, I.; Eom, J.; Lee, M. Finite element analysis of tensile testing with emphasis on necking. *Comput. Mater. Sci.* **2007**, *41*, 63–69. [CrossRef]

22.   Coppieters, S.; Ichikawa, K.; Kuwabara, T. Identification of Strain Hardening Phenomena in Sheet Metal at Large Plastic Strains. *Procedia Eng.* **2014**, *81*, 1288–1293. [CrossRef]

23.   Kuwabara, T.; Sugawara, F. Multiaxial tube expansion test method for measurement of sheet metal deformation behavior under biaxial tension for a large strain range. *Int. J. Plast.* **2013**, *45*, 103–118. [CrossRef]

24.   Hu, M.; Dong, L.M.; Zhang, Z.Q.; Lei, X.F.; Yang, R.; Sha, Y.H. Correction of Flow Curves and Constitutive Modelling of a Ti-6Al-4V Alloy. *Metals* **2018**, *8*, 1–15. [CrossRef]

25.   Tsao, L.C.; Wu, H.Y.; Leong, J.C.; Fang, C.J. Flow stress behavior of commercial pure titanium sheet during warm tensile deformation. *Mater. Des.* **2012**, *34*, 179–184. [CrossRef]

26.   Shiratori, E.; Ikegami, K. A new biaxial tensile testing machine with flat specimen. *Bull. Tokyo Inst. Technol.* **1967**, *82*, 105–118.

27.   Kulawinski, D.; Ackermann, S.; Seupel, A.; Lippmann, T.; Henkel, S.; Kuna, M.; Weidner, A.; Biermann, H. Deformation and strain hardening behavior of powder metallurgical TRIP steel under quasi-static biaxial-planar loading. *Mater. Sci. Eng. A* **2015**, *642*, 317–329. [CrossRef]

28. Hannon, A.; Tiernan, P. A review of planar biaxial tensile test systems for sheet metal. *J. Mater. Process. Technol.* **2008**, *198*, 1–13. [CrossRef]

29. Makinde, A.; Thibodeau, L.; Neale, K.W.; Lefebvre, D. Design of a biaxial extensometer for measuring strains in cruciform specimens. *Exp. Mech.* **1992**, *32*, 132–137. [CrossRef]

30. Kuwabara, T.; Ikeda, S.; Kuroda, K. Measurement and analysis of differential work hardening in cold-rolled steel sheet under biaxial tension. *J. Mater. Process. Technol.* **1998**, *80–81*, 517–523. [CrossRef]

31. Xiao, R.; Li, X.X.; Lang, L.H.; Chen, Y.K.; Yang, Y.F. Biaxial tensile testing of cruciform slim superalloy at elevated temperatures. *Mater. Des.* **2016**, *94*, 286–294. [CrossRef]

32. Welsh, J.S.; Adams, D.F. An experimental investigation of the biaxial strength of IM6/3501-6 carbon/epoxy cross-ply laminates using cruciform specimens. *Compos. Part A* **2002**, *33*, 829–839. [CrossRef]

33. Leotoing, L.; Guines, D. Investigations of the effect of strain path changes on forming limit curves using an in-plane biaxial tensile test. *Int. J. Mech. Sci.* **2015**, *99*, 21–28. [CrossRef]

© 2018 by the authors. Licensee MDPI, Basel, Switzerland. This article is an open access article distributed under the terms and conditions of the Creative Commons Attribution (CC BY) license (http://creativecommons.org/licenses/by/4.0/).

Article

Improving the Reliability of Mechanical Components That Have Failed in the Field Due to Repetitive Stress

Seongwoo Woo [1,*] and Dennis L. O'Neal [2]

[1] Reliability Association of Korea, 146, Sunyoo-ro, Yeongdeungpo-gu, Seoul 07255, Korea
[2] Dean of Engineering and Computer Science, Baylor University, Waco, TX 76798-7356, USA; Dennis_ONeal@baylor.edu
* Correspondence: twinwoo@yahoo.com; Tel.: +82-10-8602-9084; Fax: +82-2-780-2555

Received: 29 November 2018; Accepted: 25 December 2018; Published: 4 January 2019

Abstract: To improve the reliability of mechanical parts that have failed in the field, a reliability methodology for parametric accelerated life testing (ALT) is proposed. It consists of: (1) a parametric ALT plan, (2) a load analysis, (3) a tailored series of parametric ALTs with action plans, and (4) an evaluation of the final designs to ensure the design requirements are satisfied. This parametric ALT should help an engineer reproduce the fractured or failed parts in a product subjectedto repetitive loading and correct the faulty designs. As a test case, the helix upper dispenser of a refrigerator ice-maker fractured in field was studied. Using a load analysis, we discerned that the helix upper dispenser fracture was due to repetitive loads and a faulty design with a 2 mm gap between the blade dispenser and the helix upper dispenser. During the first and second ALTs, the fracture in the helix upper dispenser was reproduced. The failure modes and mechanisms found were similar to those of the failed sample in field. As an action plan, the design of the helix upper dispenser was modified by eliminating the 2 mm gap and adding enforced ribs. In the third ALT there were no problems. After three rounds of parametric ALTs, the reliability of the helix upper dispenser was guaranteed as a 10-year life with an accumulated failure rate of 1%.

Keywords: reliability design; helix upper dispenser; fracture; parametric accelerated life testing; faulty designs

---

1. Introduction

A refrigerator is designed to store fresh food. Its evaporator supplies chilled air to both the freezer and refrigerator compartments. A refrigerator that includes an ice-maker can consist of several different subsystems—cabinet, door, shelves and drawers, control system, motor or compressor, heat exchanger, water supply device, and other miscellaneous parts. The total number of components can be as many as 3000. Product reliability is targeted to have at least 10 years of B20 life that will have an accumulated failure rate of 20% (Figure 1).

Because customers have requested the inclusion of an ice-maker in a refrigerator, manufacturers have developed refrigerators with this feature. For this (intended) function, a mechanical module like the ice-maker is designed to harvest ice. As seen in Figure 2, the mechanical parts in an ice-maker should operate under the conditions subjected to it by consumers. In the field, the ice-maker might experience failures which were not anticipated in the design process. In such cases, the company may have to decide to perform a massive recall of the product. Reliability testing should reproduce any defective configurations or uses of the product which failed in field and correct them immediately. However, in this instance, there had been no proper testing methods for the failures of the mechanical product detected in field [1–3].

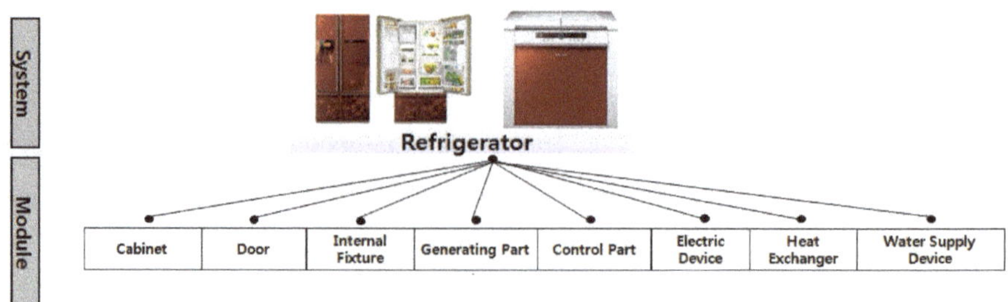

Figure 1. Breakdown of an appliance which includes a refrigeration system, with its modules.

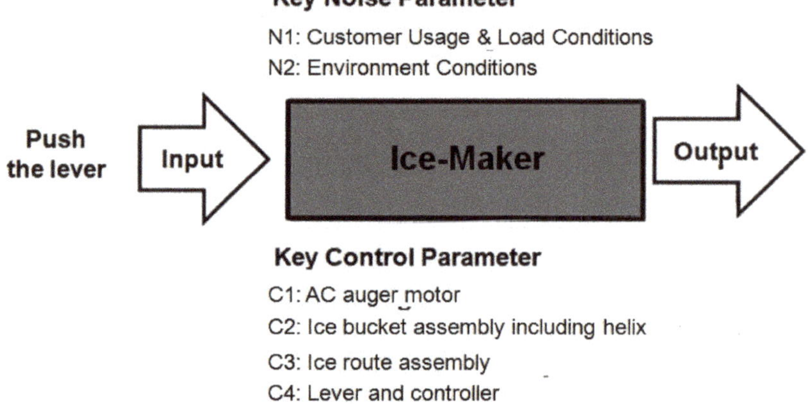

Figure 2. Robust design schematic of a mechanical system, such as an ice-maker.

Repeated loads or overloading due to daily usage may cause structural failure in a product and reduce its lifetime [4,5]. Many engineers think such possibilities can be assessed by: (1) mathematical modeling using Newtonian methods, (2) assessing the time response of the system for dynamic loads, (3) utilizing the rain-flow counting method [6,7], or (4) estimating system damage using the Palmgren-Miner's rule [8]. However, while these analytical methodologies may provide exact answers, there are many assumptions in these methods that simplify the complexities in modeling product failures due to design flaws.

Robust design methods, including the statistical design of experiments (DOE) and the Taguchi approach [9], were developed to carry out optimal design for mechanical products. In particular, Taguchi's robust design method uses parametric design to place it in a position where random "noise" does not affect the outcome. Thus, mechanical systems can be used to find out the proper design parameters and their levels [10–14]. Through utilizing interactions between control factors and noise factors, the parametric design of a mechanical system can be used to determine the proper control factors that make the design robust, regardless of the change of noise factors. In an orthogonal array, as the control factors are assigned to an inner array, the noise factors are assigned to an outer array.

However, because huge experimental computations in the Taguchi product array are required, a lot of design parameters for a mechanical structure need to be considered. As new products are introduced with faulty designs in the mechanical structure, products may be recalled and loss of brand-name value experienced.

The purpose of this paper is to present a reliability methodology of mechanical parts like the fractured helix upper dispenser which is subjected to repetitive loading in field. This reliability methodology includes: (1) a parametric accelerated life testing (ALT) plan, (2) a load analysis, (3) a tailored series of parametric ALTs with action plans, and (4) an evaluation to ensure the final design requirements of the mechanical parts were satisfied.

2. Parametric Accelerated Life Testing

2.1. Setting an Overall Parametric ALT Plan

The reliability of a mechanical system can be defined as the ability of a system or module to function under stated conditions for a specified period of time [15]. It can be illustrated in Figure 3, which includes a component called "the bathtub curve" that consists of three parts [16]. There is initially a decreasing failure rate, then a constant failure rate, and then an increasing failure rate. If a product follows the bathtub curve, it will have difficulty in succeeding in the market. Because of faulty design, the mechanical product will have higher failure rates early in its early and incur financial losses for the company. The company will then need to set goals for new products to (1) reduce early failures, (2) decrease random failures during the product operating time, and (3) increase product lifetime [17]. As the reliability of a mechanical product is improved, the traditional bathtub curve can be transformed into a straight line with the shape parameter β (the bottom curve in Figure 3).

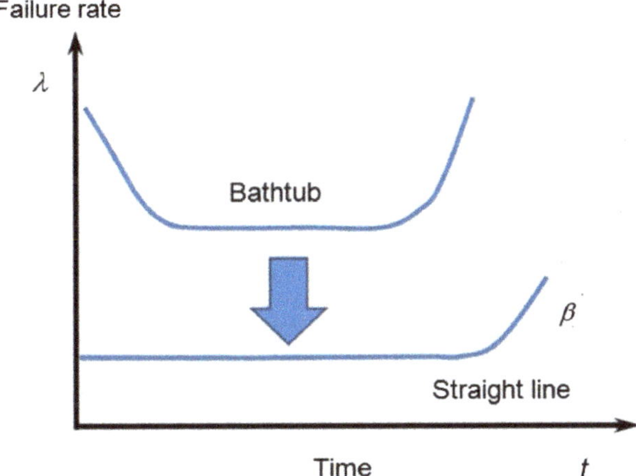

Figure 3. Bathtub curve and straight line with slope β.

Therefore, the reliability of a mechanical system might be quantified from the product lifetime L_B and failure rate λ as follows:

$$R(L_B) = 1 - F(L_B) = e^{-\lambda L_B} \cong 1 - \lambda L_{BX} \tag{1}$$

This equation is applicable below about 20 percent of the cumulative failures [18]. The reliability design of a mechanical system can be achieved by obtaining the targeted product lifetime L_B and failure rate λ after reproducing the field failure through parametric ALT and correcting the defective configuration of structures.

In targeting the reliability of a mechanical product in a parametric ALT analysis, there are three cases for modules in a mechanical product: (1) a modified module, (2) a new module, and (3) a similar module to the prior design on the basis of market demand. The ice-maker with a helix upper dispenser discussed here as a case study is a similar module with a faulty design that should be corrected because customers wanted to replace it with a new one.

The newly designed module *A* from the field data in Table 1 had a yearly failure rate of 0.35% per year and a lifetime, L_{B1}, of 2.9 years. To respond to customer claims, a new reliability target for the ice-maker was set as 10 years of B1 life with an accumulative failure rate of one percent.

Table 1. Overall parametric accelerated life testing (ALT) plan of a refrigerator.

Modules	Market Data		Expected Reliability			Targeted Reliability		
	Yearly Failure Rate, %/year	B_x Life, year	Yearly Failure Rate, %/year			B_x Life, year	Yearly Failure Rate, %/year	B_x Life, year
A	0.35	2.9	Similar	×1	0.35	2.86	0.10	10(*x* = 1.0)
B	0.24	4.2	New	×5	1.20	0.83	0.15	10(*x* = 1.5)
C	0.30	3.3	Similar	×1	0.30	3.33	0.10	10(*x* = 1.0)
D	0.31	3.2	Modified	×2	0.62	1.61	0.10	10(*x* = 1.0)
E	0.15	6.7	Modified	×2	0.30	3.33	0.15	10(*x* = 1.5)
Others	0.50	2.0	Similar	×1	0.50	2.00	0.40	10(*x* = 4.0)
Product	1.9	5.4	-	-	3.27	3.06	1.00	10(*x* = 10)

2.2. Parametric Accelerated Life Testing of Mechanical Systems

For solid-state diffusion of impurities in silicon, the junction equation *J* might be expressed as

$$J = [aC(x - a)] \cdot \exp\left[-\frac{q}{kT}\left(w - \frac{1}{2}a\xi\right)\right] \cdot v$$

$$[\text{Density}/\text{Area}] \cdot [\text{Jump Probability}] \cdot [\text{Jump Frequency}]$$
$$= -\left[a^2 v e^{-qw/kT}\right] \cdot \cosh\frac{qa\xi}{2kT}\frac{\partial C}{\partial x} + \left[2ave^{-qw/kT}\right]C\sinh\frac{qa\xi}{2kT}$$
$$= \Phi(x, t, T)\sinh(a\xi)\exp\left(-\frac{Q}{kT}\right)$$
$$= A\sinh(a\xi)\exp\left(-\frac{Q}{kT}\right) \tag{2}$$

On the other hand, the reaction process that is dependent on speed might be expressed as

$$K \;\; = K^+ - K^- = a\frac{kT}{h}e^{-\frac{\Delta E - aS}{kT}} - a\frac{kT}{h}e^{-\frac{\Delta E + aS}{kT}}$$
$$= a\frac{kT}{h}e^{-\frac{\Delta E}{kT}}\sinh\left(\frac{aS}{kT}\right) \tag{3}$$

The reaction rate *K* from Equations (2) and (3) can therefore be summarized as

$$K = B\sinh(aS)\exp\left(-\frac{E_a}{kT}\right) \tag{4}$$

If the reaction rate in Equation (4) takes an inverse function, the generalized stress model can be obtained as

$$TF = A[\sinh(aS)]^{-1}\exp\left(\frac{E_a}{kT}\right) \tag{5}$$

The range of the hyperbolic sine stress term $[\sinh(aS)]^{-1}$. in Equation (4) increases the stress as follows: (1) initially $(S)^{-1}$ has a small effect, (2) $(S)^{-n}$ has what is considered a medium effect, and (3) $(e^{aS})^{-n}$ has a large effect. Accelerated testing is usually conducted in the medium stress range. The hyperbolic sine stress term of Equation (5) in the medium range can be redefined as

$$TF = A(S)^{-n}\exp\left(\frac{E_a}{kT}\right) \tag{6}$$

The internal (or external) stress in a product is difficult to quantify and use in accelerated testing. Stresses in mechanical systems may come from efforts (or loads) like force, torque, and pressure. For a mechanical system, when replacing stress with effort, the time-to-failure (TF) can be modified as

$$TF = A(S)^{-n} \exp\left(\frac{E_a}{kT}\right) = A(e)^{-\lambda} \exp\left(\frac{E_a}{kT}\right) \tag{7}$$

From the time-to-failure in Equation (7), the acceleration factor can be defined as the ratio between the proper accelerated stress levels and typical operating conditions. The acceleration factor (AF) can be modified to include the effort concepts:

$$AF = \left(\frac{S_1}{S_0}\right)^n \left[\frac{E_a}{k}\left(\frac{1}{T_0} - \frac{1}{T_1}\right)\right] = \left(\frac{e_1}{e_0}\right)^\lambda \left[\frac{E_a}{k}\left(\frac{1}{T_0} - \frac{1}{T_1}\right)\right] \tag{8}$$

To carry out parametric ALTs, the sample size equation with the acceleration factors in Equation (8) might be expressed as [19]:

$$n \geq (r+1) \cdot \frac{1}{x} \cdot \left(\frac{L_{BX}^*}{AF \cdot h_a}\right)^\beta + r \tag{9}$$

If the reliability of the mechanical system was targeted, the number of required test cycles (or mission cycles) can be obtained for a given sample size. Through parametric ALTs, the faulty designs of a mechanical system can be identified to achieve the desired reliability target [20–22].

The estimated lifetime L_{Bx} in each ALT is approximated as

$$L_{Bx}^\beta \cong x \cdot \frac{n \cdot (h_a \cdot AF)^\beta}{r+1} \tag{10}$$

Let $x = \lambda \cdot L_{Bx}$. The estimated failure rate of the design samples λ can be described as

$$\lambda \cong \frac{1}{L_{Bx}} \cdot (r+1) \cdot \frac{L_B^\beta}{n \cdot (h_a \cdot AF)^\beta} \tag{11}$$

In each ALT, by quantifying the reliability from the multiplication of the estimated L_{Bx} life and failure rate λ, we can ensure the reliability of the final design for a mechanical system.

2.3. Case Study-Reliability Design of the Helix Upper Dispenser in an Ice-Maker

Since the customer needs ice, an ice-maker is designed to harvest ice. The primary parts in an ice-maker consist of the bucket case, helix support, helix dispenser clamp, blade dispenser, helix upper dispenser, and blade, as shown in Figure 4.

In the field, these ice-maker parts in a refrigerator cracked and fractured due to design failures under unknown customer usage conditions. Field data indicated that the damaged products may have had two structural design flaws: (1) a 2 mm gap between the blade dispenser and the helix upper dispenser, and (2) a weld line around the impact area of the helix upper dispenser. Below −20 °C the rotating blade dispenser (stainless steel) impacts the fixed helix upper dispenser (plastic) while the crushed ice is being harvested. A crack may occur in the helix dispenser (Figure 5).

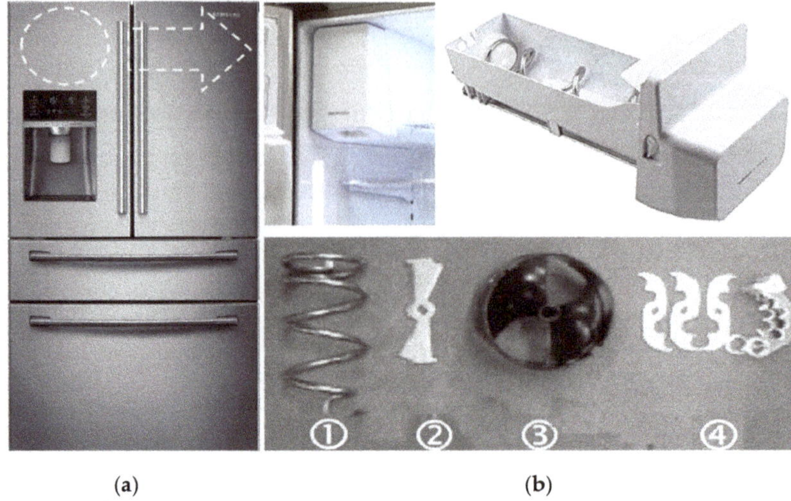

(a)　　　　　　　　　　　　　　　(b)

Figure 4. Refrigerator and ice-maker assembly. (**a**) Refrigerator; (**b**) mechanical parts of an ice-maker assembly: helix support (1), blade dispenser (2), helix upper dispenser (3), and blade (4).

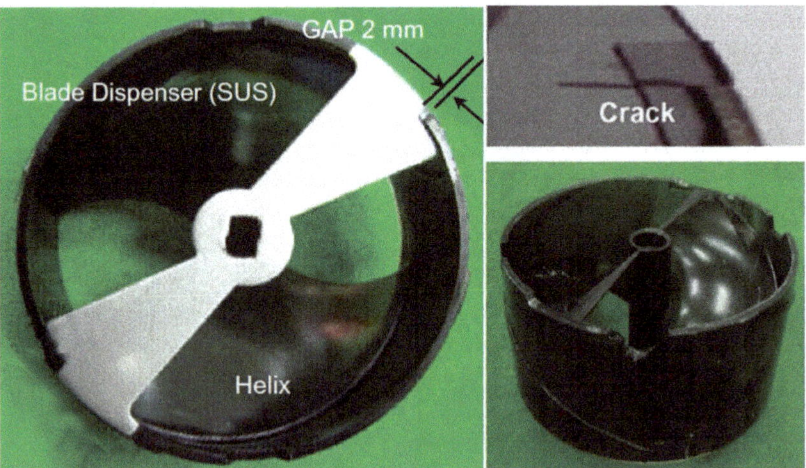

Figure 5. Damaged helix upper dispenser after use.

By utilizing the above failure analysis (and tests), fractures that started in voids inside mechanical parts such as gear have been shown to propagate to their ends [23–30]. In particular, intergranular fractures of austenitic stainless steels have been identified and discussed in a similar way to the failure of the helix dispenser [31]. A phase field visco-plastic model has also been proposed to describe the influence of the loading rate on the ductile fracture [32]. Cracks in the helix dispenser require the manufacturer to redesign the product to keep it functioning for its expected lifetime.

If repetitive loads are applied to the product structure and there are design faults, the structure will fail short of its desired lifetime. Therefore, an ice-maker's actual lifetime depends on faulty parts like the helix upper dispenser. To reproduce and correct the part(s), an engineer is required to conduct reliability testing like parametric ALT for a new design. The process consists of (1) a load analysis for the returned product, (2) the utilization of parametric ALTs with action plans, and (3) the verification of whether the reliability target of final designs has been achieved.

Ice-making involves several repetitive mechanical processes: (1) filtered water supplies the tray; (2) water freezes into ice via cold air in the heat exchanger; and (3) the ice is harvested until the bucket is full. When the customer pushes the lever by force, cubed or crushed ice is then dispensed. In this ice-making process, the ice-maker parts receive a variety of mechanical loads.

In the United States, refrigerators are designed to produce 10 cubes per use and upto 200 cubes a day. Because the ice-maker system is repetitively used in both cubed and crushed ice modes, it is continuously subjected to mechanical loads. Ice production may also be influenced by customer usage conditions such as water pressure, ice consumption, refrigerator notch settings, and the number of times the door is opened.

Figure 6 provides an overview of the schematic of an ice-maker that represents the mechanical load transfer in the ice bucket assembly using a bond graph model. To generate enough torque to crush the ice, an AC motor provides power through the gear system which is then transferred to the ice bucket assembly. Through the helix blade dispenser and the upper dispenser in the bucket, the ice is distributed by the blade. If subjected to different loads, the ice can also be crushed.

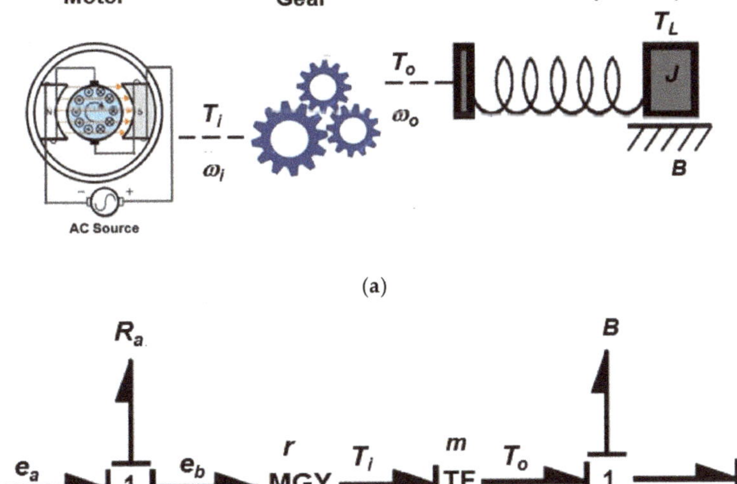

Figure 6. Design concept of an ice-maker. (**a**) Schematic diagram of auger motor and ice bucket assembly. (**b**) Bond graph modeling of auger motor and ice bucket assembly.

To derive the state equations, the bond graph model in Figure 6b can be solved at each node, that is,

$$df \times E_2/dt = 1/L_a \times eE_2 \tag{12}$$

$$dfM_2/dt = 1/J \times eM_2 \tag{13}$$

The junction from Equation (12) is

$$eE_2 = e_a - eE_3 \tag{14a}$$

$$eE_3 = R_a \times fE_3 \tag{14b}$$

The junction from Equation (13) is

$$eM_2 = eM_1 - eM_3 \tag{15a}$$

$$eM_1 = (K_a \times i) - T_{Pulse} \tag{15b}$$

$$eM_3 = B \times fM_3 \tag{15c}$$

Because $fM_1 = fM_2 = fM_3 = \omega$ and $i = fE_1 = fE_2 = fE_3 = i_a$ from Equation (14)

$$eE_2 = e_a - R_a \times fE_3 \tag{16}$$

$$fE_2 = fE_3 = i_a \tag{17}$$

If substituting Equations (16) and (17) into Equation (12), then

$$di_a/dt = 1/L_a \times (e_a - R_a \times i_a) \tag{18}$$

And from Equations (15) we can obtain

$$eM_2 = [(K_a \times i) - T_L] - B \times fM_3 \tag{19a}$$

$$i = i_a \tag{19b}$$

$$fM_3 = fM_2 = \omega \tag{19c}$$

If substituting Equations (19) into (13), then

$$d\omega/dt = 1/J \times [(K_a \times i) - T_L] - B \times \omega \tag{20}$$

We can obtain the state equation from Equations (18) and (20) as follows:

$$\begin{bmatrix} di_a/dt \\ d\omega/dt \end{bmatrix} = \begin{bmatrix} -R_a/L_a & 0 \\ mk_a & -B/J \end{bmatrix} \begin{bmatrix} i_a \\ \omega \end{bmatrix} + \begin{bmatrix} 1/L_a \\ 0 \end{bmatrix} e_a + \begin{bmatrix} 1 \\ -1/J \end{bmatrix} T_L \tag{21}$$

When Equation (21) is integrated, the output of the ac-motor and ice bucket assembly is obtained as

$$y_p = \begin{bmatrix} 0 & 1 \end{bmatrix} \begin{bmatrix} i_a \\ \omega \end{bmatrix} \tag{22}$$

From Equation (21) we know that the lifetime of the ice bucket assembly depends on the stress (or torque) due to forces required to crush the ice. The life-stress model (LS model) in Equation (7) can then be modified as

$$TF = A(S)^{-n} = AT_L^{-\lambda} = A(F_c \times R)^{-\lambda} = B(F_c)^{-\lambda} \tag{23}$$

Therefore, the AF can be derived as

$$AF = \left(\frac{S_1}{S_0}\right)^n = \left(\frac{T_1}{T_0}\right)^\lambda = \left(\frac{F_1 \times R}{F_0 \times R}\right)^\lambda = \left(\frac{F_1}{F_0}\right)^\lambda \tag{24}$$

We can carry out parametric ALT from Equation (9) until the required mission cycles that provide the reliability target of 10 years of B1 life are achieved.

The environment operating conditions of the ice bucket assembly in a refrigerator icemaker can vary from approximately -15 to $-30\,^{\circ}$C with a relative humidity ranging from 0% to 20%. Depending on customer usage, an ice dispenser is used an average of approximately 3–18 times per day. Under maximum use for 10 years, the dispenser incurs about 65,700 usage cycles. Data from the motor company specifies that normal torque is 0.69 kN-cm and maximum torque is 1.47 kN-cm. Assuming the cumulative damage exponent $\lambda = 2$, the acceleration factor is approximately 5 in Equation (24).

For 10 years of B1 life, the test cycles for a sample of ten pieces (calculated using Equation (9)) were approximately 42,000 cycles if the shape parameter was supposed to be 2.0. This parametric ALT is designed to ensure a B1 life of 10 years so that it would fail less than once during 42,000 cycles. Figure 7 shows the experimental setup of an ALT for the reproduction of the failed helix upper dispenser in the field. Figure 8 represents the duty cycles for the ice-crushing load T_L.

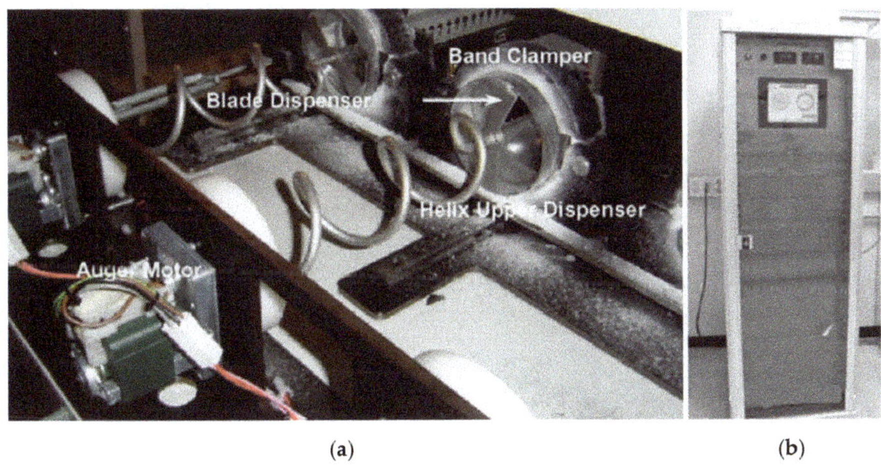

(a)   (b)

Figure 7. Equipment used in accelerated life testing and controller. (**a**) ALT Equipment; (**b**) controller.

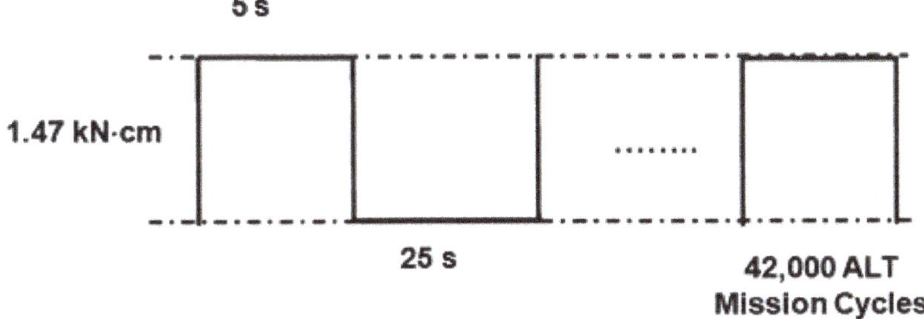

Figure 8. Duty cycles of disturbance load T_L on the band clamper.

The equipment in the chamber was designed to operate down to a temperature of about −30 °C. The controller outside can start or stop the equipment and can indicate the completed test cycles and the test periods, such as sample on/off time. To apply the maximum ice-crushing torque T_L, the helix upper dispenser and the blade dispenser were bolted together with a band clamper. As the controller gives the start signal, the augermotor rotates. At this point, the rotating blade dispenser impacts the fixed helix upper dispenser to the maximum mechanical ice-crushing torque (1.47 kN-cm).

3. Results & Discussion

In the first ALT, the helix upper dispenser fractured at 170 cycles, 5200 cycles, 7880 cycles, 8800 cycles, and 11,600 cycles. Figure 9 shows a photograph comparing the failed product from the field and that from the first ALT, respectively. Because they are similar, by parametric ALT we were able to reproduce the fractured helix upper dispenser. There was a structural design flaw—a 2 mm gap between the blade dispenser and the helix upper dispenser. As the blade dispenser (stainless steel) struck the helix upper dispenser (plastic), it cracked and fractured. Figure 10 represents the graphical analysis of the ALT results and field data on a Weibull plot. The shape parameter in the first ALT was estimated to be 2.0. For the final design, the shape parameter from the Weibull plot was confirmed to be 4.78.

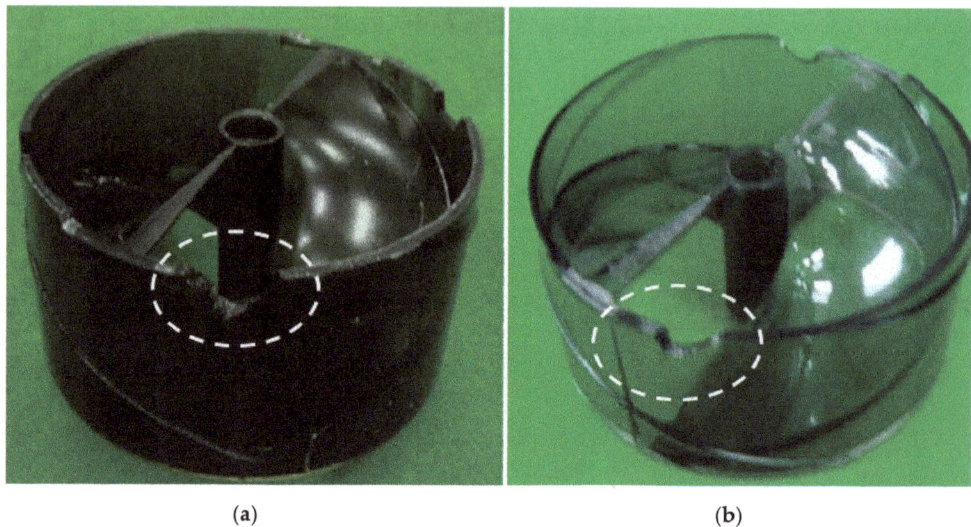

Figure 9. Failed helix upper dispensers in field and during the first ALT. (**a**) Failed product in field; (**b**) Product with crack after first ALT.

Figure 10. Field data and results of ALT on Weibull chart.

To withstand repetitive impact loads, the problematic helix upper dispenser used in field was redesigned as follows: the 2 mm gap between the blade dispenser and helix upper dispenser was eliminated (Figure 11).

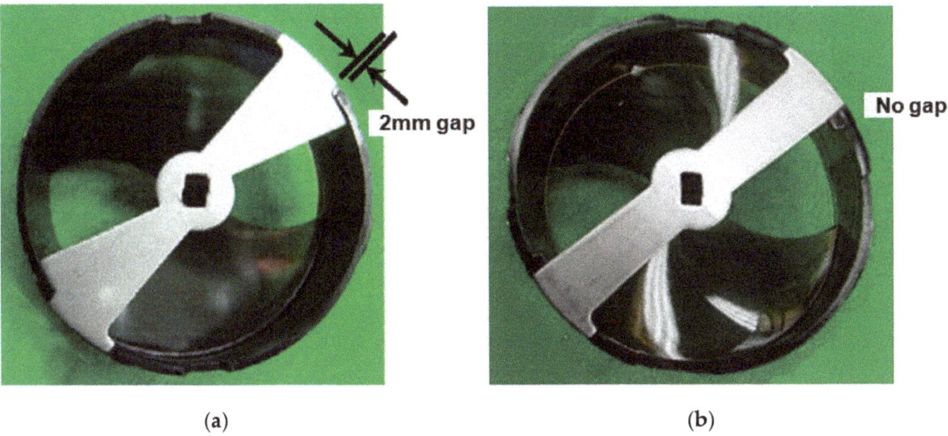

(a)   (b)

Figure 11. Redesigned helix upper dispenser. (**a**) Old design; (**b**) New design.

In the second ALT, the helix upper dispenser fractured at 17,000 cycles, 25,000 cycles, 28,200 cycles, and 38,000 cycles. When the gap between the blade dispenser and the helix upper dispenser was eliminated, the lifetime of the helix upper dispenser was extended. Because the helix upper dispenser did not have enough strength for stress, 42,000 mission cycles in the second ALT still wasn't met. As an action plan, a reinforced rib on the outside of the helix was added.

In the third ALT, there were no problems until 75,000 cycles. Over the course of three ALTs with these design changes, the helix upper dispenser was guaranteed to have 10 years of B1 life. Table 2 shows a summary of the results of the ALTs.

Table 2. Results of ALTs.

Parametric ALT	1st ALT	2nd ALT	3rd ALT
	Initial Design	Second Design	Final Design
Over the course of 42,000 cycles, the helix upper dispenser has no problems	170 cycles: 1/10 fracture 5200 cycles: 1/10 fracture 7880cycles: 2/10 fracture 8880cycles: 2/10 fracture 11,600 cycles: 4/10 fracture	17,000 cycles: 1/6 fracture 25,000 cycles: 3/6 fracture 28,000 cycles: 1/6 fracture 38,000 cycles: 1/6 fracture	42,000 cycles: 6/6 OK 75,000 cycles: 6/6 OK
Helix structure			-
Material and specification	C1: Gap of 2 mm → 0 mm	C2: Added rib on the outside of helix	-

4. Conclusions

To improve the reliability of a mechanical part that fails in the field, we have suggested a reliability methodology that includes: (1) a parametric ALT plan, (2) a load analysis, (3) a tailored series of parametric ALTs with action plans, and (4) an evaluation of the final design requirements of the mechanical part to ensure the requirements were satisfied. A helix upper dispenser fractured in field was used as a case study.

Regarding the parametric ALTs of the helix upper dispenser, based on the products that failed both in the field and in the first ALT, the helix upper dispenser fractured around a 2 mm gap between the blade dispenser and helix upper dispenser. When crushed ice was being made, the blade dispenser (stainless steel) struck the plastic helix upper dispenser. The helix upper dispenser, subjected to these repetitive stresses, fractured short of its expected lifetime. As a corrective action plan, the 2 mm gap was eliminated.

During the second ALT, the helix upper dispenser also fractured because it did not have enough strength to withstand repetitive strikes from the blade dispenser. As a corrective action plan, a reinforced rib on the outside of the helix was added. After a sequence of ALT testing, a helix upper dispenser with the proper values for the design parameters was determined to meet the reliability target of 10 years of a B1 life. These were shown to be effective in reproducing the reliability of the helix upper dispenser claimed in field and in enhancing its reliability. This reliability design methodology should be applicable to other mechanical systems such as automobiles, airplanes, construction equipment, washing machines, vacuum cleaners, and civil structures like bridges.

Author Contributions: The primary author (S.W.) was responsible for the experimental testing. Both authors contributed to the analysis and writing of the paper.

Conflicts of Interest: The authors declare no conflicts of interest.

Abbreviations

B	viscous friction coefficient
BX	time which isan accumulated failure rate of $X\%$, durability index
E_a	activation energy, eV
e	effort
e_a	applied voltage, V
e_b	counter-electromotive force
e_f	field voltage, V
f	flow
F_c	ice crushing force, kN
$F(t)$	unreliability
h	testing cycles (or cycles)
h^*	non-dimensional testing cycles, $h^* = h/L_B \geq 1$
i_a	applied current, A
i_f	field current, A
J	momentum of inertia, kg m^2
k	Boltzmann's constant, 8.62×10^{-5} eVdeg^{-1}
k_a	constant of the counter-electromotive force
L_a	electromagnetic inductance
L_B	target BX life and $x = 0.01X$, on the condition that $x \leq 0.2$
m	gear ratio
MGY	gyrator in causal forms for basic 2-ports and 3-ports
n	number of test samples
r	failed numbers
r	coefficient of gyrator
R_a	electromagnetic resistance
S	stress
t_i	test time for each sample
T	torque, kN cm
T_L	ice-crushing torque in bucket, kN cm

TF	time to failure
X	accumulated failure rate, %
x	$x = 0.01X$, on condition that $x \leq 0.2$.
Greek symbols	
ω	angular velocity in ice bucket, rad/s
η	characteristic life
λ	cumulative damage exponent in Palmgren-Miner's rule
Superscripts	
β	shape parameter in Weibull distribution
n	stress dependence, $n = -\left[\frac{\partial \ln(T_f)}{\partial \ln(S)}\right]_T$
Subscripts	
0	normal stress conditions
1	accelerated stress conditions

References

1. Magaziner, I.C.; Patinkin, M. Cold Competition: GE Wages the Refrigerator War. *Harv. Bus. Rev.* **1989**, *89*, 114–124.
2. Bertsche, B. *Reliability in Automotive and Mechanical Engineering: Determination of Component and System Reliability*; Springer: Berlin, Germany, 2010.
3. O'Connor, P.; Kleyner, A. *Practical Reliability Engineering*, 5th ed.; Wiley & Sons: Hoboken, NJ, USA, 2012.
4. Toribio, J.; Lorenzo, M.; Vergara, D.; Aguado, L. The Role of Overloading on the Reduction of Residual Stress by Cyclic Loading in Cold-Drawn Prestressing Steel Wires. *Appl. Sci.* **2017**, *7*, 84. [CrossRef]
5. Nie, B.; Zhao, Z.; Ouyang, Y.; Chen, D.; Chen, H.; Sun, H.; Liu, S. Effect of Low Cycle Fatigue Pre-damage on Very High Cycle Fatigue Behavior of TC21 Titanium Alloy. *Materials* **2017**, *10*, 1384. [CrossRef] [PubMed]
6. Matsuishi, M.; Endo, T. Fatigue of metals subjected to varying stress. *Jpn. Soc. Mech. Eng.* **1968**, *68*, 37–40.
7. Mottand Robert, L. *Machine Elements in Mechanical Design*, 4th ed.; Pearson Prentice Hall: Upper Saddle River, NJ, USA, 2004.
8. Palmgren, A.G. Die Lebensdauer von Kugellagern Zeitschrift des Vereines Deutscher Ingenieure. *Sci. Res.* **1924**, *68*, 339–341.
9. Taguchi, G. Off-line and on-line quality control systems. In Proceedings of the International Conference on Quality Control, Tokyo, Japan, October 1978.
10. Taguchi, G.; Shih-Chung, T. *Introduction to Quality Engineering: Bringing Quality Engineering Upstream*; American Society of Mechanical Engineering: New York, NY, USA, 1992.
11. Ashley, S. Applying Taguchi's quality engineering to technology development. *Mech. Eng.* **1992**, *114*, 58.
12. Wilkins, J. Putting Taguchi Methods to Work to Solve Design Flaws. *Qual. Prog.* **2000**, *33*, 55–59.
13. Phadke, M. *Quality Engineering Using Robust Design*; Englewood Cliffs: Prentice Hall, NJ, USA, 1989.
14. Byrne, D.; Taguchi, S. The Taguchi approach to parameter design. *Qual. Prog.* **1987**, *20*, 19–26.
15. *IEEE Standard Glossary of Software Engineering Terminology, IEEE Std 610.12-1990*; Standards Coordinating Committee of the Computer Society of IEEE: New York, NY, USA, 1990.
16. Klutke, G.; Kiessler, P.C.; Wortman, M.A. A critical look at the bathtub curve. *IEEE Trans. Reliab.* **2015**, *52*, 125–129. [CrossRef]
17. Duane, J.T. Learning Curve Approach to Reliability Monitoring. *IEEE Trans. Aerosp.* **1964**, *2*, 563–566. [CrossRef]
18. Kreyszig, E. *Advanced Engineering Mathematics*, 9th ed.; John Wiley and Son: Hoboken, NJ, USA, 2006; p. 683.
19. Woo, S.; O'Neal, D. Reliability Design of Mechanical Systems Subject to Repetitive Stresses. *Recent Patents Mech. Eng.* **2015**, *8*, 222–234. [CrossRef]
20. Woo, S.; O'Neal, D.; Pecht, M. Improving the reliability of a water dispenser lever in a refrigerator subjected to repetitive stresses. *Eng. Fail. Anal.* **2009**, *16*, 1597–1606. [CrossRef]
21. Woo, S.; O'Neal, D.; Pecht, M. Design of a Hinge kit system in a Kimchi refrigerator receiving repetitive stresses. *Eng. Fail. Anal.* **2009**, *16*, 1655–1665. [CrossRef]
22. Woo, S.; O'Neal, D.; Pecht, M. Failure analysis and redesign of the evaporator tubing in a Kimchi refrigerator. *Eng. Fail. Anal.* **2010**, *17*, 369–379. [CrossRef]

23. Zhao, F.; Ding, X.; Fan, X.; Cui, R.; Li, Y.; Wang, T. Contact Fatigue Failure Analysis of Helical Gears with Non-Entire Tooth Meshing Tests. *Metals* **2018**, *8*, 693. [CrossRef]

24. Hamandi, F.; Laughlin, R.; Goswami, T. Failure Analysis of PHILOS Plate Construct Used for Pantalar Arthrodesis Paper II—Screws and FEM Simulations. *Metals* **2018**, *8*, 279. [CrossRef]

25. Alqedairi, A.; Alfawaz, H.; Rabba, A.; Almutairi, A.; Alnafaiy, S.; Mohammed, M.K. Failure Analysis and Reliability of Ni–Ti-Based Dental Rotary Files Subjected to Cyclic Fatigue. *Metals* **2018**, *8*, 36. [CrossRef]

26. Cao, X.; Xu, L.; Xu, X.; Wang, Q. Fatigue Fracture Characteristics of Ti6Al4V Subjected to Ultrasonic Nanocrystal Surface Modification. *Metals* **2018**, *8*, 77. [CrossRef]

27. Nagarajan, V.; Putatunda, S.; Boileau, J. Fatigue Crack Growth Behavior of Austempered AISI 4140 Steel with Dissolved Hydrogen. *Metals* **2017**, *7*, 466. [CrossRef]

28. Zhang, W.; Wang, H.; Zhang, J.; Dai, W.; Huang, Y. Brittle Fracture Behaviors of Large Die Holders Used in Hot Die Forging. *Metals* **2017**, *7*, 198. [CrossRef]

29. Alvarez, J.; Lacalle, R.; Arroyo, B.; Cicero, S.; Gutiérrez-Solana, F. Failure Analysis of High Strength Galvanized Bolts Used in Steel Towers. *Metals* **2016**, *6*, 163. [CrossRef]

30. Persaud-Sharma, D.; Budianaky, N.; McGoron, A. Mechanical Properties and Tensile Failure Analysis of Novel Bio-absorbable Mg-Zn-Cu and Mg-Zn-Se Alloys for Endovascular Applications. *Metals* **2013**, *3*, 23–40. [CrossRef] [PubMed]

31. Hojna, A. Overview of Intergranular Fracture of Neutron Irradiated Austenitic Stainless Steels. *Metals* **2017**, *7*, 392. [CrossRef]

32. Badnava, H.; Etamadi, E.; Msekh, M. A Phase Field Model for Rate-Dependent Ductile Fracture. *Metals* **2017**, *7*, 180. [CrossRef]

© 2019 by the authors. Licensee MDPI, Basel, Switzerland. This article is an open access article distributed under the terms and conditions of the Creative Commons Attribution (CC BY) license (http://creativecommons.org/licenses/by/4.0/).

Review

A Short Review on Fracture Mechanisms of Mechanical Components Operated under Industrial Process Conditions: Fractographic Analysis and Selected Prevention Strategies

George A. Pantazopoulos

ELKEME Hellenic Research Centre for Metals S.A., 61st km Athens–Lamia National Road, 32011 Oinofyta, Viotias, Greece; gpantaz@elkeme.vionet.gr; Tel.: +30-2262-60-4463

Received: 9 January 2019; Accepted: 27 January 2019; Published: 29 January 2019

Abstract: An insight of the dominant fracture mechanisms occurring in mechanical metallic components during industrial service conditions is offered through this short overview. Emphasis is given on the phenomenological aspects of fracture and their relationships with the emergent fracture mode(s) with respect to the prevailed operating parameters and loading conditions. This presentation is basically fulfilled by embracing and reviewing industrial case histories addressed from a technical expert viewpoint. The referenced case histories reflected mainly the author's team expertise in failure analysis investigation. As a secondary perspective of the current study, selected failure investigation and prevention methodological approaches are briefly summarized and discussed, aiming to provide a holistic overview of the specific frameworks and systems in place, which could assist the organization of risk minimization and quality enhancement.

Keywords: metal components; fracture mechanisms; fractography; fracture mechanics; quality improvement

---

1. Introduction

Failure analysis (FA) is a multidisciplinary, multifaceted scientific field, connecting areas of engineering from diverse backgrounds and bodies of knowledge; from applied mechanics to electrochemistry and corrosion and from numerical modeling, to the understanding of surface science and tribology. The complexity of the nature of the subject requires the embracing of various engineering disciplines, to succeed high process performance and effective root-cause analysis, which is the core and the central objective of the failure investigation process [1]. The evolution of the failure analysis area is massive and perpetual, since it advances together with the numerous independent fields and core competencies, which are considered as the main constituents of its entire body of knowledge.

There are representative textbooks and monographs, aiming to serve the FA subject from different viewpoints and perspectives, assisting and guiding engineers and failure analysis practitioners working in a variety of industry areas. Some characteristic examples add appreciable effort in understanding the basic elementary technical disciplines, building the framework of FA, and highlighting the role of fractography are referred to in [2–6].

Another approach, which is noteworthy, is focused on the management and organizational aspects of failure analysis, aiming to present and examine its current status as a structured and disciplined procedure [7]. Without the need to mention other literature sources, when treating niche areas of FA, undoubtedly, the great variety and extent of its subject constitutes a common perception. Therefore, filtering and condensing areas of specific knowledge will be valuable for researchers and engineers working in their respective disciplines, while a rigorous review in FA is rather a futile endeavor.

The aim of this short and targeted review is, firstly, to summarize the most common fracture mechanisms of machine components, highlighting their unique identification fingerprints through the contribution of macro- and micro-fractography. The value of fractography in FA is remarkable and it was specifically referred to in an excellent historical overview [8] and in a focused presentation on machine components, presented in [9].

Apart from traditional qualitative fractography, the application of quantitative fractography aims to measure the fracture surface topographic features, revealing significant characteristics of the fracture surface, in terms of true surface areas, distances, sizes, numbers, morphologies, orientations, and positions, as well as statistical distributions of these quantities. Modern quantitative fracture image analysis systems play an important part in the progress and successful achievement of these goals, not only to accelerate the measurement procedures, but also to perform operations that would not be possible by other techniques. Although classical macro-fractography is mostly elaborated utilizing optical methods (stereo-microscopy), research studies are increasingly motivated to realize quantitative measurements using Scanning Electron Microscopy (SEM) microfractography [10].

The experimental routes together with the presentation of fracture mechanisms, decorated through special case studies, will be discussed in Sections 2 and 2.1–2.3. For the sake of simplicity and taxonomy, two classes of fracture are discussed: Instant or overload (Section 2.2) and progressive (Section 2.3) fractures. The subject is limited to the fracture of metallic components, while corrosion/environmental or surface (tribological) degradation mechanisms are out of the scope of the present work. Secondarily, in Section 3, a typical selection of methodologies used to tackle and prevent failures, aiming to enhance production efficiency and minimize machine downtime, are addressed. This section is a complementary part of the preceding section, which presents the "diagnostic" part of fracture analysis, acting as its "prognostic" inseparable twin sister – following the scheme, as the generic Equation (1) addresses:

$$\text{Component Failure Investigation} \rightarrow \text{Root-Cause Analysis} \rightarrow \text{Failure Prevention} \tag{1}$$

The contents of the short review paper concern principally the engineers and FA practitioners working in the manufacturing and metal working industry, aiming to provide:

(a)   A quick and condensed guide sharing knowledge from a technical expert point of view;
(b)   The basic methodological tools used for further preventive actions, at least in the form of their titles and not in a comprehensive and rigorous manner; and
(c)   Offer a sort of inspiration for research and continuous learning, which constitute the driving force and the backbone of improvement and sustainability.

2. Phenomenological Aspects of Fracture

The recognition of a fracture mode is a morphological identification process; fracture history is traced back to its origin, development (growth or propagation), and final ending stage. The description of the fracture process is literally connected to the construction of fracture history and main failure hypothesis. Fractographic analysis, using optical and scanning electron microscopy, constitutes an irresistible technique towards the resolution of fracture analysis problem solving. The unfolding of fracture process history is clearly stipulated through the rigorous description of the fracture surface topography; hence the preservation of fracture surfaces plays a key role in the entire FA investigation.

2.1. Experimental Procedure

Optical stereomicroscopy was performed using a Nikon SMZ 1500 (Nikon, Tokyo, Japan) stereo-microscope, using image analysis software (Image Pro Plus, Rockville, MD, USA). High-magnification fracture surface observations, utilizing a FEI XL40 SFEG scanning electron microscope (FEI, Eindhoven, The Netherlands), were realized using secondary electron (SE) and backscattered electron (BSE) imaging under 20 kV accelerating voltage conditions.

2.2. Instant (Overload) Fracture Mechanisms

This section summarizes the main types of overload fracture, which occurs instantly, once the intensity of the operating conditions exceeds the load-carrying capability of the component(s). Typically, the main overload fracture classes are categorized according to the accompanied plasticity (ductility) criterion, leading to two main groups:

$\rightarrow$ Ductile Fracture (Section 2.2.1) and

$\rightarrow$ Low/Limited Ductility Fracture (Section 2.2.2). In this case, the term "brittle fracture" is intentionally avoided and it is included with the present fracture type. Brittle fracture is an extreme case of low/limited ductility fracture where the absorbed plastic strain energy is negligible.

2.2.1. Ductile Fracture

Ductile fracture is accompanied by an appreciable amount of permanent plastic deformation, which is manifested even macroscopically by shape-geometry or cross section distortion (necking). Microscopically, voids are generated around inclusions, inclusion/matrix interfaces, and at the centre of the neck, where the hydrostatic stress is maximized and stress triaxiality dominates. The process of fracture development includes the following steps:

$\rightarrow$ Void nucleation;

$\rightarrow$ void growth; and

$\rightarrow$ void linking (coalescence).

The inclusion density affects the microvoid nucleation rate, leading to a higher number of nucleation sites and lower growth potential, resulting in a high void distribution density and lower size dimples, signifying lower overall plasticity [11]. The Gurson based model, later modified by Tvegaard and Needleman, treated the damage evolution using void failure criteria and correlating hydrostatic stress with von Mises stress and the void volume fraction number, see Equation (2) [11–13]. The examination of ductile fracture evolution is out of the scope of the present work and, therefore, only a simple reference to the previous classical GTN (Gurson-Tvegaard-Needleman) model is attempted:

$$\varphi = \frac{\sigma_{eq}^2}{\sigma_0^2} + 2\,C_1 \cdot f \cdot \cosh\left(\frac{3}{2} \cdot C_2 \cdot \frac{\sigma_h}{\sigma_0}\right) - \left(1 + C_3 \cdot f^2\right) = 0 \tag{2}$$

where σ_{eq} is the equivalent von Mises stress, σ_h is the hydrostatic stress, σ_0 is the actual yield stress, C_1, C_2, and C_3 are model constants, and f is the effective void volume fraction. When, $C_1 = C_2 = C_3 = 1$, the model is transitioned to the original Gurson model.

This model refers to a yield state, of the form, φ, as a function of the stresses (σ_{eq}, σ_0, σ_h) and void volume fraction (f), which equals 0 for a porous ductile material [13].

The microvoid coalescence gives rise to the evolution of ductile (plastic) fracture, which creates characteristic signatures and fracture surface patterns [14]. The observation of fracture surface topography, using scanning electron microscopy (SEM) reveals a specific anaglyph consisting of dimples of various size, shape, and distribution (Figure 1). The growth of coarse voids proceeds against fine ones. As void evolution advances, the un-fractured metal links between the holes (behaving as columns or webs) are consecutively strained until final failure occurs, creating lower size secondary dimples and leading to an assigned distribution of cavities, as it is also described in [15]. The variation of dimple size is affected by the void threshold stress and growth rate, which are principally influenced by the geometrical and physical characteristics of the microstructural features [15].

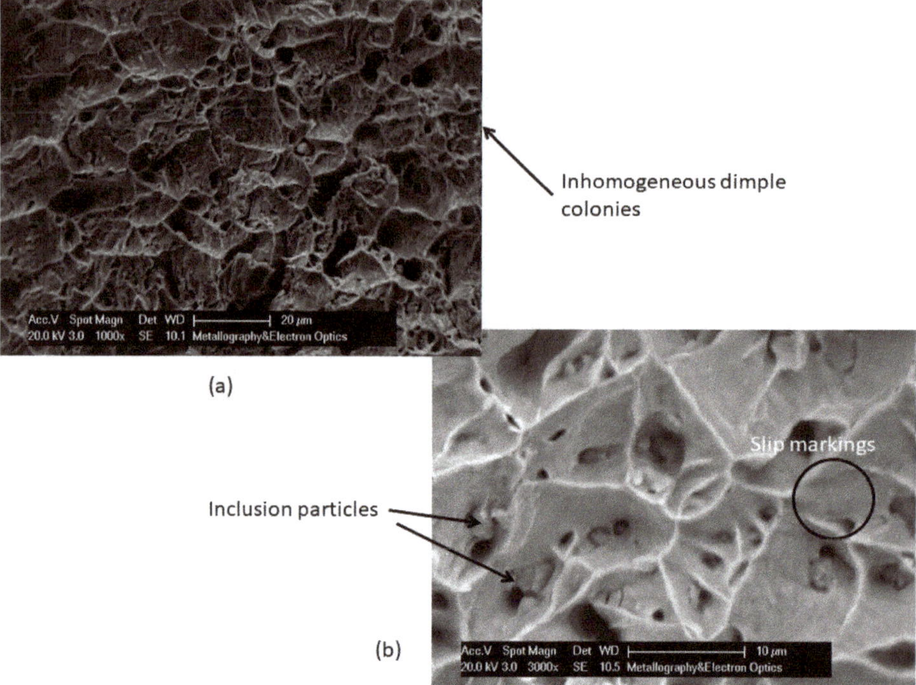

Figure 1. SEM micrographs showing: (**a**) Ductile fracture surface (SE imaging) showing inhomogeneous dimple distribution and (**b**) higher magnification topographic features showing details of dimples around non-metallic inclusions (BSE imaging). Material: Low carbon structural steel grade (C15), fractured under tensile overload.

The orientation of dimples denotes the load application (axial, shear, tear), see also [2]. The presence of shear dimples indicates high stress triaxiality conditions, impeding profuse void growth. In case of shear ductile fracture, slip band formation is restricted on inclusions, causing localized strain evolution and void nucleation [16].

2.2.2. Low/Limited Ductility Fracture

In this class of fracture, there are two principal categories:
(a) Transgranular (cleavage) fracture; and
(b) Intergranular fracture
The cleavage fracture proceeds on {100} planes in <110> directions for body-centered-cubic (bcc) metals, while it seems somehow contradictory to Griffith's thermodynamic criterion for brittle fracture, which foresees that the {110} should be the cleavage planes, showing minimum surface energy without any preferred crystallographic direction [8]. Since dislocation processes accompany cleavage, it has been postulated that preferred cleavage planes and directions are those of the lowest plasticity around the crack [8]. River-line patterns are characteristic features of transgranular (cleavage) fracture, represented topographically as plateaus connected by shear ledges, showing the direction of crack propagation [17]. The formation mechanism of river patterns is based on the change of the fracture mode from mode I with an increasing component of mode III [15]. A typical fractograph of the cleavage fracture of a steel chain link is given in Figure 2a [18]. In Figure 2b, a quasi-cleavage is shown, observed in a tool steel gripper's teeth, where minute areas of plastic deformation are also evidenced [19].

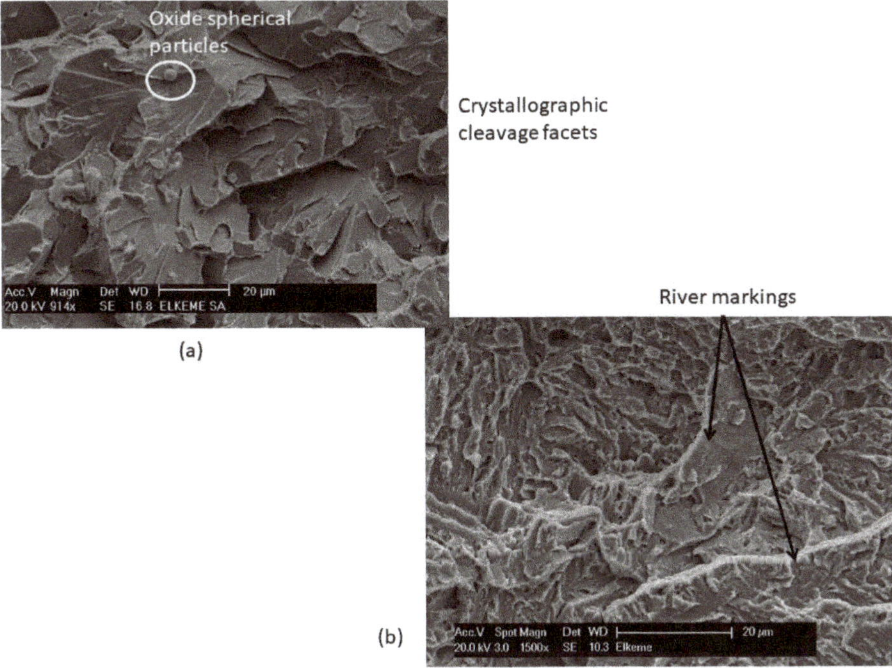

Figure 2. SEM micrographs showing: (a) Transgranular cleavage fracture surface (SE imaging) showing crystallographic facets and "occluded" river patterns (Material: Chain link, structural steel grade C40), (b) "quasi-cleavage" fracture surface (SE imaging), showing a mixed-mode transgranular fracture, accompanied with areas of localized plastic deformation (Material: Tool steel gripper W.Nr. 1.2343).

Very frequently, intergranular fractures are observed rather than cleavage, due to either the concentration of low melting-point impurity phases or the segregation of impurity elements at grain boundary areas (such as V-group elements in steels). P, As, Sb, and Sn are considered as intergranular embrittling elements, while the development of Auger electron spectroscopy (AES) in 1969 assisted in the identification of a monolayer of such impurities at grain boundaries, which promotes embrittlement [8]. The presence of minor impurities, such as Bi and Pb, could induce severe damage due to hot shortness, leading to intergranular fracture in copper and copper alloys [20–23]. Pb particles could lead to the formation of micro-voids, enlarged by growth and coalescence by an adsorption-induced de-cohesion, or emission of dislocation, which results in intergranular damage evolution, see also [24]. Typical intergranular fracture in a leaded copper alloy is shown in Figure 3.

In case of lead-free heat treated CuZn42 copper alloy, an almost entirely IG (intergranular) fracture mode was observed under static and dynamic loading conditions [25,26]. In a previous study, the absence of Pb and the room temperature testing conditions rather ruled-out the possibility of the occurrence of hot-shortness as the main cause of the identified IG fracture. According to [26], the occurrence of such an IG fracture requires a salient interpretation based on the following notions:

→ Low stress intensity factor range (ΔK);

→ coarse β-phase in relation to the developed plastic zone size; and

→ high-angle grain boundaries.

All the above parameters could be considered as a potential that could act separately or synergistically, enhancing the possibility of the IG fracture for the heat treated CuZn42 alloy.

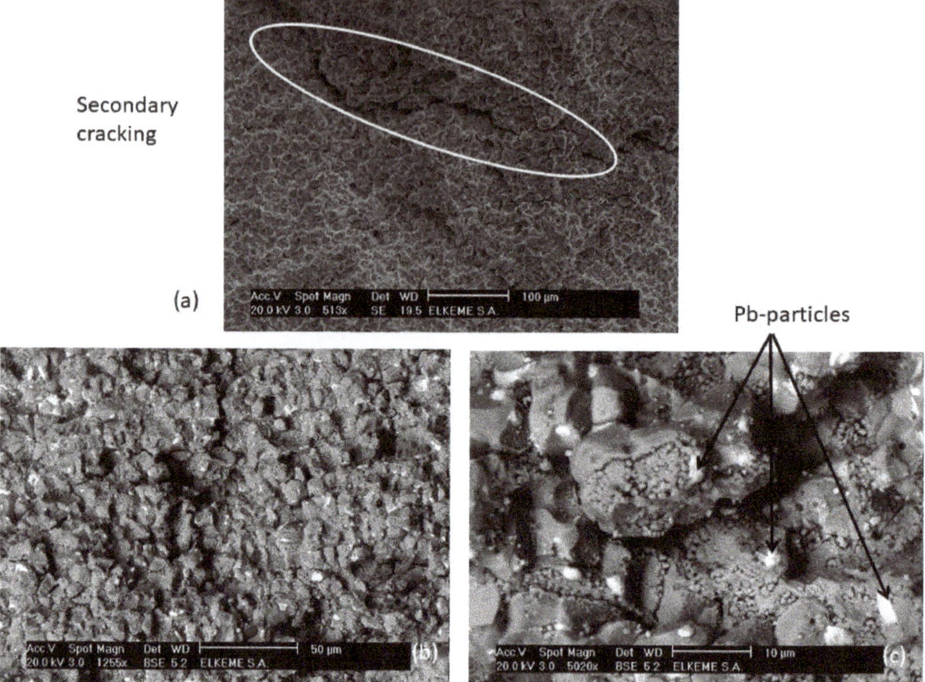

Figure 3. (a) SEM fractograph (SE imaging) showing a typical intergranular fracture mechanism and (b,c) higher magnification SEM fractographs (BSE imaging) showing details of the intergranular fracture facets; note the "bright spots", representing Pb particles, accumulated in the grain boundaries (shown by black arrows). Material: CuZn39Pb3 rod fractured during hot extrusion.

2.3. Progressive Fracture Mechanisms

This section elucidates the so-called "progressive fracture modes" or "delayed fracture modes", pertaining to two main broad categories:

→ Fatigue Fracture (Section 2.3.1); and
→ Creep Fracture (Section 2.3.2).

Although the extent of the above subjects is undoubtedly complex and enormous, special emphasis is placed on the major fractographic aspects, which are commonly observed and are considered as the "fracture fingerprints" or "signatures", in analogy to the previously presented sections.

2.3.1. Fatigue Fracture

Fatigue failure is a progressive failure process that occurs under cyclic loading and it is comprised of three distinct stages: Stage I is related to the crack nucleation at 45° to the load direction (following slip planes); Stage II is the continuous crack growth, perpendicular to the stress up to the point when the remaining cross section can no longer withstand the applied load; and, finally, Stage III, which is the instant ultimate fracture due to overload [27]. The rate of crack propagation during Stage III fracture is almost equal to half of the speed of sound in the material. The majority of the mechanical components (shaft, gears, turbine blades, bearings, rolls, etc.) are subjected to cyclic/periodic loading conditions and, hence, the fatigue failure mode is the predominant fracture mechanism. As a rule of thumb, the time necessary for the nucleation of the fatigue crack is almost equal to 80–90% of the total lifetime of the machine element.

Macro-fractographic or visual observation of fatigue fracture reveals rather a smooth surface texture, comprising various topographic features; the most significant ones are the following (Figure 4), see also [27,28]:

(a) Crack progression marks (also called beach marks or crack arrest marks). These are elliptical or semi-circular shaped marks, signifying a change of the position of the fatigue crack front. They are also related to the arrestment or decrease of crack growth due to the load interruption during machine operation, or due to the development of a compressive stress field ahead of the crack tip [27]; and

(b) Ratchet marks. They look like "shear ridges", separating successive crack fronts. The existence of ratchet marks indicates the presence of multiple crack initiation sites and high stress concentration conditions. Ratchet marks are created when cracks initiated at different positions are joined together, creating steps on the fracture surface (Figure 4a).

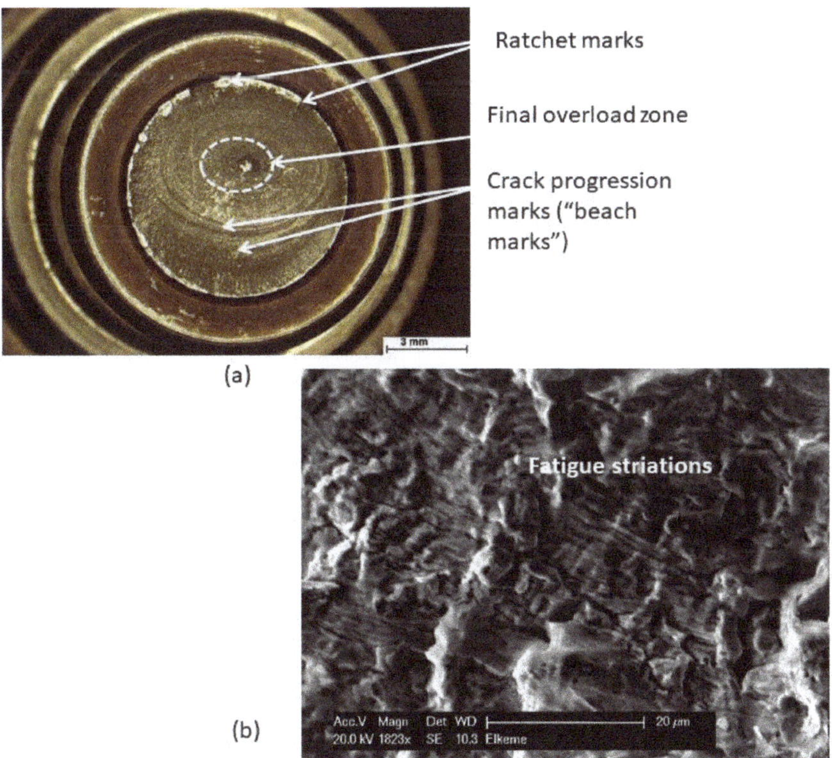

Figure 4. (a) Stereo-micrograph showing the fatigue fracture of a steel pin of a roll chain mechanism. The fractographic signs advocate the occurrence of rotating bending fatigue, under low load and high stress concentration conditions. (b) SEM fractograph (SE imaging) of the same roll showing a typical fine striation pattern. Material: Structural steel, ferritic-pearlitic microstructure, hardness: 210 HV.

The final fracture zone (overload or instant fracture zone) is rough and its pattern corresponds either to ductile (dimpled) or low energy (transgranular or intergranular) failure, depending on the loading conditions, the mechanical properties, and the geometry of the component. The size of the overload zone with respect to the entire cross section is related to the magnitude of the loading conditions at the end of the fracture process; a large overload zone (approximately >70% of the whole section) indicates severe loading conditions as opposed to the small size overload zone (approximately <40% of the whole cross section). The presence, orientation, and extent of the above macro-fractographic features are widely used in failure analysis as diagnostic tracing marks of the loading history (régime and magnitude) and stress concentration conditions [2,3]. For instance, in Figure 4a, the evidence of multiple crack initiation sites (ratchet marks) around the roll pin circumference, together with the location and size of the ultimate failure area, suggest that the component failed due to rotating bending fatigue, under low applied load and high stress concentration conditions, see also [28].

Through SEM micro-fractography (SE imaging), a fine striation-pattern, which corresponds to a microscopic fingerprint of fatigue crack propagation (Stage II), can be also detected (Figure 4b). This topographic feature is a sign of microscopic plasticity and it is generated by blunting and re-sharpening of the crack-tip, during individual load cycles, known as Laird's mechanism [27]. The striation spacing normally corresponds to the local fatigue crack propagation rate (*da/dN*).

In Figure 5, a bent deoxidized high phosphorus (DHP) copper tube, working in a refrigerating system, failed due to fatigue. The pulsating loads applied due to alternating pressure conditions and the residual stresses imposed during bending constitute synergistic parameters of the imminent fatigue fracture [29]. The stepwise crack extension fashion, denoted by the presence of ratchet steps, is a plausible indicator of stress concentration on the outer bent area.

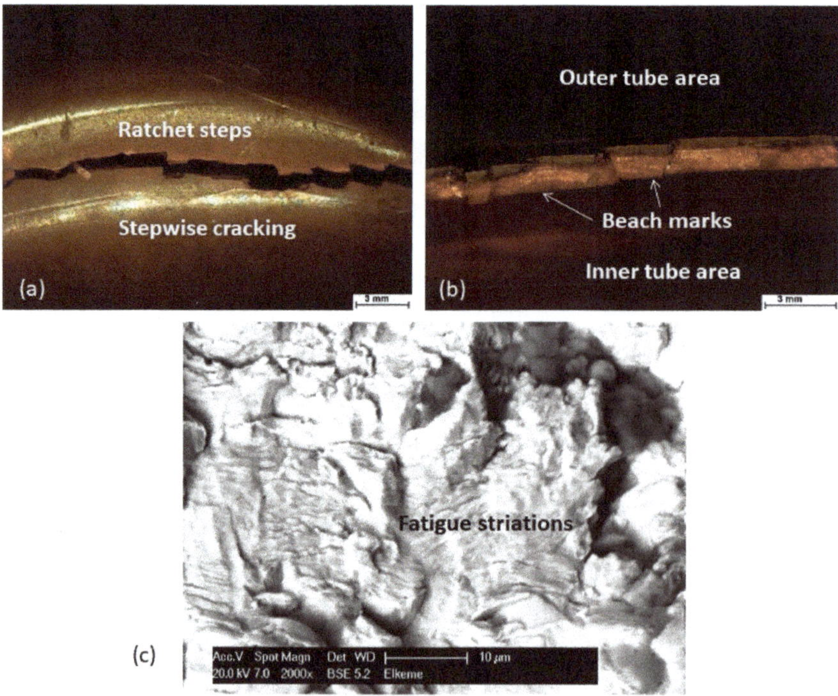

Figure 5. (**a**) Stereo-micrograph showing the fatigue crack (stepwise propagation) on the bent tube surface; (**b**) stereo-micrograph of the tube fracture surface, showing multiple crack initiation sites and beach marks; (**c**) higher magnification of the fracture surface performed by SEM (BSE imaging) where fine fatigue striations are readily resolved. Material: deoxidized high phosphorus (DHP) copper tube.

2.3.2. Creep Fracture

Metallic components show a lower lifetime when operated under high temperature conditions, e.g., in energy generation systems, power plants, gas turbines, chemical process industries [6]. Depending on the dominant environmental and stress conditions, the metallic structural parts suffer from creep-originated, environmentally induced fractures, high temperature fatigue, and thermal fatigue. Creep is realized through the nucleation and growth of transgranular or intergranular voids depending on the applied operating conditions, driven by diffusion and dislocation motion as a result of simultaneous application of stress and heat input, see also [30]. Creep is a progressive time-dependent plastic deformation, which leads to final failure with potential destructive consequences to health, safety, and the environment. It is noteworthy that remaining service life assessment techniques constitute a significant framework in failure prediction, resulting in cost minimization, due to its contribution to the maintenance planning and scheduling of high temperature process equipment, without unnecessary replacements and avoiding the risk of catastrophic property losses with severe implication to human, health, and environmental safety. Methods, using the classical Larson-Miller parameter (LMP) expressed in Equation (3) and life fraction rules using the damage accumulation concepts, are very widely applied in remaining life prediction, assisted also by finite element analysis techniques, see [31–33]:

$$LMP = T \cdot (23 + logt) \tag{3}$$

where, T is the operating temperature (K) and t the operation time (h).

Creep rupture curves present the variation of stress variation, as a function of the Larson Miller parameter. Knowledge of stress and temperature conditions could therefore lead to the estimation of the creep lifetime at a certain statistical confidence level. The life fraction rule is applied for the calculation of the remaining life of structural elements, working at high temperatures, see Equation (4):

$$\frac{t_{op}}{T_{op}} + \frac{t_{test}}{T_{test}} = 1 \tag{4}$$

where, t_{op} is the operating time, T_{op} is the total operating time under service conditions, t_{test} is the time required for rupture at the test under accelerating creep conditions, and T_{test} is the time for rupture of the fresh component under the above accelerating creep conditions.

In the case of creep failure of superheater pipes (ferritic steel 15Mo3 grade, working at 500 °C approximately), a characteristic "fish-mouth" deformation pattern is evidenced, see [34]. Macro-fractographic observations performed on the pipe ruptured lips are illustrated in Figure 6a. A typical ductile fracture surface consisting of profuse elongated dimples is directly identified, signifying the occurrence of exhaustive plastic deformation (Figure 6b). Large equiaxed cavities developed from the growth and linking of creep voids of a smaller size were readily discerned (Figure 6c). Elongated dimples corresponding to the initial grain orientation to the original pipe metal forming operation are evidenced (Figure 6b). Isolated areas of transgranular facets and ductile tearing are found in the interior of the dimples (Figure 6c). The evolution of cavitation is, therefore, a significant fingerprint of creep damage evolution across the pipe wall.

The idealized creep curve is shown in Figure 6d and it consists of three stages (primary, secondary, and tertiary creep). Following a primary creep stage (I), where strain is increased at relatively high rates, the creep rate decreases, reaching a steady state value ($d\varepsilon/dt$) at the secondary phase of creep. During the third creep stage (known as tertiary creep), the creep rate is asymptotically increased up to the final failure. Tertiary creep usually occurs under high stress–high temperature conditions. During third stage creep, an effective reduction of cross-section happens, and various microscopic degradation phenomena take place, such as massive void formation (cavitation) and microstructural changes driven by diffusion processes (e.g., intermetallic phase coarsening) and recrystallization [17].

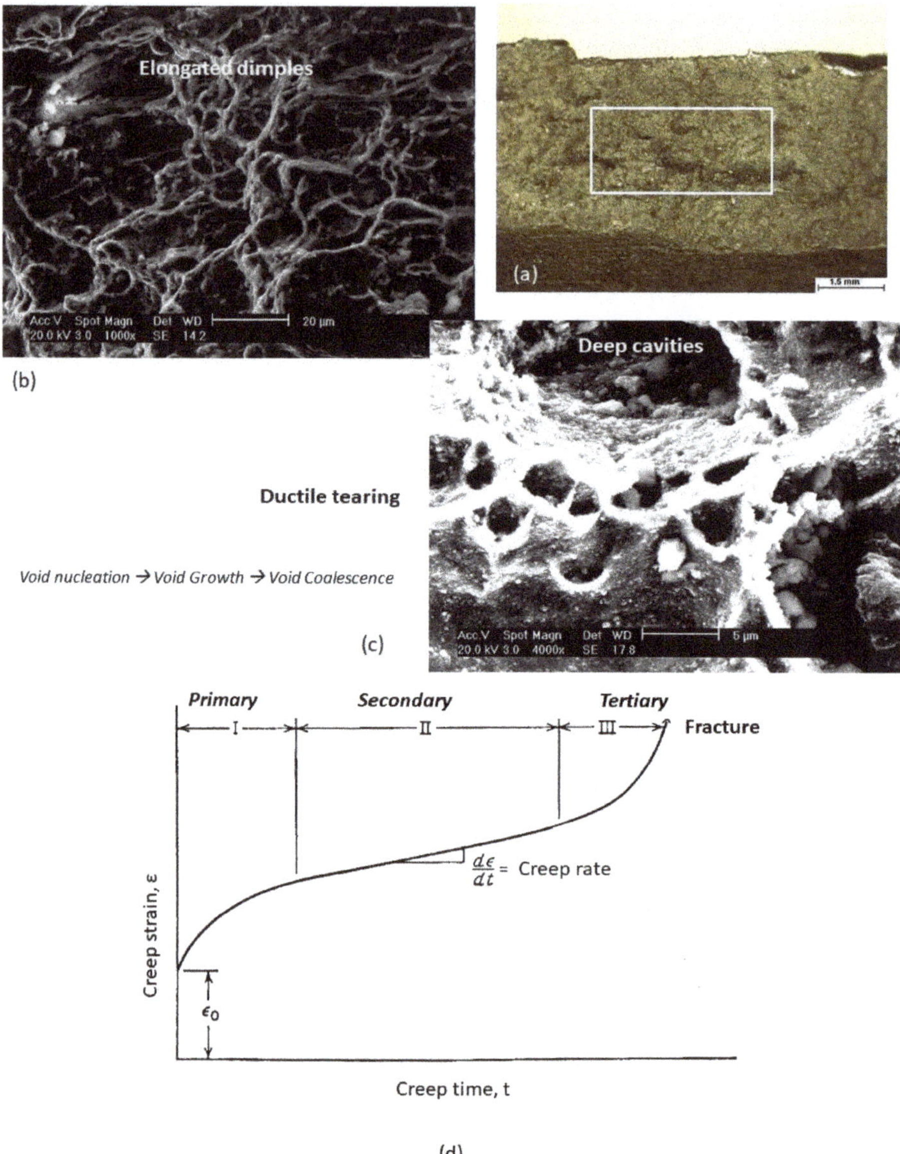

$\dfrac{d\epsilon}{dt}$ = Creep rate

ϵ_0

(d)

Figure 6. (**a**) Stereo-micrograph showing the fracture surface of the pipe-wall; (**b**) SEM micrograph (SE-imaging) showing the profuse dimpled fracture surface; (**c**) SEM micrograph (SE-imaging) showing details of the fracture surface, highlighting the presence of deep cavities and ductile tearing. (**d**) Typical diagram of creep strain evolution as a function of time/duration.

Creep and fatigue interaction is the progressive time-dependent inelastic deformation under constant temperature and variable loading conditions. The creep-fatigue process is accompanied by many different slow structure rearrangements, including dislocation motion, aging of the microstructure, and grain boundary cavity formation [35]. In a relevant study, creep-fatigue crack growth rate tests were performed on a specially designed program test cycle. In a relevant study, continuum damage mechanics were applied to estimate creep damage under multi-axial loading conditions [36]. Using the compact tension ($C(T)$) specimen, a creep-fatigue crack growth test was realized in 12Cr1MoV steel at 550 °C. Finally, the application of the creep stress intensity factor (as a creep-damage-sensitive parameter) for creep-fatigue interaction was discussed. The coupled micro-damage and macro-fracture analysis involves the definition of the creep crack-tip stress field model, induced by the application of the ductility-based model and the formulation of the damage evolution and constitutive equations regarding creep-strain rate behavior. The effect of the specimen thickness on the stress and process zone was studied and reproduced utilizing a full-field 3D finite element (FE) model [36].

3. Failure Analysis and Selected Prevention Strategies

3.1. Fracture Mechanics Approach

The main aspects of fracture mechanics, which can be considered as the corners of a triangle, are the following: Loading, material toughness, and defect crack size. If two corners of the triangle are given, the third one can be estimated.

- The fracture resistance (toughness) of the material and the crack size are both known. Then, the critical load can be estimated and a decision can be made whether further operation is safe or not;
- the loading conditions and the maximum (undetected) crack or minimum (detectable) crack size specified, which can be accurately measured by quality control, are known. Based on this information, a minimum fracture resistance (toughness) of the material can be ascertained used for material selection or during the design stage; and
- for a given fracture resistance (toughness) and loading conditions, a critical crack or defect size can be calculated and used as further information for non-destructive testing (NDT).

A review of the emergence of fracture mechanics in failure analysis projects is comprehensively attempted in [37]. It is noteworthy that the utilization of fracture mechanics in FA creates some issues related with conservatism, originating mainly from different sources, such as uncertainties and a lower number of data, critical stress intensity factor estimation (K_{Ic}), and the consideration of an extremely sharp crack tip size (zero-tip radius).

The present approach encompasses mostly the principal aspects of linear-elastic fracture mechanics (LEFM), while a limited involvement of elastic-plastic fracture mechanics (EPFM) is intended. The topic and the contribution of fracture mechanics are quite extensive and the potential utilization in FA is very promising. In this short reference, an attempt was made, focused on the following important parameters in FA investigation:

(a) Predicting the critical defect size that can be permissible to the applied loading conditions; and
(b) estimating the fracture resistance using standard methods.

The first point can provide significant input in non-destructive-testing techniques, where defect size is monitored and replacement of critical components can be precisely forecasted.

The presence of surface or internal defects, such as grooves from machining, dents/pits, inclusions, etc., are considered potential sites for fatigue crack initiation and propagation (Figure 7). Using the assumptions related for short cracks extending from minute surface flaws, see [38,39], the following expression representing the Murakami-Endo approximation, given by Equation (5), can be applied, employing the mean surface Vickers hardness of the material:

$$\sigma_{r,th} = 2.86 \cdot (HV + 120) \cdot \left(\sqrt{A}\right)^{-\frac{1}{6}} \cdot \frac{1 - R}{2}^{(0.226 + 10^{-4} \cdot HV)} \tag{5}$$

where, $\sigma_{r,th}$ is the threshold stress range (MPa), HV is the Vickers indentation hardness, A is the defect projection area normal to the maximum stress (μm^2), and R is the stress ratio (equals to $\sigma_{min}/\sigma_{max}$, for a pure tensile stress regime, $\sigma_{min} = 0$, $R = 0$).

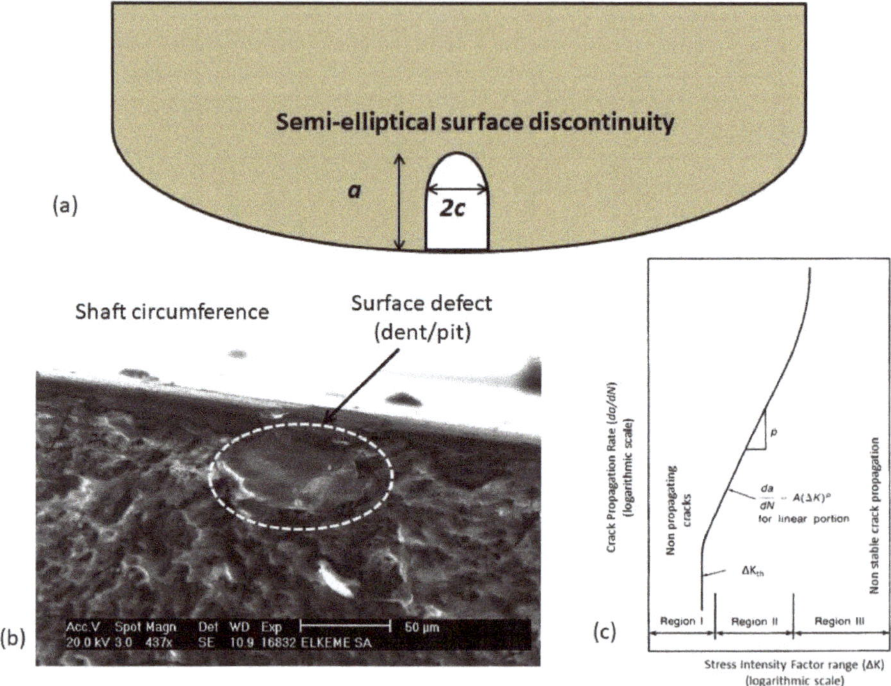

Figure 7. (**a**) Schematic showing the geometry of semi-elliptical surface imperfection, located on shaft periphery; (**b**) SEM micrograph showing the fatigue crack origin, indicating the presence of a surface defect and (**c**) fatigue crack propagation graph.

Equation (5) can be applied to a wide range of materials possessing high and low ductility behaviour (more brittle), ranging within the Vickers hardness range: $70 < \text{HV} < 720$ [39].

Under this stress range level ($\sigma_{r,th}$), the crack does not propagate or it advances at very low (rather negligible) rates. Equation (5) can be re-written using, ΔK_{th} instead of $\sigma_{r,th}$, under the principal LEFM expression, see Equation (6):

$$\Delta K_{th} = \sigma_{r,th} \cdot \sqrt{\pi \cdot a} \tag{6}$$

where, ΔK_{th} is threshold stress intensity range and α is the semi-elliptical defect size (Figure 7a).

In a recent work, the above analysis was used to analyze the level of stress required for the propagation of fatigue crack in an austenitic stainless steel propeller shaft, due to the presence of a mechanical surface defect [40]. The fatigue crack origin located on a surface defect appeared as a "dent or pit" coming most probably from localized mechanical damage (Figure 7b), see also [40]. The investigation findings suggested strongly that the propeller shaft failed due to combined rotating bending/torsional fatigue initiated from the shaft periphery and close to the keyway. The formation of surface defects is responsible for the initiation and propagation of fatigue cracks. The presence of a semi-elliptical surface defect actuates the propagation of fatigue cracks for applied tensile stress threshold above 590 MPa, for $R = 0$. For completely reversed loading conditions ($R = -1$), the threshold alternating stress is 355 MPa, which is almost approximately 30% of the shaft tensile strength (1000–1200 MPa).

The characteristic shape of crack propagation rate (*da/dN*) as a function of the stress intensity factor range (ΔK) is shown in Figure 7c. The curve is divided in three regions [17,39]:

→ In Region I, where there is a stress intensity threshold range (ΔK_{th}), below which fatigue cracks do not propagate (or propagate at quite low rates);

→ in Region II, noted as the continuous crack propagation region (linear portion of the log-log diagram), where Paris law is in effect; and

→ in Region III, where fatigue cracks propagate unstably up to the point where the maximum stress intensity factor takes the value of the critical stress intensity factor (K_{Ic}), resulting in overload fracture (intergranular, transgranular, or dimpled fracture).

The second point serves in the determination of the required or the expected fracture mechanics properties, which are foreseen by the standards or by the customer's/design requirements. The fracture toughness is often confused with impact toughness, which corresponds to the energy absorbed during dynamic loading applications, such as foreseen during the Charpy test (CVN = Charpy V–notched value), using a 45°, 2 mm notched specimen of dimensions $10 \times 10 \times 55$ mm^3 [41]. However, the most suitable testing fracture mechanics methods refer to the determination of LEFM properties, such as the critical stress intensity factor (K_{Ic}), and EPFM properties, such as J-integral and Critical-Crack-Tip-Opening-Displacement (CTOD), see [42]. In previous works, the fracture mechanics properties, such as CTOD of extruded/drawn and heat treated copper alloys, were experimentally determined [26,43]. The CTOD value represents the distance between the crack flanks, which corresponds to the extent of plastic deformation at the onset of unstable fracture propagation. For bend type specimens, the CTOD values (δ) comprise two main components, the elastic and plastic one, see Equation (7):

$$\delta = \delta_{el} + \delta_{pl} = \left[\frac{S}{B} \cdot \frac{P}{W^{1.5}} \cdot g_1\left(\frac{a_o}{W}\right) \right]^2 \cdot \frac{1 - v^2}{2YS \cdot E} + \frac{0.4\,(W - a_o)}{0.6a_o + 0.4W + z} \cdot V_p \tag{7}$$

where, S is the span distance between outer bending supports, P is the load, B is the specimen thickness, W is the specimen width, $g_1(\alpha_0/W)$ is the stress intensity function, α_0 is the average original crack length, v is the Poisson's ratio, E is the Young's modulus, V_p is the plastic component of CMOD (Crack-Mouth-Opening-Displacement), z is the knife edges' thickness, and YS is the material yield strength (actually the 0.2% proof strength is used, $R_{p0.2}$).

Apart from the critical CTOD values, the Load vs. Crack-Mouth-Opening-Displacement (CMOD) curves, denoted also as *P-V* graphs, constitute additional indicators of the fracture behaviour for tested alloys and their metallurgical conditions. The "type 6" curves, as categorized by BS 7448-1 standard [42] and the absence of abrupt load-drops ("pop-in's"), due to localized instabilities, signify adequate crack arrestment and fully plastic behaviour [26]. Therefore, the assessment of fracture mechanics properties, together with the rest of the fractographic and load-displacement behaviour evidence, provide vital information, simulating the actual material behaviour under the real service conditions, resulting in design optimization aiming to prevent failure.

Various fracture mechanics models are used to analyze the fracture toughness of notched components; the most known ones are the theory of critical distances (TCD) [44,45], strain energy density (SED) [46], and cohesive zone model (CZM) [47]. The rigorous elaboration of the above mentioned fracture mechanics models is out of the scope of the present work; rather a short highlight of the TCD is worth mentioning.

The theory of critical distances (TCD) consists of a framework of methods introducing the critical distance or length parameter [44]. This length parameter is expressed in Equation (8):

$$L = \frac{1}{\pi} \cdot \left(\frac{K_{mat}}{\sigma_0} \right)^2 \tag{8}$$

where, K_{mat} is the fracture toughness of the material, while σ_0 is the inherent strength parameter.

In case of full LEFM conditions, σ_0 coincides with the ultimate tensile strength (σ_{UTS}). A relevant study of the fracture toughness of U-notched components using the TCD is given in [45].

The parameter, L, presented as the length constant, can be calculated in various different ways [44,45]. Particularly, a number of researchers have proposed that failure can be predicted by modifying the critical stress idea so that the stress to be used is not the maximum stress (at the notch root), but the stress at a point located at a certain distance from the notch. Other researchers have used the average stress on a line starting from the notch. These two techniques, which are called as the point method (PM) and line method (LM), are very popular and they are proposed for further use in fatigue.

Cracks and notches in components and structures are frequently subjected to complex loading under states and combinations of normal and shear stresses ahead of the crack tip. Mixed-mode fracture mechanics deals with experimental studies and theoretical models for predicting the onset and path of crack propagation under the combination of Mode I (opening), Mode II (sliding), and Mode III (tearing) conditions [48]. Problems of this type are considered in case complex materials, such as welded structures, adhesive joints, and composites, in plain and reinforced concrete structures, aircrafts, bridges, etc. A mixed-mode superposition can also occur during crack branching, i.e., when a crack changes path and the classical energy balance of Griffith's theory can no longer be valid in a simple way, since cracking is not collinear, as it has been assumed previously [49].

Fast fracture and crack arrest is a highly specialized topic in fracture mechanics. The problem of dynamic fracture is considered mainly in to two basic areas: In pipelines and thick-walled pressure vessels [39]. In gas pipelines, the dynamic fracture issue rises from the rapid gas depressurization and the release of a decompression shock wave. In the case of a liquid-filled pipeline, the higher speed decompression shock-wave leads to a subsequent load release and to possible crack arrestment. The problem of crack propagation and arrestment in this type of structure (pipelines) is viewed in terms of dynamic energy balance. Dynamic fracture in thick walled pressure vessels considers the presence of partially through-the-wall cracks initiated from the inner wall surface. Unstable crack propagation becomes feasible upon thermal shock (caused due to rapid cooling of the interior of pressure vessel), which leads to the development of high tensile stress fields imposed at the inner wall.

3.2. Quality Tools and Techniques—Process Approach and FMEA

The organization of the FA as a systematic nine-step procedure is referred to in the textbook of [7]. The detailed analysis of the FA procedure aims to provide guidance to engineers and practitioners and to offer consistent results towards failure prevention and quality improvement. Such a procedure could be carefully and suitably implemented, reflecting also the goals of the hosting organization, and reviewed for the attained performance and resource availability. The complexity, cross-disciplinary nature, and team involvement necessitates a process-oriented approach, treating the transformation of "inputs" to the "desired" outputs.

The FA process could be designed as a procedure having steps and interconnected links, using the "flow-chart" visualization (Figure 8), see also [1]. The FA procedure should be both detailed and generic, applicable for the entire corpus of failure analysis investigation types of the organization, as is demonstrated by the exemplar diagram in Figure 8. It is usually embedded in a quality management system and therefore QA (quality assurance) plays a vital role in maintaining, reviewing, and revising this procedure.

The first phase (initiation of a FA project—resource allocation and planning) should involve the handling, registration, and approval stages of the FA project, together with planning/scheduling, background information, and resource allocation. The second phase (failure investigation—examination of the potential root causes-suggestion of the most plausible cause) constitutes the "core" of the technical FA investigation, containing NDT, material testing, and analysis, performance of numerical modeling, etc. together with the compilation and interpretation of the collected evidence.

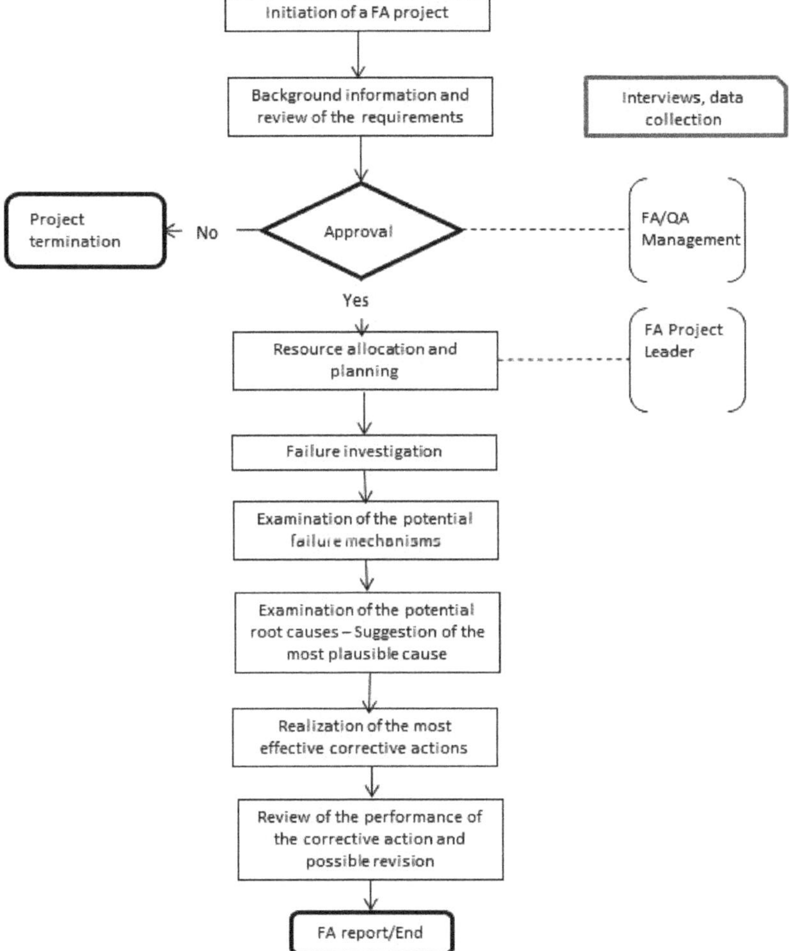

Figure 8. Flow-chart depicting the various stages of a Failure Analysis (FA) procedure.

This will lead to the determination of the failure mechanism(s), answering the question "how has the component failed?" and to the judgment of the root-cause(s), answering the question "why has the component failed via the stated failure mechanism?" Finally, in the last phase (realization of the most effective corrective actions—FA report/end), the FA team should propose and decide for the appropriate corrective actions to minimize the risk, preventing the recurrence of the failure.

In the frame of failure prevention, one of the most important techniques is Failure Mode and Effects Analysis (FMEA). This technique is dedicated to reducing risks of failure and understanding the nature of preventive actions needed to be taken as measures of continuous improvement and sustainability. The implementation of a rigorous FMEA foresees that preventive actions have to be identified prior to an incident and should be applied without delay. The principal stages of FMEA application can be summarized as the following 10 steps:

1. Determination of process parameters;
2. Determination of the possible failure modes;
3. Determination of the failure effects on the final product, system or service;

4. Determination of the root-causes;
5. Assessment of the criticality of the failure;
6. Assessment of the failure occurrence (probability);
7. Assessment of the failure detectability;
8. Determination of the risk priority number (RPN);
9. Suggestion/proposal of preventive actions;
10. Re-estimation of the RPN under the new revised conditions.

The risk priority number (RPN) is defined as the product of the three independent terms (criticality, occurrence, and detectability), see Equation (9):

$$RPN = (Criticality) \times (Occurrence) \times (Detectability) \tag{9}$$

where, criticality is the failure indicator and it is ranked within 1 to 10 (1: Low criticality, 10: High criticality), occurrence is the failure frequency indicator and it is ranked within the range 1 to 10 (1: Low frequency, 10: High frequency), and detectability is the failure detection capability and it is ranked within 1 to 10 with decreasing tendency (1: High detection capability, 10: Low detection capability).

The scheme of the FMEA flow-chart is illustrated in Figure 9. FMEA is not only an appropriate and useful risk analysis technique in the quality assurance/management field, but also in environmental health and safety (EHS). The presentation of FMEA scores is usually performed using common spreadsheets. A paradigm of a rigorous process-FMEA applied in the metal forming industry, which was used for quality improvement in brass annealing, is analytically presented in [50].

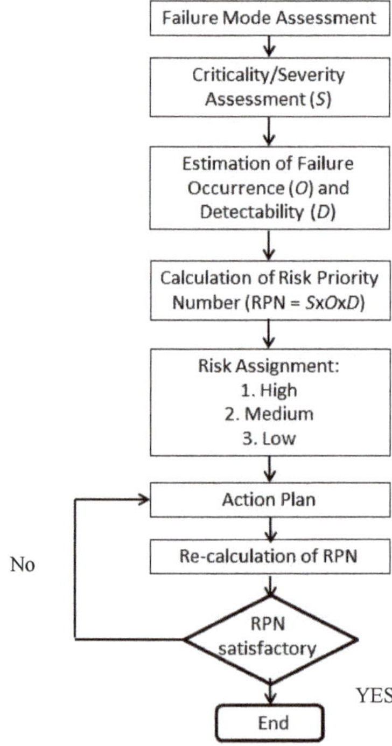

Figure 9. Representative flow-chart depicting the various stages of a Failure Mode and Effects Analysis (FMEA) procedure.

Statistical techniques using non-parametric analysis (e.g., Taguchi Design of Experiments/DOE, using Signal-to-Noise ratio) and analysis of variance (ANOVA) could be employed also to find the most influential factor(s) and succeed in process optimization [51]. The above techniques can be used in conjunction to FMEA to optimize the most critical process parameters [52]. It is noteworthy that modern trends in machine learning and artificial intelligence explore the possibilities of implementing and applying computer algorithms in the field of failure identification. The shafts' failure mode identification, using an expert system deploying three types of inference engines (rule-based, fuzzy logic, and Bayesian statistics) was presented in [53]. The knowledge was transmitted to the three inference engines through specific tables containing the most typical visual characteristics recognized in the failed shafts (e.g., beach marks, distortion, corrosion evidence), according to the technical expertise. According to [53], the best performance was recorded by the Bayesian statistics inference engine.

3.3. Systems Approach—ISO 9001:2015 and Risk Analysis

The family of the International Quality Management Standards ISO 9001 constitutes, in a broader context, a strategic decision of the industry to improve the overall performance for its customers, employees, stakeholders, and society. The adoption and conformance to the requirements of the standard contribute to the genesis of a new quality culture, which provides significant benefits to the organization towards the following areas:

1. Meeting regulatory and statutory requirements;
2. Enhancing customer satisfaction through the delivery of sound products; and
3. Addressing risks and opportunities.

The ISO 9001 quality management system is based on the following management principles [54]:

- Customer focus;
- leadership;
- people engagement;
- process approach;
- improvement;
- evidence-based decision making; and
- relationship management

The recently revised ISO 9001:2015 standard has a process-based approach, using the well-known Deming's Plan-Do-Check-Act (PDCA) cycle and risk-based thinking [54]. The application of the PDCA cycle ensures that the processes, established by the organization, are adequately organized and managed and the necessary resources are provided by the top management. Continuous improvement is also an essential element of the PDCA cycle and the new quality philosophy. Risk-based thinking provides to the organization the necessary approach to establish controls and preventive measures to mitigate risks incurred by the unplanned results of the quality management system, and to obtain maximum benefit from the opportunities that arise.

The PDCA cycle ingredients are described as follows:

1. Plan: Establish the objectives of the system and the resources to achieve the necessary results;
2. Do: Realization of what has been planned;
3. Check: Monitoring and measurement of product and process performance; and
4. Act: Taking the necessary actions to improve the results.

Risk is the effect of uncertainty and, in the present case, it reflects negative results. Risk-based thinking in the industry is a significant driver of failure prevention and quality improvement. The recognition of risks and the establishment of preventive actions to avoid failures and non-conformities increase the reliability of the supplied products and minimize the negative effects of the resulted incidents, creating a significant impact on the progress and welfare of society.

4. Epilogue

In the present short review, the basic fracture mechanisms of mechanical metallic components working in an industrial environment were summarized using fractographic analysis performed by the respective case studies. Although the subject is deemed as quite extensive, this approach was adopted from a technical expert point of view, aiming to provide condensed knowledge and guidance for component failure investigation and root-cause analysis using fracture mode identification.

Furthermore, an insight was offered to selected failure prevention strategies (or frameworks), which, in general, embrace broader aspects of industrial processes, assisting in design, failure prediction, and failure prevention. The highlighted strategies were briefly presented while they concerned various established areas in the field, such as the following:

(1) The emergence of fracture mechanics, which encourages the design of damage tolerant components;

(2) The use of a process approach and failure prevention methodologies (such as FMEA); and

(3) The adoption of a systems approach and the institution of quality management systems (such as ISO 9001:2015), which are based on a Plan-Do-Check-Act cycle and risk based thinking.

Using a presentation from "specific" to "generic", this review places emphasis and provides the necessary inspiration to FA researchers and engineers to seek knowledge following reasonable (logic paths) and determining the "cause-and-effect" relationships, which are the essential ingredients of the failure analysis procedure. As a final note, the basic principles of the failure prevention philosophy constitute an un-separable part of the entire corpus of the continuous quality improvement, which is an effective driver for industrial breakthrough change and innovation.

Funding: This research received no external funding.

Acknowledgments: The author wish to express his gratitude to ELKEME colleagues for the fruitful discussions and ELKEME Management for the encouragement and continuous support.

Conflicts of Interest: The author declares no conflict of interest.

References

1. Pantazopoulos, G.A. A process-based approach in failure analysis. *J. Fail. Anal. Prev.* **2014**, *14*, 551–553. [CrossRef]
2. Wulpi, D.J. *Understanding How Components Fail*, 2nd ed.; ASM International: Materials Park, OH, USA, 2005.
3. Sachs, N.W. *Practical Plant Failure Analysis*; CRC Press: Boca Raton, FL, USA, 2007.
4. González-Velázquez, J.L. *Fractography and Failure Analysis, Structural Integrity 3*; Springer Nature: Cham, Switzerland, 2018.
5. Broek, D. Some contributions of electron fractography to the theory of fracture. *Int. Metall. Rev.* **1974**, *19*, 135–182. [CrossRef]
6. Pantazopoulos, G.A.; Psyllaki, P.P. Progressive failures of components in chemical process industry: Case history investigation and root-cause analysis. In *Handbook of Materials Failure Analysis with Case Studies from Chemical, Concrete and Power Industries*; Makhlouf, A.S.H., Aliofkhazraei, M., Eds.; Elsevier: Oxford, UK, 2016; pp. 1–23.
7. Dennies, D.P. *How to Organize and Run a Failure Investigation*; ASM International: Materials Park, OH, USA, 2005.
8. Lynch, S.P.; Moutsos, S. A brief history of fractography. *J. Fail. Anal. Prev.* **2006**, *6*, 54–69. [CrossRef]
9. Pantazopoulos, G. Damage assessment using fractography as failure evaluation: Applications in industrial metalworking machinery. *J. Fail. Anal. Prev.* **2011**, *11*, 588–594. [CrossRef]
10. Underwood, E.E. *Quantitative Fractography*; Springer Nature: Cham, Switzerland, 1986.
11. Hosford, W.F. *Solid Mechanics*; Cambridge University Press: Cambridge, UK, 2013.
12. Gurson, A.L. Plastic Flow and Fracture Behavior of Ductile Materials Incorporating Void Nucleation, Growth and Interaction. Ph.D. Thesis, Brown University, Providence, RI, USA, 1975.
13. Tvegaard, V.; Needleman, A. Analysis of the cup-cone fracture in a round tensile bar. *Acta Metall.* **1984**, *32*, 157–169. [CrossRef]

14. Pantazopoulos, G.; Toulfatzis, A.; Vazdirvanidis, A.; Rikos, A. Analysis of the degradation process of structural steel component subjected to prolonged thermal exposure. *Metall. Microstruct. Anal.* **2016**, *5*, 149–156. [CrossRef]

15. Hull, D. *Fractography: Observing, Measuring and Interpreting the Fracture Surface Topography*; Cambridge University Press: Cambridge, UK, 1999.

16. Abbasi, S.; Esmailian, M.; Ahangarani, S. Investigation of the microstructure, micro-texture and mechanical properties of the HSLA steel, hot-rolled and quenched at different cooling rates. *Metall. Microstruct. Anal.* **2018**, *7*, 596–607. [CrossRef]

17. Dieter, G.E. *Mechanical Metallurgy*; McGraw Hill: New York, NY, USA, 1988.

18. Pantazopoulos, G.; Vazdirvanidis, A.; Toulfatzis, A.; Rikos, A. Fatigue failure of steel links operating as chain components in a heavy duty draw bench. *Eng. Fail. Anal.* **2009**, *16*, 2440–2449. [CrossRef]

19. Pantazopoulos, G.; Zormalia, S. Analysis of failure mechanism of gripping tool steel component operated in an industrial draw bench. *Eng. Fail. Anal.* **2011**, *18*, 1595–1604. [CrossRef]

20. Pantazopoulos, G. Leaded brass rods C38500 for automatic machining operations. *J. Mater. Eng. Perform.* **2002**, *11*, 402–407. [CrossRef]

21. Pantazopoulos, G. A review of defects and failures in brass rods and related components. *Pract. Fail. Anal.* **2003**, *3*, 14–22. [CrossRef]

22. Pantazopoulos, G.; Vazdirvanidis, A. Failure analysis of a fractured leaded-brass (CuZn39Pb3) extruded hexagonal rod. *J. Fail. Anal. Prev.* **2008**, *8*, 218–222. [CrossRef]

23. Pantazopoulos, G.; Vazdirvanidis, A. Fracture analysis and embrittlement phenomena of machined brass components. *Procedia Struct. Integrity* **2017**, *5*, 476–483. [CrossRef]

24. Lynch, S.P. Failures of structures and components by metal-induced embrittlement. *J. Fail. Anal. Prev.* **2008**, *8*, 259–274. [CrossRef]

25. Toulfatzis, A.; Pantazopoulos, G.; Paipetis, A. Microstructure and properties of lead-free brasses using post-processing heat treatment cycles. *Mater. Sci. Technol.* **2016**, *32*, 1771–1781. [CrossRef]

26. Toulfatzis, A.; Pantazopoulos, G.; Paipetis, A. Fracture mechanics properties and failure mechanisms of environmental-friendly brass alloys under impact, cyclic and monotonic loading conditions. *Eng. Fail. Anal.* **2018**, *90*, 497–517. [CrossRef]

27. Totten, G. Fatigue crack propagation. *Adv. Mater. Processes* **2008**, *5*, 39–41.

28. Pantazopoulos, G.; Zormalia, S.; Vazdirvanidis, A. Investigation of fatigue failure of roll shafts in a tube manufacturing line. *J. Fail. Anal. Prev.* **2010**, *10*, 358–362. [CrossRef]

29. Pantazopoulos, G.; Zormalia, S.; Vazdirvanidis, A. Failure analysis of copper tube in an industrial refrigeration unit: A case history. *Int. J. Struct. Integrity* **2013**, *4*, 55–66. [CrossRef]

30. Jones, D.R.H. *Engineering Materials 3—Materials Failure Analysis*; Pergamon Press: Oxford, UK, 1993.

31. Benac, D.J. Failure avoidance brief: Estimating heater tube life. *J. Fail. Anal. Prev.* **2009**, *9*, 5–7. [CrossRef]

32. Ilman, M.N. Analysis of material degradation mechanism and life assessment of 25Cr-38Ni-Mo-Ti wrought alloy steel (HPM) for cracking tubes in an ethylene plant. *Eng. Fail. Anal.* **2014**, *42*, 100–108. [CrossRef]

33. Quickel, G.; Taske, C.; Rollins, B.; Beavers, J. Failure analysis and remaining life assessment of methanol reformer tubes. *J. Fail. Anal. Prev.* **2009**, *9*, 511–516. [CrossRef]

34. Psyllaki, P.; Pantazopoulos, G.; Lefakis, H. Metallurgical evaluation of creep-failed superheater tubes. *Eng. Fail. Anal.* **2009**, *16*, 1420–1431. [CrossRef]

35. Shlyannikov, V.N.; Tumanov, A.V.; Boychenko, N.V.; Tartygasheva, A.M. Loading history effect on creep-fatigue crack growth in pipe bend. *Int. J. Press. Vessels Pip.* **2016**, *139-140*, 86–95. [CrossRef]

36. Shlyannikov, V.; Tumanov, A.; Boychenko, N. Creep-fatigue crack growth rate assessment using ductility damage model. *Int. J. Fatigue* **2018**, *116*, 448–461. [CrossRef]

37. Zerbst, U.; Klinger, C.; Clegg, R. Fracture mechanics as a tool in failure analysis—Prospects and limitations. *Eng. Fail. Anal.* **2015**, *55*, 376–410. [CrossRef]

38. Murakami, Y.; Endo, M. Effects of defects, inclusions and inhomogeneities on fatigue strength. *Int. J. Fatigue* **1994**, *16*, 163–182. [CrossRef]

39. Janssen, M.; Juidema, M.; Wanhill, R. *Fracture Mechanics*, 2nd ed.; SPON Press: London, UK, 2004.

40. Pantazopoulos, G.; Papaefthymiou, S. Failure and fracture analysis of austenitic stainless steel marine propeller shaft. *J. Fail. Anal. Prev.* **2015**, *15*, 762–767. [CrossRef]

41. ISO 148-1: 2009. *Metallic Materials—Charpy Pendulum Impact Test. Test Method*; ISO: Geneva, Switzerland, 2009.

42. BS 7448-Part 1: 1991. *Fracture Mechanics Tests. Determination of K_{Ic}, Critical CTOD and Critical J Values of Metallic Materials*; BSI: London, UK, 1991.

43. Toulfatzis, A.; Pantazopoulos, G.; Paipetis, A. Fracture behavior and characterization of lead-free brass alloys for machining applications. *J. Mater. Eng. Perform.* **2014**, *23*, 3193–3206. [CrossRef]

44. Taylor, D. *The Theory of Critical Distances: A New Perspective in Fracture Mechanics*; Elsevier: Oxford, UK, 2007.

45. Cicero, S.; Fuentes, J.D.; Procopio, I.; Madrazo, V.; González, P. Critical Distance Default Values for Structural Steels and a Simple Formulation to Estimate the Apparent Fracture Toughness in U-Notched Conditions. *Metals* **2018**, *8*, 871. [CrossRef]

46. Li, M.Q. Strain energy density failure criterion. *Int. J. Solids Struct.* **2001**, *38*, 6997–7003. [CrossRef]

47. Cricrì, G. Cohesive law identification of adhesive layers subject to shear load—An exact inverse solution. *Int. J. Solids Struct.* **2019**, *158*, 150–164. [CrossRef]

48. Berto, F.; Majid, A.; Marsavina, L. Mixed mode fracture. *Theor. Appl. Fract. Mech.* **2017**, *91*, 1. [CrossRef]

49. Perez, N. *Mixed Mode Fracture Mechanics, Fracture Mechanics*; Springer Nature: Cham, Switzerland, 2016; pp. 289–325.

50. Pantazopoulos, G.; Tsinopoulos, G. Process failure modes and effects analysis (PFMEA): A structured approach for quality improvement in metal-forming industry. *J. Fail. Anal. Prev.* **2005**, *5*, 5–10. [CrossRef]

51. Toulfatzis, A.; Pantazopoulos, G.; David, C.; Sagris, D.; Paipetis, A. Machinability of eco-friendly lead-free brass alloys: Cutting-force and surface-roughness optimization. *Metals* **2018**, *8*, 250. [CrossRef]

52. Mariajayaprakash, A.; Senthilvelan, T. Optimizing process parameters of screw conveyor (sugar mill boiler) through Failure Mode and Effect Analysis (FMEA) and Taguchi Method. *J. Fail. Anal. Prev.* **2014**, *14*, 772–783. [CrossRef]

53. Moreno, C.J.; Espejo, E. A performance evaluation of three inference engines as expert systems for failure mode identification in shafts. *Eng. Fail. Anal.* **2015**, *53*, 24–35. [CrossRef]

54. ISO 9001:2015. *Quality Management Systems—Requirements*; CEN—European Committee for Standardization: Brussels, Belgium, 2015.

© 2019 by the author. Licensee MDPI, Basel, Switzerland. This article is an open access article distributed under the terms and conditions of the Creative Commons Attribution (CC BY) license (http://creativecommons.org/licenses/by/4.0/).

Article

Numerical Computation of Material Properties of Nanocrystalline Materials Utilizing Three-Dimensional Voronoi Models

Panagiotis Bazios, Konstantinos Tserpes * and Spiros Pantelakis

Laboratory of Technology & Strength of Materials, Department of Mechanical Engineering & Aeronautics, University of Patras, 26500 Patras, Greece; pbazios@upatras.gr (P.B.); pantelak@upatras.gr (S.P.)
* Correspondence: kitserpes@upatras.gr; Tel.: +30-2610-969-498

Received: 15 January 2019; Accepted: 3 February 2019; Published: 8 February 2019

Abstract: Nanocrystalline metals have been the cause of substantial intrigue over the past two decades due to their high strength, which is highly sensitive to their microstructure. The aim of the present project is to develop a finite element two-phase model that is able to predict the elastic moduli and the yield strength of nanostructured material as functions of their microstructure. The numerical methodology uses representative volume elements (RVEs) in which the material microstructure, i.e., the grains and grain boundaries, is presented utilizing the three-dimensional (3D) Voronoi algorithm. The implementation of the 3D Voronoi particles was performed on the nanostructure investigation of ultrafine materials by SEM and TEM. Proper material properties for the grain interiors (GI) and grain boundaries (GB) were computed using the Hall-Petch equation and a dislocation-based analytical approach, respectively. The numerical outcomes show that the Young's Modulus of nanostructured copper increased by increasing the crystallite volume fraction, while the yield strength increased by decreasing the grain size. The numerical predictions were strongly confirmed in opposition to finite element outcomes, experimental results from the open literature, and predictions from the rule of mixtures and the Mori-Tanaka analytical models.

Keywords: finite element modeling; nanocrystalline materials; elastic moduli; yield strength

---

1. Introduction

Nanostructured (NS) metals have developed significant intrigue over the past twenty years due to their distinctive material properties [1]. Among the most investigated feature of the NS materials is their mechanical behavior [2]. It is generally well-known that along with their microcrystalline materials, NS materials are characterized by enhanced values in terms of yield stress, ultimate tensile strength (UTS), and hardness, while their plastic behavior and fracture toughness usually seem to have decreased values [3]. Due to several issues demonstrated throughout their production (for example, the thermodynamic instability because of the Gibbs free energy monotonic reduction), it is evidently accepted that experimental test campaigns for the investigation of mechanical properties of NS materials is a very tough task due to the inadequacy associated with the research community in providing sufficient quantities of material with regard to test specimens. Nonetheless, the aforementioned production obstacles are also the driving force for the implementation of numerical models suitable for associating the material microstructure (grain size, volume fraction of each phase, etc.) with the mechanical properties of nanostructured materials.

Kim et al. [4] applied the plastic strain of nanostructured metals using constitutive expression based on the dislocation density evolution along with diffusion controlled plastic flow. Estrin et al. [5] introduced a new model with crystal plasticity that incorporated terms with consideration to the fact that double-slip geometry has a diffusional behavior. Wei et al. [6] utilized the cohesive behavior of elements on the grain boundaries (GB) in order to model a GB slip mechanism. The aforementioned elements have progressive equations that utilize the slip and separation effects between adjoining grains, simulating the reversible elastic and inelastic responses. The grain interiors (GI) were defined with the approach of the single crystallite plasticity. Wei et al. [6] utilized a numerical approach to the investigation of nanostructured nickel and compared it with the experimental results of the aforementioned nanocrystalline material. It was shown that dislocation density in nickel GIs after plastic strain presented a declining trend, indicating that annihilation might be accommodated at the GB area. They detected that the non-linear response was mainly because of the elasto-plastic behavior of the GB phase. The mechanism of plasticity of GIs was infrequently detected previous to GB cracking.

Fu et al. [7] utilized, via a numerical approach, a model in which the grain of a nanocrystalline copper consisted of two phases—a GB with a definite thickness and the grain interior (GI). Typically, it has a non-linear relationship with the crystallite diameter. A number of two dimensional (2D) elements of grain sizes varying from 26 to 100 nanometers were subjected to compression. The crystal morphology continued to be constant, thus the GB thickness ratio to the grain diameter was liable for the deformation variations. The aforementioned approach investigated several grain sizes varied by ratios between GB and bulk dimensions.

Four distinctive models with differing numerical approaches and employed GI and GB phases were examined. In the first investigation [8], the GIs referred to one crystallographic orientation related to three different types of hardness orientation. The GIs were modeled isotropic and associated with the indicative stress-strain curve utilizing the experimental outcomes acquired from Diehl [9] and Suzuki et al. [10]. The GBs were modeled with perfectly plastic behavior.

A more recent work [11] used a numerical methodology by applying the separated crystallite plasticity for the GI and the Voce equation in order to simulate the hardening behavior of the GBs. In [11], the approach of separated crystallite plasticity for the GI and GB was used. The grain interior and grain boundary acquired different hardening properties. This numerical approach shows the localization evolution, which happened via the work hardening effect. In [11], another numerical methodology is presented that replaced the separate GI and GB model with a hardening approach that correlated to the evolution of dislocations. The aforementioned phenomenon is controlled by the orientation variation of maximum shear planar stress, which is calculated by the angle of shear stress.

The computations were performed with a multiphase Eulerian analytical implementation [12]. Apart from the highest strain values (which were substantially higher than experimental results), the strains were lower than the Lagrangian computations. The analytical formulas profitably predicted the Hall-Petch effect; however, the fluctuation in the experimental results was noticeable. Considering the aforementioned obstacle, it would be impossible to tune the aforementioned analytical and numerical methodologies.

GB slip was examined by taking into account the restrictive examples of perfect bonding and the ideally frictionless GB slip mechanism [8]. The limitation of ideal GB bonding increased the yield strength compared to ideal slip by 200 MPa at a proportional plastic strain of 0.2, which was close enough for the models to start acting in an almost perfectly plastic way.

It is evident from the above that the study of the material properties of nanocrystalline metals is at early stage—with upward trends—due to enhanced mechanical properties relative to commercial metallic materials and their recyclability relative to composite materials. Furthermore, it is common knowledge that there is currently an inability within the scientific and industrial communities to produce sufficient quantities of nanocrystalline that are thermodynamically stable in order to sustain their superior mechanical properties while also dealing with their low plasticity.

In the current project, a numerical methodology for predicting the mechanical properties of nanocrystalline materials utilizing a sophisticated 3D Finite Element FE model is shown. The scope combines a multidisciplinary study of nanocrystal characterization using SEM-TEM scans and the development of a numerical methodology using a representative volume element (RVE) for predicting the material properties of NS materials. For simplicity's sake, we selected the numerical implementation of cubic volume element (RVE), as it is the smallest volume in which a sampling of all microstructural heterogeneities can be considered for the investigation of macroscopic mechanical responses on the condition that it remains small enough to be considered a volume element of continuum mechanics. Thus, the indirect scope of the present thesis is to extend the scientific effort for the prediction of nanocrystalline mechanical properties from 2D models to three dimensional (3D) models by utilizing a sophisticated 3D Voronoi tessellation algorithm accompanied with a numerical work-flow, which implements only the basic mechanical properties of pure materials by combining the Hall-Petch (H-P) effect. The results of 3D models are more reliable than 2D models due to their realistic microstructure. This way, the FE approach may become a useful tool, and could become the keystone for evaluating the adequacy of nanocrystalline metals by making their implementation in the aerospace sector feasible.

This paper is an extension of a conference proceeding of the same mentioned authors of the existing paper [13]. The conference proceeding was focused on the elastic properties of nanostructured materials. In the present study, the above authors extended their investigation of plastic properties (yield strength) and the general elastoplastic response of nanocrystalline materials under tensile loading conditions.

2. The Analytical Approach

2.1. Rule of Mixtures Approach

The rule of mixtures (RoM) approach is described in its approximate configuration. The idea (Figure 1) regarding the basic approach considers that NS microstructure is a new composite-like material made of two definite phases—the GIs and the GBs. Additionally, an ideal bonding connecting the two different components is considered. In order to make the analysis less complex, triple junctions (TJs) were included in the GBs in relation to the computed volume fraction (VF). This reduction of the RoM phases to the two basic definite phases (GB and GI) should not have a considerable impact on the analytical outcomes for NS metals with a GI diameter higher of 10 nanometers because of the volume.

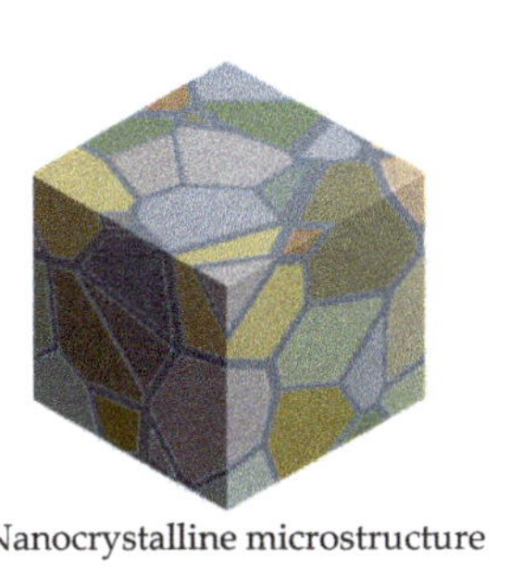

Nanocrystalline microstructure

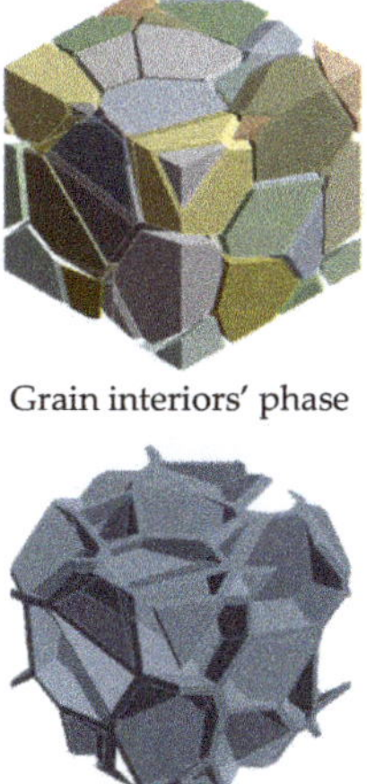

Grain interiors' phase

Grain boundaries' phase

Figure 1. Representation of rule of mixtures (RoM) approach.

Nanocrystalline materials commonly refer to the category of metals whose mean GI diameter is below 100 nanometers. Considering that many atoms settle in the GB areas, the VF of the GB component is not zero. As far as the GI diameter d and GB thickness t are concerned, the VF of the crystallites can be closed by:

$$VF_{GI} = \frac{d^3}{(d+t)^3} \tag{1}$$

as well as that of the GB phase. Furthermore, the GI phase is likewise assumed homogeneous. In a microcrystalline metal, the entire polycrystalline microstructure is filled by the crystallites, and their elastic behavior is simply the overall behavior of these grain interiors. However, for an NS metal at a grain size of 20 nanometers and grain boundary thickness equal to 1 nm, for example, the VF of the GB component is close to 14% and its contribution to the comprehensive elastic response presents some variations.

Figure 1 demonstrates a graphic of the RoM approach in an NS metal. The cubic volume of the NS metal is comprised of GBs and GIs as they were described above. It was indicated in [14,15] that the outcomes of the numerical computation plastic strain in NS metals are not precisely equal to those acquired utilizing the RoM approach. As a consequence, in this project, the subsequent simple RoM methodology based on the VFs of the two phases was utilized for the investigation of deformation response:

$$E = VF_{gi}E_{gi} + VF_{gb}E_{gb} \tag{2}$$

where the indices gi and gb correspond to grain interiors and grain boundary phase, respectively. The stress of the aforementioned definite phases of the NS material is measured utilizing the presumption that the strains of GB and GI phases are equivalent to the macroscopic applied strain.

2.2. Mori-Tanaka Approach

As it is known, the Mori-Tanaka approach belongs to the mean-field homogenization methods (MFH). The objective of mean-field homogenization is to closely and precisely calculate computations of the mean values of the stress and strain fields at the representative volume element level and in each definite component. It is imperative to indicate that the mean-field homogenization method is not able to solve the representative volume element in detail. As a result, it is incapable of computing the stress localization in both of the definite phases.

In the present project, we investigated a composite-like material made out of a matrix material filled with a definite number of inclusions (I) and holding identical material properties, elliptical morphology, and orientation. We implemented subscripts 0 for the matrix phase and 1 for the inclusions component. The VFs in the two definite components reflect this. Typically, the average strain fields over the RVE, the GB phase (matrix), and the GI phase (inclusions) are associated as follows [16]:

$$\langle \varepsilon \rangle_\omega = VF_0 \langle \varepsilon \rangle_\omega + VF_1 \langle \varepsilon \rangle_{\omega 1}. \tag{3}$$

This identity contains any stress or strain field. Any mean-field homogenization method can be described by strain concentration tensors like that of [16]:

$$\langle \varepsilon \rangle_{\omega 1} = B^\varepsilon : \varepsilon_\omega, \langle \varepsilon \rangle_{\omega 1} = A^\varepsilon : \langle \varepsilon \rangle_\omega. \tag{4}$$

The average strains over all grain interiors are identified with the average strains over the grain boundary phase by means of the primary tensor. The average deformation over the whole representative volume element is identified through the secondary tensor. The two deformation concentration tensors do not behave autonomously. In fact, the secondary term can be estimated from the primary term [16]:

$$A^\varepsilon = B^\varepsilon : [VF_1 B^\varepsilon + (1 - VF_1)I]^{-1}. \tag{5}$$

These outcomes are acceptable for any numerical methodology for either component. For any MFH approach described by a deformation concentration tensor, the stiffness of the RVE is:

$$\overline{C} = [VF_1 C_1 : B^\varepsilon + (1 - VF_1)C_0] : [VF_1 B^\varepsilon + (1 - VF_1)I]^{-1}. \tag{6}$$

An infinite part is exposed to linear displacements on its boundary, which corresponds to a uniform remote strain ε. The body consists of a GB phase (matrix) of uniform stiffness C_0 in which there is embedded an ellipsoidal inclusion (I) of uniform stiffness C_1 (Figure 2).

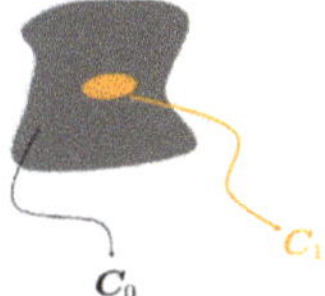

Figure 2. Illustration demonstrating an inclusion embedded in an infinite body.

Applying the solution proposed by Eshelby [17], the aforementioned issue can be figured out in closed form. It was proven that the deformation into the inclusion (I) is uniform and associated with the remote strain as follows:

$$\varepsilon(x) = H^\varepsilon(I, C_0, C_1) : E, \tag{7}$$

where H^ε is the specific (I) strain concentration tensor [17], described as follows:

$$H^\varepsilon(I, C_0, C_1) = \left\{ I + \zeta(I, C_0) : C_0^{-1} : [C_1 - C_0] \right\}^{-1}. \tag{8}$$

One more tensor that performs a noteworthy purpose is Hill's tensor, defined as:

$$P^\varepsilon(I, C_0) = \zeta(I, C_0) : C_0^{-1}. \tag{9}$$

The solution of the specific (I) issue is the cornerstone of commonly known MFH models.

The model of Mori-Tanaka [16] was proposed to solve the aforementioned analytical problem. The derivation was adapted from the implementation of the solution proposed by Eshelby [17]. It was proven that the deformation concentration tensor related to the overall deformation and inclusions to the overall matrix deformation is specified by:

$$B^\varepsilon = H^\varepsilon(I, C_0, C_1), \tag{10}$$

which is definitely the deformation concentration tensor of the single (I) problem. Benveniste [18] used the aforementioned mathematical expression in order to provide the subsequent understanding of the Mori-Tanaka (M-T) methodology. Each of the inclusions in the real representative volume element acted as if it was isolated within the real matrix. The body was infinite and exposed to the mean deformations in the real representative volume element as the far remote field deformation. This is depicted in Figure 3. Typically, the Mori-Tanaka formula is very efficient in predicting the effective material properties of composite-like materials. In principle, it truly is limited to average VF of inclusions (less than 25%), but from a practical point of view, it can provide very satisfactory predictions well beyond this particular range.

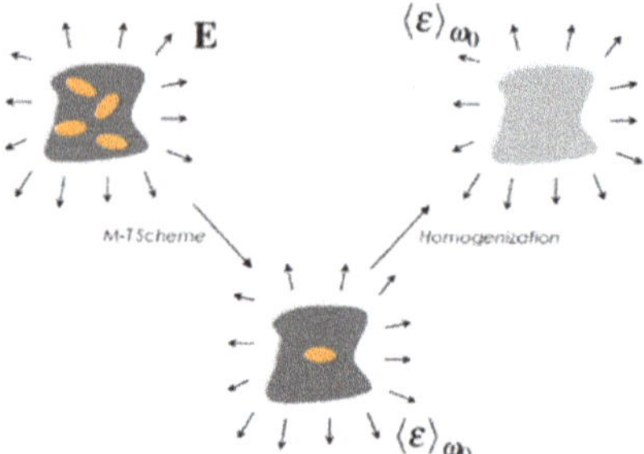

Figure 3. Representation of the Mori-Tanaka (M-T) analytical approach.

3. The Finite Element Approach

Atomistic simulation approaches give notable understanding to the microstructural behavior of nanocrystalline materials. Be that as it may, their use for foreseeing the corresponding material properties is very problematic because of fundamental limitations of the techniques. Hence, the implementation of a finite element methodology for NS materials is focused on providing an important tool for the design-by-analysis of the fundamental NS material morphologic characteristics so as to acquire the required mechanical response.

For the purpose of the tensile simulation of NS material, an FE-based model was utilized. The recommended numerical methodology was utilized in the implementation of RVEs of the NS metal. As is commonly known, RVE is the smallest numerical volume for any material in which a macro-mechanical material property can be evaluated through a multi-scale modeling approach. Due to the microscopic size of RVEs, detailed morphology of a material's microstructural characteristics applying different algorithms can be modeled.

3.1. The Representative Volume Element

The analysis objective was focused on the development of a numerical procedure via the parametric interaction representing the geometrical characteristics of nanostructured metals. The presumptions used in the present work are described extensively here.

Investigation of NS materials SEM/TEM photographs (Figure 4) from [19] showed that NS metals are made of randomly distributed polyhedral-shaped grains. For the purpose of the realistic morphology of NS metals into representative volume element, the microstructure numerical morphology was made utilizing the commonly known Voronoi algorithm. The Voronoi algorithm is a semi-analytical methodology of plane partition to multiple definite regions using the distance between points in an explicit plane subset. The aforementioned set of points is clearly defined beforehand, and for each point, there is a related area made of all points closer to that particular seed than to any other. These areas are named Voronoi cells.

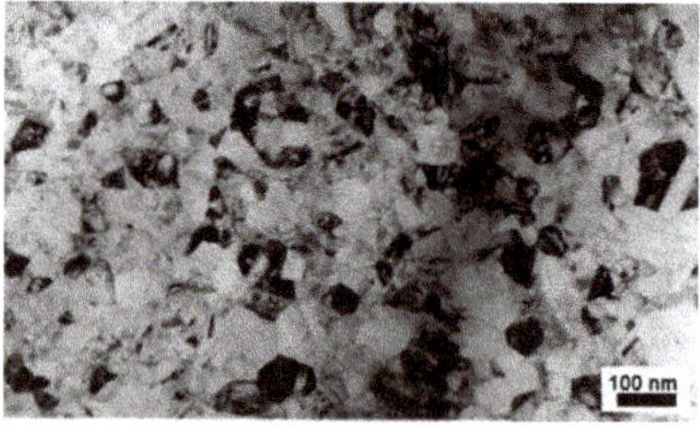

Figure 4. A TEM illustration of the nanocrystalline Copper.

Authors utilized the next assumptions for the numerical generation of the polycrystalline aggregate:

1. Crystallite generation starts at all points (p_i) in a finite set of nuclei (point seed) at the same time (T_0). The nuclei are fixed at their randomly spatial positions during the growth process; as a result, they are not able to move.
2. The velocity of the crystallite generation (following the parabolic plane curve) is assumed to be equal in all grains and all directions leading to the isotropic and uniform grain creation.
3. Crystallite growth in a direction stops as two grain boundaries contact each other. The grain overlapping is not allowed. The growth process stops when there is no further grain growth in any direction in any grain. The entire volume has been filled with grains; as a result, there are no voids.

The Voronoi method further takes for granted that there may not be more than one nucleus within each sphere of radius $r_{\min}$ (Figure 5). This is equivalent to requiring:

$$\|p_i - p_j\|_2 > r_{\min} \quad \forall\, p_i, p_j \in S: p_i \,/= p_j. \tag{11}$$

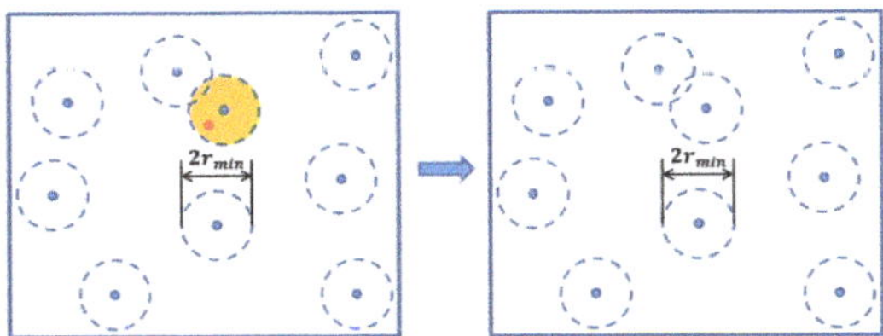

Figure 5. Implementation of point seeders.

Regarding the issue of periodicity for the numerical microstructure, Voronoi tessellation models can be periodized using their initial periodic geometry as input data. Utilizing the cell multiplication function, the final unit cell can be constructed as is illustrated in the 2D plane in Figure 6.

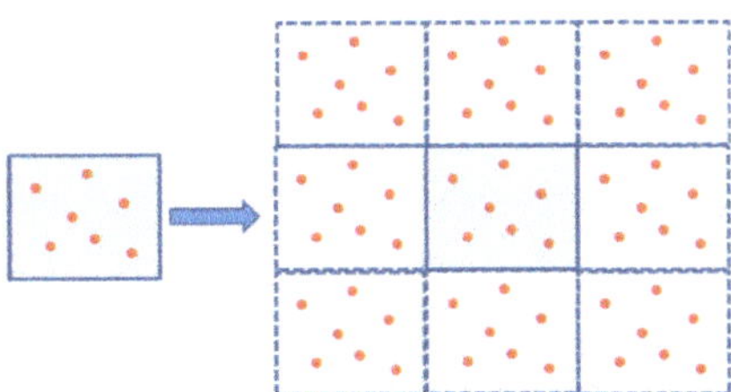

Figure 6. Creation of a periodic set of Voronoi vertices for implication of a periodic microstructure.

A detailed step-by-step procedure of the proposed meshing algorithm is shown below.

1. Create point seed S.
2. Compute Voronoi tessellation.
3. Intersect Voronoi cells with unit cell (generate input geometry).
4. Separate geometric master and slave entities.
5. Create surface mesh with constraint boundary nodes (2D meshing).
6. Surface periodic mesh: copy mesh from master to slave surfaces.
7. Create volume mesh with constraint boundary nodes (3D meshing).
8. Refine volume mesh with constraint boundary nodes.

9.   Create input file for ANSYS APDL computation (scripting).

The recommended meshing procedure was used with the DIGIMAT$^{©}$ software, which generated an ultra-fine meshed model. After the mesh generation, the model was imported to ANSYS APDL using its superior computational performance.

Initially, the point seed was generated by a Poisson process. N indicates the number of nuclei that performed in RVE. In a Poisson process, each point p_i ($i = 1, \ldots N$) is given an x-, y-, and z-coordinate through random variables X_i, Y_i, Z_i, which are uniformly distributed around the origin at $(0,0,0)$ with:

$$X_i \approx \mathcal{U}\left(\left[-\frac{w}{2}, \frac{w}{2}\right]\right), \; Y_i \approx \mathcal{U}\left(\left[-\frac{d}{2}, \frac{d}{2}\right]\right), \; Z_i \approx \mathcal{U}\left(\left[-\frac{h}{2}, \frac{h}{2}\right]\right) \tag{12}$$

where w, d and h are the width, depth, and height of the cuboid volume element, respectively. For the Voronoi tessellation, all points p_j with:

$$\|p_i - p_j\|_2 \leq r_{\min} \quad (j > i) \tag{13}$$

are removed from the point seed, and new points are added. The aforementioned procedure was repeated until the number of points satisfying the prescribed radius was obtained. The computation of Voronoi tessellation consisted of a set of vertices (P) and a list of corner vertices for each Voronoi cell. Therefore, the determination of the intersection of the Voronoi cells with the bounding box was necessary. In this work, we applied the fast and robust algorithm described in Figure 7 for the identification of such an interior point.

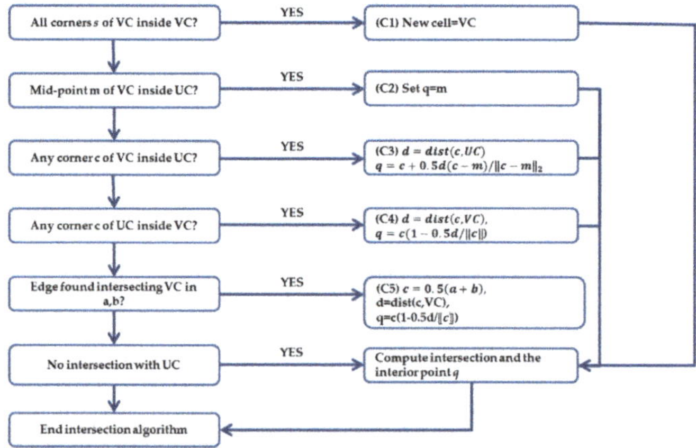

Figure 7. Algorithm for intersecting a Voronoi cell (VC) and the unit cell (UC).

In each RVE, a comprehensive 3D model of the GI and GB phases as randomly-distributed sub-volumes was developed (Figure 8). As referenced above, the VF of GI and GB phases play a significant role on the calculated mechanical response of the RVE. These VFs could be parametrically defined in the NS model.

The thickness (t) of GBs was equal for nanocrystalline NC metals and coarse-grained metals [20], and it was kept at $t = 1$ nm in the numerical case studies shown below. The difference from the VF of GBs was important only in the case of NS metals. For the simulations, the VF of the grain interior phase was defined by Equation (1).

The representative volume element was meshed utilizing tetrahedral finite elements (Figure 9). Appropriate mechanical properties at each component were given, and the representative volume elements were subjected to representative loading conditions.

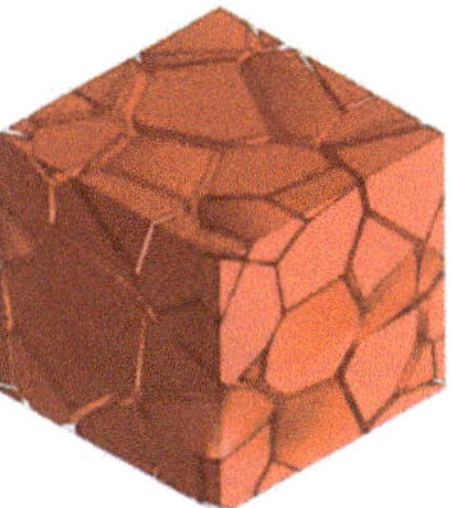

Figure 8. Voronoi unit cell.

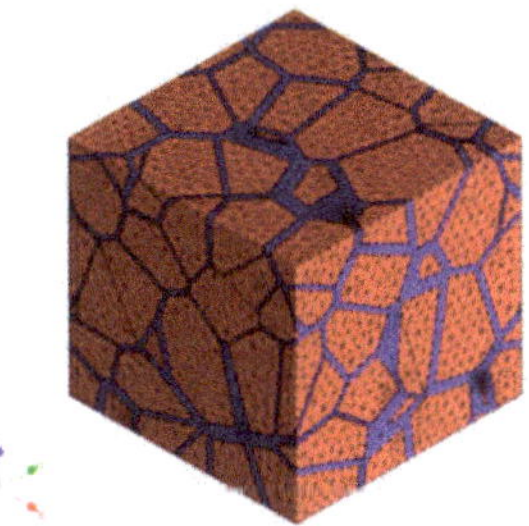

Figure 9. Meshed Voronoi unit cell.

The material properties employed for the numerical investigation of the mechanical response of the nanocrystalline copper were established through the open literature review [21–23] of bilinear stress-strain curves, defined by values of yield stresses. The aforementioned material properties presumed in the simulations are presented in Table 1.

Table 1. Material properties used in the simulations.

Phases	Young Modulus [GPa]	Yield Stress (MPa)		Poisson Ratio	Work Hardening Coefficient (MPa)
Grains	120	$\sigma_{10\,nm} = 1385$ $\sigma_{20\,nm} = 988$ $\sigma_{40\,nm} = 708$	$\sigma_{60\,nm} = 584$ $\sigma_{80\,nm} = 510$ $\sigma_{100\,nm} = 460$	0.336	315
Grain Boundaries	96	-		0.336	315

The value of σ_0 in the H-P equation was 33 MPa, and the value of k_y was 0.135 MPa m$^{1/2}$. The previously mentioned values are demonstrative for copper. The nanocrystalline copper with a grain size of about 10 nanometers led to the identical value of macroscopic flow stress.

$$\sigma_{gi} = \sigma_0 + k_y(d)^{1/2} \tag{14}$$

The Young's modulus of elasticity for the GB component, E_{gb}, was 20% lower than the value for the GI, E_{GI}. The supposition of the reduced value of the GB elastic constant was based on the outcomes of ab initio calculations founded in [8] and experimental findings reported in [9]. The yield stress of the GB phase was constantly equal to 861 MPa due to the pile-up breakdown effect. The cause behind the aforementioned effect relies on the Hall–Petch phenomenon. As the GI diameter diminished, the number of dislocations accumulated contrary to a GB phase reduced because this number was a relationship between the applied stress and the distance to the source. On the contrary, an upward stress value was required to come up with the identical number of dislocations at the pile-up. At an essential crystallite size, we could no longer utilize the approach of a pile-up mechanism in order to define the plastic flow. As the GI size decreased to the NC regime, the number of dislocations at the pile-up decreased to one after some time (see Figure 10). For even more decreased values of the GB yield stresses, total plastic strain was accommodated at the GB phase, and these stresses defined the comprehensive yield stress of the nanocrystalline aggregate. Wadsworth and Nieh and Wadsworth [24] foretold the critical GI size in several metals in which the H–P effect would collapse by considering that there is a critical value at which crystallite in a nanocrystalline specimen can no longer be adequate to hold more than one dislocation.

$$\sigma_{gb} = \frac{G * b}{d_{cr} * \pi(1 - v)} \tag{15}$$

where G, b, d_{cr}, and v are the shear modulus, the Burger's vector, the critical grain size, and Poisson's ratio, respectively.

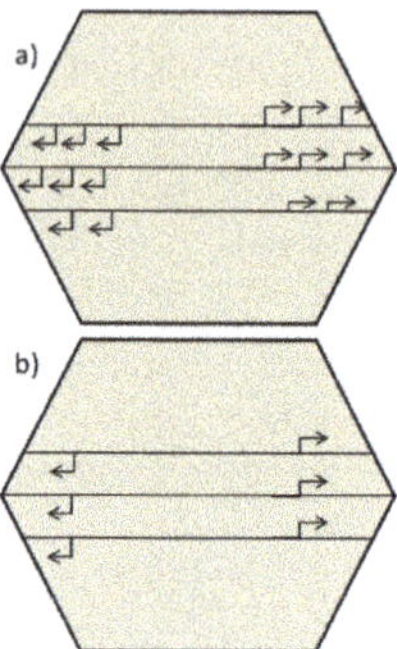

Figure 10. Breakdown of pile up mechanism: (**a**) microcrystalline level and (**b**) nanocrystalline level.

It was assumed that grain boundaries and grain interiors exhibit the elastic–plastic properties described by the following relationships:

$$\sigma_{cr/gb} = E_{cr/gb}\varepsilon \; for \; \varepsilon \leq \varepsilon_0 \tag{16}$$

$$\sigma_{cr/gb} = E_{cr/gb}\varepsilon_0 + \theta_{cr/gb}(\varepsilon - \varepsilon_0) \; for \; \varepsilon > \varepsilon_0 \tag{17}$$

where e_0 is the elastic strain at the yield point, E is the Young's modulus, and h is the work-hardening coefficient. The plasticity of grain boundaries was approached based on the Hill criterion [25], which (in general) accounts for differences in the yield strength in orthogonal directions. The equivalent stress can be described as:

$$\sigma_{gb} = \left(\frac{1}{2}\{\sigma\}^T [M]\{\sigma\} - \frac{1}{3}\{\sigma\}^T \{L\}\right)^{\frac{1}{2}} \tag{18}$$

where $[M]$ is a matrix of yield stress variations with orientation and $\{L\}$ describes the difference between tension and compression of yield strengths:

$$[M] = \begin{bmatrix} M_{11} & M_{12} & M_{13} & 0 & 0 & 0 \\ M_{12} & M_{22} & M_{23} & 0 & 0 & 0 \\ M_{13} & M_{23} & M_{33} & 0 & 0 & 0 \\ 0 & 0 & 0 & M_{44} & 0 & 0 \\ 0 & 0 & 0 & 0 & M_{55} & 0 \\ 0 & 0 & 0 & 0 & 0 & M_{66} \end{bmatrix} \; where \; M_{jj} = \frac{K}{\sigma_{+j}\sigma_{-j}}, \; j = 1\ldots6 \tag{19}$$

$$\{L\} = [L_1 \; L_2 \; L_3 \; 0 \; 0 \; 0], \; L_j = M_{jj}(\sigma_{+j} - \sigma_{-j}), \; j = 1\ldots3$$

where σ_{+j} and σ_{-j} are tensile and compressive yield strengths in the direction j, $j = x, y, z, xy, yz, xz$.

Numerical analyses were executed utilizing ANSYS software with SOLID 185 type of elements. Such 3D elements allow non-linear elastoplastic, anisotropic features using large deformations. In the analyses, it was considered that the numerical model was loaded under tensile forces. The force was conducted by the surface-boundary of the RVE in order to compute the macroscopic stress, σ_m.

Furthermore, the consequential displacement at the surface of the RVE was conducted by its length in the vertical direction to acquire the macroscopic deformation, e_m. In a numerical analysis under uniaxial tensile condition, boundary conditions (BC) utilized for the RVE are shown in Figure 11.

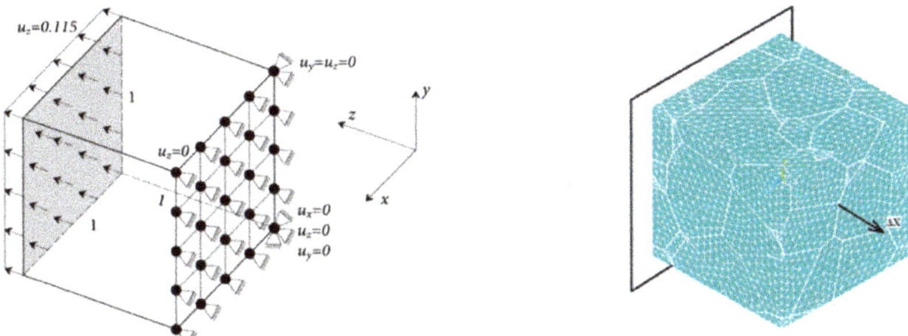

Figure 11. Applied boundary conditions (BCs) in the tensile numerical analysis.

The Young's Moduli of Elasticity, Shear Moduli, and yield strength of the nanocrystalline metal can be numerically predicted. This led us to pass over the commonly known fabrication issue of nanocrystalline materials at sufficient quantities in order to execute a considerable mechanical test campaign. Considering the aforementioned proposed numerical methodology, a reduced number of specimens for experimental tests was required for validation purposes. The implemented numerical analysis can provide the necessary tools for creating the essential nanocrystalline microstructure based on the appropriate mechanical properties.

3.2. Validation of the Proposed Numerical Approach with Experimental Tensile Test

Utilizing the Finite Element model, the overall mechanical response of the RVE was simulated for the case study of nanocrystalline copper with GI size and GB thickness equal to 54 nm and 1 nm, respectively. Figure 12 shows compares the indicative stress-strain response derived from the RVE with the experimental curve obtained from the tension test on nanocrystalline copper [26] with a grain size distribution shown in Figure 13.

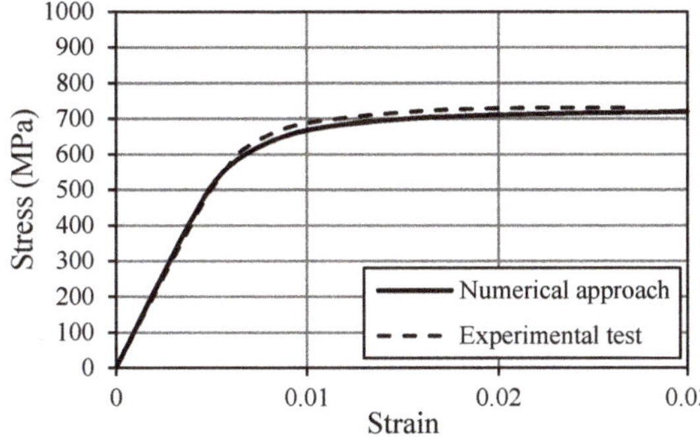

Figure 12. Comparison of the experimental and numerical outcomes of nanocrystalline copper with a grain size of 54 nm under tensile testing.

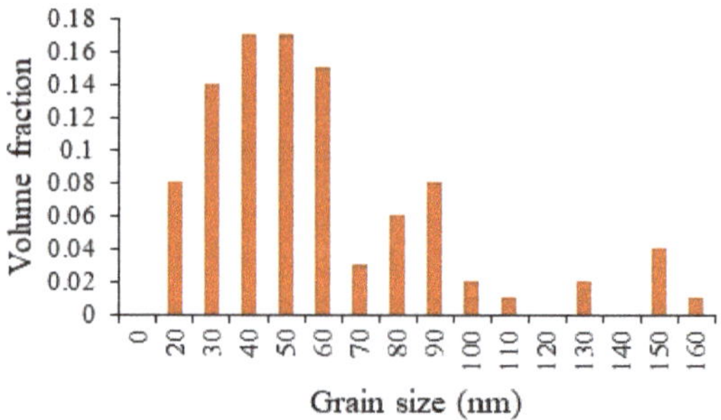

Figure 13. Grain size distribution of nanocrystalline copper used in the experimental test.

3.3. Numerical Parametric Study on Mechanical Behavior of Nanocrystalline Copper

3D Voronoi inclusion-reinforced models were randomly generated by applying the Christoffersen formula [10]. In the present work, several numerical cases were executed so as to define the importance of the size of the Voronoi inclusion on the mechanical response of these composite-like metals (NS materials) and the importance of GB thickness.

3.3.1. Young's and Shear Moduli

The numerical outcomes indicated that the importance of the Voronoi inclusion sizes on the effective mechanical behavior was trivial in the elastic regime. Considering the aforementioned fact and utilizing several numerical analyses by varying the Voronoi inclusion sizes, the effective mechanical properties of these composite-like structures were calculated for up to 95% volume fractions.

The outcomes of the proposed finite element methodology were associated with various analytical approaches, such as M-T and RoM. A few numerical analyses were also executed so as to define the importance of the GB thickness on the effective mechanical properties of nanocrystalline materials.

Several case studies were utilized in order to define the effect of the Voronoi particle sizes on effective mechanical properties of these nanocrystalline metals in each of the three material directions. Figure 14 depicts the effect of the Voronoi inclusion size on the mechanical properties. In our numerical case, we retained the GB thickness constant by varying the particle size. Effective mechanical properties were calculated at the range of 83–95% volume fraction. From Figure 14, it can be noted that there were inconsequential fluctuations on the effective mechanical properties by changing the particle size.

Figure 14 also presents the correlation between the calculated mechanical properties of the finite element technique and M-T approaches using the presumption that inclusions were modeled like spherical inclusions. Moreover, the outcomes of RoM approaches appeared with the previously mentioned examination; however, this analytical approach did not investigate the shape of the RVE inclusions. From the aforementioned cases, it can be noted that the mechanical response of the Voronoi two-component materials relied only on the volume fraction of the Voronoi inclusions. The particle sizes had a minor effect on the effective material properties in the elastic regime. The aforementioned FE results and analytical outcomes are in compliance with the FE outcomes of [21], which are referenced in Figure 14 as "Reference Paper". In addition, Figure 15 depicts the indicative strain distribution in RVE, which was accommodated in the grain boundary phase. The strain distribution accumulated at the intergranular area because the boundary phase possessed a lower value of yield strength. In case of a numerical investigation of a more brittle material with trans-granular fracture, the grains (Voronoi-shaped inclusions) will be modeled with a higher value of yield strength than those of the grain boundaries phase, strictly following the aforementioned analytical expressions of Hall-Petch for the grain phase and the Wadsworth and Nieh expression for the grain boundary phase.

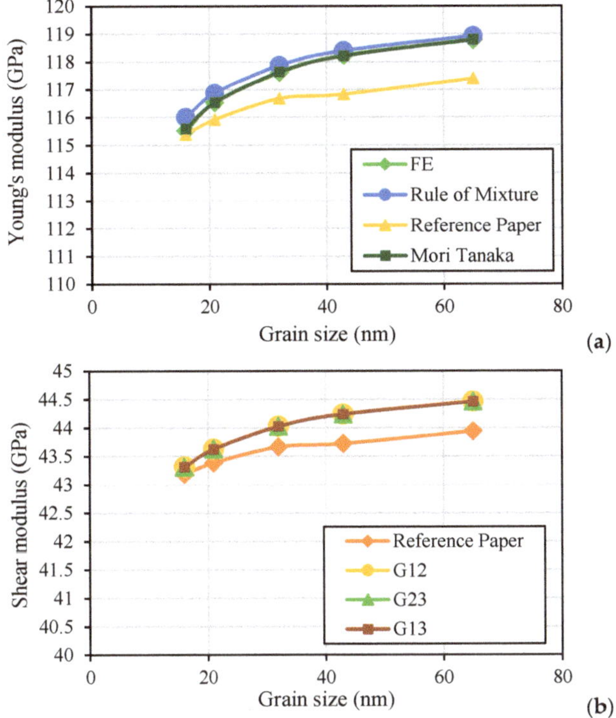

Figure 14. Comparison of FE and analytical outcomes of nanostructured materials for (**a**) Young's modulus of elasticity and (**b**) Shear modulus in each of three material directions, applying a constant value of 1 nm for the GB thickness.

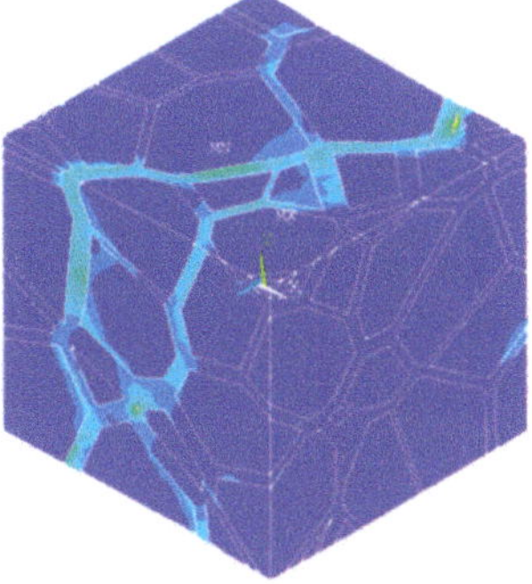

Figure 15. Illustration of strain distribution into representative volume element (RVE) subjected to uniaxial tensile loading.

3.3.2. Yield Strength

According to the investigation of the effective yield stress, it should be noted that this stress drops at the GB phase, which is more deformable than the GI phase. Thus, plastic deformation is ideally absorbed at the GB phase.

Taking into consideration one crystallite with its adjacent GB phase, the mean yield stress for an RVE can be calculated depending on the crystallite size (d), the yield stress of the crystallite (σ_i), the GB definite thickness (t), and its yield stress (σ_{gb}).

By varying d, we change the value of σ_{cr}; despite this, the other related material parameters, t (grain boundary thickness) and σ_{gb}, continue being constant. The proposed parameter definition leads to the crystallite size dependency of macroscopic stress, as discussed extensively below.

The indicative macroscopic stress–strain curves calculated from various mean grain sizes are shown in Figure 16. With regard to the above mentioned, these curves precisely revealed familiar properties of continuously yielding metals. The numerical results of parametric study are summarized in the Table 2. Furthermore, Figure 17 shows the yield strength in terms of the grain size dependency, taking into account the stress-strain curves of Figure 16. The values of yield stresses were significantly lower for crystallites following the grain size effect due to the smaller value of k in the Hall-Petch mathematical expression. A convincing relationship of the crystallite diameter was also shown, as expected for crystallite diameters varying from 10 to 60 nanometers. Outcomes received from the numerical analyses are presented in Figure 17 in the form of H-P plots. In the case of $m = 1/2$, the calculated values of the yield stress for mean crystallite size >20 nm depended linearly on $d^{1/2}$. Moreover, this was not a simple effect of the $d^{1/2}$ parameter of the H-P expression governing yield stress of single crystallites. The computed k value for nanocrystalline aggregates was smaller than the one presumed for crystallites ($k = 0.135$ MPa m^1/2). Outcomes shown in this numerical methodology clearly depicted the influence of crystallite size on the yield stress of nanocrystalline metals.

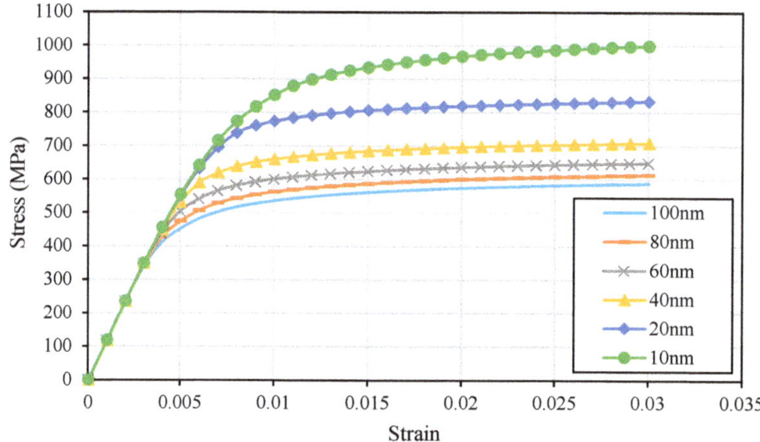

Figure 16. The numerical outcomes of nanocrystalline NC copper for the investigation of the elastoplastic behavior by varying the grain size from 100 nm to 10 nm.

Table 2. Numerical findings of parametric study.

Grain Size (nm)	Young's Modulus (GPa)	Yield Strength (MPa)	Ultimate Tensile Strength (MPa)
100	119.6	390	580
80	119.2	455	612
60	118.8	503	649
40	118.1	588	709
20	116.5	694	832
10	115.8	816	998

It should be noted that the model utilized in the current numerical investigation precisely estimated the mechanical properties of NS metals. Specifically, the crystallite size phenomenon and the Hall–Petch effect were clearly evident, which were results of the macroscopic deformation being preferentially aggregated at the GB phase.

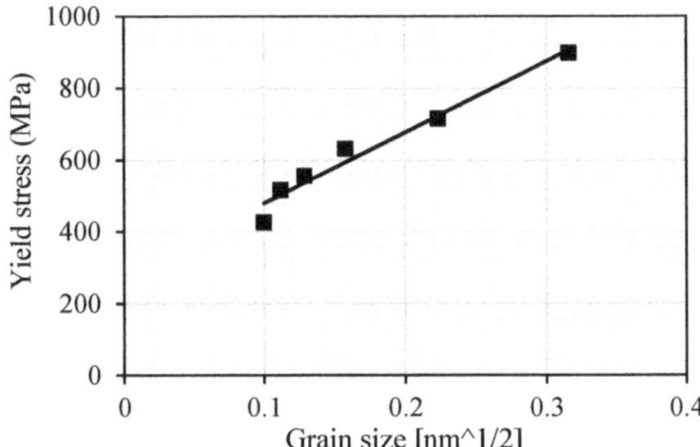

Figure 17. The numerical results of NC copper presenting the crystallite size dependency on the yield strength.

This strain aggregate at grain boundaries is intrinsic to numerical models that consist of two phases assuming a composite-like microstructure of nanocrystalline metals. In view of this methodology, this kind of material can be presented as a system with a high volume fraction of inclusions—more accurately, as an arrangement of particles with strength, which is varied by changing the definite thin GB layer. The reaction of such an arrangement to load in the plastic deformation regime cannot be evaluated by a straightforward averaging of the reaction of its constituents due to the strain partitioning.

Consequently, a presumption of the H-P effect for the yield stress of GI phase does not require the H–P expression be applied to the accumulation. This is shown by the minor presence of critical crystallite size, underneath which one can notice the crystallite size softening.

4. Conclusions

The finite element homogenization methodology was utilized for the investigation of the overall mechanical properties of composite-like materials with Voronoi-reinforced inclusions. The elastic moduli of these composite-like materials were acquired by utilizing these tools in comparison to the outcomes of different analytical approaches. Our finite element predictions were closely associated with the M-T results and RoM estimations. A few case studies were examined in order to define the importance of the Voronoi inclusion size on the overall mechanical properties. The outcomes revealed that the general mechanical properties depend mainly on the volume fraction. There were minor variations in regard to the change in size of the Voronoi particles. This explanation is valid for elastic regime for the estimation of the effective mechanical properties only. There might be some impact on plastic case and damage predictions.

Further case studies need to be done in order to define the impact of the Voronoi particle size on the elastic-plastic response of NS materials at a macro-level. This can be done by investigating different numerical approaches. A generalized numerical methodology was utilized in order to compute various effective material parameters for all required volume fractions using DIGIMAT © and an estimation on the yield strength by changing the grain size on ANSYS APDL. This numerical homogenization approach diminishes the workload and can be utilized as a principle for computing the effective material properties of composite-like materials with randomly distributed Voronoi-shaped reinforced inclusions.

Author Contributions: S.P. and K.T. planned the project, the main conceptual ideas and proof outline, while P.B. worked out the technical details, and performed the numerical analysis for the suggested material.

Funding: This research was funded by ICARUS, grant agreement No. 713514. A European Union's Horizon 2020 research and innovation programme.

Conflicts of Interest: The authors declare no conflict of interest.

References

1. Kumar, K.; Van Swygenhoven, H.; Suresh, S. Mechanical behavior of nanocrystalline metals and alloys. The Golden Jubilee Issue—Selected topics in Materials Science and Engineering: Past, Present and Future, edited by S. Suresh. *Acta Mater.* **2003**, *51*, 5743–5774. [CrossRef]
2. Meyers, M.; Mishra, A.; Benson, D. Mechanical properties of nanocrystalline materials. *Progress Mater. Sci.* **2006**, *51*, 427–556. [CrossRef]
3. Pachla, W.; Kulczyk, M.; Smalc-Koziorowska, J.; Wróblewska, M.; Skiba, J.; Przybysz, S.; Przybysz, M. Mechanical properties and microstructure of ultrafine grained commercial purity aluminium prepared by cryo-hydrostatic extrusion. *Mater. Sci. Eng. A* **2017**, *695*, 178–192. [CrossRef]
4. Kim, H.; Estrin, Y.; Bush, M. Plastic deformation behaviour of fine-grained materials. *Acta Mater.* **2000**, *48*, 493–504. [CrossRef]
5. Estrin, Y.; Gottstein, G.; Shvindlerman, L. Diffusion controlled creep in nanocrystalline materials under grain growth. *Scr. Mater.* **2004**, *50*, 993–997. [CrossRef]
6. Wei, Y.; Anand, L. Grain-boundary sliding and separation in polycrystalline metals: Application to nanocrystalline fcc metals. *J. Mech. Phys. Solids* **2004**, *52*, 2587–2616. [CrossRef]
7. Fu, H.; Benson, D.; Meyers, M. Analytical and computational description of effect of grain size on yield stress of metals. *Acta Mater.* **2001**, *49*, 2567–2582. [CrossRef]
8. Volpp, T.; Göring, E.; Kuschke, W.; Arzt, E. Grain size determination and limits to Hall-Petch behavior in nanocrystalline NiAl powders. *Nanostruct. Mater.* **1997**, *8*, 855–865. [CrossRef]
9. Diehl, J. Verfestigungskurven und Oberflächenerscheinungen. *Z. Met.* **1956**, *47*, 331–343.
10. Suzuki, H.; Ikeda, S.; Takeuchi, S. Deformation of Thin Copper Crystals. *J. Phys. Soc. Jpn.* **1956**, *11*, 382–393. [CrossRef]
11. Fu, H.; Benson, D.; André Meyers, M. Computational description of nanocrystalline deformation based on crystal plasticity. *Acta Mater.* **2004**, *52*, 4413–4425. [CrossRef]
12. Benson, D. Computational methods in Lagrangian and Eulerian hydrocodes. *Comput. Methods Appl. Mech. Eng.* **1992**, *99*, 235–394. [CrossRef]
13. Bazios, P.; Tserpes, K.; Pantelakis, S. Computation of elastic moduli of nanocrystalline materials using Voronoi models of representative volume elements. *MATEC Web Conf.* **2018**, *188*, 02006. [CrossRef]
14. Kim, H.; Estrin, Y.; Bush, M. Constitutive modelling of strength and plasticity of nanocrystalline metallic materials. *Mater. Sci. Eng. A* **2001**, *316*, 195–199. [CrossRef]
15. Kim, H.; Suryanarayana, C.; Kim, S.; Chun, B. Numerical Investigation of Mechanical Behaviour of Nanocrystalline Copper. *Powder Metall.* **1998**, *41*, 217–220. [CrossRef]
16. Mori, T.; Tanaka, K. Average stress in matrix and average elastic energy of materials with misfitting inclusions. *Acta Metall.* **1973**, *21*, 571–574. [CrossRef]
17. Eshelby, J. The determination of the elastic field of an ellipsoidal inclusion, and related problems. *Proc. R. Soc. London. Ser. A Math. Phys. Sci.* **1957**, *241*, 376–396.
18. Benveniste, Y. A new approach to the application of Mori-Tanaka's theory in composite materials. *Mech. Mater.* **1987**, *6*, 147–157. [CrossRef]
19. Hu, J.; Han, S.; Sun, G.; Sun, S.; Jiang, Z.; Wang, G.; Lian, J. Effect of strain rate on tensile properties of electric brush-plated nanocrystalline copper. *Mater. Sci. Eng. A* **2014**, *618*, 621–628. [CrossRef]
20. Van Swygenhoven, H.; Farkas, D.; Caro, A. Grain-boundary structures in polycrystalline metals at the nanoscale. *Phys. Rev. B* **2000**, *62*, 831–838. [CrossRef]
21. Dobosz, R.; Lewandowska, M.; Kurzydlowski, K. The effect of grain size diversity on the flow stress of nanocrystalline metals by finite-element modelling. *Scr. Mater.* **2012**, *67*, 408–411. [CrossRef]
22. Online Materials Information Resource-MatWeb. Available online: http://www.matweb.com/ (accessed on 23 October 2018).

23. Callister, W.; Rethwisch, D. *Materials Science and Engineering*; John Wiley & Sons: Hoboken, NJ, USA, 2012.
24. Nieh, T.; Wadsworth, J. Hall-petch relation in nanocrystalline solids. *Scr. Metall. Mater.* **1991**, *25*, 955–958. [CrossRef]
25. Hill, R. A theory of the yielding and plastic flow of anisotropic metals. *Proc. R. Soc. London. Ser. A Math. Phys. Sci.* **1948**, *193*, 281–297.
26. Cheng, S.; Ma, E.; Wang, Y.; Kecskes, L.; Youssef, K.; Koch, C.; Trociewitz, U.; Han, K. Tensile properties of in situ consolidated nanocrystalline Cu. *Acta Mater.* **2005**, *53*, 1521–1533. [CrossRef]

© 2019 by the authors. Licensee MDPI, Basel, Switzerland. This article is an open access article distributed under the terms and conditions of the Creative Commons Attribution (CC BY) license (http://creativecommons.org/licenses/by/4.0/).

Article

In Situ Observation of the Deformation and Fracture Behaviors of Long-Term Thermally Aged Cast Duplex Stainless Steels

Shilei Li [1],*, Yanli Wang [1] and Xitao Wang [2,3],*

[1] State Key Laboratory for Advanced Metals and Materials, University of Science and Technology Beijing, Beijing 100083, China; wangyl@ustb.edu.cn

[2] Collaborative Innovation Center of Steel Technology, University of Science and Technology Beijing, Beijing 100083, China

[3] Shandong Provincial Key Laboratory for High Strength Lightweight Metallic Materials, Advanced Materials Institute, Qilu University of Technology (Shandong Academy of Science), Jinan 250353, China

* Correspondence: lishilei@ustb.edu.cn (S.L.); xtwang@ustb.edu.cn (X.W.); Tel.: +86-135-5272-7287 (S.L.); +86-139-1029-7623 (X.W.)

Received: 31 January 2019; Accepted: 15 February 2019; Published: 21 February 2019

Abstract: Cast duplex stainless steel (CDSS) components suffer embrittlement after long-term thermal aging. The deformation and fracture behaviors of un-aged and thermally aged (at 400 °C for 20,000 h) CDSS were investigated using in situ scanning electron microscopy (SEM). The tensile strength of CDSS had a small increase, and the tensile fracture changed from ductile to brittle after thermal aging. Observations using in situ SEM indicated that the initial cracks appeared in the ferrite perpendicular to the loading direction after the macroscopic stress exceeded a critical value. The premature fracture of ferrite grains caused stress on the phase boundaries, leading the cracks to grow into austenite. The cleavage fracture of ferrite accelerated the shearing of austenite and reduced the plasticity of the thermally aged CDSS.

Keywords: cast duplex stainless steels; thermal aging; tensile deformation; spinodal decomposition

---

1. Introduction

Cast duplex stainless steel (CDSS), widely used in pressure water reactors (PWRs) as the primary circuit piping and the reactor coolant pump casing, is sensitive to thermal aging embrittlement after long-term service [1–6]. This embrittlement causes a degradation in the mechanical properties of CDSS, such as impact toughness, tensile properties, and fatigue properties [7–13]. Spinodal decomposition in ferrite is considered to be the primary mechanism of thermal aging embrittlement, which has been widely investigated by transition electron microscopy (TEM) [3,7,8,12–15] and atom probe tomography (APT) [2,7,14,16–18].

While it has not resulted in any reported problems, thermal aging embrittlement of CDSS components may become an issue when the service lifetimes of PWRs extend to 60 years or even 80 years. It is customary to simulate metallurgical reactions by accelerated aging at or near 400 °C because realistic aging of a component for end-of-life or life-extension conditions at service temperature cannot be produced [1]. The mechanisms of thermal aging embrittlement have been confirmed to be identical to accelerated aging and reactor operating conditions according to previous studies [2,4,11].

Thermal aging causes severe hardening in ferrite [3,7,18], leading to deformation heterogeneity between ferrite and austenite in the aged CDSS. As a result, the stress and strain repartition of the two phases may play an important role in the fracture process [19]. The hardened ferrite and stress phase boundary may act as the site of crack initiation. Studies on local approaches to fracture have been conducted to explore the fracture in duplex stainless steels [20,21]. However, less research has been conducted on in situ observations of the entire process of crack initiation, propagation, and rupture in CDSS.

In this study, an in situ scanning electron microscopy (SEM) was used to observe crack initiation and propagation in long-term thermally aged CDSS during tensile loading. The deformed microstructures near the fracture were observed by TEM and the samples were prepared using a focused ion beam (FIB). We aimed to clarify the deformation and fracture behaviors of the long-term thermally aged CDSS, including crack initiation, propagation, and final rupture.

2. Materials and Methods

The cast CDSS materials being studied were centrifugal casting Z3CN20-09M (similar to CF3) steel cut from the primary coolant water pipe in Daya Bay Nuclear Power Plant. The materials were thermally aged at 400 °C for as long as 20,000 h to simulate the behavior of CDSS after long-term operation at real service temperatures. The aging time at service temperatures, corresponding to the accelerated thermal aging experiment at 400 °C, can be calculated using an Arrhenius extrapolation [4]. In this study, the accelerated thermal aging at 400 °C for 3000 and 20,000 h was equal to real service at 290 °C for 11.0 and 73.2 years, respectively, using an activation energy of 100 kJ·mol^{-1}. The activation energy in Arrhenius's equation is taken from the activation energy for Cr self-diffusion because the spinodal decomposition in ferrite is the primary mechanism of thermal aging embrittlement. Spinodal decomposition occurs at temperatures ranging from 250–500 °C, resulting in the Cr-rich and Cr-depleted domains by Cr segregation [6].

The studied CDSS has the compositions in wt.% of 20.12 Cr, 9.73 Ni, 1.04 Si, 0.96 Mn, 0.14 Mo, 0.033 C, 0.044 N, 0.014 P, 0.0009 S, and balance Fe. The microstructure of the as-cast material was observed using electron backscatter diffraction (EBSD) in an SEM (ZEISS Supra 55, Oberkochen, Germany). The EBSD sample was prepared by electropolishing with a 60 H_3PO_4:25 H_2SO_4:15 H_2O solution.

The mechanical property changes during thermal aging were investigated by tensile tests and hardness tests. The hardness of both ferrite and austenite phases was studied by a Vickers hardness tester (Leica VMHT 30M, Wetzlar, Germany). A load of 25 g was used in the hardness tests to make sure that the indentation sizes were smaller than 10 μm in both phases. In addition, ferrite islands larger than 40 μm were selected in the hardness testing to avoid the soft substrate effect of austenite in the thermally aged specimen. Prior to hardness testing, the samples were ground with 2000-grit SiC papers, then mechanically polished using 1.5-grit diamond pastes, and finally etched by a ferric chloride solution (5g $FeCl_3$ + 100 mL HCl + 100 mL CH_3OH + 100 mL H_2O).

An in situ observation of crack initiation and propagation in the long-term thermally aged CDSS was performed at room temperature in the vacuum chamber of the SEM, equipped with a specially designed servo-hydraulic testing system (SEM-SERVO 550, Shimadzu, Japan). The specimen geometry and the stage are shown in Figure 1. This sample was ground with 2000-grit SiC papers and then electropolished with a 60 H_3PO_4:25 H_2SO_4:15 H_2O solution to distinguish ferrite from the austenite matrix after deformation. The in situ SEM tensile test was performed at room temperature using a normal strain rate of 1×10^{-4} s^{-1}. During tensile loading, images were taken when the crack initiation and propagation appeared. The strains corresponding to these images were obtained by comparing the recording time of the images and the tensile loading data.

After the in situ tensile tests were performed, the fracture surface morphologies of both un-aged and aged samples were observed by SEM. The deformed microstructure of the long-term aged specimen was observed by a field-emission TEM (FEI Tecnai F20-ST, Eindhoven, The Netherlands) operated at 200 kV. The TEM specimen was cut near the tensile fracture and thinned by an FIB-SEM system (Zeiss Auriga, Oberkochen, Germany).

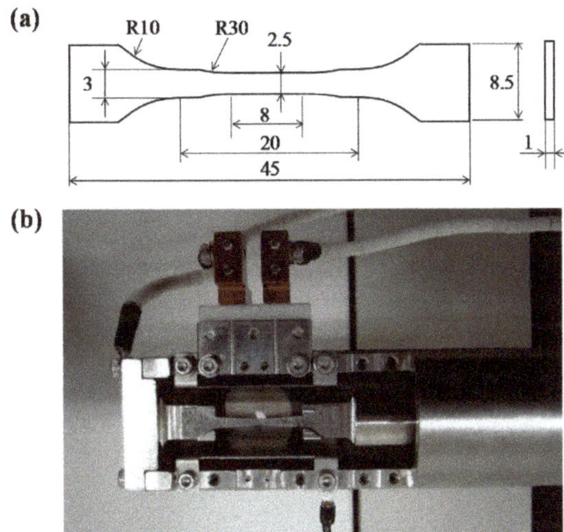

Figure 1. (**a**) The tensile specimen geometry; (**b**) the stage for in situ SEM observation (unit: mm).

3. Results and Discussions

3.1. Microstructure Observation

The EBSD orientation map is shown in Figure 2a, and the insert depicts the orientation of each grain. The phase map, shown in Figure 2b, shows the duplex structure of austenite (blue) and ferrite (red). As a cast material, austenite in the present studied CDSS has a very large grain size of several hundred microns, and the ferrite islands with much smaller sizes distribute within the austenite matrix. The ferrite islands within different austenite grains may have the same orientation, which is marked by the arrows in the same directions in Figure 2a. The volume fraction of ferrite was evaluated as 12% using optical metallography.

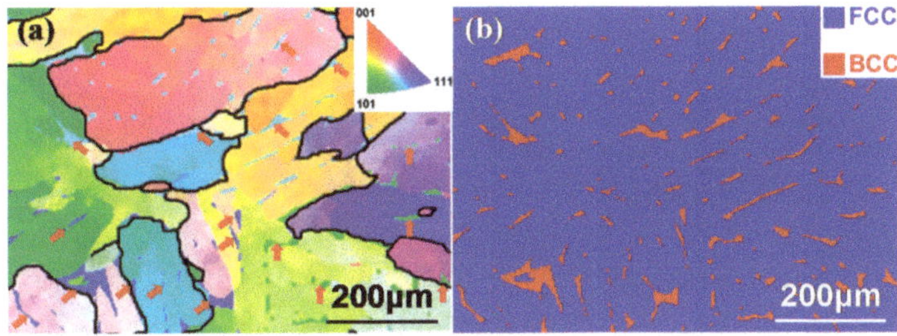

Figure 2. (**a**) Electron backscatter diffraction (EBSD) orientation map, with the insert showing the orientation of each grain; (**b**) phase map of the studied steel.

3.2. Thermal-Aging-Induced Mechanical Property Changes

The tensile properties and hardness in both phases of the un-aged and long-term thermally aged steels are shown in Table 1. Both the yield strength (YS) and ultimate tensile strength (UTS) increased by approximately 15%, and the plasticity decreased by 21% after thermal aging at 400 °C for 20,000 h. Table 1 also lists the Vickers hardness of austenite and ferrite in both the un-aged and aged CDSS. For the un-aged specimen, both the austenitic and ferritic phases had almost the same hardness. After aging at 400 °C for 20,000 h, the hardness in austenite remained unchanged, but ferrite hardness had a drastic increase of approximately 140%. This indicates that the increase in strength and decrease in ductility of CDSS after thermal aging is closely related to the severe hardening in ferrite.

The impact properties, hardness, and tensile properties of the CDSS during thermal aging were studied in our previous study, and we found that the impact energy and the hardness in ferrite significantly changed in the early stage and then became saturated, while the tensile properties had no such dramatic change after thermal aging [3,22]. Although the mechanical properties of the material aged for 3000 h were similar to those of materials aged for 20,000 h, the cleavage facets of the brittle ferrite phases were only observed on the tensile fracture surface of the material thermally aged for 20,000 h, indicating that the behaviors of crack initiation and propagation in the material aged for 20,000 h are quite different to those of the materials aged for less time [22]. To observe the crack initiation and propagation in the long-term thermally aged material during tensile loading, the material aged for 20,000 h was selected to carry out this work using an in situ SEM.

Table 1. Tensile properties and hardness in the un-aged and aged cast duplex stainless steel (CDSS).

Materials	Tensile Properties			Vickers Hardness, HV	
	Yield Strength YS (MPa)	Ultimate Tensile Strength UTS (MPa)	Elongation (%)	Austenite	Ferrite
Un-aged	237.4	523.5	54.6	217.1 ± 17.2	230.1 ± 8.9
Aged for 3000 h	288.3	674.9	40.1	211.6 ± 13.5	468.7 ± 20.9
Aged for 20,000 h	230.7	599.8	44.1	201.3 ± 10.8	556.4 ± 71.9

Figure 3a,c shows the macroscopic and microscopic morphologies of the fracture surfaces of the un-aged specimen. The homogeneous distribution of fine dimples is a typical ductile fracture characteristic, indicating that there is no difference in the deformation ability of ferrite and austenite in the un-aged CDSS. In stark contrast, the fracture surfaces of the long-term thermally aged specimen show the mixed features of brittle cleavages in ferrite and ductile shearing in austenite, as shown in Figure 3b,d. As cleavage is the mechanism of brittle trans-granular fracture, and occurs through the cleaving of the crystals along crystallographic planes, the appearance of typical cleavage facets are the flat areas on the fracture surface. Therefore, the brittle cleavage areas can be easily distinguished from the other areas by their flat feathers, as shown in Figure 3d, marked by dashed lines. Hardening in ferrite caused by thermal aging is considered to destroy the deformation compatibility between the two phases in CDSS. The deterioration of plastic deformation capacity during the aging process makes ferrite prone to brittle fracture.

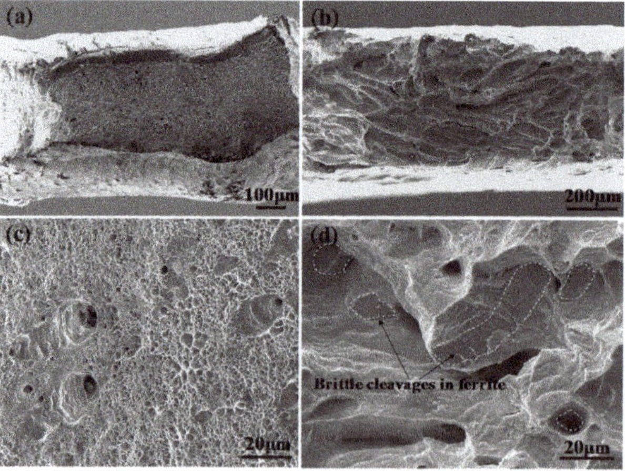

Figure 3. (a,c) Macroscopic and microscopic morphologies of the fracture surface of the un-aged CDSS (b,d) and of the long-term thermally aged CDSS.

3.3. In Situ SEM Observation of Crack Initiation in the Aged CDSS

In situ SEM observation during tensile loading was conducted in the long-term thermally aged CDSS and the microstructural evolution is shown in Figure 4. Some images are selected at specific strains (9%, 14%, 20%, 30%, and rupture, corresponding to B, C, D, E, and F in Figure 4a, respectively) to show the stages of crack initiation, propagation, and final rupture.

At the stage of low strain, no visible plastic deformation features were observed. Both ferrite and austenite appeared to be flat and smooth, and the two phases could not be distinguished from each other. When strain increased to 9%, deformation characteristics could be seen (as in Figure 4b) and the ferrite phases became slightly concave. The first crack (marked as Crack 1) in the tensile process initiated in ferrite and propagated within the ferrite grain along the direction perpendicular to stress, as shown in Figure 4c. Although the ferrite phases become very brittle after long-term thermal aging, the fracture of ferrite underwent a two-step process of crack initiation and propagation. Long-strip ferrite phases could not complete fracture-like particles after the crack initiation. Further increases in strain caused another crack in ferrite (marked as Crack 2), as shown in Figure 4d; Crack 1 and Crack 2 were arranged in parallel. After long-term thermal aging, ferrite became brittle but austenite remained ductile. During tensile loading, the austenite matrix deformed easily and bore most of the plastic deformation. As the ferrite phases became brittle and hard to deform, the phase boundaries between austenite and ferrite accumulated more stress with increasing plastic deformation. When this concentrated stress exceeded the cleavage stress of ferrite, cracks formed in the brittle ferrite. The cracks in ferrite mainly formed in the B-C-D stage, as shown in Figure 4a, and this suggests that the critical cleavage fracture stress of ferrite in the long-term thermally aged CDSS was approximately 400–500 MPa.

After the cracks extended throughout the ferrite grains, the increase in stress pulled the cracks open and caused considerably greater stress on the phase boundaries of ferrite and austenite, as shown in Figure 4e. The preferential propagation path of the primary crack for the thermally aged CDSS linked the prior cracks in ferrite. As seen from the surface morphologies after rupturing in Figure 4f, these prior cracks (e.g., Crack 1 and 2) were very close to the fracture surface.

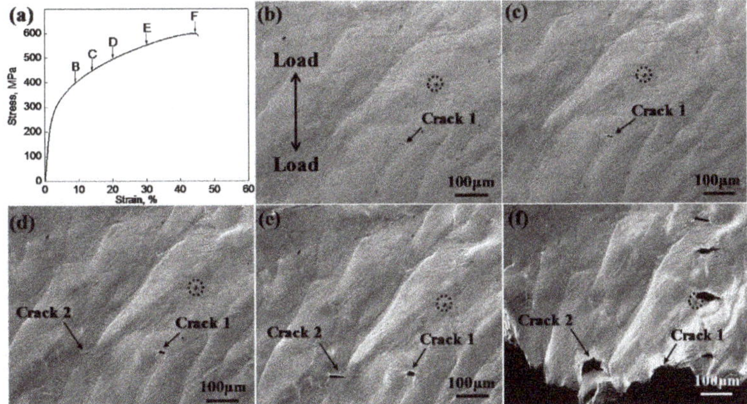

Figure 4. (a) In situ observation at different stages for strains (b) $\varepsilon = 9\%$; (c) 14%; (d) 20%; (e) 30%; and (f) rupture of CDSS thermally aged at 400 °C for 20,000 h. An inclusion is marked by a dashed circle as the position mark.

3.4. TEM Observation of Deformed Microstructures in the Aged CDSS

The transition from ductile to brittle in ferrite of the aged CDSS, as well as its severe hardening, is caused by nanoscale metallurgical reactions, including spinodal decomposition and G-phase precipitation in ferrite, which were systematically characterized by Atom probe tomography (APT) and

TEM in our previous research [3,18,23]. The deformed microstructures of the aged CDSS were observed by TEM, which enabled us to study the interactions between these precipitates and dislocations. The TEM foil of the long-term thermally aged material was prepared by an FIB system from the tensile sample near the fracture. This TEM sample contained both ferrite and austenite, and the SEM and TEM images of the overall morphology are shown in Figure 5a,b. There was no precipitation and only a few dislocations in austenite, as shown in Figure 5c. By contrast, intensive precipitates and dislocations were observed in ferrite. The dislocation-precipitate interactions (Figure 5d) indicate that the dislocation motion is hindered, which makes ferrite in the aged CDSS hard and brittle.

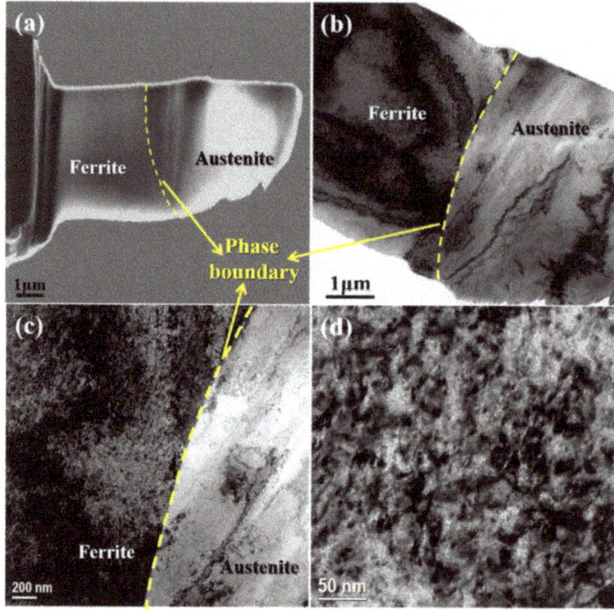

Figure 5. (a) SEM and (b) TEM images of the overall morphology; (c) TEM images of the phase boundary and (d) ferrite near the tensile fracture in the thermally aged CDSS.

The ferrite morphologies in the undeformed and deformed steel that were thermally aged at 400 °C for 20,000 h were observed and compared by HRTEM, as shown in Figure 6. G-phases can be clearly observed in the HRTEM image and the corresponding fast Fourier transform (FFT) patterns, as shown in Figure 6a,b. However, in the ferrite of the deformed sample, G-phases cannot easily be distinguished from Figure 6c,d.

It is well known that ferrite spinodally decomposes into coherent Cr-enriched and Cr-depleted domains after long-term thermal aging [2–4,7–9]. As the modulated structures had very close lattice parameters, showing the continuity of the crystal lattices, as in Figure 6a, they could not be distinguished by TEM or HRTEM. APT has been widely used to characterize the spinodal decomposition in Fe-Cr alloys and ferrite in CDSS, and the average size of the Cr-enriched and Cr-depleted domains was estimated to be approximately 7 nm for CDSS that was thermally aged for 20,000 h [18]. The Young's modulus and shear modulus of pure Cr are higher than those of pure iron, and in the alloys they increase with increasing Cr content [19]. The different moduli between Cr-enriched and Cr-depleted domains caused deformation incompatibility on the nanometer scale, forming many regions with different orientations, as shown in Figure 6c,d.

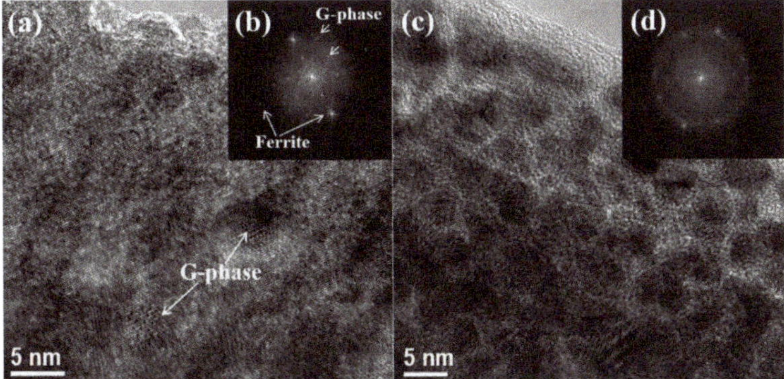

Figure 6. (**a**) HRTEM images of ferrite in the undeformed and (**c**) deformed samples of the steels thermally aged at 400 °C for 20,000 h. (**b**) and (**d**) are the corresponding FFT patterns of (**a**) and (**c**), respectively.

The lattice images of the special direction can be obtained by filtering in the corresponding FFT patterns. The region of the lattice image and its corresponding FFT pattern from the same orientation were marked in the same color, as shown in Figure 7. These regions with different orientations formed under severe deformation in the tensile process, and they had almost the same dimension as the spinodal structures of Cr-enriched and Cr-depleted domains.

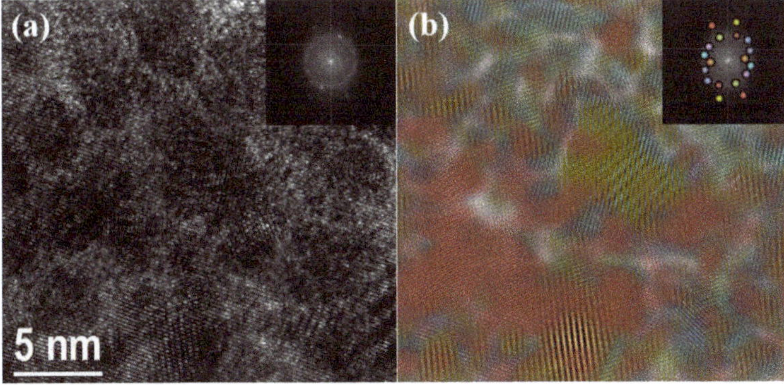

Figure 7. (**a**) HRTEM image and (**b**) the corresponding orientation distribution of ferrite in the deformed steel that was thermally aged at 400 °C for 20,000 h.

3.5. Fracture Mechanism of the Thermally Aged CDSS

After in situ SEM tensile testing, the surface morphologies of the fractured samples were also observed by SEM to investigate the crack propagation process. As shown in Figure 8a, the two phases of ferrite and austenite in the un-aged CDSS had almost the same deformation features, indicating that they had similar deformation ability. Voids initiated at inclusions, marked by solid arrows in Figure 8a, and their growth and coalescence led to ductile fracture. The halves of these voids are represented as dimples on the fracture surface, as shown in Figure 3a.

For the aged CDSS, the in situ SEM results showed the initiation and propagation of cracks within ferrite. Figure 8b shows the further extension of cracks from ferrite to austenite. The stress was released when the crack propagated throughout the ferrite grain. When the propagation of the crack encountered a phase boundary, stress concentrated on this phase boundary (marked by a solid arrow in Figure 8b), and caused the beginning of cracks in austenite. With increasing applied stress, a large number of cracks grew into austenite. When some of these cracks were connected to each other, a primary crack formed, which further caused the rupture of the aged CDSS.

After long-term thermal aging, the ferrite in CDSS decomposed into two kinds of coherent and interconnected regions with different chemical compositions and mechanical properties. Due to the deformation incompatibility of these domains, dislocations in ferrite were severely hampered, resulting in the reduction of plastic deformation ability. When the stress concentration in ferrite rose to a certain level, the initiation and propagation of cracks occurred throughout the ferrite. The local stress was released when the crack propagated throughout the ferrite grain. Further extension of the crack in ferrite was hindered by the phase boundary where the stress was most concentrated, which caused the initiation of cracks in austenite. The prior fracture of ferrite accelerated the shearing of austenite and reduced the plasticity of the thermally aged CDSS.

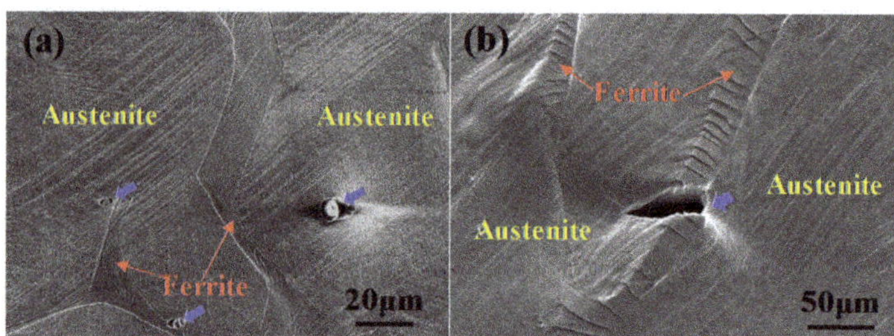

Figure 8. Deformation characteristics of austenite and ferrite phases near the tensile fracture of the (**a**) un-aged and (**b**) aged specimens.

4. Conclusions

The in situ SEM technique was used to investigate the deformation and fracture behavior of long-term thermally aged duplex stainless steel. After thermal aging at 400 °C for 20,000 h, the yield stress and ultimate tensile stress of CDSS exhibited a small increase, while the tensile fracture changed from ductile to brittle. Hardening in ferrite was caused by the spinodal decomposition into coherent and interconnected regions with different chemical compositions and mechanical properties. The deformation incompatibility of Cr-enriched and Cr-depleted domains resulted in the concentration of stress and the initiation and propagation of cracks in ferrite. The initial cracks first propagated throughout the ferrite grains and then generated stress concentrates on the phase boundaries, which caused the initiation of cracks in austenite. The aged CDSS finally ruptured by connecting these pre-existing cracks in ferrite.

Author Contributions: X.W. and Y.W. designed the research project; S.L. performed the experiments, analyzed the data, and wrote the manuscript; all the authors contributed to the discussions and commented on the manuscript.

Funding: This work was financially supported by the National Natural Science Foundation of China (NSFC) (Grant No. 51601013) and the State Key Laboratory for Advanced Metals and Materials (Grant No. 2018Z-25).

Conflicts of Interest: The authors declare no conflict of interest.

References

1. Chung, H.M. Aging and life prediction of cast duplex stainless steel components. *Int. J. Pres. Vessels Pip.* **1992**, *50*, 179–213. [CrossRef]
2. Pareige, C.; Novy, S.; Saillet, S.; Pareige, P. Study of phase transformation and mechanical properties evolution of duplex stainless steels after long term thermal ageing (>20 years). *J. Nucl. Mater.* **2011**, *411*, 90–96. [CrossRef]
3. Li, S.L.; Wang, Y.L.; Zhang, H.L.; Li, S.X.; Zheng, K.; Xue, F.; Wang, X.T. Microstructure evolution and impact fracture behaviors of Z3CN20-09M stainless steels after long-term thermal aging. *J. Nucl. Mater.* **2013**, *433*, 41–49. [CrossRef]

4. Kawaguchi, S.; Sakamoto, N.; Takano, G.; Matsuda, F.; Kikuchi, Y.; Mráz, L. Microstructural changes and fracture behavior of CF8M duplex stainless steels after long-term aging. *Nucl. Eng. Des.* **1997**, *174*, 273–285. [CrossRef]

5. Chen, Y.; Alexandreanu, B.; Chen, W.Y.; Natesan, K.; Li, Z.; Yang, Y.; Rao, A.S. Cracking behavior of thermally aged and irradiated CF-8 cast austenitic stainless steel. *J. Nucl. Mater.* **2015**, *466*, 560–568. [CrossRef]

6. Chandra, K.; Singhal, R.; Kain, V.; Raja, V.S. Low temperature embrittlement of duplex stainless steel: Correlation between mechanical and electrochemical behavior. *Mater. Sci. Eng. A* **2010**, *527*, 3904–3912. [CrossRef]

7. Yamada, T.; Okano, S.; Kuwano, H. Mechanical property and microstructural change by thermal aging of SCS14A cast duplex stainless steel. *J. Nucl. Mater.* **2006**, *350*, 47–55. [CrossRef]

8. Shiao, J.J.; Tsai, C.H.; Kai, J.J.; Huang, J.H. Aging embrittlement and lattice image analysis in a Fe-Cr-Ni duplex stainless steel aged at 400 °C. *J. Nucl. Mater.* **1994**, *217*, 269–278. [CrossRef]

9. Timofeev, B.T.; Nikolaev, Y.K. About the prediction and assessment of thermal embrittlement of Cr–Ni austenitic–ferritic weld metal and castings at the ageing temperatures 260–425 °C. *Int. J. Pres. Vessels Pip.* **1999**, *76*, 849–856. [CrossRef]

10. Miller, M.K.; Russell, K.F. Comparison of the rate of decomposition in Fe-45% Cr, Fe-45% Cr-5% Ni and duplex stainless steels. *Appl. Surf. Sci.* **1996**, *94*, 398–402. [CrossRef]

11. Chung, H.M.; Leax, T.R. Embrittlement of laboratory and reactor aged CF3, CF8, and CF8M duplex stainless steels. *Mater. Sci. Tech. Lond.* **1990**, *6*, 249–262. [CrossRef]

12. Lo, K.H.; Shek, C.H.; Lai, J.K.L. Recent developments in stainless steels. *Mater. Sci. Eng. R* **2009**, *65*, 39–104. [CrossRef]

13. Chandra, K.; Kain, V.; Bhutani, V.; Raja, V.S.; Tewari, R.; Dey, G.K.; Chakravartty, J.K. Low temperature thermal aging of austenitic stainless steel welds: Kinetics and effects on mechanical properties. *Mater. Sci. Eng. A* **2012**, *534*, 163–175. [CrossRef]

14. Takeuchi, T.; Kameda, J.; Nagai, Y.; Toyama, T.; Matsukawa, Y.; Nishiyama, Y.; Onizawa, K. Microstructural changes of a thermally aged stainless steel submerged arc weld overlay cladding of nuclear reactor pressure vessels. *J. Nucl. Mater.* **2012**, *425*, 60–64. [CrossRef]

15. Li, S.L.; Zhang, H.L.; Wang, Y.L.; Li, S.X.; Zheng, K.; Xue, F.; Wang, X.T. Annealing induced recovery of long-term thermal aging embrittlement in a duplex stainless steel. *Mater. Sci. Eng. A* **2013**, *564*, 85–91. [CrossRef]

16. Danoix, F.; Auger, P. Atom probe studies of the Fe-Cr system and stainless steels aged at intermediate temperature: A review. *Mater. Charact.* **2000**, *44*, 177–201. [CrossRef]

17. Danoix, F.; Auger, P.; Blavette, D. Hardening of aged duplex stainless steels by spinodal decomposition. *Microsc. Microanal.* **2004**, *10*, 349–354. [CrossRef]

18. Zhang, B.; Xue, F.; Li, S.L.; Wang, X.T.; Liang, N.N.; Zhao, Y.H.; Sha, G. Non-uniform phase separation in ferrite of a duplex stainless steel. *Acta Mater.* **2017**, *140*, 388–397. [CrossRef]

19. Jia, N.; Peng, R.L.; Brown, D.W.; Clausen, B.; Wang, Y.D. Tensile deformation behavior of duplex stainless steel studied by In-Situ Time-of-Flight neutron diffraction. *Metall. Mater. Trans. A* **2008**, *39*, 3134–3140. [CrossRef]

20. Hazarabedian, A.; Marini, B. Quantification of damage progression in a thermally aged duplex stainless steel. *Mater. Res.* **2002**, *5*, 113–117. [CrossRef]

21. Hazarabedian, A.; Forget, P.; Marini, B. Local approach to fracture of an aged duplex stainless steel. *Mater. Res.* **2002**, *5*, 131–135. [CrossRef]

22. Li, S.; Wang, Y.; Li, S.; Zhang, H.; Xue, F.; Wang, X. Microstructures and mechanical properties of cast austenite stainless steels after long-term thermal aging at low temperature. *Mater. Des.* **2013**, *50*, 886–892. [CrossRef]

23. Li, S.; Wang, Y.; Wang, X.; Xue, F. G-phase precipitation in duplex stainless steels after long-term thermal aging: A high-resolution transmission electron microscopy study. *J. Nucl. Mater.* **2014**, *452*, 382–388. [CrossRef]

© 2019 by the authors. Licensee MDPI, Basel, Switzerland. This article is an open access article distributed under the terms and conditions of the Creative Commons Attribution (CC BY) license (http://creativecommons.org/licenses/by/4.0/).

Article

Modeling of Ti6Al4V Alloy Orthogonal Cutting with Smooth Particle Hydrodynamics: A Parametric Analysis on Formulation and Particle Density

Adam D. Lampropoulos *, Angelos P. Markopoulos and Dimitrios E. Manolakos

Laboratory of Manufacturing Technology, School of Mechanical Engineering, National Technical University of Athens, Heroon Polytechniou 9, 15780 Athens, Greece; amark@mail.ntua.gr (A.P.M.); manolako@central.ntua.gr (D.E.M.)
* Correspondence: lambroadam@central.ntua.gr; Tel.: +30-210-772-4299

Received: 27 February 2019; Accepted: 22 March 2019; Published: 28 March 2019

Abstract: Computational modeling is a widely used method for simulation and analysis of machining processes. Smooth particle hydrodynamics (SPH) is a comparatively recently developed method that is used for the simulation of processes where high strains and fragmentation occur. The purpose of this work is the application of the SPH method for the prediction of cutting forces and chip formation mechanism in orthogonal cutting of Ti6Al4V alloy. In addition, it is examined how the final results of the simulation are influenced by the choice of the particular formulation of the SPH method, as well as by the density of the particles.

Keywords: smooth particle hydrodynamics; Titanium alloy machining; numerical simulation; cutting forces; chip formation

---

1. Introduction

Titanium alloys due to their unique mechanical properties are widely used in industrial applications. Especially, the Ti6Al4V alloy is one of the most widely used alloys and constitutes more than 50% of titanium products globally [1]. Most applications of this alloy are found in the aerospace industry, but it is also used for the production of medical devices due to its biocompatibility. However, it exhibits particular attributes as a material, such as low thermal conductivity, high strength at elevated temperatures, and low modulus of elasticity which lead to low machinability ratings [2,3]. Machining is one of the most prevalent processes for the production of industrial components and, due to the reasons mentioned above, cutting of titanium is usually accompanied by high cutting forces and high cutting temperatures. These major factors lead to excessive tool wear, shortened tool life, and poor surface quality of the final workpiece.

In practice, to increase productivity and at the same time keep quality at a high level, it is always necessary for the proper machining parameters to be chosen. There are mainly two ways that this goal can be attained. The first one involves conducting experiments to accumulate knowledge on the dependence of machining conditions upon important physical quantities (cutting forces, temperature etc.) during cutting. The second way involves the simulation of the machining process through the development of computational models. During the last decades, simulations have become increasingly popular and allow researchers to analyze, study, and understand in depth the physics of machining.

The finite element method (FEM) is one of the most frequently used numerical techniques for engineering simulations. However, there are a number of difficulties that arise when it is used for cutting simulations. Firstly, large strains lead to severe element distortions and consequently to the termination of the simulation, due to negative volume elements. Secondly, in order to model the fracture of the workpiece, a failure criterion has to be implemented into the code, so that certain elements are deleted from the grid when certain criteria are satisfied. The problem of distorted elements is usually solved with remeshing algorithms, which recreate the mesh after a user-defined number of time steps, resulting in increased computational cost. An alternative solution to this problem is based on element erosion/deletion when the stresses or strains inside an element surpass a certain threshold. The disadvantage of this approach is the fact that many elements can be artificially deleted from the grid, leading to non-physical simulations.

In [4], the orthogonal cutting is simulated by the FEM where a simple geometric criterion, combined with a mesh rezoning algorithm is used for the modeling of material separation. In [5–7], fracture is simulated by an element deletion algorithm which is based on a fracture criterion. Also, avoidance of severe element distortion is overcome by the usage of remeshing algorithms. In [8–10], material separation in the simulations is modeled by the formation of adiabatic shear bands and the usage of fracture criterions is avoided. However, the usage of remeshing algorithms still remains a necessity for the same reasons. In [11], the influence of cutting conditions on the stress at the rear surface of the tool is investigated. In [12], a mathematical model for the calculation of cutting forces is developed, based on the results of a FEM model. In [13,14], the influence of the constitutive model and the damage criteria on the prediction of cutting forces is studied. An alternative to classical FEM techniques is the use of combined Lagrangian–Eulerian formulations, where the grid is not tied on the workpiece but it is fixed in space [15–17].

During recent years, meshless numerical methods are gaining increased popularity because the particles/nodes are not strictly connected with their neighbor particles, but are relatively free to move in space, unlike the nodes of a finite element grid. In this study, the focus is on the smooth particle hydrodynamics meshless method. The SPH method was developed in the 1980s and was used for the numerical simulation of physical problems that belong to the field of astrophysics [18] and a decade later was modulated for solid mechanics problems [19]. A comprehensive analysis of the SPH method and examples regarding its applications in solid and fluid mechanics can be found in [20,21].

One of the earliest implementations of the SPH method to orthogonal cutting simulations can be found in [22], where the benefits of the method are presented in comparison to the traditional FEM approach. In [23], a 2D model is presented where friction is predicted in workpiece/cutting tool interface without the usage of a friction model, since both bodies are modeled with SPH particles. In [24], a model of the same type is used for the prediction of the variation of cutting forces caused by a worn cutting tool. In [25–30], SPH models of orthogonal, or oblique, cutting are created where the cutting tool is modeled with finite elements and the workpiece with SPH particles, which means that a friction model had to be used. In [30], the effect of subsequent cuts on the evolution of residual stresses in the workpiece is presented. Studies of SPH simulations regarding the influence of factors describing the material behavior such as the friction coefficient, the equation of state, or the usage of a damage-evolution criterion can be found in [31–33]. The influence of more intrinsic parameters to the SPH method, such as timestep and particle density, can be found in [34]. Similarly, in [35] the effect of important SPH control parameters is examined.

One of the deficiencies of the SPH method in metal cutting simulations is the inability of prediction of a realistic chip curvature, as it is mentioned in [36]. For this reason, alternative formulations are proposed to the standard SPH scheme, where this issue is resolved. One of these formulations is called renormalization and more details about its foundations and development can be found in [37,38]. Orthogonal cutting simulations with SPH can be found in the literature [23–35] for different formulations. However, there are no studies that examine the influence of the chosen formulation, on the prediction of the cutting forces. The aim of this study is the investigation of the differences in the prediction of cutting forces between two different formulations. For this purpose, numerical simulations of orthogonal cutting of Ti6Al4V titanium alloy are presented, where both formulations are used, namely the standard and the renormalized. This parametric analysis is complemented by conducting numerical simulations with different particle densities.

In Section 2, the mathematical foundations of SPH are summarized. The presentation of the theory in this section is based on [21]. In Section 3, the main formulas describing the material behavior are also provided, while in Section 4, the configuration of the numerical simulation is described. Furthermore, the way the parametric analysis was conducted, is described in detail. In Section 5, the main results of the parametric analysis are presented. For all the simulations, the commercial solver LSDYNA [39] was used.

2. SPH Foundations

In the SPH method, a function is represented by the following integral

$$f(x) = \int_{\Omega} f(x')\delta(x - x')dx' \tag{1}$$

where $\delta(x - x')$ is the Dirac delta function, which can be approximated by a smoothing function $W(x - x', h)$ that is chosen to fulfill a number of properties.

(a) The first property is the compact condition

$$W(x - x', h) = 0, \text{ if } |x - x'| > \kappa h$$

where h is the smoothing length and κ is a constant that affects the non zero area centered around point x.

(b) The second property that the smoothing function fulfills is the fact that it converges to the Dirac function

$$\lim_{h \to 0} W(x - x', h) = \delta(x - x')$$

(c) The final property that usually is fulfilled is the normalization condition

$$\int_{\Omega} W(x - x', h)dx' = 1$$

The integral representation in Equation (1) is replaced by the equation

$$< f(x) >= \int_{\Omega} f(x')W(x - x', h)dx' \tag{2}$$

where $< >$ is called the kernel approximation operator. Since the function W is zero for points outside a sphere with center at x and radius κh, it is possible to replace the problem domain Ω with the support domain $R = \left\{ x' \in R^3 : |x - x'| \leq \kappa h \right\}$. For the evaluation of the function in Equation (2), a numerical integration is carried out over a finite number of points in the neighborhood of x. In SPH, these points are represented by particles; each of them is considered to occupy a distinct amount of space and have a distinct mass. As can be seen from Figure 1, the particle X_i is positioned at the center of the support domain R of function W. For the numerical integration of Equation (2) the following sum has to be evaluated

$$< f(X_i) >= \sum_{j=1}^{N} f(X_j)W(X_j - X_i, h)\Delta V_j. \tag{3}$$

where N is the number of neighbor particles j inside the support domain and ΔV_j is the volume of each particle. Since each particle has a separate mass, the volume is substituted by the relation

$$\Delta V_j = m_j / \varrho_j \tag{4}$$

So, Equation (3) can be rewritten as

$$< f(X_i)> = \sum_{j=1}^{N} f(X_j) W(X_j - X_i, h) \frac{m_j}{\varrho_j}$$

(5)

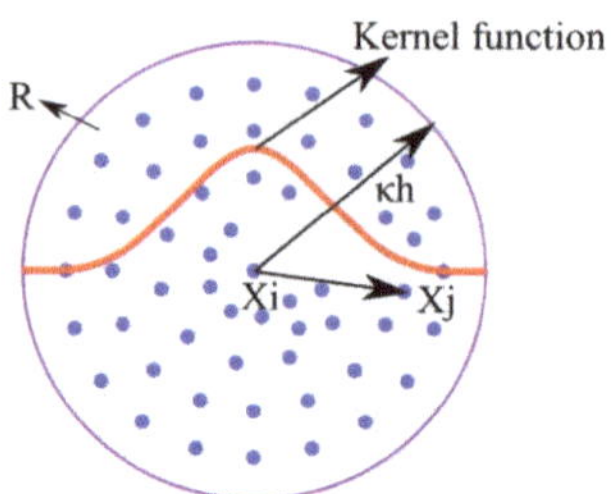

Figure 1. Support domain of kernel function centered at particle X_i.

It is worth mentioning that the density at point X_i is the weighted average of the masses of the neighbor particles, i.e.,

$$\varrho(X_i) = \sum_{j=1}^{N} m_j W(X_j - X_i, h)$$

(6)

Since many smoothing functions fulfill the aforementioned properties (a–c), there is a variety of different functions proposed over the years. A smoothing function, except properties a to c, is necessary to have a small support domain, which means that it converges to zero quickly and also it is important to have smooth derivatives. In this study, the commercial solver LS-DYNA Ver.R10.1.0.rev.123264 [39] is used for the numerical simulations, in which the cubic B-spline is chosen as a smoothing function. The B-spline was proposed in [40] and is one of the most widely used kernels, since it is numerically efficient due to its small support domain.

$$W(x - x', h) = \frac{1}{\pi h^3} \begin{cases} 1 - \frac{2}{3}u^2 + \frac{3}{4}u^3, & 0 \le u \le 1 \\ \frac{1}{4}(2 - u)^3 & 1 \le u \le 2 \\ 0 & u > 2 \end{cases}$$

(7)

where $u = |x - x'|/h$.

3. Material Modeling

3.1. Constitutive Model

For the mathematical modeling of the mechanical behavior of the workpiece material, the Johnson–Cook (J–K) constitutive relation was chosen. The Johnson–Cook model can be found in MAT_015 card [39] of LS-DYNA

$$\sigma_{eq} = (A + B\varepsilon_{eq}^n(1 + c\ln\dot{\varepsilon}_{eq}^*)^C(1 - T^{*m})$$

(8)

where, σ_{eq} is the Von Misses equivalent stress, ε_{eq} the equivalent strain, $\dot{\varepsilon}_{eq}^*$ the equivalent strain rate divided by a reference strain rate $\dot{\varepsilon}_{eq}^* = \dot{\varepsilon}_{eq}/\dot{\varepsilon}_0$ and stands for the homologous temperature $T^* = (T - T_r)/(T_m - T_r)$, where T_r is the room temperature and T_m is the melting temperature. Also, A, is the initial yield stress of the material, B, is the hardening constant, n, is the hardening exponent, C, is the strain rate constant and, m, is the thermal softening exponent. The values chosen for the parameters of the J–K model greatly affect the results of cutting simulations and the values provided in the literature for the same material vary among different authors. The values chosen in this study are provided from [41] and can be seen in Table 1. This choice is reported in [42], where it is shown that this set of parameters for the material model, provides the most accurate results for the cutting force. Also, the J–K model is used only for the workpiece since the cutting tool, due to its relatively great stiffness, is modeled as a rigid body.

Table 1. Material properties of the workpiece.

Material Parameters	Values
A (MPa)	997.9
B (MPa)	653.1
C	0.0198
m	0.7
n	0.45
$\dot{\varepsilon}_{ref}$ (s^{-1})	1
T_r (K)	298
T_m (K)	1878
Young's Modulus (E, GPa)	113.8
Density ϱ (kg/m^3)	4430
Friction coefficient (f)	0.2

3.2. Equation of State

Apart from the constitutive model, it is necessary to define the material pressure by relation to the density of the material. In LS-DYNA the simplest equation of state is the linear polynomial equation [39], where the pressure is related to the density with the relation

$$P = C_0 + C_1\mu + C_2\mu^2 + C_3\mu^3 + (C_4 + C_5\mu + C_6\mu^2)E \tag{9}$$

where $\mu = (\varrho - \varrho_0)/\varrho_0$. If $C_0 = C_2 = C_3 = C_4 = C_5 = C_6 = 0$, then C_1 is equal to the bulk modulus, since $P = K\mu$. Although the linear polynomial is considered a simplified model, it has been shown that the final results do not vary considerably when more accurate state equations are used [32].

3.3. Material Separation Modelling

For the proper simulation of cutting processes, the implementation of material separation by fracture or adiabatic shear banding is necessary. In traditional finite element codes, a fracture criterion, combined with an element deletion algorithm, is frequently used for elements located inside the material separation zone to be deleted. In the SPH method, the simulation of processes, where severe strains occur, can be modeled without the usage of a fracture criterion, since the SPH particles are loosely connected. In other words, the nodes of a finite element are strictly connected, regardless of the distance between each other; the element has to be deleted if their disconnection is necessary. On the contrary, a particle in the SPH code is affected by the neighbor particles that lie inside the support domain and there is not a predetermined connection between certain particles.

3.4. Contact and Friction Modeling

In this study, the frictional forces are considered to follow the Coulomb law

$$\tau = f\sigma_n \tag{10}$$

where τ is the frictional stress and σ_n is the normal stress. The friction coefficient f is constant over the entire contact area. In LS-Dyna, the friction coefficient can be defined in the AUTOMATIC-NODES-TO-SURFACE card. Also, on the same card, contact between the workpiece and the cutting tool is defined. A penalty type algorithm is used so that the particles of the workpiece remain in the exterior of the cutting tool's boundaries.

4. Numerical Simulation Configuration

Due to its relative simplicity, most published research is concerned with the simulation of orthogonal cutting. However, the results of a simulation need to be validated by actual experimental data and orthogonal cutting is not a widely used process in practice. Therefore, in many studies experimental data is derived from actual machining operations. Machining processes are mainly 3-dimensional and can be approximated by a 2-dimensional process to a certain extent. In this study, experimental results from reference [43] are used, where actual orthogonal cutting is conducted.

In the experimental reference, the cutting tool is moving with constant speed V into a flat strip of width W_c = 1 mm, height H = 2 mm and a length L = 10 mm. The experiment is repeated for three depths of cut h_c = 0.06, 0.04 and 0.1 mm. The material of the cutting tool is WC/Co tungsten carbide and of the workpiece Ti6Al4V titanium alloy (Table 2). The experimental conditions are summarized in Table 3.

Table 2. The chemical composition of T1-6Al-4V titanium alloy according to AMS 4928 standard [1].

Al	C	Fe	H	N	O	V	OT
5.5–6.75	0.1	0.3	0.0125	0.05	0.2	3.5–4.5	0.4

Table 3. Experimental conditions described in (data from [43]).

Machining Parameters	Values		
Cutting speed V [m/s]	0.5		
Rake angle, γ [deg]	15		
Clearance angle, α [deg]	2		
Cutting edge radius r [µm]	20		
Depth of cut, h_c [mm]	0.04	0.06	0.1

The simulation set up can be seen in Figure 2. The height of the workpiece H is three times the depth of cut (H = 3 h_c) and the width of the strip W_c is 0.01 mm. The length of the workpiece L is 1.5 mm, when the depth of cut is h_c = 0.1 mm and is reduced to 1 mm for the remaining two values of h_c. For the cutting tool and for the lower part of the workpiece up to one third its height, finite elements are used, and the rest of the workpiece is modeled with SPH particles. The nodes at the bottom and the edge of the strip are constrained in all directions.

The simulation of the experiment with the actual strip length is, from a computational standpoint, time-consuming and also unnecessary since the cutting process transitions to a steady state, where the cutting forces are stabilized long before the cutting tool reaches the end of the strip. The same applies for the width of the workpiece because all nodes are constrained to the z direction and are free to move in the x-y plane; it is redundant for the actual width of the workpiece to be modeled and scaling of the cutting forces to a factor of 100 is preferable. Finally, the SPH method is computationally more demanding than the standard FEM, and this is the reason why it is usually preferred for the area close to the path of the cutting tool to be modeled with particles. For the other parts, finite elements are considered more suitable and the mesh becomes gradually coarser at the bottom of the workpiece.

The parametric analysis is conducted based upon two parameters, the SPH formulation and the density of the particles. Initially, the distance d between the particles and the depth of cut h_c are fixed to a certain value and the simulation is carried out twice. Initially, a standard SPH formulation is used, followed by a renormalized formulation.

Consequently, the simulation with the renormalized formulation is carried out once again and the particle distance d is reduced to half of the initial value. Afterward, a different value for the depth of cut h_c is selected and the same procedure is repeated.

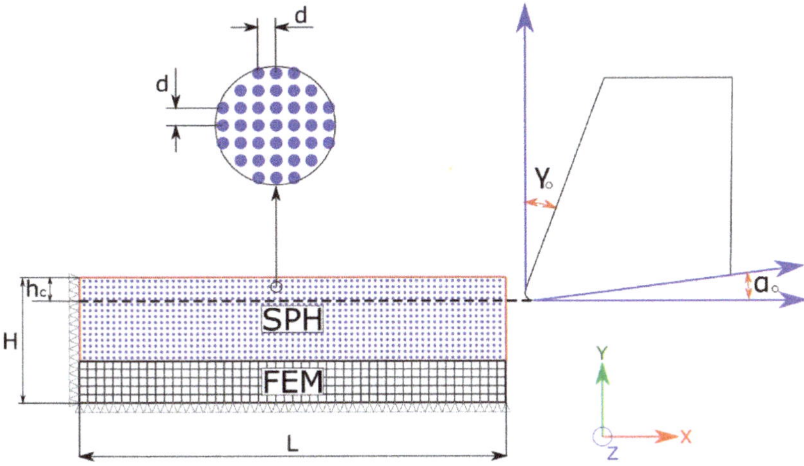

Figure 2. SPH model initial configuration.

5. Results and Discussion

In Figure 3, the equivalent stress contours are depicted for a fixed depth of cut $h_c = 0.04$ mm. In Figures 4 and 5, the stress contours are shown for depths of cut of $h_c = 0.06$ and 0.1 mm, respectively. By comparing the frames a and b for each depth of cut, an obvious observation is the chip curvature. By using the standard SPH formulation, a straight numerical chip is produced that follows the chip face of the cutting tool. The usage of the renormalized formulation leads to the production of a more realistic chip. The nonrealistic chip curvature of the standard formulation is attributed to the absence of particles in the exterior areas of the boundaries of the solution domain (workpiece) [36]. This absence leads to non-approximate calculations near the boundaries.

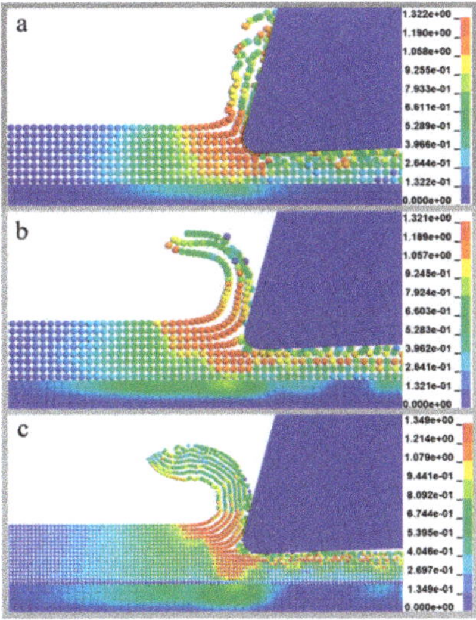

Figure 3. Equivalent stress σ_{eq} contours in GPa for 0.04 mm depth of cut: (**a**) standard formulation and d = 0.01 mm; (**b**) renormalized formulation and d = 0.01 mm; (**c**) renormalized formulation and d = 0.005 mm.

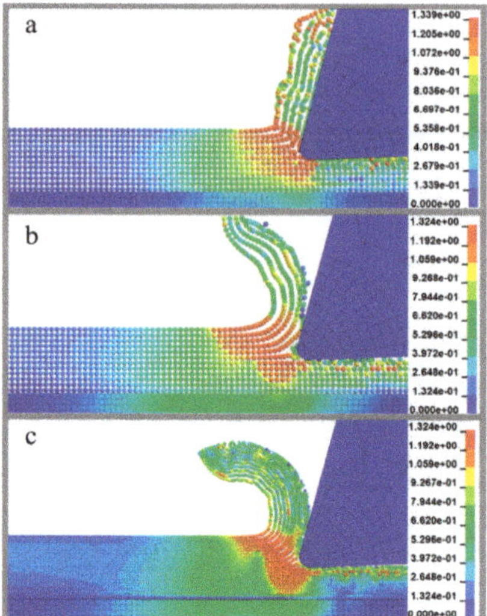

Figure 4. Equivalent stress σ_{eq} contours in GPa for 0.06 mm depth of cut: (**a**) standard formulation and d = 0.01 mm; (**b**) renormalized formulation and d = 0.01 mm; (**c**) renormalized formulation and d = 0.005 mm.

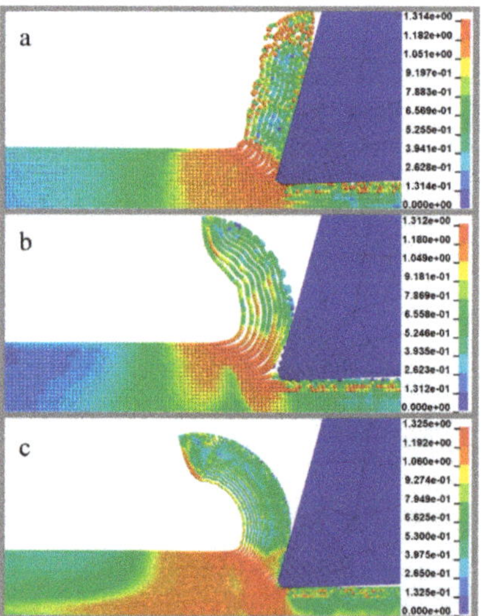

Figure 5. Equivalent stress σ_{eq} contours in GPa for 0.1 mm depth of cut: (**a**) standard formulation and d = 0.01 mm; (**b**) renormalized formulation and d = 0.01 mm; (**c**) renormalized formulation and d = 0.005 mm.

As can be seen in Figure 3b,c, the chip curvature is increased for a greater particle density and the same observation applies and for the other two values of the depth of cut. In contrast, the maximum stress is not affected considerably, regardless of the depth of cut, the formulation or the density of the particles. This can be explained by the fact that the maximum stress that is induced on the workpiece in the primary shear stress zone is dependent on the strength of the material.

In Figure 6, the plastic strain contours are depicted for each value of the depth of cut. In this figure, only the chips that are produced by the renormalized SPH are chosen, with the greater particle density. The areas with the larger values of strain can be seen at the secondary shear zone at the interface between the cutting tool and the chip, at the tertiary shear zone where the cutting tool rubs the surface of the workpiece, and also at the shear zones that are created at the primary shear zone.

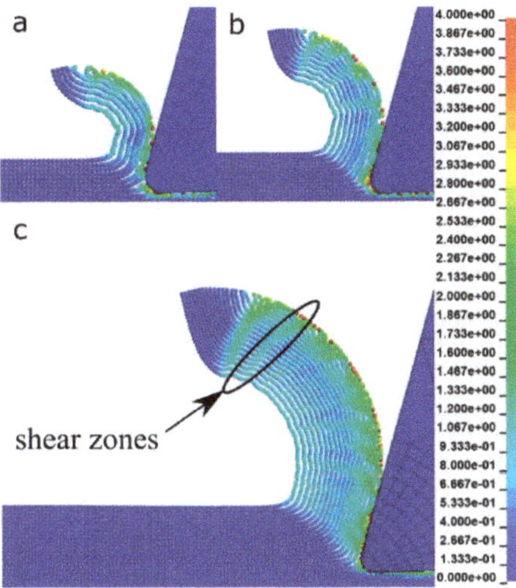

Figure 6. Equivalent plastic strain ε_{eq} contours for a depth of cut h_c: (**a**) 0.04 mm; (**b**) 0.06 mm; (**c**) 0.1 mm.

In Figures 7–9, the cutting force F_c and the feed force F_f are depicted in relation to time. Each figure corresponds to one of the three values for the depth of cut. Also, as can be seen from these figures, the forces start to stabilize at the time of 0.1 to 0.3 ms. For the values depicted in each of these figures, an average value is computed from the time that the forces remain relatively stable until the end of the simulation. The average values are compared to the experimental ones.

As shown in Figure 7a, the results of the standard SPH scheme are not satisfactory since the predicted values are less than half the values of the experimental ones. In contrast, the values predicted by the renormalized formulation, which are depicted in Figure 7b, are very close to the experimental ones. The cutting force F_c and feed force F_f are overpredicted with an error of about 7.56% and 4.39%, respectively. By comparing Figure 7b with Figure 7c, it is obvious that the forces predicted by the model with greater particle density are more stable and oscillate less. In addition, lower values are predicted for the cutting force F_c and there is also a small increase in the feed force F_f. The cutting force F_c is underpredicted with an error of about 15% and the feed force F_f is overpredicted with an error of 6.8%.

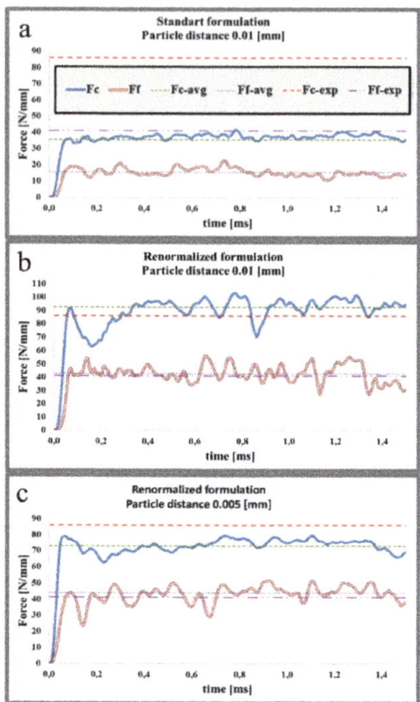

Figure 7. Cutting force F_c and feed force F_f versus time for h_c = 0.04 mm.

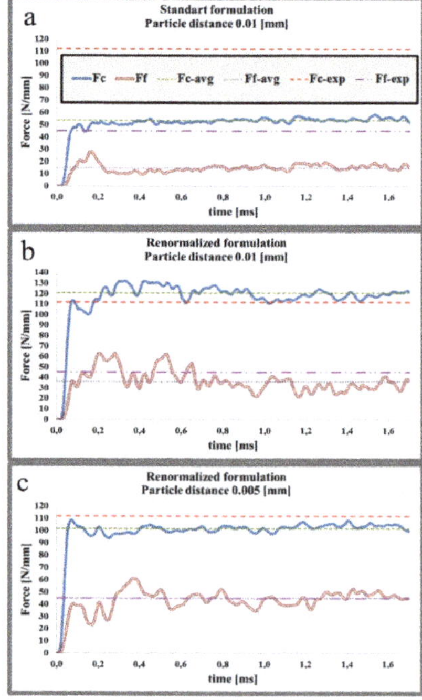

Figure 8. Cutting force F_c and feed force F_f versus time for h_c = 0.06 mm.

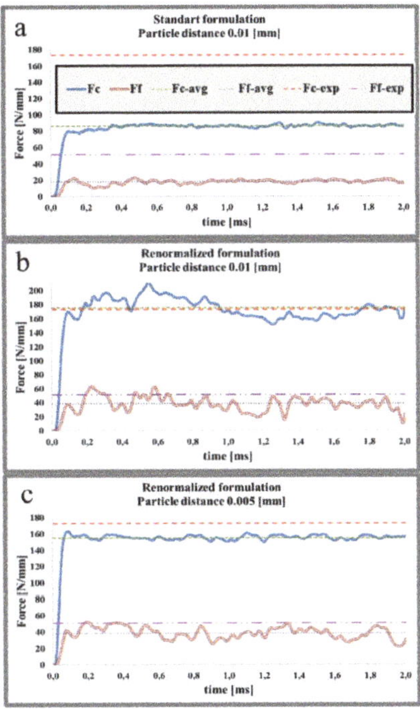

Figure 9. Cutting force F_c and feed force F_f versus time for $h_c = 0.1$ mm.

As can be seen in Figures 8a and 9a, the results produced from the standard SPH formulation remain far from the experimental values regardless of the depth of cut. By comparing Figure 8b to Figures 8c and 9b to Figure 9c, the same pattern is observed. The models with greater particle density provide relatively decreased values for the cutting force F_c and increased values for the feed force F_f. Also, especially for the cutting forces, there is less variation between the maximum and minimum values. For a depth of cut of $h_c = 0.06$ mm, the cutting force F_c is overpredicted with an error of 7.9%, the feed force F_f is underpredicted with a divergence of 20% from the experimental value. By increasing the number of particles, the predicted value of the cutting force F_c is lower than the experimental and the error is increased slightly to 9%. The predicted value for the feed force F_f is increased and almost coincides with the experimental value. For a depth of cut of $h_c = 0.01$ mm, as can be seen in Figure 9, the averaged predicted cutting force F_c is very close to the experimental value and they differentiate by less than 1.2%. However, the model which is comprised of a greater number of particles underpredicts the cutting force F_c with an error of 10.4%. For both models, as can be seen from Figure 9b,c, the feed force F_f is underpredicted and the averaged values differ from the experimental with a percentage of 25%.

The numerically predicted cutting and feed forces are gathered in Tables 4 and 5. At this point, it is obvious that only models where the renormalized formulation is used provide results close to the experimental values. Also, there is a negative correlation between cutting forces F_c and particle density. Models with smaller particle density estimate the cutting forces more accurately, which is an indication that the SPH models might converge to a value different from the experimental; still, the maximum deviation does not exceed 15%. For the feed forces F_f, a clear pattern is not visible. Although there is a positive correlation between feed force and particle density, the sensitivity is almost zero for $h_c = 0.1$ mm and there is a slight change of the predicted values for $h_c = 0.04$ mm, but for $h_c = 0.06$ mm, the deviation between the feed forces estimated by the different particle density models approaches 26%. Generally, the cutting force F_c is estimated more accurately comparing to the feed force F_f and this can be attributed to the parameters chosen for the J–K model. As mentioned in reference [42], parameters provided in the literature for the J–K model usually do not lead to accurate estimations of both cutting and feed forces as well.

Table 4. Comparison between predicted and experimental cutting forces F_c for all models.

Depth of Cut h_c [mm]	Experimental Cutting Force [N/mm]	Standard Formulation (d = 0.01 [mm]) [N/mm]	Renormalized Formulation (d = 0.01 [mm]) [N/mm]	Renormalized Formulation (d = 0.005 [mm]) [N/mm]
0.04	86	35.6	92.5	73
		−58.6%	+7.6%	−15.1%
0.06	112	53,76	120.9	101.9
		−52%	+7.94%	−9.02%
0.1	173	86	175	155
		−50.3%	+1.2%	−10.4%

Table 5. Comparison between predicted and experimental feed forces F_f for all models.

Depth of Cut h_c [mm]	Experimental Feed Force [N/mm]	Standard Formulation (d = 0.01 [mm]) [N/mm]	Renormalized Formulation (d = 0.01 [mm]) [N/mm]	Renormalized Formulation (d = 0.005 [mm]) [N/mm]
0.04	41	15.5	42.8	43.8
		−62,2%	+4.39%	+6.8%
0.06	45	14.9	35.95	45.1
		−66,9%	−20.1%	0.2%
0.1	51	18.15	38.39	38.6
		−64.4%	−24.7%	−24.3%

6. Conclusions

The orthogonal cutting of Ti6Al4V alloy was modeled by using the smooth particle hydrodynamics method. The main focus of this study was the influence of the choice between two different SPH formulations (standard and renormalized) on the prediction of cutting and feed forces. It was found that the implementation of the renormalized formulation leads to a satisfactory prediction of cutting forces. In contrast, the predictions of the standard formulations deviate from the experimental values of more than 50%. In addition, the usage of the renormalized formulation leads to a more realistic estimation of the chip geometry and this may be the main reason for the large difference in the estimation of the cutting forces between the two formulations.

This study was also complimented by the investigation of the influence of particle density in the estimation of chip geometry and cutting forces. From the results, it is evident that the curvature of the numerical chip increases when a greater number of particles is used. Also, models with more particles predict lower values of cutting forces and larger values for the feed forces.

Due to the meshless nature of this method, modeling of material separation was possible without the implementation of a fracture criterion or a remeshing algorithm. Because of these advantages, programming of the SPH method for cutting simulations is easier. Fracture criterions require calibration from experiments and element deletion algorithms may lead to an unrealistic mass loss in the simulation. Furthermore, remeshing algorithms are computationally demanding. Thus, SPH is a promising alternative to a classical FEM.

Author Contributions: Conceptualization, A.D.L., A.P.M., and D.E.M.; Formal analysis, A.D.L. and D.E.M.; Investigation, A.D.L.; Methodology, A.D.L. and A.P.M.; Resources, D.E.M.; Software, A.D.L. and A.P.M.; Supervision, D.E.M.; Validation, A.D.L.; Writing—original draft, A.D.L.; Writing—review and editing, A.P.M. and D.E.M.

Funding: This research received no external funding.

Conflicts of Interest: The authors declare no conflict of interest.

Nomenclature

A	Initial yield stress [MPa]
B	Hardening constant [MPa]
C	Strain Rate Constant
d	Distance between particles [mm]
E	Young's Modulus [GPa]
F_c	Cutting Force [N/mm]
F_f	Feed Force [N/mm]
f	Friction coefficient
H	Height of the workpiece [mm]
h	Smoothing length [mm]
h_c	Depth of cut [mm]
K	Bulk modulus [GPa]
L	Length of the workpiece [mm]
m	Thermal softening exponent
m_j	Mass of particle j
n	Hardening exponent
P	Pressure [GPa]
r	Cutting edge radius [μm]
T_r	Room temperature [K]
T_m	Melting temperature [K]
V	Cutting speed [m/s]
W_c	Width of the workpiece
W	Kernel function
X_i	Position vector of particle i
x	Position vector of a point in space
α	Clearance angle [deg]
γ	Rake angle [deg]
δ	Dirac function
ΔV_j	Volume of particle j
ε_{eq}	Equivalent strain
$\dot{\varepsilon}_{eq}$	Equivalent strain rate [1/s]
$\dot{\varepsilon}_{ref}$	Reference strain rate [1/s]
μ	Relative change of density
ϱ	Density [Kg/m^3]
ϱ_j	Density of particle j
σ_n	Normal stress [GPa]
σ_{eq}	Equivalent stress [GPa]
τ	Frictional stress [GPa]

References

1. Welsch, G.; Boyer, R.; Collings, E.W. (Eds.) *Materials Properties Handbook: Titanium Alloys*; ASM International: Materials Park, OH, USA, 1994; ISBN 978-0-87170-481-8.
2. Davim, J.P. (Ed.) *Machining of Titanium Alloys*; Springer: New York, NY, USA, 2014; ISBN 978-3-662-43901-2.
3. Ezugwu, E.O.; Wang, Z.M. Titanium alloys and their machinability—A review. *J. Mater. Process. Technol.* **1997**, *68*, 262–274. [CrossRef]
4. Mamalis, A.G.; Horváth, M.; Branis, A.S.; Manolakos, D.E. Finite element simulation of chip formation in orthogonal metal cutting. *J. Mater. Process. Technol.* **2001**, *110*, 19–27. [CrossRef]

5. Ng, E.-G.; Aspinwall, D.K. Modelling of hard part machining. *J. Mater. Process. Technol.* **2002**, *127*, 222–229. [CrossRef]

6. Guo, Y.B.; Yen, D.W. A FEM study on mechanisms of discontinuous chip formation in hard machining. *J. Mater. Process. Technol.* **2004**, *155–156*, 1350–1356. [CrossRef]

7. Hua, J.; Shivpuri, R. Prediction of chip morphology and segmentation during the machining of titanium alloys. *J. Mater. Process. Technol.* **2004**, *150*, 124–133. [CrossRef]

8. Rhim, S.-H.; Oh, S.-I. Prediction of serrated chip formation in metal cutting process with new flow stress model for AISI 1045 steel. *J. Mater. Process. Technol.* **2006**, *171*, 417–422. [CrossRef]

9. Calamaz, M.; Coupard, D.; Girot, F. A new material model for 2D numerical simulation of serrated chip formation when machining titanium alloy Ti–6Al–4V. *Int. J. Mach. Tools Manuf.* **2008**, *48*, 275–288. [CrossRef]

10. Sima, M.; Özel, T. Modified material constitutive models for serrated chip formation simulations and experimental validation in machining of titanium alloy Ti–6Al–4V. *Int. J. Mach. Tools Manuf.* **2010**, *50*, 943–960. [CrossRef]

11. Pimenov, D.Y.; Guzeev, V.I.; Koshin, A.A. Influence of cutting conditions on the stress at tool's rear surface. *Russ. Eng. Res.* **2011**, *31*, 1151–1155. [CrossRef]

12. Pimenov, D.Y.; Guzeev, V.I. Mathematical model of plowing forces to account for flank wear using FME modeling for orthogonal cutting scheme. *Int. J. Adv. Manuf. Technol.* **2017**, *89*, 3149–3159. [CrossRef]

13. Ducobu, F.; Rivière-Lorphèvre, E.; Filippi, E. Influence of the Material Behavior Law and Damage Value on the Results of an Orthogonal Cutting Finite Element Model of Ti6Al4V. In Proceedings of the 14th CIRP Conference on Modeling of Machining Operations (CIRP CMMO), Turin, Italy, 13–14 June 2013; Settineri, L., Ed.; Elsevier: Amsterdam, The Netherlands, 2013; pp. 379–384.

14. Boldyrev, I.S.; Shchurov, I.A.; Nikonov, A.V. Numerical Simulation of the Aluminum 6061-T6 Cutting and the Effect of the Constitutive Material Model and Failure Criteria on Cutting Forces' Prediction. In Proceedings of the 2nd International Conference on Industrial Engineering (ICIE-2016), Chelyabinsk, Russia, 19–20 May 2016; Radionov, A.A., Ed.; Elsevier: Amsterdam, The Netherlands, 2016; pp. 866–870.

15. Ducobu, F.; Arrazola, P.-J.; Rivière-Lorphèvre, E.; de Zarate, G.O.; Madariaga, A.; Filippi, E. The CEL Method as an Alternative to the Current Modelling Approaches for Ti6Al4V Orthogonal Cutting Simulation. In Proceedings of the 16th CIRP Conference on Modelling of Machining Operations (16th CIRP CMMO), Cluny, France, 15–16 June 2017; Outeiro, J., Poulachon, G., Eds.; Elsevier: Amsterdam, The Netherlands, 2017; pp. 245–250.

16. Ducobu, F.; Rivière-Lorphèvre, E.; Filippi, E. Finite element modelling of 3D orthogonal cutting experimental tests with the Coupled Eulerian-Lagrangian (CEL) formulation. *Finite Elem. Anal. Des.* **2017**, *134*, 27–40. [CrossRef]

17. Grissa, R.; Zemzemi, F.; Fathallah, R. Three approaches for modeling residual stresses induced by orthogonal cutting of AISI316L. *Int. J. Mech. Sci.* **2018**, *135*, 253–260. [CrossRef]

18. Gingold, R.A.; Monaghan, J.J. Smoothed particle hydrodynamics: theory and application to non-spherical stars. *Mon. Not. R. Astron. Soc.* **1977**, *181*, 375–389. [CrossRef]

19. Libersky, L.D.; Petschek, A.G. Smooth particle hydrodynamics with strength of materials. In *Advances in the Free-Lagrange Method Including Contributions on Adaptive Gridding and the Smooth Particle Hydrodynamics Method*; Trease, H.E., Fritts, M.F., Crowley, W.P., Eds.; Springer: Berlin, Germany, 1991; Volume 395, pp. 248–257, ISBN 978-3-540-54960-4.

20. Liu, G.R. *Meshfree Methods: Moving Beyond the Finite Element Method*, 2nd ed.; CRC Press: Boca Raton, FL, USA, 2009; ISBN 978-1-4200-8210-4.

21. Liu, G.R.; Liu, M.B. *Smoothed Particle Hydrodynamics: A Meshfree Particle Method*; World Scientific: Hackensack, NJ, USA, 2003; ISBN 978-981-238-456-0.

22. Heinstein, M.; Segalman, D. *Simulation of Orthogonal Cutting with Smooth Particle Hydrodynamics*; SAND-97-1961; Sandia National Labs.: Albuquerque, NM, USA, 1 September 1997.

23. Limido, J.; Espinosa, C.; Salaün, M.; Lacome, J.L. SPH method applied to high speed cutting modelling. *Int. J. Mech. Sci.* **2007**, *49*, 898–908. [CrossRef]

24. Calamaz, M.; Limido, J.; Nouari, M.; Espinosa, C.; Coupard, D.; Salaün, M.; Girot, F.; Chieragatti, R. Toward a better understanding of tool wear effect through a comparison between experiments and SPH numerical modelling of machining hard materials. *Int. J. Refract. Met. Hard Mater.* **2009**, *27*, 595–604. [CrossRef]

25. Villumsen, M.F.; Fauerholdt, T.G. Simulation of Metal Cutting using Smooth Particle Hydrodynamics. In Proceedings of the 7th LS-DYNA Anwenderforum, Bamberg, Germany, 30 September–1 October 2008; pp. 17–36.

26. Bagci, E. 3-D numerical analysis of orthogonal cutting process via mesh-free method. *Int. J. Phys. Sci.* **2011**, *6*, 1267–1282.

27. Dănuț, J. SPH Meshless Simulation of Orthogonal Cutting of AA6060-T6. *Appl. Mech. Mater.* **2015**, *808*, 258–263. [CrossRef]

28. Spreng, F.; Eberhard, P. Machining Process Simulations with Smoothed Particle Hydrodynamics. In Proceedings of the 15th CIRP Conference on Modelling of Machining Operations, Karlsruhe, Germany, 11–12 June 2015; pp. 94–99.

29. Umer, U.; Mohammed, M.K.; Qudeiri, J.A.; Al-Ahmari, A. Assessment of finite element and smoothed particles hydrodynamics methods for modeling serrated chip formation in hardened steel. *Adv. Mech. Eng.* **2016**, *8*, 1–11. [CrossRef]

30. Geng, X.; Dou, W.; Deng, J.; Yue, Z. Simulation of the cutting sequence of AISI 316L steel based on the smoothed particle hydrodynamics method. *Int. J. Adv. Manuf. Technol.* **2017**, *89*, 643–650. [CrossRef]

31. Madaj, M.; Píška, M. On the SPH Orthogonal Cutting Simulation of A2024-T351 Alloy. In Proceedings of the 14th CIRP Conference on Modeling of Machining Operations, Turin, Italy, 13–14 June 2013; Settineri, L., Ed.; Elsevier: Amsterdam, The Netherlands, 2013; pp. 152–157.

32. Olleak, A.A.; El-Hofy, H.A. SPH Modelling of Cutting Forces while Turning of Ti6Al4V Alloy. In Proceedings of the 10th European LS-DYNA Conference, Würzburg, Germany, 15–17 June 2015.

33. Takabi, B.; Tajdari, M.; Tai, B.L. Numerical study of smoothed particle hydrodynamics method in orthogonal cutting simulations – effects of damage criteria and particle density. *J. Manuf. Process.* **2017**, *30*, 523–531. [CrossRef]

34. Heisel, U.; Zaloga, W.; Krivoruchko, D.; Storchak, M.; Goloborodko, L. Modelling of orthogonal cutting processes with the method of smoothed particle hydrodynamics. *Prod. Eng.* **2013**, *7*, 639–645. [CrossRef]

35. Avachat, C.S.; Cherukuri, H.P. A Parametric Study of the Modeling of Orthogonal Machining Using the Smoothed Particle Hydrodynamics Method. In Proceedings of the ASME 2015 International Mechanical Engineering Congress and Exposition, Houston, TX, USA, 13–19 November 2015; ASME: Houston, TX, USA, 2015; p. V014T06A005.

36. Limido, J.; Espinosa, C.; Salaun, M.; Mabru, C.; Chieragatti, R.; Lacome, J.L. Metal cutting modelling SPH approach. *Int. J. Mach. Mach. Mater.* **2011**, *9*, 177–196. [CrossRef]

37. Randles, P.W.; Libersky, L.D. Smoothed Particle Hydrodynamics: Some recent improvements and applications. *Comput. Methods Appl. Mech. Eng.* **1996**, *139*, 375–408. [CrossRef]

38. Vila, J.P. SPH Renormalized Hybrid Methods for Conservation Laws: Applications to Free Surface Flows. In *Meshfree Methods for Partial Differential Equations II*; Griebel, M., Schweitzer, M.A., Eds.; Springer: Berlin, Germany, 2005; Volume 43, pp. 207–229, ISBN 978-3-540-23026-7.

39. Livermore Software Technology Corporation. *LS-DYNA KEYWORD USER'S MANUAL*; Livermore Software Technology Corporation: Livermore, CA, USA, 2017.

40. Monaghan, J.J.; Lattanzio, J.C. A refined particle method for astrophysical problems. *Astron. Astrophys.* **1985**, *149*, 135–143.

41. Seo, S.; Min, O.; Yang, H. Constitutive equation for Ti–6Al–4V at high temperatures measured using the SHPB technique. *Int. J. Impact Eng.* **2005**, *31*, 735–754. [CrossRef]

42. Ducobu, F.; Rivière-Lorphèvre, E.; Filippi, E. On the importance of the choice of the parameters of the Johnson-Cook constitutive model and their influence on the results of a Ti6Al4V orthogonal cutting model. *Int. J. Mech. Sci.* **2017**, *122*, 143–155. [CrossRef]

43. Ducobu, F.; Rivière-Lorphèvre, E.; Filippi, E. Experimental contribution to the study of the Ti6Al4V chip formation in orthogonal cutting on a milling machine. *Int. J. Mater. Form.* **2015**, *8*, 455–468. [CrossRef]

© 2019 by the authors. Licensee MDPI, Basel, Switzerland. This article is an open access article distributed under the terms and conditions of the Creative Commons Attribution (CC BY) license (http://creativecommons.org/licenses/by/4.0/).

 *metals*

Article

An Extended Iterative Identification Method for the GISSMO Model

Yue Xiao [1,2] and Yumei Hu [1,2,*]

[1]   State Key Laboratory of Mechanical Transmissions, Chongqing University, Chongqing 400044, China; shawnxiao@cqu.edu.cn

[2]   School of Automobile Engineering, Chongqing University, Chongqing 400044, China

*   Correspondence: cdrhym@163.com or cdrhym@cqu.edu.cn; Tel.: +86-136-3792-7890

Received: 22 April 2019; Accepted: 13 May 2019; Published: 15 May 2019

Abstract: This study examines an extended method to obtain the parameters in the Generalized Incremental Stress State Dependent Damage (GISSMO) model. This method is based on an iterative Finite Element Method (FEM) method aiming at predicting the fracture behavior considering softening and failure. A large number of experimental tests have been conducted on four different alloys (7003 aluminum alloy, ADC12 aluminum alloy, ZK60 magnesium alloy and 20CrMnTiH Steel), here considering tests that span a wide range of stress triaxiality. The proposed method is compared with the two existing methods. Results show that the new extended Iterative FEM method gives the good estimate of the fracture behaviors for all four alloys considered.

Keywords: fracture; iterative FEM Method; GISSMO Model; softening

---

1. Introduction

The research into lightweight and high-durability materials used in automobiles has greatly increased in recent years, due to an increased focus on preserving natural resources and reducing air pollution. One effective way to achieve weight reduction is through material development, but with this approach comes a demand for accurately predicting the flow and damage behaviors of the new material. [1,2] Flow behavior and damage evolution is, however, a challenging task.

A large amount of research has been focused on evaluating the flow and fracture properties through micromechanics-based consideration [3–11]. Among these are the porous plasticity models, the best-known being the classical model by Gurson [3] that includes the growth of voids in a metallic matrix and a yield condition that relates to porosity. The Gurson model and many of its subsequent extensions agree well with experiments; however, it fails to predict the material behavior under shear dominated loading conditions. To remedy this issue, Nahshon and Hutchinson [12] distinguished shear dominated states by the third invariant of stress. Dæhli et al. [13] extended the Gurson model by Lode-dependent void evolution, both successfully predicting fracture under low triaxiality. However, the Gurson model is computationally inefficient as the element size must scale with the dominant void sparking. Thus, the phenomenological fracture models outweigh the porous models in industrial applications, as these models often require fewer material constants to be determined through data fitting. The typical fracture models [14–17] are formalized on experimental observation, which directly related to the stress state. Identified through the stress, the triaxiality and the lode angle are proposed. Compared to the Gurson model, the Johnson-Cook (JC) model [17–19] predicts the fracture locus accounting for the strain, the strain rate and the temperature, and the model parameters in the JC model is easily to calibrate. Other fracture models in Ref. [14,15,20] predicts the fracture locus in 3D space with more accuracy. Moreover, it is worth noticing that these fracture models are uncoupled from flow behavior, which is described by the standard material model. The Generalized Incremental Stress State Dependent Damage (GISSMO) Model which is developed by Neukamm [21,22], fully describes the ductile damage accounting for material softening and fracture. The fracture behavior is described by the damage variable and the damage softening behavior is represented by an instability measure. As the GISSMO model meets the industrial requirements, it finds wide use in crashworthiness and forming process simulations [23–27].

The parameter calibration strongly influences the accuracy of a numerical prediction. The constitutive constants are classically obtained by graphical analysis [28] and linear regression methods [29] to fit experimental data. For example, the force-displacement curves obtained from experimental data is transformed into equivalent stress-strain curves, from which constitutive parameters are obtained [30]. However, these calculation methods are conducted under assumptions of relatively small strain variation, failing to predict fracture behavior in several industrial applications. Recently, an iterative FEM method [31–34] has been developed for parameter calibration that, based on trial and error, can determine a set of parameters, which leads to good agreements between experimental data and numerical results. Xiao [31] obtained the fracture parameters through the iterative FEM method to predict the fracture locus in the 3D stress state. However, fracture parameters by the iterative FEM method are difficult to predict the fracture behavior considering softening.

In this paper, an extended iterative FEM method is proposed to obtain the GISSMO parameters. The accuracy of the inverse iterative FEM method is then validated by considering four different types of alloy through an extensive experimental program that subject the materials in different stress states. FEM procedures are conducted for three sets of validated parameters obtained by three methods. The accuracy of the three methods is evaluated. These stress-strain curves and fracture modes observed throughout the tests are compared to the model predictions. Finally, a discussion on the iterative FEM method is conducted.

2. Materials and Methods

2.1. Experiment Method

Four kinds of alloy, listing 7003-T6 aluminum (Test # 1–7), ADC12 aluminum (Test # 8–11), ZK60 magnesium (Test # 12–18) and 20CrMnTiH steel (Test # 19–25), are designated as the testing material with the chemical composition shown in Table 1. These alloys are extensively used in aerospace and automobile industries. Figure 1 presents 25 different specimens for a wide range of stress states. The dimensions of tensile specimen are recommended by the GB/T228.1-2010 standard. All tests are performed in the ETM504C electronic universal testing machine (Suzhou Dymeter Automotive testing technology co., LTD, Suzhou, China) with a load range of (5–50) kN and a speed range of (0.001–500) mm/min. The deforming results are measured by the Digital Image Correlation (DIC) method, and the loading results are measured by force cell. For DIC method, signature is marked in undeformed specimen, and displacement is measured by searching the signature in deformed specimen. The crosshead speed of the testing machine is kept constant at 1.8 mm/min for smooth tensile tests and 0.2 mm/min for notched and shear tests. These tests are performed in the ambient air at the room temperature. Following experimental data is recorded: the stress-strain relation, fracture mode, and elongation.

Table 1. Chemical composition of the materials used in the study (wt.%). The single value means the maximum value of element's concentration.

Materials	Al	C	Cr	Cu	Fe	Mg	Mn	Ni	P	Pb	S	Sn	Si	Ti	Zr	Zn
7003Al	Bal.	-	-	0.1	0.35	0.5–1	0.3	-	-	-	-	-	0.3	0.2	0.05–0.25	5.5–6.5
ADC12	Bal.	-	-	1.5–3.5	1.2	0.3	0.5	0.5	-	0.1	-	0.1	9.6–12.0	-	-	1
ZK60	-	-	-	-	-	Bal.	-	-	-	-	-	-	-	-	0.45–0.9	4.8–6.2
CrMnTiH	-	0.17–0.23	1–1.35	0.3	Bal.	-	0.8–1.15	0.3	0.04	-	0.04	-	0.17–0.37	0.01–0.1	-	-

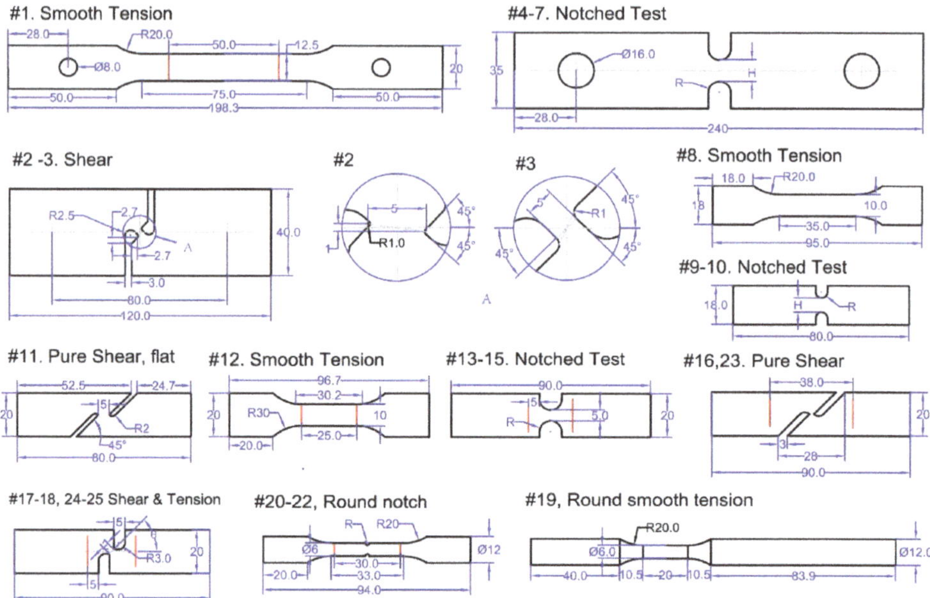

Figure 1. Schematic sketch of the specimen. specified value of R, H and θ can be found in Table 2. All dimensions are in mm.

Table 2. Summary of experiments.

Mat.	No.	Description	Gauge Length/mm
	1	Smooth tensile	50
	2	Pure Shear	19
	3	Tensile Shear, 45°, $\theta = 45°$	17.5
7003-T6	4	Notched, R5, $R = 5$ mm, $H = 10$ mm	20
	5	Notched, R10, $R = 10$ mm, $H = 10$ mm	20
	6	Notched, R15, $R = 15$ mm, $H = 8$ mm	40
	7	Notched, R20, $R = 20$ mm, $H = 8$ mm	50
	8	Smooth Tensile	35
ADC12	9	Notched, R4, $R = 4$mm, $H = 5$mm	14
	10	Notched, R8, $R = 8$mm, $H = 5$mm	18
	11	Pure shear, flat	40.6
	12	Smooth Tensile	25
	13	Notched, R1	12
	14	Notched, R5	20
ZK60	15	Notched, R10	30
	16	Pure Shear	38
	17	Tensile Shear, 30°, $\theta = 30°$	21
	18	Tensile Shear, 60°, $\theta = 60°$	21
	19	Round smooth tensile	20
	20	Round notch, R0.4, $R = 0.4$ mm, $H = 9.2$ mm	10.4
	21	Round notch, R0.8, $R = 0.8$ mm, $H = 8.4$ mm	10.8
20CrMnTiH	22	Round notch, R2, $R = 2$ mm, $H = 6$ mm	12
	23	Pure shear	38
	24	Tensile shear, 30°, $\theta = 30°$	21
	25	Tensile shear, 60°, $\theta = 60°$	21

2.2. Effective Stress and Strain Curves

Experimental obtained engineering curves cannot be directly used in parameter calibration. In small strain, engineering curves is transformed into true stress-strain curves by volume constancy law,

$$\varepsilon_{true} = \ln\left(1 + \varepsilon_{eng}\right)$$
$$\sigma_{true} = \sigma_{eng}\left(1 + \varepsilon_{eng}\right) \tag{1}$$

where σ_{true} and ε_{true} is true stress and true strain. σ_{eng} and ε_{eng} is engineering stress and strain. For some metals Equation (1) cannot compensate for the large necking effects, constitutive models are used for extrapolating the after-necking stress-strain curves. Taken smooth tension for example, the engineering curves are transformed into true stress-strain curves by volume constancy law. Then, the soften part in the true stress-strain curve is extrapolated by the power law of Ludwik [35]. The extrapolated curve is shown in Figure 2 as the yellow dash one. The effective stress-strain curve is obtained by subtracting off the elastic strain and inputting the true stress. The effective stress-strain curve is used in material model calibration. The effective strain at fracture point is input as initial value in iterative FEM method.

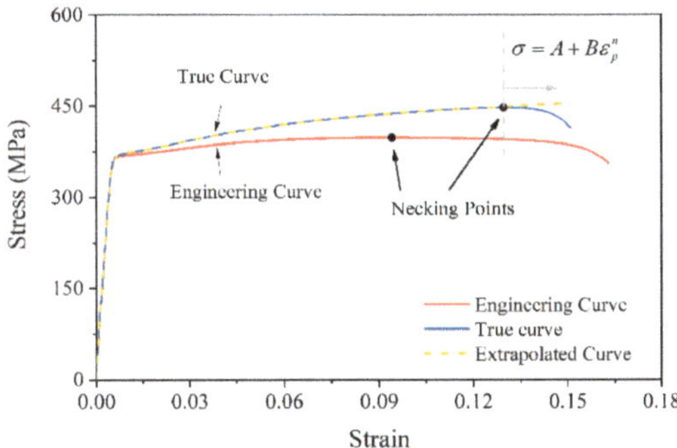

Figure 2. Stress-strain curves for experimental data processing. The red solid curve is the engineering stress-strain curves. The blue solid curve is transformed through the red one by plastic incompressibility condition. To compensate for the non-uniform deformation beyond necking, the blue solid curve is extrapolated by the power law of Ludwik. Yellow dash curve represents modified curve which could be used in calibrating constitutive model.

2.3. GISSMO Model

The GISSMO, developed by Neukamm et al. [21,22] and Basaran et al. [36], ensures flexibility for a wide range of metal as it can correctly predict damage regardless the details of the material model formulation. Besides, the GISSMO targets damage prediction that accounts for material instability, localization, and failure. Damage accumulation, which is illustrated by Johnson [17] and Xue [37], is based on an incremental formulation found from

$$\Delta D = \frac{n}{\varepsilon^f} D^{\left(1-\frac{1}{n}\right)} \dot{\varepsilon}_p, \tag{2}$$

This equation considers the nonlinear relation between plastic strain and internal damage [38,39], and the damage exponent n allows for a nonlinear accumulation of damage until failure. The equivalent plastic strain increment is denominated as $\dot{\varepsilon}_p$. ε^f represents the equivalent plastic failure strain which allows for an arbitrary definition of triaxiality dependent failure strains $\varepsilon^f(\sigma^*)$ by inputting a tabulated curve. Triaxiality is defined as p/σ_m, in which p is the average main stress and σ_m is the von Mises stress. D is the damage value, which accumulates in each element during deformation. When $D = 1$, the element is deleted.

Similar to damage accumulation, material instability is determined as

$$\Delta F = \frac{n}{\varepsilon^c} F^{\left(1-\frac{1}{n}\right)} \dot{\varepsilon}_p,$$

(3)

where ε^c is the triaxiality dependent critical strain $\varepsilon^c(\sigma^*)$ which acts as an activation for coupling damage and stress. F is the instability measure. If $F = 0$, the material is undeformed. $F = 1$ corresponds to the onset of localization.

As soon as F reaches unity, the element stress is reduced by

$$\sigma = \widetilde{\sigma}\left[1 - \left(\frac{D - D_{crit}}{1 - D_{crit}}\right)^m\right],$$

(4)

where σ is the modified stress and $\widetilde{\sigma}$ is the current stress. D_{crit} is the damage only comes to some value when F reaches unity. m is a fading exponent intended to better depict the rate of material softening.

2.4. Extended Iterative Finite Element Method

The extended Iterative FEM (EIFEM) method is a trial-and-error method, aiming at obtaining stress and strain at large strain. The EIFEM method is inspired by the method in Ref. [31], which can obtain parameters in the JC model. This paper extended the model to calibrate parameters in the GISSMO model. Five parameters need to be calibrated, including stress triaxiality σ^*, fracture strain ε^f, critical strain ε^c, fading exponent m, and damage exponent n. The proposed procedure of the extended Iterative FEM method (EIFEM) is,

1. Initial value of ε_0^f, ε_0^c, m_0 and n_0. ε_0^f is the fracture strain and ε_0^c is the necking strain at obtained effective stress-strain curve. m_0 and n_0 can be set to arbitrary value. in this paper, $m_0 = 1$, $n_0 = 3$.
2. Iteration till the shape of numerical force and displacement curve after necking coincide with experimental one. 3D FEM simulation by LS-DYNA is conducted to calculate elongation ΔL^{FEM}. In simulation, the deforming process is predicted by the JC material model and the fracture model is the GISSMO model. If numerical shape differs experimental shape, n and m are modified by $n_i = n_{i-1} - 1$ and $m_i = m_{i-1} + 0.5$ till convergence.
3. Iteration till the experimental elongation coincides with the numerical elongation $\Delta L^{EXP} = \Delta L^{FEM}$. If $\Delta L^{EXP} \neq \Delta L^{FEM}$, ε^f would be modified by $\varepsilon_{i+1}^f \times \Delta L^{EXP} = \varepsilon_i^f \times \Delta L_i^{FEM}$. If $\varepsilon_i^f = \varepsilon_i^c$ but ΔL^{FEM} still unequal to ΔL^{EXP}, ε^c is modified by $\varepsilon_{i+1}^c \times \Delta L_n^{EXP} = \varepsilon_i^c \times \Delta L_{n(i)}^{FEM}$. ΔL_n is the displacement at necking point. A satisfied pair of ε^c and ε^f is obtain when $\Delta L^{EXP} = \Delta L^{FEM}$.
4. Check if the standard deviation below 3%. The standard deviation (Std) between experimental and numerical curves is defined as

$$Std = \frac{\int_0^{l_f} \left| f_{Exp}(l) - f_{FEM}(l)\right| dl}{Max\left\{f_{Exp}(l)\right\} l_f},$$

(5)

where l is displacement. The iteration is continued until the Std below 3%. Otherwise return to Step 2 and renew the number of m, n, ε^c and ε^f till convergence. In the last iterative FEM simulation, σ^* is obtained at the fracture element right before deleting element.

2.5. Finite Element Method

The finite element (FE) analysis is conducted by the explicit mechanical solver of the commercial finite element code LS-DYNA. FE models for tensile tests are meshed through 8-node hexahedron elements by one-point integration, with single-point-constrain aligned to one side and the prescribed motion to the other side. The average mesh size is 0.2 for each specimen in gauge length. As the FEM mesh are similar between materials, the representative FEM mesh for ZK60 alloy are shown in

Figure 3. Tensile specimens all move in constants speeds according to experimental speeds. Due to the strain rate insensitivity, the velocity is set at a relatively larger value compared to experiment to reduce the calculating time. The velocity is set at 10 mm/s in this paper. The plastic flow and damage are considered separately. The GISSMO model describes the fractured and damage behavior. In LS-DYNA, the GISSMO model is embedded in the ADD_EROSION which provides the failure and erosion option. The JC material model describes the flow behavior. The JC material model, which is one of the most widely used empirical models, provides an accurate prediction of the flow behavior considering the effects of stress state and strain rate. As all tests are performed at room temperature, the temperature effect is ignored. The strain rate part is also ignored as all experiments are in quasi-static state. The JC material model [40] is simplified as follows:

$$\sigma_e = A + B\varepsilon_p{}^c, \tag{6}$$

where A, B, and c are material constants and σ_e is the flow stress. By achieving the good fit of smooth tensile experimental data, four parameter sets are calibrated and listed in Table 3. Please note that the elastic modulus is obtained from the elastic region ranging from start point to yielding point (the value stress value is A). This elastic modulus would make the numerical predicted results more accurate.

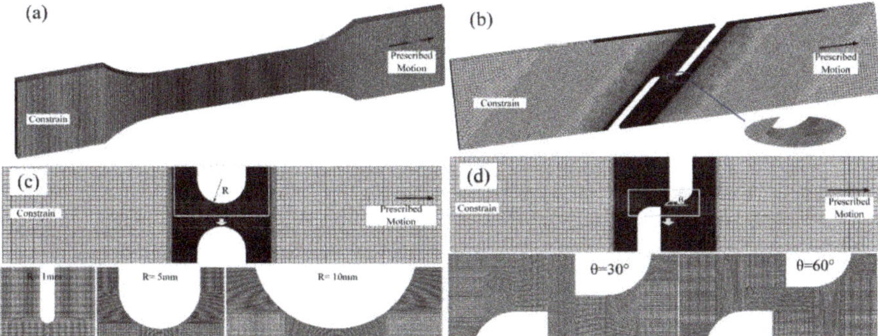

Figure 3. The representative FEM model for ZK60 alloy, (**a**) smooth tensile test, (**b**) pure shear test, (**c**) notched tensile test and (**d**) tensile and notched tests.

Table 3. Parameters for numerical simulation.

Mat.	E/GPa	A	B	c	n	m
7003-T6	66	348	252	0.44	3	2.5
ADC12	33	115	1938	0.67	2	10
ZK60	31	221	316	0.43	2	1.5
20CrMnTiH	125	944	754	0.28	3	3.5

Fracture curves $\varepsilon^f\,(\sigma^*)$, critical strain curves $\varepsilon^c\,(\sigma^*)$, the fading exponent m and the damage exponent n are identified through the numerical simulation. Although different GISSMO parameters combinations may lead to a similar numerical result, the effect could be minimized by increasing the number of tests for GISSMO model calibration. Recall the method in Section 2.4, the parameters m and n can be calibrated. The fading exponent m and the damage exponent n are listed in Table 3. $\varepsilon^f\,(\sigma^*)$ and $\varepsilon^c\,(\sigma^*)$ are cubic spline interpolations of fracture strain and critical strain.

3. Results and Discussion

3.1. Experimental Results

The original experimental results are presented in Figure 4. The engineering stress is calculated by load, which is directly obtained by force cell. The engineering strain is obtained by DIC method. Recall the method in Section 2.2, the effective stress and strain curves are obtained. Then, the triaxiality σ^*, critical strain ε^c and fracture strain ε^f are obtained based on the EIFEM method in Section 2.4. Then σ^*, ε^c and ε^f is used in numerical simulation.

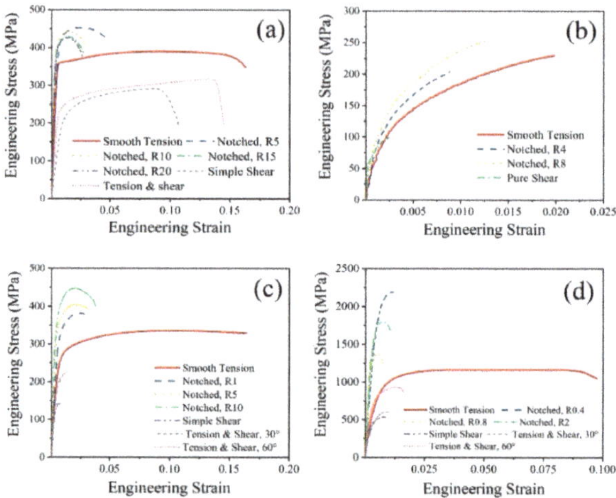

Figure 4. Engineering stress and strain curves for (**a**) 7003 aluminum alloy, (**b**) ADC12 aluminum alloy, (**c**) ZK60 magnesium alloy and (**d**) 20CrMnTiH Steel.

Table 4. Summary of experimental and numerical results.

Mat.	No.	Description	σ^*	ε^c	ε^f	Exp.	Num.	R.E. (%) [1]
	1	Smooth tensile	0.33	0.3	0.79	8.16	8.30	1.80
	2	Pure Shear	−0.01	0.3	0.83	1.94	1.84	4.87
	3	Tensile Shear, 45°	−0.08	0.05	0.73	2.55	2.49	2.33
7003-T6	4	Notched, R5	0.55	0.02	0.47	0.96	0.97	0.72
	5	Notched, R10	0.46	0.1	0.44	1.19	1.17	1.32
	6	Notched, R15	0.4	0.01	0.50	1.32	1.32	0.24
	7	Notched, R20	0.38	0.01	0.50	1.42	1.45	2.12
	8	Smooth Tensile	0.34	0.005	0.019	0.70	0.69	1.50
ADC12	9	Notched, R4	0.36	0.005	0.033	0.12	0.13	3.52
	10	Notched, R8	0.35	0.005	0.035	0.23	0.23	2.50
	11	Pure shear, flat	−0.03	0.005	0.025	0.10	0.10	1.52
	12	Smooth Tensile	0.33	0.08	0.697	4.11	4.07	0.84
	13	Notched, R1	0.41	0.10	0.262	0.36	0.35	1.51
	14	Notched, R5	0.46	0.05	0.107	0.65	0.65	0.28
ZK60	15	Notched, R10	0.48	0.03	0.289	1.13	1.09	3.39
	16	Pure Shear	0.07	0.26	0.343	0.27	0.27	0.77
	17	Tensile Shear, 30°	0.22	0.05	0.420	0.17	0.17	1.50
	18	Tensile Shear, 60°	0.30	0.09	0.587	0.31	0.31	1.10
	19	Round smooth tensile	0.33	0.03	0.17	1.95	1.94	0.51
	20	Round notch, R0.4	0.50	0.03	0.08	0.13	0.13	4.83
	21	Round notch, R0.8	0.64	0.05	0.15	0.08	0.08	1.15
20CrMnTiH	22	Round notch, R2	0.69	0.03	0.09	0.13	0.12	4.78
	23	Pure shear	−0.07	0.1	0.45	0.34	0.34	0.41
	24	Tensile shear, 30°	0.2	0.1	0.42	0.21	0.22	2.57
	25	Tensile shear, 60°	0.27	0.1	0.34	0.34	0.33	2.08

[1] R.E. denotes the relative error.

3.2. Numerical Results and Validation

Table 4 presents the experimental and numerical elongation. The relative errors are below 5% for all tests. Figure 5 shows the force and displacement curves for experiment and simulation. Figure 5a presents curves for tensile tests. Numerical results agree with the experimental results well. For 20CrMnTiH steel and 7003 aluminum alloy, the after-necking region in numerical force and displacement curves coincides with the experimental curves. The good agreements are also observed in notched and shear tests shown in Figure 5b–e.

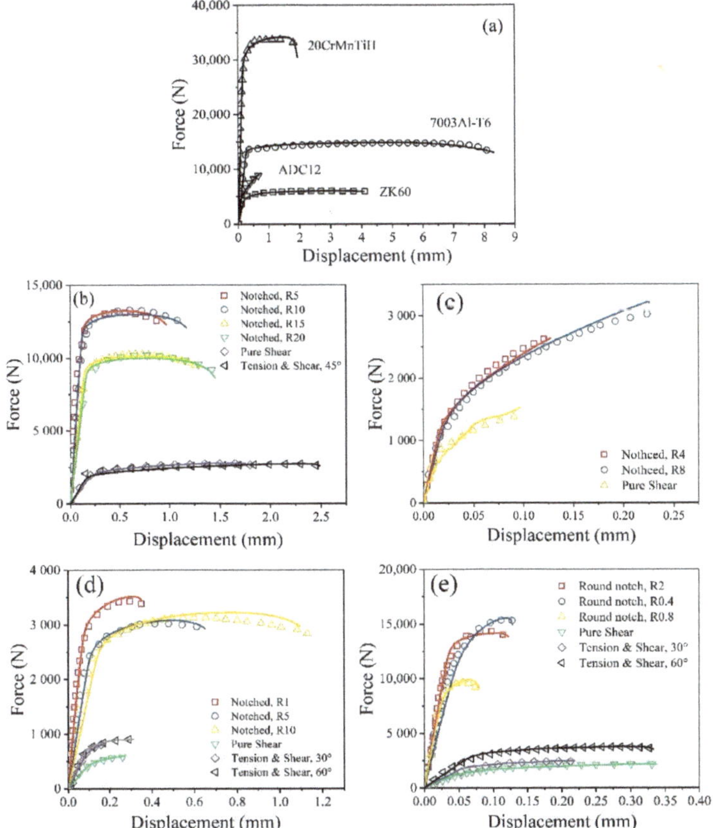

Figure 5. Numerical and experimental force and displacement curves. (**a**) Tensile tests, (**b**) notched and shear tests for 7003 aluminum alloy, (**c**) notched and shear tests for ADC12 aluminum alloy, (**d**) notched and shear tests for ZK60 magnesium alloy and (**e**) notched and shear tests for 20CrMnTiH steel. Scatters represent experimental data and solid lines represent numerical results hereinafter.

Visual comparisons between experimental and numerical predicted fracture modes are shown in Figures 6–9. Good agreements on the fracture modes are observed in those figures, indicating that the EIFEM method can reproduce the fracture behavior for different metals over a wide stress triaxiality with certain accuracy.

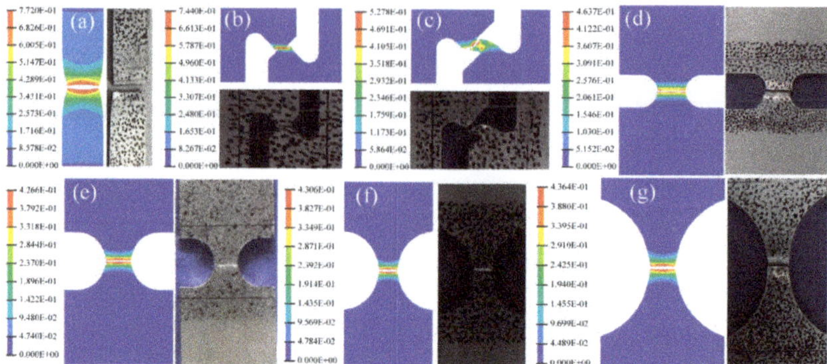

Figure 6. Fracture mode for 7003Al-T6 aluminum alloy for (**a**) smooth tensile test, (**b**) pure shear test, (**c**) tensile and shear test, (**d**) notched test for notched radii 5 mm, (**e**) notched test for notched radii 10 mm, (**f**) notched test for notched radii 15 mm and (**g**) notched test for notched radii 20 mm. Hereinafter, the numerical fracture mode is in plastic strain contour.

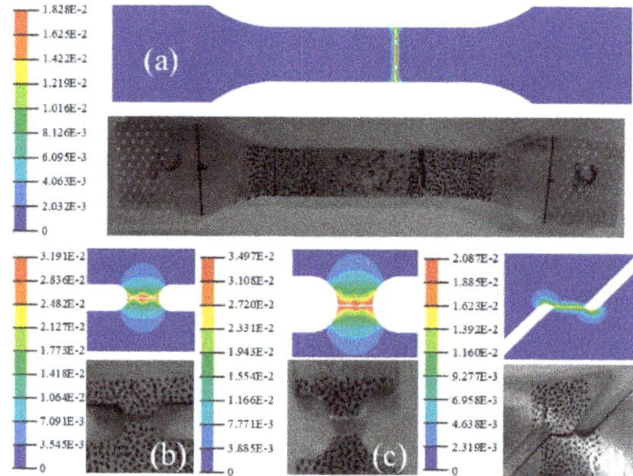

Figure 7. Fracture mode for ADC12 aluminum alloy for (**a**) smooth tensile test, (**b**) notched test for notched radii 4 mm, (**c**) notched test for notched radii 8 mm and (**d**) pure shear test.

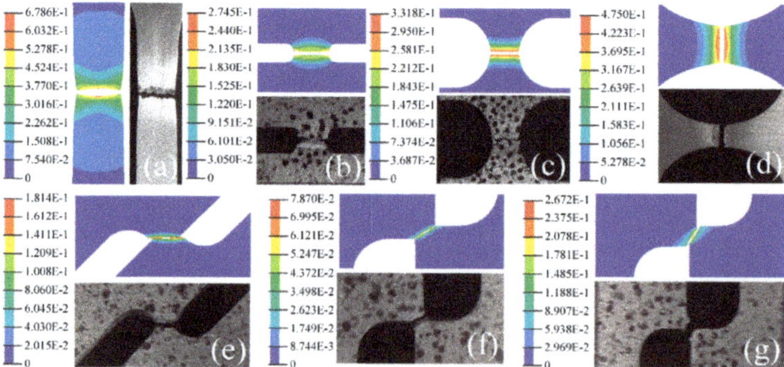

Figure 8. Fracture mode for ZK60 magnesium alloy for (**a**) smooth tensile test, (**b**) notched test for notched radii 1mm, (**c**) notched test for notched radii 5mm, (**d**) notched test for notched radii 10mm, (**e**) pure shear test, (**f**) shear and tensile test at shear angle 30° and (**g**) shear and tensile test at shear angle 60°.

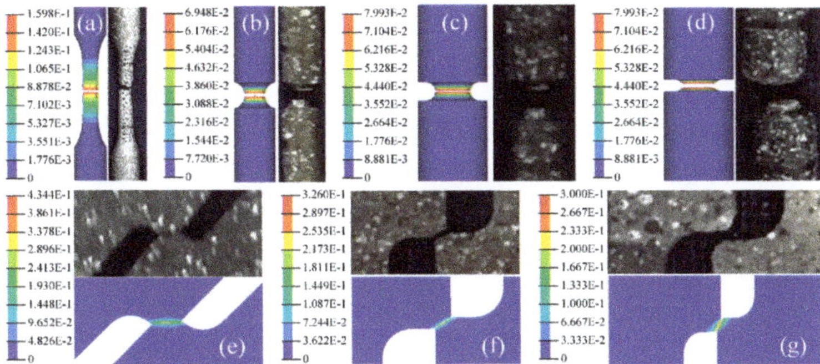

Figure 9. Fracture mode for 20CrMnTiH steel for (**a**) smooth tensile test, (**b**) notched test for notched radii 2 mm, (**c**) notched test for notched radii 0.8 mm, (**d**) notched test for notched radii 0.4 mm, (**e**) pure shear test, (**f**) shear and tensile test at shear angle 30° and (**g**) shear and tensile test at shear angle 60°.

3.3. Comparison

To further study the application range of the proposed method, the numerical results predicted by the GISSMO model are compared with other two methods. The first method is widely used in industry for rough calculation. To get the fracture strain in the first method, the experimental obtained force and displacement curves are transposed into effective stress and strain based on method in Section 2.2. Then the fracture strain is obtained at the end point in the effective stress and strain curve. Usually this fracture strain is smaller than the EIFEM obtained fracture strain. To verify which fracture strain is more accurate, the numerical simulation for first method also carried out in LS-DYNA. For better clarification, the notation in first method adds A, so the fracture strain is written as ε_A^f. The second method is a commonly used iterative FEM method as in Ref. [31]. Different value of fracture strain can be obtained by this method, and the corresponding notation adds a B, so ε_B^f. The stepwise procedure for the second method is shown in Appendix A. The numerical simulation for validating the second method is also conducted.

Numerical simulations for the aforementioned two methods carry out in LS-DYNA. The material model is John-Cook model and the fracture model is the GISSMO model, both are consistent with the EIFEM method. The method A and the method B cannot obtain the critical strain ε^c, the ε^c in the two method are set to a relatively large constant. Then the GISSMO model cannot display the soften region. The value of ε^c influence the value of ε^f. In this paper, the ε^c is set to 1 in method A and method B. m and n are set according to Table 3. The fracture strain vs. stress triaxiality for the three methods are shown in Figure 10.

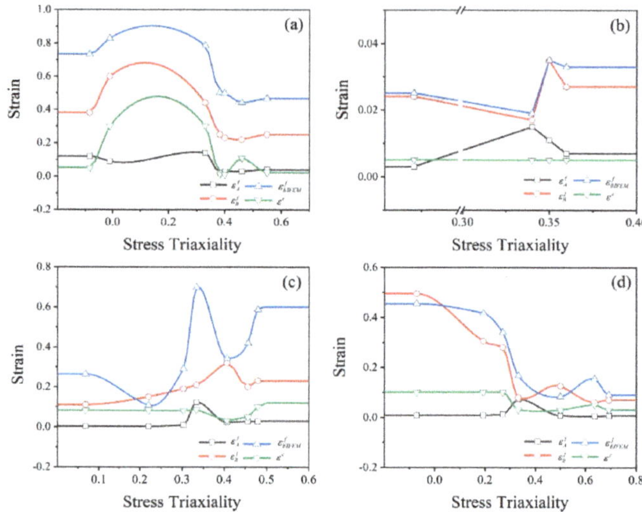

Figure 10. Strain and triaxiality curves for (**a**) 7003 aluminum alloy, (**b**) ADC12 aluminum alloy, (**c**) ZK60 magnesium alloy and (**d**) 20CrMnTiH steel.

Force and displacement relationships for method A and B is presented in Figure 11. The elongation is discussed firstly. Elongation calculated by method A is much smaller than experimental elongation. Elongations calculated by method B are closer to the experimental data. However, the relative errors for Test No. 9, 10, 17, 20 and 22 (numbered in Table 2) are beyond 5%, indicating the method B cannot accurately predict the failure behavior. Then the force and displacement curves are compared. For method A, the numerical force and displacement curves overlap the elastic region, failing to predict the harden and soften region. For method B, numerical force and displacement curves in Test No. 2, 3, 8–11,16–18, 23–25 match the experimental data. However, for other tests, the numerical curves cannot capture the soften region. The experiments well predicted by methods B include tests of ADC12 alloy and all shear tests. The experiments failing to be predicted by method B include the uniaxial tensile tests for 7003 alloy and 20CrMnTiH steel and notched tests for 7003 alloy, ADC12 alloy, and 20CrMnTiH steel. The well predicted experiments are brittle fracture (ADC12 alloy) and shear fracture, with relatively small strain and no soften effects. The bad estimated force and displacement curves are featured by obvious soften effects. So, the method B cannot predict soften effect. The reason lies in the constant value of critical strain ε^c. ε^c will influence the value of fracture strain ε^f. Usually, ε^f in condition $\varepsilon^c < \varepsilon^f$ is larger than condition $\varepsilon^c \geq \varepsilon^f$. So, the ε^f obtain by the EIFEM method is larger than ε^f by method B. Physically, the soften effect leads to the inaccurate estimate of fracture stain. Usually, $\varepsilon^f_{\text{EIFEM}} \geq \varepsilon^f_B \geq \varepsilon^f_A$. In conclusion, method A is not recommending unless for small strain conditions or very rough calculation. Method B can be safely used in deformation with little or no soften effects. The EIFEM method can accurately predict the large deformation and obvious soften conditions.

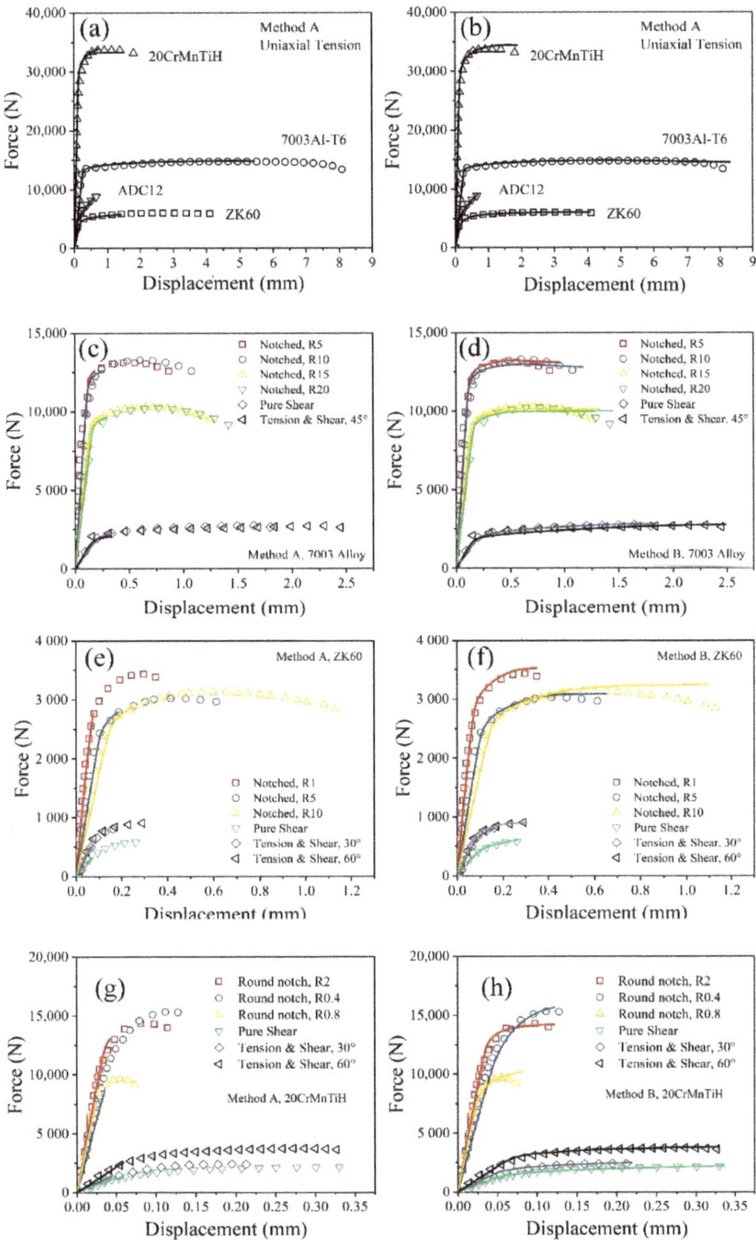

Figure 11. Force and displacement curves for (**a**) uniaxial tensile tests based on method A, (**b**) uniaxial tensile tests based on method B, (**c**) 7003Al alloy based on method A, (**d**) 7003Al alloy based on method B, (**e**) ZK60 Mg alloy based on method A, (**f**) ZK60 Mg alloy based on method B, (**g**) 20CrMnTiH based on method A and (**h**) 20CrMnTiH based on method B.

4. Conclusions

This paper validates the accuracy for an extended parameter identification method by experiments and simulations, followed by comparison with two existing methods. The EIFEM method obtains the parameters in the GISSMO fracture model. The numerical obtained elongation, force, and displacement curves and the fractured position agree well with experiment. The EIFEM method achieves the best estimate compared with other two methods. Method A is recommended to use in small strain conditions or rough calculation. Method B can be used in metal with little or no soften effects. The EIFEM method can be used in predicting the large deformation and obvious soften effects.

Author Contributions: Conceptualization, Y.X.; methodology, Y.X.; software, Y.X.; validation, Y.X.; formal analysis, Y.X.; data curation, Y.X.; writing—original draft preparation, Y.X.; writing—review and editing, Y.X. and Y.H.; visualization, Y.X.; supervision, Y.H.; project administration, Y.H.; funding acquisition, Y.H.

Funding: This research was funded by Fundamental Research Funds for the Central Universities, grant number 2018CDXYTW0031 and Graduate Scientific Research and Innovation Foundation of Chongqing, grant number CYB18060; Fundamental Research Funds for the Central Universities, grant number 0009005202001002.

Acknowledgments: Special thanks to Kim Lau Nielsen in the Technical University of Denmark for his comments and improvements to this paper. This work is also supported by State Key Laboratory of Mechanical Transmission 2017 Open Fund, grant number SKLMT-ZZKT-2017M15

Conflicts of Interest: The authors declare no conflict of interest.

Appendix A

The iterated FEM method is used to obtain stress triaxiality σ^* and fracture strain ε^f. In stepwise Step L is the gauge length and ΔL is elongation.

Step 1: Obtaining the initial value of ε_0^f. $\varepsilon_0^f = \Delta L/L$.

Step 2: Numerical simulation for tensile tests. In this step, the FEM method is performed to simulate the tensile tests. Through the first attempt of tensile simulation, σ^* can be obtained in the first fractured element at the fractured moment.

Step 3: Iteration of ε^f. The elongation $\Delta L^{(\text{FEM})}$ is not identical to the $\Delta L^{(\text{EXP})}$ under ε_0^f. Therefore, the fracture strain is modified by $\varepsilon_{i+1}^f \times \Delta L^{(\text{EXP})} = \varepsilon_i^f \times \Delta L^{(\text{FEM})}$ to reduce the difference between $\Delta L^{(\text{FEM})}$ and $\Delta L^{(\text{EXP})}$. This iteration is continued until the relative error between $\Delta L^{(\text{FEM})}$ and $\Delta L^{(\text{EXP})}$ below 3%. Through the FEM procedures, the precise values of stress triaxiality σ^* and fracture strain ε^f are identified.

References

1. Hu, M.Q.; Zhang, J.J.; Sun, B.Z.; Gu, B.H. Finite element modeling of multiple transverse impact damage behaviors of 3-D braided composite beams at microstructure level. *Int. J. Mech. Sci.* **2018**, *148*, 730–744. [CrossRef]
2. Spagnoli, A.; Terzano, M.; Brighenti, R.; Artoni, F.; Stahle, P. The fracture mechanics in cutting: A comparative study on hard and soft polymeric materials. *Int. J. Mech. Sci.* **2018**, *148*, 554–564. [CrossRef]
3. Gurson, A.L. Continuum theory of ductile rupture by void nucleation and growth: Part I Yield criteria and flow rules for porous ductile media. *J. Eng. Mater-T ASME* **1977**, *99*, 2–15. [CrossRef]
4. McClintock, F.A. A criterion for ductile fracture by the growth of holes. *Int. J. Appl. Mech.* **1968**, *35*, 363–371. [CrossRef]
5. Nguyen, N.T.; Kim, D.Y.; Kim, H.Y. A continuous damage fracture model to predict formability of sheet metal. *Fatigue Fract. Eng. Mater. Struct.* **2013**, *36*, 202–216. [CrossRef]
6. Nielsen, K.L.; Tvergaard, V. Ductile shear failure or plug failure of spot welds modelled by modified Gurson model. *Eng. Fract. Mech.* **2010**, *77*, 1031–1047. [CrossRef]
7. Rice, J.R.; Tracey, D.M. On the ductile enlargement of voids in triaxial stress fields. *J. Mech. Phys. Solids* **1969**, *17*, 201–217. [CrossRef]
8. Tvergaard, V.; Needleman, A. Analysis of the cup-cone fracture in a round tensile bar. *Acta Metall.* **1984**, *32*, 157–169. [CrossRef]

9.   Xue, L. Constitutive modeling of void shearing effect in ductile fracture of porous materials. *Eng. Fract. Mech.* **2008**, *75*, 3343–3366. [CrossRef]

10. Cricrì, G. A consistent use of the Gurson-Tvergaard-Needleman damage model for the R-curve calculation. *Fract. Struct. Integrity* **2013**, *7*, 161–174. [CrossRef]

11. Sepe, R.; Lamanna, G.; Caputo, F. A robust approach for the determination of Gurson model parameters. *Fract. Struct. Integrity* **2016**, *10*, 369–381. [CrossRef]

12. Nahshon, K.; Hutchinson, J.W. Modification of the Gurson model for shear failure. *Eur. J. Mech. A Solids* **2008**, *27*, 1–17. [CrossRef]

13. Dæhli, L.E.; Morin, D.; Børvik, T.; Hopperstad, O.S. A Lode-dependent Gurson model motivated by unit cell analyses. *Eng. Fract. Mech.* **2018**, *190*, 299–318. [CrossRef]

14. Bao, Y.; Wierzbicki, T. On fracture locus in the equivalent strain and stress triaxiality space. *Int. J. Mech. Sci.* **2004**, *46*, 81–98. [CrossRef]

15. Bao, Y.; Wierzbicki, T. A comparative study on various ductile crack formation criteria. *J. Eng. Mater-T ASME* **2004**, *126*, 314–324. [CrossRef]

16. Bao, Y.; Wierzbicki, T. On the cut-off value of negative triaxiality for fracture. *Eng. Fract. Mech.* **2005**, *72*, 1049–1069. [CrossRef]

17. Johnson, G.R.; Cook, W.H. Fracture characteristics of three metals subjected to various strains, strain rates, temperatures and pressures. *Eng. Fract. Mech.* **1985**, *21*, 31–48. [CrossRef]

18. Senthil, K.; Iqbal, M.A. Effect of projectile diameter on ballistic resistance and failure mechanism of single and layered aluminum plates. *Theor. Appl. Fract. Mech.* **2013**, *67–68*, 53–64. [CrossRef]

19. Xue, L.; Ling, X.; Yang, S.S. Mechanical behaviour and strain rate sensitivity analysis of TA2 by the small punch test. *Theor. Appl. Fract. Mech.* **2019**, *99*, 9–17. [CrossRef]

20. Benzerga, A.A.; Surovik, D.A.; Keralavarma, S.M. On the path-dependence of the fracture locus in ductile materials–analysis. *Int. J. Plasticity* **2012**, *37*, 157–170. [CrossRef]

21. Neukamm, F.; Feucht, M.; Haufe, A. Considering damage history in crashworthiness simulations. In Proceedings of the 7th European LS-DYNA Conference, Salzburg, Austria, 14–15 May 2009.

22. Neukamm, F.; Feucht, M.; Haufe, A.; Roll, K. On closing the constitutive gap between forming and crash simulation. In Proceedings of the 10th Internation LS-DYNA Users Conference, Dearborn, MI, USA, 8–10 June 2008.

23. Anderson, D.; Butcher, C.; Pathak, N.; Worswick, M.J. Failure parameter identification and validation for a dual-phase 780 steel sheet. *Int. J. Solids. Struct.* **2017**, *124*, 89–107. [CrossRef]

24. Andrade, F.X.C.; Feucht, M.; Haufe, A.; Neukamm, F. An incremental stress state dependent damage model for ductile failure prediction. *Int. J. Fract.* **2016**, *200*, 127–150. [CrossRef]

25. Effelsberg, J.; Haufe, A.; Feucht, M.; Neukamm, F.; Du Bois, P. On parameter identification for the GISSMO damage model. In Proceedings of the 12th International LS-DYNA®Users Conference, Dearborn, MI, USA, 3–5 June 2012.

26. Haufe, A.; Feucht, M.; Neukamm, F. The Challenge to Predict Material Failure in Crashworthiness Applications: Simulation of Producibility to Serviceability. In *Predictive Modeling of Dynamic Processes: A Tribute to Professor Klaus Thoma*; Hiermaier, S., Ed.; Springer: Boston, MA, USA, 2009.

27. Heibel, S.; Nester, W.; Clausmeyer, T.; Tekkaya, A.E. Influence of different yield loci on failure prediction with damage models. In *36th Iddrg Conference—Materials Modelling and Testing for Sheet Metal Forming*; Volk, W., Ed.; IOP Publishing: Munich, Germany, 2017; Volume 896.

28. Cingara, A.M.; McQueen, H.J. New formula for calculating flow curves from high temperature constitutive data for 300 austenitic steels. *J. Mater. Process. Technol.* **1992**, *36*, 31–42. [CrossRef]

29. Khoddam, S.; Lam, Y.C.; Thomson, P.F. The effect of specimen geometry on the accuracy of the constitutive equation derived from the hot torsion test. *Steel Res.* **1995**, *66*, 45–49. [CrossRef]

30. Hu, Y.; Xiao, Y.; Jin, X.; Zheng, H.; Zhou, Y.; Shao, J. Experiments and FEM simulations of fracture behaviors for ADC12 aluminum alloy under impact load. *WITH Mater. Int.* **2016**, *22*, 1015–1025. [CrossRef]

31. Xiao, Y.; Tang, Q.; Hu, Y.; Peng, J.; Luo, W. Flow and fracture study for ZK60 alloy at dynamic strain rates and different loading states. *Mater. Sci. Eng. A Struct* **2018**, *724*, 208–219. [CrossRef]

32. Gavrus, A.; Massoni, E.; Chenot, J.L. An inverse analysis using a finite element model for identification of rheological parameters. *J. Mater. Process. Technol.* **1996**, *60*, 447–454. [CrossRef]

33. Gelin, J.C.; Ghouati, O. An inverse method for determining viscoplastic properties of aluminium alloys. *J. Mater. Process. Technol.* **1994**, *45*, 435–440. [CrossRef]

34. Maniatty, A.; Zabaras, N. Method for solving inverse elastoviscoplastic problems. *J. Eng. Mech.* **1989**, *115*, 2216–2231. [CrossRef]

35. Ludwik, P. *Elemente der technologischen Mechanik*; Springer: Berlin, Germany, 2013.

36. Basaran, M.; Wölkerling, S.D.; Feucht, M.; Neukamm, F.; Weichert, D.; AG, D. An extension of the GISSMO damage model based on lode angle dependence. In Proceedings of the LS-DYNA Anwenderforum, Bamberg, Germany, 30 September–1 October 2010.

37. Xue, L. Ductile fracture modeling: theory, experimental investigation and numerical verification. Ph.D. Thesis, Massachusetts Institute of Technology, Cambridge, UK, 2007.

38. Tasan, C.C. Micro-mechanical characterization of ductile damage in sheet metal. Ph.D. Thesis, Eindhoven University of Technology, Eindhoven, The Netherlands, 2010.

39. Weck, A.G. The role of coalescence on ductile fracture. Ph.D. Thesis, McMaster University, Hamilton, ON, Canada, 2007.

40. Johnson, G.R. A Constitutive Model and Data for Metals Subjected to Large Strains, High Strain Rates and High Temperatures. In Proceedings of the International Symposium on Ballistics, Hague, The Netherlands, 19–21 April 1983; pp. 541–548.

© 2019 by the authors. Licensee MDPI, Basel, Switzerland. This article is an open access article distributed under the terms and conditions of the Creative Commons Attribution (CC BY) license (http://creativecommons.org/licenses/by/4.0/).

 *metals*

Article

A Macroscopic Strength Criterion for Isotropic Metals Based on the Concept of Fracture Plane

Jiefei Gu [1,2,*], **Puhui Chen** [2,*], **Ke Li** [3] **and Lei Su** [3]

1   Jiangsu Key Laboratory of Advanced Food Manufacturing Equipment and Technology, Jiangnan University, Wuxi 214122, China
2   State Key Laboratory of Mechanics and Control of Mechanical Structures, Nanjing University of Aeronautics and Astronautics, Nanjing 210016, China
3   School of Mechanical Engineering, Jiangnan University, Wuxi 214122, China; like@jiangnan.edu.cn (K.L.); lei_su2015@jiangnan.edu.cn (L.S.)
*   Correspondence: jfgu@jiangnan.edu.cn (J.G.); phchen@nuaa.edu.cn (P.C.); Tel.: +86-15150673608 (J.G.); +86-13813988209 (P.C.)

Received: 16 May 2019; Accepted: 30 May 2019; Published: 31 May 2019

Abstract: Although the linear Mohr–Coulomb criterion is frequently applied to predict the failure of brittle materials such as cast iron, it can be used for ductile metals too. However, the criterion has some significant deficiencies which limit its predictive ability. In the present study, the underlying failure hypotheses of the linear Mohr–Coulomb criterion were thoroughly discussed. Based on Mohr's physically meaningful concept of fracture plane, a macroscopic strength criterion was developed to explain the failure mechanism of isotropic metals. The failure function was expressed as a polynomial expansion in terms of the stresses acting on the fracture plane, and the quadratic approximation was employed to describe the non-linear behavior of the failure envelope. With an in-depth understanding of Mohr's fracture plane concept, the failure angle was regarded as a generalized strength parameter in addition to the failure stress (i.e., the conventional basic strength). The undetermined coefficients of the non-linear failure function were calibrated by the strength parameters obtained from the common uniaxial tension and compression tests. Theoretical and experimental assessment for different types of isotropic metals validated the effectiveness of the proposed criterion in predicting material failure.

Keywords: macroscopic strength criterion; isotropic metals; fracture plane; linear Mohr–Coulomb criterion; failure mechanism

---

1. Introduction

A considerable number of failure criteria for isotropic materials have been developed since the establishment of classical mechanics [1]. Among all the proposed criteria, the Mises criterion is extensively used for ductile metals. However, it cannot be applied to metallic materials which have the strength difference effect (i.e., the uniaxial tensile strength T is not equal to the uniaxial compressive strength C). The linear Mohr–Coulomb criterion is very popular due to its simplicity and general applicability. Although the criterion is frequently used to predict the failure of brittle materials such as cast iron, it can also be applied to very ductile metals [2]. In the special case of ductile materials without the strength difference effect, the criterion degenerates into the maximum shear stress criterion (also known as the Tresca criterion), which is widely-used for conventional ductile metals.

Both Mohr and Coulomb have made an important contribution to the linear Mohr–Coulomb criterion. Mohr proposed that the fracture limit of a material is determined by the stress components σ_n and τ_n on the fracture plane [3] (see Figure 1). Coulomb's assumption is based on a linear failure envelope to determine the critical combination of σ_n and τ_n [4], which gives:

$$\tau_n + \mu\sigma_n = c. \tag{1}$$

Under uniaxial loading, material failure occurs when Mohr's circle for uniaxial tension or compression is just tangent to the envelope (see Figure 2). Hence the two material-specific parameters μ and c in Equation (1) can be calibrated by the uniaxial tensile strength T and the uniaxial compressive strength C.

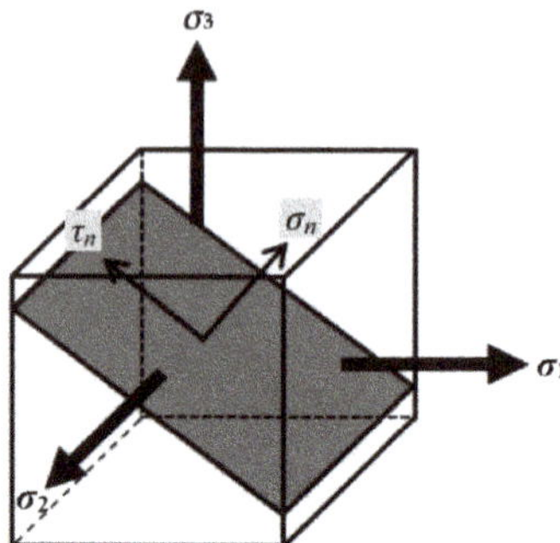

Figure 1. Stress components σ_n and τ_n on the fracture plane.

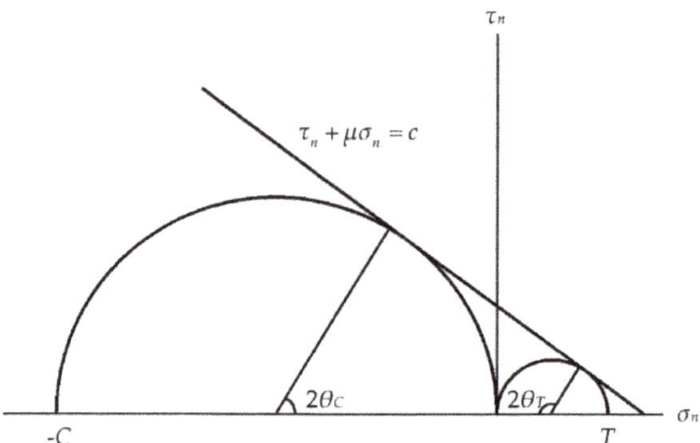

Figure 2. Coulomb's linear envelope.

In the principal stress space (σ_1, σ_2, σ_3), the linear Mohr–Coulomb criterion can be expressed in an extremely simple form of

$$\frac{\sigma_1}{T} - \frac{\sigma_3}{C} = 1, \tag{2}$$

where σ_1 and σ_3 are the maximum and minimum principal stresses respectively.

Although the classic Mohr–Coulomb criterion has been extensively used in research and engineering [5], it has several significant deficiencies which limit its predictive ability:

Firstly, the fracture angles predicted by the linear Mohr–Coulomb criterion do not always agree with the experimental observations. For example, experimental results show that typical brittle metals such as cast iron with $T/C = 1/4$ fail in tension on the plane parallel to the action plane of the applied load, i.e., the tensile fracture angle $\theta_T = 90.0°$ (see Figure 3a). However, the criterion predicts the tensile fracture angle $\theta_T = 63.4°$. Under uniaxial compression the fracture angle of cast iron should be approximately $37.0°$ (see Figure 3b), yet the predicted angle $\theta_C = 26.6°$. Moreover, the summation of the tensile and compressive failure angles is exactly $90°$ for all types of isotropic metals according to the prediction of the linear Mohr–Coulomb criterion. This conclusion does not correlate with the measured data of cast iron ($\theta_T = 90.0°$, $\theta_C = 37.0°$ (see Figure 3)) and metallic glass ($\theta_T = 50.7°$, $\theta_C = 43.0°$ (see Figure 4)).

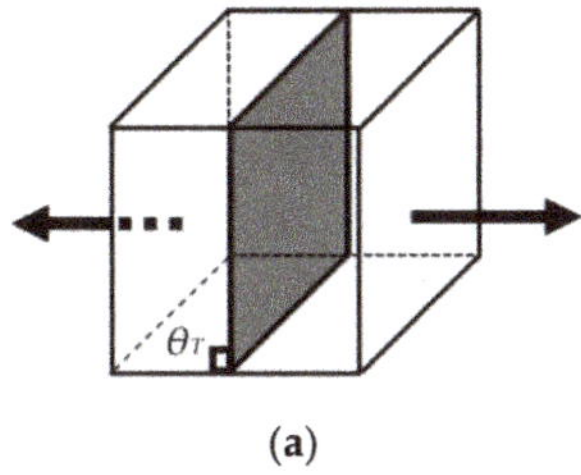
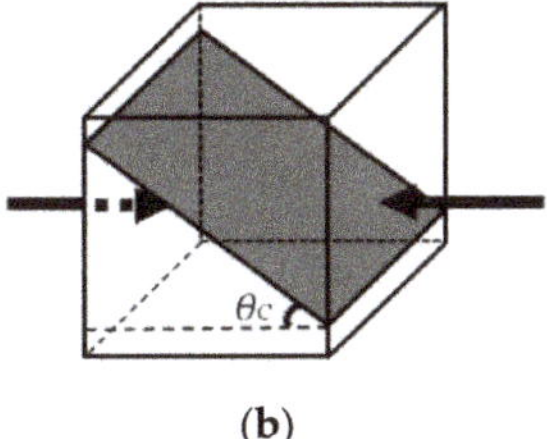

Figure 3. Fracture angles of cast iron under uniaxial tension and compression [6]. (**a**) Uniaxial tension, $\theta_T = 90.0°$; (**b**) uniaxial compression, $\theta_C = 37.0°$.

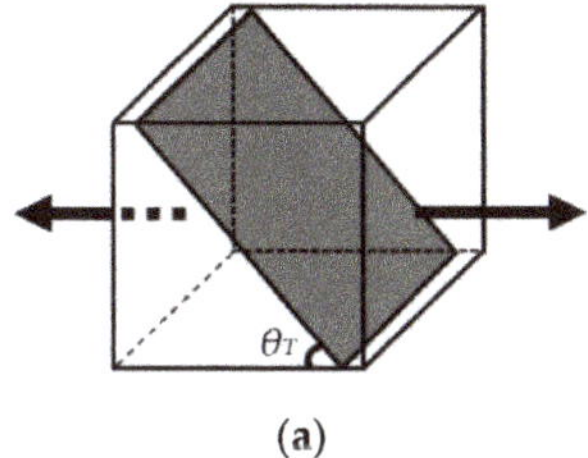
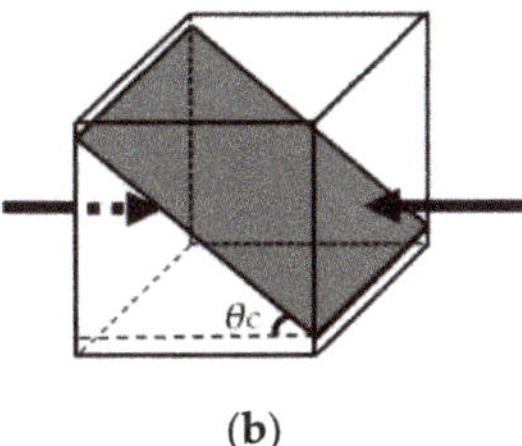

Figure 4. Fracture angles of metallic glass under uniaxial tension and compression [7]. (**a**) Uniaxial tension, $\theta_T = 50.7°$; (**b**) uniaxial compression, $\theta_C = 43.0°$.

Secondly, the linear Mohr–Coulomb criterion cannot explain the pure shear fracture behavior of cast iron with $T/C = 1/4$. The predicted pure shear strength is $S = 0.8T$ with the fracture angle $\theta_S = 26.6°$, while the measured strength is $S = T$ with the fracture angle $\theta_S = 45°$ (see Figure 5).

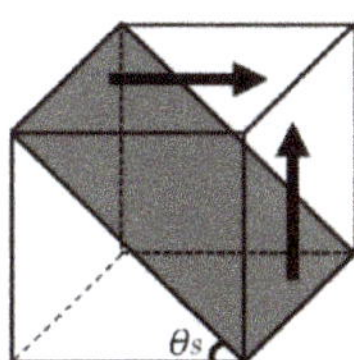

Figure 5. Fracture angle of cast iron under pure shear, $\theta_S = 45°$ [8].

Thirdly, the criterion asserts that the equi-triaxial tensile strength T_{tri} is much stronger than the uniaxial tensile strength T for brittle materials. This unphysical behavior has no supporting evidence [9].

Lastly, Christensen [9] gave a simple example which shows the inaccuracy of the linear Mohr–Coulomb criterion. Take a 3D compressive stress state given by $\sigma_1 = \sigma_2 = -\sigma$, $\sigma_3 = -2\sigma$. The criterion predicts that isotropic materials with $T/C \le 1/2$ can sustain unlimited compressive stresses, which is completely unrealistic.

The above problems reveal the inappropriateness of the linear Mohr–Coulomb criterion for isotropic metals in certain cases. Much effort has been made to modify the criterion. Paul suggested combining the linear Mohr–Coulomb criterion with the maximum normal stress criterion [10]. Yu proposed the twin-shear strength theory to replace the single shear strength theory (i.e., the linear Mohr–Coulomb criterion) [1]. Bigoni and Piccolroaz generalized the criterion using the invariants of the stress tensor [11]. However, Mohr's physically meaningful concept of fracture plane was ignored by these researchers, thus the aforementioned contradictions between the predictions and experimental results cannot be fundamentally solved.

It is worth noting that the concept of critical plane in fatigue analysis is similar to the aforementioned concept of fracture plane in static failure analysis. There are also a lot of critical plane-based fatigue failure criteria for metallic materials [12]. Brown and Miller [13] assumed the critical plane is the plane with maximum shear strain, and proposed that fatigue failure depends on the combination of the normal and shear strains acting on the critical plane. Glinka et al. [14] applied the normal and shear strain energy densities on the critical plane instead of the strains, formulating a strain energy density criterion based on the critical plane approach. The approach has been recently extended to the nanoscale by Gallo et al. [15,16].

In the present study, the linear Mohr–Coulomb criterion is modified based on an in-depth understanding of Mohr's concept of fracture plane. Not only the uniaxial strengths T and C but also the failure angles θ_T and θ_C are used as basic strength parameters to calibrate the unknown coefficients of the non-linear failure function. The macroscopic strength criterion shows good agreement with the experimental data of different types of isotropic metals, and has a better predictive ability compared with the linear Mohr–Coulomb criterion.

2. Discussion of the Linear Mohr–Coulomb Criterion

Since the proposed macroscopic strength criterion is based on Mohr's fracture plane concept, the failure hypotheses of the linear Mohr–Coulomb criterion are re-examined at first.

Material failure often originates from a specific plane [17]. Mohr proposed that the fracture limit of a material is determined by the stress components σ_n and τ_n on the fracture plane (see Figure 1). As shown in Figure 6a,b, both the plane separation driven by the normal tensile stress σ_n and the plane sliding driven by the shear stress τ_n can result in macroscopic material failure. These two stress components are correlated with the two main failure mechanisms in solids: cleavage and slip, respectively [18]. The normal compressive stress increases the difficulty of shearing along the plane, thus suppressing material failure (see Figure 6c). From the microscopic point of view, materials contain micro defects to varying degrees. The tensile normal stress is expected to open these flaws and cause them to grow, whereas under the normal compressive stress the flaws tend to have their opposite sides pressed together [6]. The shear stress drives the dislocation movement of a large number of planes of atoms [19], leading to macroscopic material distortion. Hence it can be concluded that the normal tensile stress and the shear stress promote material failure, while the normal compressive stress inhibits failure.

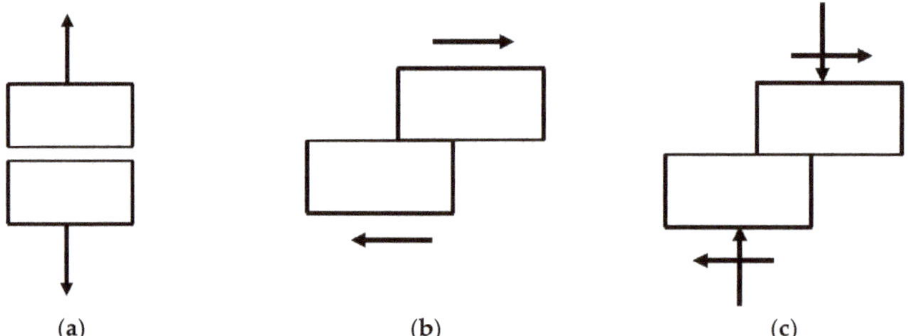

(a)   (b)   (c)

Figure 6. The effects of the normal stress σ_n and the shear stress τ_n. (**a**) Plane separation driven by the normal tensile stress; (**b**) plane sliding driven by the shear stress; (**c**) plane sliding under the combined shear stress and normal compressive stress.

The undetermined coefficients of failure criteria are usually calibrated by the maximum sustainable stresses under certain special loading conditions. The absolute value of the maximum sustainable stress is commonly referred to as "basic strength", which is calculated by dividing the applied load by the area of its own action plane [3]. The linear Mohr–Coulomb criterion takes the uniaxial tensile strength T and the uniaxial compressive strength C as the basic strengths.

It can be inferred from Mohr's fracture hypothesis that the maximum sustainable stresses actually should be the stress components acting on the fracture plane. Nevertheless, the action plane of the applied load may not be parallel to the fracture plane. For example, as observed in the uniaxial tension test, the fracture of metallic glass occurs on an inclined plane of $\theta_T = 50.7°$ (see Figure 4a). The uniaxial tensile strength T is defined as the value of the tensile failure stress (i.e., the uniaxial tensile failure load divided by the area of its action plane). However, it is the normal tensile stress component $\sigma_n = \frac{T}{2}(1 - \cos 2\theta_T)$ and the shear stress component $\tau_n = \frac{T}{2}\sin 2\theta_T$ on the fracture plane that lead to material failure. Similarly, the fracture of metallic glass occurs on an inclined plane of $\theta_C = 43.0°$ under the uniaxial compressive loading (see Figure 4b). The compressive fracture behavior of metallic glass is actually determined by the normal compressive stress component $\sigma_n = \frac{C}{2}(\cos 2\theta_C - 1)$ and the shear stress component $\tau_n = \frac{C}{2}\sin 2\theta_C$ on the fracture plane. Therefore, it is insufficient to characterize Mohr's concept solely by the conventional basic strength. Only by both the failure stress (i.e., the basic strength) and the failure angle, can the maximum sustainable stresses on the fracture plane be determined. It indicates that in strict accordance with Mohr's fracture plane concept, both the conventional strength value and the failure angle should be measured in a uniaxial test [20].

Although Coulomb's linear failure envelope is able to distinguish the different effects of the normal tensile and compressive stresses on material failure, the experimentally-determined envelopes often exhibit non-linear behavior [21]. Thus, the linear strength response is regarded as a major limitation of the classic Mohr–Coulomb criterion [19], and a non-linear form of the envelope is supposed to fit the experimental data better. Nevertheless, besides the common uniaxial tension and compression tests, additional experiments are usually required in order to determine the unknown parameters of the non-linear failure function. For example, the pure torsion test is needed in our previous research work [22]; equi-biaxial tension, equi-biaxial compression or other combined stress state tests are needed in the criterion proposed by Hu and Wang [23].

3. Formulation of the Strength Criterion

In this section, we propose a feasible method to modify the linear Mohr–Coulomb criterion. Only two common types of tests (i.e., the uniaxial tension and compression tests) are required in order to use the present criterion.

3.1. Mathematical Expression of the Failure Function

The mathematical expression of the failure function is constructed using the general approach put forward by us [22,24], which is briefly described below:

According to Mohr's fracture hypothesis, the failure function, F, should be the function of the stress components (σ_n, τ_n) on the fracture plane. Material failure occurs when $F(\sigma_n, \tau_n)$ reaches the failure index 1. Expanding F into a polynomial in terms of (σ_n, τ_n), we get:

$$F(\sigma_n, \tau_n) = \alpha\sigma_n + \beta\sigma_n^2 + \gamma\sigma_n\tau_n + \lambda\tau_n + \omega\tau_n^2 + \ldots = 1, \tag{3}$$

where $\ldots$ represents the terms of cubic and higher orders.

The quadratic form is frequently chosen as the non-linear failure function for isotropic materials due to its relatively good curve-fitting results [7,22,25,26]. In addition, much more experimental data are required to determine the unknown coefficients if cubic or higher order approximations are employed. Therefore, in the present study the failure function F is truncated at the quadratic order, i.e.,

$$F(\sigma_n, \tau_n) = \alpha\sigma_n + \beta\sigma_n^2 + \gamma\sigma_n\tau_n + \lambda\tau_n + \omega\tau_n^2 = 1. \tag{4}$$

Whether the shear stress component is positive or negative, it always makes an identical contribution to material failure. Hence the linear terms of the shear stress component, namely $\sigma_n\tau_n$ and τ_n, shall be vanished in Equation (4), leaving

$$F(\sigma_n, \tau_n) = \alpha\sigma_n + \beta\sigma_n^2 + \omega\tau_n^2 = 1. \tag{5}$$

The normal tensile stress on the fracture plane promotes material failure, while the normal compressive stress inhibits failure. It has been demonstrated that the normal tensile stress has much more pronounced effect on material failure than the normal compressive stress [7]. Therefore, unlike the linear Mohr–Coulomb criterion, the failure behaviors under the normal tensile and compressive stresses are treated separately in the present theory:

$$F(\sigma_n, \tau_n) = \alpha_C\sigma_n + \beta_C\sigma_n^2 + \omega\tau_n^2 = 1 \text{ for } \sigma_n \le 0, \tag{6}$$

$$F(\sigma_n, \tau_n) = \alpha_T\sigma_n + \beta_T\sigma_n^2 + \omega\tau_n^2 = 1 \text{ for } \sigma_n > 0, \tag{7}$$

where $\alpha_C, \beta_C, \alpha_T, \beta_T$ and ω are the undetermined parameters.

3.2. Failure Function for $\sigma_n \le 0$

We first consider the equi-triaxial compressive strength condition $\sigma_1 = \sigma_2 = \sigma_3 = -C_{tri}$. The stress components on any section plane are given by ($\sigma_n = -C_{tri}$, $\tau_n = 0$) under hydrostatic compression. Substituting ($\sigma_n = -C_{tri}$, $\tau_n = 0$) into Equation (6), we get:

$$\beta_C = \frac{1}{C_{tri}^2} + \frac{\alpha_C}{C_{tri}}. \tag{8}$$

Experiments have shown that isotropic materials can be loaded to very high values of hydrostatic pressure without failure [19], i.e., $C_{tri} \to \infty$. Hence β_C can be approximated by

$$\beta_C = 0. \tag{9}$$

Additional information can be obtained from the uniaxial compression test. As is discussed in Section 2, both the uniaxial compressive strength C and the corresponding failure angle θ_C are used as generalized strength parameters in the present theory. As shown in Figure 7, the stress components on the potential failure plane under uniaxial compression are given by:

$$\sigma_n = \frac{C}{2}(\cos 2\theta - 1), \tag{10}$$

$$\tau_n = \frac{C}{2}\sin 2\theta. \tag{11}$$

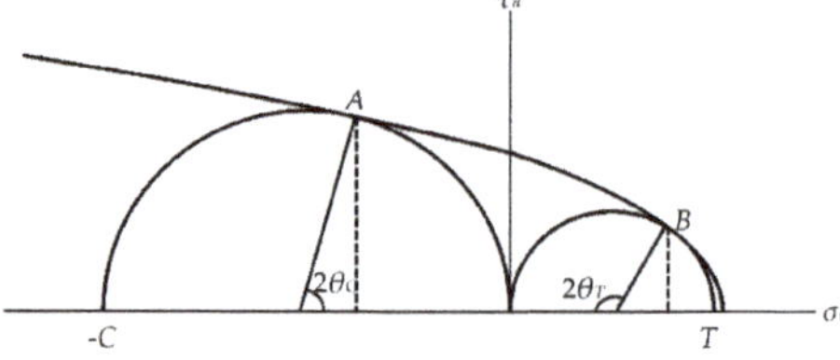

Figure 7. Failure envelope and Mohr's circles for uniaxial tension and compression.

Substituting Equations (9)–(11) into the failure function F in Equation (6), we get:

$$F(\theta) = \alpha_C \frac{C}{2}(\cos 2\theta - 1) + \omega\left(\frac{C}{2}\sin 2\theta\right)^2. \tag{12}$$

The failure function F reaches the maximum value 1 when failure occurs on the action plane oriented at $\theta = \theta_C$:

$$F(\theta_C) = \alpha_C \frac{C}{2}(\cos 2\theta_C - 1) + \omega\left(\frac{C}{2}\sin 2\theta_C\right)^2 = 1, \tag{13}$$

$$\left.\frac{dF}{d\theta}\right|_{\theta=\theta_C} = \left(-\alpha_C C + \omega C^2 \cos 2\theta_C\right)\sin 2\theta_C = 0. \tag{14}$$

As shown in Figure 7, there is no stress component on the plane oriented at $\theta = 0°$, and only the normal compressive stress which inhibits material failure acts on the plane oriented at $\theta = 90°$. Therefore, $\sin 2\theta_C \neq 0$ always holds. Solving Equation (14), we get

$$-\alpha_C + C\cos 2\theta_C \omega = 0. \tag{15}$$

The coefficients α_C and ω can be determined by solving Equations (13) and (15) together:

$$\alpha_C = \frac{4\cos 2\theta_C}{C(\cos 2\theta_C - 1)^2}, \tag{16}$$

$$\omega = \frac{4}{C^2(\cos 2\theta_C - 1)^2}. \tag{17}$$

3.3. Failure Function for $\sigma_n > 0$

The uniaxial tensile strength T and the corresponding failure angle θ_T are used to calibrate the undetermined parameters of the failure function for $\sigma_n > 0$. As shown in Figure 7, the stress components on the potential failure plane under uniaxial tension are given by:

$$\sigma_n = \frac{T}{2}(1 - \cos 2\theta), \tag{18}$$

$$\tau_n = \frac{T}{2}\sin 2\theta. \tag{19}$$

Substituting Equations (18) and (19) into the failure function F in Equation (7), we obtain:

$$F(\theta) = \alpha_T \frac{T}{2}(1 - \cos 2\theta) + \beta_T\left[\frac{T}{2}(1 - \cos 2\theta)\right]^2 + \omega\left(\frac{T}{2}\sin 2\theta\right)^2. \tag{20}$$

The failure function F reaches the maximum value 1 when failure occurs on the action plane oriented at $\theta = \theta_T$:

$$F(\theta_T) = \alpha_T \frac{T}{2}(1 - \cos 2\theta_T) + \beta_T\left[\frac{T}{2}(1 - \cos 2\theta_T)\right]^2 + \omega\left(\frac{T}{2}\sin 2\theta_T\right)^2 = 1, \tag{21}$$

$$\left.\frac{dF}{d\theta}\right|_{\theta=\theta_T} = [\alpha_T + \beta_T T(1 - \cos 2\theta_T) + \omega T\cos 2\theta_T]T\sin 2\theta_T = 0. \tag{22}$$

As shown in Figure 7, since there is no stress component on the plane oriented at $\theta = 0°$, $\theta_T \neq 0°$ always holds. If $\theta_T \neq 90°$, from Equation (22) we obtain:

$$\alpha_T + \beta_T T(1 - \cos 2\theta_T) + \omega T\cos 2\theta_T = 0. \tag{23}$$

The coefficients α_T and β_T can be determined by solving Equations (17), (21), and (23) together:

$$\alpha_T = \frac{4}{T(1 - \cos 2\theta_T)} - \frac{4T}{C^2(1 - \cos 2\theta_C)^2},$$ (24)

$$\beta_T = \frac{4}{C^2(1 - \cos 2\theta_C)^2} - \frac{4}{T^2(1 - \cos 2\theta_T)^2}.$$ (25)

If $\theta_T = 90°$, from Equation (21) we get:

$$\alpha_T T + \beta_T T^2 = 1.$$ (26)

However, Equation (22) is naturally satisfied if $\theta_T = 90°$. Therefore, in this case the uniaxial tension test actually provides only one equation for determination of the two unknown coefficients α_T and β_T. No supplementary information is available from the uniaxial tension test to establish another equation. Since extra experiments may be difficult, time-consuming, and expensive, an alternative method is proposed to construct an additional equation.

The failure envelope is commonly expected to be as smooth as possible considering the implementation of numerical methods [19]. The left derivative of the failure envelope at $\sigma_n = 0$ can be derived from Equation (6):

$$\left. \frac{d\tau_n}{d\sigma_n} \right|_{\sigma_n=0_-} = -\frac{\alpha_C}{2\omega \left. \tau_n \right|_{\sigma_n=0}},$$ (27)

and the right derivative of the failure envelope at $\sigma_n = 0$ can be derived from Equation (7):

$$\left. \frac{d\tau_n}{d\sigma_n} \right|_{\sigma_n=0_+} = -\frac{\alpha_T}{2\omega \left. \tau_n \right|_{\sigma_n=0}}.$$ (28)

Applying the smooth condition, we get:

$$\alpha_C = \alpha_T.$$ (29)

The coefficients α_T and β_T can be determined by solving Equations (17), (26) and (29) together:

$$\alpha_T = \frac{4 \cos 2\theta_C}{C(\cos 2\theta_C - 1)^2},$$ (30)

$$\beta_T = \frac{1}{T^2} - \frac{4 \cos 2\theta_C}{TC(\cos 2\theta_C - 1)^2}.$$ (31)

3.4. Function of the Failure Envelope

To sum up, the function of the failure envelope is expressed as

$$F(\sigma_n, \tau_n) = \begin{cases} \alpha_C \sigma_n + \omega \tau_n^2 = 1 & \text{for } \sigma_n \leq 0 \\ \alpha_T \sigma_n + \beta_T \sigma_n^2 + \omega \tau_n^2 = 1 & \text{for } \sigma_n > 0 \end{cases},$$ (32)

where

$$\alpha_C = \frac{4 \cos 2\theta_C}{C(\cos 2\theta_C - 1)^2},$$ (33)

$$\omega = \frac{4}{C^2(\cos 2\theta_C - 1)^2},$$ (34)

$$\alpha_T = \begin{cases} \dfrac{4}{T(1-\cos 2\theta_T)} - \dfrac{4T}{C^2(1-\cos 2\theta_C)^2} & \text{if } \theta_T \neq 90° \\[2ex] \dfrac{4\cos 2\theta_C}{C(\cos 2\theta_C - 1)^2} & \text{if } \theta_T = 90° \end{cases} \tag{35}$$

$$\beta_T = \begin{cases} \dfrac{4}{C^2(1-\cos 2\theta_C)^2} - \dfrac{4}{T^2(1-\cos 2\theta_T)^2} & \text{if } \theta_T \neq 90° \\[2ex] \dfrac{1}{T^2} - \dfrac{4\cos 2\theta_C}{TC(\cos 2\theta_C - 1)^2} & \text{if } \theta_T = 90° \end{cases} \tag{36}$$

The terms $\alpha_C \sigma_n$, $\omega \tau_n^2$, and $\alpha_T \sigma_n + \beta_T \sigma_n^2$ in Equation (32) represent the contribution of the normal compressive stress, the normal tensile stress, and the shear stress to material failure respectively.

4. Theoretical and Experimental Evaluation

4.1. Failure Modes under Uniaxial Tension and Compression

The uniaxial tension and compression tests were conducted on three different types of isotropic metals, namely the ductile metallic material, metallic glass, and brittle cast iron. The specimens were elaborately designed to avoid either material damage near the clamping end under uniaxial tension, or buckling and end effects under uniaxial compression. The tensile and compressive specimens were tested at a constant strain rate using the universal testing machine, and the failure strengths and failure angles were measured carefully. Further details about the experiments can be found in [6,7,27]. The strength parameters of the isotropic metals are listed in Table 1, while the failure angles θ_T and θ_C predicted by the linear Mohr–Coulomb criterion are listed in Table 2. Comparison between Tables 1 and 2 shows that the predicted failure angles of the three tested materials are not entirely consistent with the measured values, especially in the case of brittle cast iron.

Table 1. Strength parameters of three different types of isotropic metals.

Material Type	C/T	θ_T	θ_C
Ductile metallic material [7]	1.00	45.0°	45.0°
Metallic glass [7]	1.11	50.7°	43.0°
Brittle cast iron [6]	4.00	90.0°	37.0°

Table 2. Failure angles θ_T and θ_C predicted by the linear Mohr–Coulomb criterion

Material Type	C/T	θ_T (Predicted)	θ_C (Predicted)
Ductile metallic material	1.00	45.0°	45.0°
Metallic glass	1.11	46.5°	43.5°
Brittle cast iron	4.00	63.4°	26.6°

As shown in Table 3, since the normal compressive stress σ_n suppresses material failure under uniaxial compression, $\alpha_C \sigma_n \leq 0$ always holds. Hence all types of isotropic metals fail in the shear mode under uniaxial compression. Under uniaxial tension, because both the normal tensile stress and the shear stress promote material failure, the terms related to σ_n and τ_n are always non-negative (see Table 4). With the increase of material brittleness, the failure mode gradually transfers from shear to tension under the uniaxial tensile loading.

Table 3. Failure modes under uniaxial compression

Material Type	$\alpha_C \sigma_n$	$\omega \tau_n^2$	Failure Index	Failure Mode
Ductile metallic material	0.00	1.00		Shear
Metallic glass	−0.15	1.15	1.00	Shear
Brittle cast iron	−0.76	1.76		Shear

Table 4. Failure modes under uniaxial tension

Material Type	$\alpha_T \sigma_n + \beta_T \sigma_n^2$	$\omega \tau_n^2$	Failure Index	Failure Mode
Ductile metallic material	0.00	1.00		Shear
Metallic glass	0.10	0.90	1.00	Combination of shear and tension
Brittle cast iron	1.00	0.00		Tension

4.2. Ductile Metallic Material

For conventional ductile metallic materials such as Al-alloy, Ti-alloy and steels, material failure is usually specified by yielding. Therefore, the measured strength and failure plane actually should be the yield strength and slip plane of ductile metals. Nearly no difference between the tensile and compressive strengths can be observed for these materials [7]. In the ductile limiting case $T = C$, $\theta_T = \theta_C = 45°$, both the present criterion and the linear Mohr–Coulomb criterion degenerate into the form of

$$\tau_n = \frac{\sigma_1 - \sigma_3}{2} = \frac{T}{2},$$ (37)

see Figure 8. Equation (37) is the exact form of the Tresca criterion, which is suitable for typical ductile materials.

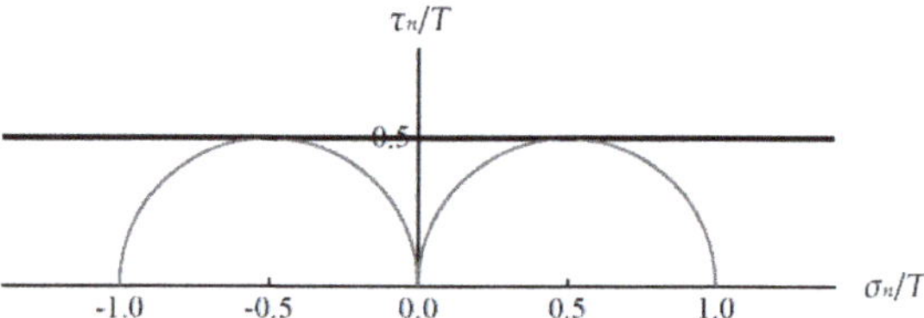

Figure 8. Failure envelope of typical ductile material. $C/T = 1$, $\theta_T = 45°$, and $\theta_C = 45°$.

4.3. Metallic Glass with Moderate Ductility

Metallic glass is a kind of high-strength isotropic material, yet with relatively lower ductility than conventional ductile metallic materials. The linear Mohr–Coulomb criterion has been applied to describe the fracture behavior of metallic glass due to its ability to characterize the T–C strength asymmetry [28]. However, as shown in Figure 9, obviously the present criterion fits the experimental data better than the linear Mohr–Coulomb criterion in the high normal tensile stress range. In the normal compressive stress range, the envelope predicted by the present criterion is similar to that predicted by the linear Mohr–Coulomb criterion. It is worth noting that the proposed failure envelope is non-smooth at the transition location $\sigma_n = 0$. This is because the proposed function of the failure envelope, Equation (32), is only an acceptable, but not perfect approximation to the "true" or "ideal" failure function.

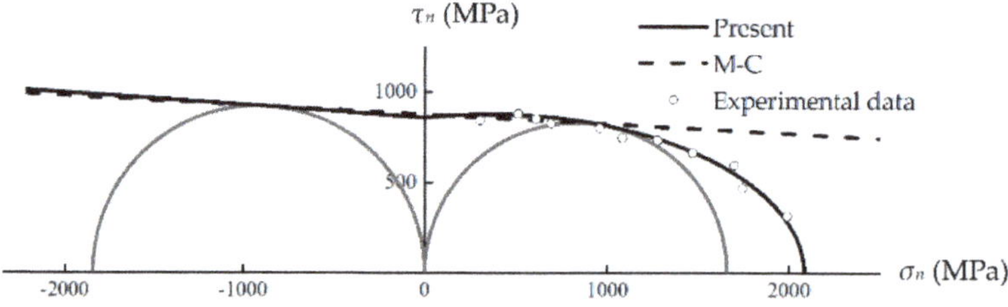

Figure 9. Failure envelopes and the experimental data of metallic glass [7]. $T = 1660$ MPa, $C = 1843$ MPa, $\theta_T = 50.7°$, and $\theta_C = 43.0°$.

4.4. Brittle Cast Iron

Cast iron is a typical brittle metal with the measured fracture angles $\theta_T = 90.0°$ and $\theta_C = 37.0°$ [6]. The failure envelopes predicted by the present criterion and the linear Mohr–Coulomb criterion are plotted in Figure 10, and Mohr's circles for uniaxial tension, uniaxial compression and pure shear are also depicted. The envelope given by the present criterion agrees with the experimental observation that fracture occurs on the plane with maximum tensile stress under uniaxial tension, i.e., $\theta_T = 90°$ (point B in Figure 10), whereas the fracture plane predicted by the linear Mohr–Coulomb criterion is incorrect. The present criterion also successfully predicts that under pure shear failure stress S, fracture occurs on the plane where the normal tensile stress $\sigma_n = S$ reaches its maximum value T (point B in Figure 10), i.e., $S = T$. However, the linear Mohr–Coulomb criterion fails to describe this brittle behavior. Besides, the linear Mohr–Coulomb criterion results in the over-valued equi-triaxial tensile strength $T_{tri} = 1.33T$, while $T_{\text{tri}} = T$ according to the present criterion.

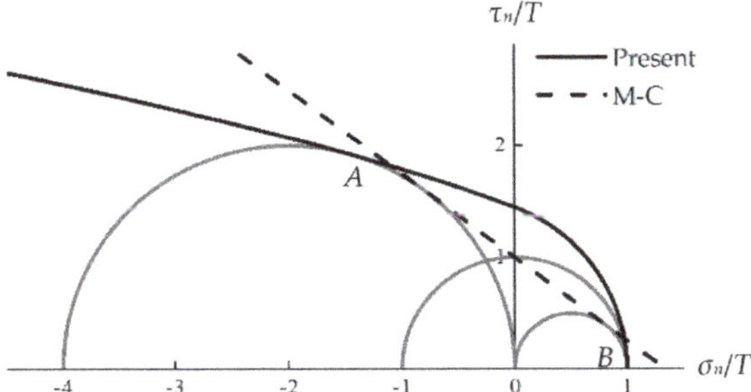

Figure 10. Failure envelopes and Mohr's stress circles of cast iron. $C/T = 4$, $\theta_T = 90.0°$, and $\theta_C = 37.0°$.

As shown in Figure 11, a limited maximum Mohr's circle is predicted by the present criterion under the stress state $\sigma_1 = \sigma_2 = -\sigma$, $\sigma_3 = -2\sigma$, indicating the failure stress σ is a finite value for cast iron with $C/T = 4$. Nevertheless, since Coulomb's linear envelope overestimates the material strength greatly in the normal compressive stress range, it results in the unrealistic prediction of infinite strength under this specific loading condition [9].

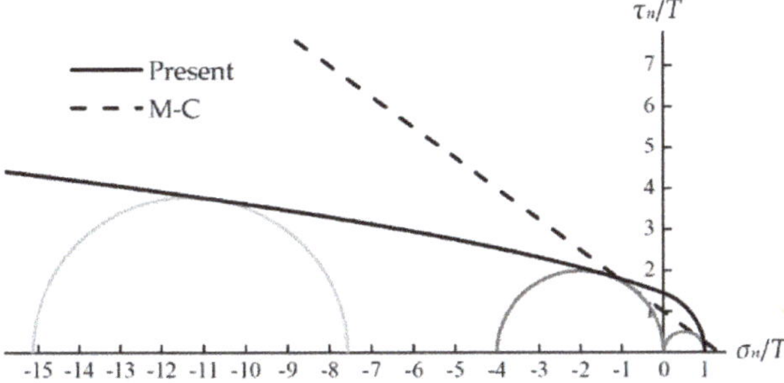

Figure 11. Maximum Mohr's stress circle under the stress state $\sigma_1 = \sigma_2 = -\sigma$, $\sigma_3 = -2\sigma$. $C/T = 4$, $\theta_T = 90.0°$, and $\theta_C = 37.0°$.

A series of experiments have been performed on cast irons subjected to combined stress loadings. As shown in Figures 12 and 13, the results predicted by the proposed criterion show good agreement with the test data, while the fit of the linear Mohr–Coulomb criterion is relatively poor.

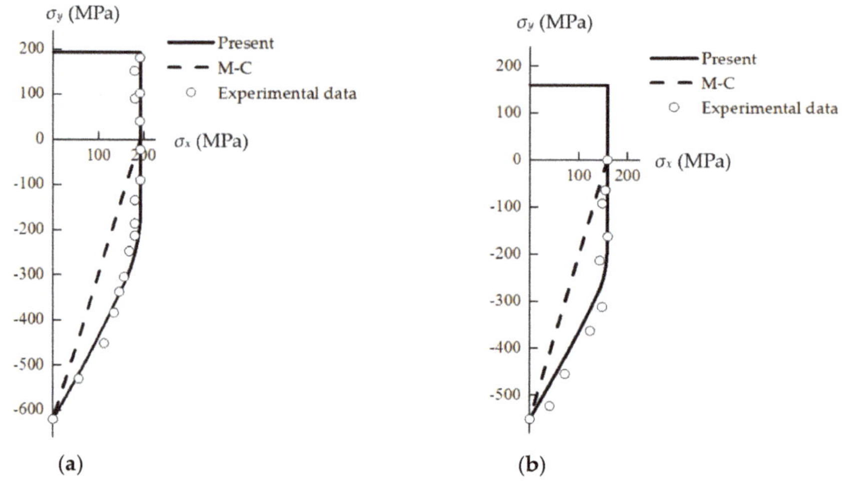

Figure 12. Failure envelopes and the experimental data of cast iron under the combined stress state σ_x-σ_y [29,30]. $\theta_T = 90.0°$, $\theta_C = 37.0°$. (**a**) $T = 193$ MPa, $C = 620.6$ MPa; (**b**) $T = 159$ MPa, $C = 551$ MPa.

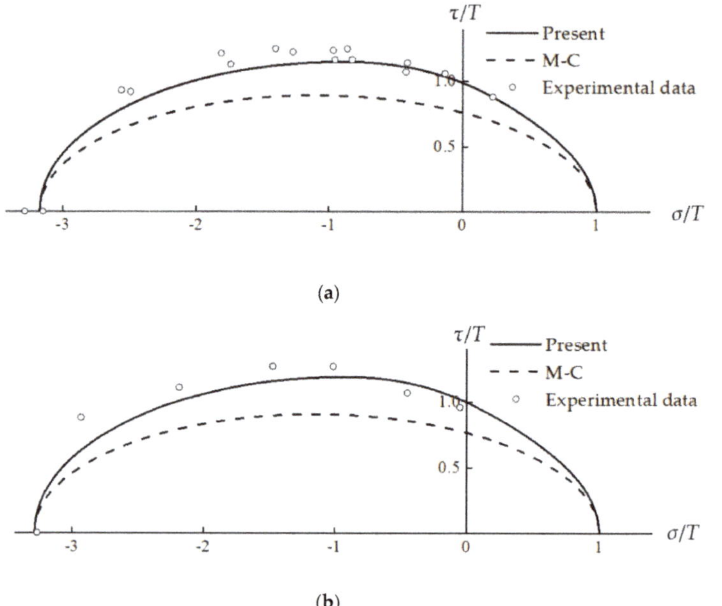

Figure 13. Failure envelopes and the experimental data of cast iron under the combined stress state σ-τ [29,30]. $\theta_T = 90.0°$, $\theta_C = 37.0°$. (**a**) $C/T = 3.18$; (**b**) $C/T = 3.29$.

5. Conclusions

In the present study, a macroscopic strength criterion for isotropic metals has been proposed to modify the linear Mohr–Coulomb criterion. It is developed on the basis of Mohr's physically meaningful concept of fracture plane, and the most notable features of the present criterion are as follows:

(1) Based on the observation that experimentally-determined envelopes often exhibit non-linear behavior, the quadratic approximation of the failure function is adopted to replace Coulomb's linear form.

(2) With an in-depth understanding of the concept of fracture plane, both the failure stress and the failure angle are used as generalized strength parameters to calibrate the undetermined coefficients of the non-linear failure function. Only two common types of tests (i.e., the uniaxial tension and compression tests) are required in order to use the criterion.

The validity of the proposed strength criterion was verified by comparing with the linear Mohr–Coulomb criterion and the experimental results of different kinds of isotropic metals. The macroscopic strength criterion has good accuracy and wide applicability.

Author Contributions: Conceptualization, J.G. and P.C.; formal analysis, J.G.; methodology, J.G.; data curation, K.L. and L.S.; funding acquisition, P.C. and K.L; writing—original draft preparation, J.G.; writing—review and editing, P.C., K.L. and L.S.

Funding: This research was funded by the National Natural Science Foundation of China (Grant No. 11572152), Key Project of Industry Foresight and Common Key Technologies of Science and Technology Department of Jiangsu Province (BE2017002-2), and Fundamental Research Funds for the Central Universities (Grant No. JUSRP51732B).

Conflicts of Interest: The authors declare no conflict of interest.

References

1. Yu, M.H. Advances in strength theories for materials under complex stress state in the 20th Century. *Appl. Mech. Rev.* **2002**, *55*, 169. [CrossRef]
2. Comanici, A.M.; Barsanescu, P.D. Modification of Mohr's criterion in order to consider the effect of the intermediate principal stress. *Int. J. Plast.* **2018**, *108*, 40–54. [CrossRef]
3. Knops, M. *Analysis of Failure in Fiber Polymer Laminates: The Theory of Alfred Puck*; Springer: Berlin, German, 2008; pp. 45–46.
4. Labuz, J.F.; Zang, A. Mohr–Coulomb Failure Criterion. *Rock Mech. Rock Eng.* **2012**, *45*, 975–979. [CrossRef]
5. Jiang, H.; Xie, Y. A note on the Mohr–Coulomb and Drucker–Prager strength criteria. *Mech. Res. Commun.* **2011**, *38*, 309–314. [CrossRef]
6. Dowling, N.E. *Mechanical Behavior of Materials/Engineering Methods for Deformation, Fracture, and Fatigue*, 4th ed.; Pearson Prentice Hall: Bergen County, NJ, USA, 2013; pp. 118, 300, 305, 311.
7. Qu, R.T.; Zhang, Z.F. A universal fracture criterion for high-strength materials. *Sci. Rep.* **2013**, *3*, 1117. [CrossRef]
8. Hibbeler, R.C. *Mechanics of Materials*, 9th ed.; Pearson Prentice Hall: Bergen County, NJ, USA, 2014; p. 528.
9. Christensen, R.M. *The Theory of Materials Failure*, 1st ed.; Oxford University Press: London, UK, 2013; pp. 8, 27.
10. Paul, B. Modification of the Coulomb-Mohr theory of fracture. *J. Appl. Mech.* **1960**, *28*, 259–268. [CrossRef]
11. Bigoni, D.; Piccolroaz, A. Yield criteria for quasibrittle and frictional materials. *Int. J. Solids Struct.* **2004**, *41*, 2855–2878. [CrossRef]
12. Karolczuk, A.; Macha, E. Critical planes in multiaxial fatigue. *Mater. Sci. Forum* **2005**, *482*, 109–114. [CrossRef]
13. Brown, M.W.; Miller, K.J. A theory for fatigue failure under multiaxial stress-strain conditions. *Proc. Inst. Mech. Eng.* **1973**, *187*, 745–755. [CrossRef]
14. Glinka, G.; Shen, G.; Plumtree, A. A multiaxial fatigue strain energy density parameter related to the critical fracture plane. *Fatigue Fract. Eng. Mater.* **1995**, *18*, 37–46. [CrossRef]
15. Gallo, P.; Sumigawa, T.; Kitamura, T.; Berto, F. Static assessment of nanoscale notched silicon beams using the averaged strain energy density method. *Theor. Appl. Fract. Mech.* **2018**, *95*, 261–269. [CrossRef]
16. Gallo, P.; Yan, Y.; Sumigawa, T.; Kitamura, T. Fracture behavior of nanoscale notched silicon beams investigated by the theory of critical distances. *Adv. Theory Simul.* **2018**, *1*, 1700006. [CrossRef]
17. Qu, R.T.; Zhang, Z.J.; Zhang, P.; Liu, Z.Q.; Zhang, Z.F. Generalized energy failure criterion. *Sci. Rep.* **2016**, *6*, 23359. [CrossRef]
18. Andrianopoulos, N.P.; Manolopoulos, V.M. Can Coulomb criterion be generalized in case of ductile materials? An application to Bridgman experiments. *Int. J. Mech. Sci.* **2012**, *54*, 241–248. [CrossRef]

19. Barsanescu, P.; Sandovici, A.; Serban, A. Mohr–Coulomb criterion with circular failure envelope, extended to materials with strength-differential effect. *Mater. Des.* **2018**, *148*, 49–70. [CrossRef]

20. Li, N.; Gu, J.F.; Chen, P.H. Fracture plane based failure criteria for fibre-reinforced composites under three-dimensional stress state. *Compos. Struct.* **2018**, *204*, 466–474. [CrossRef]

21. Baker, R. Nonlinear Mohr Envelopes Based on Triaxial Data. *J. Geotech. Geoenviron.* **2004**, *130*, 498–506. [CrossRef]

22. Gu, J.; Chen, P. A failure criterion for isotropic materials based on Mohr's failure plane theory. *Mech. Res. Commun.* **2018**, *87*, 1–6. [CrossRef]

23. Hu, W.; Wang, Z.R. Multiple-factor dependence of the yielding behavior to isotropic ductile materials. *Comput. Mater. Sci.* **2005**, *32*, 31–46. [CrossRef]

24. Gu, J.; Chen, P. A failure criterion for homogeneous and isotropic materials distinguishing the different effects of hydrostatic tension and compression. *Eur. J. Mech. A-Solids* **2018**, *70*, 7015–7022. [CrossRef]

25. Abbo, A.J.; Lyamin, A.V.; Sloan, S.W.; Hambleton, J.P. A C2 continuous approximation to the Mohr–Coulomb yield surface. *Int. J. Solids Struct.* **2011**, *48*, 3001–3010. [CrossRef]

26. Zambrano-Mendoza, O.; Valkó, P.P.; Russell, J.E. Error-in-variables for rock failure envelope. *Int. J. Rock Mech. Min. Sci.* **2003**, *40*, 137–143. [CrossRef]

27. Qu, R.T.; Eckert, J.; Zhang, Z.F. Tensile fracture criterion of metallic glass. *J. Appl. Phys.* **2013**, *109*, 869. [CrossRef]

28. Ming, Z.; Mo, L. Comparative study of elastoplastic constitutive models for deformation of metallic glasses. *Metals* **2012**, *2*, 488–507.

29. Grassi, R.C.; Cornet, I. Fracture of gray cast-iron tubes under biaxial stresses. *J. Appl. Mech.* **1949**, *16*, 178–182.

30. Mair, W.M. Fracture criteria for cast iron under biaxial stresses. *J. Strain Anal.* **1968**, *3*, 254–263. [CrossRef]

© 2019 by the authors. Licensee MDPI, Basel, Switzerland. This article is an open access article distributed under the terms and conditions of the Creative Commons Attribution (CC BY) license (http://creativecommons.org/licenses/by/4.0/).

Article

Enhanced Ductility of a W-30Cu Composite by Improving Microstructure Homogeneity

Zehui Hao [1], Jinxu Liu [1,2,*], Jin Cao [1], Shukui Li [1,2,3], Xingwei Liu [1], Chuan He [1,2] and Xinying Xue [4]

[1] School of Materials Science and Engineering, Beijing Institute of Technology, Beijing 100081, China; 2120161132@bit.edu.cn (Z.H.); 3120160524@bit.edu.cn (J.C.); bitleesk@bit.edu.cn (S.L.); 7520190032@bit.edu.cn (X.L.); chuan.he@bit.edu.cn (C.H.)

[2] China National Key Laboratory of Science and Technology on Materials under Shock and Impact, Beijing Institute of Technology, Beijing 100081, China

[3] State Key Laboratory of Explosion Science and Technology, Beijing Institute of Technology, Beijing 100081, China

[4] Xi'an Modern Control Technology Research Institute, Xi'an 710065, China; xinying_xue@163.com

* Correspondence: liujinxu@bit.edu.cn; Tel.: +86-(010)-689-1393-7802

Received: 3 May 2019; Accepted: 31 May 2019; Published: 3 June 2019

Abstract: Due to its negligible solubility, it is difficult to obtain a W-30Cu composite with a homogenous microstructure. However, with a selected W skeleton, a homogeneous phase distribution was achieved for a W-30Cu composite in the present study. By detailed characterization of the mechanical performance and microstructure of the W-30Cu composite, as well as the stress distribution state under a loading condition, the effects of microstructure homogeneity on the mechanical properties and failure mechanisms are identified. The mechanisms by which the ductility and strength depend on microstructure homogeneity contain the effects on plastic deformation and stress coordination of the Cu phase network. The dominant factors for the high ductility and strength of W-30Cu composites are proposed.

Keywords: W-30Cu; microstructure homogeneity; dynamic compression strength; ductility; failure mechanism

---

1. Introduction

Tungsten–copper composites have been widely applied in electrical contact materials and heat sink materials due to their excellent properties of high electrical conductivity and good resistance to high temperature [1–3]. With the development of military technology, W-Cu composites have shown good potential in warhead materials that demand materials with good homogeneity and excellent dynamic mechanical properties (especially ductility) [4,5]. Particularly, W-30Cu composite shows an enormous potential in the military field due to its excellent comprehensive mechanical performance [6,7]. Compared with W-20Cu composite, the W-30Cu composite has a higher plastic deformation capacity and proper yield strength [8]. However, with the increase of copper content (particularly, from 20 wt.% to 30 wt.%), it would be more challenging to fabricate a W-Cu composite with a homogeneous microstructure via the conventional liquid phase sintering method due to the negligible solubility of the W-Cu system [9,10]. In general, to achieve homogenous distribution of the microstructure, the infiltration sintering method is commonly applied in the W-Cu system [11,12]. Ibrahim and Aziz [13] mentioned that subsequent infiltration treatment could increase the density and microstructure homogeneity of the W-Cu composite via liquid phase sintering. Further, to improve the sintering properties of the W-Cu composite, pre-treatments of powders such as mechanical alloying [14,15], oxidation-reduction method [16], and preparation of Cu-coated W powders [17,18] have been investigated extensively.

It is known that W-W contiguity can be influenced significantly by Cu content and plays a dominant role in the mechanical performance of the W-Cu composite [19,20]. Since W-W contiguity will decrease according to the increase of Cu content, a considerable transformation of fracture behavior from brittle to ductile may occur in a W-Cu composite [21]. Furthermore, Deng et al. [22] fabricated a W-30Cu composite with a homogenous copper network structure by spark plasma sintering using core-shell powders. They confirmed that the W-Cu composite shows a homogeneous microstructure and high thermal conductivity due to the formation of a fine copper network structure and low W-W contiguity. However, detailed effects of microstructure homogeneity on mechanical properties for the W-30Cu composite are relatively deficient in previous studies.

In the present work, by inserting molten copper into the W-Cu skeleton and W skeletons, respectively, W-30Cu composites with different microstructure characteristics were fabricated. It is noteworthy that W powders with optimized particle sizes were used. Microstructure parameters were proposed to quantitatively describe the distribution features of the W and Cu phases. ANSYS LS-DYNA software was applied to simulate the stress distribution of the specimens during the deformation process, and high-speed photograph technology was used to record deformation behavior of specimens. The effects of microstructure homogeneity on mechanical properties and failure mechanisms under tensile and compressive tests were analyzed.

2. Materials and Methods

2.1. W-Cu and W Skeleton

For the first method, W and Cu powders with compositions of W-15 wt.% Cu were ball milled for 6 hours in a WC-Co container with WC-Co balls at a speed of 200 rpm. The weight ratio of ball to powders was 3:1. The milled powders were compacted at 250 MPa for 15 min by cold isostatic pressing to obtain a W-Cu skeleton. The W-30Cu compact was produced by infiltrating molten copper into the W-Cu skeleton. According to the processing steps (ball milling, compacting, and infiltrating (BCI)) the W-30Cu compact composite was named as the BCI composite. For the second method, to obtain a W skeleton with high porosity at a low compacting pressure, the particle size of W powders was first sifted and controlled within a narrow range. To obtain the W skeleton with proper porosity corresponding to the W-30Cu composite, the selected tungsten powders were compacted at various pressures and times by cold isostatic pressing without any pre-sintering treatment. Eventually, the tungsten powders were compacted at 100 MPa for 15 min to obtain a W skeleton. Another W-30Cu compact composite was produced by infiltrating molten copper into the W skeleton. According to the processing steps (sifting, compacting, and infiltrating (SCI)), the W-30Cu composite was named as the SCI composite. The isothermal infiltration processes of the W-Cu skeleton and W skeleton were performed for 2 h in dry H_2 at 1400 °C with a heating rate of 10 °C/min. Both desirable skeletons were infiltrated by oxygen-free molten copper.

The morphology and size distribution characteristics of two kinds of skeletons are given in Figure 1. It is seen that W particles possess a polyhedral morphology and a concentrated size distribution. The W-Cu skeleton is composed of copper and tungsten powders. Most of the copper particles with irregular shapes contain a broad particle size distribution and distribute around the tungsten powders. These large-sized copper particles can be converted into a copper pool after melting, which may promote the rearrangement of the tungsten particles, but the residual copper pool will cause copper phase aggregation after solidification. In a pure W skeleton, the W powder particles mainly exhibit a polygon shape with a diameter of 5–8 μm. The large-sized tungsten particles constitute the framework of the W skeleton, and the small-sized tungsten particles fill the gaps between the large particles. This is owing to the fact that fine tungsten particles are able to enter into the pores of the W skeleton by pressure applied during the compacting process [23]. The W particles can be rearranged randomly by the capillary forces of molten copper, which are favorable to the homogenous microstructure of W-30Cu composite.

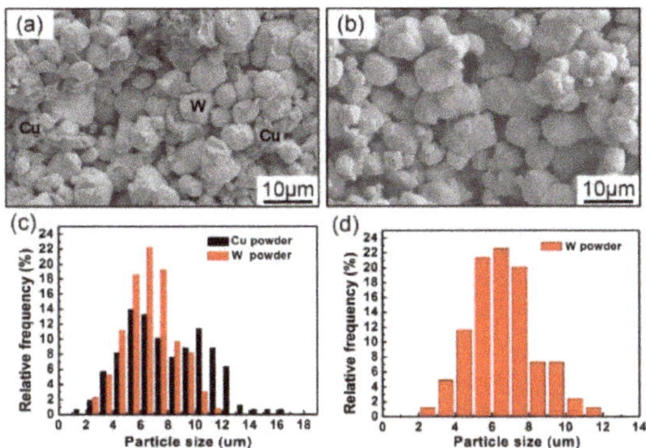

Figure 1. Microstructure morphology of skeletons: (**a**) W-Cu skeleton; (**b**) W skeleton. Size distribution characteristics of skeletons: (**c**) W-Cu skeleton; (**d**) W skeleton.

2.2. Experimental Method

The density of the W-30Cu composite was measured by the Archimedes method. The theoretical density was calculated by using inductively coupled plasma optical emission spectrometer (ICP-OES) method. Scanning electron microscopy (SEM) (Hitachi High–Technologies Corporation, Tokyo, Japan) and the Image-Pro Plus software (V6.0 For Windows, Media Cybernetics Corporation, Rockville, MD, USA) were employed to analyze the microstructure characteristics of the two kinds of composites. The SEM micrographs with different magnifications were applied to measure the W-W contiguity and phase distribution features.

The W-W contiguity $C_{W\text{-}W}$ was calculated using the following equation:

$$C_{W\text{-}W} = 2L_{W\text{-}W}/(2L_{W\text{-}W} + L_{W\text{-}Cu}) \tag{1}$$

where the $L_{W\text{-}Cu}$ and $L_{W\text{-}W}$ represent the length of the W/Cu and W/W interfaces respectively and were counted through the Image-Pro Plus software [24].

To quantitatively describe the distribution feature of W and Cu phase, a statistical parameter, i.e., the distribution homogeneity D_U, was proposed and calculated according to the following expression:

$$\frac{1}{D_U} = \sqrt{\frac{\sum_{i=1}^{n}\left(\frac{A_{Cu}}{A_W} - \overline{\frac{A_{Cu}}{A_W}}\right)^2}{n}} \tag{2}$$

where A_{Cu} and A_W represented the area of the Cu phase and W phase in SEM micrographs respectively, which are counted through the Image-Pro Plus software for each W-30Cu composite. n is the amount of statistical SEM micrographs. To make the statistic results more accurate, the value of n was at least greater than 10. The ratio of A_{Cu} to A_W was a constant if the phase distribution is completely homogenous, implying that the value of D_U tends to be infinite. Obviously, the larger the distribution homogeneity D_U, the more homogenous the distribution of the W and Cu phases is.

Quasi-static tensile tests under different temperature conditions were conducted. The tensile specimens were designed with dumbbell shapes, where the size of the reduced section was Φ5 mm × 25 mm. Quasi-static tensile tests were carried out using the CMT5150 universal testing machine with a strain rate of 10^{-3} s^{-1}. At least three specimens were prepared for tensile tests under different temperatures, respectively.

Furthermore, to clearly identify the effect of the microstructure homogeneity on the mechanical properties in the W-30Cu composite, a hollow specimen was designed for compressive tests. The geometry of the compressive specimen was designed as a hollow shaft of length L = 5 mm, with an outer diameter of 5 mm and inner diameter of 3 mm. Quasi-static compression tests were performed using CMT4150 universal testing machine with a strain rate of 10^{-3} s^{-1}, while the dynamic compression tests were carried out using split hopkinson pressure bar (SHPB) (Beijing Institute of Technology, Beijing, China) with a strain rate of 4500 s^{-1}. At least three specimens were tested for each condition. Furthermore, the deformation behavior of specimens under dynamic compression tests was captured via high-speed photograph technology. The Photron FASTCAM Mini AX200 high-speed camera (Photron, Tokyo, Japan) was used in the present work. In order to obtain more photos in the same time on the premise of adequate pixel resolution, 10,000 frames per second (fps) at 768 × 768 pixel resolution was chosen.

ANSYS LS-DYNA software (V15.0, Livermore Software Technology Corporation, Livermore, CA, USA) was utilized to simulate the stress distribution during the deformation process. The fracture surfaces of the specimens were observed by SEM to study the failure mechanism after compressive and tensile tests.

2.3. Simulation Model

According to the actual structure of the Split Hopkinson Pressure Bar device, the geometric models of the bullet, input bar, and output bar were established. The element type of bullet, input bar, and output bar are solid elements, while the element type of hollow specimens is a shell element. It is noted that the mapped mesh method is applied to reduce the calculating error in the mesh partition. The total number of elements reaches 325,136. To simplify the calculation of simulation, the bullet, input bar, and output bar were modeled through a rigid material model. The deformation process of compressive specimens was modeled through the material model "MAT_PIECEWISE_LINEAR_PLASTICITY". In this model, the stress–strain curve obtained from dynamic compression tests was selected for data fitting in our simulation. The tested material parameters used for the specimen simulation model are given in Table 1. Young's modulus and the material density of the W-30Cu composite applied in this simulation were obtained from the experimental results. Poisson's ratio is an empirical parameter. The initial condition of the simulation was set as follows: a bullet with a speed of 25 m/s strikes the input bar, corresponding to the bullet velocity in the actual experiment.

Table 1. Material parameters for specimen simulation model.

Quantity	Symbol	Value	Unit
Young's modulus	E	195	GPa
Poisson's ratio	M	0.35	-
Material density	ρ	14,330	Kg/m^3

3. Results and Discussion

3.1. Microstructure Characteristics

Quantitative characterizations of microstructure for W-30Cu composites are presented in Table 2. The distribution homogeneity D_U of the SCI composite is 9.09 which is much higher than that of the BCI (0.7). Further, the contiguity $C_{W\text{-}W}$ of the SCI composite is 0.38. However, the $C_{W\text{-}W}$ of BCI is 0.49. The results imply that, compared with BCI composite, the phase distribution is more homogeneous in the SCI composite. A tungsten skeleton in the form of "isolated island" morphology distributes homogeneously in the continuous copper network. In the SCI composite, polyhedral W grains have a homogeneous distribution feature in the Cu matrix, which possesses a continuous network structure, as shown in Figure 2a. By contrast, W grains with aggregated distribution exist in the Cu matrix in the BCI composite, as shown in Figure 2b, which is attributed to the large-sized copper powders in the initial W-Cu skeleton. The copper network shows a phenomenon of "interruptions", where tungsten particles aggregate together.

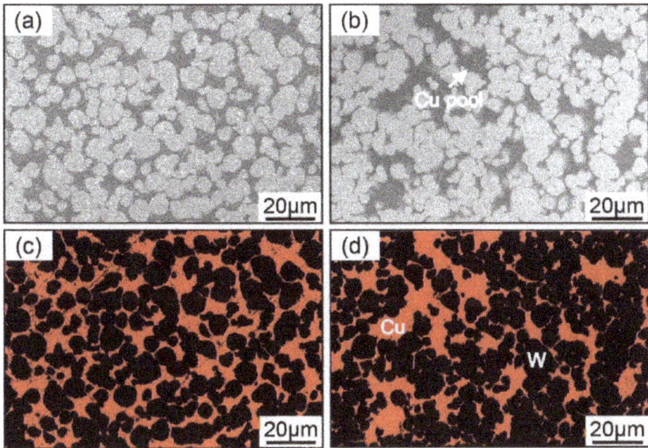

Figure 2. Microstructure of W-30Cu composites prepared by different process: (**a**) and (**c**) microstructure of the sifting, compacting, and infiltrating (SCI) composite; (**b**) and (**d**) microstructure of the ball milling, compacting, and infiltrating (BCI) composite.

Table 2. Microstructure parameters of W-30Cu composites.

Quantity	SCI Composite	BCI Composite
Relative density	98.5%	97.6%
Distribution homogeneity D_U	9.09	0.70
W-W contiguity C_{W-W}	0.38	0.49

3.2. Tensile Test

The true stress–strain curves under tensile conditions for the W-30Cu composites as a function of temperature are plotted in Figure 3a,b. It can be seen that temperature plays an important role in the tensile mechanical properties. The tensile properties of the W-30Cu composites exhibit remarkable temperature dependence and follow the same trend: with an increase of testing temperature, the strength value reaches its maximum at room temperature and then decreases, while the percent elongation reaches its maximum at 300 °C and then sharply decreases. As depicted in Figure 3c in detail, the strength of the SCI composite is higher than that of the BCI composite within a given temperature range from 25 °C to 600 °C. In Figure 3d, the tensile percent elongation as a function of temperature is given for the W-30Cu composites. It is noted that the SCI composite exhibits a higher value in tensile elongation within the given testing temperature range. At room temperature, tensile elongation of the SCI composite is 10.2%, while the tensile elongation of the BCI composite is only 6.51%. Considering the microstructure characteristics of W-30Cu composites, these results indicate that a homogeneous microstructure can lead to a significant advantage in tensile mechanical performance.

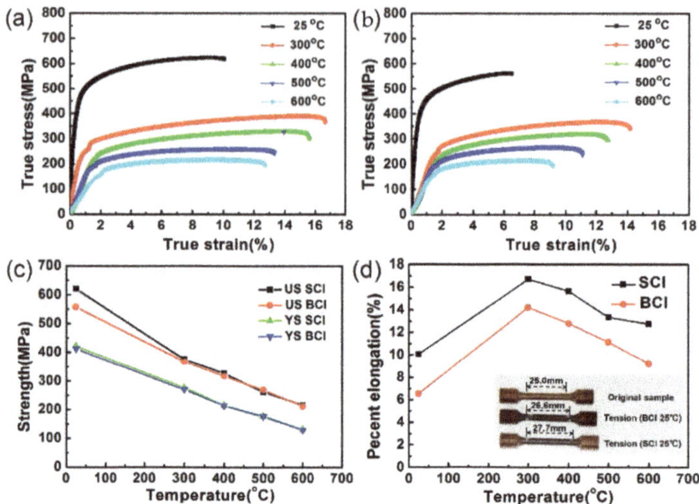

Figure 3. Quasi-static tensile tests of W-30Cu composites: (**a**) true stress–strain curve of the SCI composite; (**b**) true stress–strain curve of the BCI composite; (**c**) tensile strength-temperature curve; (**d**) percent elongation–temperature curve.

Figure 4 presents the typical fracture surfaces of the W-30Cu composites after tensile testing at 25 °C, 300 °C and 500 °C. At room temperature, as shown in Figure 4a,b, the deformation of the Cu phase with the typical dimples and intergranular fractures of W particles dominate the failure of the W-30Cu composites. However, the cleavage that occurred in the W-W interfaces can be observed as a result of the presence of W phase agglomeration, as shown in Figure 4b, while the W phase is surrounded by continuous well-deformed copper in Figure 4a. The deformation of the copper phase in the SCI composite is more prominent and particularly visible. Furthermore, with an increase of the testing temperature, an appreciable number of micro-cracks appear on the fracture surface in the BCI composite (Figure 4d). The fracture surface shows uneven morphology, with a valley in the depth (Figure 4f). It is noticeable that either visible micro-cracks or uneven surface provide direct evidence for the inhomogeneous deformation of the BCI composite at room temperature, as well as high temperatures. The deformation degree of the Cu phase represents a crucial factor affecting the tensile ductility of the W-30Cu composite. As a consequence, the SCI composite with superior tensile ductile performance can be attributed to its continuous copper network and homogenous phase distribution features. Although the difference in microstructure homogeneity can lead to a difference in ductility, the two kinds of composites exhibit similar temperature dependence. As shown in Figure 4c,d, as a result of the dimples fraction increasing and reaching the maximum at 300 °C, the fracture surfaces shows transition phenomena from a rough surface at room temperature to a smooth surface at 300 °C. As we know that copper displays features of "intermediate temperature embrittlement" [25], the ductility of the Cu phase may have been weakened as the testing temperature increased above 400 °C, which may have led to stress concentrations at W/Cu interfaces, thus destroying the W/Cu interfaces, as shown in Figure 4e,f. In previous studies [26,27], the Cu phase can cause local crack deflection and generate an extensive plastic deformation at low temperatures ($T < 400$ °C). However, the degradation in ductility of the Cu phase is the main reason for early failure of W-Cu composites at high temperatures ($T > 400$ °C).

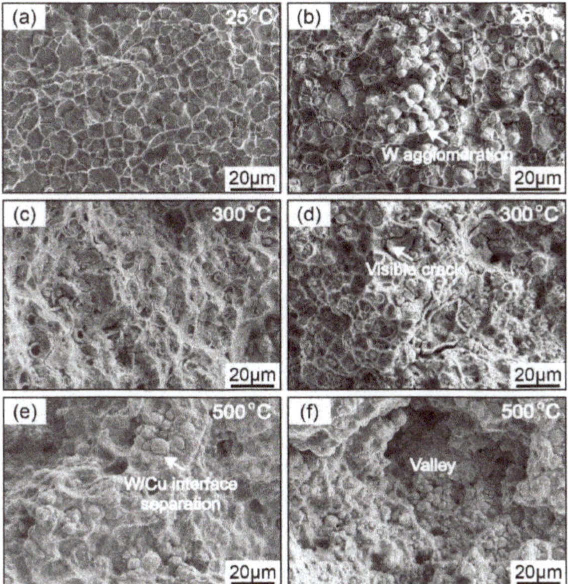

Figure 4. SEM micrographs of the fracture surfaces of W-30Cu composites after tensile testing at different temperatures: (**a**) SCI composite at 25 °C; (**b**) BCI composite at 25 °C; (**c**) SCI composite at 300 °C; (**d**) BCI composite at 300 °C; (**e**) SCI composite at 500 °C; (**f**) BCI composite at 500 °C.

3.3. Compression Test

To study the stress distribution and deformation behavior of the two kinds of composites, a hollow cylindrical specimen was designed for compression tests in this work. The true stress–strain curve of W-30Cu composites under dynamic compression conditions is plotted in Figure 5a. It can be found that the SCI composite exhibits a higher failure strain of 23.6%. In contrast, the failure strain of the BCI composite is 13.9%. Furthermore, the strength of the SCI composite exhibits a constant stage, while the strength of the BCI composite reaches its maximum at a strain of 12% and then drops off rapidly. The different mechanical behaviors of the two composites indicate that a homogeneous microstructure can contribute to homogenous plastic deformation under dynamic compression conditions.

To visually describe the stress distribution under dynamic compression conditions, stress nephograms at different simulation moments were obtained via ANSYS LS-DYNA software (V15.0, Livermore Software Technology Corporation, Livermore, CA, USA). In Figure 5b, it can be found that the stress exhibits the same level in the whole specimen when the simulation time reaches 0.14 ms. The value of stress is mainly in the order of 970MPa. As shown in Figure 5c, the stress gradually increases with the increase of simulation time. Particularly, the maximum value of stress is observed around the center of the inner surface before failure. Thus, for hollow specimens, stress waves rapidly pass and reflect through the whole specimen due to its thin-wall structure. The stress reaches a high value in a short time and the stress distribution is relatively homogeneous, which demands that the W-Cu composites should contain high strength and deformation capabilities.

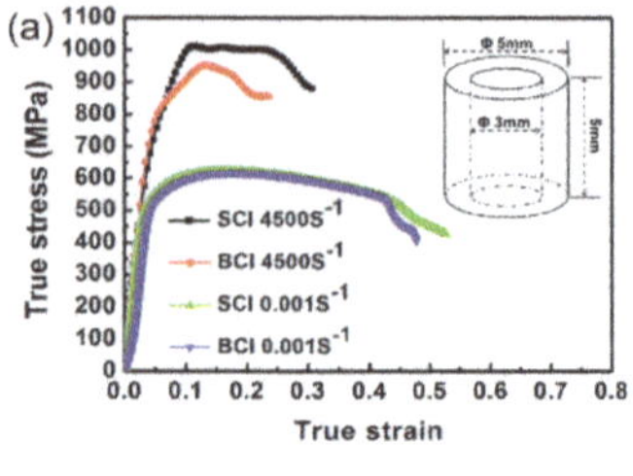

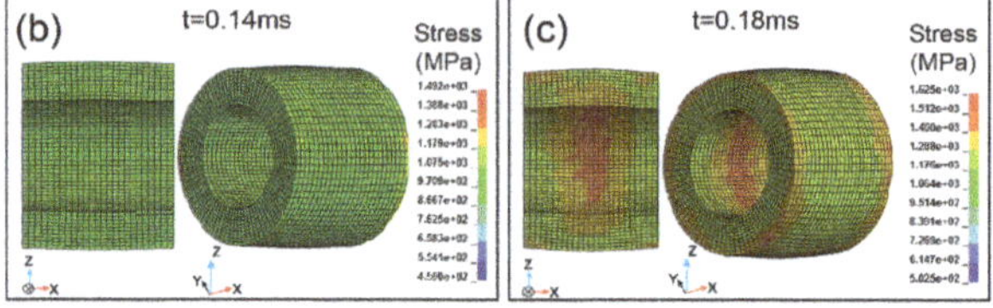

Figure 5. Mechanical performance and finite element modelling under compressive tests for W-30Cu composites at room temperature: (**a**) true stress–strain curve; (**b**) stress distribution at 0.14 ms; (**c**) stress distribution at 0.18 ms.

For a better representation of evolution in material failure, as shown in Figure 6, the deformation behaviors of the W-30Cu composites are captured by using high-speed photograph technology. Under dynamic compression conditions, the SCI composite presents obvious plastic deformation in Figure 6a. By contrast, in Figure 6b, the BCI composite shows rigid characteristics, which correspond to the sharp decrease of stress values in Figure 5a. Furthermore, a significant amount of tiny debris can be found in the SCI composite under dynamic compression, as shown in Figure 6c, while several fragments appear in the BCI composite in Figure 6d. These results indicate that, for the SCI composite, homogeneous phase distribution can facilitate continuous plastic deformation and high dynamic compressive strength.

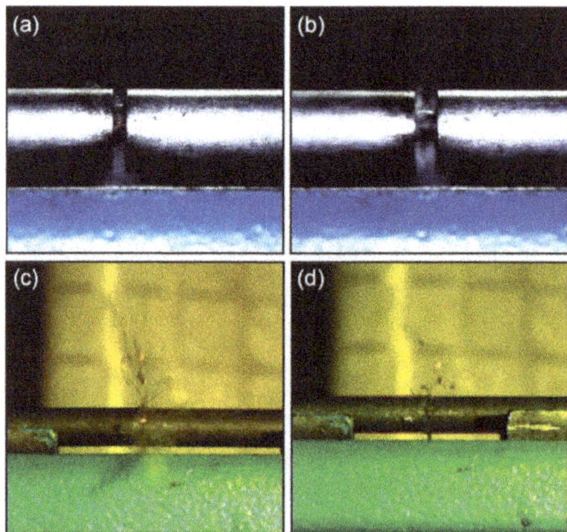

Figure 6. High speed photographs of W-30Cu composites under dynamic compression conditions: (**a**) the SCI composite with plastic deformation; (**b**) the BCI composite without plastic deformation; (**c**) the destructed SCI composite with tiny debris; (**d**) the destructed BCI composite with several fragments.

To study the failure mechanism of the W-30Cu composites, the fracture surfaces after dynamic compression testing are investigated as shown in Figure 7. For the SCI composite, as shown in Figure 7a, the deformation of the Cu phase is obvious and continuous. Moreover, the micro-cracks are generated at the interface of the W/W or W/Cu and are isolated in the well-deformed Cu phase. In contrast, for the BCI composite, the micro-cracks propagate along the interface of the W/W at the location of the W-phase agglomeration. Indeed, the Cu matrix plays a significant role in the plastic deformation and coordination of stress concentration. Simultaneously, the W particles show slight elongation in the W-Cu composites under dynamitic compression conditions [28].

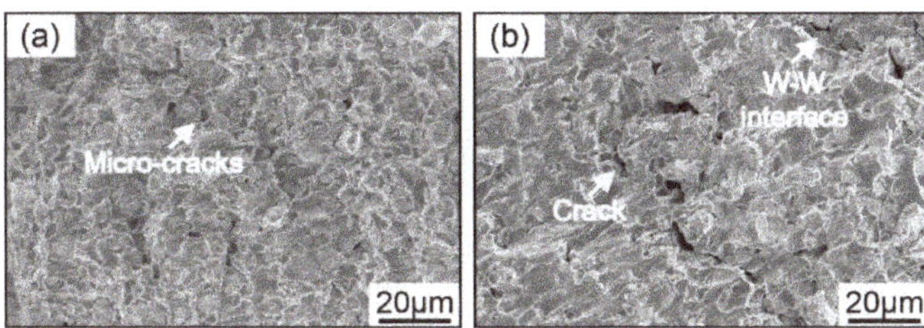

Figure 7. SEM micrographs of the fracture surfaces of W-30Cu composites under dynamic compression conditions: (**a**) the SCI composite; (**b**) the BCI composite.

Combining the results on the deformation behaviors and fracture surface morphology of W-30Cu composites under dynamic compression conditions, the relationship between microstructure homogeneity and compressive mechanical performance is established. Under the dynamic compression conditions of the hollow specimens, the stress with a high value distributes homogeneously within the whole specimen. In general, interfaces of W/W or W/Cu are readily propagated by micro-cracks due to their poor bonding strength. As for the SCI composite, due to its homogenous Cu phase distribution, micro-cracks nucleate and propagate randomly. Furthermore, it is known that the nucleation and propagation of micro-cracks are mainly prohibited and compensated by the coordinated deformation of a Cu phase network. The number of micro-cracks increases throughout the whole specimen until the rupture of the Cu phase dominates the fracture mechanism of the W-Cu composite, finally leading to the failure of the W-30Cu composite as a result of producing significant amounts of tiny debris. On the contrary, in the BCI composite, the micro-cracks nucleate intensively at the location of the W phase agglomeration. Due to the inhomogeneous distribution of the copper phase, the micro-cracks cannot be prohibited timely by the copper phase, leading micro-cracks to coalesce and then propagate along the W-W interface. Consequently, the BCI composite specimen breaks into several fragments. These results indicate that the excellent advantage of the SCI composite in ductility is caused by its homogeneous phase distribution and continuous Cu phase network.

4. Conclusions

In the present study, by infiltrating molten Cu phase into W skeleton and W-Cu skeleton, W-Cu composites with different homogeneity of phase distribution were prepared. The effects of microstructure homogeneity on the mechanical properties and failure mechanisms under tensile and compressive conditions were investigated. The relationship between microstructure characteristics and the failure of W-30Cu composites was studied in detail. The related conclusions are mentioned below:

1. The selected tungsten powders were compacted by isostatic cool pressing at 100 MPa to obtain a W skeleton. The W-30Cu composite was prepared via infiltrating molten copper into the W skeleton. The prepared W-30Cu composite exhibits homogeneous phase distribution feature and continuous Cu phase network structure.

2. Quasi-static tensile tests of W-30Cu composites with a different homogeneity of phase distribution were investigated. It was found that tensile elongation of the SCI composite is 10.2% while the

BCI composite is only 6.51% at room temperature. It was confirmed that the homogeneous phase distribution feature and continuous network structure are responsible for a high capacity of plastic deformation in the W-30Cu composites. Further, as a function of temperature, both the SCI and BCI composite display a similar ductile evolution with an increase in temperature. In addition, their tensile elongation reaches a maximum of 16.7% and 14.1% for the SCI composite and the BCI composites, respectively, at 300 °C.

3. Hollow cylindrical specimens were applied to study the effect of microstructure homogeneity on strength and ductility of the W-30Cu composites under a dynamic compression condition. The SCI composite exhibits an excellent plastic deformation capacity due to its homogenous distribution feature and continuous network structure. The micro-cracks nucleate randomly instead of concentrating on the location of the W phase agglomeration due to the homogeneous phase distribution feature, and the micro-cracks can be prohibited and compensated by the coordinated deformation of the continuous copper network.

Author Contributions: Conceptualization, methodology, and formal analysis, Z.H. and J.L.; investigation and software, Z.H. and J.C.; writing—original draft preparation, Z.H.; writing—review and editing, Z.H., X.L. and C.H.; resources and funding acquisition, S.L. and X.X.

Funding: This research is supported by the funding of the National Key R&D Program of China (2017YFB0306000) and The National Science Foundation of China (No. 51571033). This work was supported in part by the National Natural Science Foundation of China under Grant (No. 11521062).

Acknowledgments: The China National Key Laboratory of Science and Technology on Materials under Shock and Impact is acknowledged.

Conflicts of Interest: The authors declare no conflict of interest.

References

1. Li, J.; Deng, N.; Wu, P.; Zhou, Z. Elaborating the Cu-network structured of the W–Cu composites by sintering intermittently electroplated core-shell powders. *J. Alloys Compd.* **2019**, *770*, 405–410. [CrossRef]
2. Meng, Y.; Shen, Y.; Chen, C.; Li, Y.; Feng, X. Effects of Cu content and mechanical alloying parameters on the preparation of W–Cu composite coatings on copper substrate. *J. Alloys Compd.* **2014**, *585*, 368–375. [CrossRef]
3. Muller, A.; Ewert, D.; Galatanu, A.; Milwich, M.; Neu, R.; Pastor, J.Y.; Siefken, U.; Tejado, E.; You, J.H. Melt infiltrated tungsten–copper composites as advanced heat sink materials for plasma facing composites of future nuclear fusion devices. *Fusion Eng. Des.* **2017**, *124*, 455–459. [CrossRef]
4. Wang, F.; Guo, W.; Liu, J.; Li, S.; Zhou, J. Microstructural evolution and grain refinement mechanism of pure tungsten under explosive loading condition. *Int. J. Refract. Met. Hard Mater.* **2014**, *45*, 64–70. [CrossRef]
5. Elshenawy, T.; Elbeih, A.; Li, Q.M. A modified penetration model for copper-tungsten shaped charge jet with non-uniform density distribution. *Cent. Eur. J. Energetic Mater.* **2016**, *13*, 927–943. [CrossRef]
6. Wang, F.; Jiang, J.; Men, J.; Bai, Y.; Wang, S.; Li, M. Investigation on shaped charge jet density gradient for metal matrix composites: Experimental design and execution. *Int. J. Impact Eng.* **2017**, *109*, 311–320. [CrossRef]
7. Xi, B.; Liu, J.; Li, S.; Lv, C.; Guo, W.; Wu, T. Effect of interaction mechanism between jet and target on penetration performance of shaped charge liner. *Mater. Sci. Eng. A* **2012**, *553*, 142–148. [CrossRef]
8. Guo, W.; Wang, Y.; Liu, K.; Li, S.; Zhang, H. Effect of Copper Content on the Dynamic Compressive Properties of Fine-grained Tungsten Copper Alloys. *Mater. Sci. Eng. A* **2018**, *727*, 140–147. [CrossRef]
9. Johnson, J.L.; Brezovsky, J.J.; German, R.M. Effect of liquid content on distortion and rearrangement densification of liquid-phase-sintered W-Cu. *Metall. Mater. Trans. A* **2005**, *36*, 1557–1565. [CrossRef]
10. Elsayed, A.; Li, W.; Kady, O.A.E.; Daoush, W.M.; Olevsky, E.A.; German, R.M. Experimental investigations on the synthesis of W–Cu nanocomposite through spark plasma sintering. *J. Alloys Compd.* **2015**, *639*, 373–380. [CrossRef]
11. Hamidi, A.; Arabi, H.; Rastegari, S. Tungsten-copper composite production by activated sintering and infiltration. *Int. J. Refract. Met. Hard Mater.* **2011**, *29*, 538–541. [CrossRef]

12. Ahangarkani, M.; Zangeneh-Madar, K.; Borji, S. Microstructural study on the effect of directional infiltration and Ni activator on tensile strength and conductivity of W-10wt%Cu composite. *Int. J. Refract. Met. Hard Mater.* **2018**, *71*, 340–351. [CrossRef]

13. Ibrahim, H.; Aziz, A.; Rahmat, A. Enhanced liquid-phase sintering of W–Cu composites by liquid infiltration. *Int. J. Refract. Met. Hard Mater.* **2014**, *43*, 222–226. [CrossRef]

14. Li, C.; Zhou, Y.; Xie, Y.; Zhou, D.; Zhang, D. Effects of milling time and sintering temperature on structural evolution, densification behavior and properties of a W-20wt.%Cu alloy. *J. Alloys Compd.* **2018**, *731*, 537–545. [CrossRef]

15. Qiu, W.T.; Pang, Y.; Xiao, Z.; Li, Z. Preparation of W-Cu alloy with high density and ultrafine grains by mechanical alloying and high pressure sintering. *Int. J. Refract. Met. Hard Mater.* **2016**, *61*, 91–97. [CrossRef]

16. Wang, C.P.; Lin, L.C.; Xu, L.S.; Xu, W.W.; Song, J.P.; Liu, X.J.; Yu, Y. Effect of blue tungsten oxide on skeleton sintering and infiltration of W–Cu composites. *Int. J. Refract. Met. Hard Mater.* **2013**, *41*, 236–240. [CrossRef]

17. Zhou, Q.; Chen, P. Fabrication of W–Cu composite by shock consolidation of Cu-coated W powders. *J. Alloys Compd.* **2016**, *657*, 215–223. [CrossRef]

18. Zhang, Q.; Liang, S.; Zhuo, L. Fabrication and properties of the W-30wt%Cu gradient composite with W@WC core-shell structure. *J. Alloys Compd.* **2017**, *708*, 796–803. [CrossRef]

19. Hiraoka, Y.; Inoue, T.; Hanado, H.; Akiyoshi, N. Ductile-to-Brittle Transition Characteristics in W–Cu Composites with Increase of Cu Content. *Mater. Trans.* **2005**, *46*, 1663–1670. [CrossRef]

20. Ahangarkani, M.; Borji, S.; Zangeneh-madar, K.; Valefi, Z.; Ahangarcani, M. Mutual relationship between material removal rate and W-W interfacial features during ultra-high temperature erosion of infiltrated W-10wt.%Cu composite. *Int. J. Refract. Met. Hard Mater.* **2018**, *75*, 191–201. [CrossRef]

21. Tejado, E.; Muller, A.V.; You, J.-H.; Pastor, J.Y. The thermo-mechanical behaviour of W-Cu metal matrix composites for fusion heat sink applications: The influence of the Cu content. *J. Nucl. Mater.* **2018**, *498*, 468–475. [CrossRef]

22. Deng, N.; Zhou, Z.; Li, J.; Wu, Y. W-Cu composites with homogenous Cu-network structure prepared by spark plasma sintering using core-shell powders. *Int. J. Refract. Met. Hard Mater.* **2019**, *82*, 310–316. [CrossRef]

23. Montealegre-Meléndez, I.; Arévalo, C.; Perez-Soriano, E.W.; Neubauer, E.; Rubio-Escudero, C.; Kitzmantel, M. Analysis of the Influence of Starting Materials and Processing Conditions on the Properties of W/Cu Alloys. *Materials* **2017**, *10*, 142. [CrossRef] [PubMed]

24. Zheng, L.; Liu, J.; Li, S.; Wang, G.; Guo, W. Investigation on preparation and mechanical properties of W–Cu–Zn alloy with low W-W contiguity and high ductility. *Mater. Des.* **2015**, *86*, 297–304. [CrossRef]

25. Laporte, V.; Mortensen, A. Intermediate temperature embrittlement of copper alloys. *Int. Mater. Rev.* **2009**, *54*, 94–116. [CrossRef]

26. Tejado, E.; Muller, A.V.; You, J.-H.; Pastor, J.Y. Evolution of mechanical performance with temperature of W/Cu and W/CuCrZr composites for fusion heat sink applications. *Mater. Sci. Eng. A* **2018**, *712*, 738–746. [CrossRef]

27. Zivelonghi, A.; You, J.-H. Mechanism of plastic damage and fracture of a particulate tungsten-reinforced copper composite: A microstructure-based finite element study. *Comput. Mater. Sci.* **2014**, *84*, 318–326. [CrossRef]

28. Guo, W.; Liu, J.; Li, S.; Wang, Y.; Ji, W. Microstructural evolution and deformation mechanism of the 80W-20Cu alloy at ultra-high strain rates under explosive loading. *Mater. Sci. Eng. A* **2013**, *572*, 36–44.

© 2019 by the authors. Licensee MDPI, Basel, Switzerland. This article is an open access article distributed under the terms and conditions of the Creative Commons Attribution (CC BY) license (http://creativecommons.org/licenses/by/4.0/).

Article

Comparative Study of Chip Formation in Orthogonal and Oblique Slow-Rate Machining of EN 16MnCr5 Steel

Katarina Monkova [1,2,*], Peter Pavol Monka [1], Adriana Sekerakova [1], Lumir Hruzik [3], Adam Burecek [3] and Marek Urban [4]

[1]   Faculty of manufacturing technologies with the seat in Presov, Technical University of Kosice, Sturova 31, 080 01 Presov, Slovakia; peter.pavol.monka@tuke.sk (P.P.M.); adriana.sekerova@gmail.com (A.S.)
[2]   Faculty of Technology, UTB Tomas Bata University in Zlin, Vavreckova 275, 760 01 Zlin, Czech Republic
[3]   Faculty of Mechanical Engineering, VSB-Technical University of Ostrava, 17. listopadu 15/2172, 708 00 Ostrava-Poruba, Czech Republic; lumir.hruzik@vsb.cz (L.H.); adam.burecek@vsb.cz (A.B.)
[4]   Faculty of Mechanical Engineering, West Bohemia University in Pilsen, Univerzitni 22, 301 00 Pilsen, Czech Republic; maarc@kto.zcu.cz
*   Correspondence: katarina.monkova@tuke.sk; Tel.: +421-55-602-6370

Received: 23 May 2019; Accepted: 18 June 2019; Published: 20 June 2019

Abstract: In today's unmanned productions systems, it is very important that the manufacturing processes are carried out efficiently and smoothly. Therefore, controlling chip formation becomes an essential issue to be dealt with. It can be said that the material removal from a workpiece using machining is based on the degradation of material cohesion made in a controlled manner. The aim of the study was to understand the chip formation mechanisms that can, during uncontrolled processes, result in the formation and propagation of microcracks on the machined surface and, as such, cause failure of a component during its operation. This article addresses some aspects of chip formation in the orthogonal and oblique slow-rate machining of EN 16MnCr5 steel. In order to avoid chip root deformation and its thermal influence on sample acquisition, that could cause the changes in the microstructure of material, a new reliable method for sample acquisition has been developed in this research. The results of the experiments have been statistically processed. The obtained dependencies have uncovered how the cutting tool geometry and cutting conditions influence a chip shape, temperature in cutting area, or microhardness according to Vickers in the area of shear angle.

Keywords: slow-rate machining; chip formation; shape; temperature; microhardness HV

1. Introduction

Machining is a major manufacturing process in the engineering industry. The quality of a product is largely dependent on the accuracy and consistency of the machining processes used for the production of the parts. Although in the last few decades, some entirely new machining processes have been developed—such as ultrasonic machining, thermal metal removal processes, electrochemical material removal processes, and laser machining processes, which differ from the conventional machining processes—conventional metal cutting operations are still the most widely used fabrication processes, and this is the reason why it is still essential to develop a fundamental understanding of metal cutting processes.

In general, machining consists of both cutting and abrasive processes that are mostly complimentary. In spite of the fact that conventional metal cutting processes are chip-forming processes, chip control has been overlooked in the manufacturing processes for a long time. However, along with the automation of manufacturing processes, machining chip control becomes an essential issue in machining operations in order to carry out the manufacturing processes efficiently and smoothly, especially in today's unmanned machining systems. It can be said that the material removal from a workpiece using machining is based on a degradation of material cohesion realized in a controlled way.

Chip removal produces technically and scientifically interesting material responses. In single point cutting, the process concentrates the mechanical power of a machine tool into creating the small volumes of the workpiece by forcing a hard-cutting tool with a small edge radius through an outer layer of the workpiece material. The tribological conditions at the tool–chip interface can result in macroscopic welding of the material to the cutting tool, and dissolution of the tool elements into the chip. The geometry of the deformation zones and the resulting microstructures in the workpiece depend on the geometrical and tribological interaction of the cutting tool with the workpiece material. Further understanding of fundamentals of chip formation is motivated by the desire for higher material removal rates, longer tool life, tighter tolerances, and improved quality of machined surfaces.

The goal of this article is to study some aspects of chip formation in the orthogonal and oblique slow-rate machining of EN 16MnCr5 steel. For the experimental comparative study, planing operation has been selected, by which the main sliding motion was performed by a workpiece. The main reason why this technology has been chosen is because chip formation is visible to the naked eye, in contrast to inward flanging (as another type of slow-rate machining), where the chip-forming process is hidden within the workpiece. For this reason, it has also been possible to measure the temperature in the cutting zone that is generated by the planing tool. The next factor was that restoring the cutting ability of the planing tools produced from high-speed steel by sharpening—according to the required parameters—was considerably easier when compared to turning tools.

Chip flow is only a part of chip space movement. To understand the chip formation mechanism, it is necessary to study other parameters that have an effect on the chip formation, and there are still many challenging issues to deal with.

Through the experimental measurements carried out in this study, different information has been gathered to help obtain a thorough understanding of the chip formation process at relatively low cutting speeds.

2. State of the Art

Chip machining is a complex process in which several mechanisms which are working simultaneously and interacting with each other. This process is greatly affected by material properties, cutting conditions, tool geometry, and machine tool dynamics. In any machining operation, the material is removed from the workpiece in the form of chips, the nature of which differs from operation to operation. As the form and dimensions of a chip from any process can reveal a lot of information about the nature and loyalty of process, the analysis of chips in the process of chip formation, by tear formation, is very important.

A lot of work has been conducted on chip flow angle research during the last few decades, and there are many methods for calculating of the chip flow angle. The investigation of chip flow began with modeling over plane rake face tools. Merchant [1] and Shaffer and Lee [2] have used plasticity theory to attempt to obtain a unique relationship between the chip shear plane angle, the tool rake angle, and the friction angle between the chip and the tool. Palmer [3] presented shear zone theory by allowing for variation in the flow stress for a work-hardening material. Von Turkovich [4] investigated the significance of work material properties and the cyclic nature of the chip formation process in metal cutting. Okushima [5] considered that chip flow is not influenced by cutting speed and chip flow should be the summation of elemental flow angles over the entire length of the cutting edge. Slip-line field theory is widely applied in chip formation research and some slip-line field models are presented [6–8]. Another chip flow model was presented by Young [9], assuming Stabler's flow rule, with validity for infinitesimal chip width, and the directions of elemental friction forces summed up to obtain the direction of chip flow.

During machining, material is removed from the workpiece, where it flows out in the form of chips. After flowing out, the chip curls either naturally or through contact with obstacles. If the material strain exceeds the material breaking strain, the chip will break. Chip flow, chip curl, and chip breaking are three main areas of chip control research [10].

According to Toulfatzis et al. [11], an improved chip breaking capability is directly connected with lower cutting tool wear rates. The width of the chips varies according to the different depths of cut employed during the various machining experiments. Chip segmentation is of pivotal importance since it facilitates the machining ergonomics and scrap removal without damaging workpiece surface quality and ensures the safety of the working personnel.

To study the chip characteristics, it is necessary to specify the type of machining by which the chip is forming. There are two different types of cutting: orthogonal (Figure 1a) and oblique (Figure 1b). Orthogonal cutting is a type of metal cutting in which the cutting edge of the wedge shape cutting tool is perpendicular to the direction of tool motion. In this cutting, the cutting edge is wider than the width of the cut. This cutting type is also known as 2D cutting because the force developed during the cutting can be plotted on a plane or can be represented by a 2D coordinate.

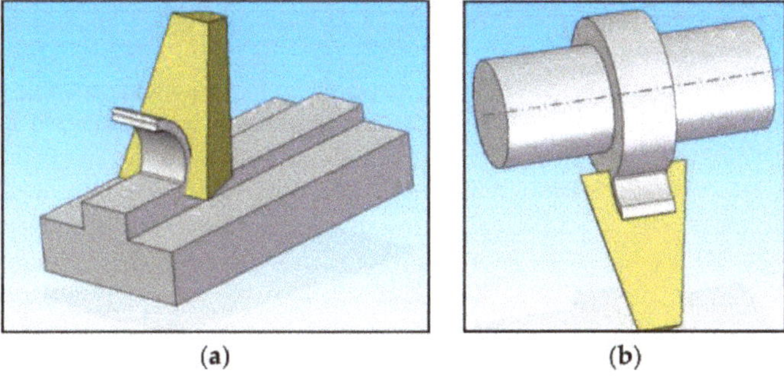

(a)   (b)

Figure 1. The principle of orthogonal cutting in (**a**) planing and (**b**) turning.

However, orthogonal cutting is only a particular case of oblique cutting and, as such, any analysis of orthogonal cutting can be applied to oblique cutting. Oblique cutting (Figure 2) is a common type of three-dimensional cutting used in the machining process [12].

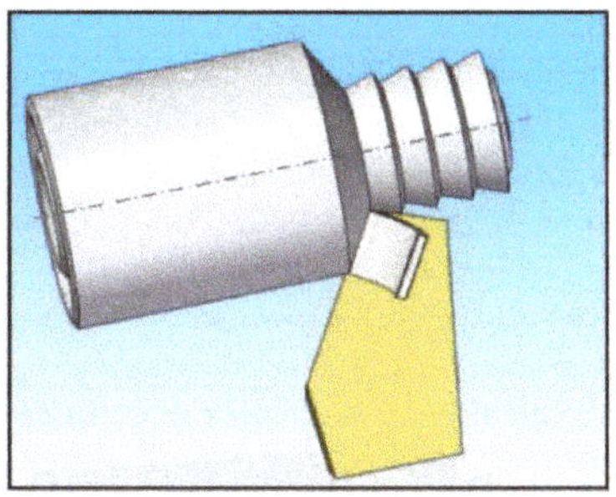

Figure 2. The principle of oblique cutting.

In this type of machining, the cutting edge of the wedge shape makes an angle, except for the right angle, to the direction of tool motion. This will affect the cutting conditions and is also known as 3D cutting because the cutting force developed during the cutting process cannot be represented by 2D coordinates and 3D coordinates must be used to represent it.

One of the most important parameters of oblique cutting is the chip flow angle. Stabler [13] stated that the chip flow angle is very close to the angle of obliquity. This rule has been accepted by many researchers and it has been considered to be a good predictor of the chip flow angle, e.g., as stated in [14–16]. However, Stabler's rule does not take into account the mechanics of the cutting process, such as the influence of the shear angle and friction. In some special cases, it can cause larger errors. Luk [17] investigated the effect of cutting parameters on a chip flow angle and their study has been verified by Lin and Oxley [18] within their experimental investigation. Also, Russel and Brown [19] confirmed the influence of the normal rake angle on the chip flow. Shamoto and Altintas [20] developed a model for shear angle prediction in oblique cutting where the shear angle is specified by two components of the resultant force and by the chip flow angle. Moufki together with his colleagues [21] calculated the chip flow angle supposing that the friction force is collinear to the chip flow direction on the tool rake face. The effect of nose radius on the chip flow angle has been studied by investigators Usui [22] and Wang [23] using an iterative energy minimization method.

The chip characteristics during the milling process by varying the feed rates and the types of materials used were investigated by Prasetyo [24].

Chip morphology and microstructure were also studied by Hernández [25] in dry machining, where the influence of cutting speed and feed rate on various geometric chip parameters was carried out when cutting Ti6Al4V alloy.

Various methods by many researchers have been used to study chip geometry. Some of them analyzed chip geometry from an analytical point of view, resulting in the formulation of some theoretical models [26–28]. However, in these cases, many simplifications had to be made with regard to the complexity of the chip formation process and, hence, some of the real, practically obtained results have not correspond to the predictions of various parameters to set up a chip geometry, and the results were inaccurate [29–31]. Next, analyses of the chip formation used numerical models (finite element method, FEM) to simulate the chip generation process [32–35]. These models require a very good and precise definition of the boundary conditions; otherwise, the models are incomplete and vague [36,37].

Many of the studies that analyze the influence of one or two cutting parameters on chip geometry can be found in which the measured data were processed using four major statistical methods: regression, factor analysis, stochastic processes, and contingency table analysis [38–42]. Salem [43] and his coauthors have studied the chip formation at the machining of a hardened alloy X160CrMoV12-1 to obtain the optimal cutting conditions and to observe the different chip formation mechanisms. For the sake of simplicity, ANOVA (ANalysis Of VAriance) was used in this study to determine the influence of cutting parameters.

Currently, research of chip formation at slow-rate machining has been concerned with the influence of one or two factors on microhardness or temperature at chip root. Only a few studies have focused on chip formation in EN 16MnCr5 steel machining. The novelty of the presented study lies in the following: This research is more complex, while still considering the mutual connections among the four factors influencing chip forming in the cutting of EN 16MnCr5 steel in combination with observing and comparing the significance and influence of each individual factor in orthogonal and oblique cutting. For this reason, a three-level planned experiment was used for statistical evaluation of the obtained data. Based on the experimental study, several parametric models have also been developed that allow for the prediction of different temperatures or microhardness evolution at a chip root as a function of input parameters (cutting speed, cutting depth, and two geometry angles and of a tool λ_s and γ_o). Hereby, a new method of obtaining chip roots has been designed. The new method can be used to prevent changes in the microstructure of material by demonstrating a non-deformed and thermally uninfluenced chip root.

3. Materials and Methods

3.1. Cutting Tools, Machined Material, and Measuring Equipment

In the presented research, chip formation in orthogonal and oblique slow-rate machining has been experimentally investigated. For this comparative study, the technology of planing has been selected, in which a main sliding motion is performed on the workpiece. The machining process was carried out using the planer machine of HJ8A type (KOVOSCIT MAS Machine Tools, Sezimovo Ústí. Czech Republic).

Various combinations of cutting parameters, cutting speed v_c, cutting depth a_p, tool angles λ_s, and γ_o, were used in the experiments. The range of values was chosen based on industrial requirements. It is necessary to point out that lowest value of cutting speed was based on the speed limitations of the machine, where the highest value corresponds to 60% of the machine power. The range of cutting depth values a_p was given by the planing machine, while the limitations were connected with a maximal cross-section of a chip. The values of rake angle γ_o were positive, in order to achieve the lowest possible specific cutting resistance values. The maximum angle value γ_o was selected in view of achieving sufficient bending strength of the cutting wedge. The angle of tool cutting edge inclination λ_s was chosen from zero up to a value that is four times higher than is commonly used in practice, in order to make the extent of the dependence under investigation large enough.

The planing necking tool type 32x20 ON 36550 HSS00 (PILANA Tools Ltd, Hulin, Czech Republic) was used at orthogonal cutting and straight roughing tool 32x20 ON 36500 HSS00 was used in oblique machining. Both types of cutting tools included brazed tips from high-speed steel with three different types of cutting-edge inclinations $\lambda_s = 0°$, $10°$, and $20°$. The used cutting tools and their geometries are presented in Table 1.

Table 1. Cutting tools and their geometries as used in the experimental study.

Tool Angle	Planing Necking Tool	Straight Roughing Tool
κ_r—tool cutting edge angle	0°	60°
$\kappa_r{'}$—tool minor (end) cutting edge angle	-	20°
ε_r—tool included angle	-	100°
γ_o—angle of tool orthogonal rake	8°	3°
α_o—angle of tool orthogonal clearance	15°	15°
λ_s—angle of tool cutting edge inclination	0° / 10° / 20°	0° / 10° / 20°

The angle of tool orthogonal rake γ_o and the angle of tool orthogonal clearance α_o have been varied in the 2nd and 3rd phase of experiments to obtain a better view of chip formation and more reliable results. Changes in the angles' values are organized in Table 2.

Table 2. Changes in the angles' values.

Changes in Angles			
2nd phase	γ_o	12°	7°
	α_o	11°	11°
3rd phase	γ_o	16°	11°
	α_o	7°	7°

In order to verify the input angles of the tool orthogonal rake γ_o, preliminary input tests were performed. The tests were carried out using 3D measuring equipment RAPID CNC THOME (Zimmer Maschinenbau GmbH, Kufstein, Austria) that is shown in Figure 3, where details of the measuring process are also presented.

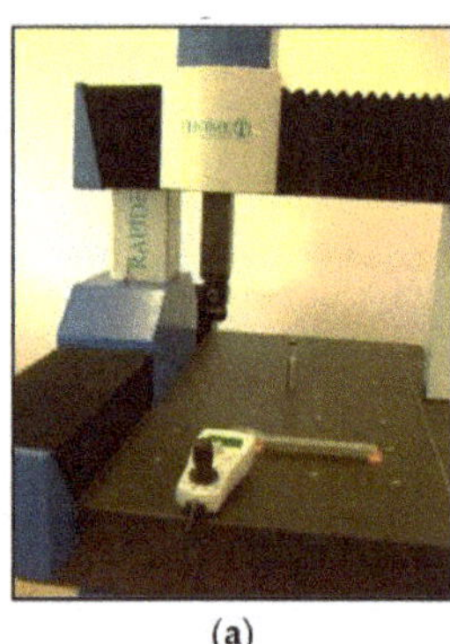
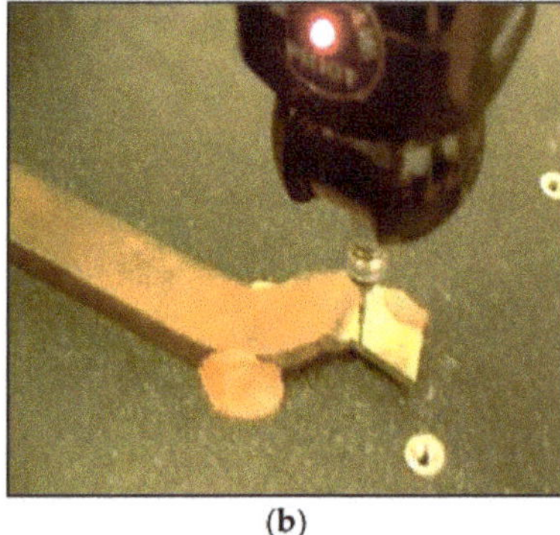

(a)   (b)

Figure 3. Preliminary tests of input angles of tool orthogonal rake γ_o, (**a**) Overall view on the testing equipment RAPID CNC THOME, (**b**) Detail view on the measuring of a tool orthogonal rage angle.

In the next experiment, the angle of tool orthogonal rake γ_o was measured for each of the cutting tools used. The protocols from measurements confirmed the values listed in Tables 1 and 2, while the deviation of all measured values did not exceed 5% and the average angles of tool orthogonal rake for planing necking tools and straight roughing tools were $\gamma_o = 8.138 = 8°2'0''$ or $\gamma_o = 3.159 = 3°2'1''$, respectively.

The 1.7131 steel (EN 16MnCr5) was selected as a machined material; the chip formation of which has been subjected to some research. The alloyed carbon steel contains smooth deformable calcium aluminates encapsulated in manganese sulfide as an alternative to tough alumina oxide inclusions. It is suitable for cementing and for die forging; it is easily hot-formable and, after annealing, also cold-formable and easily machinable and weldable. This grade of steel is generally used for elements with a required core tensile strength of 800–1100 Nmm^{-2} and a good carrying resistance, e.g., piston bolts, camshafts, levers, and other automobile and mechanical engineering add-ons. The chemical composition of this steel, as given by European EN standards, has been verified by spectral analysis at the FMT TU Kosice with the seat in Presov, and is presented in Table 3.

Table 3. Chemical composition of 1.7131 steel (EN 16MnCr5).

Steel	C (%)	Mn (%)	Si (%)	Cr (%)	P (%)	S (%)
EN 16MnCr5	0.14–0.19	1.10–1.40	0.17–0.37	0.80–1.10	max 0.035	max 0.035

The infrared thermometer, UNI-T UT305C, was used to measure the temperature based on the principle of infrared radiation emitted from a target surface.

Vickers microhardness was measured with a MICRO-VICKERS HARDNESS TESTER CV-403DAT (MetTech Ltd., Calgary, AB, Canada), which has the possibility to magnify the view 200–600×.

Etched specimens of chips were observed by means of Platinum USB digital microscope UM019 (Shenzhen Handsome Technology Co., Ltd., Shenzhen, China) with magnification 25–220×.

3.2. Design of the Composite Plan of the Experiment

The planned experiment, unlike the unplanned one, provides the maximum amount of information and performs the task very efficiently, e.g., in obtaining constants and exponents in empirical exponential dependencies that create a mathematical model [44].

In this study, the planned experiment at three levels (lower, basic, and upper) was implemented for the test preparation and the statistical method using a regression function has been used for data evaluation. The description of the experimental plan within this part of the article is given, due to a better understanding of measured data processing.

Based on the [44], the basic equations for statistical processing can be written in matrix:

$$Y = X\,b, \tag{1}$$

where

Y—column vector of measured quantities,
X—matrix of independent variables,
b—coefficient of a regression function.

The system of the normal equation (2) and a vector of the regression function coefficients (3) according to the matrix inversion can be respectively expressed in following way:

$$X^T\,Y = X^T\,X\,b, \tag{2}$$

$$b = X^T\,X^{-1}\,X^T\,Y. \tag{3}$$

It is necessary to consider that the complete three-level plan has a large scale of measurements expressed by $N = 3^k$, where k is a number of variables and N is a number of measures (e.g., considering 5 variables within an experiment, where 243 measurements should be performed in total because $N = 3^5 = 243$). [45]

A reduced number of measurements for the dependencies described by functions of the second order,

$$y = b_0 x_0 + \sum_{j=1}^{N} b_j x_j + \sum_{\substack{u,j=1 \\ u \neq j}}^{N} b_j x_j x_u + \sum_{j}^{N} b_{jj} x_j^2, \tag{4}$$

can be achieved by means of the so-called second level compositional non-rotational plan [46], while the symbols in Equation (4) have the following meanings: x_j is a variable (in the case of presented research it is one of the cutting parameters that will be varied), j, u are indexes that define a parameter, and b_j is a j-th correlation coefficient.

The composition plan, in this case, consists of [46]:

1. A core of plan that can be

   - two-level 2^k plan for $k < 5$, or as
   - shortened replica 2^{k-p} for $k \geq 5$, where p is a level of significance (Grubbs' test);

2. The star points α with coordinates: $(\pm\alpha, 0, ..., 0)$; $(0, \pm\alpha, 0, ..., 0)$; ...; $(0, 0, ..., 0, \pm\alpha)$;

3. The measurements done at the basic level; or in the middle of the plan at $x_1 = x_2 = ... = x_k = 0$ (the number of measurements in the middle of the plan is n_0).

The total number of measurements is then [47]:

$$N = 2^k + 2k + n_0, \text{ if } \quad k < 5, \text{ or}$$
$$N = 2^{k-p} + 2k + n_0, \text{ if } \quad k \geq 5. \tag{5}$$

In practical implementation, $n_0 = 1$ [48] is chosen, with no boundary. The matrix of the orthogonal composition plan for k, α, and n_0 is given in Table 4. In its general form, it is not orthogonal because the relations on the left sides of Equations (6) and (7) differ from zero:

$$\sum_{i=1}^{N} x_{oi}x_{ji}^2 \neq 0, \tag{6}$$

$$\sum_{i=1}^{N} x_{ji}^2 x_{ui}^2 \neq 0. \tag{7}$$

Table 4. General form of the matrix of the composition plan.

N	x_o	x_1	x_2	$...$	x_k	Description
	+1	−1	−1	$...$	−1	
	+1	+1	−1	$...$	−1	
	+1	−1	+1	$...$	−1	
2^k ($k < 5$) or	+1	+1	+1	$...$	−1	a core of the plan
2^{k-p} ($k > 5$)	+1	−1	−1	$...$	+1	
	+1	+1	−1	$...$	+1	
	+1	−1	+1	$...$	+1	
	+1	+1	+1	$...$	+1	
	+1	−α	0	$...$	0	
	+1	+α	0	$...$	0	
$2k$	+1	0	−α	$...$	0	the star points of the plan
	+1	0	+α	$...$	0	
	+1	0	0	$...$	−α	
	+1	0	0	$...$	+α	
	+1	0	0	$...$	0	
n_0	+1	0	0	$...$	0	the measurements in the middle of the plan
	+1	0	0	$...$	0	

The matrix is converted to orthogonal shape by quadratic variables exchanging [49]:

$$x'_j = x_j^2 - \frac{1}{N}\sum_{i=1}^{N} x_{ji}^2 = x_j^2 - \overline{x}_j^2, \tag{8}$$

This is why

$$\sum_{i=1}^{N} x_{oi}x'_{ji} = \sum_{i=1}^{N} x_{ji}^2 - N\overline{x}_j^2 = 0, \tag{9}$$

$$\sum_{i=1}^{N} x'_{ji}x_{ui} \neq 0. \tag{10}$$

The regression function correlation coefficients in (4) are independent because of the orthogonality of the experimental matrix, and they are specified by the following relations, (11)–(14):

$$b_j = \frac{\sum_{i=1}^{N} x_{ji}y_i}{\sum_{i=1}^{N} x_{ji}^2} = \frac{\sum_{i=1}^{N} x_{ji}y_i}{2^k + 2\alpha^2}, \tag{11}$$

$$b_{uj} = \frac{\sum_{i=1}^{N} x_{ji} y_i}{\sum_{i=1}^{N} x_{ji}^2} = \frac{\sum_{i=1}^{N} x_{ji} y_i}{2^k}, \tag{12}$$

$$b_{jj} = \frac{\sum_{i=1}^{N} x'_{ji} y_i}{\sum_{i=1}^{N} \left(x'_{ji}\right)^2}, \tag{13}$$

$$b'_o = \frac{1}{N} \sum_{i=1}^{N} x_{oi} y_i, \tag{14}$$

Hence, the second stage regression function (4) is then given by Equation (15):

$$y = b_0 + b_1 x_1 + b_2 x_2 + \ldots + b_k x_k + b_{12} x_1 x_2 + b_{(k-1)k} x_{(k-1)} x_k + b_{11}\left(x_1^2 - \overline{x}_1^2\right) + b_{kk}\left(x_k^2 - \overline{x}_k^2\right), \tag{15}$$

where the constant member of the regression function is corrected by quadratic variables (8) in the form of

$$b_0 = b'_o - b_{11} \overline{x}_1^2 - \cdots - b_{kk} \overline{x}_k^2. \tag{16}$$

Using Grubbs' testing criteria, the outliers from the measured values have been specified for every group of measurements. The following equations, (17)–(19), have had to be kept.

$$H_i = \frac{\left|T_{ik} - \overline{T}_i\right|}{S_{Ti}} < H_p(m), \tag{17}$$

while

$$\overline{T}_i = \frac{\sum_{k=1}^{m} T_{ik}}{m}, \tag{18}$$

$$S_{Ti} = \sqrt{\frac{1}{m-1} \cdot \sum_{k=1}^{m} \left(T_{ik} - \overline{T}_i\right)^2}, \tag{19}$$

where

m—a number of evaluated measurements within the Grubbs´ test;
T_{ik}—measured value of k-th issue in the i-th group, $k = 1, 2, 3; i = 1, 2, \ldots, 24, 25$;
$\overline{T}_i$—average value of measured issues of the i-th group; calculation according to the equation;
S_{Ti}—standard deviation of measured issue of the i-th group;
$H_p(m)$—critical value of Grubbs´ testing criteria for m values ($m = 3$), where p is a level of significance and usually it is $H_p(m) = 0.05$.

The calculation of the regression coefficients was performed using MATLAB calculation software (The MathWorks, Inc., Natick, MA, USA), while the significance of the coefficients of the function $y = \log T$ was tested according to Student's test criterion.

The adequacy of regression function was assessed according to the Fisher–Snedecor test criterion $F < F0.05$ (f_1, f_2), where the degrees of freedom $f_1 = Nq$ (q is a number of significant coefficients) and $f_2 = N(m-1)$.

3.3. Process of Obtaining Samples

To track changes in the zone of chip forming, the machining process must be stopped immediately, thus interrupting tool and workpiece contact. A reliable method to achieve the immediate stop of the machining process has been developed in this research and is based on observation of the chip end produced upon interrupted cutting, e.g., in planing or face milling. As the tool leaves the cutting zone, the end of the chip is "torn off", as shown in Figure 4.

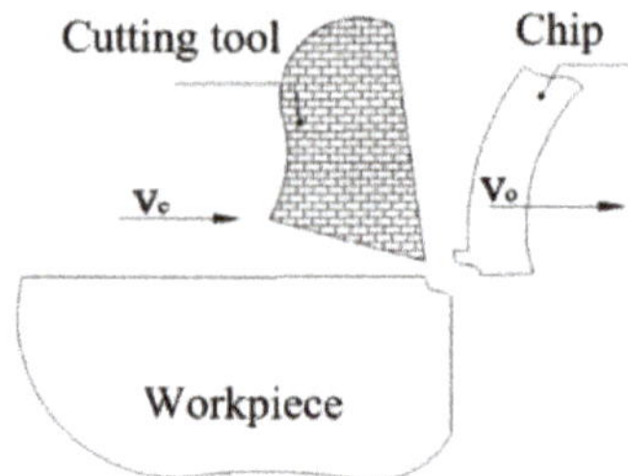

Figure 4. An imprint of cutting wedge on the chip at the interrupted cut.

The reasons mentioned above led to the design of a new method for chip root acquisition by modifying the end of the workpiece according to Figure 5. The goal of the proposed method and the special workpiece design was to avoid deformation and thermal influence of the material which could cause the changes in the microstructure of a material. For this reason, a cooling medium JCK PS was used, at a concentration of 5%, for all operations within the workpiece preparation.

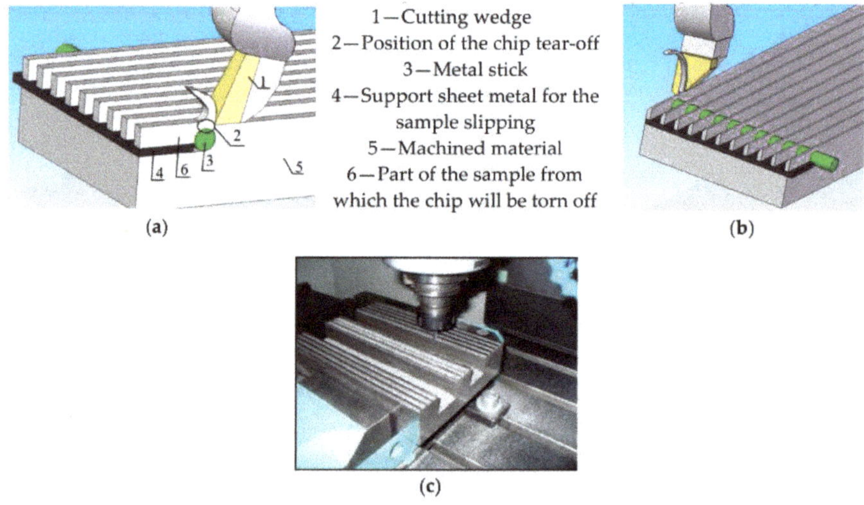

Figure 5. The principle of the designed method for instantaneous contact interruption between a tool and workpiece showing (**a**) orthogonal cutting, (**b**) oblique cutting, (**c**) real workpiece.

At the point of departure of tool "1" from the engagement, a groove has been cut where the sheet metal insert "4" has been put in to prevent deformation of the specimen during the rupture. A hole of 8 mm diameter has been drilled behind the groove and a metal rod "3" has been inserted therein, which prevents the hole from deforming. When the tool passes above the metal rod, the section "2" becomes narrower and the material ruptures, which is similar to the tensile test. Sample "6" is rapidly thrown up in the direction of tool movement at a rate greater than cutting speed and on the sample; and the plastic deformation state corresponding to the actual cutting speed is captured.

The experiment, in which chips were obtained, was carried out on a planer without the use of cooling, since the cutting length of the tool path was about 200 mm and, thus, the tool and workpiece were not overheated.

Within the experimental study, the variables according to the Table 5 have been taken into account, while the codes for the specific values of individual variables are referred to in Table 6.

Table 5. The variables and the ranges of individual values.

Variables	Symbols		−1	−α	0	+α	+1
Cutting speed (mmin^{-1})	v_c	x_1	6	8.25	10.5	12.75	15
Cutting depth (mm)	a_p	x_2	0.2	0.25	0.3	0.35	0.4
Angle of tool orthogonal rake – orthogonal cutting (°)	γ_o	x_3	8	10	12	14	16
Angle of tool orthogonal rake – oblique cutting (°)	γ_o	(x_3)	3	5	7	9	11
Angle of tool cutting edge inclination (°)	λ_s	x_4	0	5	10	15	20

Table 6. The codes for specific values of individual variables.

Code	Specific Values of Variables
−1	x_{min}
−α	$[(x_{max} + x_{min})/2] - [(x_{max} - x_{min})/2\alpha^2]$
0	$(x_{max} + x_{min})/2$
+α	$[(x_{max} + x_{min})/2] + [(x_{max} - x_{min})/2\alpha^2]$
+1	x_{max}

4. Results and Discussion

4.1. Types of Chips

During machining, different types of chips have been created. Cutting speed was not observed to have a significant impact on the chip shape, while the depth of the cut affected the radius of curvature of the chip (when increasing the thickness of the cut layer, the radius of curvature of the chip was increased). Figure 6 shows an example of the dependence of chip shapes on cutting speed v_c and cutting depth a_p at tool angles for $\gamma_o = 16°$, $\lambda_s = 0°$, $\kappa_r = 0°$.

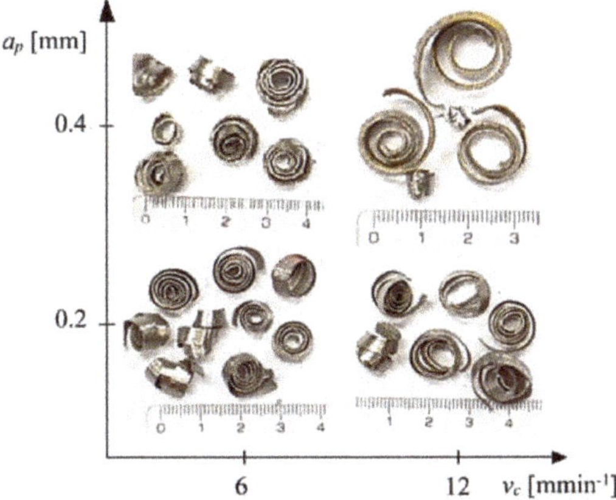

Figure 6. The shapes of chips and their dependence on cutting speed v_c and cutting depth a_p at tool angles for $\gamma_o = 16°$, $\lambda_s = 0°$, $\kappa_r = 0°$.

As for tool geometry, the angle of the orthogonal tool rake γ_o at the minimum values caused formation of a crumbly chip and the angle of inclination of the main cutting edge λ_s has influenced the chip shape in the way that is graphically presented in Figure 7.

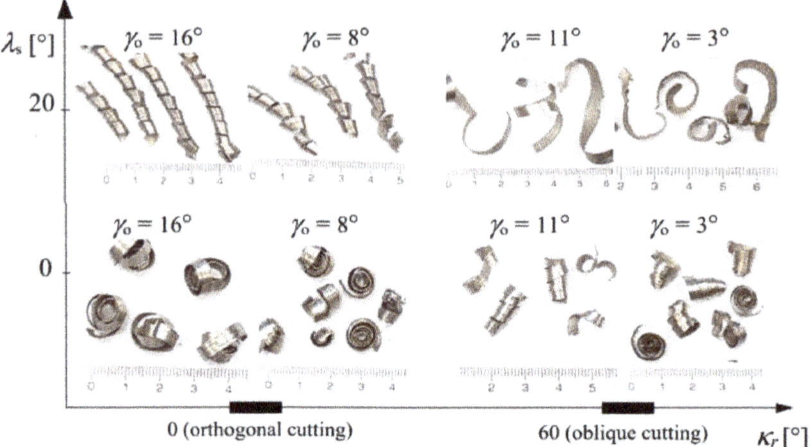

Figure 7. The shapes of chips of EN 16MnCr5 steel and their dependence on tool cutting edge angle κ_r, angle of tool cutting edge inclination λ_s, and angle of tool orthogonal rake γ_o at $a_p = 0.2$ mm, $v_c = 15$ mmin^{-1}.

A shorter chip was created at a zero angle of inclination of main cutting edge when comparing it with the angle of inclination of the main cutting edge of 20°. At the same time, at a 20° angle of inclination of the cutting edge, the chip obtained a so-called chamfer along the edges, which is adequate to the angle of inclination of the main cutting edge and, thus, the chip was compressed in the direction of λ_s inclination and acquired the character of a spiral chip. In orthogonal cutting, at which $\kappa_r = 0°$ and $\lambda_s = 0°$, a spiral flat chip was formed, which in some cases passed into a spiral conical chip. For $\kappa_r = 0°$ and $\lambda_s = 20°$, a conical–helical long chip was constituted. In oblique cutting, where $\kappa_r = 60°$ and $\lambda_s = 0°$, a conical–helical short chip was generated. At $\kappa_r = 60°$ and $\lambda_s = 20°$, the chip shape changed to a coiled chip, while at a rake angle $\gamma_o = 11°$, it was a tubular coiled chip and at $\gamma_o = 3°$ it was a strip coiled chip.

The shrinkage factor K and the segment ratio r_c at the planing can be calculated with Equations (20) and (21), respectively, where h_c is the chip width, a_p is cutting depth, γ is the tool rake angle, and Φ is the shear angle [1,12,50,51]:

$$K = \frac{h_c}{a_p} = \frac{\cos(\Phi - \gamma)}{\sin \Phi}, \tag{20}$$

$$r_c = \frac{1}{K}. \tag{21}$$

The values of the shrinkage factor K achieved within the experimental study are organized in Table 7.

Table 7. Shrinkage factor K achieved within the experimental study.

Type of Cutting	Shear Angle Φ (°)	Shrinkage Factor K
Orthogonal cutting	31.7	1.75
	42.2	1.3
Oblique cutting	30.2	1.77
	42.4	1.21

4.2. Temperature Measuring

The temperature in a cutting zone was measured by means of the infrared thermometer UNI-T UT305C. The values of measured temperature with the coded matrix (explained in the Design of the Composite Plan of the Experiment, Section 3.2.) are shown in Table 8.

Table 8. Experimentally obtained values of temperature in cutting zone for orthogonal and oblique machining of EN 16MnCr5 steel.

No.	v_c (m·min^{-1})	a_p (mm)	γ_o (°)	λ_s (°)	T (°C) Orthogonal Cutting	Oblique Cutting
1.	−1	−1	−1	−1	86.8	63.8
2.	+1	−1	−1	−1	58.7	63.0333
3.	−1	+1	−1	−1	72.5333	55.2333
4.	+1	+1	−1	−1	66.4667	48.8
5.	−1	−1	+1	−1	78.5333	40.5333
6.	+1	−1	+1	−1	50.3333	41.1333
7.	−1	+1	+1	−1	69.9	46.5
8.	+1	+1	+1	−1	51.5	48
9.	−1	−1	−1	+1	72.2333	50
10.	+1	−1	−1	+1	68.3333	96.6
11.	−1	+1	−1	+1	73	62.8667
12.	+1	+1	−1	+1	38.9667	72.8333
13.	−1	−1	+1	+1	61.3667	62.8333
14.	+1	−1	+1	+1	70.8	60.8333
15.	−1	+1	+1	+1	62.6	61.7
16.	+1	+1	+1	+1	72.3667	67.9333
17.	−α	0	0	0	103	45.1
18.	+α	0	0	0	90.6667	95.5333
19.	0	−α	0	0	85.3333	74.5667
20.	0	+α	0	0	93.3	62.9333
21.	0	0	−α	0	94	99.2333
22.	0	0	+α	0	74.8	61.3
23.	0	0	0	−α	91.1333	57.3333
24.	0	0	0	+α	72.3	84.7667
25.	0	0	0	0	79.3	62.8333

Measured data were statistically processed. According to the Grubbs' testing criterion, it could be stated that the measured values have not been burdened by grave mistakes.

Homogeneity of the variance of measured temperatures were tested according to Cochrane´s criterion by the relation

$$G = \frac{S^2_{max}}{\sum_{i=1}^{N} S^2_i} = \frac{0.025877}{0.111027} = 0.23307. \tag{22}$$

The critical value of Cochrane's criterion $G_{0,05}(f_1, f_2)$, given by [22], is $G_{0.05}(f_1, f_2) = 0.2705$, where the degrees of freedom were $f_1 = N = 25$ and $f_2 = m - 1 = 3 - 1 = 2$.

Since $G < G_{0,05}(f_1, f_2) \rightarrow$ the criterion has been satisfied and it means that the variations within parallel measurements are homogeneous.

The coefficients of regression functions have been calculated using the software MATLAB and their significance was tested according to the Student's criterion. The regression functions (23) and (24) have been built. According to the Fisher–Snedecor-tested criterion, they describe the experiment adequately, while the reliability of the relations is $R^2 = 0.91$ and 0.94, respectively:

a)   For orthogonal cutting:

$$y = 3.2502x_0 - 3.7533x_1 + 3.1462x_2 + 4.2035x_3 - 0.1865x_4 - 0.1553x_1x_2$$
$$+0.3467x_1x_3 + 0.0385x_1x_4 + 0.4106x_2x_3 - 0.0202x_2x_4 \qquad (23)$$
$$+0.0304x_3x_4 + 1.6486x_1^2 + 3.1542x_2^2 - 2.8264x_3^2 - 0.042x_4^2$$

b)   For oblique cutting:

$$y = 2.9212x_0 - 10.4247x_1 - 7.2462x_2 + 4.4859x_3 + 0.1295x_4 - 0.4264x_1x_2$$
$$-0.2621x_1x_3 + 0.0625x_1x_4 + 0.5108x_2x_3 + 0.0251x_3x_4 \qquad (24)$$
$$+5.4295x_1^2 - 6.6225x_2^2 - 2.6781x_3^2 + 0.0676x_4^2$$

Based on the measured data, the following dependencies have been evaluated for both orthogonal and oblique cuttings:

- The dependency of temperature T on cutting speed v_c and on the cutting depth a_p;
- The dependency of temperature T on cutting speed v_c and angle of tool orthogonal rake γ_o;
- The dependency of temperature T on cutting speed v_c and on the angle of tool cutting edge inclination λ_s.

The dependencies are presented in Figures 8–10.

It can be seen from the dependence of the temperature on the cutting speed and on the cutting depth at orthogonal machining (Figure 8a) that the cutting depth has an incomparably stronger influence than the cutting speed. By contrast, in oblique cutting (Figure 8b), the impact of cutting depth is milder, so the effect of cutting speed is more noticeable. Maximum temperature values were reached at a maximum cutting depth during orthogonal cutting and at a minimum cutting depth during oblique cutting.

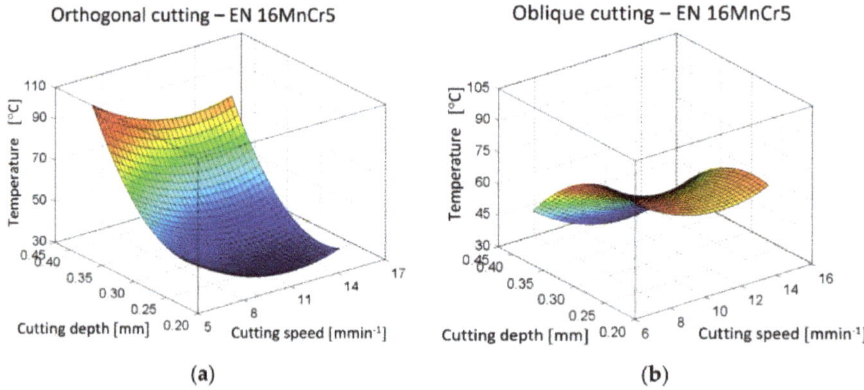

Figure 8. The dependence of temperature T on cutting speed v_c and cutting depth a_p, (**a**) Orthogonal cutting; (**b**) oblique cutting.

The angle of tool orthogonal rake γ_o at the machining of manganese chromium steel has a similar character in both cutting methods (Figure 9), while the highest values of temperature have been measured at mean values of γ_o. The influence of cutting speed also has the same character at both orthogonal and oblique cutting, and the maximum temperature has been reached at the minimal value of cutting speed.

The dependence of the cutting speed and tool cutting edge inclination λ_s on the temperature is almost identical in both orthogonal and oblique cutting (Figure 10). The influence tool cutting edge inclination is almost imperceptible in both cases compared to the impact of speed.

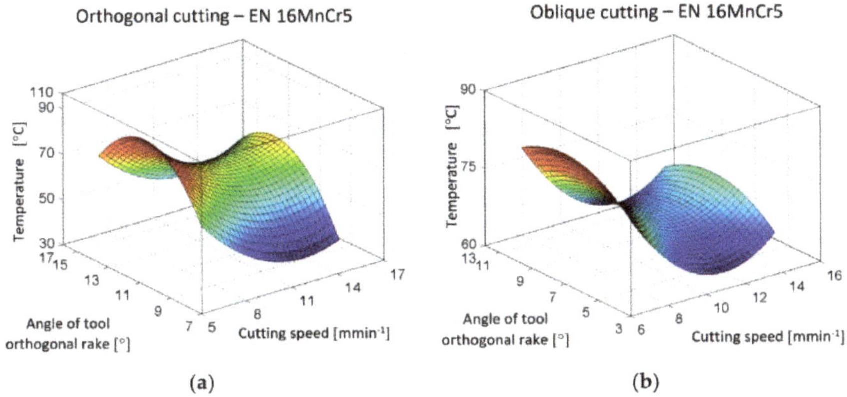

Figure 9. The dependence of temperature T on cutting speed v_c and angle of tool orthogonal rake γ_o. (**a**) Orthogonal cutting; (**b**) oblique cutting.

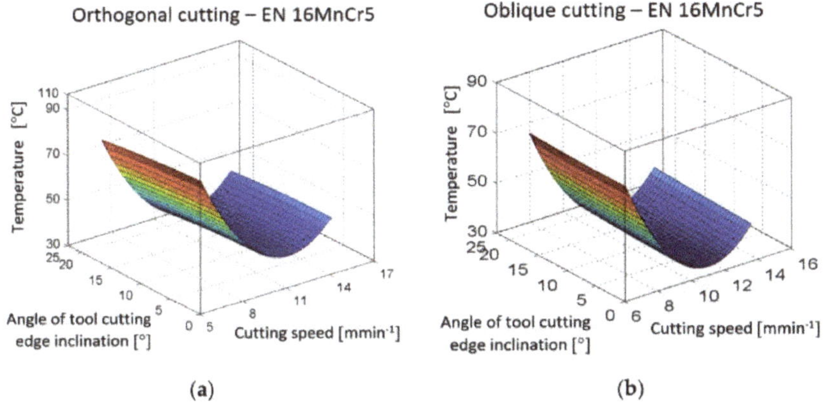

Figure 10. The dependence of temperature T on cutting speed v_c and angle of tool cutting edge inclination λ_s. (**a**) Orthogonal cutting; (**b**) oblique cutting.

4.3. Measurement of Microhardness HV According to Vickers

The next evaluated parameter was microhardness according to Vickers. The measurements were carried out on samples of chip root at a load of 200 g during intervals of 10 s, according to the standard STN EN ISO 6507-1. Vickers microhardness evaluation was performed based on the experimental plan design, similarly as at the temperature measuring. The character of the matrix for experimental composition plan was designed for both cutting methods, where four matrix shapes have been evaluated resulting in four dependencies pictured by means of surface plots. The hardness of the material was measured in the area of shear angle Φ that is shown in Figure 11; also, a diamond body imprint with magnification 600× is displayed here in detailed view. The obtained values are organized in Table 9.

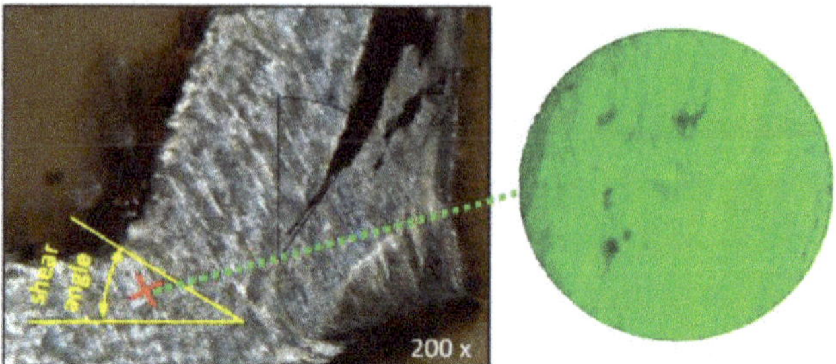

Figure 11. Diamond body imprint with magnification 600×.

Table 9. Experimentally obtained values of microhardness HV measured in the area of the shear angle at chips.

No.	v_c (mmin^{-1})	a_p (mm)	γ_o (°)	λ_s (°)	*HV* (kgmm^{-2}) Orthogonal Cutting	Oblique Cutting
1.	−1	−1	−1	−1	218.567	337.267
2.	+1	−1	−1	−1	268.833	232.667
3.	−1	+1	−1	−1	308.4	400.1
4.	+1	+1	−1	−1	253.933	281.767
5.	−1	−1	+1	−1	240.9	293.9
6.	+1	−1	+1	−1	238.533	221.2
7.	−1	+1	+1	−1	330.533	270.733
8.	+1	+1	+1	−1	204.6	216.267
9.	−1	−1	−1	+1	216.767	211.067
10.	+1	−1	−1	+1	319.367	229.7
11.	−1	+1	−1	+1	293.733	233.167
12.	+1	+1	−1	+1	196.667	363.533
13.	−1	−1	+1	+1	184.9	245.767
14.	+1	−1	+1	+1	213.933	190.367
15.	−1	+1	+1	+1	160.7	192.667
16.	+1	+1	+1	+1	165,367	216.833
17.	−α	0	0	0	239.9	273.233
18.	+α	0	0	0	230.2	237.9
19.	0	−α	0	0	270.867	254.9
20.	0	+α	0	0	321.533	318.433
21.	0	0	−α	0	301.333	329.567
22.	0	0	+α	0	271.967	194.767
23.	0	0	0	−α	265.667	167.1
24.	0	0	0	+α	448.367	279.7
25.	0	0	0	0	301.2	305.533

Similarly as for measurements of temperature and share angle, the regression functions (25) and (26) for the evaluation of microhardness of chip root have been defined. The reliabilities R^2 of the dependencies for both types of machining (orthogonal and oblique) were 0.92 and 0.93, respectively. Regression functions are specified by the following equations:

a)   For orthogonal cutting:

$$y = -5.0009x_0 + 17.2807x_1 + 5.2179x_2 0.575x_3 + 0.1881x_4 - 1.5626x_1x_2$$
$$-0.1588x_1x_3 + 0,0397x_1x_4 - 0.2036x_2x_3 - 0.0623x_2x_4 \tag{25}$$
$$-0.0376x_3x_4 - 9.2084x_1^2 + 3.3134x_2^2 - 0.4774x_3^2 + 0.0915x_4^2$$

b) For oblique machining:

$$y = -1.2673x_0 + 7.7671x_1 + 9.0532x_2 + 5.5376x_3 + 0.2281x_4 + 0.7224x_1x_2$$
$$-0.2077x_1x_3 + 0.0846x_1x_4 - 0.6964x_2x_3 - 3.6942x_1^2 + 8.2721x_2^2 \qquad (26)$$
$$-3.8696x_3^2 + 0.1162x_4^2$$

Based on the relations (23) and (24) formulated above, the dependencies of the microhardness HV on individual variables were plotted. They are presented in Figures 12–16.

The influence of cutting depth and cutting speed was the most significant effect of the four variable parameters which changed the microhardness HV. In oblique cutting (Figure 12b), the effect of the cutting depth is more pronounced than the effect of the cutting speed compared to its effect in orthogonal cutting (Figure 12a). Maximum microhardness values were achieved at maximum cutting depth and at medium and higher cutting speeds.

The effect of the rake angle on orthogonal cutting (Figure 13a) is almost imperceptible compared to the cutting speed. In oblique cutting (Figure 13b), the angle of tool orthogonal rake has the same significant impact as the cutting speed, and maximum microhardness HV was achieved at its mean values.

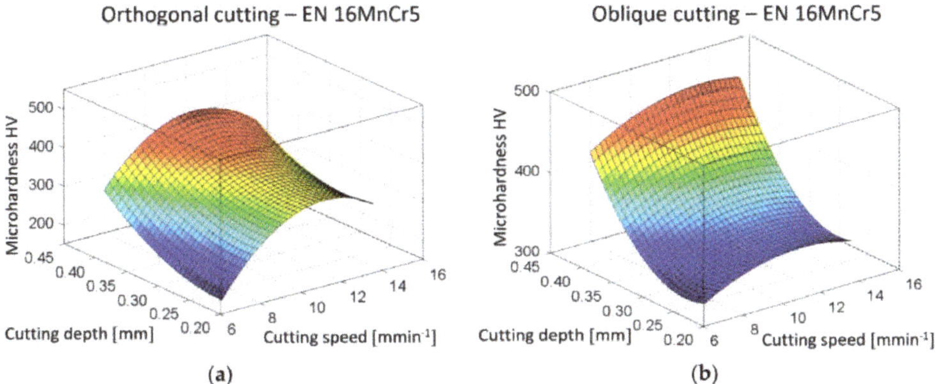

Figure 12. Dependency of microhardness HV on the cutting speed v_c and on the cutting depth a_p. (a) Orthogonal cutting; (b) oblique cutting.

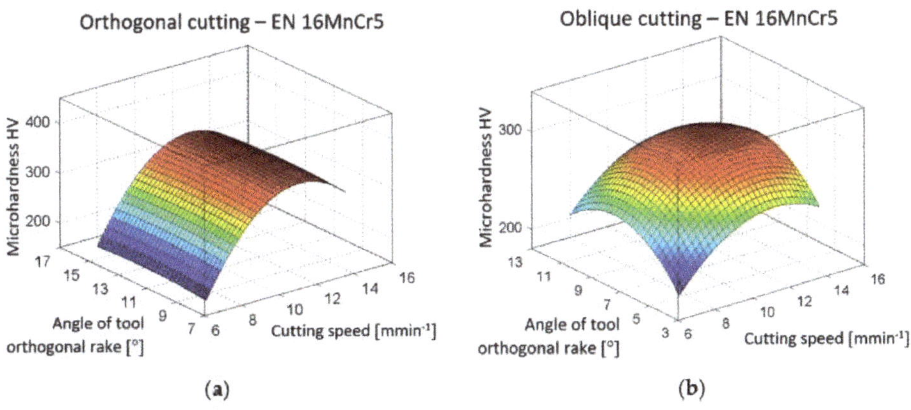

Figure 13. Dependency of microhardness HV on the cutting speed v_c and on the orthogonal rake angle γ_o. (a) Orthogonal cutting; (b) oblique cutting.

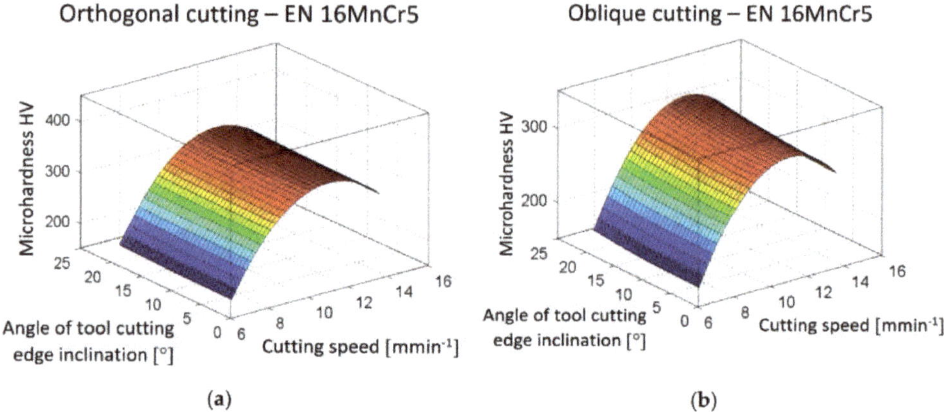

Figure 14. Dependency of microhardness HV on the cutting speed v_c and on the angle of tool cutting edge inclination λ_s. (**a**) Orthogonal cutting; (**b**) oblique cutting.

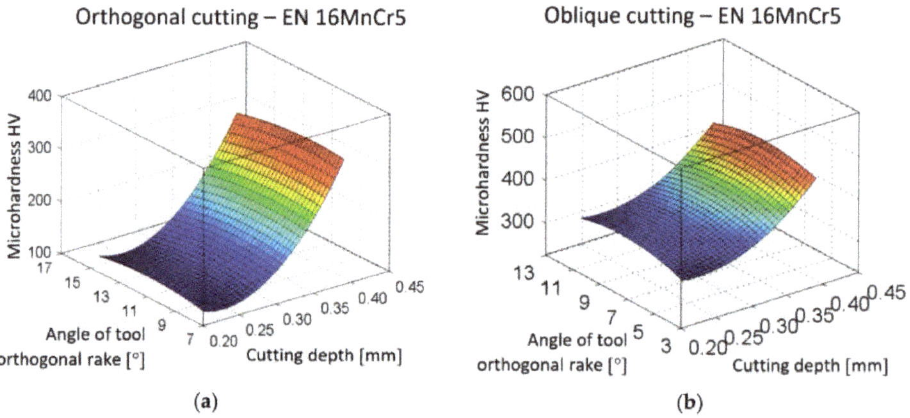

Figure 15. Dependency of microhardness HV on the cutting depth a_p and on the orthogonal rake angle γ_o. (**a**) Orthogonal cutting; (**b**) oblique cutting.

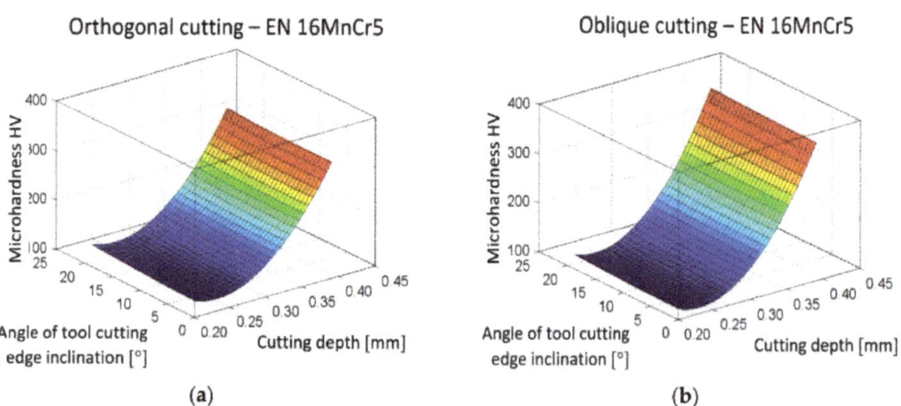

Figure 16. Dependency of microhardness HV on the cutting depth a_p and on the angle of tool cutting edge inclination λ_s. (**a**) Orthogonal cutting; (**b**) oblique cutting.

In the interaction of the angle of the tool cutting edge inclination and the cutting speed, the influence of the inclination angle on a change of the microhardness HV appears to be zero compared to the cutting speed in both cutting methods (Figure 14).

The interaction of the cutting depth and, also, the angle of the tool orthogonal rake have the same nature regarding the effect on microhardness in both cutting methods (Figure 15). The angle of tool orthogonal rake is less pronounced with respect to the cutting depth and the maximum microhardness values were measured at mean rake angle values and maximum cutting depth. Almost the same statement can be made for the effect of the cutting depth and the angle of tool cutting edge inclination on microhardness HV, as shown in Figure 16.

5. Conclusions

The chip formation in cutting processes at relatively low cutting speeds depends on the conditions of the cutting, the cutting tool and, to a large extent, on the machined material. EN 16MnCr5 manganese chromium steel was chosen as the material for this study. The experiments were carried out on specially modified workpieces for orthogonal and oblique machining (planing), in which the tool geometry and cutting conditions were changed. When changing individual parameters, the shape of chip forming was observed, the temperature was measured in the cutting area, and the Vickers HV microhardness was measured in the shear angle area. The results were statistically processed, and statistical regression dependence was constructed from all measured values.

The influence of individual parameters on the chip shape during machining of EN 16MnCr5 steel can be summarized in the following ways:

- The effect of the change in cutting speed in a given experiment on the chip shape appeared to be insignificant.
- The angle of the orthogonal tool rake caused a crumbly chip formation at minimum values, rather than at the maximum values.
- The cutting depth affected the radius of curvature of the chip. When increasing the thickness of the cut layer, the radius of chip curvature increased.

Based on the experiments, it could be stated that the most noticeable influence on the shape change of the chips is achieved by the angle of inclination of the main cutting edge λ_s and the tool cutting edge angle κ_r. At a zero angle of inclination of the main cutting edge, a shorter chip was produced than at an angle of inclination of the main cutting edge of 20°. At the same time, at a 20° angle of inclination of the cutting edge, the chip obtained a so-called chamfer along the edges, which is adequate to the angle of inclination of the main cutting edge, thus compressing the chip in the inclination direction of λ_s leading to the shape obtaining characteristics of a spiral chip.

By measuring the temperature in the cutting area (for both orthogonal and oblique cutting), the most significant influencing parameter for temperature change was found to be the depth of the cut. This was followed by cutting speed and a forehead angle, whereas the influence of the angle of inclination of the main cutting edge appeared to be insignificant. The maximum temperature during orthogonal cutting was reached at the maximum cutting depth, but in oblique cutting, the maximum temperature was reached at minimum cutting depth. In both cases, cutting at maximum temperature was characterized by mean values of the angle of tool orthogonal rake.

When measuring Vickers HV microhardness in the shear angle area, the significance of parameters was different. In orthogonal cutting, the highest microhardness was achieved at maximum cutting speed, while in oblique cutting, the maximum microhardness HV was achieved at maximum cut depth values.

Author Contributions: Conceptualization, K.M. and A.S.; methodology, K.M., P.P.M. and A.S.; software, L.H. and A.B.; validation, L.H. and A.B.; investigation, K.M. and P.P.M.; resources, K.M. and M.U.; data curation, M.U.; writing—original draft preparation, K.M.; writing—review and editing, P.P.M.; supervision, K.M.

Funding: This research was funded by Ministry of Education of the Slovak Republic by grant KEGA 007TUKE-4/2018.

Acknowledgments: The article was prepared thanks to the direct support of the Ministry of Education of the Slovak Republic by grant KEGA 007TUKE-4/2018 and thanks to European Regional Development Fund in the Research Centre of Advanced Mechatronic Systems project, project number CZ.02.1.01/0.0/0.0/16_019/0000867 within the Operational Programme Research, Development and Education.

Conflicts of Interest: The authors declare no conflict of interest.

Nomenclature

γ_o	angle of tool orthogonal rake (°)
λ_s	angle of tool cutting edge inclination (°)
α_o	angle of tool orthogonal clearance (°)
κ_r	tool cutting edge angle (°)
$\kappa_r{'}$	tool minor (end) cutting edge angle (°)
ε_r	tool included angle (°)
Φ	shear angle (°)
a_p	depth of cut (mm)
v_c	cutting speed (mmin^{-1})

References

1. Merchant, M.E. Mechanics of the metal cutting process. I. orthogonal cutting and a type 2 chip. *J. Appl. Phys.* **1945**, *16*, 267. [CrossRef]
2. Lee, E.H.; Shaffer, B.W. The theory of plasticity applied to a problem of machining. *J. Appl. Mech.* **1951**, *73*, 405.
3. Palmer, W.B.; Oxley, P.L.B. Mechanics of metal cutting. *Proc. Inst. Mech. Eng.* **1959**, *173*, 623. [CrossRef]
4. Von Turkovich, B.F. Dislocation theory of shear stress and strain rate in metal cutting. In Proceedings of the 8th International Machine Tool Design and Research Conference, Manchester, UK, September 1967; pp. 531–542.
5. Okushima, K.; Minato, K. On the behaviors of chips in steel cutting. *Bull. JSME.* **1959**, *2*, 58–64. [CrossRef]
6. Usui, E.; Hoshi, K. Slip-line fields in metal machining which involve centered fans. In Proceedings of the International Production Engineering Research (ASME Conference), Pittsburgh, PA, USA, September 1963; pp. 61–71.
7. Johnson, W.; Sowerby, R.; Haddow, J.B. *Plane-Strain Slip-line Fields: Theory and Bibliography*; Edward Arnold Ltd.: London, UK, 1970.
8. Fang, N.; Jawahir, I.S.; Oxley, P.L.B. A universal slip-line model with non-unique solutions for machining with curled chip formation and a restricted contact tool. *Int. J. Mech. Sci.* **2001**, *43*, 570–580. [CrossRef]
9. Young, H.T.; Mathew, P.; Oxley, P.L.B. Allowing for nose radius effects in predicting the chip flow direction and cutting forces in bar turning. *Proc. Inst. Mech. Eng.* **1987**, *201*, 213–226. [CrossRef]
10. Zhou, L.; Guo, H.; Rong, Y. Machining chip breaking prediction. In Proceedings of the SME 4th International Machining & Grinding Conference, Troy, MI, USA, 7–10 May 2001; p. 601.
11. Toulfatzis, A.I.; Pantazopoulos, G.A.; David, C.N.; Sagris, D.S.; Paipetis, A.S. Final heat treatment as a possible solution for the improvement of machinability of pb-free brass alloys. *Metals* **2018**, *8*, 575. [CrossRef]
12. Vasilko, K. New experimental dependence of machining. *Manuf. Technol.* **2014**, *14*, 111–116.
13. Stabler, G.V. The chip flow law and its consequences. In *Advances in Machine Tool Design and Research*; Tobias, S.A., Koenigsberger, F., Eds.; The Macmillan Press Ltd.: London, UK, 1964; pp. 243–251.
14. Moufki, A.; Dudzinski, D.; Molinari, A.; Rausch, M. Thermo-viscoplastic modelling of oblique cutting: Forces and chip flow predictions. *Int. J. Mech. Sci.* **2000**, *42*, 1205–1232. [CrossRef]
15. Lin, Z.C.; Lin, Y.Y.A. Study of an oblique cutting model. *J. Mater. Process. Technol.* **1999**, *86*, 119–130. [CrossRef]
16. Becze, C.E.; Elbestawi, M.A.A. Chip formation based analytical force model for oblique cutting. *Int. J. Mach. Tools Manuf.* **2002**, *42*, 529–538. [CrossRef]
17. Luk, W.K. The direction of chip flow in oblique cutting. *Int. J. Proc. Res.* **1972**, *10*, 67–76. [CrossRef]

18. Lin, G.C.I.; Oxley, P.L.B. Mechanics of oblique machining: predicting chip geometry and cutting forces from workmaterial properties and cutting conditions. *Proc. Inst. Mech. Eng.* **1972**, *186*, 813–820. [CrossRef]

19. Russell, J.K.; Brown, R.H. The measurement of chip flow direction. *Int. J. Mach. Tool Des. Res.* **1966**, *6*, 129–138. [CrossRef]

20. Shamoto, E.; Altintas, Y. Prediction of Shear Angle in Oblique Cutting with Maximum Shear Stress and Minimum Energy Principles. *J. Manuf. Sci. Eng.* **1999**, *121*, 399–407. [CrossRef]

21. Moufki, A.; Molinari, A.; Dudzinski, D. Modelling of orthogonal cutting with a temperature dependent friction law. *J. Mech. Phys. Solids.* **1998**, *46*, 2103–2138. [CrossRef]

22. Usui, E.; Hirota, A.; Masuko, M. Analytical prediction of three dimensional cutting process. Part 1: basic cutting model, an energy approach. *J. Eng. Ind.* **1977**, *100*, 222–228. [CrossRef]

23. Wang, J. Development of a chip flow model for turning operations. *Int. J. Mach. Tools Manuf.* **2001**, *41*, 1265–1274. [CrossRef]

24. Prasetyo, L.; Tauviqirrahman, M. Study of chip formation feedrates of various steels in low-speed milling process. *IOP Conf. Ser. Mater. Sci. Eng.* **2017**, *202*, 012097. [CrossRef]

25. Sánchez Hernández, Y.; Trujillo Vilches, F.J.; Bermudo Gamboa, C.; Sevilla Hurtado, L. Experimental parametric relationships for chip geometry in dry machining of the Ti6Al4V alloy. *Materials* **2018**, *11*, 1260. [CrossRef]

26. Sutter, G.; List, G. Very high speed cutting of Ti-6Al-4V titanium alloy—Change in morphology and mechanism of chip formation. *Int. J. Mach. Tools Manuf.* **2013**, *66*, 37–43. [CrossRef]

27. Joshi, S.; Tewari, A.; Joshi, S.S. Microstructural characterization of chip segmentation under different machining environments in orthogonal machining of Ti6Al4V. *J. Eng. Mater. Technol.* **2015**, *137*, 011005. [CrossRef]

28. Saez-de-Buruaga, M.; Soler, D.; Aristimuño, P.X.; Esnaola, J.A.; Arrazola, P.J. Determining tool/chip temperatures from thermography measurements in metal cutting. *Appl. Therm. Eng.* **2018**, *145*, 305–314. [CrossRef]

29. Calamaz, M.; Coupard, D.; Girot, F. A new material model for 2D numerical simulation of serrated chip formation when machining titanium alloy Ti-6Al-4V. *Int. J. Mach. Tools Manuf.* **2009**, *48*, 275–288. [CrossRef]

30. Bai, W.; Sun, R.; Roy, A.; Silberschmidt, V.V. Improved analytical prediction of chip formation in orthogonal cutting of titanium alloy Ti6Al4V. *Int. J. Mech. Sci.* **2017**, *133*, 357–367. [CrossRef]

31. Joshi, S. Dimensional inequalities in chip segments of titanium alloys. *Eng. Sci. Technol. Int. J.* **2018**, *21*, 238–244. [CrossRef]

32. Zhang, Y.; Outeiro, J.C.; Mabrouki, T. On the selection of Johnson-Cook constitutive model parameters for Ti-6Al-4V using three types of numerical models of orthogonal cutting. *Procedia CIRP* **2015**, *31*, 112–117. [CrossRef]

33. Wang, B.; Liu, Z. Shear localization sensitivity analysis for Johnson-Cook constitutive parameters on serrated chips in high speed machining of Ti6Al4V. *Simul. Model. Pract. Theory* **2015**, *55*, 63–76. [CrossRef]

34. Ducobu, F.; Rivière-Lorphèvre, E.; Filippi, E. Material constitutive model and chip separation criterion influence on the modeling of Ti6Al4V machining with experimental validation in strictly orthogonal cutting condition. *Int. J. Mech. Sci.* **2016**, *107*, 136–149. [CrossRef]

35. Kamiński, M.; Świta, P. Structural stability and reliability of the underground steel tanks with the stochastic finite element method. *Arch. Civ. Mech. Eng.* **2015**, *15*, 593–602. [CrossRef]

36. Calamaz, M.; Coupard, D.; Girot, F. Numerical simulation of titanium alloy dry machining with a strain softening constitutive law. *Mach. Sci. Technol.* **2010**, *14*, 244–257. [CrossRef]

37. Yameogo, D.; Haddag, B.; Makich, H.; Nouari, M. Prediction of the cutting forces and chip morphology when machining the Ti6Al4V alloy using a microstructural coupled model. *Procedia CIRP* **2017**, *58*, 335–340. [CrossRef]

38. Salguero, J.; Gerez, J.; Batista, M.; Garófano, J.E.; Marcos Bárcena, M. A study of macrogeometrical deviations in the dry turning of UNS R56400 Ti alloy. *Appl. Mech. Mater.* **2012**, *152–154*, 613–617. [CrossRef]

39. Trujillo, F.J.; Sevilla, L.; Marcos, M. Experimental parametric model for indirect adhesionwear measurement in the dry turning of UNS A97075 (Al-Zn) alloy. *Materials* **2017**, *10*, 152. [CrossRef] [PubMed]

40. Batista, M.; Calamaz, M.; Girot, F.; Salguero, J.; Marcos, M. Using image analysis techniques for single evaluation of the chip shrinkage factor in orthogonal cutting process. *Key Eng. Mater.* **2012**, *504–506*, 1329–1334. [CrossRef]

41. Veiga, C.; Davim, J.P.; Loureiro, A.J.R. Review on machinability of titanium alloys: The process perspective. *Rev. Adv. Mater. Sci.* **2013**, *34*, 148–164.

42. Nouari, M.; Makich, H. On the physics of machining titanium alloys: interactions between cutting parameters, microstructure and tool wear. *Metals* **2014**, *4*, 335–358. [CrossRef]

43. Salem, S.B.; Bayraktar, E.; Boujelbene, M.; Katundi, D. Effect of cutting parameters on chip formation in orthogonal cutting. *J. Achiev. Mater. Manuf. Eng.* **2012**, *50*, 7–17.

44. Filippov, A.V.; Filippova, E.O. Determination of cutting forces in oblique cutting. *Appl. Mech. Mater.* **2015**, *756*, 659–664. [CrossRef]

45. Davim, J.P.; Silvia, J.; Baptista, A.M. Experimental cutting model of metal matrix composites (MMCs). *J. Mater. Process. Technol.* **2007**, *183*, 358–362. [CrossRef]

46. Kiran, K.R.; Manohar, B.; Divakar, S. A central composite rotatable design analysis of lipase catalyzed synthesis of lauroyl lactic acid at bench-scale level. *Enzyme Microb. Technol.* **2001**, *29*, 122–128. [CrossRef]

47. Obeng, D.P.; Morrell, S.; Napier-Munn, T.J. Application of central composite rotatable design to modeling the effect of some operating variables on the performance of the three-product cyclone. *Int. J. Miner. Process.* **2005**, *76*, 181–192. [CrossRef]

48. Zhang, H.T.; Liu, P.D.; Hu, R.S. A three-zone model and solution of shear angle in orthogonal machining. *Wear* **1991**, *143*, 29–43. [CrossRef]

49. Harrell, F.E. *Regression Modeling Strategies: with Applications to Linear Models, Logistic Regression, and Survival Analysis*; Springer: New York, NY, USA, 2001; p. 568.

50. Kouadri, S.; Necib, K.; Atlati, S.; Haddag, B.; Nouari, M. Quantification of the chip segmentation in metal machining: Application to machining the aeronautical aluminium alloy AA2024-T351 with cemented carbide tools WC-Co. *Int. J. Mach. Tools Manuf.* **2013**, *64*, 102–113. [CrossRef]

51. Atlati, S.; Haddag, B.; Nouari, M.; Zenasni, M. Segmentation intensity ratio as a new parameter to quantify the chip segmentation phenomenon in machining ductile metals. *Int. J. Mach. Tools Manuf.* **2011**, *51*, 687–700. [CrossRef]

© 2019 by the authors. Licensee MDPI, Basel, Switzerland. This article is an open access article distributed under the terms and conditions of the Creative Commons Attribution (CC BY) license (http://creativecommons.org/licenses/by/4.0/).

Article

Aging Phenomena during In-Service Creep Exposure of Heat-Resistant Steels

G. N. Haidemenopoulos [1,2,*], K. Polychronopoulou [1,3], A. D. Zervaki [2], H. Kamoutsi [2], S. I. Alkhoori [1], S. Jaffar [1], P. Cho [1] and H. Mavros [1]

[1] Department of Mechanical Engineering, Khalifa University, Abu Dhabi 127788, UAE
[2] Department of Mechanical Engineering, University of Thessaly, Volos 38334, Greece
[3] Center for Catalysis and Separations, Khalifa University, Abu Dhabi 127788, UAE
*   Correspondence: hgreg@mie.uth.gr

Received: 17 June 2019; Accepted: 17 July 2019; Published: 19 July 2019

Abstract: An investigation of aging phenomena during creep exposure has been conducted for HP-Nb cast reformer tubes for several exposure conditions. Aging was manifested by carbide precipitation, carbide coarsening, and carbide transformation. The transformation of primary M_7C_3 to the more stable $M_{23}C_6$ carbide takes place at high exposure temperature (910 °C and above). The primary MC carbides transform to the Ni-Nb silicide or G-phase during creep exposure. The presence of Ti in the steel prevented the transformation of MC carbides to the G-phase. Morphological changes like needle to globular transitions, rounding of carbide edges, and carbide coarsening take place during creep exposure. The room-temperature tensile elongation and ultimate tensile strength are significantly reduced during creep exposure. The above aging phenomena are precursors to creep damage.

Keywords: creep; steam reforming; carbides; G-phase; aging; cast reformer tubes

1. Introduction

Catalytic reforming is routinely employed in oil refineries for the production of petroleum byproducts. The process takes place in tubes suspended in a furnace at a high temperature. The tubes are fed with hydrocarbon, mostly methane, and steam mixture in the presence of a catalyst according to the reaction network:

$$C_nH_m + nH_2O \rightarrow nCO + (\frac{m}{2} + n)H_2, \tag{1}$$

$$CO + H_2O \leftrightarrow CO_2 + H_2. \tag{2}$$

The first is the steam reforming reaction, which, for the case of methane, is strongly endothermic (ΔH_r = 206 KJ/mol) and takes place at a high temperature of the order of 900 °C. The second is the water-gas shift reaction and is mildly exothermic (ΔH_r = −41 KJ/mol). The internal pressure in the tubes generates stresses, causing creep of the tube material. However, prior to creep damage, several aging phenomena take place, which gradually changes the microstructure of the material. The reformer furnace tubes are centrifugally cast and are manufactured from heat-resistant steels, containing Ni and Cr. The current state-of-the-art is to use modified HP-alloys (25Cr-35Ni), with Nb and/or Ti additions. Niobium is added for the precipitation of NbC, which due to its thermal stability contributes to an increased creep resistance [1,2]. Centrifugal casting results in austenitic dendritic grains oriented in the radial direction from the inside diameter (ID) towards the outside diameter (OD) of the tube. The material is strengthened by a network of primary interdendritic carbides. The design life of these tubes is 100,000 h (11.4 years) according to API recommended practice [3]. Premature failures of reformer tubes are frequently observed and are caused by various mechanisms such as creep, carburization, and thermal shock. Evaluation of creep damage in reformer tubes has been reported by several investigators, either after creep testing [2,4,5], after a certain operational time [6–10], or after premature tube rupture [11,12]. In addition, specific studies of microstructural evolution after long-term aging at high temperatures have been performed [13–15]. The present work contributes to the study of the evolution of carbides and other intermetallic compounds with aging during creep exposure. These phenomena lead to a decrease of ductility and may lead to cracking from thermal stresses during shut-down and start-up operations or during weld repairs.

It is important to note that most of the published data are concerned with studies involving laboratory creep exposure. Results based on specimens taken from real service are rather rare. It is not common to retrieve specimens during a shut-down of a reformer furnace. Therefore, the value of the present work is that the analysis was based entirely on specimens of reformer tubes extracted after a certain operational period, before the end of operational life, and for different temperatures.

2. Materials and Methods

Sections of reformer tubes with an internal diameter of 103 mm and a thickness of 15 mm were retrieved from an oil refinery after 11 years of service. These sections correspond to isothermal operation at three different temperatures 830, 880, and 910 °C and are named as conditions L, M, and H respectively. Another section came from a tube that operated for 8 years at 910 °C and was then nipped for the remaining 3 years, during which the temperature was 950 °C. This section corresponds to condition X. The material of these sections is an HP25Cr-35Ni alloy with Nb addition (HP-Nb steel). The creep strain was estimated by measurements of the outer diameter of the tubes to be of the order of 0.2%. It was also possible to retrieve a section of a reformer tube from another refinery after 5 years of service at 910 °C. This condition is named H5 and the chemical composition of H5 is similar to the previous one with the most important difference being the addition of Ti. This tube sampling enabled the study of the effect of temperature on aging phenomena by comparing sections L, M, H, and X. At the same time, the effect of time and Ti addition could also be performed by comparing sections H and H5. Chemical analysis of the tube materials was performed by optical emission spectrometry. The composition of the alloys and creep exposure conditions are shown in Table 1.

Table 1. Composition (wt%) of the alloys and creep exposure conditions.

Code	T (°C)	t (y)	C	Si	Mn	P	S	Cr	Ni	Nb	Mo	Ti	Fe
L	830	11											
M	880	11	0.49	2	0.80	0.01	0.01	25.7	33.6	0.93	0.18	-	B
H	910	11											
X	910/950	8/3											
H5	910	5	0.41	1.6	1.0	0.02	0.01	24	34.7	0.96	0.07	0.2	B

Characterization of microstructure, including aging phenomena, carbide transformation, and carbide morphology was performed by means of light optical metallography (LOM) and scanning electron microscopy equipped with an energy dispersive X-ray analysis system (EDX). Metallography was performed on thickness specimens from all tube sections. Specimen preparation for metallography involved cutting with a Struers Labotom-3 machine (Struers, Ballerup, Denmark), grinding with SiC papers with 240, 400, 800, 1200, and 2400 grit, polishing with 6, 3, and 1 µm diamond paste and finishing with colloidal silica suspension of 0.04 µm. The etching was performed by swabbing with Kalling's reagent ($CuCl_2$ in a mixture of hydrochloric acid and ethanol). Optical metallography was performed on an Olympus inverted microscope (Olympus, Tokyo, Japan) at magnifications 50–1000×. Scanning electron microscopy (SEM) was performed on etched specimens on a field emission JEOL JSM 7610F (JEOL, Tokyo, Japan) and Quanta 3D microscopes (ThermoFisher Scientific, Hillsboro, OR, USA)equipped with an energy dispersive X-ray analysis (EDX) detector (Oxford Instruments, Abingdon, UK). Imaging was performed with secondary electron mode (SE) and in certain cases with backscattered electron (BSE) mode. Room temperature uniaxial tensile testing was performed according to ISO 6892 specification [16].

3. Results

3.1. Tensile Properties

Room temperature tensile testing results are shown in Figure 1. The results indicate a decrease of the ultimate tensile strength (UTS) and a considerable reduction in tensile elongation (strain to failure) with exposure temperature, as a result of aging. Condition H5 also exhibits considerable reduction in elongation, however the elongation is higher than the elongation of condition H. Similar reductions in tensile ductility have been reported by Pan et al. [11] after 36,000 h at 910 °C (4.12 years), and Monobe et al. [13] after exposure for 200 h at 720–780 °C. The aging phenomena and carbide transformations are discussed in the following section.

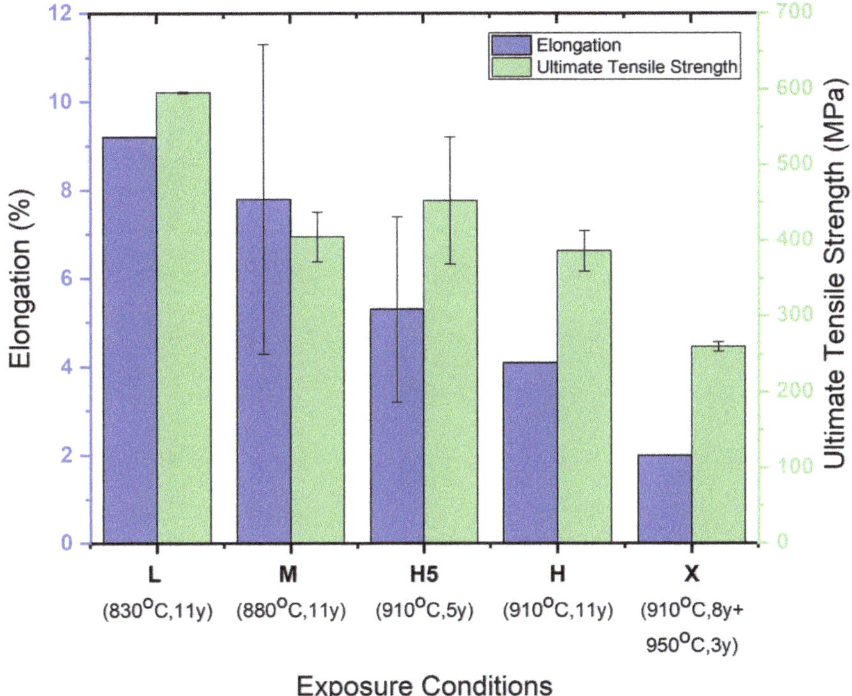

Figure 1. Room-temperature tensile properties vs. creep exposure conditions.

3.2. Microstructure Analysis

Carbides and other phases have been identified in this work with SEM/EDX analysis. Carbides were identified by the M:C ratio and the G-phase by the Ni:Nb:Si ratio. It should be noted that the EDX composition analysis of carbides is semi-quantitative, since it depends on the spot size relative to the carbide size. In order to avoid matrix contribution to the EDX spectrum, a low voltage was used during the EDX analysis. The analysis, therefore, provides a valid indication of the nature of the carbides.

The microstructure of the alloy consists of an austenitic matrix, forming dendrites, and a network of primary carbides. The dendrite length varies from 0.3 to 6.5 mm with an average length of 1.4 mm as depicted in Figure 2. The dendrites are oriented in the radial direction from the inner diameter (ID) to the outer diameter (OD) of the tube. Some dendrites are very long with a length being a significant fraction of the tube thickness. In the as-cast condition, the austenite matrix is free of precipitates, as reported by Alvino et al. [6].

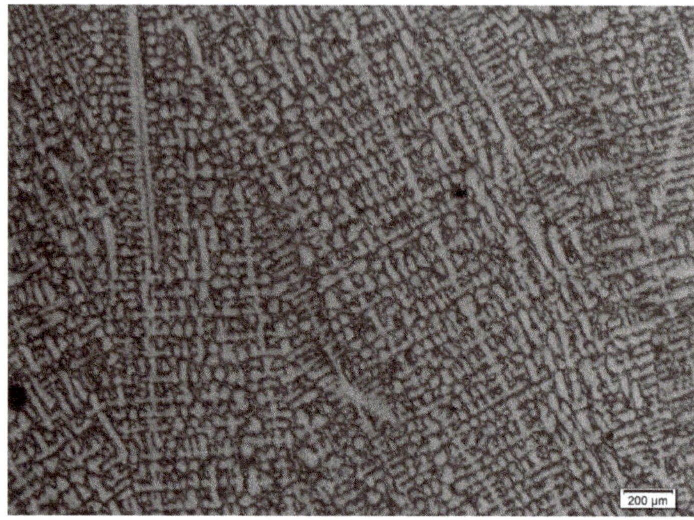

Figure 2. Microstructure of HP-Nb steel with austenite dendritic structure and carbide network in the interdendritic regions.

The carbides are located either in interdendritic or intradendritic regions and are either Cr-rich of M_7C_3 or $M_{23}C_6$ type and Nb-rich MC carbides. $M_{23}C_6$ are mostly grain boundary film-like carbides. M_7C_3 carbides exhibit a "Chinese script" morphology. Long-term exposure resulted in aging, manifested by the secondary precipitation in the austenitic matrix and coarsening of carbides. Such intragranular precipitation and carbide coarsening have also been observed by Attarian et al. [4] in HP-Nb steel after aging at 982 °C for 100 h. The results are presented for each exposure condition below.

Condition L (830 °C/11y): The microstructure is depicted in Figure 3a, consisting of M_7C_3 carbides. Identification of M_7C_3 carbides was based on SEM/EDX analysis of Figure 4. The numbers in Figure 4 correspond to EDX composition analysis, which is presented in Table 2. These carbides exhibit a "Chinese script" morphology. In addition to this morphology, M_7C_3 carbides appear with a needle morphology (Figures 3a and 4b).

Condition M (880 °C/11y): The microstructure is depicted in Figure 3b. The "Chinese script" morphology of M_7C_3 carbides is preserved with considerable rounding of edges. However, the needle-shaped M_7C_3 carbides have undergone a morphological transformation and exhibit a globular shape. This morphological transformation has been observed in directionally solidified HP-Ni steel [4]. The NbC carbides, originally present in the as-cast microstructure, transformed to a Ni-Nb silicide, the G-phase, with nominal composition $Ni_{16}Nb_7Si_6$, as depicted in Figure 5a,b. The numbers in Figure 5 correspond to EDX composition analysis, which is presented in Table 2. This transformation has been observed, in Ti-free alloys, by Pan [11] and by Soares et al. [17]. The G-phase is hard and brittle and, therefore, prone to cavity formation when the material is subjected to temperature excursions [11,17]. However, it should be noted that no cavities formed on the G-phase were detected at this operating condition. In addition, no $M_{23}C_6$ carbides were detected in condition M, indicating that the transformation of M_7C_3 to $M_{23}C_6$ carbides had not taken place.

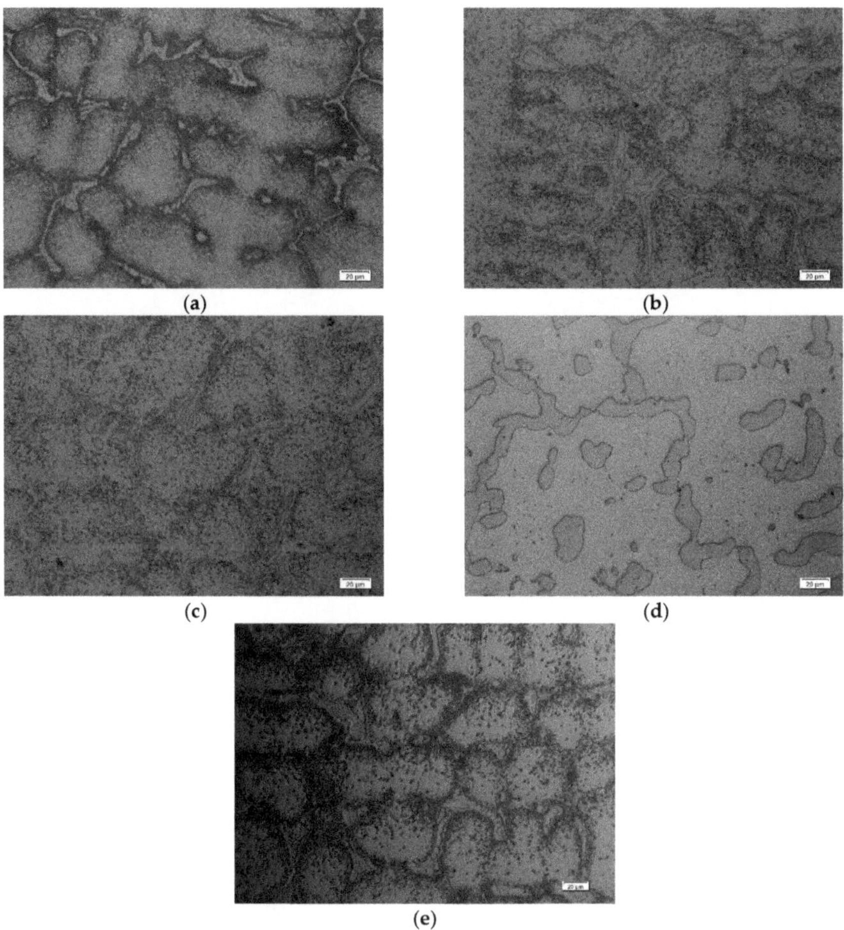

Figure 3. Microstructure revealed by metallography: (**a**) Condition L, (**b**) condition M, (**c**) condition H, (**d**) condition X, and (**e**) condition H5.

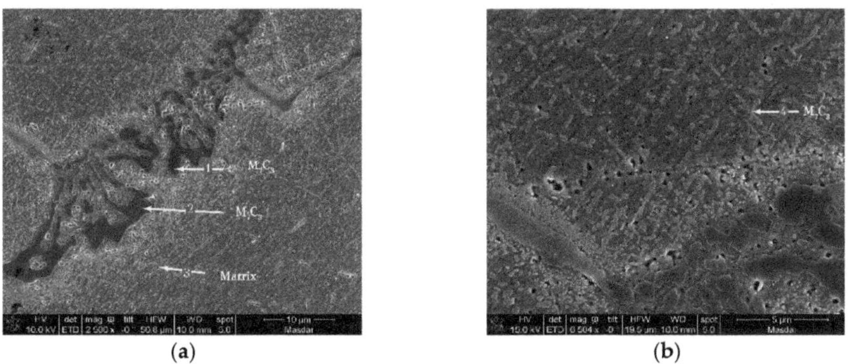

Figure 4. SEM image depicting M_7C_3 carbides in condition L. (**a**) "Chinese script" morphology; (**b**) needle morphology. Numbers indicate EDX composition analysis in Table 2.

Table 2. EDX analysis (at%).

Conditions		Cr	Ni	Fe	Nb	Ti	Si	C	M	M:C	Phases
Figure 4 (L)	1	54.3	4	10	-	-	-	31.6	68.3	2.16	M_7C_3
	2	55.8	4	8.5	-	-	-	31.6	68.3	2.16	M_7C_3
	3 matrix	23.8	29.6	38.5	0.3	0.5	-	4	-	-	-
	4 needles	19.4	21.5	22.2	-	-	-	31.8	63.1	1.98	M_7C_3
Figure 5 (M)	1	62.7	3.72	7.2	-	-	-	25.4	73.62	2.87	M_7C_3
	2	-	46	-	16.4	-	21.8	-	-	-	G
	3	-	54.17	-	19.1	-	26.7	-	-	-	G
	4	59.2	3	6	-	-	-	32	68.2	2.13	M_7C_3
	5	-	55	-	19.6	-	26	-	-	-	G
Figure 6 (H)	1	68	4	7	-	-	-	21	79	3.76	$M_{23}C_6$
	2	67.5	3.8	5.5	-	-	-	23.3	76.8	3.30	$M_{23}C_6$
	3	67.15	4.2	5.9	-	-	-	22.8	77.25	3.38	$M_{23}C_6$
	4	67.1	4.3	5.8	-	-	-	22.7	77.2	3.4	$M_{23}C_6$
Figure 7 (X)	1	59	3	11	5	-	-	22	78	3.54	$M_{23}C_6$
	2	59.2	3.7	10	4.3	-	-	22.6	77.2	3.41	$M_{23}C_6$
	3	46.7	20.1	3.9	4.6	-	-	21.5	75.3	3.50	$M_{23}C_6$
Figure 8a (H5)	1	-	-	-	52.2	1.7	-	46.1	53.9	1.17	MC
	2	-	-	-	50	2.7	-	47.3	52.7	1.11	MC
	3	-	-	-	51.3	2.5	-	46.2	53.8	1.16	MC
	4	-	-	-	41.4	5	-	53.6	46.4	0.86	MC
	5	53.4	7.8	16.8	-	-	-	22	78	3.54	$M_{23}C_6$
	6	59	4.7	15.3	-	-	-	21	79	3.76	$M_{23}C_6$
	matrix	25.4	33.4	34.7	-	-	-	5.2	-	-	-
	matrix	25.9	33.2	34.6	-	-	-	5	-	-	-
Figure 8c (H5)	1	57.4	1.4	7.6	-	-	-	33.6	66.4	1.97	M_7C_3
	2	28.1	22.2	29.2	-	-	-	20.5	79.5	3.87	$M_{23}C_6$
	3	33.1	19.6	25.3	-	-	-	21.3	78	3.66	$M_{23}C_6$
	4	-	-	-	41.3	2.7	-	56	44	0.78	MC

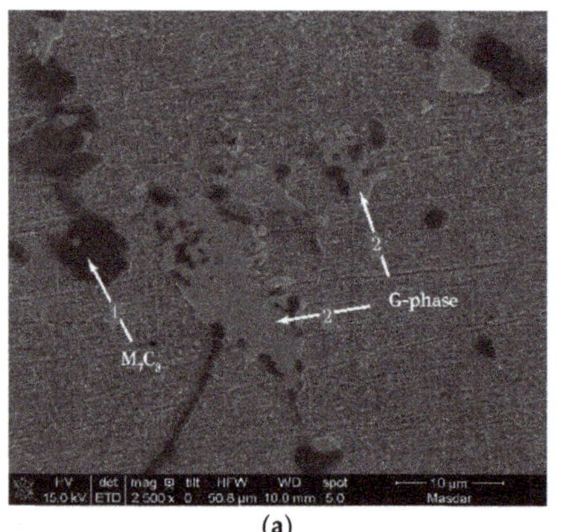

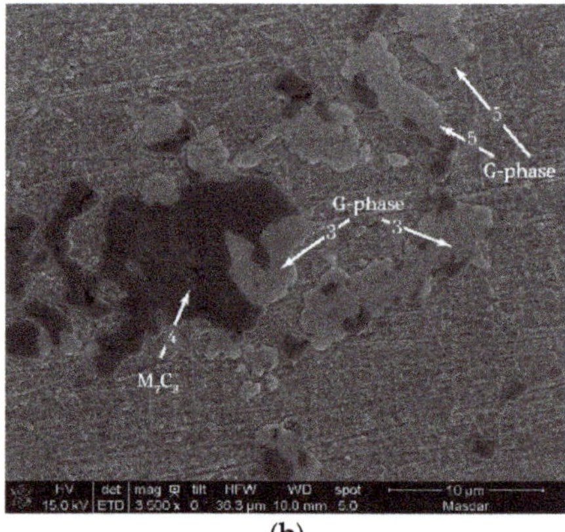

(a)　　　　(b)

Figure 5. SEM images, (**a**,**b**), depicting M_7C_3 carbides and G-phase in condition M. Numbers indicate EDX composition analysis in Table 2.

Condition H (910 °C/11y): The microstructure is depicted in Figure 3c. Most of the M_7C_3 carbides have transformed into the more stable $M_{23}C_6$ carbides as shown in Figure 6a,b. The numbers in Figure 6 correspond to EDX composition analysis, which is presented in Table 2. Significant coarsening of the $M_{23}C_6$ carbides has also taken place driven by the reduction in interfacial energy.

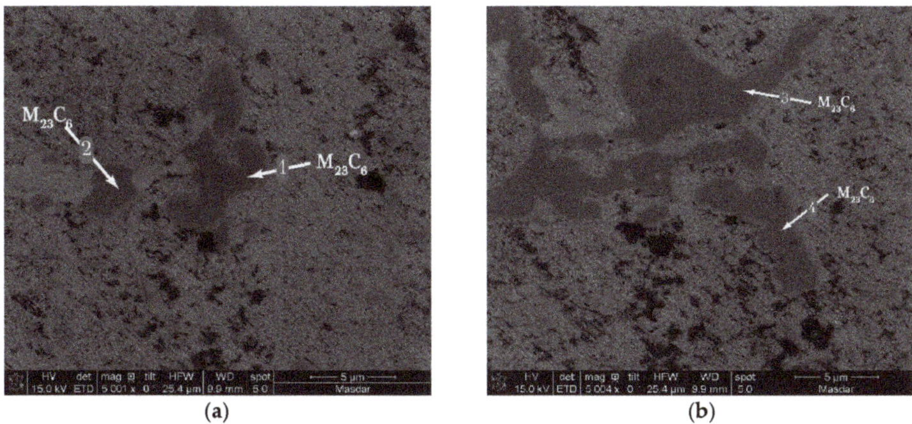

(a)   (b)

Figure 6. SEM images, (**a**,**b**), depicting $M_{23}C_6$ carbides in condition H. Numbers indicate EDX composition analysis in Table 2.

Condition X (910 °C/8y + 950 °C/3y): The microstructure is depicted in Figure 3d. All carbides are of the $M_{23}C_6$ type and significant coarsening has taken place due to exposure at high temperature (950 °C). Nb has also been incorporated in the $M_{23}C_6$ carbides as indicated in Figure 7a,b. The numbers in Figure 7 correspond to EDX composition analysis, which is presented in Table 2.

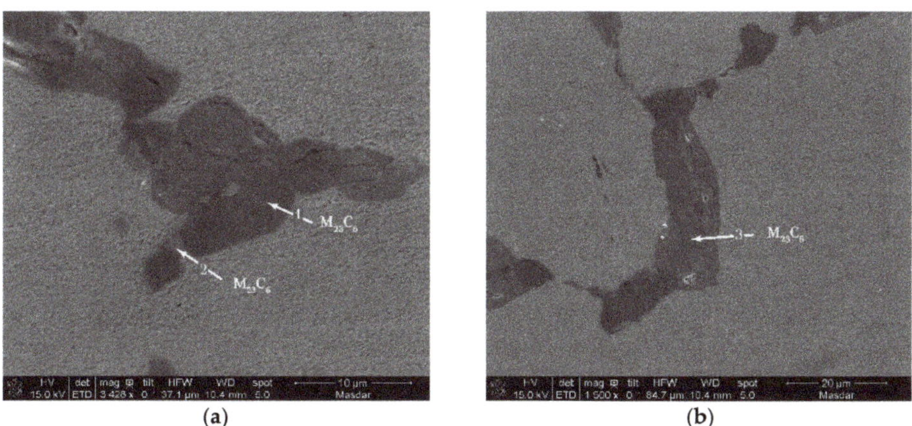

(a)   (b)

Figure 7. SEM images, (**a**,**b**), depicting coarse $M_{23}C_6$ carbides in condition X. Numbers indicate EDX composition analysis in Table 2.

Condition H5 (910 °C/5y): It is noted that the H5 condition concerns the material containing Ti. The optical micrograph of the microstructure is depicted in Figure 3e. The microstructure of H5 material was investigated with SEM operating in the SE and BSE modes. Typical microstructures are depicted in Figure 8. The numbers in Figure 8 correspond to EDX composition analysis, which is presented in Table 2. MC carbides (Nb-rich), appear white, while $M_{23}C_6$, Cr-rich carbides appear dark grey in the BSE images due to the difference in atomic number between Nb and Cr. SEM-BSE imaging has been used by other researchers for carbide identification in cast heat-resistant steels [6,15]. The results indicate that MC carbides contain Ti, which according to Swaminathan et al. [18], stabilizes the MC carbide against transformation to the G-phase. This explains the fact that no G-phase was detected in the H5 material. Most probably, the transformation of the primary M_7C_3 carbides to

$M_{23}C_6$ is partial, contrary to the H condition where the transformation was complete. This partial transformation is depicted in Figure 8c with the associated EDX analysis of carbides in Table 2.

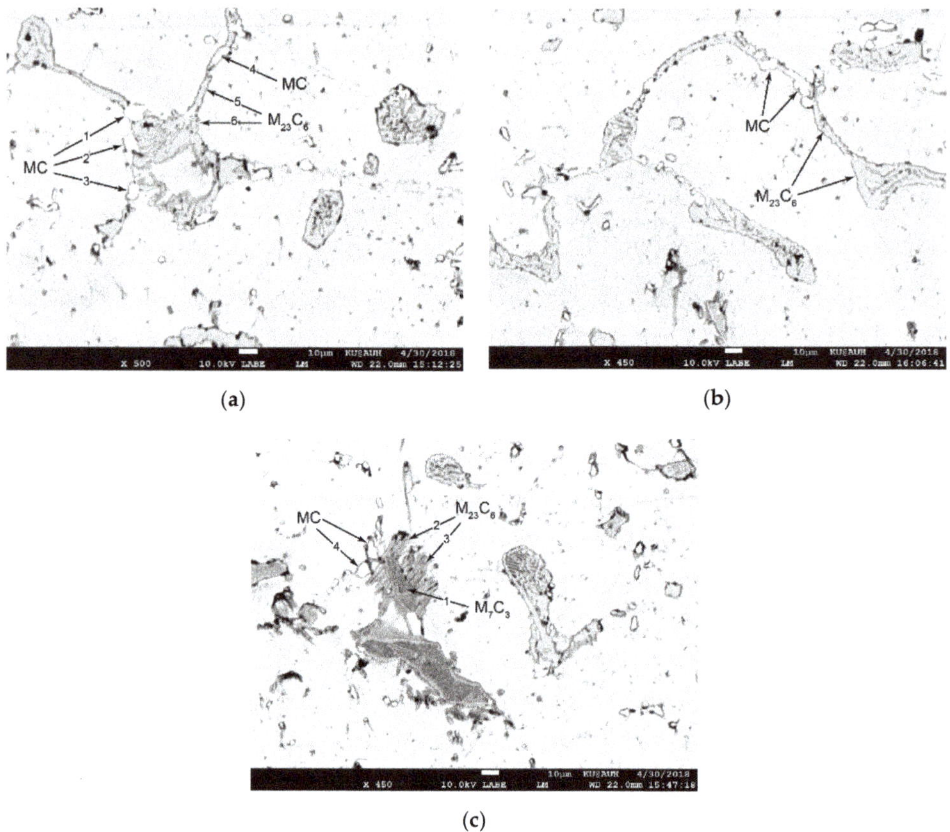

Figure 8. SEM-BSE images depicting: (**a**,**b**) MC and $M_{23}C_6$ carbides in condition H5; (**c**) transformation of M_7C_3 to $M_{23}C_6$ carbide after creep exposure of 5 years. Numbers correspond to EDX composition analysis in Table 2.

4. Discussion

Creep resistance in cast steel reformer tubes depends, among other factors, on the thermal stability of the carbides. During operation at high temperatures, the carbides undergo phase changes as well as morphological changes, which degrade the creep resistance of the material. Phase changes involve the transformation of primary M_7C_3 to the more thermodynamically stable, at those temperatures of operation, $M_{23}C_6$. Morphological changes involve the coarsening of carbides driven by the reduction of interfacial energy [4,6]. Finally, an aging reaction, manifested as secondary precipitation of carbides within the austenitic matrix takes place during creep exposure [6]. In the present work, the transformation of M_7C_3 to the more stable $M_{23}C_6$ was observed to take place only at 910 °C and above for 11 years of operation period (conditions H and X). In these conditions the transformation is complete. In an intermediate operation period of five years (condition H5), the M_7C_3 to $M_{23}C_6$ transformation is partial. Significant coarsening of the carbides takes place at high temperatures (conditions H and X) while at lower temperatures the changes are morphological, i.e., the rounding of edges or the change from needle-shaped carbides to a globular morphology (condition M). Another interesting aging phenomenon during creep exposure is the transformation of the primary MC carbides to the Ni-Nb silicide of G-phase manifested at intermediate temperatures (condition M). At higher temperatures, the G-phase dissolves and the Nb is incorporated in the $M_{23}C_6$ carbide. Similar observations regarding the type and morphologies of various carbides in creep-exposed HP-Nb steels have been reported by Allesio et al. [2] and Almeida et al. [15]. Finally, the presence of NbC carbides and the corresponding absence of G-phase in the H5 condition in the Ti-containing material validated previous observations that Ti stabilizes the MC carbide and acts against the transformation of the MC carbides to the G-phase [11,17]. As a result of aging the room temperature tensile elongation (strain to failure) is reduced considerably. The reduction is higher the higher the temperature of exposure. This reduced ductility makes the material prone to cracking due to expansions and contractions, which take place during shut-down/start-up of the reformer furnace.

5. Conclusions

An investigation of aging phenomena during creep has been conducted for HP-Nb cast reformer tubes. Based on the results presented above, the following conclusions can be drawn:

Aging was manifested by carbide precipitation, carbide coarsening and transformation from M_7C_3 to $M_{23}C_6$ carbides.

The MC carbides transform to the G-phase during creep exposure making the material more prone to cavity formation. The presence of Ti in the steel of condition H5 prevented the transformation of MC carbides to the G-phase.

Morphological changes occur as needle to globular transitions, rounding of edges, and carbide coarsening take place during creep exposure.

As a result of aging phenomena, the room-temperature tensile ductility is reduced, making the material prone to cracking during shut-down/start-up operations.

Author Contributions: Conceptualization, G.N.H.; methodology, K.P.; mechanical testing H.K. and A.D.Z.; metallography S.I.A., P.C., H.M. and S.J.; SEM/EDX, K.P.; paper preparation and editing, G.N.H.

Funding: This research received no external funding.

Acknowledgments: The assistance of Roba Khaled Saab of KU with SEM/EDX is greatly appreciated.

Conflicts of Interest: The authors declare no conflict of interest.

References

1. Barbabela, G.D.; de Almeida, L.H.; da Silveira, T.L.; Le May, I. Role of Nb in modifying the microstructure of heat-resistant cast HP steel. *Mater. Charact.* **1991**, *26*, 193–197. [CrossRef]
2. Alessio, D.; Gonzalez, G.; Pirrone, V.F.; Iurman, L.; Moro, L. Variation of Creep Properties in HP Steel by Influence of Temperature. *Proc. Mater. Sci.* **2012**, *1*, 104–109. [CrossRef]
3. API. *Calculation of Heater Tube Thickness in Petroleum Refineries: API Recommended Practice 530*, 3rd ed.; API: Washington, DC, USA, 1988.
4. Attarian, M.; Taheri, A.K. Microstructural evolution in creep aged of directionally solidified heat resistant HP-Nb steel alloyed with tungsten and nitrogen. *Mater. Sci. Eng. A* **2016**, *659*, 104–118. [CrossRef]
5. Ray, A.K.; Roy, N.; Raj, A.; Roy, B.N. Structural integrity of service exposed primary reformer tube in a petrochemical industry. *Int. J. Pres. Ves. Pip.* **2016**, *137*, 46–57. [CrossRef]
6. Alvino, A.; Lega, D.; Giacobbe, F.; Mazzocchi, V.; Rinaldi, A. Damage characterization in two reformer heater tubes after nearly 10 years of service at different operative and maintenance conditions. *Eng. Fail. Anal.* **2010**, *17*, 1526–1541. [CrossRef]
7. Bonaccorsi, L.; Guglielmino, E.; Pino, R.; Servetto, C.; Sili, A. Damage analysis in Fe–Cr–Ni centrifugally cast alloy tubes for reforming furnaces. *Eng. Fail. Anal.* **2014**, *36*, 65–74. [CrossRef]
8. Gong, J.-M.; Tu, S.-T.; Yoon, K.-B. Damage assessment and maintenance strategy ofhydrogen reformer furnace tubes. *Eng. Fail. Anal.* **1999**, *6*, 143–153. [CrossRef]
9. Lee, J.H.; Yang, W.J.; Yoo, W.D.; Cho, K.S. Microstructural and mechanical property changes in HK40 reformer tubes after long term use. *Eng. Fail. Anal.* **2009**, *16*, 1883–1888. [CrossRef]

10. Ray, A.K.; Kumar, S.; Krishna, G.; Gunjan, M.; Goswami, B.; Bose, S.C. Microstructural studies and remnant life assessment of eleven years service exposed reformer tube. *Mater. Sci. Eng. A* **2011**, *529*, 102–112. [CrossRef]

11. Pan, J.-H.; Chen, Z.; Fan, Z.-C.; Wu, Y.-C. An experimental investigation on manifold failure and material deterioration. *Int. J. Pres. Ves. Pip.* **2018**, *162*, 1–10. [CrossRef]

12. Ray, A.K.; Sinha, S.K.; Tiwari, Y.N.; Swaminathan, J.; Das, G.; Chaudhuri, S.; Singh, R. Analysis of failed reformer tubes. *Eng. Fail. Anal.* **2003**, *10*, 351–362. [CrossRef]

13. Monobe, L.S.; Schön, C.G. Characterization of the cold ductility degradation after aging in centrifugally cast 20Cr32Ni+Nb alloy tube. *Int. J. Pres. Ves. Pip.* **2009**, *86*, 207–210. [CrossRef]

14. Kenik, E.A.; Maziasz, P.J.; Swindeman, R.W.; Cervenka, J.; May, D. Structure and phase stability in a cast modified-HP austenite after long-term ageing. *Scr. Mater.* **2003**, *49*, 117–122. [CrossRef]

15. De Almeida, L.H.; Ribeiro, A.F.; Le May, I. Microstructural characterization of modified 25Cr–35Ni centrifugally cast steel furnace tubes. *Mater. Charact.* **2002**, *49*, 219–229. [CrossRef]

16. *ISO 6892-1:2016. Metallic Materials—Tensile testing—Part 1: Method of Test at Room*; International Organization for Standardization: Geneva, Switzerland, 2016.

17. De Almeida Soares, G.D.; de Almeida, L.H.; da Silveira, T.L.; Le May, I. Niobium additions in HP heat-resistant cast stainless steels. *Mater. Charact.* **1992**, *29*, 387–396. [CrossRef]

18. Swaminathan, J.; Guguloth, K.; Gunjan, M.; Roy, P.; Ghosh, R. Failure analysis and remaining life assessment of service exposed primary reformer heater tubes. *Eng. Fail. Anal.* **2008**, *15*, 311–331. [CrossRef]

© 2019 by the authors. Licensee MDPI, Basel, Switzerland. This article is an open access article distributed under the terms and conditions of the Creative Commons Attribution (CC BY) license (http://creativecommons.org/licenses/by/4.0/).

Article

Failure-Analysis Based Redesign of Furnace Conveyor System Components: A Case Study

Beatriz González-Ciordia [1,2], **Borja Fernández** [1,2], **Garikoitz Artola** [3], **Maider Muro** [3], **Ángel Sanz** [4] **and Luis Norberto López de Lacalle** [2,*]

[1] Gestamp ES&AM, AIC-Automotive Intelligence Center, Barrio Boroa, 2, 483401 Amorebieta-Etxano, Spain
[2] Department of Mechanical Engineering, University of the Basque Country (UPV/EHU), Plaza Ingeniero Torres Quevedo 1, 48013 Bilbao, Spain
[3] Fundación Azterlan, Aliendalde Auzunea 6, 48200 Durango, Spain
[4] GHI Hornos Industriales, Aperribai 4, 48960 Galdakao, Spain
[*] Correspondence: norberto.lzlacalle@ehu.eus; Tel.: +34-946-014-024

Received: 17 June 2019; Accepted: 20 July 2019; Published: 25 July 2019

Abstract: Any manufacturing equipment designed from scratch requires a detailed follow-up of the performance for the first units placed in service during the production ramp-up, so that lessons learned are immediately implemented in next deliveries and running equipment is accordingly updated. Component failure analysis is one of the most valuable sources of improvement among these lessons. In this context, a failure-assessment based design revision of the conveying system of a newly developed press hardening furnace is presented. The proposed method starts with a forensic metallurgical analysis of the failed components, followed by an investigation of the working conditions to ensure they match the forensic observations. The results of this approach evidenced an initially unforeseen thermo-mechanical damage produced by a combination of thermal distortions, material ageing, and mechanical fatigue. Once the cause–effect relationship for the failure is backed up by evidence, an improved design is proposed. As a conclusion, a new standard design for the furnace entrance set of rollers in hot stamping lines was established for roller hearth furnaces. The solution can be extended to similar applications, ensuring the same issues will not arise thanks to the lessons learned.

Keywords: hot stamping; press hardening; austenitizing furnace; high temperature fatigue; thermal distortion; conveying system; refractory steels; furnace component failure

---

1. Introduction

Press hardening is a fast-growing technology, mainly exploited for the safety and weight improvements that it introduces in the automotive industry [1–3]. Austenitizing the steel to be hardened is a must in the manufacturing process, which turns the austenitizing furnace into a key piece of equipment. Despite that several heating alternatives have been studied, such as resistance heating [4], induction heating [5,6], or direct contact heating [7], radiant heating is the leading industrial solution for mass production. The amount of published works related to dedicated furnaces for press hardening is, though, scarce and mostly focused on roller hearth furnaces [8].

Roller hearth furnaces carry the steel sheet blanks through the furnace by means of ceramic rollers that stand the nominal working temperatures required for press hardening furnaces, which range from 890 °C to 950 °C. These ceramic rollers can show a variety of porosity levels and chemical compositions, ranging from high density fused silica to very porous aluminosilicates. The main drawback of this conveying system lies in the brittle nature of the ceramics and its impact in furnace maintenance costs and production dead-times arising from for broken roller replacement. These maintenance requirements are increased by the predominance AlSi coated steel sheet in the press hardened products. The AlSi coating damages the ceramics rollers in two ways (Figure 1):

- It can partially melt and adhere to the roll, generating a build-up that leads to format trajectory deviation inside the furnace. Therefore, the formats do not exit the furnace in a proper position for robotic transferring to the stamping press.
- The molten AlSi can infiltrate in the pores and micro-cracks in the roll surfaces. When the furnace is cooled down for a stop or an energy saving condition, such as a weekend stand-by, the infiltrated metal solidifies and acts as a wedge, which leads to roller fracture.

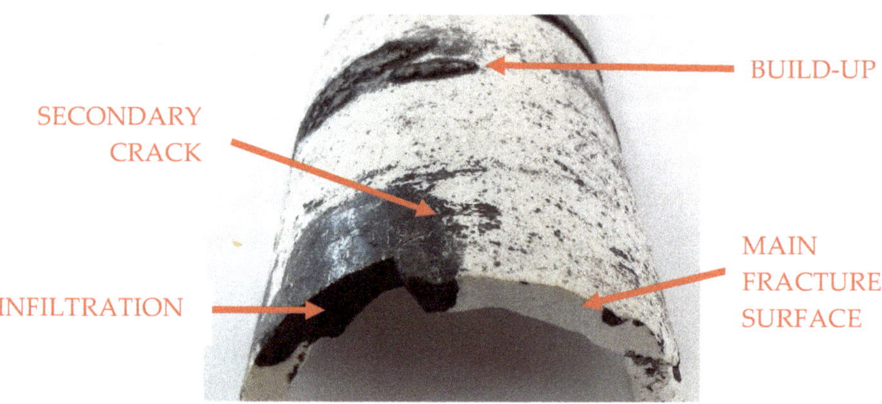

Figure 1. Picture of a cracked roller showing both aluminum build-up due to adhesion and aluminum infiltration across the full thickness of the roller wall.

The rolling beam sheet blank conveying concept is an advantageous alternative to the roller hearth furnace system as it allows avoiding these two issues. More specifically, the rolling beam design overcomes the limitations imposed by the roller hearth thanks to the following improvements [9]:

- Blanks can be loaded widthwise, leading to shorter production lines.
- Transport system-to-blank contact is reduced, minimizing conveying system contamination problems due to AlSi coating melting.
- Blank location accuracy at the exit point of the blanks from the furnace is improved as a result of the fixed stroke of the rolling beams.
- Overall equipment efficiency is increased owing to lower maintenance needs and less stops are needed for correcting blank positioning at the furnace exit.

On the drawback side, roller hearth furnaces are a mature option for austenitizing that have worked successfully for years in press hardening, while roller beam furnaces have been in use for less than five years now. As the novel roller beam sheet metal format conveying system was designed from scratch, close follow-up of the furnaces deployed first is a must to ensure a quick maturation of the equipment. During this sort of follow-up, component failure analysis has proven to be a source of major information to introduce design and operation improvements in many applications [10–12]. In the specific case of the furnaces, studies on elements such as radiant tubes [13], internal supports [14], gas conductions [15], and product transportation systems [16] are representative examples.

In the following sections, a failure analysis-based improvement of the steel rollers in the conveying system of a roller beam press hardening furnace is presented. These rollers are the elements transmitting movement from the drives to the beams that convey the blanks inside the furnace. After an operating time of 1800 h at a nominal temperature of 930 °C, a roller from the front section of the furnace (initial

2 m of the chamber) failed in the fillet between the end bell and the journal that connected the roller to the drive (Figure 2a). The fillet fractured across the diameter where it met the journal and showed clear fatigue marks (Figure 2b).

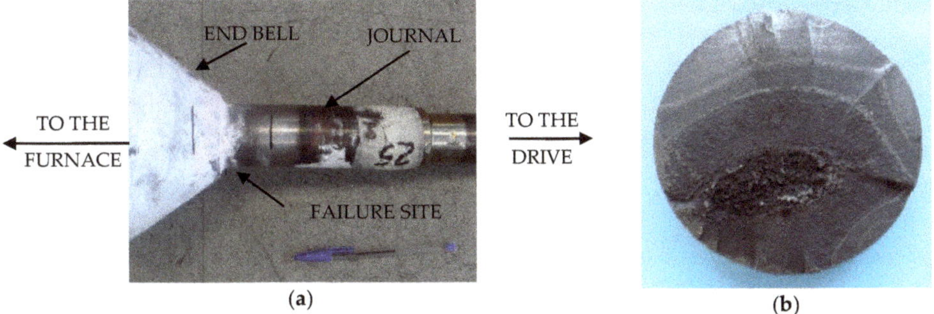

Figure 2. General description of the failure site: (**a**) overview of the fracture location in the drive end of the roll; (**b**) detail of the fracture surface.

The journal, end-bell, and fillet design criteria followed the furnace maker's standard, which was backed-up by successful experiences with components that were predicted to show analogous working conditions in other types of furnaces and should fail due to creep. More specifically, the applied criteria for design was a maximum 1% creep after 100,000 h under operating conditions. This was translated to a maximum working stress of 125 MPa for the failure site [17]. This case study focuses on determining the root causes behind the failure of only 2 rollers out of the 36 identical rollers that were installed in the furnace and in redesigning the conveying system to mature the robustness of the equipment.

2. Materials and Methods

The studied rollers were built in the stainless-steel grade AISI 304 and were comprised by two bell and journal ends (Figure 3), welded to a cylindrical body (tube) that crossed the entire furnace chamber width. The tube ends were welded to the Ø180 mm of the conical bells. One of the journals was fit into a bearing adapted to accommodate the roller expansion during furnace warm-up, while the other was fixed to the motor that applied the thrust movements involved in rolling beam format conveying.

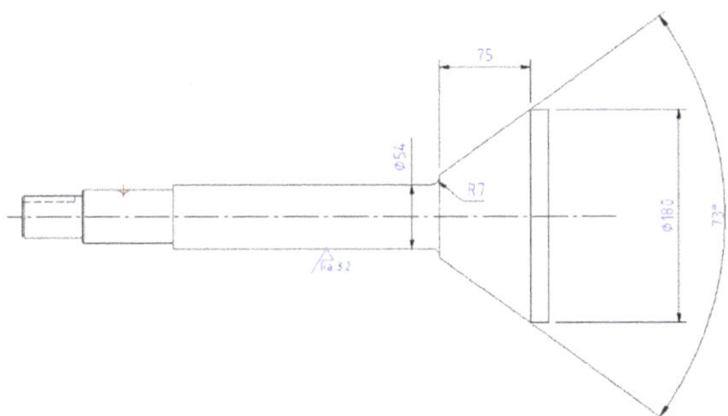

Figure 3. Most relevant dimensions of the original bell and journal design (dimensions in mm).

The materials employed for the study corresponded to two rollers (identified as BR1 and BR2) that were found collapsed during furnace maintenance stops, both showing the fracture in the motor side journal (Figure 4). Both rollers were in the first section of the furnace, where the formats enter the austenitizing chamber. Following the detection of the two broken rollers, all journal fillet radiuses, both motor side and expansion side, were penetrant dye inspected for the presence of cracks in all the rollers of the furnace. Despite no cracks being detected, five rolls were randomly selected and disassembled from the furnace to confirm the absence of cracks in the fillets by metallographic cross sectioning and the absence of cracking was confirmed. The failure was thus specific to the two affected rollers, which were studied more in depth.

Figure 4. Sample of broken roll 1 (BR1) that was inspected.

The failure analysis on the cracked rollers was performed in the following steps:

1. Metallurgical analysis of the fractured component and proposal of the failure mechanism.

    a. Chemical analysis to verify that the material was correctly supplied.
    b. Tensile testing of a specimen from the area of the journal that works the coldest in service, to ensure there was not an initial mechanical property problem from the beginning.
    c. Microstructural inspection in the neighborhood of the fracture and in a position far from the fracture (cold area), to check the phase evolution of the material during service in the cracked area compared with a section thermally unaltered in service.
    d. Fractographic inspection of the fracture surface to determine the nature of the damage that the roll suffered and its trajectory.
    e. Verification of the concordance of the observations against the expected failure mode by design of the roller.

2. Investigation of the deviations between the expected working conditions and the actual working conditions of the rolls that would fit with the failure mechanism.

    a. In situ measurement of the temperature in the failure site with the furnace running.
    b. Verification of the stress in the failure site by studying the load profile during the movement sequence in the running furnace.
    c. Proposal of the failure mechanism taking into account the deviations of the actual working condition from the loads and temperatures expected from design.

3. Redesign of the roll to avoid future failures.

3. Results and Discussion

3.1. Metallurgical Analyses

Starting with the verification of the chemical composition, it was performed by employing spark emission spectrometry (Spectrolab + Spark Analyzer Vision V2, Model M10; Spectro Analytical Instruments GmbH, Boschstr. 10, Kleve, Germany). This allowed evaluating whether or not there was any mistake in the material supply in terms of satisfying the roller drawing indications. Table 1 shows that there was no unexpected deviation on the materials, and they agreed on the AISI 304 that was chosen by design.

Table 1. Specification and actual measured chemical composition of the rollers in % weight (expanded uncertainty $K = 2$ of the measurements <0.03%).

Reference	C	Mn	Si	S	P	Cr	Ni
AISI 304 Specification [18]	0.08	<2	<1	<0.03	<0.045	18–20	8–10.5
Broken Roll 1 (BR1)	0.05	1.04	0.62	<0.02	<0.02	19.58	9.14
Broken Roll 2 (BR2)	0.06	1.05	0.58	<0.02	<0.02	19.61	9.15

Room temperature tensile properties were also checked following the work of [19] in a 100 kN universal testing machine (Zwick Z100, ZwickRoell GmbH & Co. KG, Ulm, Germany) to validate that the material was in a proper mechanical condition when the component was installed. Ø10 mm cylindrical specimens were machined from the coldest end of the journal. This location was chosen to ensure that the was no thermal modification in the material because of the time the roller spent in service. The results are shown in Table 2 and allow discarding the presence of intrinsically faulty raw material, as the purchasing specifications were met.

Table 2. Purchasing specification and actual tensile properties for the BR1 and BR2 samples (indicated tolerance corresponds to expanded measurement uncertainty $K = 2$).

Reference	Yield Strength $Rp0.2$ (MPa)	Tensile Strength Rm (MPa)	Elongation E (%)	Reduction in Area RA (%)
Purchasing specification from design	>210	>560	>58	-
Tensile specimen from BR1	248 ± 5	602 ± 5	61.0 ± 1.5	75.0 ± 1.5
Tensile specimen from BR2	250 ± 5	610 ± 5	60.0 ± 1.5	74.0 ± 1.5

The metallographic analyses of the materials were performed by means of light microscopy (Leica Microsystems, Wetzlar, Germany). The inspections showed that the fracture area was submitted to high temperatures, in excess of 550 °C, as extensive sigma phase evolution and intergranular carbide precipitation were observed (Figure 5 left). The material did not show this microstructure in origin, as evidenced by the inspection of the cold journal zone fitting the motor (Figure 5, right).

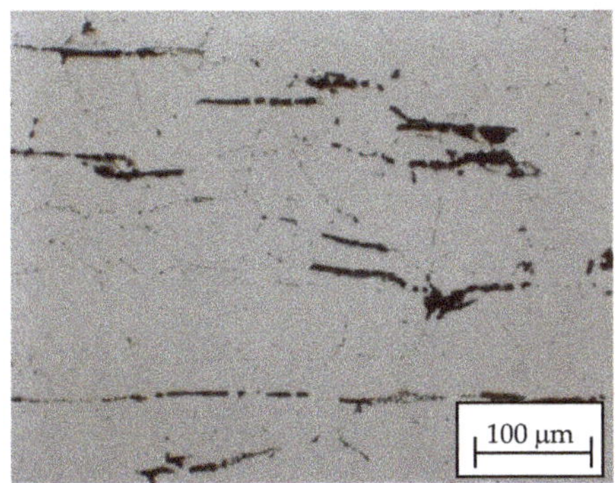

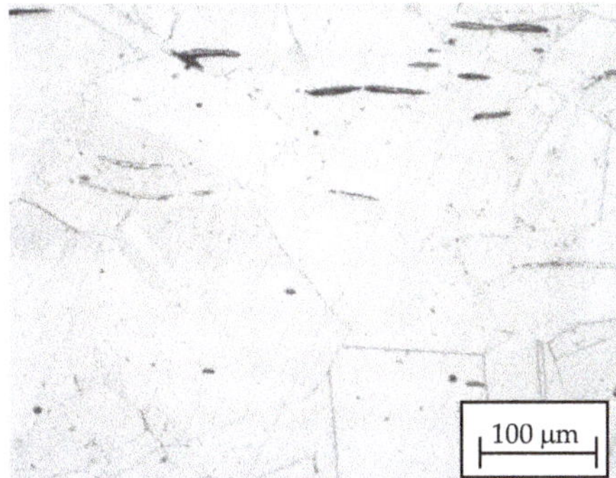

Figure 5. Micrographs of the broken journals. Left. Carbide precipitation in grain boundaries and sigma phase nucleation in the fractured zone. Right. Regular solution annealed austenitic microstructure free of carbides and sigma phase in cold zone of the journal.

Regarding macroscopic inspection of the fracture surfaces of BR1 and BR2 (Figure 6), both samples looked very similar, indicating that the mechanical damage progress was analogous. In both cases, fractures showed several fatigue crack nucleation sites. The fatigue crack growing pattern left lighter color growing marks, which are interpreted as crack stops during bank-holidays and preventive maintenance furnace stand-by periods. The final ductile collapse of the rollers is displaced from the geometrical section of the roll because of the different loads that the rollers must bear depending on their angular position.

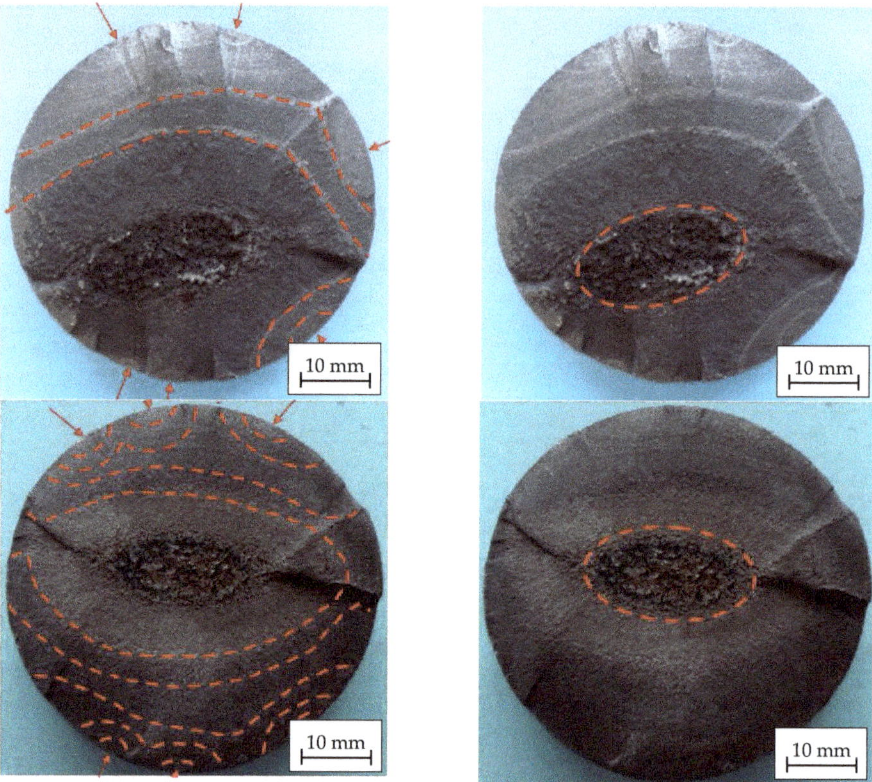

Figure 6. Fracture surface of broken roll 1 (BR1) (top) and BR2 (bottom). Left. Several fatigue nucleation sites marked by arrows and crack stop lines marked with discontinuous lines. Right. Final ductile collapse surface of the roll.

Scanning electron microscopy (FEG-SEM, Ultra Pluss, Carl Zeiss AG, Oberkochen, Germany) inspection confirmed the macroscopic observations. Most of the surface showed oxidation, which meant that some features were hidden, but fatigue nucleation, fatigue crack growth, and ductile fracture patterns (Figure 7) were found. In consequence, the failure mechanism was undoubtedly a fatigue failure, which was probably enhanced by thermal deterioration of the mechanical properties. This deterioration consisted of carbide precipitation and sigma phase evolution coinciding in the fractured section.

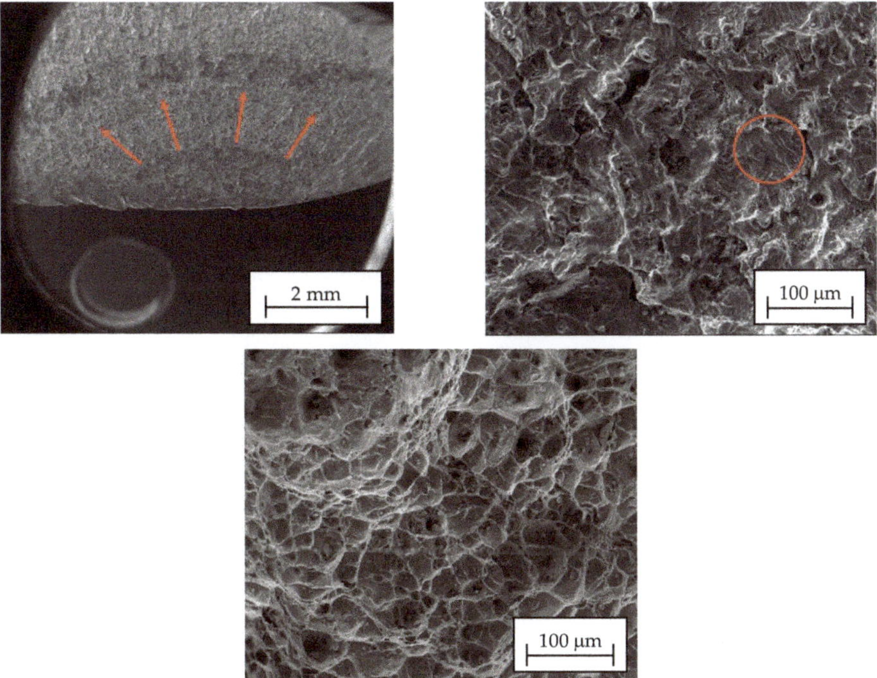

Figure 7. Scanning electron microscopy (SEM) micrographs of the three main fractographic features observed in BR1 and BR2. Top left. Fatigue nucleation site with the crack growth direction marked by arrows. Top right. Fatigue lines on the oxidized surface of the fatigue cracked area. Bottom. Ductile dimples in the final collapse zone.

It is important to note that the bearing journal fillet was dimensioned supposing an average working temperature of 500 °C, and thus material ageing was not expected in this section. On top of that, the design of the roller was directed to avoid creep as the forecast failure mode, but fatigue was the actual cause. Thus, higher loads than expected from design must have been involved in the roller fracture. The next section explains the investigations that led to verifying that the failed section happened to be working at higher temperature and under higher loads than designed, and why the rest of the rollers appeared in perfect condition during penetrant dye inspection.

3.2. Investigation of the Deviations

As the metallurgical observations pointed to an excess of temperature in the failure area, a direct temperature measurement was performed on the furnace during service, both in the failed roller's position (on new rollers used to substitute the BR1 and BR2) and in three further randomly selected rollers along the furnace.

The verification of the working temperature of the rollers in the position where the journal met the fillet radius was performed by drilling an inclined Ø5 mm hole through the outer furnace steel sheet cover and the refractory insulation, so that it was possible for a straight K type thermocouple to be guided to contact the area of interest. The incision angle was calculated in such a way that a maximum deviation of ±2 mm from the measuring junction of the thermocouple to the failure position would occur. The thermocouple was driven through the hole until contact was made and then retracted 1 mm to avoid friction while the roll was rotating. Values obtained during service ranged from 530 °C to 570 °C in all positions, confirming that the actual average working temperature was higher than designed. In consequence, the creep strength of the steel dropped from 125 MPa to 75 MPa. Nevertheless, all the measurements were above the expected values both in failed and sound sites, pointing to the loading condition as the key factor causing the failure.

Regarding the working loads, careful inspection during service allowed identifying an unexpected vibration during the conveying movement, right when the tip of the first walking beam contacted the rolls in the back and forth trajectory. This beam worked half stroke outside the furnace, as it dragged the formats from the feeding platform to the inside of the furnace and was directly heated by the ceiling burners. While the bottom of the beam was cooled down by the contact with the cold rollers, the top of the beam was unable to cool down at the same pace and a thermal distortion was generated. It was found that the beam was bent as shown in Figure 8. This caused the vibration that was observed when the leading edge of the beam hit the rollers. It also meant the presence of a gap between the beam and its expected contact points. Consequently, the weight of the beam was not evenly distributed between three to four rollers, but between two instead. Complex deformation of machines in other processes was also studied [20], but in their case, the main cause was mechanical deformation due to process forces.

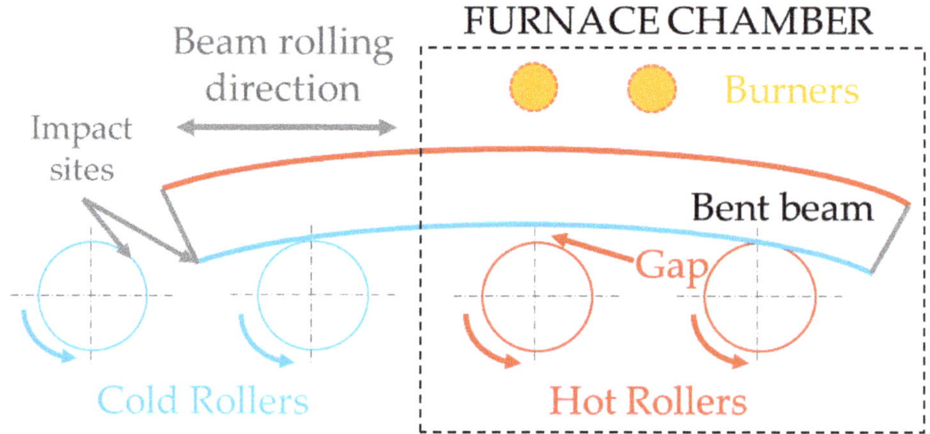

Figure 8. Scheme of the effect of temperature differences between the top and the bottom of the beam. The rollers marked as "hot rollers" correspond to the fractured ones.

It must be also remarked that the rollers and the applied loads were not symmetrical (Figure 9); the distance between the bearing reaction point and the fillet radius at the expansion side was 30% shorter than in the motor side, leading to lower stress and explaining why cracking was observed only in one side of the rollers.

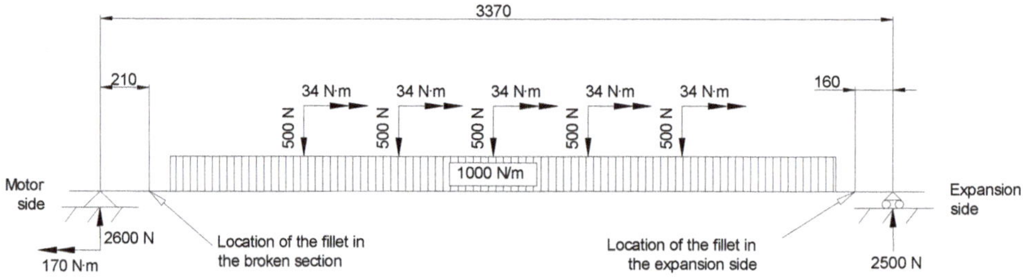

Figure 9. Loading scheme of the roller when the beams are sitting on four rollers at the same time (dimensions in mm).

The investigations in the working furnace confirmed that the broken sections were subjected to temperatures 10% higher than design temperatures and loads between 50% and 100% higher than design loads, depending on the expected number of contacts of the beam in each point of the stroke. More specifically, 50% for the positions where the beams were sitting on two rollers instead of three (3/2 higher load than expected) and 100% for the positions where the beams were sitting on two rollers instead of four (4/2 higher load than expected).

Considering the load scenario from Figure 9 with a classical approach [21], the maximum bending and torsional stresses applied on the broken Ø54 mm should have been 35 MPa (Equation (1)) and 6 MPa (Equation (2)), which gives maximum principal stress of 36 MPa (Equation (3)). No thermal stresses were considered because of the installation of a slider in the expansion side of the roller, which allowed the unrestrained expansion of the component. This meant a creep design safety factor over 3.

$$\sigma_{bending} = \frac{4 \cdot M}{\pi \cdot R^3} = \frac{4 \cdot 2600 \cdot 0.21}{\pi \cdot 0.027^3} = 35\ MPa \tag{1}$$

$$\tau_{torsion} = \frac{2 \cdot T}{\pi \cdot R^3} = \frac{2 \cdot 170}{\pi \cdot 0.027^3} = 6\ MPa \tag{2}$$

$$\sigma_{principal\ max} = \frac{\sigma_{bending}}{2} + \sqrt{\left(\frac{\sigma_{bending}}{2}\right)^2 + \tau_{torsion}^2} = 36\ MPa \tag{3}$$

The beam sitting problems, though, doubled the peak stress, and adding the stress concentration factors due to fillet radii on 1.55 (taking r/d = 0.13 and D/d = 1.28 from the bending stepped round bar with a fillet chart on [22]) makes the peak stress 112 MPa. Introducing the surface finish (1.1) and scale factors (1.25) [23], the final peak stress reaches 154 MPa.

This would have been admissible if the temperature of the journal had stayed close to 450 °C, as the dynamic hardening gives a fatigue limit plateau of approximately 225 MPa for the AISI 304 [24,25]. The fatigue strength of AISI 304, though, drops when working over 450 °C. It can be estimated from AISI 304 fatigue curves in the work of [17] that a reasonable fatigue strength for the broken journal would be close to 165 MPa. Despite it not being possible to perform fatigue tests on the journal left overs to verify this information, these values of stress and strength fit with the observed failure.

3.3. Roller Redesign

The main criterion for journal and sleeve redesign was minimizing the stress in the failure area, while keeping a geometry compatible with the existing insulation sleeves that mated the modified roller section. This strategy allowed standardizing a new roller geometry without any major changes in further components of the furnace. The intervention to redesign the rollers comprised three working lines:

1. The insulation of the area was improved and a temperature reduction of 50 °C was achieved in the affected area, slowing down the carbide precipitation kinetics and arresting sigma phase embrittlement.
2. The thermal differences leading to beam bending were evened with radiation deflectors and a proper sitting of the beam on three or four rollers, depending on the stroke's position, was achieved. Working loads were corrected this way, reducing the stress in the fillet radius.
3. The design of the roll was made more robust by increasing the journal diameter from Ø54 mm to Ø64 mm and by reducing the stress concentration factor thanks to a fillet of R20 mm and tangent to the cone instead of a truncated R7 mm fillet (Figure 10).

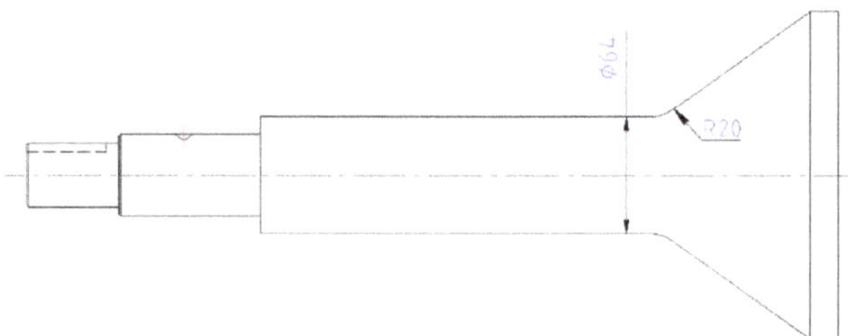

Figure 10. Final journal design that substituted the original geometry (dimensions in mm).

These three actions led to an improved loading condition in which the following occurred:

- The service loads were reduced by between 33% and 50% depending on the beam's position (33% if the beam passed from sitting in two rollers to sitting on three rollers and 50% in the case in which the beams passed from sitting on four rollers to sitting on four rollers).
- The generation of over temperatures in the fillet was avoided, so that the reduction of the hot strength was avoided.
- The stress in the failure area was reduced by 40%. The main factor contributing to this improvement is an increased inertia when changing from Ø54 mm to Ø64 mm, as both the bending and torsion stresses in the failure area are inversely proportional to $\varnothing^3$: $54^3/64^3 = 0.60$.

This can be expressed in terms of principal stress (Equations (4)–(6)) to build the following new scenario:

$$\sigma_{bending} = \frac{4 \cdot M}{\pi \cdot R^3} = \frac{4 \cdot 2600 \cdot 0.21}{\pi \cdot 0.032^3} = 21 \; MPa, \tag{4}$$

$$\tau_{torsion} = \frac{2 \cdot T}{\pi \cdot R^3} = \frac{2 \cdot 170}{\pi \cdot 0.032^3} = 3 \; MPa, \tag{5}$$

$$\sigma_{principal \; max} = \frac{\sigma_{bending}}{2} + \sqrt{\left(\frac{\sigma_{bending}}{2}\right)^2 + \tau_{torsion}^2} = 21 \; MPa. \tag{6}$$

- The stress concentration factor was reduced to 1.25 ($r/d = 0.31$ and $D/d = 1.12$ from the bending stepped round bar with a fillet chart in the work of [22]). Also introducing the scale (1.25) and roughness (1.1) leaves a peak stress of $1.25 \times 1.25 \times 1.1 \times 21 = 39$ MPa.

Adding up all the actions, the peak stress for the $R = -1$ fatigue condition in which the rollers work changed from 154 MPa in initial service to 39 MPa in the redesigned condition. The four-fold reduction of stress led to an improvement effect fatigue wise, as reflected in the fact that the modified rolls were in production for double the initial roll failure time without any trace of cracking after penetrant dye inspection. The new design was implemented in all the furnaces of the same family. This result emphasizes the importance of performing mechanical calculations that must consider all factors involved in service when designing new equipment.

4. Conclusions

The metallurgical study of the fractured rollers and the investigations on the actual operating conditions of the furnace allowed reaching the following conclusions:

- The supply condition of the broken steel was correct and clearly different from its metallurgical condition on the fracture site.
- The roller was designed against creep, but the failure was caused by mechanical fatigue crack growth combined with ductile collapse when the resistant section was severely reduced.

- The actual metallurgical condition of the steel did not fit with the design working parameters that were expected.
- The investigations performed on the running furnace confirmed that the working temperature on the failure site exceeded design temperature and that working loads even doubled design load at some positions of the stroke.
- The actual working parameters perfectly fit the observations in the fractured component, both regarding microstructural and fractographical evolutions.
- Actions against the root cause, leveling the contact of the rollers with the beam, adjusting the thermal barriers, and reducing the stress in the failure site were successfully implemented and the service times of the rollers working in these conditions doubled the failure times of the original broken rollers without showing any crack traces in liquid penetrant inspections.

This case study shows the importance of investigating the deviations between the expected working condition and the actual working condition when a failure analysis is performed. The operating temperature and loads are critical in furnace components. In terms of temperatures, it becomes critical to assess the carbide and sigma phase precipitation kinetics when dealing with embrittlement susceptible austenitic stainless steels. Regarding loads, not only the functional loads must be considered, but also thermal borne stresses, even those generated by faulty contacts caused by thermal expansion and bending. Design-wise, the recommendation of analyzing the loading scenarios due to deviations from expected service conditions for mechanical design reliability of new equipment is strongly underlined. When resistant section must be dimensioned for a rotary steel part that is susceptible to creep, it must be taken into account that the rotation compensates the creep and fatigue becomes a design factor that must also be considered.

The results and method followed in the above sections are of general application to production lines in cases of hot-stamping lines, and can be considered, along with conclusions of modelling given in the work of [26], as a key aspect of the high-production process currently in full use in automotive white-body assembly.

Author Contributions: The present work was conceptualized by B.G.-C., B.F., A.S., M.M., and G.A.; the failure analysis methodology was put in place by B.G.-C., B.F., M.M., and G.A.; hypothesis and result validation was performed by L.N.L.d.L.; working condition investigation was carried out by B.G.-C., B.F., and A.S.; writing—original draft preparation was performed by B.G.-C. and G.A.; writing—review and editing was carried out by L.N.L.d.L.; supervision was carried out by L.N.L.d.L.

Funding: This research was funded by the Basque Government, with budget from the Departamento de Industria, Innovación, Comercio y Turismo and from Fondo de Desarrollo Regional (FEDER) in the frame of the Etorgai Program supporting the LABEBERRI Project (exp. ER-2013/00032). Thanks are also due to funds from Excellence groups of the Basque university system n IT1337-19, and to projects Spanish Ministry of Science DPI2016-74845-R and RETOS NewMINE RTC-2017-6039.

Conflicts of Interest: The authors declare no conflict of interest.

References

1. Karbasian, H.; Tekkaya, A.E. A review on hot stamping. *J. Mater. Process. Technol.* **2010**, *210*, 2103–2118. [CrossRef]
2. Neugebauer, R.; Schieck, F.; Polster, S.; Mosel, A.; Rautenstrauch, A.; Schönherr, J.; Pierschel, N. Press hardening—An innovative and challenging technology. *Arch. Civil. Mech. Eng.* **2012**, *12*, 113–118. [CrossRef]
3. Mori, K.; Bariani, P.F.; Behrens, B.-A.; Brosius, A.; Bruschi, S.; Maeno, T.; Merklein, M.; Yanagimoto, J. Hot stamping of ultra-high strength steel parts. *CIRP Ann. Manuf. Technol.* **2017**, *66*, 755–777. [CrossRef]
4. Mori, K.; Maki, S.; Tanaka, Y. Warm and hot stamping of ultra high tensile strength steel sheets using resistance heating. *CIRP Ann. Manuf. Technol.* **2005**, *54*, 209–212. [CrossRef]
5. Kolleck, R.; Veit, R.; Merklein, M.; Lechler, J.; Geiger, M. Investigation on induction heating for hot stamping of boron alloyed steels. *CIRP Ann. Manuf. Technol.* **2009**, *58*, 275–278. [CrossRef]

6. Pedraza, J.P.; Landa-Mejía, R.; García-Rincón, O.; García, C.I. The effect of rapid heating and fast cooling on the transformation behavior and mechanical properties of an advanced high strength steel (AHSS). *Metals* **2019**, *9*, 545. [CrossRef]

7. Rasera, J.N.; Daun, K.J.; Shi, C.J.; Souza, M.D. Direct contact heating for hot forming die quenching. *Appl. Therm. Eng.* **2016**, *98*, 1165–1173. [CrossRef]

8. Oh, J.; Han, U.; Park, J.; Lee, H. Numerical investigation on energy performance of hot stamping furnace. *Appl. Thermal. Eng.* **2019**, *147*, 694–706. [CrossRef]

9. Saez, A.; Angulo, D.; Berasategui, J. Advanced technology for hot stamping. *Heat Process.* **2018**, *3*, 77–80.

10. Pantazopoulos, G.A. A short review on fracture mechanisms of mechanical components operated under industrial process conditions: Fractographic analysis and selected prevention strategies. *Metals* **2019**, *9*, 148. [CrossRef]

11. Álvarez, J.A.; Lacalle, R.; Arroyo, B.; Cicero, S.; Gutiérrez-Solana, F. Failure analysis of high strength galvanized bolts used in steel towers. *Metals* **2019**, *6*, 163. [CrossRef]

12. Zhang, W.; Wang, H.; Zhang, J.; Dai, W.; Huang, Y. Brittle fracture behaviors of large die holders used in hot die forging. *Metals* **2017**, *7*, 198. [CrossRef]

13. Al-Meshari, A.; Al-Rabie, M.; Al-Dajane, M. Failure analysis of furnace tube. *J. Fail. Anal. Preven.* **2013**, *13*, 282–291. [CrossRef]

14. Babakr, A.M.; Al-Ahmrai, A.; Al-Jumayiah, K.; Habiby, F. Failure investigation of a furnace tube support. *J. Fail. Anal. Preven.* **2009**, *9*, 16–22. [CrossRef]

15. Acevedo, F.; Sáenz, L. Study of embrittlement in austenitic stainless steel AISI 304H with 15 years in service exposed to high temperature. *Dyna* **2018**, *93*, 428–434. [CrossRef]

16. Yang, Y.P.; Mohr, W.C. Finite element creep-fatigue analysis of a welded furnace roll for identifying failure root cause. *J. Mat. Eng. Perform.* **2015**, *24*, 4388–4399. [CrossRef]

17. Davies, J.R. *ASM Specialty Handbook—Heat-Resistant Materials*; ASM Materials, Materials Park: Russell, OH, USA, 1997.

18. *ASTM A479/A479M–18 Standard Specification for Stainless Steel Bars and Shapes for Use in Boilers and Other Pressure Vessels*; ASTM International: West Conshohocken, PA, USA, 2013.

19. *ASTM E8/E8M–16 Standard Test Methods for Tension Testing of Metallic Materials*; ASTM International: West Conshohocken, PA, USA, 2013.

20. Pozo, D.; López de Lacalle, L.D.; López, J.M.; Hernández, A. Prediction of press/die deformation for an accurate manufacturing of drawing dies. *Int. J. Adv. Manuf. Technol.* **2008**, *37*, 649–656. [CrossRef]

21. Gere, J.M. *Mechanics of Materials*; Brooks Cole: Hampshire, UK, 2003.

22. Peterson, R.E. *Stress Concentration Factors*; John Wiley and Sons: New York, NY, USA, 1974.

23. Shigley, J.E.; Mischke, C.R.; Brown, T.H., Jr. *Standard Handbook of Machine Design*; McGraw-Hill: New York, NY, USA, 2004.

24. Milella, P.P. *Fatigue and Corrosion in Metals*; Springer Verlag: Milano, Italia, 2013.

25. Nikitin, I.; Scholtes, B.; Maier, H.J.; Altenberg, I. High temperature fatigue behavior and residual stress stability of laser-shock peened and deep roller austenitic steel AISI 304. *Scr. Mater.* **2004**, *50*, 1345–1350. [CrossRef]

26. Borja, F.; Beatriz, G.; Artola, G.; López de Lacalle, N.; Angulo, C. A Quick Cycle Time Sensitivity Analysis of Boron Steel Hot Stamping. *Metals* **2019**, *9*, 235. [CrossRef]

© 2019 by the authors. Licensee MDPI, Basel, Switzerland. This article is an open access article distributed under the terms and conditions of the Creative Commons Attribution (CC BY) license (http://creativecommons.org/licenses/by/4.0/).

Article

Validation of a New Quality Assessment Procedure for Ductile Irons Production Based on Strain Hardening Analysis

Giuliano Angella [1],* **and Franco Zanardi** [2]

[1] Department of Chemical Sciences and Materials Technology, Institute of Condensed Matter Chemistry and Technologies for Energy (ICMATE), National Research Council of Italy, via R. Cozzi 53, 20125 Milano, Italy
[2] Zanardi Fonderie S.p.A., via Nazionale 3, 37046 Minerbe (VR), Italy
* Correspondence: giuliano.angella@cnr.it; Tel.: +39-02-6617-3327

Received: 19 June 2019; Accepted: 24 July 2019; Published: 27 July 2019

Abstract: A mathematical procedure based on the analysis of tensile flow curves has been proposed to assess the microstructure quality of several ductile irons (DIs). The procedure consists of a first diagram for the assessment of the ideal microstructure of DIs, that is, the matrix where mobile dislocations move, and a second diagram for the assessment of the casting integrity because of potential metallurgical discontinuities and defects in DIs. Both diagrams are based on the dislocation-density-related constitutive Voce equation that is used for modeling the tensile plastic behavior of DIs. The procedure stands on the fundamental assumption that the strain hardening behavior of DIs is not affected by the nature and the density of the potential metallurgical discontinuities and defects, which are expected to affect only the elongations to fracture. However, this fundamental assumption is not obvious, and so its validity was evaluated through tensile testing Isothermed Ductile Irons (IDIs) 800, showing a wide scatter of elongations to rupture. The analysis of the strain hardening behaviors supported by strain energy density calculations of IDIs tensile tests proved that the fundamental assumption was valid and the quality assessment procedure could be applied to IDIs. A modified Voce equation was also introduced to improve the fitting of the experimental tensile flow curves and the strain energy density calculations.

Keywords: ductile irons; tensile tests; mechanical properties; constitutive equations; quality assessment

---

1. Introduction

Due to the demand of improving mechanical properties, new chemical compositions and production routes of Ductile Irons (DIs) have been explored to obtain different microstructures. Thus, new classes of advanced DIs have been produced, that are, for instance, the alloyed SiBoDur [1] and High Silicon Strengthened (HSiS) DIs, i.e., with silicon content above 3.5–4.0 wt% [2–4], and the Austempered DIs (ADIs) [5–8] and Isothermed DIs (IDIs) [8] that are produced through heat treatments. The current international standards were originally produced for classifying conventional DIs, where silicon content is almost constant (1.8–2.8 wt%) and the pearlite to ferrite ratio, changing because of alloying elements like copper, is the key microstructure parameter that increases yield and tensile strengths, and reduces elongations to rupture. Thus, the current DIs classification is based on the minimal tensile mechanical properties, but unfortunately this classification cannot give the correct picture of the advanced DIs [9]. A specific classification approach to properly weigh up these new generations of advanced DIs is needed, which should be capable of taking into account increasing alloying elements content and new production routes, mainly through heat treatments resulting in complex microstructures, like ausferrite in ADIs and perferrite in IDIs.

Microstructure in DIs and any castings may enclose metallurgical discontinuities and defects that cause high variability in mechanical properties. To minimize this variability metallurgical discontinuities and defects have to be easily evaluated, hopefully based on tensile data [10]. Microstructure is a generic term that involves different structural features at different scales. However, only those structural features that produce effects because of the application of some external stimulus are meaningful to define the microstructure of materials. Under the experimental conditions to test metallic materials in tension, in ferritic-pearlitic DIs or perferritic IDIs, bcc ferrite deforms plastically through dislocation multiplication and storage, and motion of mobile dislocations. Therefore, as the correlation between microstructure and tensile plastic behavior is our concern, dislocation dynamics has to be considered. At room temperature, dislocations move on crystallographic planes along specific crystallographic directions of bcc ferrite, and interact with other structural features that they encounter on these planes, and are obstacles to their motion, i.e., other dislocations, grain boundaries, and second phases boundaries, like α/cementite interfaces in pearlite. The set of the structural features involved in this scenario results in what is called as the matrix. The matrix and the loading conditions (loading mode, strain rate, and temperature) determine the plastic flow curves of metallic materials. However, metallurgical discontinuities and defects that in DIs are irregular morphology graphite, shrinkages, slag inclusions, and gas holes, may be present in the matrix and affect the tensile flow curves, the extent of which depends on their nature and density. The microstructure results in matrix with metallurgical discontinuities and defects. In the contest of cast products, quality mainly assumes the meaning of microstructure integrity, that is, the absence of metallurgical discontinuities and defects. The microstructure is of high quality or ideal, when the nature and density of defects in the matrix are below a threshold and do not affect the flow curve; in other words, the ideal microstructure is the matrix. Otherwise, when defects affect the flow curve to some extent with respect to the flow curve of the ideal microstructure, the microstructure is defected or of lower quality.

A framework to classify grade (conventional and heat treated) and quality level of DIs based on plastic properties has been proposed by Zanardi et al. [9], where a Material Quality Index (MQI) was defined as

$$\mathrm{MQI} = R_m{}^2 \cdot A_5/(8200 + 3R_m), \tag{1}$$

with R_m the ultimate tensile strength, and A_5 the elongation to rupture in tensile test. Accordingly, new generation DIs have MQI larger than 360, while in conventional ferritic-pearlitic DIs MQI ranges between about 100 and 190 according to the pearlite content. Indeed, a first MQI was first proposed by Siefer and Ortis, and developed later by Crews in 1974 [10], defining $\mathrm{MQI} = R_m{}^2 \cdot A_5$. However, an effective microstructure quality assessment procedure should be capable of cataloguing univocally the grade of the matrix (the ideal microstructure), depending on chemical composition and production route, and within that, of classifying the integrity of the material, which these approaches based on MQIs cannot do. An innovative mathematical procedure [11,12] to assess the microstructure quality of DIs has been recently proposed and it is indeed capable of addressing these two issues: Cataloguing different grades of DIs and classifying their integrity. The procedure comes into two diagrams. The first diagram is for the assessment of DIs matrix, that is, the ideal microstructure. This first diagram is based on the correlation between matrix and plastic behavior, which can be described by using the dislocation-density-related constitutive Voce equation. In fact, the Voce equation parameters obtained from the best fits of the experimental tensile flow curves are used as coordinates of the matrix assessment diagram. Conventional DIs with different ferrite–pearlite ratios and advanced DIs with high silicon content or ausferrite and perferrite, are univocally identified by lining on distinct positions on it [11,12]. Thus, the diagram can classify the different microstructures of the modern DIs, overtaking the limitations of the current classification standards based on the minimal mechanical properties and MQIs. The second diagram is for the material integrity assessment, and is also based on Voce formalism. Metallurgical discontinuities and defects can cause premature rupture that occurs before the geometrical instability of tensile specimens, that is, localized deformation (necking), and so in the integrity assessment diagram the experimental elongations to rupture (e_r) and uniform elongations (e_u) where necking occurs, are compared [11,12].

A similar approach has been published to assess the integrity of Al casting alloys for aerospace applications, based on the concept of the target properties [13,14] by using Voce equation and Strain Energy Density (SED) that is the energy per unit volume stored during tensile plastic deformation until rupture. In the reported tests on Al casting alloys, the strain hardening behavior was significantly affected by metallurgical discontinuities and defects, because of the pour integrity typical in Al castings, which made it difficult to define a specimen free of major defects, and in fact the target properties approach has been discarded for a statistical approach based on the ductility potential [15,16]. Indeed, the two diagrams procedure here proposed to assess the microstructure quality of DIs stands on the fundamental assumption that the strain hardening behavior of DIs is not affected by the nature and the density of the potential metallurgical discontinuities and defects, which are expected to only affect the elongations to fracture. However, this is not obvious, as already reported in Al castings [13–16], and if the assumption were not valid, the integrity assessment diagram for DIs should be abandoned. This paper is focused on evaluating whether, in DIs, the metallurgical discontinuities and defects only affected the elongations to rupture, and did not the strain hardening behavior. As a consequence of this validation, the mathematical procedure proposed for the material quality assessment of DIs [11,12] should be validated.

For this investigation, the heat treated IDIs 800 (Ultimate Tensile Strength—UTS—of 800 MPa) produced in Zanardi S.p.A. was used, intentionally selected in a wide range of elongations to rupture before and after necking, which was not typical for this material grade that is usually characterized by an improved uniformity of mechanical properties, thanks to the absence of any alloying and for a short time for the solid state transformation during quenching after partial austenitization. Isothermed Ductile Irons (IDIs) [17] belong to a new generation of DIs that are very attractive for the good combination of outstanding mechanical properties and low production costs. IDI 800s were produced from conventional GJS 400 heats (with a C content of about 3.6wt% and Si 2.46–2.66, Mn 0.10–0.15, Cu 0.01–0.15, Ni < 0.06, Mo < 0.01, and Sn < 0.01, Fe balance) that were cast with different cooling rates in Zanardi Fonderie S.p.A. through four different mold geometries, namely, cylindrical Lynchburg mold with 25 mm diameter (here after L25) produced complying with the standard UNI EN 1563, and three different Y moulds with thickness 25, 50 and 75 mm (here after Y25, Y50 and Y75) complying with the standard ASTM A 536-84. Then, from room temperature the GJS 400 heats were heated up at 815 °C in the intercritical range Ac_1–Ac_3 for 150 min [17] to be partially austenitized, leaving an opportune fraction of pro-eutectoid ferrite, and then, after quenching in salt bath above the Ms, the austenite transformed into pearlite. The resulting microstructure is called as perferrite (that is a neologism [9]) and consists of alternated ferrite and pearlite, which is different from the bull-eye structure of conventional pearlitic-ferritic DIs, where ferrite is surrounded by pearlite. The microstructure of alternated ferrite and pearlite confers, to IDI 800, excellent elongations to rupture, and in fact the minimum elongation to rupture is 6% for IDI 800, while it is 2.5% for conventional pearlitic-ferritic grades with UTS of 800 MPa [9]. An example of perferrite is reported in Figure 1, where pearlite is bright (volume fraction typically higher than 60%), ferrite is gray and graphite is black (volume fraction typically 10%). Microstructure observations of IDI 800 were performed through a high resolution Scanning Electron Microscope (SEM) SU-70 by Hitachi (Hitachi High-Technologies Corporation, Tokyo, Japan) after conventional metallographic polishing procedure and final etching with Nital 2%.

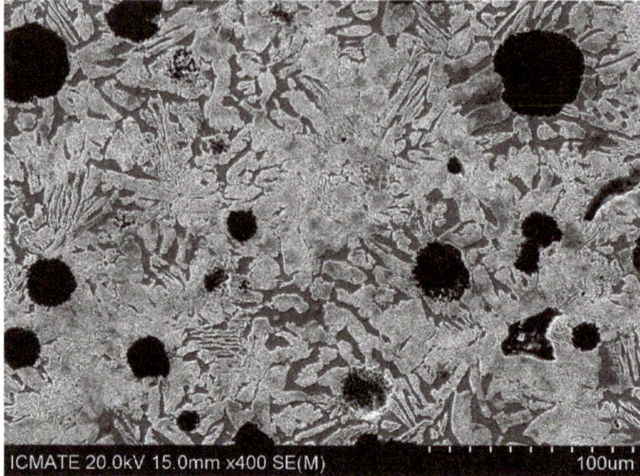

Figure 1. Scanning electron microscopy (SEM) micrograph with secondary electron signal of typical spheroidal graphite perferritic microstructure of IDI 800 after metallographic polishing and etching with 2% Nital, consisting of nodular graphite (black) embedded in alternated ferrite (gray) and pearlite (bright).

The wide range of elongations of the 20 tensile-tested samples of the investigated IDI 800 was associated either to a wide range of density of possible metallurgical discontinuities and defects, and to the ideal matrix when uniform elongations were achieved. The strain hardening behaviors of the tensile curves of IDIs 800 were analyzed, and the fundamental assumption of the quality assessment procedure was quantitatively analyzed on strain hardening comparison and SED considerations.

2. Methods

2.1. Mechanical Properties and Plastic Behaviour

Tensile tests were carried out at the strain rate of 10^{-4} s^{-1} on round specimens made of 20 different IDI 800 samples with an initial gauge in the diameter d_o = 12.5 mm and length l_o = 50 mm complying with ASTM E8-8M. Engineering tensile flow curves S vs. e with $S = F/A_o$ and $e = (l - l_o)/l_o$ were obtained, where F, A_o and l were the instantaneous load, the initial sectional area and instantaneous gauge length, respectively. The mechanical properties that are conventionally used to qualify metallic materials are Yield Stress (YS), Ultimate Tensile Stress (UTS), and elongation to rupture (e_r). These parameters are correlated with the material microstructure. However, they are not enough to assess the microstructure quality, since the uniform elongation e_u is also relevant. In fact, plastic deformation lastly induces failure, and in tensile testing strain localization in the form of necking occurs: The elongation at necking is called uniform elongation (e_u). Metallurgical discontinuities and defects are stress raisers that can nucleate cracks and produce premature failure before necking, causing $e_r < e_u$, and so e_u is a parameter that can give indications of the quality of the material microstructure.

A rigorous method to find the uniform elongation e_u and compare it to the experimental elongation to rupture e_r is based on two steps. First, true stress-true plastic strains (σ vs. ε_p) have to be found, where $\sigma = S \cdot (1 + e)$ and $\varepsilon_p = \varepsilon - \sigma/E = \ln(1 + e) - \sigma/E$, with S and e the engineering stress and elongation, respectively, and E the experimental Young modulus obtained from tensile tests. Secondly, an appropriate constitutive equation [18–20] has to be used to model the experimental flow curves to predict, at best, the uniform strain ε_u, that is, the plastic strain at which the Considère's condition $d\sigma/d\varepsilon_p = \sigma$ is fulfilled, where $d\sigma/d\varepsilon_p$ is the strain hardening rate. Note that the uniform strain ε_u is univocally linked to the uniform elongation e_u, as well as, the strain to rupture ε_r is related to the elongation to rupture e_r. However, since height and shape of tensile flow curves could be affected by the nature and the density of defects, a straightforward comparisons between e_r and e_u might be meaningless, and the effects of defects on strain hardening behavior of DIs, that is the increase of strength with straining, necessarily has to be analyzed.

2.2. Voce Equation

Voce equation is

$$\sigma = \sigma_V + (\sigma_0 - \sigma_V) \cdot \exp(-\varepsilon_p/\varepsilon_c), \tag{2}$$

where σ is the true flow stress, ε_p the true plastic strain, σ_V is the saturation stress that is achieved asymptotically with straining, ε_c is the characteristic transient strain that defines the rate with which σ_V is approached, and σ_0 is the back-extrapolated stress to $\varepsilon_p = 0$. Voce equation has been successfully applied to a wide number of metallic materials, like copper [18,19], austenitic stainless steels [21,22], aluminum alloys [13–16], ferritic-pearlitic DIs, IDIs, and ADIs [8,9,11,12], and also solid solution and precipitation strengthened nickel based superalloys [23].

The differential form of the Voce equation is used to investigate the strain hardening behavior of IDIs:

$$\frac{d\sigma}{d\varepsilon_p} = \frac{\sigma_V}{\varepsilon_c} - \frac{\sigma}{\varepsilon_c} = \Theta_0 - \frac{\sigma}{\varepsilon_c}, \tag{3}$$

where Θ_0 and ε_c are constants. Through fitting Equation (3) to the linear regions at high stresses in the differential experimental data ($d\sigma/d\varepsilon_p$ vs. σ), the Voce parameters Θ_0 and ε_c are found, and, as a consequence, σ_V. Finally, σ_0 is found from the best fitting of Voce equation to the experimental tensile flow curves.

In Figure 2, an example of fitting procedure on a typical IDI 800 is reported (sample No. 18 in Table 1). In Figure 2a, where data $d\sigma/d\varepsilon_p$ ($= \Theta$) vs. σ are reported, the slope of the best linear fit at high stresses is $1/\varepsilon_c$, whilst Θ_0 is the intercept with $\sigma = 0$ according to Equation (3). In Figure 2b, the corresponding tensile flow curve with the best fitting Voce equation with the parameters found in Figure 2a are reported. The saturation stress is $\sigma_V = \Theta_0 \cdot \varepsilon_c$, and σ_0 is found through minimizing the variance between Voce equation and experimental data. The first diagram of the material quality assessment procedure is for the assessment of the IDI ideal matrix and is based on plotting $1/\varepsilon_c$ vs. Θ_0.

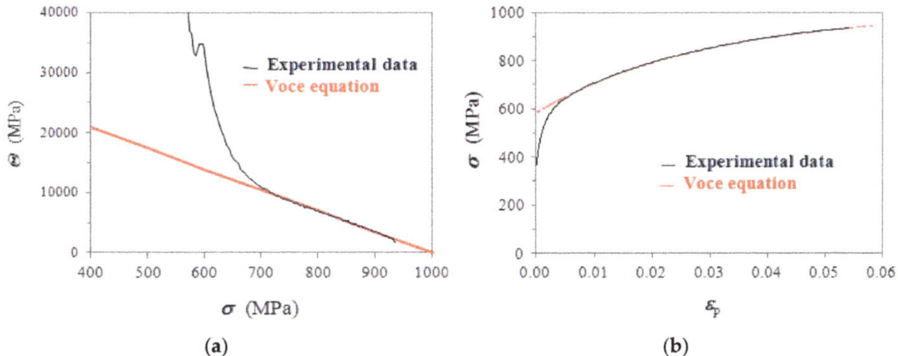

Figure 2. (**a**) Strain hardening analysis with typical IDI 800 (sample No. 18 in Table 1): differential data ($d\sigma/d\varepsilon_p = \Theta$ vs. σ) with the best fit of the differential form of Voce equation (Equation (3)) at high stresses; (**b**) experimental tensile flow curve with the best fitting Voce equation (Equation (2)).

For the definitions of the second diagram concerning the casting integrity, which is for the assessment of metallurgical discontinuities and defects, the uniform strain $\varepsilon_{u,p}$ is defined according to Voce formalism as follows:

$$\varepsilon_{u,p} = \varepsilon_c \cdot \ln\left[\frac{\varepsilon_c + 1}{\varepsilon_c} \cdot \frac{\sigma_V - \sigma_0}{\sigma_V}\right], \tag{4}$$

It is important to bear in mind that $\varepsilon_{u,p}$ was determined from Voce parameters, independently of the fact that necking was actually achieved or not. Indeed, if necking was achieved, then $\varepsilon_{u,p}$ could be compared to the actual uniform elongation $\varepsilon_{u,exp}$, found experimentally at necking. However, if necking did not occur, then $\varepsilon_{u,p}$ was the ideal value that was not achieved because of metallurgical discontinuities or defects. Then, the uniform elongation is

$$e_u(\%) = \left[\exp\left(\varepsilon_{u,p} + \sigma_u / E\right) - 1\right] \cdot 100, \tag{5}$$

where σ_u is the stress at $\varepsilon_{u,p}$, E the experimental Young modulus, and σ_u/E is the elastic component of strain. The second diagram concerning the casting integrity was built by plotting the experimental elongations to rupture e_r vs. e_u, that is the ideal uniform elongations.

Table 1. Summary of the Voce parameters for the quality assessment procedure of IDI 800 and strain energy density at uniform strains (SED$_U$).

No.	Mould	Θ_o (MPa)	$1/\varepsilon_c$	$\varepsilon_{u,p}$	e_u (%)	e_r/e_u	SED$_U$ (MJ/m^3)
1	Y75	28,780.5	28.8	0.0822	8.95	1.17	70.4
2	L25	28,892.7	31.4	0.0847	9.15	0.82	65.7
3	Y25	29,386.7	31.1	0.0832	9.01	0.79	66.2
4	Y25	29,648.3	30.9	0.0840	9.10	1.12	68.1
5	Y25	29,758.1	31.2	0.0828	8.98	1.20	67.1
6	Y50	30,224.5	30.1	0.0804	8.76	0.72	68.8
7	Y50	30,340.6	30.3	0.0804	8.75	1.12	69.4
8	L25	30,391.0	30.4	0.0807	8.79	1.09	68.8
9	Y50	31,316.7	32.0	0.0828	8.97	0.97	68.9
10	Y25	31,531.7	32.9	0.0801	8.68	0.91	65.6
11	Y50	31,870.6	33.3	0.0803	8.69	0.72	65.3
12	L25	32,115.0	31.8	0.0782	8.52	1.00	67.9
13	Y50	32,338.6	33.3	0.0806	8.73	0.78	66.5
14	Y50	32,539.3	33.0	0.0819	8.87	0.72	68.3
15	L25	33,911.0	33.1	0.0766	8.34	0.96	67.9
16	Y75	34,114.7	33.9	0.0783	8.50	0.94	67.1
17	Y75	34,127.6	33.8	0.0804	8.72	0.95	69.3
18	Y50	34,719.9	34.7	0.0776	8.42	0.74	66.7
19	Y25	37,007.8	36.4	0.0778	8.42	0.78	66.9
20	Y25	38,694.2	37.1	0.0764	8.28	1.12	69.0

2.3. Strain Energy Density (SED) at Uniform Strain

In order to compare the strain hardening behavior of two samples quantitatively, the only comparison of the Voce parameters coming from strain hardening analysis ($1/\varepsilon_c$ and Θ_o) might not be enough, since the yielding where $d\sigma/d\varepsilon_p$ varied rapidly, could not be easily compared in the $d\sigma/d\varepsilon_p = \Theta$ vs. σ plots. Thus, we proposed that a third parameter should be involved in strain hardening comparisons, which was the plastic SED at the uniform strain $\varepsilon_{u,p}$, hereafter called as SED$_U$, that is the volumetric plastic strain energy (MJ/m^3) stored in the tensile sample up to the uniform strain $\varepsilon_{u,p}$, defined as

$$SED_U = \int_0^{\varepsilon_{u,p}} \sigma\left(\varepsilon_p\right) d\varepsilon_p, \tag{6}$$

where $\sigma(\varepsilon_p)$ was the Voce equation (see Equation (2)). $\varepsilon_{u,p}$ depended on Voce parameter σ_0 (see Equation (4)), that is close to the yielding proof stress $\sigma_{0.2\%}$. SED_U had the advantage to be an integral quantity involving the whole plastic flow curve, and furthermore SED_U was particularly useful to compare the plastic behaviors of ideal matrix, where $\varepsilon_{u,p}$ was indeed achieved ($e_r \geq e_u$) and defected material where $\varepsilon_{u,p}$ was not ($e_r < e_u$).

In Figure 2b, Voce equation matches very well the experimental flow curve only at high strains. The reasons of this mismatch could be attributed to a transient in the dislocation microstructure evolution and are debated [19]. However, for the determination of SED_U in Equation (6) proper modeling of the tensile flow curve was also necessary at small strains. A negative Ludwigson-kind correction Δ was proposed to the Voce equation, which was meant to also match the experimental flow curve at small strains:

$$\Delta = -\sigma_1 \cdot \exp(n_0 + n_1 \cdot \varepsilon^{n_2}), \tag{7}$$

with $\sigma_1 = 1$ MPa (inserted for dimensional consistency), and n_0, n_1 and n_2 constants obtained from fitting. So the modified Voce equation used to model the IDI 800 experimental flow curves was finally

$$\sigma = \Delta + \left[\sigma_V + (\sigma_o - \sigma_V) \cdot \exp\left(-\varepsilon_p / \varepsilon_c\right)\right], \tag{8}$$

In Figure 3 an example of the best fitting with Equation (8) of the same tensile flow curve in Figure 2b (sample No. 18 in Table 1) is reported. The match is excellent all over the strains. For instance, at $\varepsilon_p = 0.002$ the error was 8.1% with Voce equation, while the error was reduced at −0.2% with the modified Voce equation (Equation (8)).

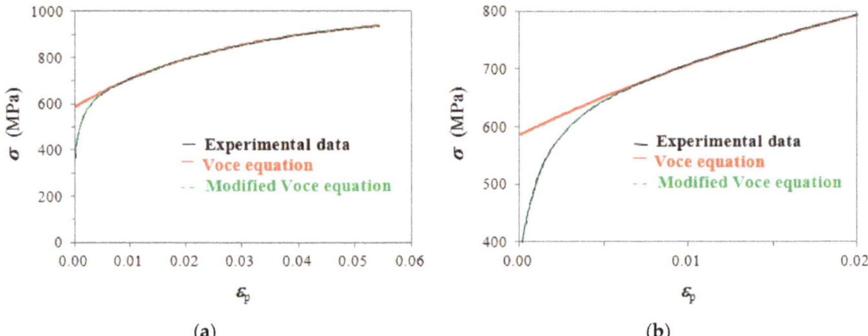

Figure 3. (a) Typical experimental tensile flow curve of IDI 800 (the same in Figure 2b) with the best fitting modified Voce equation (in Equation (7) $n_0 = 5.38$, $n_1 = -543.9$ and $n_2 = 0.96$); (b) close-up at small strains to underline the excellent match between experimental flow curve and modified Voce equation.

3. Results

3.1. Microstructure Quality Assessment Procedure

The Voce parameters Θ_0 and $1/\varepsilon_c$ resulting from the strain hardening analysis of the IDI 800 tensile flow curves are reported in Table 1, with increasing Θ_0 in order to help the data interpretation. Uniform plastic strains $\varepsilon_{u,p}$ (Equation (4)), uniform elongations e_u (Equation (5)) and e_r/e_u ratios are reported for rupture conditions analysis, and SED_U values (Equation (6)) calculated using the modified Voce equation (Equation (8)) data for strain hardening comparison purposes. The mold types are also reported for indication of the cooling rates of the GJS 400 heats from which the IDIs 800 were produced, the fastest with L25 mould and the slowest with Y75 mold. The nodule counts measured complying on the standard ASTM E 2567 resulted to be 420 nodules/mm^2 in L25, 280 in Y25, 190 in Y50 and 150 in Y75, and pearlite ranged between nil and 5%, which was typical of fully ferritic grade GJS 400. In Table 1, there was no correlation between molds (cooling rates) and plastic behavior, namely Voce parameters. This is distinctive of IDIs because of their low alloying elements contents. In fact, in IDIs, there are no major segregations that may be particularly significant and detrimental in thicker walls when the alloying elements content is higher. However, the aim of the work was not to find a stringent correlation between Voce parameters and microstructure that was not investigated through quantitative techniques, but was to evaluate the goodness of the quality assessment procedure [11,12] on a IDI grade that could present a wide range of final elongations. This variability might be associated either to a wide range of density of possible metallurgical discontinuities and defects, or to an ideal matrix when uniform elongations were achieved. Thus, the set of 20 samples of IDIs 800 was intentionally selected in a wide range of elongations to rupture before and after necking, which was not typical for this material grade that is usually characterized by an improved uniformity of mechanical properties, thanks to the absence of any alloying and to the short time for the solid state transformation during quenching after partial austenitization.

The matrix assessment diagram for IDI 800 is reported in Figure 4a, where GJS 400 data (different from the heats used to produce the investigated IDI 800) are also reported for comparison. GJS 400 was also produced in Zanardi Fonderie S.p.A. with L25, Y25, Y50, and Y75 molds, and presented a microstructure consisting of ferrite with a pearlite volume fraction spanning from nil to about 4% depending on the cooling rate. Each material lies on specific regions of the diagram along specific lines, identifying univocally the material in the diagram. The matrix quality assessment diagram is capable of classifying the different microstructures of the modern DIs produced with different chemical compositions (mainly the Si content) and through different industrial routes and heat treatments, like IDIs and ADIs [11,12]. In fact, the current classification standards are only based on the minimal tensile properties (yield stress, ultimate tensile stress, and ductility) that are sufficient to classify only the conventional DIs produced through a single process, where the only key microstructure parameter is the pearlite volume fraction. The integrity assessment diagram is reported in Figure 4b for IDI 800, where GJS 400 data are not reported since the ratios e_r/e_u were always bigger than 1.5. For each tensile curve the experimental plastic elongations e_r with the ideal uniform plastic elongations e_u (Equation (5)) were compared, where e_u represented the ideal behavior of metallic materials, since necking was achieved. However, because of metallurgical discontinuities or defects premature ruptures could occur, resulting in plastic elongations $e_r < e_u$. Thus, in Figure 4b, the data points below the ideal behavior line ($e_r = e_u$) identify premature ruptures because of metallurgical discontinuities and defects, while data points lining on it or above (elongations beyond the necking) identify samples with metallurgical integrity. IDI 800 shown a wide range of possible rupture conditions for the selected set of test specimens IDI 800, so IDI 800 represented the optimal material for the present investigation.

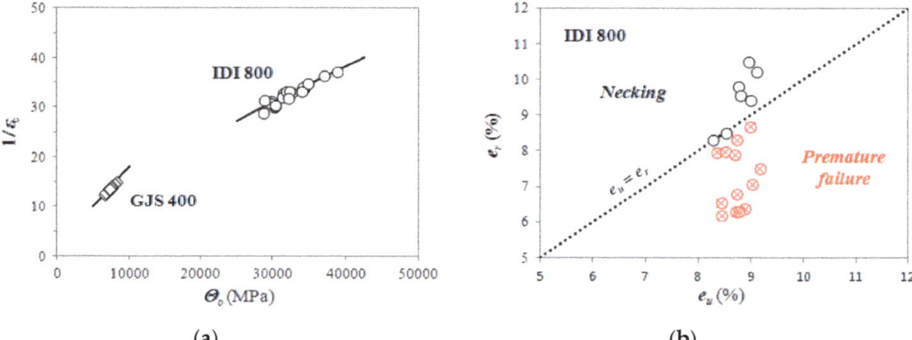

Figure 4. (a) Matrix assessment diagram for IDI 800 and GJS400; (b) integrity assessment diagram for IDI 800.

3.2. Strain Hardening Comparison

In Figure 5a,c,e, strain hardening plots Θ vs. σ of typical IDI 800 flow curves with similar Voce parameters and different e_r/e_u ratios are reported for comparison. The Voce parameters were selected in order to span the full possible range of Θ_0. Indeed, to compare the strain hardening behavior, a third parameter was necessary, because yielding of different samples could be different even if the Voce parameters matched, and for this purpose SED$_U$ were calculated in Table 1 and was reported in the plots of Figure 5. The dot lines passing through the origin in Figure 5a,c,e represent the Considère's criterion, that is, the achievement of the uniform strain $\varepsilon_{u,p}$, i.e., the ideal behavior (necking). In Figure 5b,d,f, the engineering flow curves S vs. e corresponding to the strain hardening data in Figure 5a,c,e are reported. The differential data with similar Voce parameters and SED$_U$ overlap whatever e_r/e_u ratios they have, and so no significant change in the strain hardening behavior could be observed, even if significantly different rupture elongations were found: In Figure 5b e_r spans from a minimum of ≈6.5% to a maximum of ≈10%, while in Figure 5d,f from ≈6% to ≈8%. In other words, the tensile engineering flow curves matched utterly from yielding until the premature ruptures in the defected matrixes.

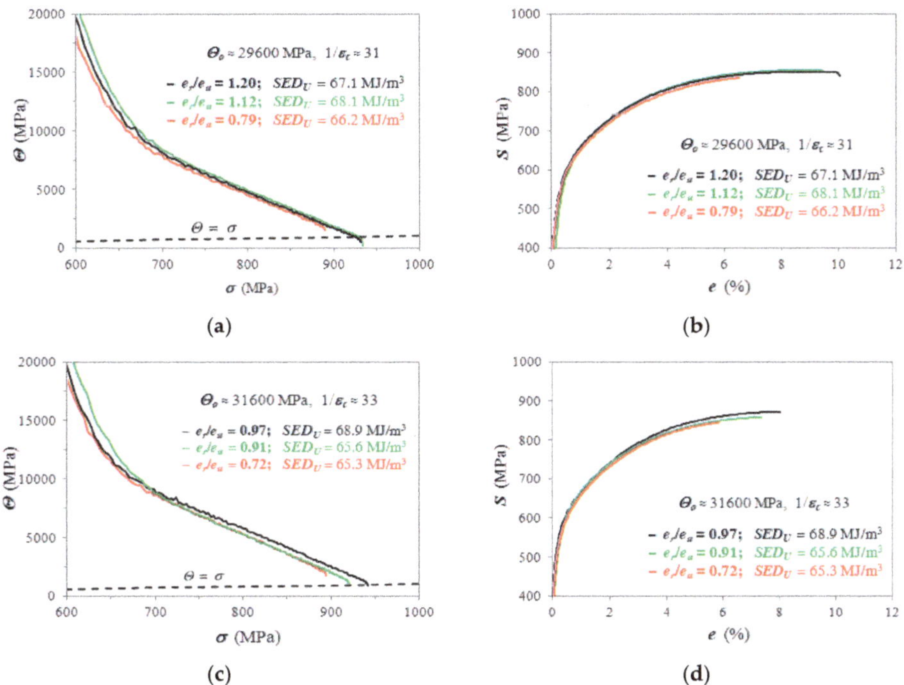

Figure 5. *Cont.*

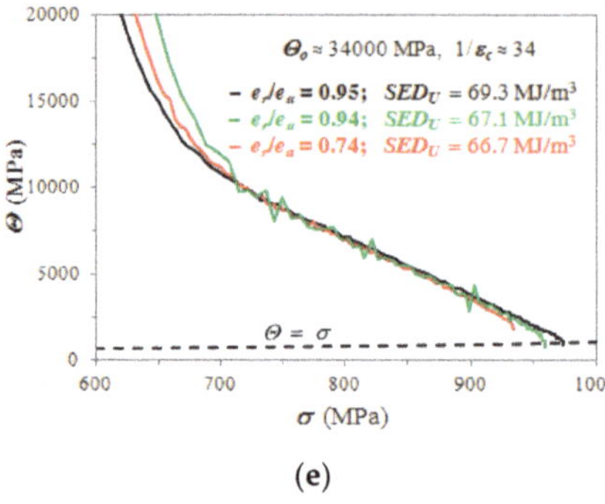
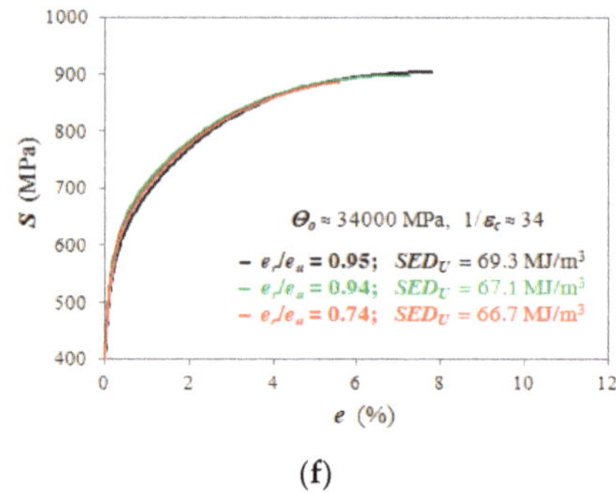

Figure 5. (**a**) Strain hardening data of typical IDI 800 with $\Theta_o \approx 29{,}600$ MPa and $1/\varepsilon_c \approx 31$; (**b**) engineering flow curves corresponding to the differential data in (**a**); (**c**) with $\Theta_o \approx 31{,}600$ MPa and $1/\varepsilon_c \approx 33$; (**d**) engineering flow curves corresponding to (**c**); (**e**) with $\Theta_o \approx 34{,}000$ MPa and $1/\varepsilon_c \approx 34$; (**f**) engineering flow curves corresponding to (**e**). The dot lines identified in (**a**)–(**f**) identify the Considère's criterion $\Theta = \sigma$.

4. Discussion

Through the matrix assessment diagram in Figure 4a, different grades of DIs with different silicon content and different production routes involving heat treatments could be classified unambiguously. This is true for ferritic, ferritic-pearlitic DIs, ADIs 800 and 1000, and IDIs 800 and 1000 [11,12]. This capability is fundamental for a new classification approach needed for the appearance on the market of new advanced DIs, since the actual classifications of DIs are based on minimal tensile mechanical properties, and they are shaped on a single production route with similar silicon content, i.e., between 1.8–2.8 wt%, where pro-pearlite elements, like, for instance, copper, determine the pearlite to ferrite ratio and so yielding, tensile strength, and ductility. The integrity assessment diagram gives hint on the presence of metallurgical discontinuities and defects that reduce the material ductility. Indeed, the integrity assessment diagram is a graphical representation of the quality index e_r/e_u that has some similarity with the approach presented by Caceres on Al castings [24], who has defined a relative ductility parameter q as

$$q = e_r/e_{r,mdf}, \tag{9}$$

where e_r has the usual meaning, and $e_{r,mdf}$ is the elongation to rupture of a major-defect-free specimen (subscript *mdf*). For $e_{r,mdf}$ the strain hardening exponent n in the constitutive Hollomon equation, $\sigma = K \cdot \varepsilon^n$, was used, where K is the strength coefficient (MPa). Equation (9) is the ratio of extrinsic to intrinsic elongations, where extrinsic elongation is measured in the defected matrix, while the intrinsic one is from the ideal matrix free of defects. There are two major differences between Caceres' approach and the here-proposed one. Firstly, the Voce equation has been proved to be more reliable on fitting tensile flow curves and determining the uniform elongations [8,19,20]. In Reference [8], it has been reported that theoretical uniform strains $\varepsilon_{u,p}$ and the experimental $\varepsilon_{u,exp}$ matched very well in IDIs and ferritic-pearlitic DIs with errors between experimental and theoretical values between 1% and 4%, if Voce formalism was used, in opposition to other constitutive equations, like Hollomon and Hollomon-type equations, where deviations between 50 and 100% were found. Secondly, in the here-proposed approach, the $e_{r,mdf}$ is calculated by the Voce parameters found from the tensile flow curve itself, by replacing e_u (see Equations (4) and (5)) to $e_{r,mdf}$. However, this replacing is only correct if the metallurgical discontinuities and defects only affect the experimental elongations to rupture e_r, and do not influence the strain hardening behavior.

Indeed, both the matrix assessment and the integrity assessment diagrams are based on the assumption that the metallurgical discontinuities and defects affect only the experimental elongations to rupture e_r, and the Voce parameters $1/\varepsilon_c$, Θ_o and the uniform elongations e_u should not be affected.

So $1/\varepsilon_c$ and Θ_o should become characteristic parameters of the DI matrix. In other words, the strain hardening behaviors of ideal and defected materials should be similar even if they have different elongations to rupture. This assumption was validated in the present work on IDI 800 (see Figure 5a,c,e), where, even if tensile flow curves had different e_r/e_u values, the strain hardening data were similar. In Figure 5b,d,f the corresponding engineering tensile flow curves of the ideal ($e_r/e_u \geq 1$) and defected materials ($e_r/e_u < 1$) could almost overlap from yielding until the failure (of the defected materials). Further support to the strain hardening consistency was that the SED_U values were comparable for the different flow curves reported in Figure 5. Indeed, the range of the SED_U values along the whole range of the IDI 800 reported in Table 1, resulted in being quite narrow, spanning from 65.3 to 70.4 MJ/m^3, and having a mean value of 67.7 MJ/m^3 with a mean absolute variance of 1.2 MJ/m^3. Thus, the SED_U could be considered almost constant for the analyzed IDI 800, indicating that metallurgical discontinuities and defects caused premature failures, but did not significantly affect strain hardening behaviors.

The fact that this assumption was fulfilled in IDI 800 is not obvious in metallic alloys produced through casting. For instance, in Al alloys castings for airspace applications this assumption has not been satisfied because of the nature and the high density of metallurgical defects [25]. As reported in Reference [26] for samples of Al206-T76 produced through the same casting route, the strain hardening behaviors of the flow curves (reported in Figure 8a of Reference [26]), with "different levels of quality index (as indicated by their elongations)" have significantly different Voce parameters $1/\varepsilon_c$ and Θ_o. The engineering flow curves corresponding to the strain hardening data shown in Figure 8a of Reference [26], are re-constructed and here reported in Figure 6a, where the dash parts of the engineering flow curves are beyond the rupture (e_r) until the achievement of the uniform elongations e_u. In Figure 6a, the strain hardening behaviors of Al206-T76 casting samples [26] are strongly affected by metallurgical discontinuities and defects, since the Voce parameters change dramatically, and the engineering flow curves do not overlap at all. For instance, in the blue flow curve with $\Theta_o = 9{,}746$ and $1/\varepsilon_c = 21.8$, it was found $e_r/e_u \approx 0.94$, which could indicate a quite sound material with a SED_U of 39.8 MJ/m^3. However, the black flow curve ($e_r/e_u \approx 1.00$) is significantly higher with significant larger SED_U of 68.1 MJ/m^3. Indeed, the plastic behavior in the highly defected Al206-T76 changed so dramatically that the SED_U values of the re-constructed curves varied significantly from 68.1 ($e_r/e_u \approx 1.00$) to 23.2 MJ/m^3 ($e_r/e_u \approx 0.42$), with a mean value of 41.8 MJ/m^3 and a mean absolute variance of 13.2 MJ/m^3. This data scatter was significantly wider than in IDI 800, where the mean absolute variance of SED_U values was 1.7 MJ/m^3. Thus, from the e_r/e_u values reported in Figure 6a, it was evident that these ratios could not be used as an indication of the microstructure quality of Al206-T76, because the strain hardening behavior was dramatically affected by metallurgical discontinuities and defects. Conversely, in the IDI 800 results reported in Figure 5, even the widest changes of e_r/e_u did not produce any significant changes of SED_U. In Figure 6b, the SED_U values vs. $1/\varepsilon_c$ of IDI 800 and the Al206-T76 are reported for comparison.

From Figure 6b, other considerations arose. It was evident that the range of Voce parameters $1/\varepsilon_c$ in IDI 800 was very narrow if compared to Al206-T76. Even more significant, there was no large scatter in the SED_U values of IDI 800 that appeared constant if compared to Al206-T76, resulting in no evident relationship between SED_U and $1/\varepsilon_c$. Indeed, SED_U and $1/\varepsilon_c$ are mathematically related through Equations (4) and (6), so this result meant that the differences of plastic behaviors were caused by random microstructure features that affected the tensile flow curves (σ_V and σ_0), or even by contributions that could be associated to measurement errors that were random by nature. This suggested that the nature and density of defects in the IDI 800 matrix was below a significant threshold and did not affect the flow curve shapes, but only affected the elongations to rupture that could be caused by single defects in the matrix according to the fracture mechanics approach. Conversely, in the specimens Al206-T76 the SED_U and $1/\varepsilon_c$ were strongly related, since the nature and density of defects in the samples were so high that the defects contributed significantly to the plastic behaviors of the different samples, affecting σ_V and σ_0. In the latter case, a direct comparison between e_r/e_u should not make any sense. From Figure 6b the SED_U seemed to be a promising tool to discriminate between sound and defected castings, and potential applications could be carried out, for instance, on comparing DIs produced in thin walls and heavy sections (same chemical compositions with different production routes), or DIs produced with increasing alloying elements contents causing segregations and/or carbides precipitations (same production routes with different chemical compositions).

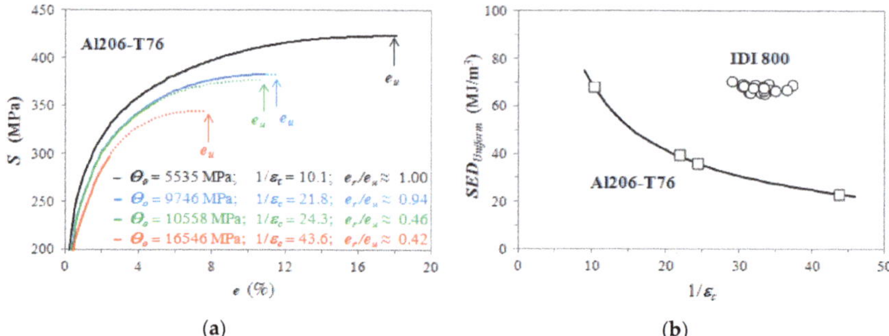

Figure 6. (**a**) Re-constructed engineering flow curves from the strain hardening data of Al206-T76 reported in Figure 8a in Reference [26]; the dashed parts of the flow curves concern the achievement of the uniform strains e_u for the calculation of SED$_U$; (**b**) SED$_U$ vs. $1/\varepsilon_c$ for IDI 800 and Al206-T76 from the flow curves in (**a**).

These results proved that, with IDI 800, the material quality assessment procedure [11,12] based on the matrix assessment and the integrity assessment diagrams are valuable. These results suggested that the procedure should be possible for any DI, where the difference of density between melt metal and floating scrap is so high that the concentration of metallurgical discontinuities and defects could be usually kept low with a proper foundry practice. However, even if the nominal compositions and pouring conditions of DIs are the same, from different foundries the density of defects may be significantly different, and so the fundamental assumption of the material quality assessment proposed here has to validated every time for every material. Thus, the capability of a foundry to adopt the procedure on a material could be itself an indication of the quality of the foundry practice.

5. Conclusions

A mathematical procedure to assess the microstructure quality of DIs [11,12] has been proposed and comes into two diagrams. A first matrix assessment diagram is for the classification of DIs ideal microstructure, that is, the matrix, identifying DIs with different silicon content and DIs produced through different routes and heat treatments. A second integrity assessment diagram is for the evaluation of the presence of potential metallurgical discontinuities and defects in DIs through comparing the rupture elongations (e_r) of DIs to the uniform elongations (e_u) during tensile tests. Both diagrams are based on the formalism of the dislocation-density-related constitutive Voce equation that is related to dislocation dynamics. However, the procedure is based on the fundamental assumption that the strain hardening behaviors of DIs are not affected by the nature and the density of the metallurgical discontinuities and defects, and so they only affect the elongations to fracture. Strain Energy Density at the uniform strains (SED$_U$) calculations were used to have quantitative support to the strain hardening analysis.

- Twenty samples of IDIs 800 were intentionally selected in a untypical wide range of elongations to rupture to evaluate this assumption;
- The strain hardening behaviors and the SED$_U$ values of the tested IDI 800 resulted in being independent of rupture conditions, and so the fundamental assumption was proven to be true;
- Interestingly SED$_U$ analysis seemed to be a promising tool to discriminate between sound and defected castings, like, for instance, comparing DIs produced in thin walls and heavy

sections, or DIs produced with increasing alloying elements contents causing segregations and carbides precipitations;

- The results proved that the microstructure quality assessment procedure was valid for IDI 800.

The procedure should also be valid in other DIs, even if indeed the fundamental assumption of the material quality assessment has to validated every time for every material, and for the same DI also for every foundry. Indeed, since metallurgical discontinuities and defects depend on the foundry practice, the capability of a foundry to adopt the proposed quality assessment procedure is itself an indication of the quality of the foundry practice.

Author Contributions: Material production, F.Z.; tensile testing, F.Z.; scanning electron microscopy G.A.; data formal analysis, G.A.; method conceptualization, G.A.; original draft preparation, G.A. and F.Z.; review and editing, G.A.

Acknowledgments: Davide Della Torre and Tullio Ranucci are warmly thanked for their experimental support.

Conflicts of Interest: The authors declare no conflict of interest.

References

1. Menk, W. A new high strength high ductile nodular iron. *Mater. Sci. Forum* **2018**, *925*, 224–230. [CrossRef]
2. Hammersberg, P.; Hamberg, K.; Borgström, H.; Lindkvist, J.; Björkegren, L. Variation of tensile properties of high silicon ductile irons. In Proceedings of the 11th International Symposium on the Science and Processing of Cast Iron, SPCI-XI 2017, Jonkoping, Sweden, 4–7 September 2017; Volume 925, pp. 280–287.
3. González-Martínez, R.; de la Torre, U.; Lacaze, J.; Sertucha, J. Effects of high silicon contents on graphite morphology and room temperature mechanical properties of as-cast ferritic ductile cast irons. Part I—Microstructure. *Mater. Sci. Eng. A* **2018**, *712*, 794–802. [CrossRef]
4. González-Martínez, R.; de la Torre, U.; Ebel, A.; Lacaze, J.; Sertucha, J. Effects of high silicon contents on graphite morphology and room temperature mechanical properties of as-cast ferritic ductile cast irons. Part II—Mechanical properties. *Mater. Sci. Eng. A* **2018**, *712*, 803–811. [CrossRef]
5. Cekic, O.E.; Sidjanin, L.; Rajnovic, D.; Balos, S. Austempering kinetics of Cu-Ni alloyed austempered Ductile Iron. *Met. Mater. Int.* **2014**, *20*, 1131–1138. [CrossRef]
6. Donnini, R.; Fabrizi, A.; Bonollo, F.; Zanardi, F.; Angella, G. Assessment of Microstructure Evolution of an Austempered Ductile Iron during Austempering Process through Strain Hardening Analysis. *Met. Mater. Int.* **2017**, *23*, 855–864. [CrossRef]
7. Górny, M.; Angella, G.; Tyrala, E.; Kawalec, M.; Paz, S.; Kmita, A. Role of Austenitization Temperature on Structure Homogeneity and Transformation Kinetics in Austempered Ductile Iron. *Met. Mater. Int.* **2019**, *25*, 956–965. [CrossRef]
8. Angella, G.; Zanardi, F.; Donnini, R. On the significance to use dislocation-density-related constitutive equations to correlate strain hardening with microstructure of metallic alloys: The case of conventional and austempered ductile irons. *J. Alloy Compd.* **2016**, *669*, 262–271. [CrossRef]
9. Zanardi, F.; Bonollo, F.; Bonora, N.; Ruggiero, A.; Angella, G. A contribution to new material standards for Ductile Irons and Austempered Ductile Irons. *Int. J. Met.* **2017**, *11*, 136–147.
10. Ductile Iron Data for Design Engineers/Section III Engineering Data/Part 1 Tensile Properties/Relationships between Tensile Properties. Available online: www.ductile.org (accessed on 14 June 2019).
11. Donnini, R.; Zanardi, F.; Vettore, F.; Angella, G. Evaluation of microstructure quality in ductile irons based on tensile behaviour analysis. *Mater. Sci. Forum* **2018**, *925*, 342–349. [CrossRef]
12. Angella, G.; Zanardi, F. Microstructure Quality Assessment of Isothermed Ductile Irons through Tensile Tests. In Proceedings of the 73rd World Foundry Congress "Creative Foundry", WFC 2018–Proceedings, Krakow, Poland, 23–27 September 2018; pp. 265–266.
13. Tiryakioğlu, M.; Campbell, J.; Staley, J.T. Evaluating Structural Integrity of Al-7%Si-Mg Alloys via Work Hardening Characteristics: I. Concept of Target Properties. *Mater. Sci. Eng. A* **2004**, *368*, 205–211. [CrossRef]
14. Tiryakioğlu, M.; Staley, J.T.; Campbell, J. Evaluating Structural Integrity of Al-7%Si-Mg Alloys via Work Hardening Characteristics: II. A New Quality Index. *Mater. Sci. Eng. A* **2004**, *368*, 231–238. [CrossRef]

15. Tiryakioğlu, M.; Alexopoulos, N.D.; Campbell, J. On the Ductility Potential of Cast Al-Cu-Mg (2009) Alloys. *Mater. Sci. Eng. A* **2004**, *506*, 23–26. [CrossRef]

16. Tiryakioğlu, M.; Campbell, J. Quality index for aluminum alloy castings. *Int. J. Met.* **2014**, *8*, 39–42. [CrossRef]

17. Zanardi Fonderie–Products: IDI. Patented by Zanardi Fonderie S.P.A., Italy. Available online: http://zanardifonderie.com/IDI (accessed on 14 June 2019).

18. Estrin, Y. Dislocation theory based constitutive modelling: Foundations and applications. *J. Mater. Process. Technol.* **1998**, *80*, 33–39. [CrossRef]

19. Angella, G.; Zanardi, F. Comparison among Different Constitutive Equations on Investigating Tensile Plastic Behavior and Microstructure in Austempered Ductile Iron. *J. Cast. Mater. Eng.* **2018**, *2*, 14–23. [CrossRef]

20. Reed-Hill, R.E.; Crebb, W.R.; Monteiro, S.N. Concerning the analysis of tensile stress-strain data using log $d\sigma/d\varepsilon p$ versus log σ diagrams. *Met. Trans. A* **1973**, *4*, 2665–2667. [CrossRef]

21. Angella, G.; Wynne, B.P.; Rainforth, W.M.; Beynon, J.H. Strength of AISI 316L in torsion at high temperature. *Mater. Sci. Eng. A* **2008**, *475*, 257–267. [CrossRef]

22. Angella, G.; Donnini, R.; Ripamonti, D.; Maldini, M. Combination between Voce formalism and improved Kocks-Mecking approach to model small strains of flow curves at high temperatures. *Mater. Sci. Eng. A* **2014**, *594*, 381–388. [CrossRef]

23. Serafini, A.; Angella, G.; Malara, C.; Brunella, M.F. Mechanical and Microstructural Characterization of AF955 (UNS N09955) Nickel-Based Superalloy after Different Heat Treatments. *Mater. Sci. Eng. A* **2018**, *49*, 5339–5352. [CrossRef]

24. Caceres, C.H. A Rationale for the Quality Index of Al-Si-Mg Casting Alloys. *Int. J. Cast. Met. Res.* **1998**, *10*, 293–299. [CrossRef]

25. Fiorese, E.; Bonollo, F.; Timelli, G.; Arnberg, L.; Gariboldi, E. New Classification of Defects and Imperfections for Aluminum Alloy Castings. *Int. J. Met.* **2015**, *9*, 55–66. [CrossRef]

26. Tiryakioğlu, M.; Staley, J.T., Jr.; Campbell, J. The effect of structural integrity on the tensile deformation characteristics of A206-T71 alloy castings. *Mater. Sci. Eng. A* **2008**, *487*, 383–387. [CrossRef]

© 2019 by the authors. Licensee MDPI, Basel, Switzerland. This article is an open access article distributed under the terms and conditions of the Creative Commons Attribution (CC BY) license (http://creativecommons.org/licenses/by/4.0/).

Article

Research on Chip Shear Angle and Built-Up Edge of Slow-Rate Machining EN C45 and EN 16MnCr5 Steels

Katarina Monkova [1,2,*], **Peter Pavol Monka** [1], **Adriana Sekerakova** [1], **Jozef Tkac** [1], **Martin Bednarik** [2], **Juraj Kovac** [3] and **Andrej Jahnatek** [4]

[1] Faculty of Manufacturing Technologies with the Seat in Presov, Technical University of Kosice, Sturova 31, 08001 Presov, Slovakia
[2] Faculty of Technology, UTB Tomas Bata University in Zlin, Vavreckova 275, 76001 Zlin, Czech Republic
[3] Faculty of Mechanical Engineering, Technical University of Kosice, Letna 9, 04200 Kosice, Slovakia
[4] Jadrová Energetická Spoločnosť Slovenska, a. s., Tomasikova 22, 82102 Bratislava, Slovakia
* Correspondence: katarina.monkova@tuke.sk; Tel.: +421-55-602-6370

Received: 5 August 2019; Accepted: 28 August 2019; Published: 30 August 2019

Abstract: One of the phenomena that accompanies metal cutting is extensive plastic deformation and fracture. The excess material is plastically deformed, fractured, and removed from the workpiece in the form of chips, the formation of which depends on the type of crack and their propagation. Even in case of the so-called 'continuous' chip formation there still has to be a fracture, as the cutting process involves the separation of a chip from the workpiece. Controlling the chip separation and its patterning in a suitable form is the most important problem of the current industrial processes, which should be highly automated to achieve maximal production efficiency. The article deals with the chip root evaluation of two EN C45 and EN 16MnCr5 steels, focusing on the shear angle measuring and built-up edge observation as important factors influencing the machining process, because a repeated formation and dislodgement of built-up edge unfavorably affects changes in the rake angle, causing fluctuation in cutting forces, and thus inducing vibration, which is harmful to the cutting tool. Consequently, this leads to surface finish deterioration. The planing was selected as a slow-rate machining operation, within which orthogonal and oblique cutting has been used for the comparative chips' root study. The planned experiment was implemented at three levels (lower, basic, and upper) for the test preparation and the statistical method, and regression function was used for the data evaluation. The mutual connections among the four considered factors (cutting speed, cutting depth, tool cutting edge inclination, and rake angle) and investigated by the shear angle were plotted in the form of graphical dependencies. Finally, chips obtained from both steels types and within both cutting methods were systematically processed from the microscopic (chip root) and macroscopic (chip pattern) points of view.

Keywords: shear angle; chip root; shape; built-up edge; slow-rate machining

---

1. Introduction

Almost every mechanical component in use has undergone a machining operation at some stage of its manufacturing process. Considering that the economics of the metal cutting process significantly affect the overall cost of the final products, there is a strong drive to reduce the time and cost of machining operations. This is one of the reasons why chip machining technology underwent huge developments in the last several decades, which has been displayed by its massive use in industrial practice. Despite the opinion of some experts [1,2] that the manufacturing of components by machining is not efficient and that it is necessary to substitute it with chip-less methods, this technology is irreplaceable by other technological methods in many cases. The demanded accuracy and the quality of a machined surface are in many cases not achievable by mechanical working or casting processes, because the surfaces of the parts are adversely affected. This is why machining is needed. [3] Controlling the chip separation and its patterning in a suitable form is the most important problem of the current industrial processes, which should be highly automated to achieve maximal production efficiency.

One of the phenomena that accompanies metal cutting is extensive plastic deformation and fracture. The excess material is plastically deformed, fractured, and removed from the workpiece in the form of chips, the formation of which depends on the type of crack and their propagation. Even in the case of so-called 'continuous' chip formation there still has to be a fracture, as the cutting process involves the separation of a chip from the workpiece. Chips at the machining are formed due to tearing and shearing, the workpiece material adjacent to the tool face is compressed and a crack runs ahead of the cutting tool and towards the body of the workpiece. Cutting takes place intermittently and there is no movement of the work-piece material over the tool face. In chip formation by shear, there is general movement of the chip over the tool face.

As the tool advances into the work-piece, the metal ahead of the tool is severely stressed. The cutting tool causes internal shearing action in the metal, such that the metal below the cutting edge flows plastically in the form of a chip. Firstly, compression of the metal under the tool edge takes place, followed by the separation of the metal when the compression limit of that metal has been exceeded. Plastic flow takes place in a localized region called the shear plane, which extends from the cutting edge obliquely up to the uncut surface ahead of the tool. When the metal is sheared the crystals are elongated, the direction of elongation being different than that of the shear.

The angle made by a plane of shear with the direction of tool travel is known as the shear angle, Φ. Its value depends on the material being cut and the cutting conditions. If the shear angle is small, the path of the shear will be long, chips will be thick, and the force required to remove the layer of metal of given thickness will be high, and vice versa.

2. State of the Art

The classic shear plane model is based on a free body diagram of the chip, which is held in equilibrium by two equal, opposite, and collinear resultant forces, one acting on a shear plane, and the other on the rake face of the tool. This was developed by Merchant [4] in the early 1940s. Figure 1 depicts the famous Merchant's circle, which shows the relationship between various force and velocity components acting on the chip. The individual symbols are defined as follows: v_c—chip velocity, v_s—shearing velocity, v_w—workpiece velocity, R_s is the resultant force on the tool with its projections in shear and in frictional directions, α, Φ, and β are the rake, shear, and friction angles, respectively, L_c is the contact length, and h and h_c are uncut chip thickness and chip thickness, respectively.

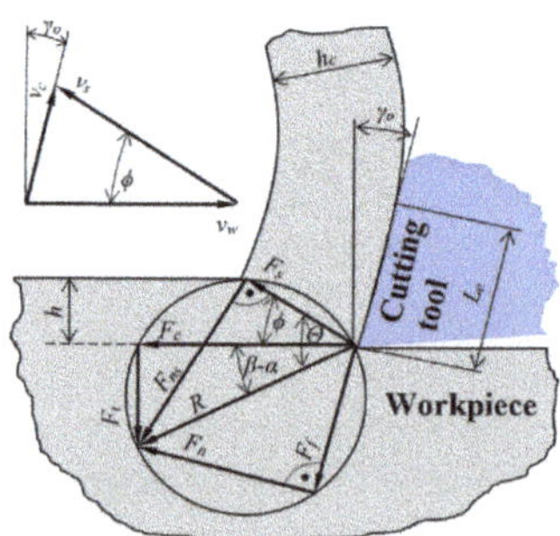

Figure 1. Force and velocity components in Merchant's model.

The basis of the scientific research related to orthogonal machining provided by Merchant has been further developed by many scientists, such as Weiner [5], Wallace and Boothroyd [6], and Muraka et al. [7]. Later, the involvement of computer aid enabled the development of new numerical models of orthogonal machining. The related research has been presented by Li et al. [8], Shi et al. [9], Filice et al. [10], Bagci [11], Priyadarshini et al. [12], and other researchers.

To study chip characteristics, it is necessary to specify the type of machining by which the chip forms. There are two different types of cutting: orthogonal and oblique. Orthogonal cutting is a type of metal cutting in which the cutting edge of the wedge shape cutting tool is perpendicular to the direction of tool motion. In this cutting, the cutting edge is wider than the width of a cut. This cutting is also known as 2D cutting, because the force developed during cutting can be a plot on a plane or can be represented by a 2D coordinate. However, orthogonal cutting is only a particular case of oblique cutting, such that any analysis of orthogonal cutting can be applied to oblique cutting. Oblique cutting is a common type of three-dimensional cutting used in the machining process [13].

The characterization of chip morphology is defined by analyzing the chip, both on a micro and macro level. On a micro-level, the chip is characterized as chip shape and at the macro level as chip curl. The chip shape characterizes the chip's cross-section in a plane perpendicular to the rake face and the cutting plane. Continuous chips, segmented chips, and discontinuous chips could be defined as the most important classification of chip shape and this is influenced to a large extent by workpiece material behavior, cutting process parameters, and cutting tool macro geometry [14]. Mathematical formulations employing the above defined geometrical parameters were later developed by Nakayama et al. to define the chip in 3D practical applications [15]. A more systematic study on chip curl was developed by various research groups over several decades [14,16–18]. In addition, chip curl research has been well-reviewed and documented [19,20] and has resulted in a more quantifiable chip form classification. Among all these fundamental influences, from a cutting tool geometry viewpoint, the cutting tool macro geometry influences chip curl to a large extent. 2D orthogonal cutting would primarily result in an up curled chip, with side curling to a very small extent. In 3D oblique cutting, chips are a combination of both side curl and up curl. Chip curl is also influenced to a large extent by the nose configuration and its effect is dependent on the relationship between the nose radius and process parameters selected [21]. Chip segmentation is of pivotal importance, since it facilitates the machining ergonomics and scrap removal without damaging workpiece surface quality and ensuring the safety of the working personnel [22].

Noteworthy studies within the chip formation field are the works of Shaw et al. [23], Luk [24], and Okushima and Minato [25]. Palmer and Oxley [26] used experimental flow fields obtained by tracing the path of individual grains in low speed cutting to determine the extent of the deformation zone and the shear angle. They showed that the plastic deformation zone has a substantial thickness and the streamlines form smooth curves from the workpiece into the chip. They also developed slip-line field solutions from the experimental results and calculated the stress distribution in the deformation zone. Oxley and Welsh [27] developed a parallel-sided shear zone theory, which allows for strain hardening of the material. In this model, the shear flow stress changes from the initial yield stress at the lower boundary of the shear zone to a higher value at its upper boundary.

Along with an evolution of new materials that should be machined and continual cutting tools improvements, the topic of chip formation and its morphology is still relevant, so many researchers are presently addressing this issue. Prasetyo and Tauviqirrahman [28] developed more comprehensive models by using a power-law flow stress equation, whose coefficients are dependent on strain rate and temperature. These effects were combined using the concept of velocity-modified temperature, and it was suggested that the flow stress for a particular material at a given strain can be assumed to be a unique function of this parameter. Hegab with his research team [29] investigated the effects of two types of nano-cutting fluids on tool performance and chip morphology during the turning of Inconel 718. Similarly, Adekunle et al. [30] dealt with the chip morphology and behavior of tool temperature during the turning of AISI 301steel using different biodegradable oils. Anthony [31] analyzed a cutting force and chip morphology during the hard turning of AISI D2 steel. Their research confirmed the results of other studies, within which it was found that depth of the cut, followed by cutting speed, influence cutting force considerably. Umer et al. studied chip morphology predictions while machining hardened tool steel using finite element and smoothed particles hydrodynamics methods [32]. The topic of chip morphology and its formation modeling during metal cutting operations was also the subject of a study by Mabrouki et al. [33].

In chip machining, the chip forming process is often accompanied by the formation of a built-up edge (BUE) in the face of the tool and it is necessary to take this phenomenon into account. Its creation changes the geometry of the cutting tool and thus the conditions of the cutting process. It affects the removal of the chip from the cutting zone, its twisting, and shaping due to the material sticking to the tool face and thereby causing the tool face angle to change during the cutting process. During chip formation, the material layers of the workpiece become highly plastically deformed, and after separation by friction on the tool face this forms a so-called braked layer between the exiting chip and the tool face. The height of this layer varies considerably, depending on the cutting speed. At very low cutting speeds, the material is not sufficiently heated and plastic to cause such phenomena. However, when machining with cutting tools from high-speed steel at a cutting speed of 10 to 30 m·min^{-1}, these layers can be welded under pressure and thus lead to an increase in the tool face. Figure 2a–d shows the succession of the built-up-edge using the slip lines [34,35].

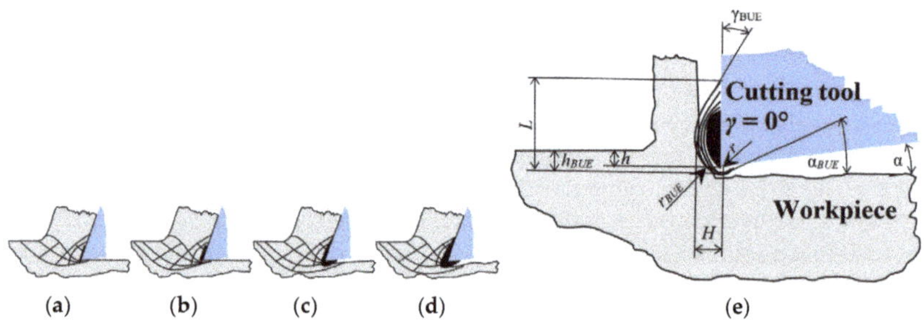

Figure 2. The succession of the built-up-edge specified by the field of slip lines: (**a**) the formation of a core of built-up edge; (**b**) formation of a microfracture on a machined surface; (**c**) separation of the free chip in front of the rake surface and grow up of the built-up edge, (**d**) crushing of built-up edge by a crack propagation on a machined surface; (**e**) model of the machining with a built-up edge.

The tool angles of rake and clearance increase plastically due to the build-up edge (BUE) appearing, significantly increasing the corner radius (from r to r_{BUE}), depending on the size of the build-up edge. Since the built-up edge plastically extends the cutting edge, it increases the thickness of the cutting layer (from h to h_{BUE}) (Figure 2e) [36].

The important features of the built-up edge are the variability of its dimensions over time and space (in the line of the cutting edge) and the high hardness that will allow it to assume the function of the cutting edge. There is some range of cutting speed at which the built-up edge occurs. Outside this cutting speed range, no built-up edge can be observed [37,38].

As can be seen from the research works presented above, a great deal of research has been devoted toward understanding the mechanics of metal cutting, with the objective of obtaining more effective cutting tools and more efficient manufacturing process plans. Traditionally, these objectives have been achieved by experimentation and numerical simulation. In spite of extensive research in this field, the basic mechanics of the cutting process and the interplay of many factors which lead to its great variety are not yet totally understood, so the search for more effective models continues on analytical, experimental, and numerical fronts.

This study is an original contribution to the chip root research of two EN C45 and EN 16MnCr5 steel types as one of the possibilities to develop knowledge within a chip fracturing that can minimize cutting-tool wear and poor machinability, restricting the quality and productivity of the machining process. It aims to contribute to the analysis of the chip formation process at orthogonal and oblique cutting with relatively slower speeds, focusing on the chip shear angle and built-up edge investigation, as well as to the finding out the impact of selected machining factors on this process. The planing operation has been selected due to the visibility of the chip formation to the naked eye. This type of machining has also allowed authors to develop a new reliable method for sample acquisition in order to avoid chip root deformation and its thermal influence on sample acquisition, which could cause changes in the microstructure of the material.

To the best of our knowledge, no relevant studies have been published concerning chip root observations and the influence of cutting parameters (cutting speed, cutting depth, tool cutting edge inclination, and rake angle affected by forming built-up edge) on the shear angle of both EN C45 and EN 16MnCr5 steels types within such deep and complex comparative research. In addition, the mutual connections among the four considered factors influencing shear angle in orthogonal and oblique cutting have been plotted in the form of graphical dependencies.

3. Materials and Methods

3.1. Cutting Tools, Machined Material, and Measuring Equipment

Chip formation at orthogonal and oblique slow-rate machining has been experimentally investigated within this research. For this study, the technology of planing was selected, at which the main sliding motion performs a workpiece. The machining process was carried out using the planer machine of KOVOSVIT MAS Machine Tools (Sezimovo Ústí, Czech Republic).

The planing necking tool type 32 × 20 ON 36550 HSS00 (PILANA Tools Ltd., Hulín, Czech Republic) was used for orthogonal cutting and a straight roughing tool 32 × 20 ON 36500 HSS00 was used in oblique machining. Both types of cutting tools included brazed-tips from high-speed steel with three different types of cutting edge inclinations: $\lambda_s = 0°$, 10°, and 20°. The angle of tool orthogonal rake γ_o and angle of tool orthogonal clearance α_o were varied in the second and third phase of the experiments to get a better view of the chip formation and to obtain more reliable results. Pictures of the cutting tools are presented in Figures 3 and 4 and their geometries are arranged in Table 1.

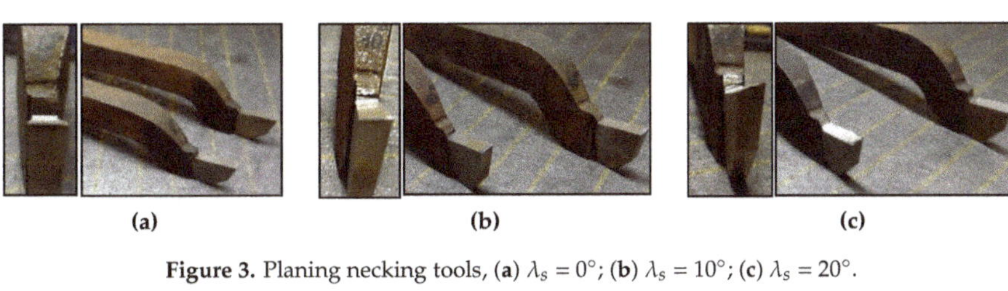

Figure 3. Planing necking tools, (**a**) $\lambda_s = 0°$; (**b**) $\lambda_s = 10°$; (**c**) $\lambda_s = 20°$.

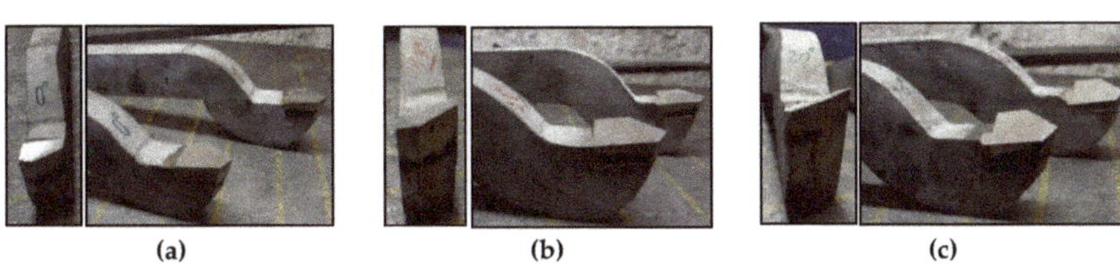

Figure 4. Straight roughing tool, (**a**) $\lambda_s = 0°$; (**b**) $\lambda_s = 10°$; (**c**) $\lambda_s = 20°$.

Table 1. Changes in the angle values.

Type of Tool Angle	Planing Necking Tool	Straight Roughing Tool
Tool cutting edge angle κ_r	0°	60°
Tool minor (end) cutting edge angle $\kappa_r{}'$	-	20°
Tool included angle ε_r	-	100°
Angle of tool orthogonal rake γ_o	8°/12°/16°	3°/7°/11°
Angle of tool orthogonal clearance α_o	15°/11°/7°	15°/11°/7°

In order to verify the input angles of the tool orthogonal rake γ_o, preliminary input tests were performed. The tests were carried out using 3D measuring equipment RAPID CNC THOME (Zimme Maschinenbau GmbH, Kufstein, Austria), as shown in Figure 5, along with detail from the measuring process.

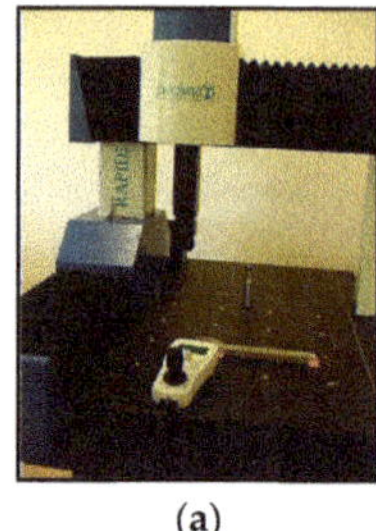
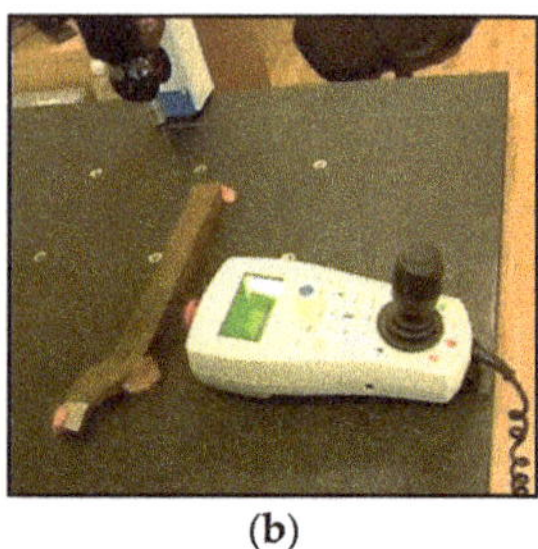

(a)   (b)

Figure 5. Preliminary tests of input angles of tool orthogonal rake γ_o, (**a**) overall view of the testing equipment RAPID CNC THOME, (**b**) detailed view of the measuring of a tool orthogonal rake angle.

The angle of tool orthogonal rake γ_o was measured for every cutting tool used in the next experiments. The protocols from the measurements confirmed the values listed in Table 1, while the deviation of all measured values did not exceed 5% and the average angles of tool orthogonal rake for planing necking tools and straight roughing tool were $\gamma_o = 8.138° = 8°2'0''$ or $\gamma_o = 3.159 = 3°2'1''$, respectively.

The steels 1.7131 (EN 16MnCr5) and 1.0503 (EN C45) were selected as a machined material, the chip formation of which has been subjected to research. The chemical composition of these steels types is presented in Table 2.

The steel 1.7131 (EN 16MnCr5) contains smooth deformable calcium aluminates, encapsulated in manganese sulphide, as an alternative to tough alumina oxide inclusions. It is suitable for cementing and for die forging; it is well machinable, well weldable, and, after annealing, also well formable. This grade of steel is generally used for elements with a required core tensile strength of 800–1100 N·mm^{-2} and good carrying resistance, e.g., piston bolts, camshafts, levers, and other automobile and mechanical engineering add-ons.

The steel 1.0503 (EN C45) is an unalloyed medium carbon engineering steel which offers moderate tensile strengths, wear resistance, and good machinability. This material is capable of through hardening by quenching and tempering on limited sections and can also be flame or induction hardened to a surface hardness of min 55 HRC. C45 is generally supplied in an untreated or normalized condition, with a typical tensile strength range of 570–700 MPa and Brinell hardness range of 170–210.

Table 2. Chemical composition of the steels.

Steel	C (%)	Mn (%)	Si (%)	Cr (%)	Ni (%)	Cu (%)	P (%)	S (%)
EN C45	0.42–0.50	0.50–0.80	0.17–0.37	max 0.25	max 0.30	max 0.30	max 0.040	max 0.040
EN 16MnCr5	0.14–0.19	1.10–1.40	0.17–0.37	0.80–1.10	-	-	max 0.035	max 0.035

The infrared thermometer UNI-T UT305C (manufacturer UNI-TREND Technology, China Co., Ltd., Dongguan, China), based on a principle of infrared radiation emitted from the target surface, was used to measure the temperature.

Vickers microhardness was measured with the MICRO—VICKERS HARDNESS TESTER CV—403DAT (MetTech Ltd., Calgary, AB, Canada), which has the possibility of magnifying a view 200 times and 600 times.

Etched specimens of chips were observed by means of the Platinum USB digital microscope UM019 (Shenzhen Handsome Technology Co., Ltd., Shenzhen, China) with magnification 25–220×.

3.2. Methods

In order to maximize the amount of "information" that was obtained by a given experimental effort, it is necessary to design the experimental plan well. Many experimental designs (full factorial design in two/three levels, fractional factorial design, Plackett–Burman design, Doehlert matrix, A Box–Wilson Central Composite Design, Box Behnken designs, and others) have been recognized as useful techniques to optimize process variables. The influence of four factors (cutting speed, cutting depth, rake angle, and angle of cutting edge inclination) on chip formation (shear angle) in a slow rate machining process was investigated in this study. In this case, the most powerful tool of the planned experiment appeared to be how the basic principle was a measurement of each factor's influence on three levels [36,39,40].

The description of the experimental plan within this part of the article is given due to a better understanding of measured data processing.

Based on [41], the basic equations for statistical processing can be written in the matrix (1):

$$Y = X \cdot b, \tag{1}$$

where Y—column vector of measured quantities, X—matrix of independent variables, b—coefficient of a regression function.

The system of normal Equation (2) can be expressed in the following way:

$$[X^T \cdot Y] = [X^T \cdot X] \cdot b. \tag{2}$$

The vector "b" in relationship (2) is specified by the least-squares' method of the matrix regression analysis (3):

$$b = [X^T \cdot X]^{-1} \cdot X^T \cdot Y. \tag{3}$$

It is necessary to consider that the complete three-level plan had a large scale of measurements expressed by $N = 3^k$, where k is a number of variables (in this case factors) and N is a number of measures (e.g., considering four variables within an experiment, a total of 81 measurements should be performed, because $N = 3^4 = 81$) [36].

A reduced number of measurements for the dependencies described by functions of the second order can be achieved by means of the so-called second level compositional non-rotational plan [42], while the symbols in Equation (4) have the following meanings: x_j is a variable (in the case of presented research it is one of the cutting parameters that will be varied), j, u are indexes that define a parameter, b_j is a j-th correlation coefficient:

$$y = b_0 x_0 + \sum_{j=1}^{N} b_j x_j + \sum_{\substack{u,j=1 \\ u \neq j}}^{N} b_{uj} x_j x_u + \sum_{j=1}^{N} b_{jj} x_j^2. \tag{4}$$

The composition plan, in this case, consisted of [36,42]:

1. a core of plan that can be

   - two-level 2^k plan for $k < 5$, or as
   - shortened replica $2^{k\text{-}p}$ for $k \geq 5$, in which p is the linear effects associated with k—interaction effects,

2. the star points α with coordinates: $(\pm\alpha, 0, ..., 0)$; $(0, \pm\alpha, 0, ..., 0)$; ...; $(0, 0, ..., 0, \pm\alpha)$,
3. the measurements that were done on a basic level—in the middle of the plan at $x_1 = x_2 = ... = x_k = 0$ (the number of measurements in the middle of the plan is n_0).

The total number of measurements is then [43]:

$$N = 2^k + 2k + n_0, \text{ if} \qquad k < 5, \text{ or}$$
$$N = 2^{k-p} + 2k + n_0, \text{ if} \qquad k \geq 5. \tag{5}$$

In practical implementation, $n_0 = 1$ was chosen, with no boundary [36,44]. The matrix of the orthogonal composition plan for k, α, and n_0 is given in Table 3. In its general form, it is not orthogonal, because the relationships on the left sides of Equations (6) and (7) are different from zero:

$$\sum_{i=1}^{N} x_{oi} x_{ji}^2 \neq 0, \tag{6}$$

$$\sum_{i=1}^{N} x_{ji}^2 x_{ui}^2 \neq 0. \tag{7}$$

Table 3. The general form of the matrix of the composition plan.

N	x_o	x_1	$x_2 \ldots$	x_k	Description
	+1	−1	−1 ...	−1	
	+1	+1	−1 ...	−1	
	+1	−1	+1 ...	−1	
2^k ($k < 5$) or 2^{k-p} ($k > 5$)	+1	+1	+1 ...	−1	the core of the plan
	+1	−1	−1 ...	+1	
	+1	+1	−1 ...	+1	
	+1	−1	+1 ...	+1	
	+1	+1	+1 ...	+1	
	+1	−α	0 ...	0	
	+1	+α	0 ...	0	
$2k$	+1	0	−α ...	0	the star points of the plan
	+1	0	+α ...	0	
	+1	0	0 ...	−α	
	+1	0	0 ...	+α	
	+1	0	0 ...	0	
n_0	+1	0	0 ...	0	the measurements in the middle of the plan
	+1	0	0 ...	0	

The matrix is converted to orthogonal shape by quadratic variables exchanging [36,45]:

$$x'_j = x_j^2 - \frac{1}{N} \sum_{i=1}^{N} x_{ji}^2 = x_j^2 - \bar{x}_j^2. \tag{8}$$

Which is why:

$$\sum_{i=1}^{N} x_{oi} x'_{ji} = \sum_{t=1}^{N} x_{ji}^2 - N\bar{x}_j^2 = 0, \tag{9}$$

$$\sum_{i=1}^{N} x'_{ji} x_{ui} \neq 0, \tag{10}$$

The regression function correlation coefficients in (4) are independent because of the orthogonality of the experimental matrix, and they are specified by the following relations (11)–(14):

$$b_j = \frac{\sum_{i=1}^{N} x_{ji} y_i}{\sum_{i=1}^{N} x_{ji}^2} = \frac{\sum_{i=1}^{N} x_{ji} y_i}{2^k + 2\alpha^2}, \tag{11}$$

$$b_{uj} = \frac{\sum_{i=1}^{N} x_{ji} y_i}{\sum_{i=1}^{N} x_{ji}^2} = \frac{\sum_{i=1}^{N} x_{ji} y_i}{2^k}, \tag{12}$$

$$b_{jj} = \frac{\sum_{i=1}^{N} x'_{ji} y_i}{\sum_{i=1}^{N} \left(x'_{ji}\right)^2}, \tag{13}$$

$$b'_o = \frac{1}{N} \sum_{i=1}^{N} x_{oi} y_i. \tag{14}$$

Hence, the second stage regression function (4) is then given by equation (15):

$$y = b_o + b_1 x_1 + b_2 x_2 + \ldots + b_k x_k + b_{12} x_1 x_2 + b_{(k-1)k} x_{k-1} x_k + b_{11}\left(x_1^2 - \overline{x}_1^2\right)$$
$$+ b_{kk}\left(x_k^2 - \overline{x}_k^2\right), \tag{15}$$

where the constant member of the regression function is corrected by quadratic variables (8) in the form of:

$$b_o = b'_o - b_{11}\overline{x}_1^2 - b_{22}\overline{x}_2^2 - \ldots - b_{kk}\overline{x}_k^2. \tag{16}$$

Using Grubbs' testing criteria, the outliers from the measured values were specified for every group of measurements. The following Equations (17)–(19) have had to be kept as:

$$H_i = \frac{|T_{ik} - \overline{T}_i|}{S_{T_i}} < H_p(m), \tag{17}$$

while

$$\overline{T}_i = \frac{\sum_{k=1}^{m} T_{ik}}{m}, \tag{18}$$

$$S_{T_i} = \sqrt{\frac{1}{m-1} \cdot \sum_{k=1}^{m}\left(T_{ik} - \overline{T}_i\right)^2}, \tag{19}$$

where m = the number of evaluated measurements within the Grubbs' test, T_{ik} = the measured value of the k-th issue in the i-th group, $k = 1, 2, 3; i = 1, 2, \ldots, 24, 25, \overline{T}_i$ = the average value of the measured issues of the i-th group; calculation according to the equation, S_{T_i} = the standard deviation of the measured issue of the i-th group, according to $H_p(m)$, which is the critical value of Grubbs' testing criteria for m values ($m = 3$), where p is the level of significance, and usually it is $H_p(m) = 0.05$.

The calculation of the regression coefficients was performed using the MATLAB 2016 software (The MathWorks, Inc., Natick, MA, USA) while the significance of the coefficients of the function $y = logT$ was tested according to the Student's test criterion.

The adequacy of regression function was assessed according to the Fisher–Snedecor test criterion $F < F0.05\ (f_1, f_2)$, where the degrees of freedom $f_1 = Nq$ (q is a number of significant coefficients) and $f_2 = N(m-1)$.

The methodology of the orthogonal compositional non-rotational plan enabled a decreased number of experiments, from 81 to 25. In this methodology, a reduction based on p-value (so-called shortened replica) or Fisher's coefficient is recommended only when the number of experiment factors is equal to or greater than five.

Within the experimental study, the variables according to Table 4 were taken into account, while the codes for the specific values of individual variables are referred in Table 5.

To track changes in the zone of chip forming, the machining process had to be stopped immediately, thus interrupting tool and workpiece contact. This is based on the observation of the chip end produced in interrupted cutting, e.g., in planing or face milling.

A reliable method to stop the machining process immediately was developed. Its principle lies in a chip root acquisition by modifying the end of the workpiece, according to Figure 6. The goal of the proposed method and the special workpiece design was to avoid deformation and a thermal influence of the material, which could cause changes in the microstructure of a material. Due to this reason, for all operations within the workpiece preparation, a cooling medium 5% emulsion from fully synthetic oil JCK PS (manufacturer JCK, Ltd., Prešov, Slovakia) was used.

Table 4. The variables and ranges of individual values.

Variables	Symbols		$-\alpha$	-1	0	$+1$	$+\alpha$
Cutting speed (m·min^{-1})	v_c	x_1	6	8.25	10.5	12.75	15
Cutting depth (mm)	a_p	x_2	0.2	0.25	0.3	0.35	0.4
Angle of tool orthogonal rake—orthogonal cutting (°)	γ	x_3	8	10	12	14	16
Angle of tool orthogonal rake—oblique cutting (°)	γ	(x_3)	3	5	7	9	11
Angle of tool cutting edge inclination (°)	λ_s	x_4	0	5	10	15	20

Table 5. The codes for the specific values of individual variables.

Code	Specific Values of Variables
$-\alpha$	x_{min}
-1	$[(x_{max} + x_{min})/2] - [(x_{max} - x_{min})/2\alpha^2]$
0	$(x_{max} + x_{min})/2$
$+1$	$[(x_{max} + x_{min})/2] + [(x_{max} - x_{min})/2\alpha^2]$
$+\alpha$	x_{max}

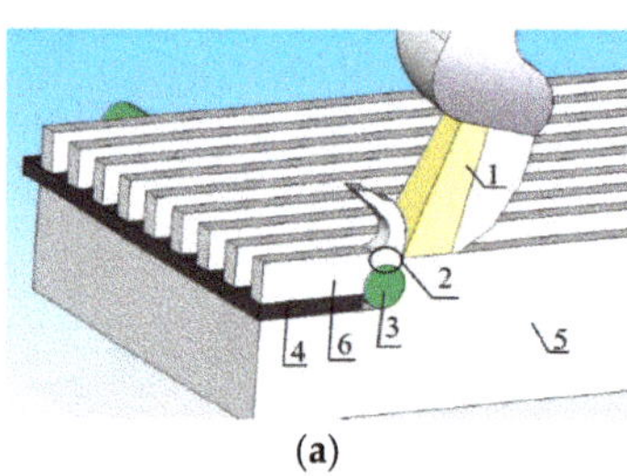

1—Cutting wedge
2—Position of the chip tear-off
3—Metal Stick
4—Support sheet metal for the sample slipping
5—Machined material
6—Part of the sample from which the chip will be torn-off

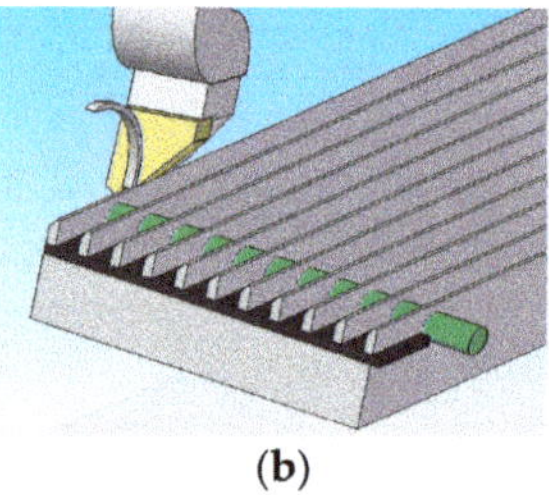

(a) (b)

Figure 6. The principle of the designed method for instantaneous contact interruption between the tool and the workpiece; (**a**) orthogonal cutting, (**b**) oblique cutting.

At the point of departure of the tool "1" from the engagement, a groove was cut, where the sheet metal insert "4" was inserted to prevent deformation of the specimen during rupture. A hole of 8 mm diameter was drilled behind the groove, and a metal rod "3" was inserted therein, which prevented the hole from deforming. When the tool passed above the metal rod, the section "2" became narrower and the material ruptured, similar to in the tensile test. Sample "6" was rapidly thrown up in the direction of tool movement at a rate greater than cutting speed, and on the sample, the plastic deformation state corresponding to the actual cutting speed was captured.

The experiment was carried out on a planer without the use of cooling, since the cutting length of the tool path was about 200 mm, and thus the tool and workpiece were not overheated.

The shear angle Φ was measured at a chip root. The samples were ground five times, polished, and etched. Also, the shear angle, i.e., angle of the boundary between the deformed and undeformed material, and angle of built up edge were measured five times at each sample. An example of an investigated sample is presented in Figure 7.

The angle of the tool cutting edge at orthogonal cutting was $\kappa_r = 0°$ and at oblique cutting it was $\kappa_r = 60°$. At the same time, the angle of the tool of cutting edge inclination λ_s was also changed. Thus, it is very important to evaluate the shear angle correctly. Figure 8 shows a chip root sample, and shows how the change of the angle of main cutting edge inclination λ_s would affect the individual cross-sections of the sample.

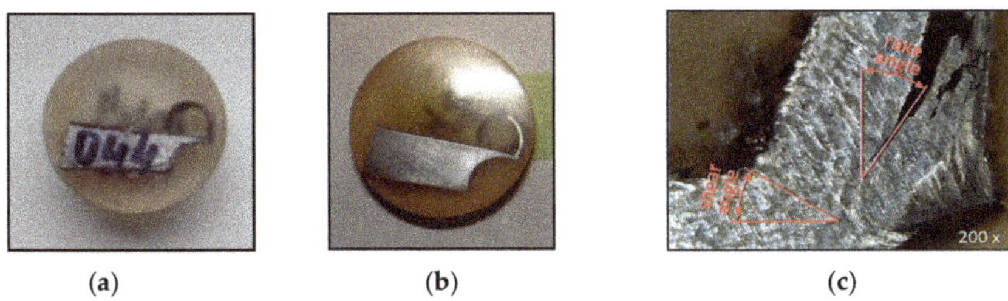

(a)   (b)   (c)

Figure 7. An example of investigated sample (**a**) before grinding, (**b**) after grinding and polishing, (**c**) shear and rake angle measuring.

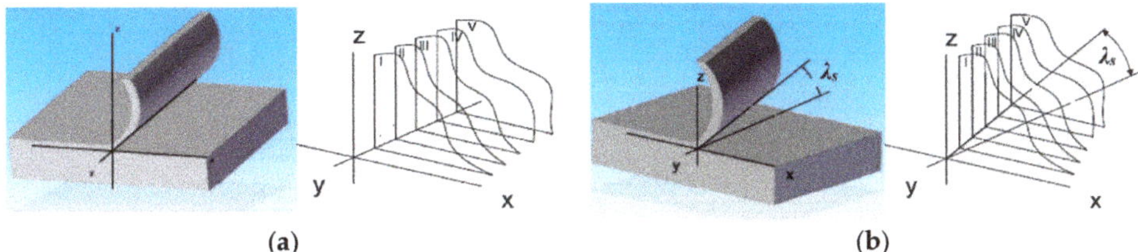

(a)   (b)

Figure 8. A sample of the chip root for orthogonal cutting with the presentation of individual facet planes, (**a**) $\lambda_s = 0°$; (**b**) $\lambda_s > 0°$.

It is not very difficult to solve the issue of orthogonal cutting. However, for oblique cutting, the angle of the tool main cutting edge κ_r and also the angle of the tool cutting edge inclination λ_s have to be taken into account. The figures below show the chip root sample and successive cross-sections of the sample at $\lambda_s = 0°$ (Figure 9a) and $\lambda_s > 0°$ (Figure 9b). This approach was also used for chip root evaluation within the research.

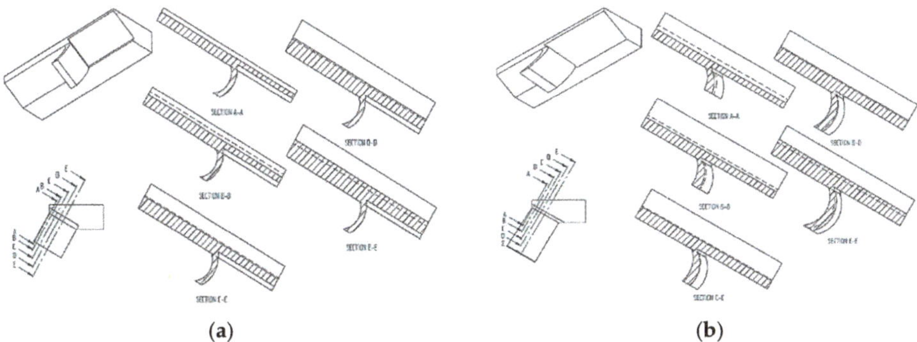

(a)   (b)

Figure 9. Cross-sections of a chip root at oblique cutting at the angle (**a**) $\lambda_s = 0°$, (**b**) $\lambda_s > 0°$.

4. Results and Discussions

4.1. Types of Chips

4.1.1. A Chip Root of EN C45 Steel

For the orthogonal cutting of EN C45 steel, the sample at cutting speed $v_c = 15$ m·min^{-1}, depth of cut $a_p = 0.2$ mm, rake angle $\gamma = 8°$, and angle of main cutting edge inclination $\lambda_s = 0°$, was selected as a representative sample. Since it was orthogonal cutting, the tool cutting edge angle was $\kappa_r = 0°$. The views (magnified 200×) on an etched sample in all five cross-sections are in Figure 10.

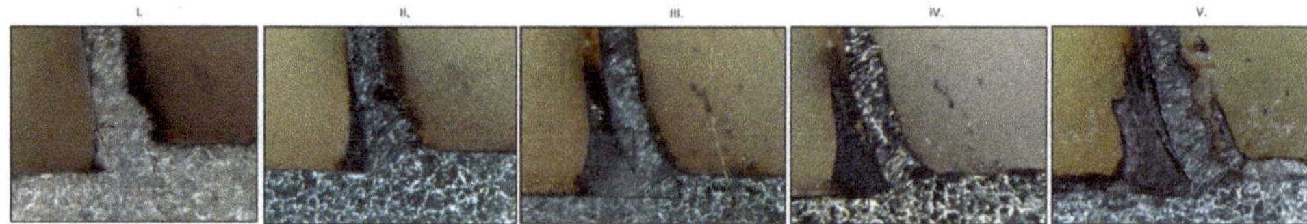

Figure 10. Successive chip roots (magnified 200×) of carbon steel EN C45 at orthogonal cutting with the parameters $v_c = 15$ m·min^{-1}, $a_p = 0.2$ mm, $\gamma = 8°$, and $\lambda_s = 0°$.

In the first cross-section, the measured shear angle Φ was 34°. A second cut was made on the sample after the removal of an approximately 0.5 mm layer. The structure was more clearly visible due to better etching of the sample and the shear measured angle was $\Phi = 37°$. This angle is 3° greater than on the first cut. The evaluation of the clearly observable built-up edge shows that the tool rake angle of 8° was increased to 19°. The created built-up edge on the tool assumed the function of the cutting tool, and thus it can be claimed that the rake angle increased significantly. At the third cut, it can be stated that the shear angle increased by 2° and at the same time the rake angle increased by the built-up edge of 20°, so the rake angle was 28°. On the fourth cut of the chip root sample, the shear angle was retained but an increase in the rake angle to 33° was observed through the built-up edge. On the built-up edge in the third cut, there was glued material, and the length of contact of the built-up edge and the outgoing chip was longer than that observed on the fourth cut. Based on this, it can be stated that the length of the contact of the tool face with the outgoing chip does not have to be the same over the entire width of the cut, and thus the friction coefficient will also be different. On the fifth cut of the chip root, the shear angle increased by 3° and, at the same time, the rake angle increased. At this cross-section, it was shown that the length of contact of the outgoing chip with the tool face, in this case through the face of the formed built-up edge, was not constant over the entire width of the cut. For this reason, the friction coefficient, which affects the heat generated in the cutting area, also changed.

As a representative sample for the chip root observation at the oblique slow-rate cutting of EN C45 steel, a sample obtained from machining with the following parameters was selected: $v_c = 6$ m·min^{-1}, $a_p = 0.2$ mm, $\gamma = 3°$, $\lambda_s = 0°$. Individual shapes of the chip roots, magnified 200×, are presented in Figure 11.

Figure 11. Successive chip roots (magnified 200×) of carbon steel EN C45 at oblique cutting with the parameters $v_c = 6$ m·min^{-1}, $a_p = 0.2$ mm, $\gamma = 3°$, $\lambda_s = 0°$.

In the first cut, the measured shear angle Φ was 35°. In the second chip root cut, the shear angle was 4° less, so it was $\Phi = 31°$. In both figures of the first and second chip root cross-section, there was no clearly noticeable built-up edge. The third cut showed an even smaller shear angle of $\Phi = 26°$ and at the same time, it can be stated that the rake angle was also reduced compared to the second cut. A slight increase in shear angle to 28° in the fourth chip root cut was caused by a slight increase in the rake angle. Similarly, in the fifth and final cut, an increase in shear angle to 32° was also observed. As mentioned above, the size of the shear angle is closely related to the size of the rake angle. The built-up edge directly on this sample was not clearly observable, but it occurred at the cutting edge of the tool in all cases of cutting.

Based on the shear angles measured in all five cross-sections, the average value was computed. The average values of the shear angles obtained by the machining of EN C45 steel with variable machining conditions, that were characterized by the matrix for statistical processing, are organized in Table 6.

The data were statistically processed and according to the Grubbs' testing criterion [36], it could be stated that the measured values were not burdened by grave mistakes. The coefficients of regression functions were calculated using the software MATLAB 2016 and their significance was tested according to the Student's criterion.

Table 6. Experimentally obtained values of temperature in cutting zone for orthogonal and oblique machining of EN C45 steel.

v_c (m·min^{-1})	a_p (mm)	γ (°)	λ_s (°)	Φ (°)	
				Orthogonal Cutting	Oblique Cutting
-1	-1	-1	-1	39.8	30.4
$+1$	-1	-1	-1	38.8	32.4
-1	$+1$	-1	-1	42.0	33.6
$+1$	$+1$	-1	-1	40.6	37.2
-1	-1	$+1$	-1	36.6	39.0
$+1$	-1	$+1$	-1	42.2	32.0
-1	$+1$	$+1$	-1	40.2	34.0
$+1$	$+1$	$+1$	-1	42.0	37.8
-1	-1	-1	$+1$	35.0	31.4
$+1$	-1	-1	$+1$	36.4	45.6
-1	$+1$	-1	$+1$	42.4	26.6
$+1$	$+1$	-1	$+1$	37.8	30.2
-1	-1	$+1$	$+1$	37.8	34.0
$+1$	-1	$+1$	$+1$	39.8	45.0
-1	$+1$	$+1$	$+1$	40.4	35.6
$+1$	$+1$	$+1$	$+1$	39.0	34.6
$-\alpha$	0	0	0	38.2	34.2
$+\alpha$	0	0	0	29.8	39.8
0	$-\alpha$	0	0	39.0	39.2
0	$+\alpha$	0	0	38.0	33.8
0	0	$-\alpha$	0	39.2	38.6
0	0	$+\alpha$	0	38.2	42.6
0	0	0	$-\alpha$	40.8	44.6
0	0	0	$+\alpha$	40.8	37.8
0	0	0	0	38.0	27.6

The regression functions (20) and (21) were built. According to the Fisher–Snedecor-tested criterion [38], they described the experiment adequately, while the reliability of the relationship was $R^2 = 0.92$ and 0.94, respectively:

(a) For orthogonal cutting:

$$y = -0.3947x_0 + 8.4952x_1 + 5.5938x_2 - 1.6685x_3 + 0.0176x_4 - 0.3475x_1x_2 \\ + 0.1452x_1x_3 - 0.0118x_1x_4 - 0.1465x_2x_3 - 4.5093x_1^2 + 4.6121x_2^2 + 0.9621x_3^2. \tag{20}$$

(b) For oblique cutting:

$$y = 1.5410x_0 - 0.6329x_1 - 3.5076x_2 - 1.7228x_3 - 0.2007x_4 - 0.2010x_1x_2 \\ - 0.2384x_1x_3 + 0.0361x_1x_4 + 0.1180x_2x_3 - 0.0771x_2x_4 + 0.0091x_3x_4 + 0.4437x_1^2 \\ - 3.1096x_2^2 + 1.3852x_3^2 - 0.0434x_4^2. \tag{21}$$

Based on measured data, the dependencies of the shear angle Φ on the following combinations pairs of parameters were evaluated for both orthogonal and oblique cuttings:

- cutting speed v_c and cutting depth a_p;
- cutting speed v_c and rake angle γ;
- cutting speed v_c and the angle of tool cutting edge inclination λ_s;
- cutting depth a_p and rake angle γ;
- cutting depth a_p and the angle of tool cutting edge inclination λ_s;
- rake angle γ and the angle of tool cutting edge inclination λ_s.

The dependences were plotted and they are presented in Figures 12–17.

For the orthogonal cutting of EN C45 carbon steel (Figure 12a), the cutting depth had a slightly more pronounced effect than the cutting speed, while for oblique cutting (Figure 12b), the cutting depth had a much more significant effect on the shear angle than the cutting speed. Maximum shear angle values were achieved at the maximum cutting depth for orthogonal cutting, and vice versa for oblique cutting at minimum cutting depth and higher cutting speeds.

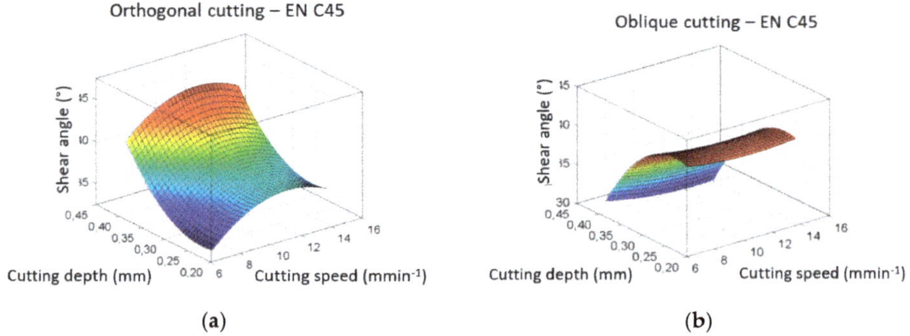

Figure 12. The dependence of shear angle Φ on cutting speed v_c and on cutting depth a_p; (**a**) orthogonal cutting; (**b**) oblique cutting.

The interaction of cutting speed and rake angle was totally opposite in both cutting methods. In orthogonal cutting (Figure 13a), the cutting speed had a more significant effect compared to the rake angle and the maximum shear angle values were reached at extreme rake angle and maximum cutting speed values. With respect to oblique cutting (Figure 13b), the effect of the rake angle was more pronounced than the effect of the cutting speed and the maximum shear angle values were reached at the minimum rake angle values.

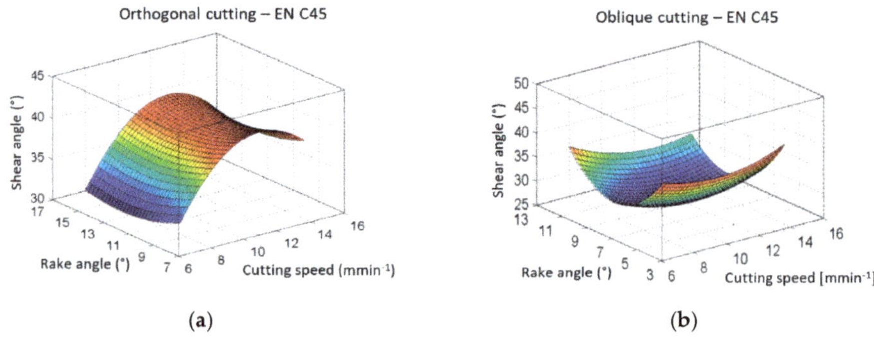

Figure 13. The dependence of shear angle Φ on cutting speed v_c and on rake angle γ; (**a**) orthogonal cutting; (**b**) oblique cutting.

The influence of the angle of tool cutting edge inclination on the change in shear angle compared to the cutting speed was not noticeable for orthogonal cutting (Figure 14a). For oblique cutting (Figure 14b), as was already mentioned above, the effect of the cutting speed was less significant, and therefore the effect of the angle of the tool cutting edge inclination on the shear angle change is also observable in the graph. For orthogonal cutting, maximum shear angle values were achieved at higher cutting speeds (from the used speed range), and within oblique cutting, they were achieved at minimum cutting speeds and minimum angles of the tool cutting edge inclination.

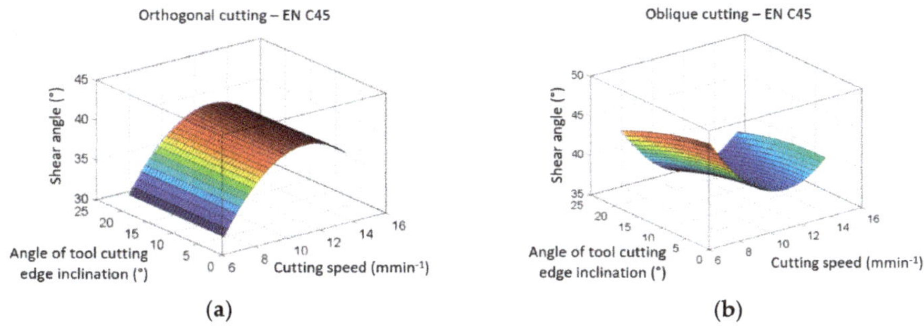

Figure 14. The dependence of shear angle Φ on cutting speed v_c and on the angle of tool cutting edge inclination λ_s; (**a**) orthogonal cutting; (**b**) oblique cutting.

The depth of the cut layer had a significant influence on the shear angle for both types of machining. For orthogonal cutting (Figure 15a), the rake angle affected the change in shear angle slightly more than for oblique cutting (Figure 15b). For orthogonal cutting, the maximum values of the shear angle were achieved at the maximum values of the cutting depth, and for oblique cutting, maximum values of shear angle were observed at the maximum rake angle.

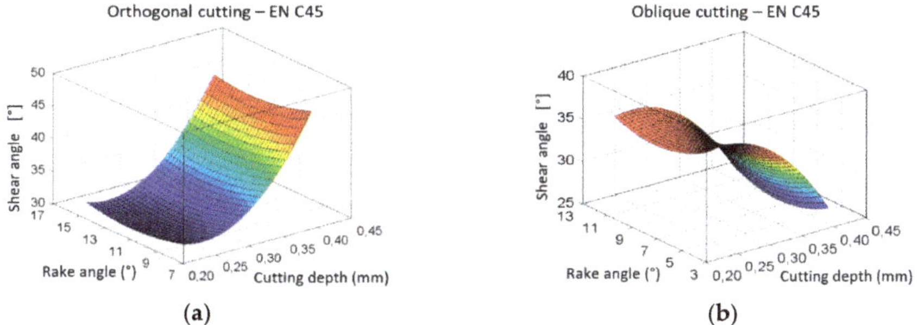

Figure 15. The dependence of shear angle Φ on cutting depth a_p and on rake angle γ; (**a**) orthogonal cutting; (**b**) oblique cutting.

A common feature of both cutting methods, within the interactions of the parameters presented in Figure 16, was an invisible effect of the angle of the tool cutting edge inclination on the shear angle. As is clear from these graphs, the maximum shear angle values were achieved for orthogonal cutting at the maximum cutting depth and within oblique cutting at the minimum depth of cut.

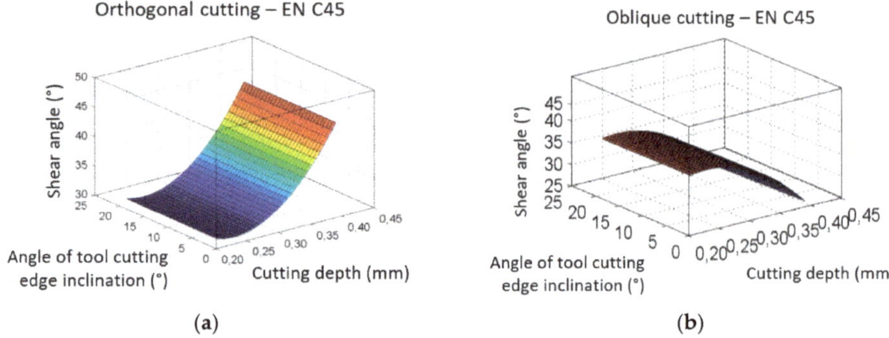

Figure 16. The dependence of shear angle Φ on cutting depth a_p and on the angle of tool cutting edge inclination λ_s; (**a**) orthogonal cutting; (**b**) oblique cutting.

The effect of the rake angle and the angle of the tool cutting edge inclination on the shear angle had the same character in both cutting methods (Figure 17). Maximum shear angle values were achieved at minimum rake angle values, and the lowest values of the shear angle Φ were achieved at mean rake angle values.

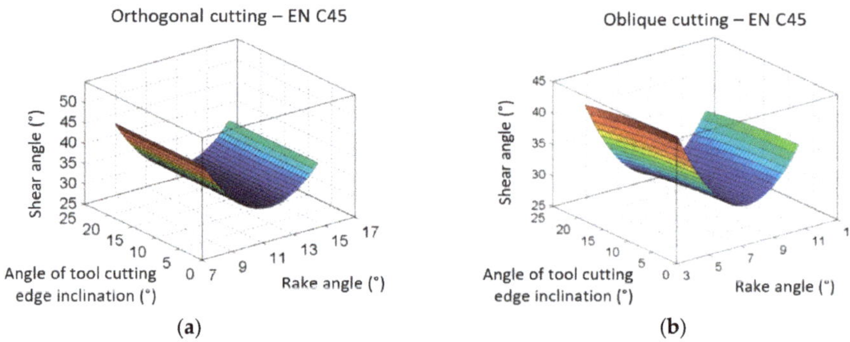

Figure 17. The dependence of shear angle Φ on rake angle γ and on the angle of tool cutting edge inclination λ_s; (**a**) orthogonal cutting; (**b**) oblique cutting.

4.1.2. A Chip Root of EN 16MnCr5 Steel Observation

For the orthogonal cutting of EN 16MnCr5 steel, a chip root obtained in machining with the following cutting parameters was selected as a representative sample for observation: cutting speed $v_c = 15$ m·min^{-1}, cutting depth $a_p = 0.4$ mm, rake angle $\gamma = 8°$, and angle of the main cutting edge inclination $\lambda_s = 0°$. All five cross-sections of the chip root with magnification 200× are shown in Figure 18.

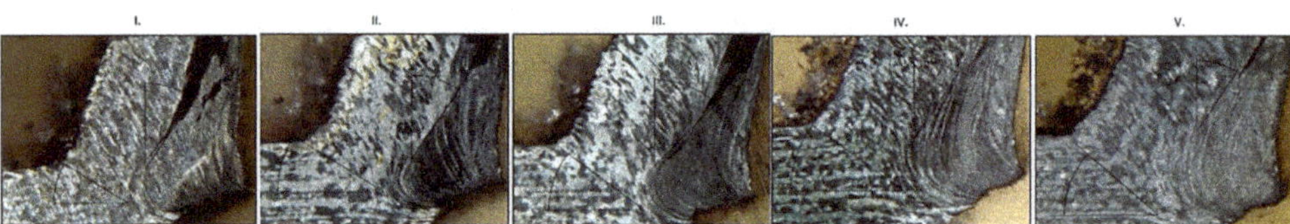

Figure 18. Successive chip roots (magnified 200×) of carbon steel EN 16MnCr5 steel for orthogonal cutting with the parameters $v_c = 15$ m·min^{-1}, $a_p = 0.4$ mm, $\gamma = 8°$, $\lambda_s = 0°$.

On the first cut of the sample, the shear angle was $\Phi = 31°$ and the rake angle of the formed built-up edge was 34°. On the second cut of the chip root, the shear angle increased by only one negligible degree and the rake angle of the formed built-up-edge remained unchanged at 34°. However, in this cross-section, the length of a contact surface between the built-up edge rake and the outgoing chip was several times greater than in the first section, resulting in a different coefficient of friction for both cases mentioned. In the third cut, a greater shear angle of $\Phi = 37°$ was measured, as well as a greater rake angle, until 44°. Again, the length of the contact between the outgoing chip and the built-up edge was different, and the stalled layers of hardened material were visible on the built-up edge. In some cases, the Vickers microhardness HV of the built-up edge was also measured at a load of 200 g in a duration of 10 s. While in the area of primary deformation the hardness was about 200 HV, in the built-up edge it was about 600 HV. The shear angle in the fourth section was 36°. The rake angle (of the built-up edge) was 33°, and the chip was adhered to the face of the tool built-up-edge, while the contact length between the face of the built-up edge and outgoing chip was greater compared to the third cut. In the fifth cut of the chip root of EN 16MnCr5 steel, the shear angle was the largest and the built-up edge appeared almost the same as in the fourth cut. The built-up edge on the cutting edge of the tool over the entire width of the cut significantly increased the rake angle, and thus influenced the shear angle in the individual cross-sections. At the same time, the built-up edge considerably increased the radius of the cutting edge round, indicating that the tool geometry was different during cutting than what was chosen to perform the experiment. An important element is the length of the contact surface between the face of the built-up edge and outgoing chip. It is shown in Figure 18 that the length of this contact surface was not constant over the entire width of the cut. Since the built-up edge was not stable, it could "tear-off" from the tool face and so change the shape of the tool during cutting, due to the same reason the rake angle cannot be considered constant even in the individual parts of the cutting width.

The sample obtained at oblique machining of EN 16MnCr5 steel with cutting parameters $v_c = 15$ m·min^{-1}, the cutting depth $a_p = 0.4$ mm, the rake angle $\gamma = 3°$ and the angle of the main cutting edge inclination $\lambda_s = 0°$, was selected as the representative sample for a description. The chip-roots in cross-sections of this sample magnified 200× are presented in Figure 19.

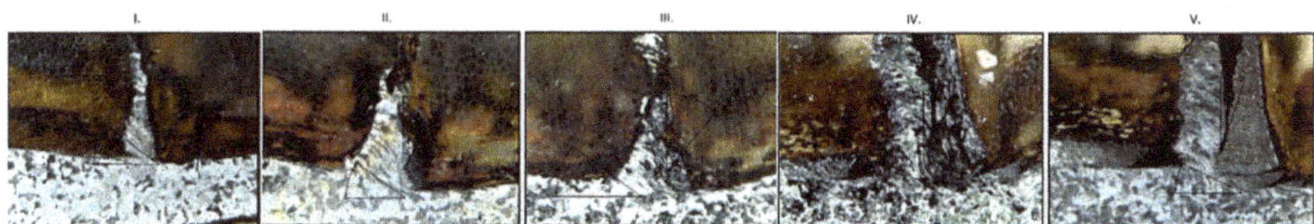

Figure 19. Successive chip roots (magnified 200×) of carbon steel EN 16MnCr5 steel for orthogonal cutting, with the parameters $v_c = 15$ m·min^{-1}, $a_p = 0.4$ mm, $\gamma = 3°$, $\lambda_s = 0°$.

On the first cut of the sample chip root, the built-up edge was not noticeable, but according to the shape of the outgoing chip, the rake angle appeared negative, while the rake angle of the straight roughing tool was positive, at $\gamma = 3°$. The cutting angle Φ was 36°. On the second cut, the built-up edge was already visible and the shear angle was reduced by 1° compared to its size on the first cut to 35°, which is a negligible change. On the third cut of the chip root, the rake angle increased and, at the same time, the shear angle slightly increased, to the value of 36°. The chip root in the fourth cut pointed to how the radius of the cutting edge round increased and, at the same time, the rake angle reduced. For this reason, the shear angle was also reduced to 29°. In the fifth cut, both the rake angle and the shear angle ($\Phi = 26°$) were smaller. In this case, it can be seen how the machined surface was damaged by the build-up "tearing off", which left from the cutting area.

The matrix of the experimental compositional plan for the evaluation of shear angle for the EN 16MnCr5 steel machining was designed in the same way as it was for EN C45 steel, for both cutting methods. The experimentally obtained values of the shear angle Φ, as well as the encoded matrix of individual cutting condition parameters, are shown in Table 7.

Table 7. The measured shear angle values Φ for both orthogonal and oblique cutting of EN 16MnCr5 steel.

v_c (m·min^{-1})	a_p (mm)	γ (°)	λ_s (°)	Φ (°)	
				Orthogonal Cutting	**Oblique Cutting**
−1	−1	−1	−1	43.0	35.4
+1	−1	−1	−1	39.2	42.8
−1	+1	−1	−1	35.6	24.2
+1	+1	−1	−1	35.4	31.4
−1	−1	+1	−1	37.2	30.8
+1	−1	+1	−1	38.2	33.8
−1	+1	+1	−1	42.2	31.8
+1	+1	+1	−1	39.0	41.2
−1	−1	−1	+1	37.4	41.0
+1	−1	−1	+1	37.8	34.4
−1	+1	−1	+1	40.4	41.2
+1	+1	−1	+1	39.6	34.6
-1	−1	+1	+1	35.8	37.6
+1	−1	+1	+1	41.6	35.6
−1	+1	+1	+1	38.2	38
+1	+1	+1	+1	35.2	34.4
$-\alpha$	0	0	0	40.2	35.8
$+\alpha$	0	0	0	33.8	40.4
0	$-\alpha$	0	0	34.8	39.2
0	$+\alpha$	0	0	38.4	37.2
0	0	$-\alpha$	0	37.4	35
0	0	$+\alpha$	0	36.4	41.6
0	0	0	$-\alpha$	41.8	36.0
0	0	0	$+\alpha$	32.4	39.8
0	0	0	0	33.2	42.4

Using the software MATLAB, the following regression functions, (22) and (23), were defined for both types of machining, while their reliabilities R^2 were 0.91 and 0.93, respectively:

(a) for orthogonal cutting:

$$y = 2.7826x_0 + 1.1743x_1 + 4.7878x_2 - 0.7749x_3 - 0.1338x_4 - 0.2652x_1x_2$$
$$+0.0469x_1x_3 + 0.0088x_1x_4 + 0.1330x_2x_3 + 0.0130x_2x_4 - 0.0085x_3x_4 - 0.6994x_1^2 \quad (22)$$
$$+4.2200x_2^2 + 0.5177x_3^2 - 0.0510x_4^2;$$

(b) for oblique machining:

$$y = 0.2562x_0 + 0.0132x_1 - 5.6702x_2 - 0.7684x_3 + 0.1414x_4 + 0.2051x_1x_2$$
$$+0.0752x_1x_3 - 0.0652x_1x_4 + 0.5759x_2x_3 + 0.0315x_2x_4 - 4.5505x_2^2 + 0.6638x_3^2 \quad (23)$$
$$+0.0169x_4^2.$$

For both types of machining, the simultaneous interaction of two of the four parameters (cutting depth a_p, cutting speed v_c, rake angle γ, and angle of tool cutting edge inclination λ_s) and their effect on the shear angle Φ was evaluated. The obtained dependencies are presented in Figures 20–25.

For the planing of manganese chromium steel EN 16MnCr5, the cutting depth had a more pronounced effect than the cutting speed on the shear angle change within both cutting methods.

In orthogonal cutting (Figure 20a) the maximum shear angle was reached at the maximum depth of cut and in oblique cutting (Figure 20b) at the minimum depth of cut.

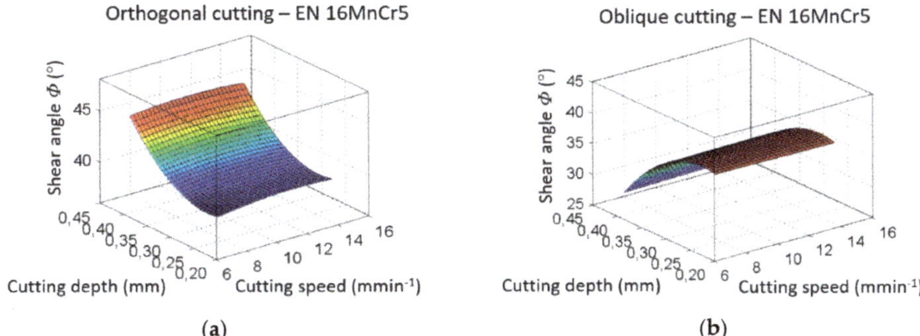

Figure 20. Dependence of shear angle Φ on cutting speed v_c and cutting depth a_p; (**a**) orthogonal cutting; (**b**) oblique cutting.

The effect of cutting speed compared to the rake angle was more pronounced in orthogonal cutting (Figure 21a), where maximum shear angle values were achieved at a higher cutting speed and at the boundary rake angle values. In oblique cutting (Figure 21b), maximum values were reached at minimum rake angle values.

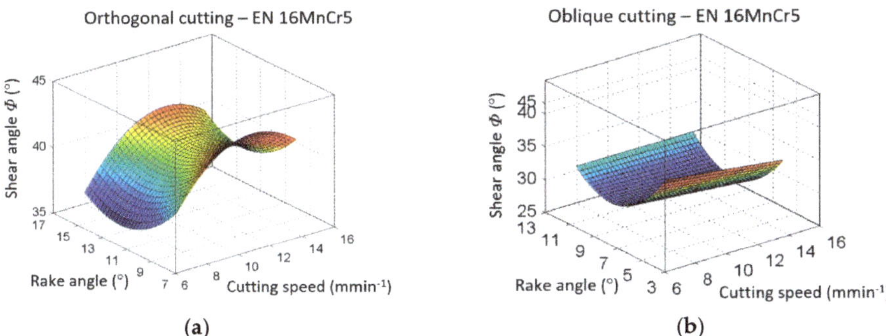

Figure 21. Dependence of shear angle Φ on cutting speed v_c and rake γ; (**a**) orthogonal cutting; (**b**) oblique cutting.

The graphs in Figure 22 show a demonstrable influence of the angle of the tool cutting edge inclination on the change of shear angle compared to the cutting speed. In orthogonal cutting (Figure 22a), the cutting speed had a greater impact, but in oblique cutting (Figure 22b), the angle of tool cutting edge inclination had a more pronounced effect. In orthogonal cutting, the maximum shear angle values were achieved at a minimum angle of tool cutting edge inclination and at higher cutting speed values. In oblique cutting, maximum shear angle values were visible at a minimum cutting speed and maximum angle of tool cutting edge inclination.

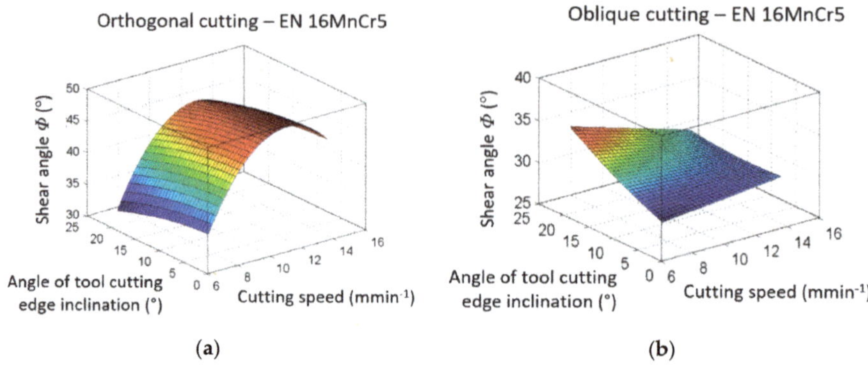

Figure 22. Dependence of shear angle Φ on cutting speed v_c and angle of tool cutting edge inclination λ_s; (**a**) orthogonal cutting; (**b**) oblique cutting.

Compared to cutting depth, the rake angle had only a slight effect on changes in the shear angle within both types of machining (Figure 23). The influence of cutting depth on the shear angle values have already been mentioned above. The angle of the tool cutting edge inclination (Figure 24) had a very similar slight effect to the rake angle in Figure 23, so the dependencies were almost of the same character.

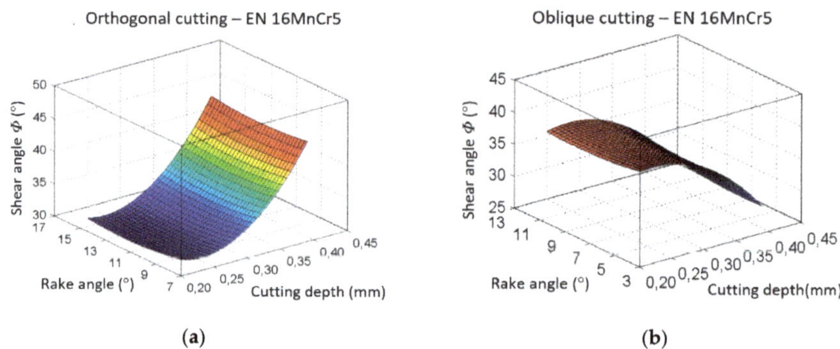

Figure 23. The dependence of shear angle Φ on cutting depth a_p and rake angle γ; (**a**) orthogonal cutting; (**b**) oblique cutting.

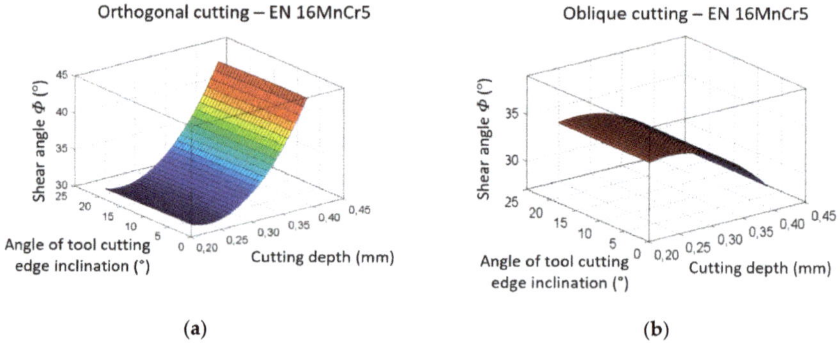

Figure 24. The dependence of shear angle Φ on cutting depth a_p and angle of tool cutting edge inclination λ_s; (**a**) orthogonal cutting; (**b**) oblique cutting.

The interaction of the shear angle with both the cutting edge inclination and rake angle is presented in Figure 25. It can be observed from the graphs that the dependencies had a similar character for both types of machining, while the angle of tool cutting edge inclination had an only slight effect on the shear angle compared to the rake angle. The minimal values were achieved at the mean values of the rake angle.

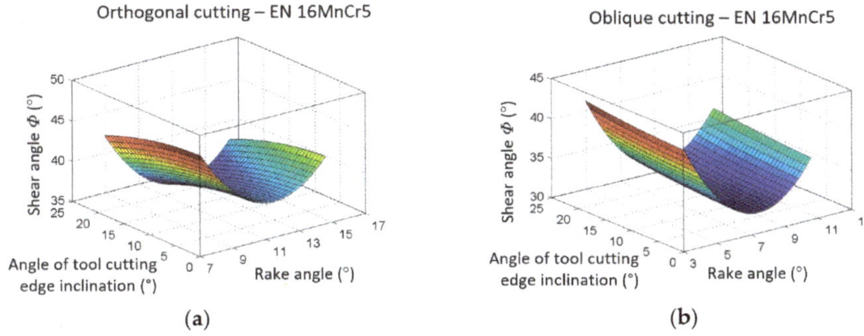

Figure 25. The dependence of shear angle Φ on rake angle γ and angle of tool cutting edge inclination λ_s; (**a**) orthogonal cutting; (**b**) oblique cutting.

4.1.3. Comparison of the Chips of EN C45 and EN 16MnCr5 Steels

A comparison of some of the chip roots made from both types of investigated steel, i.e., EN C45 and EN 16MnCr5, is presented in Table 8. All chips were formed by machining procedures, with the same cutting parameters obtained for orthogonal and oblique cutting. It can be seen that the behavior of the chip formation strongly depended not only on the materials but also on the cutting conditions.

Table 8. Comparison of chip roots of EN C45 and EN 16MnCr5 steels with the same cutting parameters.

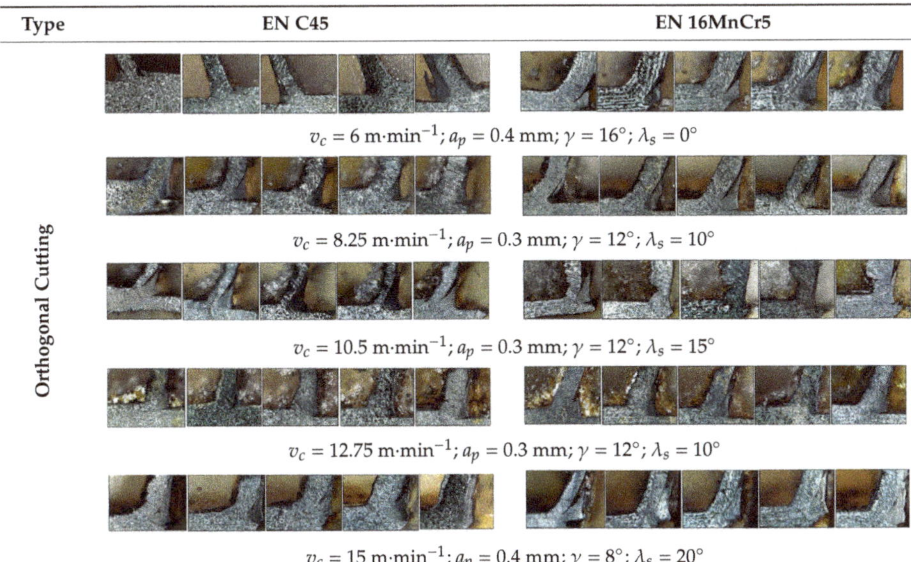

Type	EN C45	EN 16MnCr5
Orthogonal Cutting	$v_c = 6$ m·min^{-1}; $a_p = 0.4$ mm; $\gamma = 16°$; $\lambda_s = 0°$	
	$v_c = 8.25$ m·min^{-1}; $a_p = 0.3$ mm; $\gamma = 12°$; $\lambda_s = 10°$	
	$v_c = 10.5$ m·min^{-1}; $a_p = 0.3$ mm; $\gamma = 12°$; $\lambda_s = 15°$	
	$v_c = 12.75$ m·min^{-1}; $a_p = 0.3$ mm; $\gamma = 12°$; $\lambda_s = 10°$	
	$v_c = 15$ m·min^{-1}; $a_p = 0.4$ mm; $\gamma = 8°$; $\lambda_s = 20°$	

Table 8. *Cont.*

Type	EN C45	EN 16MnCr5
Oblique Cutting	$v_c = 6$ m·min^{-11}; $a_p = 0.4$ mm; $\gamma = 11°$; $\lambda_s = 0°$	
	$v_c = 8.25$ m·min^{-1}; $a_p = 0.3$ mm; $\gamma = 7°$; $\lambda_s = 10°$	
	$v_c = 10.5$ m·min^{-1}; $a_p = 0.3$ mm; $\gamma = 7°$; $\lambda_s = 15°$	
	$v_c = 12.75$ m·min^{-1}; $a_p = 0.3$ mm; $\gamma = 7°$; $\lambda_s = 10°$	
	$v_c = 15$ m·min^{-1}; $a_p = 0.4$ mm; $\gamma = 3°$; $\lambda_s = 0°$	

Except for the input value of the tool orthogonal rake, the formed built-up edge influenced also the shear angle and the shrinkage factor K, along with the segment ratio r_c. These can be calculated with Equations (24) and (25), where h_c is the chip width, a_p is cutting depth, γ is the tool rake angle, and Φ is the shear angle [4,45,46]:

$$K = \frac{h_c}{a_p} = \frac{\cos(\phi - \gamma)}{\sin \phi}; \tag{24}$$

$$r_c = \frac{1}{K}. \tag{25}$$

The values of the shrinkage factor K achieved within the experimental study are organized in Table 9.

Table 9. Shrinkage factor K achieved within the experimental study.

Type of Steel	Type of Cutting	Shear Angle Φ (°)	Shrinkage Factor K
EN C45	Orthogonal cutting	29.8	1.92
		42.4	1.22
	Oblique cutting	26.6	2.05
		39.8	1.31
EN 16MnCr5	Orthogonal cutting	31.7	1.75
		42.2	1.3
	Oblique cutting	30.2	1.77
		42.4	1.21

All these real cutting parameters that entered into the machining process influenced the chip forming during machining. The summary overview of the effects of these parameters on the chip shapes is graphically presented in Figures 26 and 27.

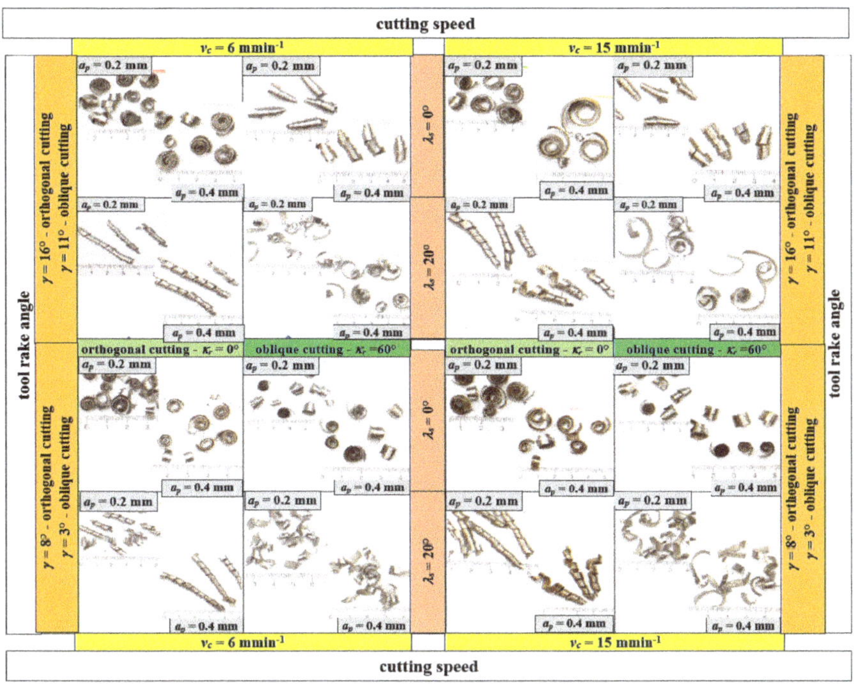

Figure 26. The shapes of chips of EN C45 steel and their dependence on cutting parameters.

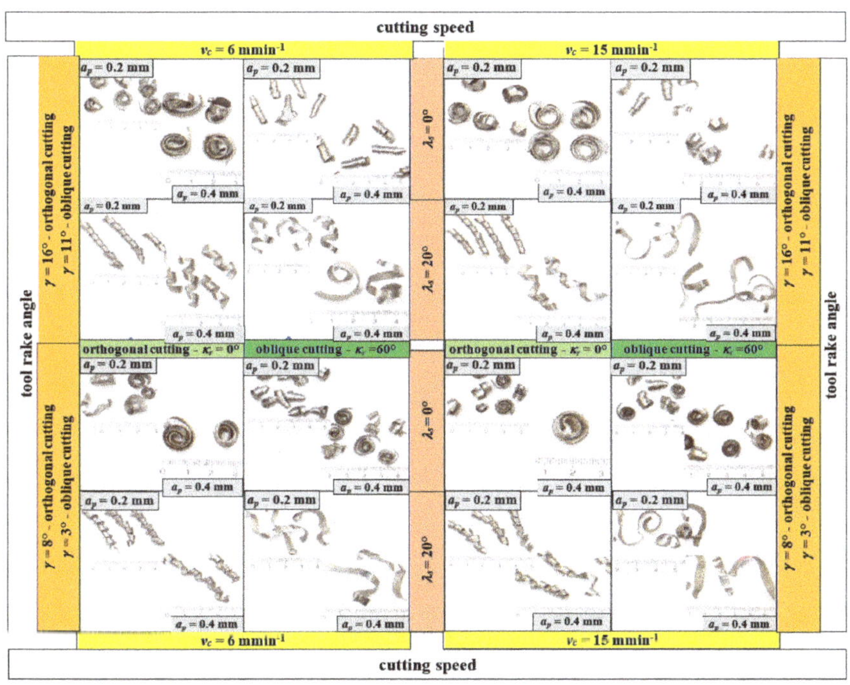

Figure 27. The shapes of chips of EN 16MnCr5 steel and their dependence on cutting parameters.

In the slow-rate machining of both steels, a shorter chip was created at a zero angle of inclination of main cutting edge when compared with the angle of inclination of the main cutting edge of 20°. At the same time, at a 20° angle of inclination of the cutting edge, a chamfer was formed along the edges of the chip, which corresponded to the angle of inclination of the main cutting edge and, thus, the chip was compressed in the direction of λ_s inclination and acquired the character of a spiral chip. In orthogonal cutting, at which $\kappa_r = 0°$ and $\lambda_s = 0°$, a spiral flat chip was formed, which in some cases passed into a spiral conical chip. For $\kappa_r = 0°$ and $\lambda_s = 20°$, a conical–helical long chip was constituted. In oblique cutting, where $\kappa_r = 60°$ and $\lambda_s = 0°$, a conical–helical short chip was generated. At $\kappa_r = 60°$ and $\lambda_s = 20°$, the chip shape changed to a coiled chip, while at rake angle $\gamma_o = 11°$, it was a tubular coiled chip and at $\gamma_o = 3°$ it was a strip coiled chip.

5. Conclusions

Within the presented study, the shear angle for orthogonal and oblique slow rate machining was evaluated, depending on selected cutting parameters—cutting speed v_c; cutting depth a_p; rake angle γ; angle of tool cutting edge inclination λ_s—and the chip roots were observed, focusing on the formed built-up edge.

To evaluate the interaction of the parameters correctly, the compositional planned experiment was designed and a new reliable method for the chip root obtaining was developed. Experimentally obtained data were statistically processed, while the regression functions for both types of machining and for both steel types were defined and the dependencies of the selected parameters on the shear angle were plotted.

Based on the shear angle measurements, the individual dependencies were determined and the significance of the influences of the individual factors was evaluated. In both cutting methods, the cutting depth proved itself to be the parameter with the most significant influence on the shear angle change. For orthogonal cutting, the cutting speed was the second most significant parameter, and the rake angle was the third while for oblique cutting, the rake angle was the second and the tool cutting edge inclination angle was placed third. Maximum shear angle values were measured on the chip samples obtained in the orthogonal cutting of both steel types when the cutting speed and cutting depth were maximal and the rake angle was minimal. In the oblique cutting of both materials, this was at minimum values of the cutting speed, the cutting depth, and the rake angle.

In the preliminary research of the authors [13], where the changes in temperature and microhardness of a chip root were evaluated, the angle of tool cutting edge inclination appeared to be an insignificant parameter. However, regarding changes in the chip shape, the angles of tool cutting edge inclination λ_s and tool cutting edge κ_r were the most important parameters.

A major problem with respect to the accuracy and reproducibility of the experiment results was the build-up on the tool rake. This built-up edge increased the tool rake angle by 30° and, due to this, it adheres to the face of tool and it assumes the function of the tool. When measuring the rake angle on successive cuts of the chip root samples, the built-up edge was observed on every sample. Along with the shear angle, the size of the newly originated rake angle was observed. As the rake angle increased, the shear angle also increased, causing a more favorable shear direction in the plastic zone, and thus the chip formation process became simpler [47]. It is disturbing that the build-up edge was unstable and, as has been proven by the chip root cuts, the shape of the build-up edge and its influence on the tool geometry was not constant over the entire width of the cut. Also, the length of the contact face (taken at the moment as the tool face) with the outgoing chip was not constant over the entire cutting width. Consequently, there was a clear change in the friction coefficient in the contact region of the face and the outgoing material, and thus there is a presumption that there is an indirect dependence between the formation of the built-up edge or its adherence to the outgoing material and between the temperature generated in the cutting area by changing the friction coefficient.

Author Contributions: Conceptualization, K.M. and A.S.; methodology, P.P.M. and A.S.; software, J.T. and J.K.; validation, M.B., J.K. and A.J.; investigation, K.M. and P.P.M.; resources, K.M. and J.T.; data curation, P.P.M. and M.B.; writing—original draft preparation, K.M.; writing—review and editing, K.M.; supervision, K.M.

Funding: This research was funded by the Ministry of Education of the Slovak Republic by grant KEGA 007TUKE-4/2018.

Acknowledgments: The article was prepared thanks to the direct support of the Ministry of Education of the Slovak Republic by grant KEGA 007TUKE-4/2018.

Conflicts of Interest: The authors declare no conflict of interest.

Nomenclature

γ_o	angle of tool orthogonal rake (°)
λ_s	angle of tool cutting edge inclination (°)
α_o	angle of tool orthogonal clearance (°)
κ_r	tool cutting edge angle (°)
κ_r'	tool minor (end) cutting edge angle (°)
ε_r	tool included angle (°)
Φ	shear angle (°)
a_p	depth of cut (mm)
v_c	cutting speed (m·min^{-1})
BUE	built-up edge

References

1. Jurko, J.; Panda, A.; Behun, M. Prediction of a new form of the cutting tool according to achieve the desired surface quality. *Appl. Mech. Mater.* **2013**, *268–270*, 473–476. [CrossRef]
2. Jerzy, J.; Kuric, I.; Grozav, S.; Ceclan, V. Diagnostics of CNC machine tool with R-Test system. *Acad. J. Manuf. Eng.* **2014**, *12*, 56–60.
3. Zetek, M.; Zetkova, I. Increasing of the Cutting Tool Efficiency from Tool Steel by Using Fluidization Method. *Procedia Eng.* **2015**, *100*, 912–917. [CrossRef]
4. Merchant, M.E. Mechanics of the metal cutting process. I. orthogonal cutting and a type 2 chip. *J. Appl. Phys.* **1945**, *16*, 267. [CrossRef]
5. Weiner, J.H. Shear plane temperature distribution in orthogonal machining. *Trans. ASME* **1955**, *77*, 1331–1341.
6. Wallace, P.W.; Boothroyd, G. Tool forces and tool-chip friction in orthogonal machining. *J. Mech. Eng. Sci.* **1964**, *6*, 74. [CrossRef]
7. Muraka, P.D.; Barrow, G.; Hinduja, S. Influence of the process variables on the temperature distribution in orthogonal machining using the finite element method. *Int. J. Mech. Sci.* **1979**, *21*, 445–456. [CrossRef]
8. Li, K.; Gao, X.-L.; Sutherland, J.W. Finite element simulation of the orthogonal metal cutting process for qualitative understanding of the effects of crater wear on the chip formation. *J. Mater. Process. Tech.* **2002**, *127*, 309–324. [CrossRef]
9. Shi, G.; Deng, X.; Shet, C. A finite element study of the effect of friction in orthogonal metal cutting. *Finite Elem. Anal. Des.* **2002**, *38*, 863–883. [CrossRef]
10. Filice, L.; Micari, F.; Rizzuti, S.; Umbrello, D. A critical analysis on the friction modeling in orthogonal machining. *Int. J. Mach. Tools Manuf.* **2007**, *47*, 709–714. [CrossRef]
11. Bagci, E. 3-D numerical analysis of orthogonal cutting process via mesh-free method. *Int. J. Phys. Sci.* **2011**, *6*, 1267–1282.
12. Priyadarshini, A.; Pal, S.K.; Samantaray, A.K. A Finite Element Study of Chip Formation Process in Orthogonal Machining. *Int. J. Manufact. Mater. Mech. Eng.* **2011**, *1*, 19–45. [CrossRef]
13. Monkova, K.; Monka, P.P.; Sekerakova, A.; Hruzik, L.; Burecek, A.; Urban, M. Comparative Study of Chip Formation in Orthogonal and Oblique Slow-Rate Machining of EN 16MnCr5 Steel. *Metals* **2019**, *9*, 698. [CrossRef]
14. Devotta, A.; Beno, T.; Löf, R.; Espes, E. Quantitative characterization of chip morphology using computed tomography in orthogonal turning process. *Procedia CIRP* **2015**, *33*, 299–304. [CrossRef]

15. Nakayama, K.; Arai, M. Comprehensive Chip Form Classification Based on the Cutting Mechanism. *CIRP Ann. Manuf. Technol.* **1992**, *41*, 71–74. [CrossRef]
16. Jawahir, I.S. A survey and future predictions for the use of chip breaking in unmanned systems. *Int. J. Adv. Manuf. Technol.* **1988**, *3*, 87–104. [CrossRef]
17. Kharkevich, A.; Venuvinod, P. Basic geometric analysis of 3-D chip forms in metal cutting. Part 1 determining up-curl and side-curl radii. *Int. J. Mach. Tools* **1999**, *39*, 751–769. [CrossRef]
18. Spaans, C. *The Fundamentals of Three-Dimensional Chip Curl, Chip Breaking and Chip Control*; TU Delft: Delft, The Netherlands, 1971.
19. Kluft, W.; Konig, W.; Van Luttervelt, C.A.; Nakayama, K.; Pekelharing, A.J. Present knowledge of chip control. *Ann. CIRP* **1979**, *28*, 441–455.
20. Jawahir, I.S.; Luttervelt, C.A. Recent Developments in Chip Control Research and Applications. *CIRP Ann. Manuf. Technol.* **1993**, *42*, 659–693. [CrossRef]
21. Lee, E.H.; Shaffer, B.W. The Theory of Plasticity Applied to Problem of Machining. *J. Appl. Mech.* **1951**, *13*, 405–413.
22. Toulfatzis, A.I.; Pantazopoulos, G.A.; David, C.N.; Sagris, D.S.; Paipetis, A.S. Final heat treatment as a possible solution for the improvement of machinability of pb-free brass alloys. *Metals* **2018**, *8*, 575. [CrossRef]
23. Shaw, M.C.; Cook, N.H.; Finnie, I. Shear Angle Relationship in Metal Cutting. *Trans. ASME* **1953**, *75*, 273–288.
24. Luk, W.K. The direction of chip flow in oblique cutting. *Int. J. Proc. Res.* **1972**, *10*, 67–76. [CrossRef]
25. Okushima, K.; Minato, K. On the behaviors of chips in steel cutting. *Bull. JSME.* **1959**, *2*, 58–64. [CrossRef]
26. Palmer, W.B.; Oxley, P.L.B. Mechanics of Metal Cutting. *Proc. Inst. Mech. Eng.* **1959**, *173*, 623–654. [CrossRef]
27. Oxley, P.L.B.; Welsh, M.J.M. Calculating the Shear Angle in Orthogonal Metal Cutting from Fundamental Stress, Strain Rate Properties of the Work Material. In *Proceedings of the 4th International Machine Tool Research Conference*; Pergamon Press: Oxford, UK, 1963; pp. 73–86.
28. Prasetyo, L.; Tauviqirrahman, M. Study of chip formation feedrates of various steels in low-speed milling process. *IOP Conf. Ser. Mater. Sci. Eng.* **2017**, *202*, 012097. [CrossRef]
29. Hegab, H.; Umer, U.; Soliman, M.; Kishawy, H.A. Effects of nano-cutting fluids on tool performance and chip morphology during machining Inconel 718. *Int. J. Adv. Manuf. Technol.* **2018**, *96*, 3449–3458. [CrossRef]
30. Adekunle, A.S.; Adedayo, S.M.; Ohijeagbon, I.O.; Olusegun, H.D. Chip morphology and behaviour of tool temperature during turning of AISI 301 using different biodegradable oils. *J. Prod. Eng.* **2015**, *18*, 18–22.
31. Anthony, X.M. Analysis of cutting force and chip morphology during hard turning of AISI D2 steel. *J. Eng. Sci. Technol.* **2015**, *10*, 282–290.
32. Umer, U.; Qudeiri, J.A.; Ashfaq, M.; AL-Ahmari, A. Chip morphology predictions while machining hardened tool steel using finite element and smoothed particles hydrodynamics methods. *J. Zhejiang Univ.-Sci. A (Appl. Phys. Eng.)* **2016**, *17*, 873–885. [CrossRef]
33. Mabrouki, T.; Courbon, C.; Zhang, Y.; Rech, J.; Nélias, D.; Asad, M.; Hamdi, H.; Belhadi, S.; Salvatore, F. Some insights on the modelling of chip formation and its morphology during metal cutting operations. *Comptes Rendus Mécanique* **2016**, *344*, 335–350. [CrossRef]
34. Vasilko, K. New experimental dependence of machining. *Manuf. Technol.* **2014**, *14*, 111–116.
35. Saglam, H.; Yaldiz, S.; Unsacar, F. The effect of tool geometry and cutting speed on main cutting force and tool tip temperature. *Elsevier Mater. Des.* **2007**, *28*, 101–111. [CrossRef]
36. Beno, J. *Theory of Metal Cutting*; Vienal Publisher: Kosice, Slovakia, 1999; p. 255. ISBN 80-7099-429-0.
37. Davim, J.P.; Silvia, J.; Baptista, A.M. Experimental cutting model of metal matrix composites (MMCs). *Elsevier J. Mater. Process. Technol.* **2007**, *183*, 358–362. [CrossRef]
38. Pastucha, P.; Majstorovic, V.; Kučera, M.; Beno, P.; Krile, S. Study of Cutting Tool Durability at a Short-Term Discontinuous Turning Test. *Lect. Notes Mech. Eng.* **2019**, 493–501. [CrossRef]
39. Vychytil, J.; Holecek, M. The simple model of cell prestress maintained by cell incompressibility. *Math. Comput. Simul.* **2010**, *80*, 1337–1344. [CrossRef]
40. Zhang, H.T.; Liu, P.D.; Hu, R.S. A three-zone model and solution of shear angle in orthogonal machining. *Wear* **1991**, *143*, 29–43. [CrossRef]
41. Obeng, D.P.; Morrell, S.; Napier-Munn, T.J. Application of central composite rotatable design to modeling the effect of some operating variables on the performance of the three-product cyclone. *Int. J. Miner. Process.* **2005**, *76*, 181–192. [CrossRef]

42. Kiran, K.R.; Manohar, B.; Divakar, S. A central composite rotatable design analysis of lipase catalyzed synthesis of lauryl lactic acid at bench-scale level. *Enzyme Microb. Technol.* **2001**, *29*, 122–128. [CrossRef]

43. Harrell, F.E. *Regression Modeling Strategies: With Applications to Linear Models, Logistic Regression, and Survival Analysis*; Springer: New York, NY, USA, 2001; p. 568.

44. Filippov, A.V.; Filippova, E.O. Determination of cutting forces in oblique cutting. *Appl. Mech. Mater.* **2015**, *756*, 659–664. [CrossRef]

45. Kouadri, S.; Necib, K.; Atlati, S.; Haddag, B.; Nouari, M. Quantification of the chip segmentation in metal machining: Application to machining the aeronautical aluminum alloy AA2024-T351 with cemented carbide tools WC-Co. *Int. J. Mach. Tools Manuf.* **2013**, *64*, 102–113. [CrossRef]

46. Atlati, S.; Haddag, B.; Nouari, M.; Zenasni, M. Segmentation intensity ratio as a new parameter to quantify the chip segmentation phenomenon in machining ductile metals. *Int. J. Mach. Tools Manuf.* **2011**, *51*, 687–700. [CrossRef]

47. Shamoto, E.; Altintas, Y. Prediction of Shear Angle in Oblique Cutting with Maximum Shear Stress and Minimum Energy Principles. *J. Manuf. Sci. Eng.* **1999**, *121*, 399–407. [CrossRef]

© 2019 by the authors. Licensee MDPI, Basel, Switzerland. This article is an open access article distributed under the terms and conditions of the Creative Commons Attribution (CC BY) license (http://creativecommons.org/licenses/by/4.0/).

Article

A Study on the Failure of AISI 304 Stainless Steel Tubes in a Gas Heater Unit

Abbas Bahrami [1] and Peyman Taheri [2,*]

[1] Department of Materials Engineering, Isfahan University of Technology, Isfahan 84156-83111, Iran
[2] Department of Materials Science and Engineering, Delft University of Technology, Mekelweg 2, 2628 CD Delft, The Netherlands
* Correspondence: p.taheri@tudelft.nl; Tel.: +31-15-278-2275

Received: 7 August 2019; Accepted: 31 August 2019; Published: 3 September 2019

Abstract: This paper investigates a failure in convection section tubes of a gas heater unit in a petrochemical plant. Tubes are made of AISI 304 stainless steel. The failure is reported after 5 years of service at working temperature 500 °C. The failure is in the form of circumferential cracks in the vicinity of the weld. Various characterization techniques, including optical and electron microscopes as well as energy-dispersive X-ray spectroscopy (EDS), were used to study the failure. Results showed that that the damage has initiated in the heat affected zone (HAZ) area parallel to the weld/base metal interface. Cracks have propagated alongside grain boundaries, resulting in an intergranular fracture. The main cause of failure was concluded to be attributable to the grain boundary sensitization and intergranular grain boundary attack due to improper welding and long time exposure of tubes to high temperature. Possible mitigation strategies to minimize similar failures will be discussed.

Keywords: convection tubes; AISI 304 stainless steel; failure analysis; sensitization

---

1. Introduction and Case Background

This paper investigates the root cause analysis of a failure in convection section tubes in a gas heater unit in a petrochemical plant. This unit is used to heat up the reformed gas from a reformer unit. The gas composition is up to 70% H_2, 20%CO, and the rest is a mixture of CH_4, CO_2, H_2O, and N_2. The gas enters into the heating unit at 250 °C and exits the unit at 570 °C. The flue gas at the fireside is from coal firing. The temperature of the flue gas at the inlet of the gas heating unit is 1050 °C. This temperature decreases to 450 °C at the outlet of the unit. Tubes at the first four rows are made of heat resistant HP40 alloy. Tubes at the 5th and 6th rows, where the failure is observed, are made of AISI 304 stainless steels. The temperature at 5th and 6th rows is approximately 500 °C. Austenitic stainless steel AISI 304 is known to have an excellent combination of high temperature mechanical properties, high temperature corrosion resistance, and superior structural stability [1–6], making it a popular choice for convection tubes for temperature range 500 to 700 °C. Tubes in similar working conditions are reported to be failed due to oxidation, erosion, creep, thermal fatigue and stress relaxation cracking [7–17]. Depending on the working condition, either of these mechanisms or a combination of them can control the damage. In this case, the installation is designed to continuously work. There is a shutdown in the system every 3 months for cleaning and maintenance. In one of the shutdowns after 5 years of service, a temperature abnormality was reported, which is normally taken as an indication for leakage. None-destructive testing (NDT) evaluation confirmed the leakage in the vicinity of weldments. This paper investigates root cause of observed failure in this failure. Results will be used to propose mitigation strategies. Mitigation measures are effective only when there is a good understanding of correlation between microstructure, mechanical properties, environmental parameters, and service conditions.

2. Experimental Methods

Samples were taken from damaged tubes. According to the specification, the alloy is AISI 304. Inductively-coupled plasma atomic energy spectroscopy (ICPAES) was used to make sure that the alloy has standard chemical composition. Optical and Scanning Electron Microscopes (SEM Philips XL30, Philips, Eindhoven, The Netherlands) together with energy-dispersive X-ray spectroscopy (EDS, Ametek, PA, USA) were used to study the microstructure of failed specimens. Macro images were taken by Stereo Microscope Nikon SMZ 800 (Nikon, Tokyo, Japan). For metallography, samples were prepared by grinding the surface with 180 to 1200 grinding papers, followed by polishing with diamond paste (3 and 1 micron). Microstructure of samples were studied in both as-polished and after etching. The former is useful when it comes to analyzing sensitized grain boundaries. To make sure that there is a minimum trace of polishing particles on the surface, samples were washed with acetone. Etching was performed with Nitric acid 60% (Voltage was 3 V and samples were kept for 45 s in etchant). The fracture surface was studied by means of a scanning electron microscope (SEM).

3. Results and Discussion

3.1. Visual Examinations

Figure 1 shows an example of a failed tube together with stereomicroscopic image of crack. As can be seen, the crack has propagated circumferentially parallel to the welding line. The crack has propagated in HAZ area in the vicinity of the weld. stereomicroscopic image shows that the crack has propagated alongside grain boundaries. There is no defect in the form of crack or cavities inside the weld, inferring that the weld itself has a perfect quality. In some areas there are some deposits on the surface. Other than these mentioned features, there is hardly any other abnormality in the form of bulging, thinning, or localized deformation in any form. The fact that there is no bulging or localized deformation infers that creep is not a major issue.

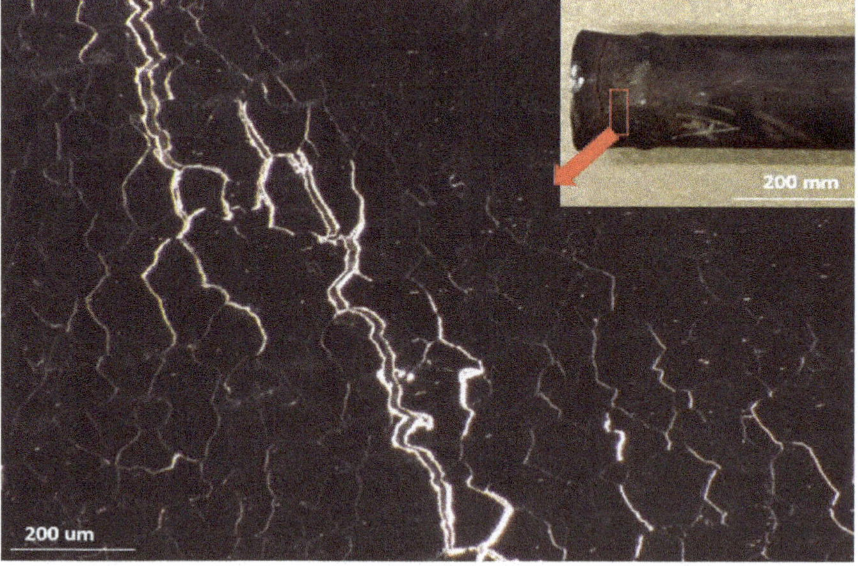

Figure 1. Macro and stereomicroscopic images of failed specimen.

3.2. Chemical Analysis

The chemical composition of the alloy was checked with ICPAES to make sure that the chemical composition of the alloy is within the standard range. Table 1 compares the chemical composition of the alloy with the standard values, showing that the alloy has a standard chemical composition.

Table 1. Chemical composition of the alloy, compared with standard values (in wt%).

AISI 304	C	Si	Mn	P	S	Cr	Ni	Fe
Measured	-	0.7	1.40	0.011	0.006	18.10	10.00	Bal.
Standard	-	0.75	2.00	0.045	0.030	18.00–20.00	8.00–10.50	Bal.

3.3. Analysis of Surface Deposits/Structure

Figure 2 shows grain structure and deposits on the surface of the tube near the damaged area. Two main features are noteworthy to mention. The first noticeable observation is the so-called "grain falling-out" phenomenon (see Figure 2a). It appears that grain boundaries at the surface are heavily oxidized, resulting in the weakening of grain boundary and detachment of a whole grain from the surface. This is also confirmed by the SEM image (Figure 2b). It is also noticeable that some areas on the surface are covered a black deposit. The EDS analysis of these deposits are given in Figure 2c, showing that elements like C, Cl, Mg, Na, and Ca are present in the composition of these deposits. Given that the outer surface of the tube is the fire-side, one can conclude that these deposits are typical fly ashes, coming from the fuel. In this case fly ash corrosion appears to have a significant attribution to the observed failure. Fly ash is a residue generated due to coal combustion and contains fine particles. In this case ash is brought to the system via flue gas. The composition of fly ash depends on the coal and the combustion condition. In most cases, fly ash contains ferric oxide, aluminum oxide, silicon dioxide, and calcium oxide [18]. Fly ash particles are normally collected by filter bags and/or electrostatic filter. Yet, there is a chance that very small residues are not captured by filters. These residues are then precipitated on the outer surface of the tube. Fly ash residues can accelerate surface corrosions. Elements like Ca and Na can diffuse into grain boundaries, resulting in the weakening of grain boundaries and ultimately grain detachment on the surface. Needless to mention that elements like Cl can also severely corrode stainless steels.

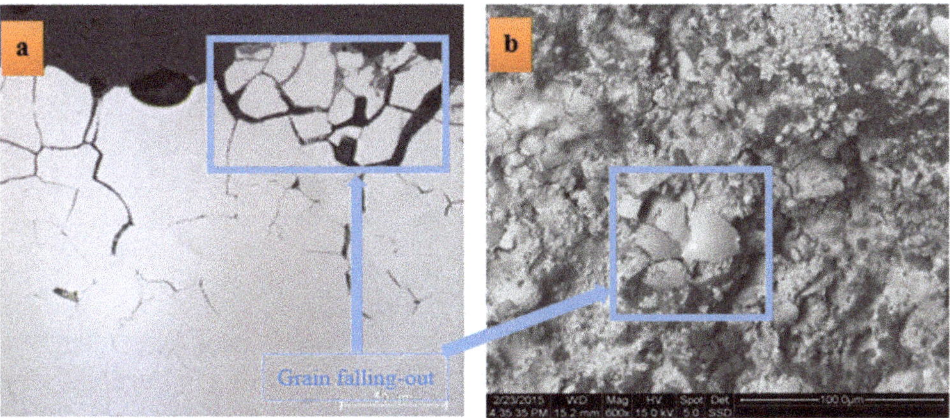

Figure 2. *Cont.*

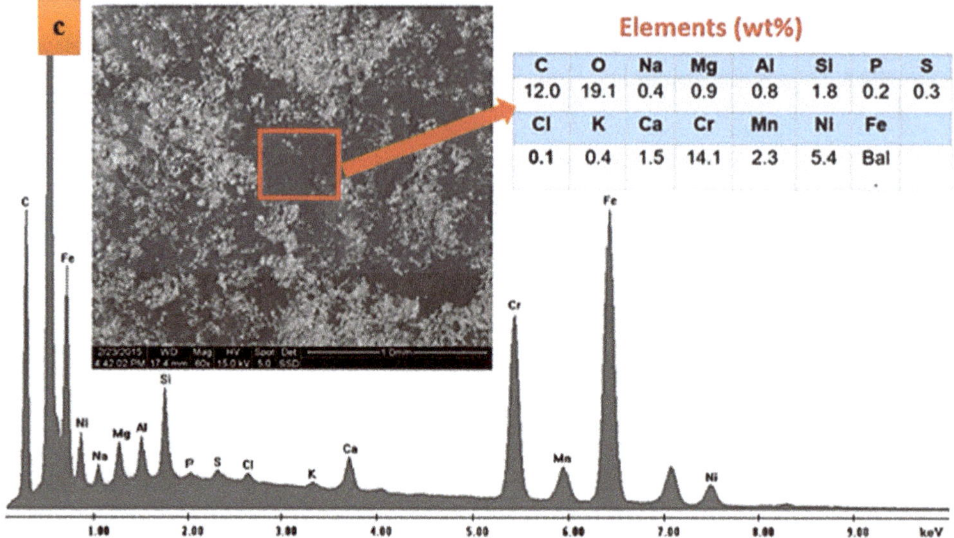

C	O	Na	Mg	Al	Si	P	S
12.0	19.1	0.4	0.9	0.8	1.8	0.2	0.3

Cl	K	Ca	Cr	Mn	Ni	Fe	
0.1	0.4	1.5	14.1	2.3	5.4	Bal	

Figure 2. (**a**) Optical microscope image of grain falling-out at the surface, (**b**) SEM image of grain falling-out at the surface, and (**c**) SEM/energy-dispersive X-ray spectroscopy (EDS) analyses of deposits on the surface.

3.4. Microstructure Analysis

Figure 3 shows optical microscope and SEM images of HAZ structure in the as-polished condition. Figure 3a,b show that grain boundaries are mostly covered with a rather thick black phase layer. Figure 3c,d depict that the thickness of the grain boundary phase layer can be as high as 10 μm. The grain boundary phase appears to be rather brittle. This is clearly seen in Figure 3c in which the grain boundary phase is broken. The EDS analyses of grain boundary, given in Figure 3c,d, show that the grain boundary phase is rich in oxygen, carbon, and chromium. It appears that grain boundaries are sensitized and are heavily oxidized. Grain boundary sensitization takes place when chromium carbide precipitates at grain boundaries. Chromium carbide precipitation at grain boundary creates a chromium-depleted zone in the vicinity of sensitized grain boundaries. Given that chromium is the major element when it comes to the corrosion resistance in stainless steels, this narrow chromium-depleted region adjacent to the grain boundaries makes grain boundaries preferential corrosion spots. A continuous network of carbides on grain boundaries and intergranular corrosion makes grain boundaries weak and brittle. Cracks can therefore more easily propagate alongside sensitized grain boundaries. Sensitization of grain boundaries of austenitic stainless steels takes place in temperature range 510 to 780 °C [19]. Any thermal exposure into this temperature range during welding or service could potentially result in the sensitization of grain boundaries in stainless steels. The temperature and time required to cause sensitization and intergranular corrosion depends on the chemistry of the alloy, and more specifically on the carbon content of the alloy. Obviously, the higher the carbon content of the alloy, the higher is the risk of sensitization. For this reason, low carbon stainless steel grades such as AISI 304L and AISI 316L alloys are known to have an excellent resistance against sensitization during welding. In higher alloyed stainless steels such as alloys AISI 904L, grain boundary susceptibility to sensitization is also essentially not a concern [19]. In this case, the observed sensitization has possibly its root in the welding practice. The service temperature is on the lower side of the sensitization temperature range and service exposure has possibly not the major contribution to the sensitization. If sensitization was due to the service exposure, one should have seen sensitization in the base metal as well, which obviously is not the case.

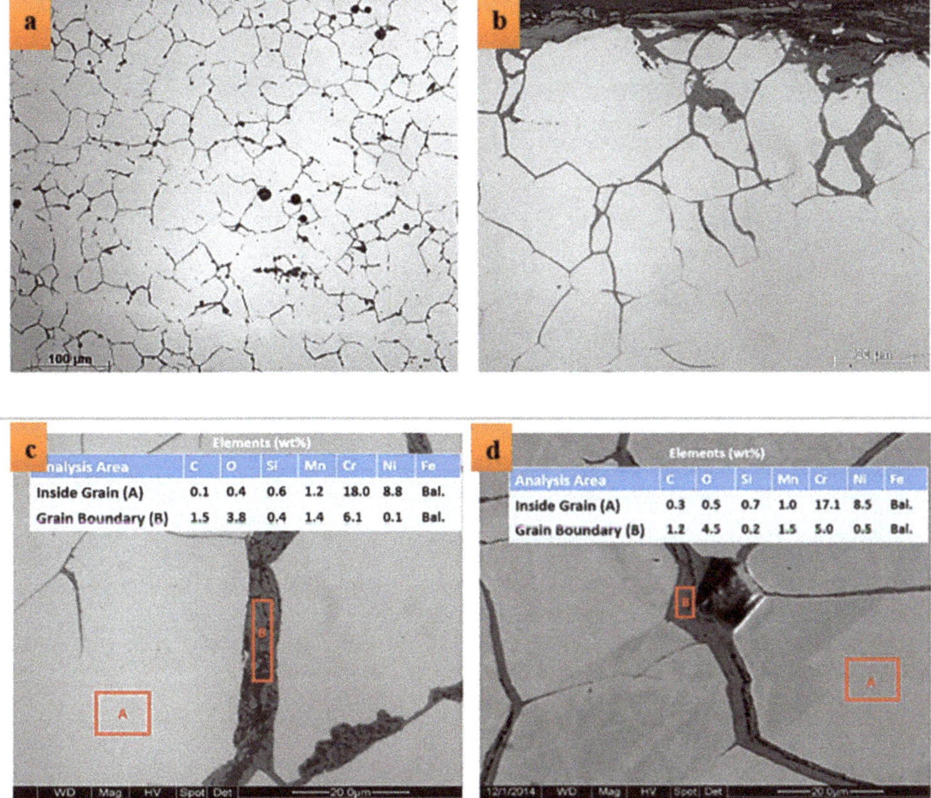

Figure 3. Optical microscope images of grain structure in the heat affected zone (HAZ) area of failed specimens; (**a**,**b**) an overall view, and (**c**,**d**) SEM/EDS analyses of grain boundary phase and inside grains.

3.5. Assessment of Grain Boundary Sensitization

ASTM A262 [20] is a standard intergranular corrosion assessment method that can be used to evaluate grain boundary sensitization in austenitic stainless steels. The ASTM A262 standard contains five different practices. When stainless steels are exposed to high temperature service for a long time, carbide precipitates on grain boundaries. This creates Cr-depleted areas in the vicinity of grain boundaries, making the alloy susceptible to intergranular corrosion. This standard is a popular assessment method for interganular corrosion susceptibility of stainless steels. This standard contains five different practices, which each being useful for different conditions and materials. Practice A is the simplest method, which is based on the etching of alloys in Oxalic acid. Obtained etch structures can be categorized into the following groups [20]:

Step Structure: steps are formed between grains, no ditches at grain boundaries;
Dual Structure: Some ditches at grain boundaries in addition to steps; and
Ditch Structure: One or more grains completely surrounded by ditches.

The formation of the latter etch structure is an indication of grain boundary sensitization. Figure 4 shows the etch structure of AISI 304 stainless steel in the base metal (Figure 4a) and in the HAZ (Figure 4b). A clear difference is observed between these structures. While the former is a typical step structure, the latter is a clear ditch structure, inferring that HAZ has become severely sensitized during welding and service exposure.

Figure 4. Optical microscope images of grain structure in (**a**) base metal and (**b**) HAZ area of failed specimen, etched according to standard ASTM A262, Practice A.

4. Concluding Remarks and Preventive Measures

This paper investigates a failure in convection section tubes of a gas heater unit in a petrochemical plant. Tubes are made of AISI 304 stainless steel. The failure is reported after 5 years of service at working temperature 500 °C. The failure is in the form of circumferential cracks in the vicinity of the weld. The following conclusions can be drawn, based on the obtained results:

- Damage has initiated at HAZ area in the vicinity of the weld. Cracks have propagated alongside grain boundaries.
- Grain boundaries in the HAZ area are covered with a rather thick grain boundary phase. Further from the HAZ area in the base metal, there is hardly any indication of such grain boundary phase.
- Results show that grain boundaries are rich in chromium and are sensitized. Sensitization of grain boundaries is associated with the formation of a narrow Cr-depleted region in the vicinity of grain boundaries. This makes grain boundaries susceptible against corrosion.
- The test method, proposed in ASTM A262, confirms the susceptibility of grain boundaries in the HAZ area to intergranular attack.
- Some deposits on the outer surface of tubes are observed. These deposits are rich in C, Cl, Mg, Na, Si, Al, and Ca, indicating that these residues on the surface are fly ashes.
- Carbon content has a very decisive influence on the susceptibility of grain boundaries. Superior resistance to grain boundary susceptibility can be achieved by reducing the carbon content values less than 0.03 wt%. It is highly recommended to use AISI 304L or other low-carbon stainless steel grades instead of AISI 304.
- Stabilizing elements like Nb and Ti can also enhance the resistance to sensitization. These elements form stable carbides at high temperature. This way less carbon is available for the formation of chromium carbides at grain boundaries. Commonly used stabilized austenitic stainless steel grades are AISI 321 and AISI 347. Replacement of currently used AISI 304 alloy with stabilized grades is also highly recommended.

Author Contributions: Conceptualization, A.B.; methodology, A.B.; formal analysis, A.B.; investigation, A.B. and P.T.; writing—original draft preparation, A.B.; writing—review and editing, P.T.; visualization, A.B.; supervision, P.T.; project administration, P.T.

Funding: This research received no external funding.

Acknowledgments: Authors would like to thank Delft University of Technology for support in publishing this manuscript.

Conflicts of Interest: The authors declare no conflict of interest.

References

1. Ghalambaz, M.; Abdollahi, M.; Eslami, A.; Bahrami, A. A case study on failure of AISI 347H stabilized stainless steel pipe in a petrochemical plant. *Case Stud. Eng. Fail. Anal.* **2017**, *9*, 52–62. [CrossRef]
2. Mousavi Anijdan, S.H.; Madaah-Hosseini, H.R.; Bahrami, A. Flow stress optimization for 304 stainless steel under cold and warm compression by artificial neural network and genetic algorithm. *Mater. Des.* **2007**, *28*, 609–615. [CrossRef]
3. Jafarian, H.; Eivani, A.R. Texture development and microstructure evolution in metastable austenitic steel processed by accumulative roll bonding and subsequent annealing. *J. Mater. Sci.* **2014**, *49*, 6570–6578. [CrossRef]
4. Kiani Khouzani, M.; Bahrami, A.; Eslami, A. Metallurgical aspects of failure in a broken femoral HIP prosthesis. *Eng. Fail. Anal.* **2018**, *90*, 168–178. [CrossRef]
5. Lavvafi, H. Effects of Laser Machining on Structure and Fatigue of 316LVM Biomedical Wires. Ph.D. Thesis, Case Western Reserve University, Cleveland, OH, USA, 2013.
6. Bahrami, A.; Mousavi Anijdan, S.H.; Taheri, P.; Yazdan Mehr, M. Failure of AISI 304H stainless steel elbows in a heat exchanger. *Eng. Fail. Anal.* **2018**, *90*, 397–403. [CrossRef]
7. Lavvafi, H.; Lewandowski, M.E.; Schwam, D.; Lewandowski, J.J. Effects of surface laser treatments on microstructure, tension, and fatigue behavior of AISI 316LVM biomedical wires. *Mater. Sci. Eng. A* **2017**, *688*, 101–113. [CrossRef]
8. Natesan, K.; Park, J.H. Fireside and steamside corrosion of alloys for USC plants. *Int. J. Hydrog. Energy* **2007**, *32*, 3689–3697. [CrossRef]
9. Ahmad, J.; Rahman, M.M.; Zuhairi, M.H.A. High operating steam pressure and localized overheating of a primary superheater tube. *Eng. Fail. Anal.* **2012**, *26*, 344–348. [CrossRef]
10. Ananda Rao, M.; Sankara Narayanan, T.S.N. Failure investigation of a boiler bank tube from a 77 × 2 MW coal based thermal power plant in the northwest region of India. *Eng. Fail. Anal.* **2012**, *26*, 325–331. [CrossRef]
11. Almazrouee, A.; Singh, R.K. Role of oxide notching and degraded alloy microstructure in remarkably premature failure of steam generator tubes. *Eng. Fail. Anal.* **2011**, *11*, 2288–2295. [CrossRef]
12. Movahedi-Rad, A.; Plasseyed, S.S.; Attarian, M. Failure analysis of superheater tube. *Eng. Fail. Anal.* **2015**, *48*, 94–104. [CrossRef]
13. Psyllaki, P.P.; Pantazopoulos, G.; Lefakis, H. Metallurgical evaluation of creep-failed superheater. *Eng. Fail. Anal.* **2009**, *16*, 1420–1431. [CrossRef]
14. Khodamorad, S.H.; Alinezhad, N.; Haghshenas Fatmehsari, D.; Ghahtan, K. Stress corrosion cracking in Type.316 plates of a heat exchanger. *Case Stud. Eng. Fail. Anal.* **2016**, *5–6*, 59–66. [CrossRef]
15. Corleto, C.R.; Argade, G.R. Failure analysis of dissimilar weld in heat exchanger. *Case Stud. Eng. Fail. Anal.* **2017**, *9*, 27–34. [CrossRef]
16. Corte, J.S.; Rebello, J.M.A.; Areiza, M.C.L.; Tavares, S.S.M.; Araujo, M.D.A. Failure analysis of AISI 321 tubes of heat exchanger. *Eng. Fail. Anal.* **2015**, *56*, 170–176. [CrossRef]
17. Laurent, M.; Estevez, R.; Fabregue, D.; Ayax, E. Thermomechanical fatigue life prediction of 316L compact heat exchanger. *Eng. Fail. Anal.* **2016**, *68*, 138–149. [CrossRef]
18. Fly Ash. Available online: https://www.corrosionpedia.com/definition/1624/fly-ash (accessed on 1 August 2019).
19. Corrosion: Intergranular Corrosion. Available online: http://www.ssina.com/corrosion/igc.html (accessed on 1 August 2019).
20. ASTM A262-15. *Standard Practices for Detecting Susceptibility to Intergranular Attack in Austenitic Stainless Steels*; ASTM International: West Conshohocken, PA, USA, 2015.

© 2019 by the authors. Licensee MDPI, Basel, Switzerland. This article is an open access article distributed under the terms and conditions of the Creative Commons Attribution (CC BY) license (http://creativecommons.org/licenses/by/4.0/).

Article

Prediction of Bake Hardening Behavior of Selected Advanced High Strength Automotive Steels and Hailstone Failure Discussion

Mária Mihaliková [1], Kristína Zgodavová [1,*], Peter Bober [2] and Andrea Sütőová [1]

[1] Faculty of Materials, Metallurgy and Recycling, Technical University of Košice, 04200 Košice, Slovakia; maria.mihalikova@tuke.sk (M.M.); andrea.sutoova@tuke.sk (A.S.)

[2] Faculty of Electrical Engineering and Informatics, Technical University of Košice, 04200 Košice, Slovakia; peter.bober@tuke.sk

* Correspondence: kristina.zgodavova@tuke.sk; Tel.: +421-903-750-590

Received: 7 August 2019; Accepted: 13 September 2019; Published: 18 September 2019

Abstract: The purpose of the present study is three-fold. Firstly, it attempts to describe the bake hardening (BH) behavior of selected interstitial free (IF) and dual phase (DP) steels. Secondly, it predicts the BH behavior of the IF DX 51D and DP 500 HCT 590X plates of steel, and thirdly studies material failure prevention in scholarly sources. The research is aimed at investigating the increasing steel strength during the BH of these two high-strength sheets of steel used for outer vehicle body parts. Samples of steel were pre-strained to 1%, 2%, and 5% and then baked at 140–220 °C for 10 to 30 min. The BH effect was determined from three factors: pre-strain, baking temperature, and baking time. Research has shown that increasing the yield strength by the BH effect is predictable. Therefore, the number of experiments could be reduced for the investigation of BH effect for other kinds of IF and DP steels. The literature study of the hailstone failure reveals that the knowledge of BH steels behavior helps to calculate the steel supplier´s failure mode effect analysis (FMEA) risk priority number.

Keywords: bake hardening; dent resistance; failure study; polynomial regression; yield strength; automotive steels

---

1. Introduction

In Europe, there were about 25,000 fatal road accidents in 2016, of which passenger accidents accounted for more than 25.6% [1]. The European Union's environmental policy and high customer demands for car quality are forcing manufacturers to adhere to strict quality, safety, and environmental management standards. Therefore, becoming a new supplier in the automotive chain in today's busy market is extremely complex [2,3].

According to [4] many industrial applications, such as car bodies require steels with good formability and high strengths, taking into account safety, fuel efficiency, environmental performance, manufacturability, durability, and other quality properties.

According to the Eurostat 2018 study [1], such steels contribute significantly to reducing the number of fatal road accidents. The goal of car manufacturers is to increase the use of high-strength steels with minimized-thickness because it is necessary for preventing material failure and avoiding cost increases.

However, these requirements are often contradictory since the increase in strength must be achieved without compromising formability.

Due to a variety of requirements, different automotive parts utilize various steel types and grades to achieve the required properties [5]. For example, outer body parts should possess good surface quality, dent resistance, and also a good hemming ability.

Bake hardened steels have good formability before stamping and enhanced strength after baking. Hence, in recent years bake hardened steel has been widely used in vehicles components, thus leading to a reduction in vehicle weight and improved safety. Bake hardening steels derive their increase in strength from a strain aging [6].

Considering that the investigated bake hardenable steels are planned to be used for the automotive outer body parts, there is a high probability that during use, various damages will occur due to static or dynamic stone impact, e.g., hailstones. It is therefore worth considering for further research the increased number of hailstorms that have recently occurred in the world. There were 4611 hailstorms in the USA [7] and Europe per year. A report by Munich RE [8] shows that damage from individual events can exceed billions of dollars as the risk has increased in the past decades. Hail damage to properties, such as cars, becomes substantial when the diameter of hailstones approaches and exceeds 50 mm. Forecasts for the year 2019 say that world-wide hailstorms might cause record damage [9]. Therefore, steel sheets are required to have high dent resistance, which is closely related to high strength.

The aim of our research is to find out how to control a particular grade of interstitial free (IF) and dual phase (DP) steel by changing input parameters during bake hardening (BH) and to explore the possibility of generalizing this knowledge to other grades of IF and DP steels by microscopic analysis and predict their behavior by multi-polynomial regression. Hailstone failure of bake hardenable steels is studied in scholarly sources, and inclusion of increased frequency of occurrence and severity of hailstones into the failure mode effect analysis (FMEA) at the steel-maker company (supplier) is recommended.

2. Materials and Methods

Conventional bake hardening steels were initially developed and patented in 1977 to propose an increased strength in cold-formed sheet metal parts that require good formability [10].

BH effect can be observed in many different grades of steels, and this phenomenon is discussed in detail, e.g., in [11]. Bake hardening occurring in steel materials corresponds to an increase of the flow stress after a pre-strain followed by heat treatment (or annealing) within a specific temperature range [12]. Bake hardening uses the deformation aging process to increase yield strength. The maximum is reached at a temperature range between 150–220 °C according to [13,14].

The pre-strain is a permanent elongation of the sample. The sample was loaded to a given pre-strain elongation and unloaded at room temperature. The value of pre-strain is the percent of the length before elongation.

Bake hardening increases the yield strength by the deformation aging process of the formed part for increased dent resistance without a reduction in formability [12]. Typical applications are automotive outer body panels where increased dent resistance is required.

For press-formed car body structural components, the paint-baking treatment is the last process cycle which is performed at low-temperature. Paint-baking gives the car not only an aesthetic appearance but also positively affects the strength of the material and increases the dent resistance via the BH effect [15–17].

Two low carbon sheets of steel referred to as IF DX 51D and DP 500 HCT 590X that exhibit bake hardening effect were used in this study. According to the metallurgical designation, we can classify IF DX 51D and DP 500 HCT 590X as advanced high-strength steels (AHSS). AHSS have carefully selected chemical compositions and multiphase microstructures resulting from precisely controlled heating and cooling processes. Table 1 shows the chemical composition of the investigated steels.

Table 1. Chemical compositions of investigated steels (wt.%).

Steel	C	Si	Mn	P	S	Ti	Al_{TOT}	Cr + Mo	Nb + Ti	V	B
IF DX 51D	0.15	0.55	-	0.04	0.015	0.3	-	-	-	-	-
DP 500 HCT 590X	0.12	0.5	1.6	0.1	0.045	0.3	0.015–1.5	1.4	0.15	0.2	0.005

Most automotive components are subject to low stress during the final shaping operation before baking. Therefore, BH behavior experiments are performed by exposing the samples to a pre-strain range of 1–5% at 20 °C. Applying pre-forming to samples increased the density of dislocations [18,19]. After that, the sample was baked for a defined time at a specified temperature.

Bake hardening effects can be evaluated using a standardized concept of a bake hardening index. The value of BH_2 index for pre-strain 2% is given according to DIN EN 10325:2006 [20] by Equation (1):

$$BH_2 = Re_{L,t} \text{ (or } Rp_{0.2,t}) - Rp_{0.2,r} \tag{1}$$

where $Re_{L,t}$—is the yield strength of lower point when the stress-strain shows a sharp yielding point for the sample after baking; $Rp_{0.2,t}$—the yield strength of the crossing point of 0.2% offset line with the tensile test curve when the stress-strain does not show a sharp yielding point; $Rp_{0.2,r}$—the yield strength of the relevant pre-strain for the sample before baking.

The dislocations generated by the preformation are anchored by free soluble atoms that diffuse into the dislocation core during baking. The consequence of this process is an increased yield strength (YS) [20].

In general, the BH index is 30–40 MPa for IF steels [21] and 30–60 MPa for DP steels [22] after 2% pre-strain and baking at a temperature of 170 °C throughout 20 min. The magnitude of this phenomenon depends on the baking temperature and time [12,23].

3. Results

3.1. Bake Hardening Behavior of Investigated Steels

The mechanical properties of investigated steels before and after baking were tested using a tensile test according to [24]. The sample thickness for DP steel was 1.5 mm and was 0.55 mm for IF steel. The geometry of the samples used for the tensile test is given in Figure 1.

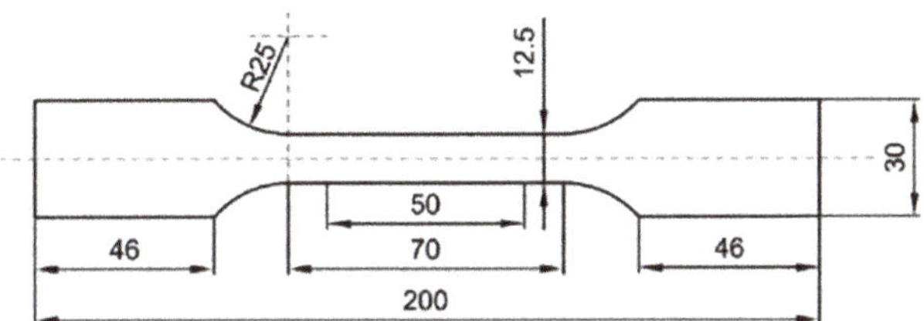

Figure 1. Sample geometry for tensile testing (unit: mm).

Table 2 shows the mechanical properties of investigated steels before baking.

Table 2. Mechanical properties of as-received steels.

Steel	YS (MPa)	UTS (MPa)	A80 (%)
IF DX 51D	220	350	36
DP 500 HCT 590X	340	590	20

YS—yield strength; UTS—ultimate tensile strength; A80—elongation 80 mm.

The microstructure of both materials was obtained by metallographic analysis. Samples were taken from cold rolled galvanized sheets in the rolling direction. Samples were produced for both types of steel by a standard metallographic procedure according to the internal procedure of the steel-maker [25]. Samples of tested steels were observed on the Olympus GX 71 microscope (Tokyo, Japan). Figure 2a and Supplementary Video S1 show the microstructure of the IF DX 51D steel before baking. The microstructure is formed by ferrite grains. Steel is purely ferritic without interstices as carbon (C) and nitrogen (N). The microstructure of the DP 500 HCT 590X steel before baking is represented in Figure 2b and Supplementary Video S1. The higher volume fraction of small martensite islands and small ferrite grains produces higher BH values. Solute carbon content in ferrite controls the speed of BH response [19].

(a)   (b)

Figure 2. The microstructure of investigated steels: (**a**) Interstitial free (IF) DX 51D steel. (**b**) Dual phase (DP) 500 HCT 590X steel.

For tensile tests, a multiaxial material testing machine Zwick Z050 (Ulm, Germany) with optical strain measurement ARAMIS 5M and the rate-controlled device was used. First, three groups of samples were pre-strained to 1%, 2%, and 5%. Then, each sample was subjected to a baking process at the unique combination of temperature (140 °C, 170 °C, 190 °C, 220 °C) and time (10, 15, 20, and 30 min). After cooling down to room temperature the tensile test was performed, and measured values were recorded. Tables 3 and 4 show the selected measured values at a typical temperature 170 °C and time 20 min. The whole set of measured values of the BH index used in a model calculation in Section 3.2 is listed in Appendix A.

Table 3. Measured values of IF DX 51D, 170 °C/20 min.

Specimen	a_0 (mm)	b_0 (mm)	Re_L (MPa)	$Rp_{x,r}$ (MPa)	A80 (%)	BH_x (MPa)	Pre-Strain
E1_170/20	0.55	20.06	352	310	20.7	42	1%
E2_170/20	0.55	20.04	374	339	15.6	35	2%
E5_170/20	0.53	20.04	433	342	19.6	91	5%

a_0—sample thickness; b_0—sample length; Re_L—yield strength of lower point; $Rp_{x,r}$—yield strength of pre-deformation 1%, 2%, 5% before baking; A80—80 mm elongation; BH_x—bake hardening index.

Table 4. Measured values of DP 500 HCT 590X, 170 °C/20 min.

Specimen	a_0 (mm)	b_0 (mm)	Re_L (MPa)	$Rp_{x,r}$ (MPa)	A80 (%)	BH_x (MPa)	Pre-Strain
K1_170/20	1.47	20.09	486	418	23.8	68	1%
K2_170/20	1.47	20.01	548	477	22.9	71	2%
K5_170/20	1.46	19.77	630	567	19.9	63	5%

a_0—sample thickness; b_0—sample length; Re_L—yield strength of lower point; $Rp_{x,r}$—yield strength of pre-deformation 1%, 2%, 5% before baking; A80—80 mm elongation; BH_x—bake hardening index.

3.2. Multi-Polynomial Regression Model

The regression analysis was used for estimating the relationships between bake hardening values and independent variables: pre-strain, baking temperature, and baking time. The second-order multi-polynomial regression model was chosen from several tested models because it had the lowest value of the residual sum of squares [26,27]. The number of measurements taken was 48 for each type of steel, which is the number of a full factorial experiment. The model is given by Equation (2):

$$Y = C_0 + C_1X_1 + C_2X_2 + C_3X_3 + C_{12}X_1X_2 + C_{13}X_1X_3 + C_{23}X_2X_3 + C_{11}X_1{}^2 + C_{22}X_2{}^2 + C_{33}X_3{}^2 \quad (2)$$

where Y is BH (MPa), X_1 is pre-strains (%), X_2 is baking temperature (°C), and X_3 is baking time (min). The coefficients C_1 to C_{33} of the model are listed in Table 5.

Table 5. Summary of regression analysis results.

Coefficient	IF DX 51D	DP 500 HCT 590X
C_0	−67.2078	31.5670
C_1	−18.9102	−14.5249
C_2	1.4452	−0.1713
C_3	−2.5467	1.2129
C_{12}	0.0656	0.0320
C_{13}	0.2182	0.0066
C_{23}	0.0021	0.0006
C_{11}	4.3385	1.0295
C_{22}	−0.0045	0.0021
C_{33}	0.0765	−0.0218
Residual Sum of Squares	1926.8700	204.4950
Coefficient of Determination R^2	0.9528	0.9897

Table 6 contains the values of the coefficient of determination (R^2) for models which were calculated in the way mentioned above for the different number of measurements. The values were taken from the existing set of measurements. The number of measurements can be reduced three times from 48 to 16, and the R^2 value is still close to the original one. It must be said that good result are achieved by choosing a set of measurements where all distinct values of independent variables (pre-strain, baking temperature, and baking time) are presented.

Table 6. Coefficient of determination (R^2) for calculated models.

No. of Measurement	Coefficient of Determination R^2	
	IF DX51D	DP 500 HCT 590X
12	0.9486	0.9832
14	0.9472	0.9826
16	0.9498	0.9893
18	0.9511	0.9892
20	0.9507	0.9890

The coefficient of determination, R^2, in Table 6 is the proportion of the variance in the variable BH that is predictable by the model. The analysis shows that the model for DP 500 HCT 590X steel is more precise. Nevertheless, it can be concluded that the increased yield strength by the BH effect is predictable and can be described by the second-order multi-polynomial regression model with ten coefficients C_1 to C_{33}. The values of coefficients in Table 5 are valid for given sample geometry and thickness.

3.3. Yield Strength and a Dent Resistance—Findings from the Literature Study

There are extensive crash studies concerning the safety of outer vehicle body parts in the literature [28,29]. Less known and respectively less published are failure studies, which are related to smaller, non-life-threatening outer damage of sheets due to static and dynamic impacts of small hard objects (up to 100 mm), which reduce aesthetic properties of vehicles and require costly repair. More extensive dent resistance tests were done by [30,31]. A key distinction that is sometimes missing in dent resistance studies is the nature of the dent.

There are two types of denting, static and dynamic, according to F. Gatto & D. Morri [32]. Static denting refers to a gradually applied load over a small area typified by a hand pushing on a vehicle bonnet. Dynamic denting occurs under impact loading typified by a hailstone. The key difference between the two is the nature of load application. A static dent indicates a slowly applied force. A dynamic dent is driven by inertia and impact energy.

Burley et al. [33] confirmed that dynamic dent resistance is affected by factors such as panel density, modulus of elasticity, and curvature. Thomas D. presented the test of the AA6111 steel sheet. Results of this research were presented as "dynamic dent models" in [30].

Mehmet E. Uz [31], simulated the hailstone impact on G300 and G550 steels in laboratory conditions and investigated the dependence of dent size on steel grade, steel thickness, yield strength, and the size, speed, and impact force of ice balls. Experimentation reveals that the dent depth was inversely proportional to thickness and yield strength, while the dent diameter was found to be proportional to yield stress. As the yield strength of the steel sheet increased, the dent depth decreased for these specific materials.

In the literature [34–36] there are applications of failure studies in steel-maker organizations. The detailed procedure, according to [37,38] aims to guide quality engineers and practitioners and to offer consistent results towards failure prevention and quality improvement.

One of the most popular engineering techniques for failure prevention is a FMEA method [39]. FMEA is used for reducing the risks of failure and understanding the nature of preventive actions needed to be taken as measures of continuous improvement.

This method can reveal the risks at the early stage of product planning, reduce time, and save investment. For failure risk assessment the risk priority number (RPN) can be used and defined as the result of the three independent factors Equation (3):

$$RPN = S \times O \times D \tag{3}$$

where S is the Severity of the failure effect; O is the occurrence likelihood factor of the failure cause and D is for the detection likelihood of the occurred failure cause, failure or failure effect, respectively.

4. Discussion

4.1. Bake Hardening Behavior

The minimum BH index value for IF DX 51D steel is 12.2 MPa at 1% pre-strain, 220 °C, and 15 min, and the maximum value is 98 MPa at 5%, 220 °C, and 20 min. The minimum BH index value for DP 500 HCT 590X steel is 33.5 MPa at 5% pre-strain, 140 °C and 10 min, and the maximum value is 109.2 MPa at 1%, 220 °C, 30 min, as can be seen in Appendix A.

The BH effect for the IF DX 51D steel is less sensitive to temperature, which is evident in Figure 3. In contrast, the BH effect for the DP 500 HCT 590X steel is quite sensitive to temperature but almost insensitive to pre-strain (Figure 4). It could be stated that BH effect for IF DX 51D steel is dependent on pre-deformation as opposed to DP 500 HCT 590X steel, where BH effect increases with the rise of temperature. That is consistent with the findings in literature sources [19,23], which state that increasing of the BH effect significantly depends on the type of material and its structure.

4.2. Prediction of the Bake Hardening Behavior

Prediction of BH behavior is visualized in Figures 3 and 4. It is evident that different steels behave differently, which is apparent from the Supplementary Video S1. However, the assumption made about the predictability of bake hardening behavior allows for calculation of the model coefficients for different steels from a reduced number of measurements. In this case, the model coefficients are calculated by solving a system of linear equations.

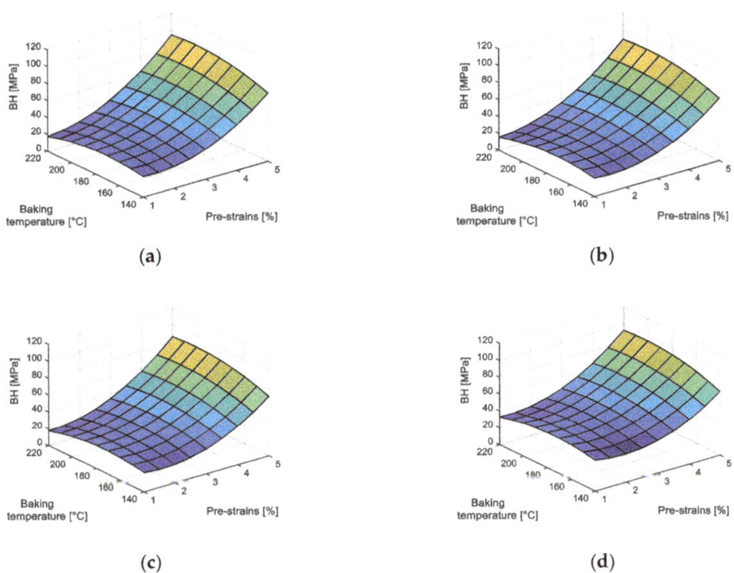

Figure 3. Bake hardening behavior of the IF DX 51D steel. (**a**) Baking time 10 min; (**b**) baking time 15 min; (**c**) baking time 20 min; (**d**) baking time 30 min.

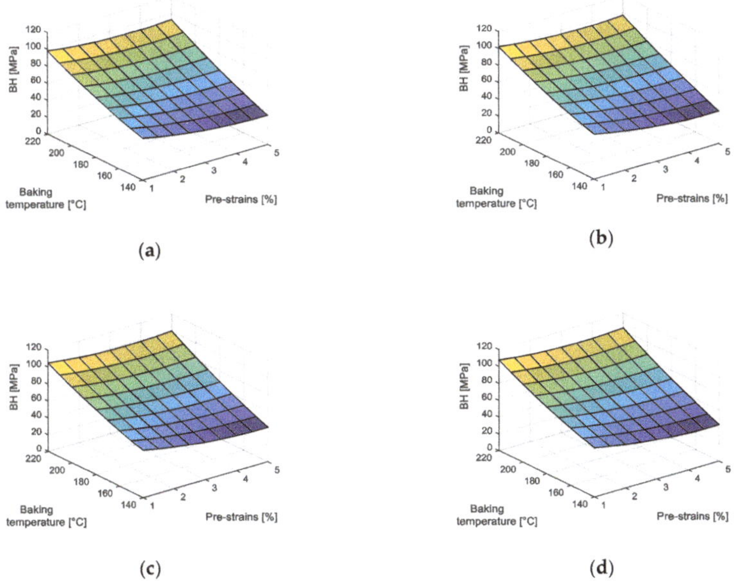

Figure 4. Bake hardening behavior of the DP 500 HCT 590X steel. (**a**) Baking time 10 min; (**b**) baking time 15 min; (**c**) baking time 20 min; (**d**) baking time 30 min.

4.3. Failure Caused by a Hailstone

According to the literature [6,12,34], BH steels have a high dent resistance. Nevertheless, we know cases when car roofs have been damaged during hailstorms as well as due to the stone impacts to the hood, doors, and mudguards. Illustration can be found in Supplementary Video S1.

According to our multi-polynomial regression model, it is possible to predict the yield strength only for sheet thicknesses shown in Tables 3 and 4. By measuring for different thicknesses, the same procedure could make the model generalizable for other thicknesses of the examined materials and thus contribute to the customer's decision on a particular sheet thickness with respect to dynamic dent resistance.

Then, implementation of failure analysis (FA) as a systematic procedure of an organization will help to identify the dent risks concerning of the outer vehicle body damage.

Since the studies mentioned above have shown dependence between yield strength and dent size caused by hailstones, the knowledge of work hardening and bake hardening effects can be used for estimating the RPN number by steel-maker organization.

Our further research will focus on the simulated hailstone in laboratory conditions and investigate steel grade dependence, steel thickness, yield stress and size, speed, and impact strength of ice balls on samples on outer vehicle body parts for IF DX 51D steel and DP 500 HCT 590X steel. FMEA will reveal the risks of hailstone damage for investigated steels at the early stage of steel research.

5. Conclusions

The subjects of our study were bake hardenable automotive body steels, for which the vehicle producer required good formability and high strengths. On the base of the producer request, we investigated the behavior of two steel types during bake hardening.

We examined the effect of pre-strain, baking temperature, and baking time on the bake hardening process, which is used for the strengthening of body sheets after the painting process.

The following conclusions can be drawn from the experimental part:

- IF DX 51D steel is created only by ferritic phase. DP 500 HCT 590X steel contains ferrite as the primary phase, and the secondary strengthened phase is formed mainly by martensite.
- Measured higher ultimate tensile strength of the primary material DP 500 HCT 590X steel is 590 MPa and elongation A80 is 20%, and for IF DX51D steel the ultimate tensile strength is 350 MPa and elongation A80 is 36%.
- The result of the regression analysis shows that the increase of yield strength by the bake hardening effect of investigated steels is predictable, therefore a reduced number of measurements are needed to describe BH behavior for another grade of IF and DP steels or for different steel sheet thickness.
- The literature review shows that there is a direct relationship between the dent depth, which is inversely proportional with thickness, and yield strength. Therefore, knowledge of BH behavior should be taken into account in the process FMEA of the steel-maker organization.

Supplementary Materials: The following are available online at http://www.mdpi.com/2075-4701/9/9/1016/s1, Video S1: Bake hardening behavior of the IF DX51D and DP 500 HCT 590X steels.

Author Contributions: Conceptualization, K.Z.; data curation, M.M., K.Z., P.B.; formal analysis, K.Z., P.B.; founding acquisition, K.Z.; investigation, M.M.; methodology, M.M., K.Z., P.B.; project administration, K.Z.; resources, M.M., K.Z., P.B., A.S.; visualization, M.M., K.Z., P.B.; writing—original draft, M.M., K.Z., P.B.; writing—review & editing, K.Z., P.B., A.S.

Funding: This research was funded by MINISTERSTVO ŠKOLSTVA, VEDY, VÝSKUMU A ŠPORTU SLOVENSKEJ REPUBLIKY, grant number KEGA 043TUKE-4/2019 "Improving material engineering and integrated management systems study programs for Industry 4.0".

Conflicts of Interest: The authors declare no conflict of interest. The funders had no role in the design of the study, in the collection, analyses, interpretation of data, in the writing of the manuscript, or in the decision to publish the results.

Appendix A

Table A1. Measured values of BH index.

Pre-Strain	Temperature (°C)	Time (min)	BHx (MPa)			
			IF DX 51D	a_0	DP 500 HCT 590X	a_0
1%	140	10	26.0	0.55	54.0	0.47
2%	140	10	30.5	0.55	40.5	0.47
5%	140	10	80.5	0.53	33.5	0.46
1%	170	10	38.0	0.55	64.0	0.47
2%	170	10	32.0	0.55	60.0	0.47
5%	170	10	84.0	0.53	52.0	0.46
1%	190	10	32.2	0.55	79.1	0.47
2%	190	10	30.1	0.55	72.1	0.47
5%	190	10	90.8	0.53	66.8	0.46
1%	220	10	24.2	0.55	102.1	0.47
2%	220	10	26.1	0.55	92.1	0.47
5%	220	10	95.8	0.53	92.8	0.46
1%	140	15	14.0	0.55	54.0	0.47
2%	140	15	19.5	0.55	43.5	0.47
5%	140	15	77.5	0.53	38.5	0.46
1%	170	15	26.0	0.55	64.0	0.47
2%	170	15	21.0	0.55	63.0	0.47
5%	170	15	81.0	0.53	57.0	0.46
1%	190	15	20.2	0.55	79.1	0.47
2%	190	15	19.1	0.55	75.1	0.47
5%	190	15	87.8	0.53	71.8	0.46
1%	220	15	12.2	0.55	102.1	0.47
2%	220	15	15.1	0.55	95.1	0.47
5%	220	15	92.8	0.53	97.8	0.46
1%	140	20	16.0	0.55	56.0	0.47
2%	140	20	21.0	0.55	49.0	0.47
5%	140	20	81.0	0.53	44.0	0.46
1%	170	20	42.0	0.55	68.0	0.47
2%	170	20	35.0	0.55	71.0	0.47
5%	170	20	91.0	0.53	63.0	0.46
1%	190	20	22.0	0.55	81.0	0.47
2%	190	20	21.0	0.55	80.0	0.47
5%	190	20	93.0	0.53	76.0	0.46
1%	220	20	14.0	0.55	104.0	0.47
2%	220	20	17.0	0.55	100.0	0.47
5%	220	20	98.0	0.53	102.0	0.46
1%	140	30	25.0	0.55	60.0	0.47

Table A1. *Cont.*

Pre-Strain	Temperature (°C)	Time (min)	BHx (MPa)			
			IF DX 51D	a_0	DP 500 HCT 590X	a_0
2%	140	30	38.0	0.55	49.0	0.47
5%	140	30	72.0	0.53	42.0	0.46
1%	170	30	51.0	0.55	72.0	0.47
2%	170	30	52.0	0.55	71.0	0.47
5%	170	30	82.0	0.53	61.0	0.46
1%	190	30	38.3	0.55	86.2	0.47
2%	190	30	43.7	0.55	82.2	0.47
5%	190	30	84.7	0.53	76.2	0.46
1%	220	30	30.3	0.55	109.2	0.47
2%	220	30	39.7	0.55	102.2	0.47
5%	220	30	89.7	0.53	102.2	0.46

a_0—sample thickness. Shaded parts are minimum and maximum values of BHx.

References

1. Eurostat. Persons Killed in Road Accidents by Type of Vehicle (CARE Data). 2018. Available online: http://appsso.eurostat.ec.europa.eu/nui/show.do?dataset=tran_sf_roadve&lang=en (accessed on 10 March 2019).
2. PennState. Automotive Manufacturing Industry Analysis. Available online: http://php.scripts.psu.edu/users/l/a/law5039/assign5.html (accessed on 6 May 2019).
3. Supply Chain Dive. How Suppliers are Innovating to Keep Pace with the Auto Industry. Available online: https://www.supplychaindive.com/news/auto-series-supplier-innovation-digitization-OEM/516585/ (accessed on 20 February 2019).
4. Momeni, A.; Dehghani, K.; Abbasi, S.; Torkan, M. Bake Hardening of a Low Carbon Steel for Automotive Applications. *Metalurgija* **2007**, *13*, 131–138.
5. Pereloma, E.; Timokhina, I. Effect of bake hardening on the performance of automotive steels. In *Bake Hardening of Automotive Steels; Design, Metallurgy, Processing and Applications, Effect of Bake Hardening on the Performance of Automotive Steels*; Elsevier Ltd.: Amsterdam, The Netherlands, 2017; pp. 259–288.
6. Total Materia. Dent Resistant Steels. Available online: https://www.totalmateria.com/page.aspx?ID=CheckArticle&site=kts&NM=452 (accessed on 10 February 2019).
7. NOAA. Storm Prediction Center, National Weather Service. Available online: https://www.spc.noaa.gov/climo/online/monthly/2018_annual_summary.html (accessed on 2 July 2019).
8. ESSL. Major Hailstorms of 2018 across Europe, European Severe Storms Laboratory. Available online: https://www.essl.org/cms/major-hailstorms-of-2018-across-europe/ (accessed on 15 June 2019).
9. CBS News. Are Hailstorms Getting Worse in US? Why 2019 Could Produce Record Damage. Available online: https://www.cbsnews.com/news/hail-damage-costs-this-year-could-hit-new-annual-high-in-u-s/ (accessed on 2 July 2019).
10. Kantereit, H. *Bake Hardening Behavior of Advanced High Strength Steels under Manufacturing Conditions*; SAE Technical Paper; SAE International: Troy, MI, USA, 2011. [CrossRef]
11. Das, S.; Singh, S.B.; Mohanty, O.N. Bake Hardening. In *Encyclopedia of Iron, Steel, and Their Alloys*, 1st ed.; Colás, R., Totten, G.E., Eds.; CRC Press: Boca Raton, FL, USA, 2016; Volume 1, pp. 304–311.
12. Thuillier, S.; Zang, S.-L.; Troufflard, J.; Manach, P.-I.; Jegat, A. Modeling Bake Hardening Effects in Steel. *Metals* **2018**, *8*, 594. [CrossRef]
13. Seath, P. Study of Bake-Hardening Behaviour of Ultra-Low Carbon BH 220 Steel at Different Strain Rates. Master's Thesis, National Institute of Technology Rourkela, Department of Metallurgical and Materials Engineering, Rourkela, India, 2014.

14. Petrov, A. Vplyv BH-efektu na zmenu vlastnosti vybraných automobilových plechov (in Slovak). Bake Hardening Effect on the Change of Properties of Selected Automotive Plates. Master's Thesis, Technical University of Košice, Košice, Slovakia, 2018.
15. Ramazani, A.; Mukherjee, K.; Prahl, U.; Bleck, W. Modelling the effect of microstructural banding on the flow curve behaviour of dual-phase (DP) steels. *Comput. Mater. Sci.* **2012**, *52*, 46–54. [CrossRef]
16. Kuang, C.; Zhang, S.; Li, H.; Wang, I.; Liu, H. Effects of pre-strain and baking parameters on the microstructure and bake-hardening behavior of dual-phase steel. *Int. J. Miner. Metall. Mater.* **2014**, *21*, 766. [CrossRef]
17. Ramazani, A.; Bruehl, S.; Gerber, T.; Bleck, W.; Prahl, U. Quantification of bake hardening effect in DP600 and TRIP700 steels. *Mater. Des.* **2014**, *57*, 479–486. [CrossRef]
18. Elsen, P.; Hougardy, H. On the mechanism of bake-hardening. *Mater. Technol.* **1993**, *64*, 431–436. [CrossRef]
19. Ji, D.; Zhang, M.; Zhu, D.; Luo, S.; Lin, L. Influence of microstructure and pre-straining on the bake hardening. *Mater. Sci. Eng. A* **2017**, *708*, 129–141. [CrossRef]
20. Deutsche Norm. *Steel-Determination of Yield Strength Increase by the Effect of Heat Treatment (Bake-Hardening-Index);* English Version (DIN EN 10325:2006); Deutsches Institut für Normung: Berlin, Germany, 2016.
21. Liu, Z.; Li, W.; Shao, X.; Kang, Y.; Li, Y. An Ultra-low-Carbon Steel with Outstanding Fish-Scaling Resistance and Cold Formability for Enameling Applications. *Metall. Mater. Trans. A* **2019**, *50*, 1805–1815. [CrossRef]
22. Kvackaj, T.; Mamuzic, I. Development of bake hardening effect by plastic deformation and annealing. *Metalurgija* **2006**, *45*, 51–55.
23. Palkowski, H.; Anke, T. Bake Hardening of Hot Rolled Multiphase Steels under Biaxial Pre-strained. *Steel Res. Int.* **2006**, *77*, 675–679. [CrossRef]
24. International Standard Organization. *Metallic materials—Tensile testing—Part 1: Method of Test at Room Temperature (ISO 6892-1:2016);* ISO: Geneva, Switzerland, 2016.
25. Slovak Technical Standard. Metallography of steel. In *Terminology;* Slovak Version (STN 420003:1973); UNMS: Bratislava, Slovakia, 1973.
26. Paris, A.S.; Tarcolea, C.; Croitoru, S.M.; Majstorović, V.D. Statistical Study of Parameters in the Process of Orthogonal Cutting Surface Hardness. In Proceedings of the 4th International Conference on the Industry 4.0 Model for Advanced Manufacturing, Belgrade, Serbia, 3–6 June 2019; Monostori, L., Majstorovic, V., Hu, S., Djurdjanovic, D., Eds.; Springer Nature Switzerland AG: Cham, Switzerlan, 2019; pp. 68–77.
27. Plura, J. Procedures and Methods of Quality Planning and their Use for Forming Process Optimization. In *Engineering the Future;* Dudas, L., Ed.; IntechOpen Limited: London, UK, 2010; pp. 257–279.
28. WorldAutoSteels. Facing the Challenge for Crash Safety, Study by Ducker Worldwide. Available online: https://www.worldautosteel.org/why-steel/safety/facing-the-challenge-for-crash-safety/ (accessed on 12 March 2019).
29. Evin, E.; Tomas, M.; Katalinic, B.; Wessely, E.; Kmec, J. Design of Dual Phase High Strength Steel Sheets for Autobody. In *DAAAM International Scientific Book;* Katalinic, B., Tekic, Z., Eds.; DAAAM International: Vienna, Austria, 2013; pp. 767–786.
30. Thomas, D. The Numerical Prediction of Panel Dent Resistance Incorporating Panel Forming Strains. Master's Thesis, University of Waterloo, Waterloo, ON, Canada, 2001.
31. Uz, M.E. Examining Dent Formation Caused by Hailstone Impact. *Shock Vib.* **2019**, *2019*, 6175206. [CrossRef]
32. Gatto, F.; Morri, D. Forming properties of some aluminium alloys sheets for car—Body use. In Proceedings of the 12th Biennial Congress, International Deep Drawing Research Group, Margherita Ligure, Italy, 24–28 May 1982.
33. Burley, C.E.; Niemeier, B.A.; Koch, G.P. *Dynamic Denting of Autobody Panels;* Technical Paper 760165; SAE: Troy, MI, USA, 1976. [CrossRef]
34. Lu, H.; Ma, M.; Jou, J.; Li, Z. A Research Progress of Dent Resistance for Automotive Body Panes. Available online: http://www.paper.edu.cn/scholar/showpdf/MUT2IN4IOTD0Uxzh (accessed on 15 March 2019).
35. Dennies, D. *How to Organize and Run a Failure Investigation;* ASM International: Materials Park, OH, USA, 2005.
36. Jumbad, V.; Salunke, J.J.; Satpute, M.A. FMEA Methodology for Quality Improvement. *Int. J. Eng. Res. Technol.* **2016**, *5*, 122–126.
37. Monka, P.; Monková, K.; Modrak, V.; Hric, S.; Pastucha, P. Study of a tap failure at the internal threads machining. *Eng. Fail. Anal.* **2019**, *100*, 25–36. [CrossRef]

38. Pantazopoulos, G. A Short Review on Fracture Mechanisms of Mechanical Components Operated under Industrial Process Conditions: Fractographic Analysis and Selected Prevention Strategies. *Metals* **2019**, *9*, 148. [CrossRef]
39. Pačaiová, H.; Sinay, J.; Nagyová, A. Development of GRAM—A risk measurement tool using risk based thinking principles. *Measurement* **2016**, *100*, 288–296. [CrossRef]

 © 2019 by the authors. Licensee MDPI, Basel, Switzerland. This article is an open access article distributed under the terms and conditions of the Creative Commons Attribution (CC BY) license (http://creativecommons.org/licenses/by/4.0/).

 *metals*

Article

Creep Failure of Reformer Tubes in a Petrochemical Plant

Abbas Bahrami [1] and Peyman Taheri [2],*

[1] Department of Materials Engineering, Isfahan University of Technology, Isfahan 84156-83111, Iran;
     a.n.bahrami@cc.iut.ac.ir
[2] Delft University of Technology, Department of Materials Science and Engineering, Mekelweg 2,
     2628 CD Delft, The Netherlands
*   Correspondence: p.taheri@tudelft.nl; Tel.: +31-15-278-2275

Received: 27 August 2019; Accepted: 20 September 2019; Published: 21 September 2019

Abstract: This paper investigates a failure in HP-Mod radiant tubes in a petrochemical plant. Tubes fail after 90,000 h of working at 950 °C. Observed failure is in the form of excessive bulging and longitudinal cracking in reformer tubes. Cracks are also largely branched. The microstructure of service-exposed tubes was evaluated using optical and scanning electron microscopes (SEM). Energy-dispersive X-ray spectroscopy (EDS) was used to analyze and characterize different phases in the microstructure. The results of this study showed that carbides are coarsened at both the inner and the outer surface due to the long exposure to a carburizing environment. Metallography examinations also revealed that there are many creep voids that are nucleated on carbide phases and scattered in between dendrites. Cracks appeared to form as a result of creep void coalescence. Failure is therefore attributed to creep due to a long exposure to a high temperature.

Keywords: reformer tubes; HP-Mod; failure analysis; creep

1. Introduction and Case Background

Hydrogen is produced in the so-called "steam reforming" process, in which a steam/hydrocarbon mixture goes through vertical reforming tubes. Hydrocarbon is then transformed to hydrogen and carbon monoxide. Creep-resistant centrifugally cast high Cr/high Ni tubes are widely used in reformer units in petrochemical plants for hydrogen production [1]. HK40 (Cr25Ni20), HP40 (Cr25Ni35), and HP-Mod (Cr25Ni35Nb) are examples of currently used heat resistant alloys with superior high temperature corrosion and creep resistance. Reformer tubes experience a severe working condition, i.e., the working temperature is as high as 1000 °C and pressures are up to 3.5 MPa [2,3]. These tubes are expected to have a service life over 100,000 h, with strains no more than 3%, at temperatures and internal pressures up to 980 °C and 35 bar, respectively [4]. Hoop stress, originated from internal pressure, together with constant high temperature service exposure results in creep stress approximately equal to 30 MPa. This in turn results in a creep deformation rate of approximately 10^{-10} s^{-1} [3]. It is therefore not surprising that reformer tubes are considered very critical components when it comes to the safe operation and the integrity of installations in a petrochemical/reforming plant. In addition, from an economic standpoint, these tubes are strategically important. Obviously, shutdowns in a petrochemical plant due to failures in reformer tubes are very costly [5]. Moreover, tube replacement operations require an extensive dismantling of the furnace components. A total replacement operation can last a few days and can easily cost millions of dollars for a petrochemical plant. With that said, one can imagine that understanding and mitigating or postponing failures in reformer tubes are very important. This paper investigates a failure in reforming tubes, made of HP-Mod alloy, after 90,000 h of service exposure at approximately 950 °C. The results, obtained from this root cause failure analysis, can be used in assessing and mitigating similar failures in reforming units.

2. Experimental Methods

Samples for microstructure and failure analyses were taken from failed tubes. According to the specification, the alloy is HP-Mod. The chemical composition of samples was measured, using inductively-coupled plasma atomic energy spectroscopy (ICPAES). Optical and scanning electron microscopes (SEM Philips XL30, Philips, Eindhoven, The Netherlands) were used to investigate the microstructure of failed specimens. Energy-dispersive X-ray spectroscopy (EDS, Ametek, PA, USA) was used to characterize different phases in the microstructure. Fractography was performed on the fracture surface of failed specimens.

3. Results and Discussion

3.1. Visual Examinations

Figure 1 shows an example of failed reformer tubes together with an image, showing how tubes are installed in a reforming unit. As can be seen, cracks have propagated longitudinally. Extensive branching is also observed overall on the outer surface. It is also noticeable that in some parts tubes are largely deformed in such a way that a clear bulging is observed. Cracks are often seen in areas of localized deformation/bulging.

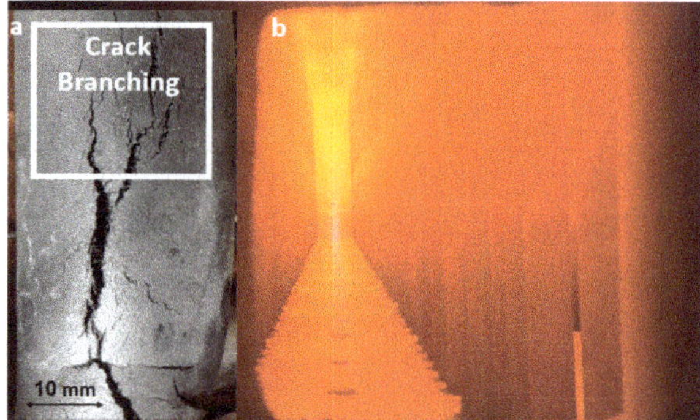

Figure 1. (a) Example of failed tubes and (b) an image of tubes in service.

3.2. Microstructure and Chemical Analyses

The chemical composition of the alloy was measured to make sure that the chemical composition of the failed tubes falls within the standard range. Table 1 shows the chemical composition of a failed specimen, showing that the alloy has a typical composition of a HP-Mod alloy.

Table 1. Chemical composition of failed specimens.

Composition	C	Si	Mn	Cr	Ni	Nb	Fe
wt%	0.41	1.2	0.9	25.5	35.6	1.3	Bal.

Optical and SEM micrographs of service-exposed tubes are shown in Figure 2. As can be seen, there are two types of morphologies for primary eutectic phases in the matrix. Primary eutectic phases can be categorized into "Chinese script" morphology (see Figure 2a) and a rather continuous grain boundary phase morphology (see Figure 2b). EDS analyses show that the eutectic phase is a mixture of dark and white phases, with the former ones being chromium carbides (Figure 2c), and more specifically $M_{23}C_6$ and M_7C_3 carbides (M = Fe, Cr) ($M_{23}C_6$ carbides reportedly gradually transform to M_7C_3 carbides during the service due to carbon diffusion/carburization [6,7]), with the latter being niobium carbide (Figure 2d).

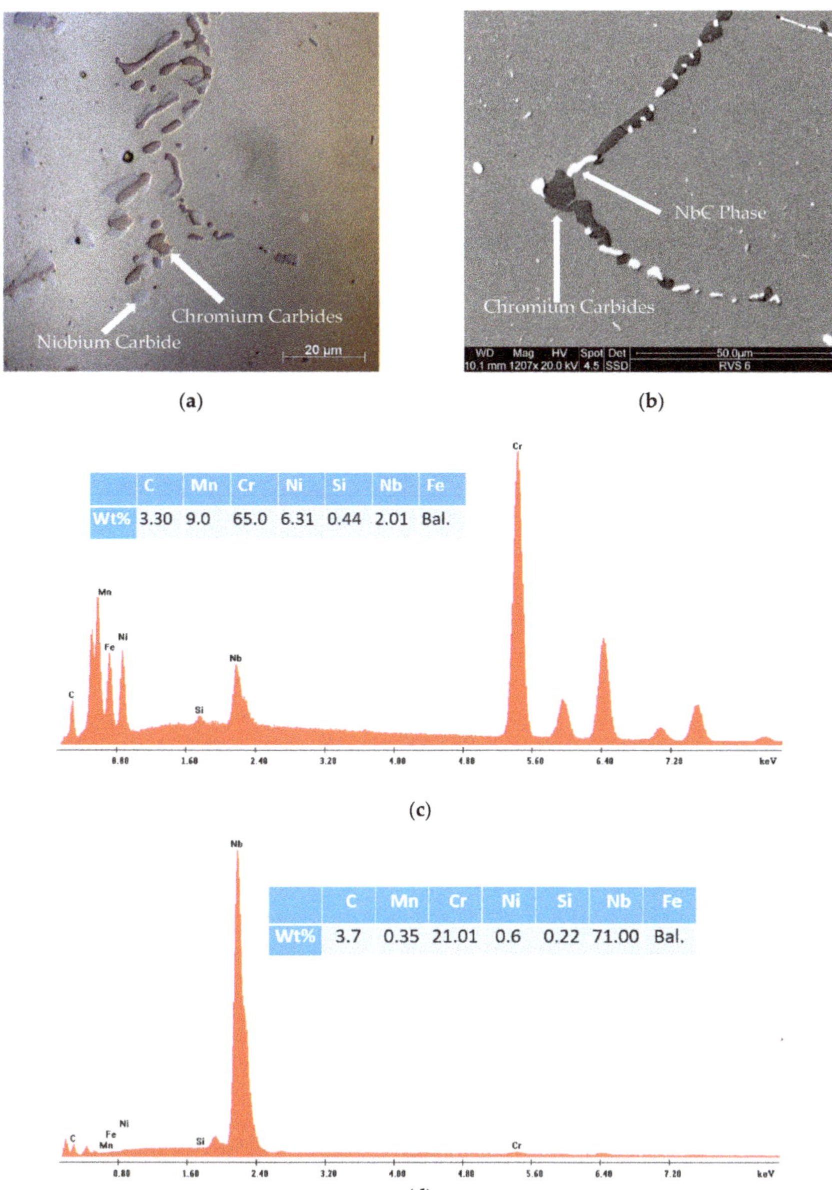

(a) **(b)**

	C	Mn	Cr	Ni	Si	Nb	Fe
Wt%	3.30	9.0	65.0	6.31	0.44	2.01	Bal.

(c)

	C	Mn	Cr	Ni	Si	Nb	Fe
Wt%	3.7	0.35	21.01	0.6	0.22	71.00	Bal.

(d)

Figure 2. (**a**) Optical and (**b**) SEM images of the microstructure in failed tubes, together with EDS analyses of the (**c**) black phase (chromium carbide) and (**d**) white phase (NbC).

Figure 3 shows a stereomicroscope image of the cross section of the failed tube. Based on this image, the grain structure can be divided into an elongated dendritic structure from the outer surface towards the half thickness, and an equiaxed grain structure from the half thickness towards the inner surface. This is a typical grain structure in centrifugally cast tubes in which solidification starts from the outer surface, followed by dendrite growth towards the center of tube. Equiaxed grains are then formed at the latest stage of solidification. In addition, it is noticeable that the examined cross-section exhibits a large number of voids. In the dendritic area, voids have more directionality and are more oriented alongside dendrite growth direction, while those in the equiaxed grain area are randomly distributed. Figure 4a,b compares the microstructure of the degraded sample with that in the as-cast condition, clearly showing how voids are formed due to longtime service exposure. In addition, Figure 4 shows optical images of voids in both dendritic structure and equiaxed structure zones. It is seen that voids in the former zone have a clear directionality and are fully connected in some areas (see Figure 4b,c). It appears that cracks are formed as a result of void coalescence. On the contrary, voids in the equiaxed structure are more randomly distributed, are more isolated and are less connected (see Figure 4c,d). It is postulated that in the case that creep prevails, embrittlement can take place, resulting in a decreased rupture ductility [8,9]. Creep and rupture strengths are considered to be extremely important design parameters [10]. Deterioration of rupture ductility due to creep largely influences crack growth rates and high temperature damage tolerance of reformer tubes. It has been hypothesized [11] that the mechanism of creep-induced embrittlement is linked to the generation and coalescence of intergranular voids, which are controlled by the formation of a denuded zone in the vicinity of the precipitates. The denuded zone around the precipitates is formed due to the segregation and diffusion of alloying elements at the precipitate/matrix interfaces. Our results show that, in this case, a network of voids has nucleated and grown during creep deformation. It appears that the failure mechanism in this case is due to the nucleation of creep voids and their evolution into microcracks, and finally macrocracks. Buchheim et al. [12] postulated that creep cracking due to long-term high temperature service can take place in the piping components and can grow due to temperature fluctuations, particularly during abrupt shutdowns. Pourmohammad et al. [5] also discussed the negative implications of sudden shutdowns for the lifetime of radiant tubes. When it comes to the creep failure of reformer tubes, the first stage is always related to the void nucleation [13–16]. Figure 5 shows that creep voids have nucleated on both chromium carbide and niobium carbide particles. Given that $M_{23}C_6$ and NbC carbides are distributed in the inter-dendritic regions [2], voids are also nucleated and distributed in-between dendrites. Figure 6 shows that, in some areas, voids are formed in between carbide particles. The working temperature of reformer tubes is high enough to cause carbide coalescence and coarsening. Fine carbide precipitates obviously enhance the creep resistance of HP-Mod alloys. Yet, in the coalesced and coarsened carbide structure, the positive attribution of carbide particles to the creep resistance of HP-Mod alloys can reverse. Coarse carbide particles can act as creep void nucleation sites. Carbide coalescence and coarsening reactions are diffusion-dependent, inferring that a slight increase in service temperature can significantly accelerate both carbide coalescence and coarsening, and this in turn results in faster nucleation of creep voids. This clearly adversely affects the creep lifetime of HP-Mod alloys. In comparison to chromium carbides, niobium carbides have a comparatively higher melting point, giving them more stability during high temperature service conditions. Rampat et al. [2] postulated that "since most of the NbC precipitates are out of the matrix and remain on the grain and/or cell boundaries, higher activation energies (via higher stresses and temperatures) are required for slip and the consequential creep voids". The results of this study depict that both chromium and niobium carbides are nucleation sites for creep voids. Examples from both cases are shown in Figure 5. Creep-induced embrittlement and distribution of voids in the microstructure lead to an easier crack propagation and transition of fracture modes. This is evidently seen in Figure 7. As can be seen in this figure, the fracture surface close to the outer surface in the dendritic region has a directional fibrous appearance. Voids are also clearly seen on the fracture surface. There is no indication of dimples anywhere on the fracture surface, supporting the argument of creep-induced embrittlement in failed tubes [17]. However, the pictures do not show ordinary intergranular or interdendritic fractures. It is likely that the fracture surface was altered by the interaction with the steam/hydrocarbon/hydrogen atmosphere.

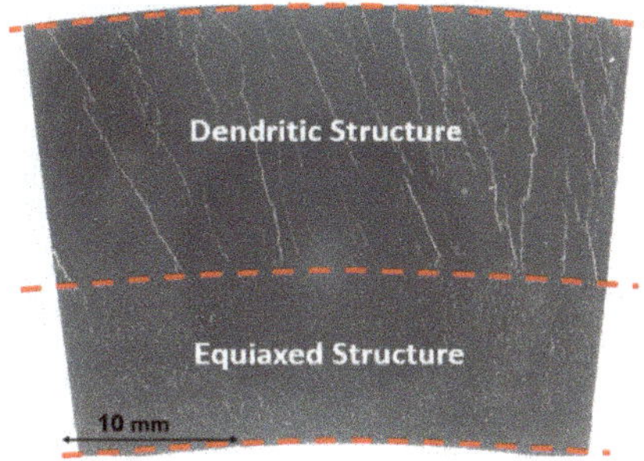

Figure 3. Stereomicroscope image of the cross section of the failed tube.

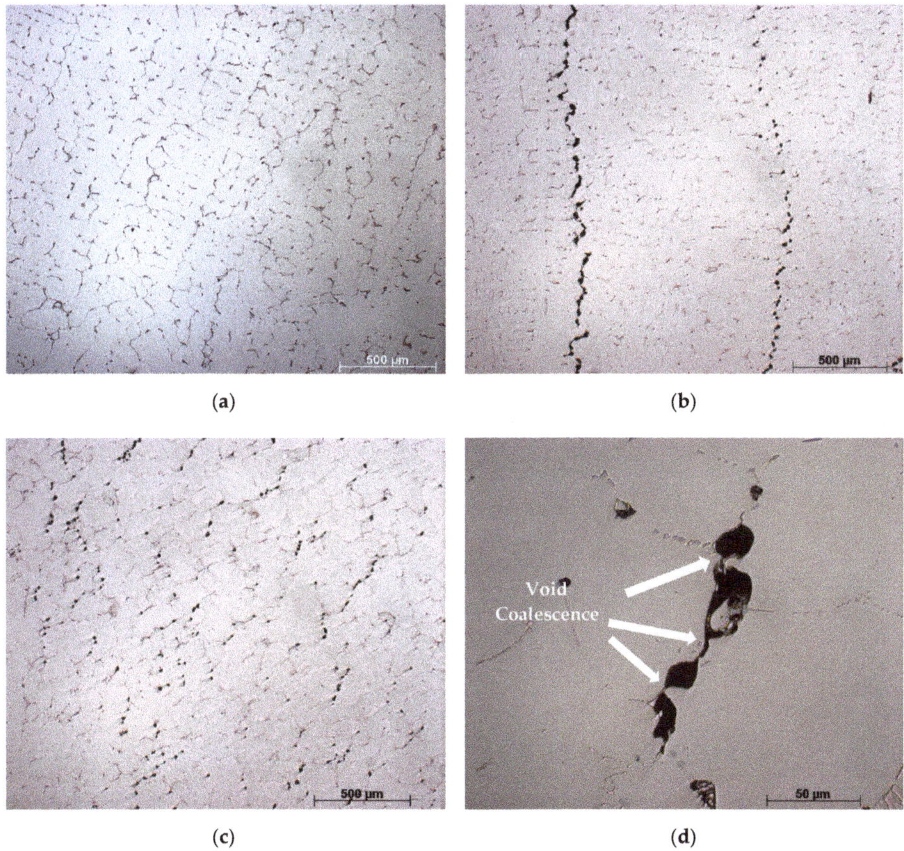

Figure 4. (**a**) The as-cast microstructure and voids in failed samples in (**b**) the dendritic structure region and (**c**,**d**) the equiaxed grain region.

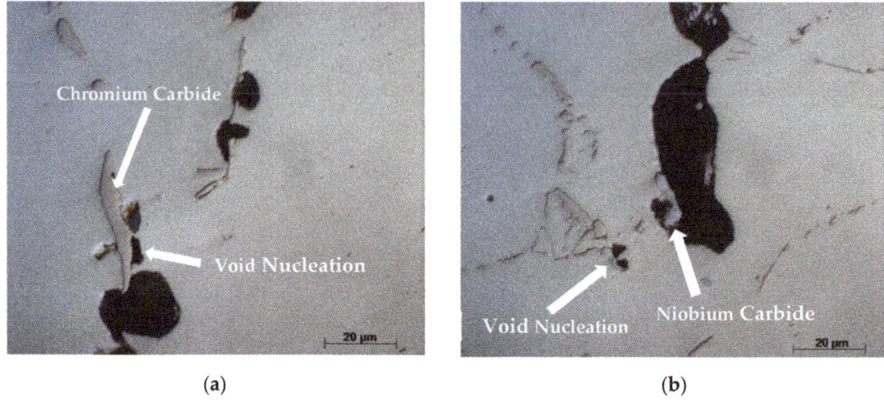

(**a**)   (**b**)

Figure 5. Void nucleation on (**a**) chromium carbide and (**b**) niobium carbide particles.

Figure 6. Creep voids are often nucleated in-between particles.

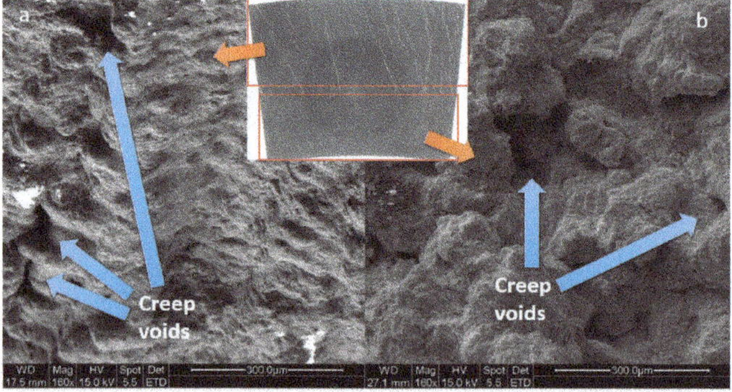

Figure 7. Fracture surface in (**a**) dendritic and (**b**) equiaxed grain regions.

4. Concluding Remarks

This paper investigated the root cause analysis of failure in HP-Mod radiant tubes in a petrochemical plant. Failure was reported after 90,000 h of working at 950 °C. Longitudinal cracks with lots of branching were observed on failed tubes in such a way that these tubes could not be used again due to crack and excessive creep deformation. The results of failure analysis showed that:

- Tubes have two typical grain structures: (i) dendritic microstructure, starting from the outer surface towards the half-thickness, and (ii) equiaxed grain structure, starting from the half thickness towards the inner surface. This is the typical microstructure of centrifugally cast heat resistant alloys.

- Two types of carbides were present in the microstructure: (i) chromium-rich carbides and (ii) niobium carbides. In most cases, carbides had a "Chinese script" morphology. Carbides were overall coalesced and coarsened to a large extent.

- Creep voids were observed through the thickness of failed tubes. Voids in the dendritic structure region were oriented more alongside dendrites and were seen more often in the inter-dendritic regions. On the contrary, voids in the equiaxed grain structure were more randomly distributed without any directionality.

- Creep voids were nucleated on both chromium and niobium carbides. Cracks were formed due to the growth and coalescence of voids.

- The root cause analysis of the concerned failure showed that tubes failed due to creep failure.

- When it comes to creep failure, the most important factor that controls the lifetime of the components during a high temperature service is temperature. Improper control of temperature and overheating can dramatically decrease the creep lifetime of reformer tubes.

Author Contributions: Conceptualization, A.B.; methodology, A.B.; formal analysis, A.B.; investigation, A.B. and P.T.; writing—original draft preparation, A.B.; writing—review and editing, P.T.; visualization, A.B.; supervision, P.T.; project administration, P.T.

Funding: This research received no external funding.

Acknowledgments: Authors would like to thank Delft University of Technology for support in publishing this manuscript.

Conflicts of Interest: The authors declare no conflict of interest.

References

1. Jones, D.R.H. Creep failures of overheated boiler, superheater and reformer tubes. *Eng. Fail. Anal.* **2004**, *11*, 873–893. [CrossRef]
2. Rampat, K.; Maharaj, C. Creep embrittlement in aged HP-Mod alloy reformer tubes. *Eng. Fail. Anal.* **2019**, *100*, 147–165. [CrossRef]
3. Guglielmino, E.; Pino, R.; Servetto, C.; Sili, A. Creep damage of high alloyed reformer tubes. In *Handbook of Materials Failure Analysis with Case Studies from the Chemicals, Concrete and Power Industries*; Butterworth-Heinemann: Oxford, UK, 2016; pp. 69–91.
4. Swaminathan, J.; Guguloth, K.; Gunjan, M.; Roy, P.; Ghosh, R. Failure analysis and remaining life assessment of service exposed primary reformer heater tubes. *Eng. Fail. Anal.* **2008**, *15*, 311–331. [CrossRef]
5. Pourmohammad, H.; Bahrami, A.; Eslami, A.; Taghipour, M. Failure investigation on a radiant tube in an ethylene cracking unit. *Eng. Fail. Anal.* **2019**, *104*, 216–226. [CrossRef]
6. Tari, V.; Najafizadeh, A.; Aghaei, M.H.; Mazloumi, M.A. Failure analysis of ethylene cracking tube. *J. Fail. Anal. Prev.* **2009**, *9*, 16–32. [CrossRef]
7. Otegui, J.L.; De Bona, J.; Fazzini, P.G. Effect of coking in massive failure of tubes in an ethylene cracking furnace. *J. Eng. Fail. Anal.* **2015**, *48*, 201–209. [CrossRef]
8. American Petroleum Institute. *API 942B Material, Fabrication, and Repair Considerations for Austenitic Alloys Subject to Embrittlement and Cracking in High Temperature 565 °C to 760 °C Refinery Services*; API: Washington DC, USA, 2017.

9. Barcik, J. Mechanism of σ-phase precipitation in Cr–Ni austenitic steels. *Mater. Sci. Technol.* **1988**, *4*, 5–15. [CrossRef]

10. Zieliński, A.; Dobrzański, J.; Purzyńska, H.; Golański, G. Properties, structure and creep resistance of austenitic steel Super 304H. *Mater. Test.* **2015**, *57*, 859–865. [CrossRef]

11. Villanueva, D.M.E.; Junior, F.C.P.; Plaut, R.L.; Padilha, A.F. Comparative study on sigma phase precipitation of three types of stainless steels: austenitic, superferritic and duplex. *Mater. Sci. Technol.* **2006**, *22*, 1098–1104. [CrossRef]

12. Buchheim, G.; Becht, C.; Nikbin, K.; Dimopolos, V.; Webster, G.; Smith, D. Influence of aging on high-temperature creep crack growth in type 304H stainless steel. In *Nonlinear Fracture Mechanics: Volume I Time-Dependent Fracture*; ASTM International: West Conshohocken, PA, USA, 1988.

13. Song, R.; Wu, S. Microstructure evolution and residual life assessment of service exposed Cr35Ni45 radiant tube alloy. *J. Eng. Fail. Anal.* **2018**, *88*, 63–72. [CrossRef]

14. Khodamorad, S.H.; Haghshenas Fatmehsari, D.; Rezaiea, H.; Sadeghipour, A. Analysis of ethylene cracking furnace tubes. *J. Eng. Fail. Anal.* **2012**, *21*, 1–8. [CrossRef]

15. Alvino, A.; Lega, D.; Giacobbe, F.; Mazzocchi, V.; Rinaldi, A. Damage characterization in two reformer heater tubes after nearly 10 years of service at different operative and maintenance conditions. *J. Eng. Fail. Anal.* **2010**, *17*, 1526–1541. [CrossRef]

16. Guan, K.; Xu, H.; Wang, Z. Analysis of failed ethylene cracking tubes. *J. Eng. Fail. Anal.* **2005**, *12*, 420–431. [CrossRef]

17. Monobe, L.S.; Shiguenobu, L.; Schön, C.G. Microstructural and fractographic investigation of a centrifugally cast 20Cr32Ni+Nb alloy tube in the 'as cast' and aged states. *J. Mater. Res. Technol.* **2013**, *2*, 195–201. [CrossRef]

© 2019 by the authors. Licensee MDPI, Basel, Switzerland. This article is an open access article distributed under the terms and conditions of the Creative Commons Attribution (CC BY) license (http://creativecommons.org/licenses/by/4.0/).

Review

An Introduction to Wear Degradation Mechanisms of Surface-Protected Metallic Components

Pandora P. Psyllaki

Department of Mechanical Engineering, University of West Attica, 250 Thivon avenue & P. Ralli, 12244 Egaleo, Greece; psyllaki@uniwa.gr; Tel.: +30-210-538-1292

Received: 1 September 2019; Accepted: 25 September 2019; Published: 28 September 2019

Abstract: Despite the fact that ceramics and polymers have found numerous applications in several mechanical systems, metals and metallic alloys still remain the main materials family for manufacturing the bulk of parts and components of engineering assemblies. However, in cases of components that are serving as parts of a tribosystem, the application of surface modification techniques is required to ensure their unhampered function during operation. After a short introduction on fundamental aspects of tribology, this review article delves further into four representative case studies, where the inappropriate application of wear protection techniques has led to acceleration of the degradation of the quasi-protected metallic material. The first deals with the effects of the deficient lubrication of rolling bearings designed to function under oil lubrication conditions; the second is focused on the effects of overloading on sliding bearing surfaces, wear-protected via nitrocarburizing; the third concerns the application of welding techniques for producing hardfacing overlayers intended for the wear protection of heavily loaded, non-lubricated surfaces; the fourth deals with the degradation of thermal-sprayed ceramic coatings, commonly used as wear-resistant layers.

Keywords: surface modification techniques; degradation of protective layers; lubrication; nitrocarburizing; hardfacings; thermal-sprayed coatings

---

1. Introduction

The term "tribosystem" or "tribopair" is commonly used to describe the conjugated function of two components that are moving relevant to each other [1]. Although in general, one element of the tribosystem could be in the fluid state, e.g., natural gas moving in a pipeline, in the majority of mechanical systems both elements are in the solid state and their coupled operation serves to transfer motion, power, or mechanical loading; thus, a typical tribosystem is schematically presented in Figure 1a.

A tribosystem is defined by (a) the geometry of the participating elements, (b) their relevant geometry that defines the contact as either conformal or contra-formal, (c) the microgeometry of the surfaces in contact (roughness, Figure 1b), and (d) the construction materials of the two elements and the degree of their hardening. Each such tribopair can perform under various operational parameters that combine the normal load applied (F), the relevant speed (v), and the intervention of a third medium (solid or fluid) at the area between the two in-contact surfaces [1]. Finally, the operation of the tribosystem is carried out in a particular surrounding environment that is characterized by the level of relevant humidity (% RH) and temperature, or even by the application of vacuum. Depending on the type of the relevant motion, the tribosystems in mechanical engineering can be generally divided into rolling or sliding bearings.

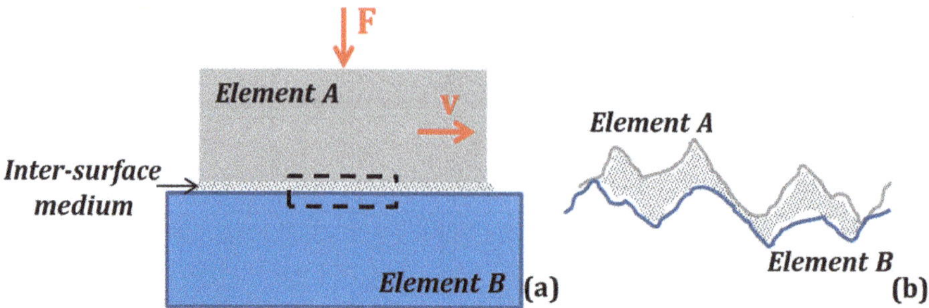

Figure 1. (**a**) Schematic representation of a tribosystem and (**b**) magnification of the inter-surface area marked in (**a**).

The resistance to the relevant motion between the two elements is reflected in the friction coefficient value (μ), which is equal to the ratio of the normal load applied to the friction force that appears ($\mu = \frac{F}{T}$). The higher the friction coefficient, the higher is the energy consumption for keeping the motion ongoing [2]. Under these conditions, the operation of the tribosystem leads simultaneously to material removal from the two conjugated surfaces, i.e., material losses [3]. In the meantime, the application of a normal load at the contact surface leads to a distribution of applied pressure that results in a distribution of normal and shear stresses within the upper surface layers of the two solids in contact; the former are maximized on the contact area, while the latter are maximized at a depth below the contact area of the two solids [1–4]. This fact could result in crack initiation and propagation during the operation of a tribosystem.

Depending on the particular application of each tribo-assembly, the requirement could be for either low or high friction coefficient values, i.e., cutting tools or brakes, respectively, either low or high wear values, i.e., brakes or workpiece, respectively. In all cases, the stable operation of the tribosystem requires a constant friction coefficient and uniform wear rates during function, as well as normal and shear stresses lower than the yield point of the construction materials of the two conjugated elements and significant durability when operating in the fatigue regime. The combined effect of the friction force, the normal and shear stresses, and the adhesion forces developed between the two elements define the wear micro-mechanisms that are activated on the contact area, leading to material removal from both conjugated surfaces [3].

The topography and the mechanical properties of the surface are of crucial importance for the operation, since they determine the stress distribution, the ease of relevant motion, and the wear rate and mechanisms that affect the surface and sub-surface crack initiation and propagation, energy consumption, and material losses, respectively. In order to enhance the performance of such bearing surfaces, several modification techniques are available, involving liquid or solid compounds that alter the distribution of surface stresses, the relevant motion, and/or material removal, but not the mechanical properties of the bulk metal of the entire structure. Even though such surface modification techniques do have a beneficial effect on engineering surfaces, their inappropriate application or their erroneous selection for a particular application, render their involvement in the tribosystem a rather detrimental factor for the integrity of the entire structure.

In order to overcome undesirable premature failure during the operation of a tribosystem consisting of two conjugated metallic solids, it is recommended to modify the properties of their surfaces exposed to the action of the above-mentioned facts. The most widely applied techniques that result in mainly the reduction of the friction force—thus, reduction of the friction coefficient and consequently diminution of the energy demands for motion sustainability—consist of the application of a lubricant medium between the two conjugated surfaces [2]. The main categories of lubrication, representative media, and common applications for each one are presented in Table 1.

Table 1. Main categories of lubrication and representative examples.

Category	Representative Lubricants	Main Applications
Solid-State Lubrication	Graphite; Phyllomorphous Minerals	Dies and Molds for Metal Forming and Shaping
Liquid-State Lubrication	Oil-Based; Water-Based	Engines and Tool-Machines for Metals Machining
Gas-State Lubrication	Air	Dental Equipment

To ensure their effectiveness, both solid and liquid lubricants should, mainly, exhibit effective adherence to the contact surfaces of the conjugated solids, coherence under the normal and shear stresses and compositional stability at the temperature range to be developed during operation of the tribosystem.

During the last decades, several surface modification techniques [4–7] were developed aiming to reduce the wear rate of the material of the main construction by several orders of magnitude, not necessarily leading to a decrease of the friction coefficient. The majority of these techniques are based on the creation of a surface overlayer on the initial material; the former being much more wear resistant than the latter. In this way, the service lifetime of the elements of the tribopair is prolonged and, in some cases, this "protective" surface layer can be easily replaced without any adverse effect on the metal of the main structure. In general, the thickness of such overlayers is smaller compared to the dimensions of the main structure; however, these are efficient to expand significantly the service lifetime of the elements, without crucial dimensional changes of the main components. The main categories of these techniques, representative surface layers, and respective applications are presented in Table 2.

Table 2. Main categories of coatings techniques and representative examples.

Technique	Representative Wear-Resistant Surface Layers	Main Applications
Thermochemical Treatments	Nitriding and nitrocarburizing	Forming, Cutting tools, Dies, Gears, Shafts, Clutches
Welding; High-power laser	Carbide-reinforced Fe-based composites	Heavily-loaded surfaces
Thermal Spraying	Oxides (Al_2O_3; TiO_2, YSZ); WC-, Cr_3C_2 based CerMets	Medical implants Heavily-loaded surfaces
Physical or Chemical Vapor Deposition (PVD or CVD)	TiN; TiC; BN; Diamond; DLC	Cutting tools
Metallization/Plating	Cr-, Ni- coatings	Automotive parts

In particular, the typical coatings obtained via PVD, CVD, and plating (electrolytic or electroless) contribute not only to wear resistance but also to the reduction of the friction coefficient. Despite the fact that both approaches, i.e., lubrication and coatings deposition, contribute significantly to the protection of the metal of the basic construction, their inadequate application or their use under non-recommended operating conditions would lead to the acceleration of degradation of the metallic structure to be protected and to premature catastrophic failure of the elements of the tribosystem. Four relevant case studies that emphasize the degradation of metallic components of representative wear-protected elements due to the failure of the protection are discussed below, the main aim being to present, in a unified and coherent way, the detrimental effects on the performance of tribo-elements that should have exhibited prolonged lifetime after the application of a surface protection technique.

2. Synopsis of Experimental Techniques

It is generally admitted that in cases of severe degradation, the root cause of failure could be accurately determined by observing the failed system at low magnification or even by optical inspection, whilst observations at higher magnifications or complementary diagnostic techniques can be applied in order to further elucidate mechanisms that are unclear, vague, and not widely known. For this purpose and on a case-by-case basis, the experimental techniques applied in the following representative examples comprised:

(1) Visual inspection and stereomicroscopic (Leica MS 5,Leica Microsystems GmbH, Wetzlar, Germany) and optical (Olympus BX60, Olympus Corporation, Tokyo, Japan) observations for determining the material's failure areas that could provide information on the root cause, as well as areas that merit further laboratory examination.

(2) Scanning electron microscopy (SEM, JEOL 840A, Jeol Ltd, Tokyo, Japan) coupled with elemental micro-analysis (EDS, OXFORD INCA 300, Oxford Instruments, Abingon, UK) and X-ray diffraction (XRD, Siemens D-800, Siemens AG, Munich, Germany) analysis of selected areas, for revealing the material's flaws or transformations that accelerated failure.

(3) Ball-on-disc tribological measurements (Centre Suiss d' Electronique et de Microtechnique, CSEM SA, Neuchâtel, Switzerland), for evaluating the performance of wear-resistant protective coatings before service.

3. Case Study I: Failure of Lubricated Rolling Bearings

3.1. Fundamentals of Liquid Lubrication

Already since 1902, Stribeck [8] proposed a model for lubricated friction that describes the behavior of a tribosystem in terms of friction coefficient, taking into account the combined factor $\left(\eta\frac{v}{P}\right)$ of the operational parameters, where (η) is the viscosity of the lubricant, (v) the rotation speed, and (P) the load per unit of the projected bearing area.

As can be observed in Figure 2, depending on the operational parameters, each particular lubricated tribosystem exhibits a wide range of friction coefficient values that correspond to three different lubrication regimes [2]:

- Within regime I, also known as boundary lubrication, the friction coefficient value tends to be similar to that under non-lubricating conditions. The thickness of the lubrication film is very thin, the load is taken on exclusively by the two solids and is transferred from the one to the other trough the contact between their surface asperities; thus, the rheological characteristics of the lubricant do not intervene and the lubrication efficacy depends only on the physicochemical characteristics of the liquid. These exactly "quasi-dry" lubrication conditions result in severe wear of the two solids.

- Within regime II, also known as mixed or partial elasto-hydrodynamic lubrication, the friction coefficient drops down to very low values. The thickness of the lubrication film is moderate, and the load is born by both the solid and the liquid film, depending on the thickness of the latter. The consecutive wear is moderate compared to that in regime I.

- Within regime III, also known as hydrodynamic lubrication, the friction coefficient values are slightly higher, the thickness of the lubrication film is significant enough to result in complete separation of the two solids and the load is transferred from the one to the other through the lubricant that should take on the entire load; thus, lubrication efficacy is depended only on the rheological characteristics of the liquid. In this case, the wear of the two non-in-contact solids is negligible.

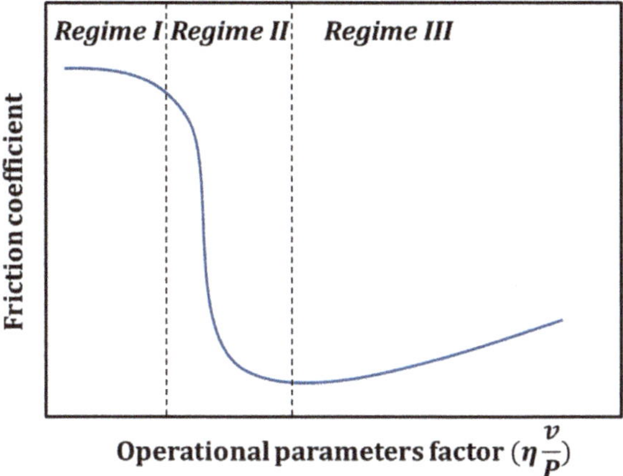

Figure 2. Typical Stribeck curve for liquid lubrication and associated regimes.

Since 1902, this model has been further improved by emphasizing the effect of the surface microgeometry (average roughness) of the two conjugated solids [9,10] and the thickness of the lubrication film, and the distinction between the three regimes is based on the lambda ratio (λ) that is given by Equation (1):

$$\lambda = \frac{h_{min}}{\sqrt{R_{a,1}^2 + R_{a,2}^2}}$$

(1)

where h_{min} is the minimum film thickness for assumed smooth bearing surfaces, and R_a is the average roughness of solids (1) and (2).

Based on this approach, boundary lubrication should be expected for (λ) values lower than one and hydrodynamic lubrication for (λ) lambda values higher than three [2]. Depending on the load and the relevant speed of motion applied to the tribosystem, as well as the viscosity of the lubricant used, one should have in mind that the low friction coefficient value could suddenly increase to very high ones and the wear from negligible could turn to a severe one. Special attention should be given to the thermal stability of the rheological characteristics of the liquid that could alter the lubrication mechanisms, as well as its additives that could chemically attack the metallic components, leading to local corrosion [11] that could be the initiation locus of cracking [12].

3.2. Catastrophic Failure of Rolling Bearings

The catastrophic failure of a rolling-element bearing that should operate in the hydrodynamic lubrication regime [13] is presented as an example of metal degradation due to failure of the protective means.

The sub-assembly failed after six months of continuous operation in a centrifugal pump that was supporting the recirculation of monoethanolamine (MEA) in a petroleum refinery. Unhampered performance of the assembly is crucial since such a cleaning process of several products is essential before their further treatment; thus, interruption of the pump operation has a negative financial effect on the entire production. The temperature at the cleaning unit did not exceed 45 °C, whilst the speed was 2899 rpm, ~24% lower than the maximum value recommended by the bearing' s supplier.

After the removal of all the organic residuals from the metallic surface, visual inspection revealed the catastrophic failure of the outer ring of the sub-assembly (Figure 3). In particular, 30% of its periphery was separated into three fragments that presented irregular fracture surfaces (Figure 4) and signs of oxidation on them (Figure 5).

Figure 3. Rolling bearing as-received (**a**) and after the removal of organic residuals (**b**).

Figure 4. Stereoscopic images of a fragment of the failed outer ring, exhibiting geometric irregularities.

Figure 5. Stereoscopic images of a fragment of the failed outer ring, exhibiting signs of oxidation on the fracture surface.

Disassembling of the rolling bearing revealed that failure was not limited to the ring, but it has also affected the rolling elements (balls) that exhibited material removal in the form of craters of smaller or higher dimensions (Figure 6).

The morphology of both the fracture surfaces of the outer ring and the rolling balls advocates for insufficient lubrication during operation that resulted in impact loading that led eventually to the detrimental catastrophic failure of the sub-assembly and the operation interruption of the whole cleaning unit.

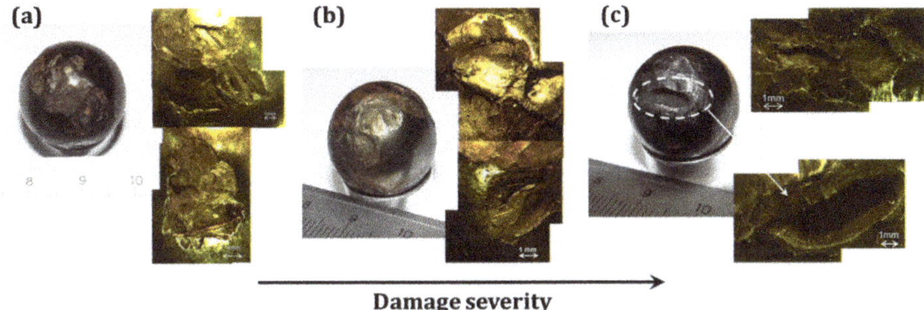

Figure 6. Representative failed rolling balls exhibiting different levels of damage severity.

4. Case Study II: Failure of Nitrocarburized Sliding Bearings

4.1. Fundamentals of Nitrocarburizing

Several engineering components, such as gears and shafts, are commonly manufactured by steel of different grades and hardening degrees, depending on the type and the level of mechanical loading to which they would be subjected during their final application. In the case of these metals serving as parts of a tribosystem, their surface reinforcement against sliding wear is a requirement not only for decreasing their wear rate, but also for altering the wear micro-mechanisms that could lead to the secondary activation of severe degradation of the entire structure via oxidation and/or crack initiation and propagation. Such a surface modification technique widely applied at an industrial scale is nitrocarburizing and, especially, Tufftriding, often referred to as liquid nitrocarburizing.

Tufftriding is a thermochemical technique governed by the diffusion of nitrogen and carbon in the ferrous matrix [5,14,15]. After a first step of preheating for moisture removal, the steel components are immersed for about 6 h in a bath of cyanide salts, with a chemical composition of 60% KCN, 24% KCl, and 16% K_2CO_3. The treatment temperature remains constant at 580 °C, lower than that of austenitization; thus, the technique is characterized as ferritic nitriding. Under these conditions, two chemical reactions take place:

$$2KCN + O_2 \rightarrow 2KCNO \tag{2}$$

$$8KCNO \rightarrow 2K_2CO_3 + 4KCN + CO_2 + (C)_{Fe} + (N)_{Fe} \tag{3}$$

During the first reaction, potassium cyanide is oxidized to potassium oxycyanide, whilst during the second one, atomic nitrogen and carbon are produced and adsorbed by the solid surface. The desirable surface layer, enriched in nitrogen and carbon, is formed via simultaneous chemical reaction of the produced species with the ferrous matrix and nitrogen in-depth diffusion, as described by Fick' s second law:

$$\frac{\partial C}{\partial t} = D\frac{\partial^2 C}{\partial x^2} \tag{1}$$

After their rinsing with water to remove salt residues mechanically attached to the steel surface, the components are subjected to a stress relief heat treatment.

These two simultaneous mechanisms lead to the consecutive formation of two distinct layers [16–18] on the treated surface, that both consist the so-called case depth (Figure 7):

(a) The outermost one is the product of the chemical reaction between the diffusing atoms and the iron of the base steel and consists mainly of ε-carbonitride Fe_{2-3} (C,N). This layer, known as white or compound layer, is the feature that provides high wear resistance to the underlying material. Although its thickness does not exceed 10 µm, the compound layer contributes to the decrease of specific wear even by an order of magnitude [17].

(b) The layer underneath is formed via the in-depth nitrogen diffusion that leads, primarily, to α-(Fe,N) solid solution with a nitrogen content decreasing with increasing depth. In previous works [18], it was found that the thickness of this diffusion layer is strongly affected by the exact chemical composition of the steel grade, its level of prior thermal hardening, or cold working percentage. Although this layer does not affect the wear resistance, it results in a field of compressive stresses that have a positive effect on the fatigue performance of the treated components.

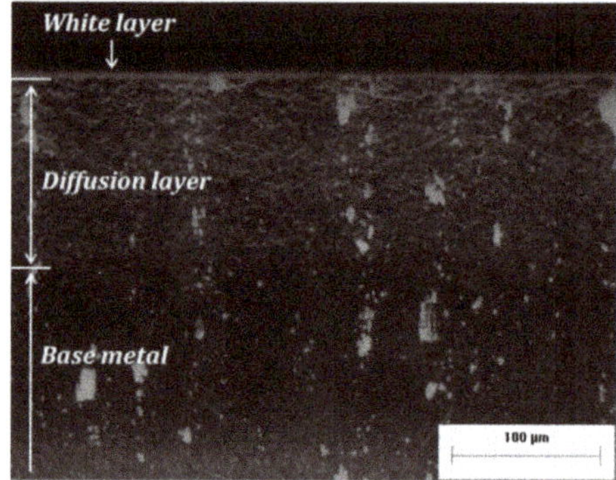

Figure 7. Representative cross-section of a tool steel subjected to Tufftriding (optical micrograph).

Compared to other gas nitriding techniques, Tufftriding is a widely applied one, mainly due to its low cost, easy application, and high reproducibility. Nowadays, the scientific interest of engineers applying this technique is focused on the waste neutralization and their management optimization strategies, since from a materials point of view, the reliability of the technique is indisputable.

4.2. Degradation of Nitriding Layers and Acceleration of Base Metal Failure

In a previous work [19] concerning the comparative study of the same steel grades with and without surface nitrocarburizing surface treatment, it was demonstrated that Tufftriding prevents galling failure (Figure 8), i.e., motion suspension of a metal–metal tribopair due to strong adhesion phenomena having occurred at the contact surface during dry sliding, typically observed in the case of sliding bearings. Moreover, nitrocarburizing protects the metallic surface from extensive oxidation (Figure 9a) that can lead to secondary tribo-corrosion mechanisms that would increase the wear rate of the metallic component.

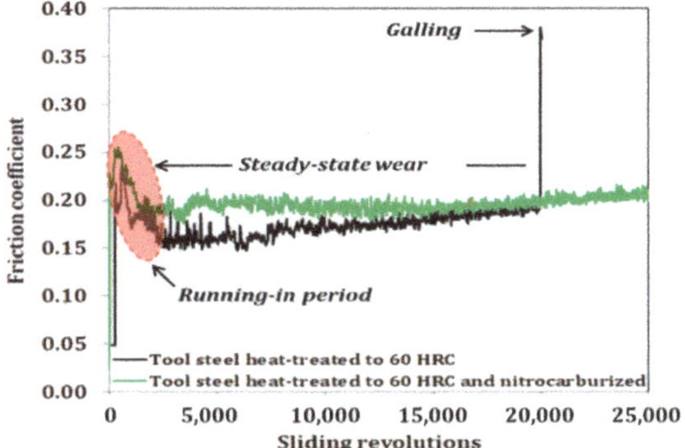

Figure 8. Evolution of the friction coefficient of a tool steel grade with and without nitrocarburizing surface treatment.

However, overloading can lead to cracking and exfoliation of the protective white layer (Figure 9b) creating cavities on the worn surface. In dry sliding under ambient atmosphere, these cavities act as loci of moisture accumulation that promote oxidation of the underlying diffusion zone, leading at the same time to extensive internal oxidation [16]. Even if nitrocarburizing has a positive effect on the dry sliding performance, a designer should take the operational limits (maximum applied load and sliding speed) per steel grade and hardening level into account. In cases of nitrocarburizing of a steel of low hardness, the "quasi-protective" layer would fail under milder operational parameters, accelerating the degradation of the underlying base metal. Since it is rather costly and time-consuming to test experimentally a material before proposing it as the suitable one for a specific application, the development of a well-structured artificial neural network (ANN) can provide wear maps that predict the areas of safe application [20] or areas of wear mechanism alteration [21].

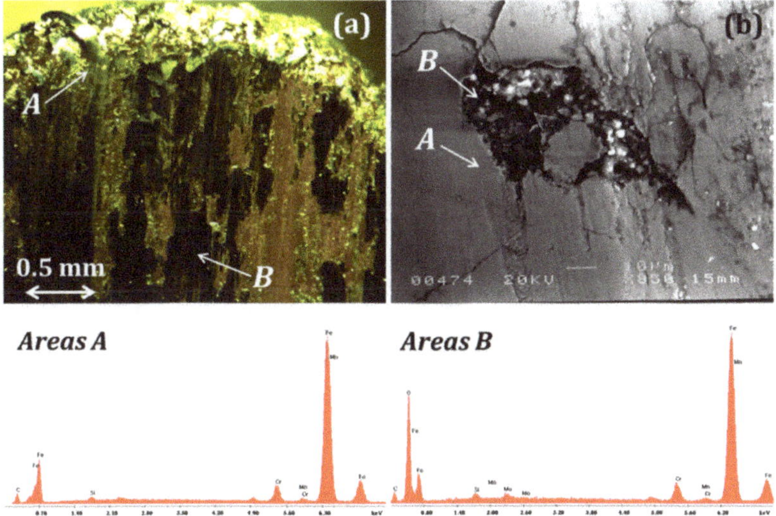

Figure 9. Scanning electron microscopy (SEM) micrographs and EDS point analysis on characteristic areas of the worn surfaces before surface treatment (a) and after liquid nitrocarburizing (b).

5. Case Study III: Failure of Wear-Resistant Hardfacing Overlayers

5.1. Hardfacing Overlayers Elaborated via Melting and Resolidification

In many applications, the increased resistance to surface loading is often achieved via the deposition of composite layers, consisting of a metallic matrix reinforced by ceramic particulates. The latter are usually oxide or carbide particles that are dispersed homogeneously in the metal. Such overlayers are deposited with a thickness of ~300 μm up to several mm, and they exhibit high hardness and a low wear coefficient. These properties render them ideal protection for surfaces of conjugated elements that operate in tribo-pairs under severe loading in demanding industrial environments, whilst their exact values are defined by both the type of the reinforcing particles as well as their percentage and size distribution.

The common methodology for depositing composite overlays reinforced by carbide particles onto steel base metal is based on the fundamentals of arc welding and is typically applied in filling cavities on metallic surfaces that have been subjected to extreme abrasive wear. In this particular case, the overlayers are referred as "hardfacings"—not simply as "surface coatings"—a term that emphasizes their inherent properties (high hardness, low wear coefficient), beneficial for applications where wear resistance is the prime requirement. In conventional arc welding techniques, the material to be deposited is provided as a consumable bulk rod that contains iron, carbon, and strongly carbide-forming elements. During hardfacing, the feedstock material at its molten state is deposited on the surface of the ferrous base metal, part of which is also subjected to melting. The relevant weight percentage of the material provided and the base metal, when both exist in the liquid state of total miscibility, is called dilution, and its exact value is depended on the particular welding technique used and the operational parameters applied. During cooling, this melt is re-solidified leading to the desirable hard overlayer, the microstructure of which is governed by the chemical composition of the melt [22] and the cooling rate imposed. This processing results in the dispersion of carbides—primarily of the strongly carbide-forming elements contained in the feedstock rod—within a Fe-based matrix, having a modified chemical composition compared to the initial one of the base metal. Typical examples of such a structure are presented in Figure 10.

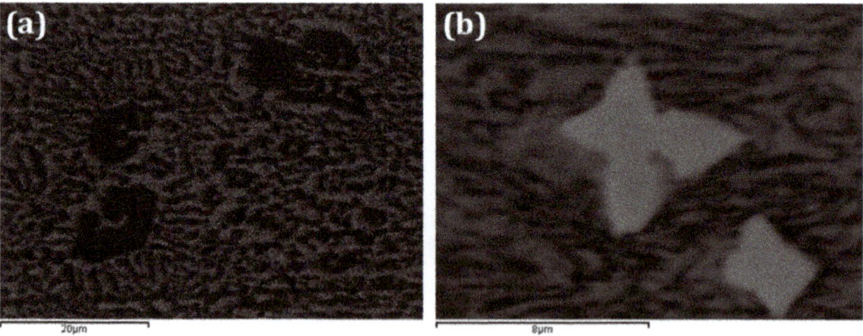

Figure 10. SEM micrographs of primary (**a**) chromium and (**b**) niobium carbides formed on an AISI 4130 steel via arc welding technique.

In a variation of hardfacing via the arc welding technique known as flux core arc welding (FCAW), the carbide powder comprises the core of a tubular metallic-shell rod, which is used as feeding material, so that the carbide particles are introduced in their solid state in the metallic melt and not precipitated thereof as primary carbides during solidification, as previously described. In both cases, the temperature at the vicinity of the metallic pool should exceed 1550 °C, typical of the melting temperature of common steels. Hardfacings deposited via arc welding find wide industrial applications, because of the low cost of the needed facilities and the occupation of non-highly-skilled personnel;

however, the elaborated overlayers often exhibit microstructure flaws due to the rather empirical and random way of application. These flaws, commonly pores and cracks (Figure 11), are typical of all processes based on the gradual solidification of melts in contact with non-molten solids, deteriorating the elements' performance and reducing their service life. Especially in the case of manual deposition, the non-precise control of the treatment parameters increases the possibility of such flaws. Moreover, the occasionally prolonged exposure of the workpieces to high thermal loads results in extended heat-affected zones and could cause distortion of the entire structure and/or detrimental transformations of the base metal microstructure.

Besides conventional arc welding, the plasma transferred arc (PTA) [23,24] and laser beams [25,26] are applied as non-conventional techniques for elaborating hard overlayers on steel components. Nevertheless, the small diameter of the high-energy beam limits the melting area to a diameter lower than 3 mm, a fact that in turn imposes multiple-step processing of larger surfaces, intensifying the negative effects of high thermal loading. Recently, concentrated solar energy (CSE) was proven [27] an attractive alternative for elaborating such hardfacing layers free of pores and cracks on large dimensional surfaces.

In the following paragraph, the case of hardfacing TiC-based overlayers deposited via automated arc melting technique is presented as a good practice example of the development of wear-resistant composite hardfacings.

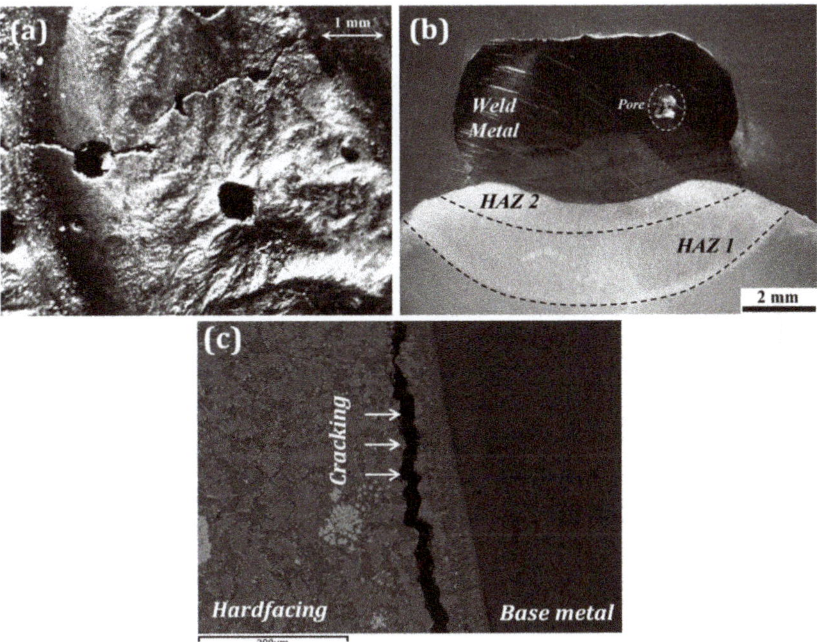

Figure 11. Microstructure flaws in hardfacing overlayers deposited manually via arc welding technique: (**a**) stereoscopic image of top view, revealing the development of surface pores and cracks; (**b**) optical image of a cross-section, revealing the existence of a sub-surface isolated macro-pore; and (**c**) SEM image of a cross-section at the vicinity of the base metal, indicating the development within the overlayer of a crack parallel to the interface.

5.2. Performance of TiC-Based Composite Overlayers Produced via FCAW Technique

The microstructure characteristics and the XRD analysis of both the core of the commercial wire fed for the elaboration of the conventional FCAW hardfacing, as well as of the deposits obtained are compared in Figure 12. The particles' agglomerates (Figure 12a) consist of TiC, which is the compound of interest in the present study, and CaF_2 and Si, which are the fluxing agents, whilst traces of ferrite were also detected originating from the wire's shell (Figure 12b). The deposit obtained (Figure 12c) is characterized by the uniform dispersion of fine TiC particulates in the metallic matrix, with rare presence of small size pores, and consists of only ferrite and TiC (Figure 12d).

Figure 12. (**a**) SEM micrograph and (**b**) X-ray diffraction (XRD) spectrum, respectively, of the core powder of the wire used for flux core arc welding (FCAW) deposits. (**c**) SEM micrograph and (**d**) XRD spectrum, respectively, of the obtained hardfacing layers.

The tribological in-service performance of these FCAW deposits against an Al_2O_3 ball is presented in Figure 13. The friction coefficient reached immediately the steady-state stage (Figure 13a), acquiring a constant value of 0.78 ± 0.02, which was practically the same for both applied loads. The wear coefficient reached the steady-state conditions after 20,000 sliding revolutions (Figure 13b), tending towards $(5.45 \pm 0.31) \times 10^{-7}$ and $(6.84 \pm 0.49) \times 10^{-7}$ mm$^3 \times$ N$^{-1} \times$ laps^{-1} for 5 and 10 N, respectively.

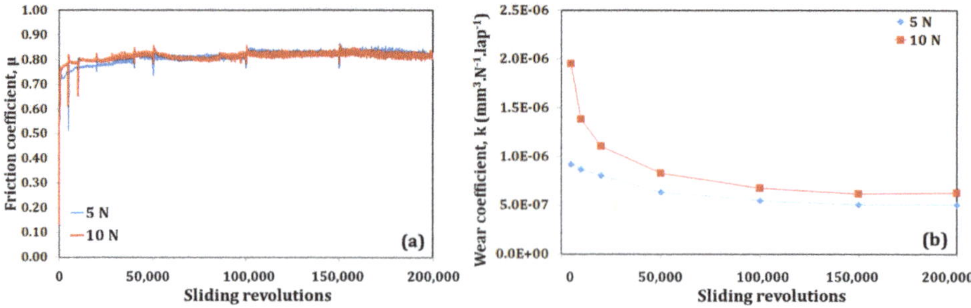

Figure 13. Representative ball-on-disk testing results of the FCWA hardfacing layers (counterbody: Al_2O_3 ball): evolution of (**a**) friction coefficient and (**b**) wear coefficient vs. sliding revolutions.

SEM observations of the worn surface revealed a wear track of ~840 µm width (Figure 14a), within which polishing lines and oxidation areas of the metallic matrix (Figure 14b) can be clearly observed, together with the emerged fine carbides dispersion (Figure 14c). Compared to the deposits obtained via the novel CSE technique previously mentioned [27], the friction coefficient values, as well as the wear coefficient and the associated micro-mechanisms, exhibit significant differences originating from the size of the reinforcing TiC particles and their dispersion in the matrix:

(a)   The conventional FCAW deposits, enhanced by TiC particles of finer size, exhibit a rather metallic nature, with high friction coefficient values, tending towards those of a typical tool steel [17] and the relevant wear micro-mechanisms, involving polishing and oxidation of the metallic matrix.

(b)   The solar surface layers, enhanced by TiC particles of larger size, exhibit the typical nature of a ceramic-particle-reinforced metal (CPRM) composite, with the ceramic phase playing the predominant role in the low friction coefficient values. The wear micro-mechanism is a clear superposition of the plastic deformation of the metallic matrix and the micro-fragmentation of the carbides, leading finally to exfoliation (Figure 15).

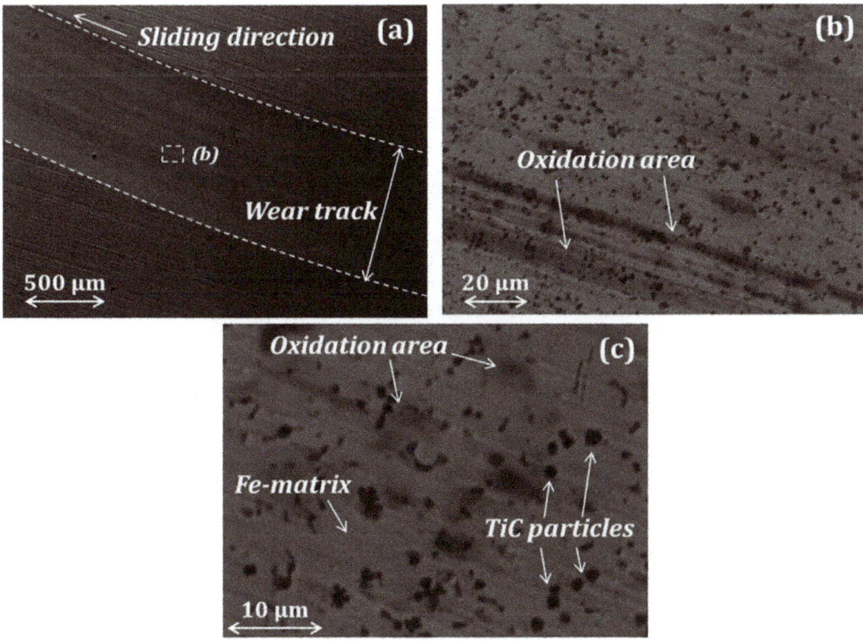

Figure 14. Top view SEM micrographs of the worn surface of the FCAW hardfacing layers, after ball-on-disk testing for 200,000 sliding revolutions, applying a normal load of 10 N: (**a**) entire width of the wear track, (**b**) and (**c**) successive magnifications within the wear track, demonstrating extensive oxidation and abrasion scars along the sliding direction.

Figure 15. Top view SEM micrographs of the worn surface of the "solar" hardfacing layers, after ball-on-disk testing for 200,000 sliding revolutions, applying a normal load of 10 N.

The comparative results of the tribological performance of the FCAW deposits and those concerning non-conventional CSE-obtained hardfacing layers are summarized in Table 3.

Table 3. Ball-on-disk tribological performance of conventional FCAW deposits, compared to CSE overlayers [27].

Applied Load	Friction Coefficient, μ		Wear Coefficient, k (mm^{-3}.N^{-1}.laps^{-1})	
	FCAW Deposits	**CSE Overlayers**	**FCAW Deposits**	**CSE Overlayers**
5 N	0.78 ± 0.02	0.43 ± 0.02	$(5.45 \pm 0.31) \times 10^{-7}$	$(3.26 \pm 0.41) \times 10^{-7}$
10 N	0.78 ± 0.02	0.46 ± 0.02	$(6.84 \pm 0.49) \times 10^{-7}$	$(4.18 \pm 0.42) \times 10^{-7}$

6. Case Study IV: Failure of Wear-Resistant Thermal-Sprayed Ceramic Coatings

The last surface modification family of techniques widely applied at an industrial scale for the wear protection of metallic components is thermal spraying [7,28]. Ceramic particles or bulk metallic material in the form of rods are introduced in an area of high temperature, where they are heated even to their melting temperatures and are accelerated toward the surface to be coated. These totally or partially molten particles arriving at the target transmit their thermal and kinetic energy almost instantaneously and are solidified rapidly providing the desirable coating. Nowadays, several thermal spraying techniques are available, each one characterized by the ratio of thermal to kinetic energy attributed to the material to be deposited, the means implemented to create the high temperature area, and the atmosphere surrounding the accelerating particles. Ceramic materials, due to their high tribological performance and their characteristic mechanical resistance to compressive loading [29], also attract interest for their application as protective coatings. Typically, atmospheric plasma spraying (APS) is applied to deposit oxide layers [30,31], whilst the high-velocity oxyfuel (HVOF) technique is used to form carbide [32,33] wear-resistant layers.

Due to the peculiarities of the deposition technique, ceramic plasma-sprayed coatings, having a thickness of 200–400 µm, exhibit stratified structure with interfacial roughness, which promotes the formation of porosity (Figure 16a) and mechanical anchorage to the metallic surface. In order to enhance the latter prior to thermal spraying, it is necessary to prepare the metallic solid via sandblasting. This pre-processing could result in sand residues (Figure 16b) on the surface leading to poor adhesion to the metallic substrate. In particular in the case of carbide coatings, their deposition under non-recommended parameters of spraying technique causes their in-flight decomposition/decarburization [34–39].

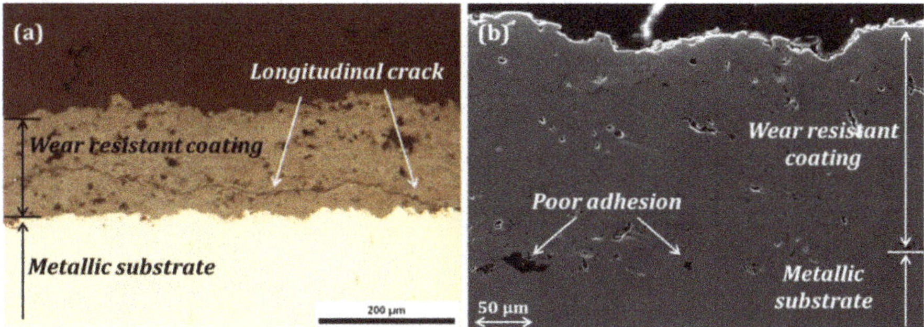

Figure 16. Microstructure flaws developed during plasma spray deposition of (**a**) a titania coating (optical micrograph), where a longitudinal crack propagated parallel to the coating/substrate interface and (**b**) a WC-Co coating (SEM micrograph) of poor adhesion to the metallic substrate, due to the presence of residues of the pre-deposition sandblasting preparation of the metallic surface.

Despite all the above microstructure flaws possibly present in a thermal-sprayed coating, their tribological performance could be acceptable when operating under mild conditions [40]. Under more severe service conditions, these structure flaws accelerate their degradation (Figure 17a) via vertical cracking (Figure 17b), providing the route to the moisture or corrosive gases, to penetrate until the metallic surface that would be hence oxidized.

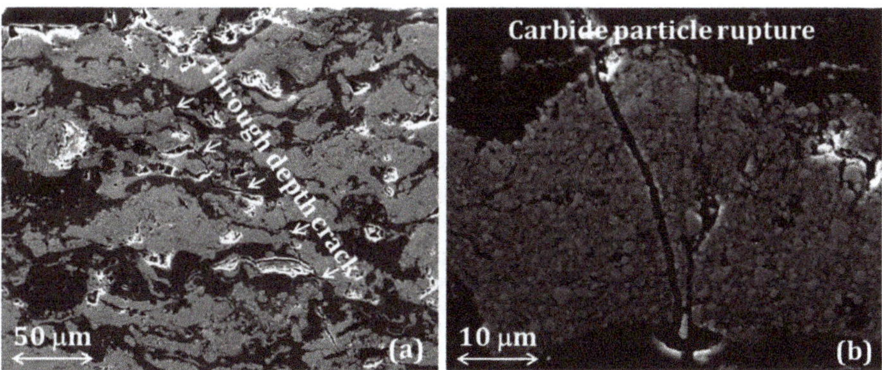

Figure 17. Degradation of a composite plasma-sprayed coating: (**a**) after tribological service a through-depth crack was developed below the contact area and (**b**) propagated through the carbide reinforcing particles resulting in their multi-rupture.

7. Epilogue

In general, the operation of metallic parts and components as elements of a tribosystem requires their further surface enhancement. Several techniques are available for facilitating relevant motion of two solids in contact, which is reflected in low friction coefficient values, and/or for prolonging their lifetime by using surface layers of lower wear rates. In this review, four cases of common wear protective techniques were presented in order to emphasize that the wrong application of the "quasi-protective" layer on the contact surfaces does not prevent but rather accelerate the degradation of the metallic component.

When using an oil lubricant, a strict schedule of periodic inspection of the sum-assembly should be followed in parallel to the periodic oil drain (oil discharge), while its selection should be made taking into account the compatibility of the lubricant' s additives to the metal of the solid.

In the case of implemented diffusion coatings, as the nitrocarburized ones, the bulk hardness of the protected component should be carefully selected, since it affects the good performance of the wear-resistant white layer.

In the case of hardfacing overlayers elaborated via arc welding techniques, the formation of structure flaws associated to solidification of liquid metals (pores and cracks) could be prevented by the proper selection of processing parameters and, if possible, of an automated technique.

Finally, in the case of thermal-sprayed ceramic coatings the selection of the proper technique is of crucial importance to avoid structural flaws that eventually lead to the rapid degradation of the underlying metallic part.

Funding: This research received no external funding.

Acknowledgments: The author would like to thank all the under-graduate and post-graduate students as well as the Ph.D. candidates having carried/carrying out their theses in the Tribology Laboratory of our Department; their work allowed collecting the experimental knowledge briefly presented in this review article.

Conflicts of Interest: The author declares no conflict of interest.

References

1. Williams, J.A. *Engineering Tribology*, 1st ed.; Oxford University Press Inc.: New York, NY, USA, 1994.
2. *Friction, Lubrication and Wear Technology*; ASM Handbook; ASM International: Materials Park, OH, USA, 1992; Volume 18.
3. Zum Gahr, K.-H. *Microstructure and Wear of Materials*, 1st ed.; Elsevier: Amsterdam, The Netherlands, 1987.
4. Holmberg, K.; Matthews, A. *Coatings Tribology*, 1st ed.; Elsevier: Amsterdam, The Netherlands, 1994.
5. *Heat Treating*; ASM Handbook; ASM International: Materials Park, OH, USA, 1991; Volume 4.
6. *Surface Engineering*; ASM Handbook; ASM International: Materials Park, OH, USA, 1994; Volume 5.
7. Fauchais, P.L.; Heberlein, J.V.R.; Boulos, M.I. *Thermal Stray Fundamentals: From Power to Part*, 1st ed.; Springer: New York, NY, USA, 2014.
8. Stribeck, R. Die Wesentlichen Eigenschaften der Gleit-und Rollenlager. *Z. Ver. Deul. Zng.* **1902**, *46*, 180.
9. Avitzur, B. Boundary and hydrodynamic lubrication. *Wear* **1990**, *139*, 49–76. [CrossRef]
10. Avitzur, B. Modelling the effect of lubrication on friction behavior. *Lubr. Sci.* **1990**, *2*, 99–132. [CrossRef]
11. Vrčeka, A.; Hultqvista, T.; Baubetb, Y.; Marklunda, P.; Larssona, R. Micro-pitting damage of bearing steel surfaces under mixed lubrication conditions: Effects of roughness, hardness and ZDDP additive. *Tribol. Int.* **2019**, *138*, 239–249. [CrossRef]
12. Ancellotti, S.; Fontanari, V.; Dallago, M.; Benedetti, M. A novel experimental procedure to reproduce the load history at the crack tip produced by lubricated rolling sliding contact fatigue. *Eng. Fract. Mech.* **2018**, *192*, 129–147. [CrossRef]
13. Stolarski, T.A.; Tobe, S. *Rolling Contacts*, 1st ed.; John Wiley & Sons, Ltd.: Hoboken, NJ, USA, 2000.
14. Pye, D. *Practical Nitriding and Ferritic Nitrocarburizing*; ASM International: Materials Park, OH, USA, 2005.
15. Pantazopoulos, G.A. Tufftriding and Tennifer Surface Treatment. In *Encyclopedia of Tribology*; Wang, Q.J., Chung, Y.W., Eds.; Springer: Boston, MA, USA, 2013.
16. Psyllaki, P.; Kefalonikas, G.; Pantazopoulos, G.; Sideris, J.; Antoniou, S. Microstructure and tribological behaviour of liquid nitrocarburized tool steels. *Surf. Coat. Tech.* **2002**, *162*, 67–78. [CrossRef]
17. Karamboiki, C.-M.; Mourlas, A.; Psyllaki, P.; Sideris, J. Influence of microstructure on the sliding wear behavior of nitrocarburized tool steels. *Wear* **2013**, *303*, 560–568. [CrossRef]
18. Pantazopoulos, G.; Psyllaki, P. An overview on the tribological behaviour of nitro-carburised steels for various industrial applications. *Tribol. Ind.* **2015**, *37*, 299–308.
19. Psyllaki, P.; Stamatiou, K.; Iliadis, I.; Mourlas, A.; Asteris, P.; Vaxevanidis, N. Surface treatment of tool steels against galling failure. *MATEC* **2018**, *188*, 04024. [CrossRef]
20. Cavaleri, L.; Asteris, P.G.; Psyllaki, P.P.; Douvika, M.G.; Skentou, A.D.; Vaxevanidis, N.M. Prediction of surface treatment effects on the tribological performance of tool steels using artificial neural networks. *Appl. Sci.* **2019**, *9*, 2788. [CrossRef]
21. Rasool, G.; Stack, M.M. Wear maps for TiC composite based coatings deposited on 303 stainless steel. *Tribol. Int.* **2014**, *74*, 93–102. [CrossRef]

22. Chaidemenopoulos, N.; Psyllaki, P.; Pavlidou, E.; Vourlias, G. Aspects on carbides transformations of Fe-based hardfacing deposits. *Surf. Coat. Tech.* **2019**, *357*, 651–661. [CrossRef]
23. Bourithis, L.; Papaefthymiou, S.; Papadimitriou, G.D. Plasma transferred arc boriding of a low carbon steel: Microstructure and wear properties. *Appl. Surf. Sci.* **2002**, *200*, 203–218. [CrossRef]
24. Bourithis, L.; Papadimitriou, G.D. The effect of microstructure and wear conditions on the wear resistance of steel metal matrix composites fabricated with PTA alloying technique. *Wear* **2009**, *266*, 1155–1164. [CrossRef]
25. Pantelis, D.; Tissandier, A.; Manolatos, P.; Ponthiaux, P. Formation of wear resistant Al-SiC composite by laser melt-particle injection process. *Mater. Sci. Tech.* **1995**, *11*, 299–303. [CrossRef]
26. Pantelis, D.; Michaud, H.; de Freitas, M. Wear behaviour of laser surface hardfaced steels with tungsten carbide powder injection. *Surf. Coat. Tech.* **1993**, *57*, 123–131. [CrossRef]
27. Mourlas, A.; Pavlidou, E.; Vourlias, G.; Rodríguez, J.; Psyllaki, P. Concentrated solar energy for in-situ elaboration of wear-resistant composite layers. Part I: TiC and chromium carbide surface enrichment of common steels. *Surf. Coat. Tech.* **2019**, *377*, 124882. [CrossRef]
28. Aranke, O.; Algenaid, W.; Awe, S.; Joshi, S. Coatings for automotive gray cast iron brake discs: A review. *Coatings* **2019**, *9*, 552. [CrossRef]
29. Basu, B.; Kalin, M. *Tribology of Ceramics and Composites*, 1st ed.; John Wiley & Sons Inc.: Hoboken, NJ, USA, 2011.
30. Pantelis, D.; Psyllaki, P.; Alexopoulos, N. Tribological behaviour of plasma-sprayed Al_2O_3 coatings under severe wear conditions. *Wear* **2000**, *237*, 197–204. [CrossRef]
31. Psyllaki, P.; Jeandin, M.; Pantelis, D. Microstructure and wear mechanisms of thermal-sprayed alumina coatings. *Mater. Lett.* **2001**, *47*, 77–82. [CrossRef]
32. Kekes, D.; Psyllaki, P.; Vardavoulias, M.; Vekinis, G. Wear micro-mechanisms of composite WC-Co/Cr-NiCrFeBSiC coatings. Part II: Cavitation erosion. *Tribol. Ind.* **2014**, *36*, 375–383.
33. Kekes, D.; Psyllaki, P.; Vardavoulias, M. Wear micro-mechanisms of composite WC-Co/Cr-NiCrFeBSiC coatings. Part I: Dry sliding. *Tribol. Ind.* **2014**, *36*, 361–374.
34. Verdon, C.; Karimi, A.; Martin, J.-L. A study of high velocity oxy-fuel thermally sprayed tungsten carbide based coatings. Part 1: Microstructures. *Mat. Sci. Eng. A Struct.* **1998**, *246*, 11–24. [CrossRef]
35. Liao, H.; Normand, B.; Coddet, C. Influence of coating microstructure on the abrasive wear resistance of WC/Co cermet coatings. *Surf. Coat. Tech.* **2000**, *124*, 235–242. [CrossRef]
36. Sahraoui, T.; Fenineche, N.-E.; Montavon, G.; Coddet, C. Structure and wear behaviour of HVOF sprayed Cr_3C_2-NiCr and WC-Co coatings. *Mater. Des.* **2003**, *24*, 309–313. [CrossRef]
37. Morks, M.F.; Gao, Y.; Fahim, N.F.; Yingqing, F.U.; Shoeib, M.A. Influence of binder materials on the properties of low power plasma sprayed cermet coatings. *Surf. Coat. Tech.* **2005**, *199*, 66–71. [CrossRef]
38. Lee, C.W.; Han, J.H.; Yoon, J.; Shin, M.C.; Kwun, S.I. A study on powder mixing for high fracture toughness and wear resistance of WC-Co-Cr coatings sprayed by HVOF. *Surf. Coat. Tech.* **2010**, *204*, 2223–2229. [CrossRef]
39. Venter, A.M.; Oladijo, O.P.; Luzin, V.; Cornish, L.A.; Sacks, N. Performance characterization of metallic substrates coated by HVOF WC-Co. *Thin Solid Films* **2013**, *549*, 330–339. [CrossRef]
40. Psyllaki, P.; Mourlas, A.; Vourlias, G.; Pavlidou, E.; Vardavoulias, M. Influence of microstructure flaws on the tribological performance of Cr-based thermal-sprayed ceramic coatings. *Ceram. Int.* **2019**, *45*, 19360–19369. [CrossRef]

© 2019 by the author. Licensee MDPI, Basel, Switzerland. This article is an open access article distributed under the terms and conditions of the Creative Commons Attribution (CC BY) license (http://creativecommons.org/licenses/by/4.0/).

Article

Finite Element Analysis of Forward Slip in Micro Flexible Rolling of Thin Aluminium Strips

Feijun Qu [1], Jianzhong Xu [2] and Zhengyi Jiang [1,3,*

[1]   School of Mechanical, Materials, Mechatronic and Biomedical Engineering, University of Wollongong, Northfields Avenue, 2522 Wollongong, Australia; fq348@uowmail.edu.au

[2]   State Key Laboratory of Rolling and Automation, Northeastern University, Shenyang 110004, China; xujz@ral.neu.edu.cn

[3]   School of Materials and Metallurgy, University of Science and Technology Liaoning, Anshan 114051, China

*   Correspondence: jiang@uow.edu.au

Received: 17 August 2019; Accepted: 19 September 2019; Published: 29 September 2019

Abstract: This study delineates a novel finite element model to consider a pattern of process parameters affecting the forward slip in micro flexible rolling, which focuses on the thickness transition area of the rolled strip with thickness in the micrometre range. According to the strip marking method, the forward slip is obtained by comparison between the distance of the bumped ridges on the roll and that of the markings indented by the ridges, which not only simplifies the calculation process, but also maintains the accuracy as compared with theoretical estimates. The simulation results identify the qualitative and quantitative variations of forward slip with regard to the variations in the reduction, rolling speed, estimated friction coefficient and the ratio of strip thickness to grain size, respectively, which also locate the cases wherein the relative sliding happens between the strip and the roll. The developed grain-based finite element model featuring 3D Voronoi tessellations allows for the investigation of the scatter effect of forward slip, which gets strengthened by the enhanced effect of every single grain attributed to the dispersion of fewer grains in a thinner strip with respect to constant grain size. The multilinear regression analysis is performed to establish a statistical model based upon the simulation results, which has been proven to be accurate in quantitatively describing the relationship between the forward slip and the aforementioned process parameters by considering both correlation and error analyses. The magnitudes of each process parameter affecting forward slip are also determined by variance analysis.

Keywords: finite element analysis; forward slip prediction; strip marking method; multilinear regression; micro flexible rolling; thickness transition area; 3D Voronoi modelling

---

1. Introduction

Strip rolling technology has been further developing rapidly in recent decades thanks to the wide application of its products in a variety of fields, including manufacturing, construction and energy, wherein higher quality and productivity have been of great interest to the researchers and engineers in the field of metal forming [1–3]. For instance, Jiang et al. [3,4] investigated the rolling force, intermediate force, roll edge contact force and wear condition, as well as the shape, profile and surface roughness of the rolled strip in response to various process parameters like reduction, rolling speed, initial strip thickness, using a combined numerical and experimental approach. Xie et al. [5] carried out analytical and experimental investigations to identify the edge crack initiation and propagation during cold rolling of low carbon steel strip with the aid of Atomic Force Microscopy and Scanning Electron Microscopy, from which they found a lower friction coefficient, as well as a finer surface finish, would not only prevent the microcracks, but delay the crack-initiation process in rolled strip. Jiang et al. [6] simulated the cold strip rolling process with consideration of friction variation along the rolling and transverse directions based on finite element method, whereof the numerical results had revealed the significant influence of friction on the rolling force, rolled strip shape and profile, which were confirmed by the measured values that the developed approach was capable of improving the accuracy of the conventional rolling model. Moreover, an elastic-plastic finite element model was established by Jiang and Tieu [7] to analyse the strip deformation in the roll bite zone taking tension into account, which was of great help to quantify the effects of tension and rolling speed on the hump value at elastic entry zone, as well as determining the mixed lubricating film in the roll bite and the elastic recovery of final rolled strips. Nonetheless, the forward slip is also a common phenomenon during strip rolling, which reflects the difference between the strip speed and the roll speed at the exit of roll bite. It is always a significant parameter that helps understand and determine the friction, and also assists in tension control, as well as in preventing the occurrence of skidding [8].

The forward slip is normally analytically predictable or can be measured in both industrial and experimental circumstances [9,10]. For instance, Heydari vini and Farhadipour [11] computed the forward slip as a function of process parameters and strip geometry in cold rolling. Pawelski [12] introduced an explicit analytical method for determining the forward slip during cold strip rolling, which reproduced most predictions of classic strip theory, except for extreme cases, such as in the foil rolling regime or for skin pass rolling. Bayoumi [13] developed a kinematic analytical approach to predict the forward slip in hot strip rolling based on formulating a velocity field in the roll bite zone that considers the effect of interfacial friction on the distribution of the axial velocity and longitudinal stresses across the strip thickness. This method could yield accurate results, compared with those of the finite element simulation, and was suitable for use in online control because of the drastic reduction in computational time. On the other hand, Tieu, Jiang et al. [14] discussed two promising experimental methods, which are the strip marking method and the laser Doppler method, respectively, and Liu [10] applied these two methods to measure the forward slip in cold rolling of aluminium alloy under lubricated condition, while the results successfully verified the effectiveness of both methods for determining the forward slip in the laboratory rolling mill. Li et al. [8] analysed the uncertainty of forward slip measurement using the laser Doppler method and proposed an improved laser Doppler velocimetry system by mixing the Doppler signals with a reference frequency signal before feeding them to the signal processors. Rolling tests of mild steel strips using the improved system showed a fivefold accuracy enhancement of forward slip measurements under different rolling conditions and surface roughness. Yuen [15] described an alternative approach to determine the forward slip in hot strip rolling, which instead compared the distance the head end of the workpiece had travelled with that the periphery of the work roll had travelled during the same time period.

Extensive investigations have shown variability in forward slip under different reductions, rolling speeds, roll diameters, tensions, lubrication, etc., however, the microstructural effects still remain to be identified, particularly with the micro-thin strips. Among most studies, the microstructural effects have been characterised by the ratios of specimen thickness (T) to grain size (D) to find their interactive relationship with the micro deformation behaviours. For example, according to Fang et al. [16], the plastic deformation behaviour along with the fracture mode of the material produces an obvious change when the value of the T/D ratio varies around 1. Lee et al. [17] revealed that both the maximum blank holder force and the limit drawing ratio increased with increasing T/D ratio during micro deep drawing of 304 stainless steel foils. They also recommended that the T/D ratio be greater than 10 for better formability and steady deep drawing behaviour. In addition, Anand and Kumar [18] demonstrated the influence of T/D ratio on both yield stress and flow stress using ultra-thin brass sheets, which emphasises on the modification in conventional constitutive equations when applied for ultra-thin sheets. Raulea et al. [19] conducted a series of uniaxial tensile tests to identify variation in the yield and tension strength when polycrystalline and single crystal aluminium sheets were adopted respectively, as well as a sequence of blanking experiments to analyse the variation in product shape and process force for both cases.

In such an engineering problem, forward slip can be considered as a dependent variable, which typical value changes for when any one of the independent variables, i.e., process and material parameters, is altered. To estimate the relationships between these variables, regression analysis can be carried out to explore the forms of these relationships, and this methodology has been applied in a wide variety of prediction and forecasting situations [20]. For instance, Vorkov et al. [21] employed multiple linear regression with workpiece and tooling dimensions as the independent variables to deliver the prediction of contact points position between the punch and the workpiece in large radius air bending using high-strength steels. Schmid et al. [22] predicted both major and minor strains based on ultimate elongation and thickness of the material by linear regression so as to further determine the whole forming limit curve geometry by performing the interpolation. Strano and Colosimo [23] also conducted the empirical determination of forming limit diagram in sheet metal processes utilising logistic regression model, which allowed the deduction of the probability of failure as a function of both the principal planar strains by taking into account both the failed and safe experimental data points in the analysis. Additionally, Hubbard et al. [24] established a two-parameter Weibull regression model to estimate strain intensity in different strain modes using surface roughness data, which suggested the critical strain localisation and/or failure, as well as the active deformation mechanisms in sheet metal forming of aluminium alloy 5754.

Moreover, flexible rolling technology has been devised by Kopp et al. [25,26] to meet the increasing demand for lightweight structures in the automotive industry, which alters the roll gap during the rolling passes to obtain the part with varied thickness distribution along its length direction as per the load requirements at respective locations. Liu et al. [27,28] also utilised finite element method to analyse the influences of process parameters, such as reduction, friction coefficient and workpiece horizontal velocity on the rolling force, forward slip and workpiece thickness profile so as to advance the numerical investigation of flexible rolling. Therefore, the conventional strip marking method cannot be simply applied to determine the forward slip in this newly developed technique as rolling phase may change prior to one circumferential rotation of the roll; alternatively, two parallel ridges have been conceived to ensure that two line markings fall within either thickness transition zone or rolling phase with an invariable thickness, so that the forward slip can be evaluated effectively by comparing the distances between the pairs of markings and ridges with respect to each rolling phase.

In this context, both analytical and numerical methods are utilised to determine the forward slip of thin strips, which occurs within the thickness transition area during micro flexible rolling. The novel finite element model developed based on strip marking method is capable of producing numerical estimates of forward slip of mimicked aluminium alloy 5052 strips in an efficient and accurate manner as compared with theoretical estimates obtained from the mathematical model. The variation of forward slip in response to the variations in different process variables, such as reduction, rolling speed, estimated friction coefficient and the T/D ratio is identified with a qualitative evaluation of the scatter centred around it, which is due to the anisotropic nature of individual grains constructed using 3D Voronoi tessellation technique. The relative slip between the strip and the roll caused by the relocation of a neutral point outside the roll bite is also discovered during the rolling phase with thinner thickness, together with an estimated friction coefficient of 0.08 or T/D ratio of 2. Furthermore, the multilinear regression analysis followed by correlation and error estimations has been employed to derive an equation model so as to generate reasonably accurate forecasts of forward slip of thin aluminium alloy 5052 strips within the thickness transition zone in micro flexible rolling with different process parameters. At last, analysis of variance is utilised to assess the significance of each process parameter and quantify their respective contributions to the forward slip.

2. Analytical Modelling of Forward Slip in Micro Flexible Rolling

2.1. Determination of Neutral Angle

Figure 1 depicts the forces acting on a unit width of the workpiece during the micro flexible rolling phases to form transition zones with thicker and thinner thicknesses, respectively, wherein the work rolls are assumed not to sustain any elastic deformation. This study is focusing on force analysis within thickness transition zones, which represent the main characteristics in flexible rolling. Therefore, Figure 1a,b just compare forces that act on the unit width of the workpiece during micro flexible rolling phases to form the transition zones with thicker and thinner thicknesses, respectively, while rolling phase with an invariable thickness is not considered in the current study, as it is the same as the flat rolling process.

The neutral angle can be determined by considering the horizontal forces acting on the workpiece of unit width, as well as neglecting the lateral spreading in the deformation zone. This is because according to [29], the rolling process can be treated as a two-dimensional deformation when the workpiece has a very large width compared with its thickness; this suits our case so that we have neglected the lateral deformation of the material.

According to [30,31], the following equations can be obtained by solving the horizontal forces on the unit width of the workpiece in equilibrium.

For the rolling phase with thicker thickness,

$$\sum x = -\int_{\theta}^{\alpha} p_x \sin \varphi R d\varphi - \int_{\theta}^{\gamma} \tau_x' \cos \varphi R d\varphi + \int_{\gamma}^{\alpha} \tau_x \cos \varphi R d\varphi = 0, \tag{1}$$

where x refers to the horizontal axis of the two-dimensional figure in Cartesian coordinates, p_x is the radial pressure, τ_x' and τ_x are the tangential pressures in the forward and backward slip zones, respectively, R is the roll radius, φ is the arc at arbitrary contact point and H denotes the initial strip thickness. As can be seen from Figure 1, point C is the intersection of the circular arc of the upper roll and the central line of the rolls. Thus, tangential pressures are not perpendicular to OC, but directed tangent to the circular arc at which the neutral plane meets.

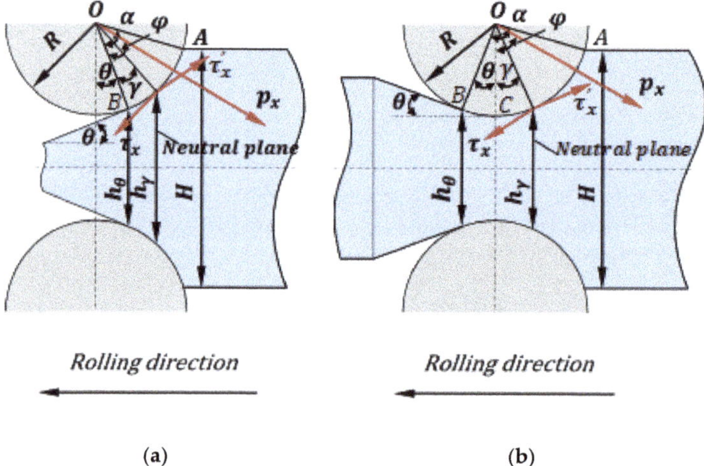

Figure 1. Forces acting on the unit width of the workpiece within thickness transition zones: (**a**) Rolling phase to form transition zone with thicker thickness; and (**b**) rolling phase to form transition zone with a thinner thickness (for the deformation area below arc *AC*) [32].

Suppose (1) stress is uniformly distributed along the arc of contact between the rolls and the rolled workpiece, namely p_x is regarded as a constant, and (2) friction coefficient μ remains invariable along the arc of contact and friction between the rolls and the workpiece obeys the Coulomb's law given by $\tau_x = \mu p_x$ [32]. Then upon integration and rearrangement, Equation (1) yields the following,

$$\sin \gamma = \frac{\sin \alpha + \sin \theta}{2} - \frac{\cos \theta - \cos \alpha}{2\mu}, \tag{2}$$

where α is the angle of bite, θ is the tilt angle for thickness transition area, γ is the arc at the neutral plane and μ defines the estimated friction coefficient along the arc of contact.

For the rolling phase with thinner thickness,

$$\sum x = \int_0^\theta p_x \sin \varphi' R d\varphi' - \int_0^\alpha p_x \sin \varphi R d\varphi - \int_0^\theta \tau_x' \cos \varphi' R d\varphi' \\ - \int_0^\gamma \tau_x' \cos \varphi R d\varphi + \int_\gamma^\alpha \tau_x \cos \varphi R d\varphi = 0, \tag{3}$$

After integration and rearrangement, this equation becomes,

$$\sin \gamma = \frac{\sin \alpha - \sin \theta}{2} - \frac{\cos \theta - \cos \alpha}{2\mu}. \tag{4}$$

2.2. Determination of Forward Slip

It is postulated that the volume rates of material flow are the same at the exit of the roll bite and the neutral plane. Thus, there exists the relationship,

$$v_\theta h_\theta = v_\gamma h_\gamma, \tag{5}$$

where v_θ, v_γ are the horizontal velocities and h_θ, h_γ are the thicknesses of the workpiece at the exit and the neutral plane, respectively.

Using

$$v_\gamma = v \cos \gamma, \tag{6}$$

and

$$h_\gamma = h_\theta + D(\cos \theta - \cos \gamma), \tag{7}$$

The following equation can be obtained,

$$\frac{v_\theta}{v} = \frac{v_\theta \cos \gamma}{v_\gamma} = \frac{h_\gamma \cos \gamma}{h_\theta} = \cos \gamma + \frac{D(\cos \theta - \cos \gamma) \cos \gamma}{h_\theta}, \tag{8}$$

where D is the roll diameter, while v is the circumferential velocity of the roll.

Then the forward slip can be calculated by,

$$S_{h_\theta} = \frac{v_\theta - v}{v} = \frac{v_\theta}{v} - 1 = \cos \theta - 1 + \frac{D(\cos \theta - \cos \gamma) \cos \gamma}{h_\theta}. \tag{9}$$

3. Numerical Estimation of Forward Slip in Micro Flexible Rolling Based on Strip Marking Method

In most finite element simulations of sheet material forming processes, constitutive modelling plays a significant role in determining the stress and strain distributions, as well as their variation with respect to different locations in the formed part, which includes the mechanical behaviour of the material, yield criterion, hardening rule, flow rule and so forth [33,34]. In the current engineering application of micro flexible rolling, the material is assumed to be elastoplastic and undergo isotropic hardening after it yields according to von Mises criterion, the flow rule being modelled by Levy-Mises with normality condition. Moreover, this problem can be considered as plane-strain state as the workpiece is assumed to be very wide compared with its thickness [29].

A half symmetry 3D finite element model has been developed in ABAQUS/CAE (also known as Complete Abaqus Environment, which is a backronym with a root in Computer-Aided Engineering) to numerically estimate the forward slip in micro flexible rolling in accordance with the strip marking method, as displayed in Figure 2. The roll was simplified as a rigid cylindrical shell in order to eliminate its elastic deflection and save computational resources, and a finer mesh was assigned to the strip than that for the roll so that the bumped ridges on the roll would leave an image on the strip, rather than penetrating into it after the simulation process was completed [35]. Then the average forward slip may be worked out by Equation (10) [10],

$$\overline{S} = \frac{l' - l}{l} \times 100\%,\tag{10}$$

where l' and l are the distances between the pairs of markings and ridges, respectively.

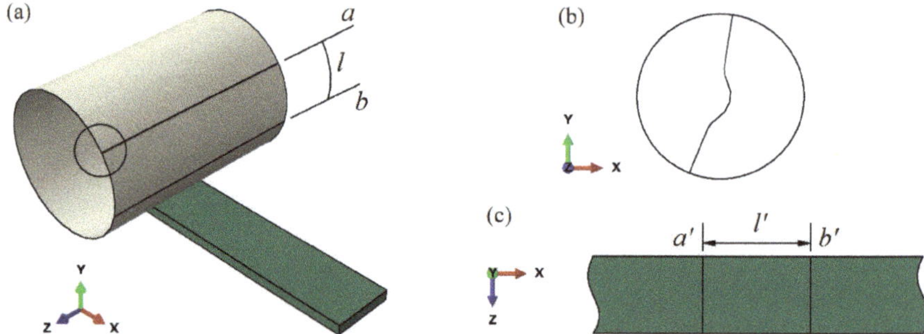

Figure 2. Schematic of strip marking method for numerical estimation of forward slip in micro flexible rolling: (**a**) Finite element analysis model with two parallel ridges running axially along roll surface; (**b**) enlarged profile view of the ridge; and (**c**) rolled strip with two line markings indented by the ridges.

As this study intends to reveal the variation of forward slip in the thickness transition area when process variables like reduction, rolling speed, friction coefficient, as well as the ratio of strip thickness to grain size change, a frictional interface has been defined between the outer surface of the roll and the top surface of the strip, and the 3D Voronoi tessellation has also been applied in the strip to mimic the polycrystalline structure of the material to be rolled.

The rolling speed is a process parameter in our work, while the numerical model has been set to contain general, static analysis steps, namely initial contact, one type of reduction, thickness transition and another type of reduction. The boundary conditions for different analysis steps are listed in Table 1, where RP refers to the geometric centre of the roll, s refers to the bottom surface of the workpiece, v_x, v_y and v_z are the velocity components in directions x, y and z, respectively, ω_x, ω_y and ω_z are the angular velocity components of the roll rotating around the axes x, y and z, respectively, Δh is the reduction amount, d is the diameter of the roll and l_c is the length of the zone of contact between the roll and the workpiece [36].

Table 1. Boundary conditions for different analysis steps.

Step	Position	Boundary Conditions
Initial contact	RP	$v_x = v_y = v_z = \omega_x = \omega_y = 0$ $\omega_z \neq 0$
	S	$v_y = 0$ $v_x \neq 0$
One type of reduction	RP	$v_x = v_y = v_z = \omega_x = \omega_y = 0$ $\omega_z \neq 0$
	S	$v_y = 0$
Thickness transition	RP	$v_x = v_z = \omega_x = \omega_y = 0$ $v_y = \pm \dfrac{\Delta h \omega_z d}{2l_c}$ $\omega_z \neq 0$
	S	$v_y = 0$
Another type of reduction	RP	$v_x = v_y = v_z = \omega_x = \omega_y = 0$ $\omega_z \neq 0$
	S	$v_y = 0$

The bilinear isotropic hardening material model was selected for the workpiece in which the basic material data were referred to Reference [37], namely aluminium alloy 5052 with Young's modulus E of 70 GPa, initial yield stress σ_{s0} of 195 MPa and tangent modulus E_{TAN} of 292 MPa; however, different heterogeneity coefficients have been employed to alter the tangent modulus to obtain different types of mechanical properties. Figure 3a presents seven types of mechanical properties distinguished by dissimilar heterogeneity coefficients, which have been assigned to Voronoi polyhedrons to substantially reflect the inhomogeneity of individual grains that constitute the strip, as exhibited in Figure 3b. Note that ξ, E_{TAN} and σ_{s0} in Figure 3a denote heterogeneity coefficient, tangent modulus and initial yield stress for the selected material, respectively.

According to [35], the concise instructions to create a Voronoi tessellated strip and subsequently associate its internal grains with a variety of properties are as follows:

(1) The strip is divided into several unit cubes, each containing a single nucleus.
(2) The 3D Voronoi function in computational software MATLAB is in use to create Voronoi polyhedrons that take those nuclei as their mass centres.
(3) Different types of mechanical properties are assigned to the Voronoi polyhedrons by means of MATLAB code to reflect the grained inhomogeneity in a real material.
(4) MATLAB finally generates a Python file containing all the geometrical and topological information of the generated Voronoi tessellation together with property attribution of each individual polyhedron.
(5) The Python script is imported to ABAQUS/CAE to complete the setup of grain aggregate with a range of mechanical properties.

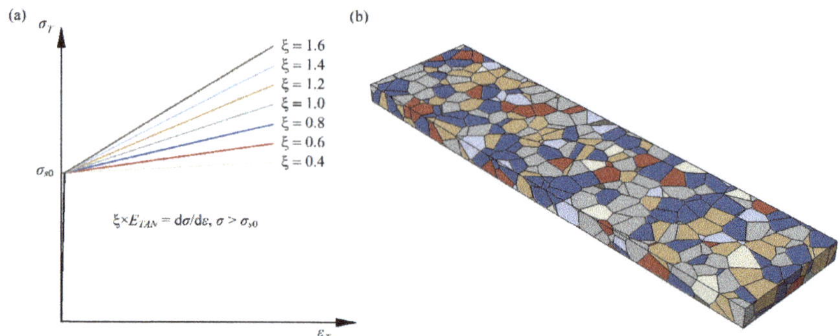

Figure 3. Voronoi tessellation-based 3D construction of the strip with consideration of grained inhomogeneity: (**a**) Bilinear stress-strain curves with different tangent moduli governed by corresponding heterogeneity coefficient; and (**b**) mimicked microstructure of the strip where grains are equipped with various mechanical properties.

The strip was meshed with 10-node modified quadratic tetrahedron elements (C3D10M) with a fine mesh size of 0.04 mm to ensure a good convergence, as well as improving the mesh quality near grain boundaries, whilst the roll was coarsely meshed with 4-node 3D bilinear rigid quadrilateral elements (R3D4) by utilising mesh size of 0.16 mm in such a way as to prevent the nodes on the roll surface from penetrating into the workpiece [35,38].

In general, the optimised model setting includes the simplification of the roll as a rigid cylindrical surface and the determination of mesh sizes for both roll and strip, in order to reduce the model complexity, save the computational time and boost both computational accuracy and efficiency.

4. Results and Discussion

Figure 4 presents the micro flexible rolled strip with a reduction of 20% to 50%, wherein red dots indicate the locations of the marks transferred from the roll within the thickness transition area. Furthermore, the strip was rolled at a speed of 20 cm/min and under dry friction (with an equivalent friction coefficient of 0.13 [38]), and both initial strip thickness and average grain size were set as 250 μm, which gives the T/D ratio of 1 for this case. In the current study grain size of 250 μm has been used, due to two main reasons, whereof one is that three types of T/D ratio can be considered, which means the scattering effect resulted from the difference between properties of single grains can be evaluated and compared when approximately one (T/D ratio is less than or equal to 1) or two layers of grains (T/D ratio is equal to 2) constitute the strip; and the other is that a bigger grain size is helpful to lessen the computation time for creating the Voronoi tessellation and performing the simulation tasks, while the grained inhomogeneity which is a key factor leading to scattering effect can still be reflected in spite of a large grain size.

It is clear from Figure 4 that the effective stress distribution on the profile exhibits a highly non-uniform pattern because of the coexistence of both elastic and plastic states of stress in the inhomogeneous granular material subjected to the applied rolling force [39].

Based on Equations (4) and (9), the theoretical value of the forward slip in this case is equal to 2.94%, whereas the numerical estimate turns out to be 3.16% (averaged result from three repeated trials to improve the accuracy of length measurement after rolling) in accordance with Equation (10), which error is 7.48% with respect to the theoretical estimate and can be calculated as follows:

$$Error = \frac{|Theoretical\ Estimate - Numerical\ Estimate|}{Theoretical\ Estimate} \times 100\%. \tag{11}$$

As the developed finite element model is capable of evaluating the forward slip by simple calculations with measurements which produce a reasonable level of accuracy, it is utilised to identify how different process parameters affect the forward slip within the thickness transition area in micro flexible rolling.

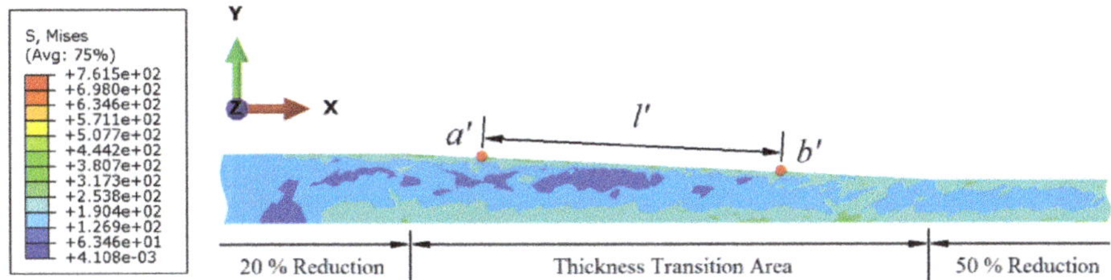

Figure 4. Profile of the micro flexible rolled sample with a reduction of 20% to 50%.

4.1. Effect of Reduction

Reduction combinations of 20% to 40%, 20% to 45%, and 20% to 50% have been selected for the rolling phase with thinner thickness, and likewise, 50% to 30%, 50% to 25%, and 50% to 20% for the rolling phase with thicker thickness by maintaining the length of thickness transition area, but altering the tilt angle θ. Figure 5 represents the relationship between the forward slip and the reduction for the rolling phase with thinner thickness. It is evident that the forward slip increases as the reduction increases, due to the fact that the increasing reduction gives rise to the increase in the displacement volume in the thickness direction, which consequently leads to the growth of material flow in the rolling direction, and thus, the increase in forward slip [40].

In Figure 6, the reduction combination is located on the abscissa, while the forward slip is given on the ordinate for the rolling phase with thicker thickness. As can be seen, the forward slip undergoes a decline because of progressively reduced material flow in the rolling direction with the decreasing reduction. Compared with Figure 5, forward slip in this stage presents generally higher values probably because the reduction starts with a greater value of 50%, which introduces a larger amount of forward slip at the beginning of this stage. Moreover, the scatter of average forward slip has been noted for each reduction combination in Figures 5 and 6. This being the case the strip comprises of approximately a single layer of grains; consequently, the property of each individual grain has a prominent effect on the overall deformation of the whole strip, which results in the scatter effect of forward slip with respect to different grain compositions.

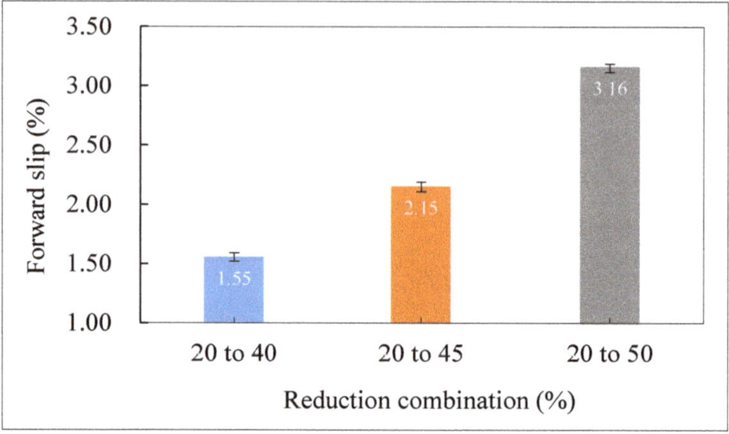

Figure 5. Forward slip versus reduction for the rolling phase with thinner thickness.

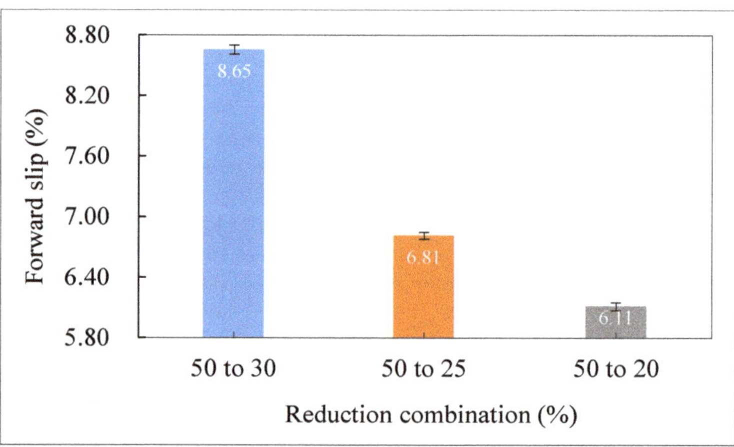

Figure 6. Forward slip versus reduction for the rolling phase with thicker thickness.

4.2. Effect of Rolling Speed

The effects of rolling speed on forward slip are depicted in Figures 7 and 8 for the rolling phases with reduction combinations of 20% to 50% and 50% to 20%, respectively, wherein the friction coefficient is 0.13 and T/D ratio remains 1 for all cases.

As can be noticed in both figures that the increase in rolling speed can lead to the increase in forward slip, regardless of the rolling phase with thinner or thicker thickness, which is primarily due to the fact that a higher rolling speed results in a larger residual friction force regarding a constant friction coefficient in the stable rolling stage, which, therefore, speeds up the plastic flow of the material in the rolling direction, causing an increase in forward slip eventually [40]. Furthermore, the scatter effect of forward slip resulted from the inhomogeneous microstructure throughout the strip has also been observed, whereof the average value varies from 6.11% to 8.02% for the rolling phase starting with a bigger reduction of 50% as the rolling speed increases from 20 to 30 cm/min (Figure 8). This change is generally greater than that of 3.16% to 4.32% for the rolling phase starting with a smaller reduction of 20% (Figure 7), which indicates once again that the reduction affects the forward slip in a significant manner.

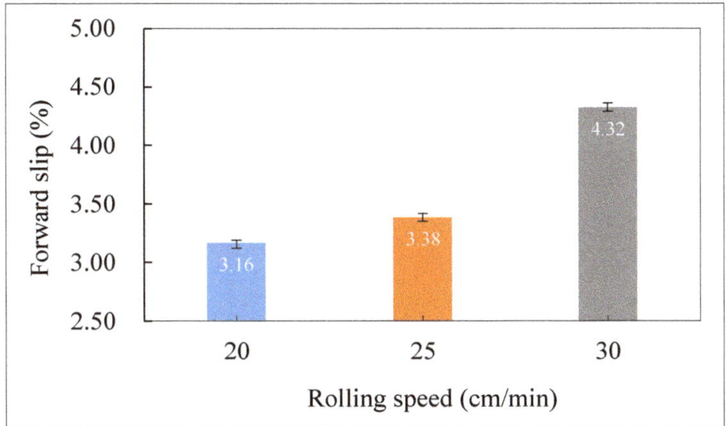

Figure 7. Variation of forward slip with rolling speed for the rolling phase with thinner thickness.

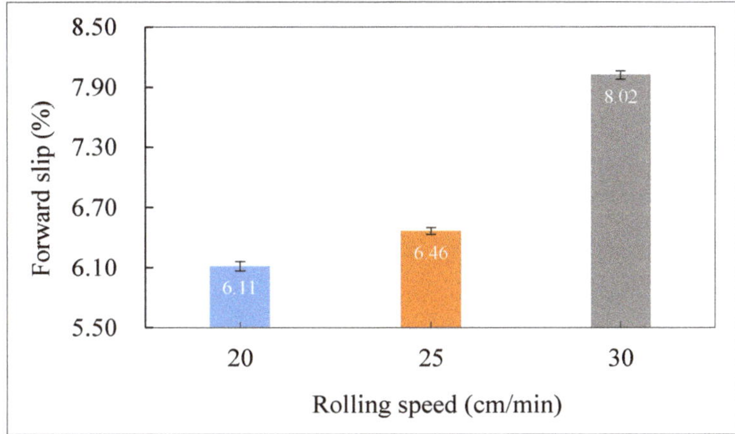

Figure 8. Variation of forward slip with rolling speed for the rolling phase with thicker thickness.

4.3. Effect of Estimated Friction Coefficient

To examine how lubrication condition affects the forward slip in micro flexible rolling four types of estimated friction coefficients have been selected to represent dry friction (with estimated friction coefficients of 0.13, 0.18, and 0.23, respectively) and rolling with lubrication (with estimated friction coefficient of 0.08) in accordance with previous investigations conducted by Qu et al. [36,38]. In this section, all cases were performed with rolling speed of 20 cm/min and T/D ratio of 1.

As can be seen in Figure 9, the average value of forward slip has undergone an overall decline from 7.86% to −3.42% as the estimated friction coefficient decreases from 0.23 to 0.08, and where this fact can be explained by the reduced residual friction force owing to the decrease in estimated friction coefficient, which brings a continued drop in forward slip, provided all other variables remain unchanged [40]. It is also quite noticeable that the forward slip experiences a significant decrease or can even be thought of as reversing from positive value of 3.16% to negative value of −3.42% when the mimicked condition varies from dry rolling to that with the presence of lubricant; this phenomenon occurs because the neutral point has taken relocation outside the roll bite, which means the strip sliding takes place to a certain extent even though it succeeds to feed in the roll bite.

For the rolling phase with reduction of 50% to 20% (Figure 10), the forward slip displays an almost identical downward trend, but stays positive, indicating that the strip sliding no longer exists during the rolling phase with thicker thickness in spite of the low friction coefficient of 0.08 reached. As is also found through comparison of Figures 9 and 10, the forward slip starts to decline from similarly high levels of 7.86% and 9.01%, respectively, which reveals that the contribution of high friction coefficient to forward slip overwhelms that of increased reduction.

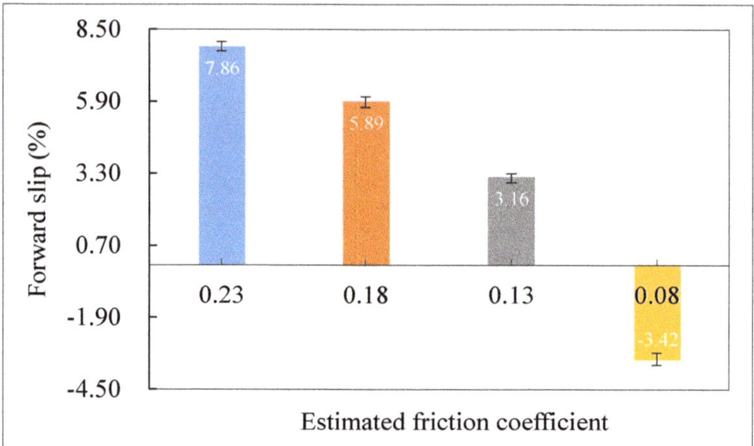

Figure 9. Dependence of forward slip on estimated friction coefficient for the rolling phase with thinner thickness.

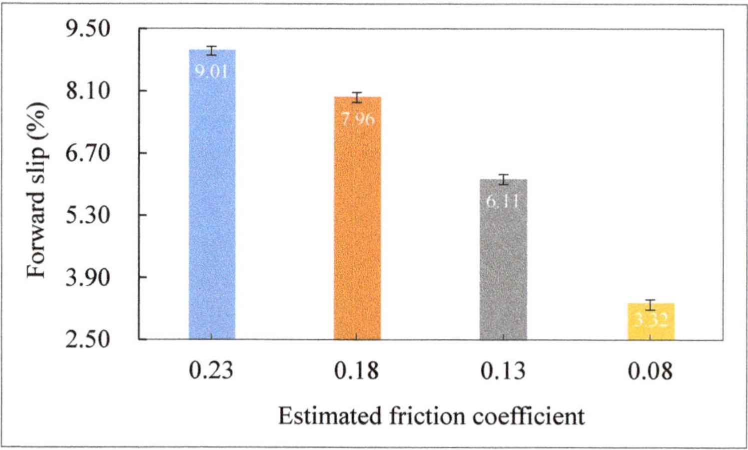

Figure 10. Dependence of forward slip on estimated friction coefficient for the rolling phase with thicker thickness.

4.4. Effect of Initial Strip Thickness (or T/D Ratio)

Figures 11 and 12 exhibits the tendency of forward slip to decrease with increasing initial strip thickness, but with a constant rolling speed of 20 cm/min and friction coefficient of 0.13 for the rolling phases with thinner (reduction of 20% to 50%) and thicker (reduction of 50% to 20%) thicknesses, respectively. Firstly, as is clearly seen in Figure 11 that the increasing initial strip thickness has driven a drastic decline in the forward slip of 3.16% to −6.89%, wherein the negative value indicates that the neutral point has moved outside the roll bite again for a larger initial strip thickness of 500 μm during the rolling phase with thinner thickness; the strip keeps feeding in the roll bite, along with a certain amount of sliding though.

While in the case with thicker thickness, as illustrated in Figure 12, there is a progressive and approximate linear decrease in forward slip from 8.99% to 3.46% as initial strip thickness increases from 100 to 500 μm during this rolling phase without strip sliding; the primary reason for the observed trends is that the elongation of the material in the rolling direction decreases with increasing strip thickness, which is equivalent to decreasing the reduction that reduces the material flow along rolling direction, and the forward slip is then declined as a result [41]. In addition, a smaller initial strip thickness exerts a greater influence on the forward slip than that of the reduction, since it reaches quite close values of 8.87 and 8.99 for the rolling phases with thinner and thicker thickness, respectively, no matter whether the reduction starts at 20% or 50% in relation to the initial strip thickness of 100 μm.

Another noticeable finding observed from both Figures 11 and 12 is that the scatter of average forward slip is smaller at the initial strip thickness of 500 μm than that for both smaller strip thicknesses of 100 and 250 μm, which is attributable to approximately two layers of grains dispersed in the thicker strip of 500 μm (with respect to the average grain size of 250 μm, namely the *T/D* ratio reaches 2), weakening the effect of each single grain and leading to a more uniform deformation within the material, as compared with the other two types of thinner strips (with *T/D* ratio less than or equal to 1) that produce larger discrepancies across simulation results.

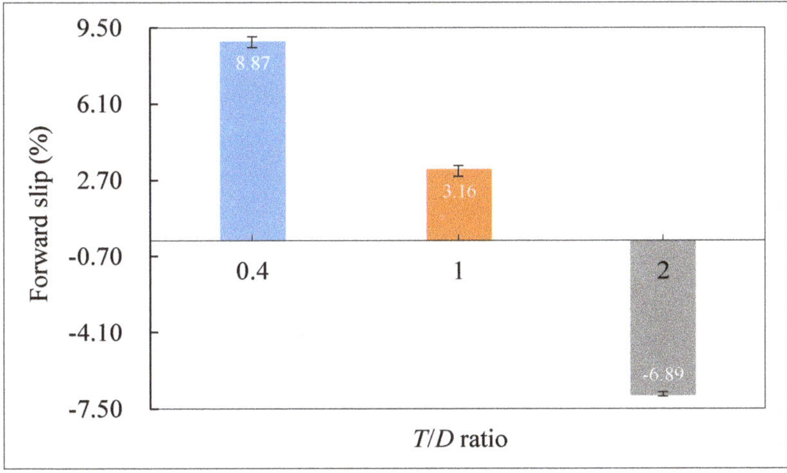

Figure 11. Tendency of forward slip with *T/D* ratio for the rolling phase with thinner thickness.

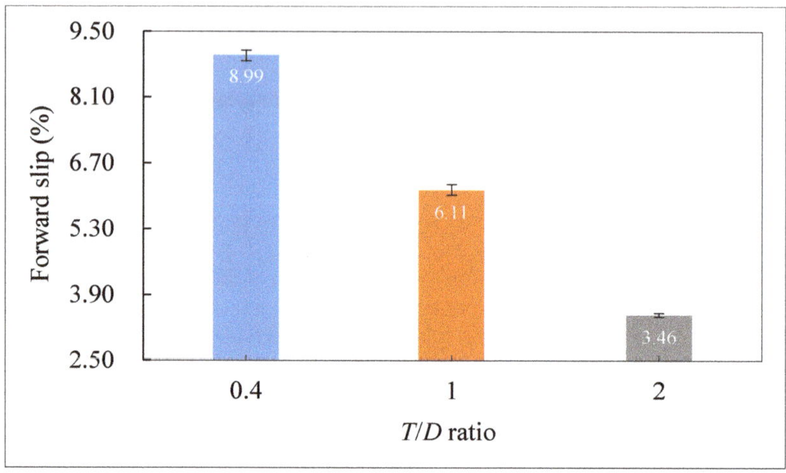

Figure 12. Tendency of forward slip with *T/D* ratio for the rolling phase with thicker thickness.

4.5. Development of Prediction Model for forward Slip in Micro Flexible Rolling Based on Regression Analysis

The multilinear regression analysis is a powerful technique for establishing a statistical model to predict the unknown value of a dependent variable from the known values of a set of independent variables in a quantitative manner [35]. In the present study, the forward slip is taken as the dependent variable y that is influenced by four types of process parameters taken as the independent variables x_1–x_4, to wit, reduction, rolling speed, estimated friction coefficient and the ratio of initial strip thickness to average grain size, respectively.

According to [35,42], the multilinear regression prediction model can be constructed as follows:

$$\hat{y}_i = a_0 + a_1 x_{i1} + a_2 x_{i2} + a_3 x_{i3} + a_4 x_{i4} + e_i, \text{ for } i = 1, 2, \ldots, n, \tag{12}$$

where n is the number of observations, a_0 is the intercept term, namely the predicted value of the dependent variable $\hat{y}_i$ when all of the independent variables x_{i1}–x_{i4} is equal to zero, a_1–a_4 are the regression coefficients estimated by least-squares method and e_i is the difference between the observed value y_i and predicted value $\hat{y}_i$, which is normally distributed with mean 0 and variance σ, i.e., $e_i \sim N(0, \sigma^2)$, and where σ can be calculated as,

$$\sigma = \sqrt{\frac{\sum e_i^2}{n - p - 1}}, \tag{13}$$

where p is used for the number of independent variables.

In a regression model, the effects of lower-order terms become conditional ones, not main effects by adding interaction terms, which means that the effect of one variable is conditional on the value of the other and it goes into full effect only when the other term in the interaction equals 0. Therefore, the interaction terms have not been taken into account in the current regression prediction model so as to make quantitative estimation of the main effects of reduction, rolling speed, estimated friction coefficient and *T/D* ratio on forward slip, respectively, since the value changes of these process parameters are independent of each other.

In order to maintain the integrity of data, simulation results from both cases with thinner and thicker thicknesses have been utilised jointly to establish one single model, wherein input variable x_1, viz. reduction, takes positive sign for the case with thinner thickness and negative for the other one. Thus, based upon 60 groups of values of independent and dependent variables, the regression model can be written in the following form,

$$\begin{cases} \hat{y}_i = 0.489503 - 7.0693x_{i1} + 0.179244x_{i2} + 51.06542x_{i3} - 6.45104x_{i4} + e_i, \; for \; i = 1,2,\ldots,60 \\ e_i \sim N(0, 3.29633). \end{cases} \tag{14}$$

Equation (14) gives the quantitative definition of the prediction model of forward slip as a function of independent variables. For instance, the first observation of values of x_{11}–x_{14} is 0.2, 20, 0.13 and 1, respectively, and the difference ε_1 is −1.328 obtained from the prediction model. By substituting the values of x_{11}–x_{14} and ε_1 into Equation (14), the predicted forward slip $\hat{y}_1$ is calculated to be 2.848%. This procedure can be applied to predict the forward slip for the other observations of values of x_{11}–x_{14}.

Moreover, the fitted regression model has also been validated by evaluating the strength of the relationship between the simulated and predicted values of forward slip under different process conditions, as well as assessing its prediction reliability in terms of Pearson correlation coefficient (r) and root mean square error ($RMSE$), which are defined according to [43,44], as,

$$r = \frac{\sum_{i=1}^{N}(y_i - \overline{y})(\hat{y}_i - \overline{\hat{y}})}{\sqrt{\sum_{i=1}^{N}(y_i - \overline{y})^2} \sqrt{\sum_{i=1}^{N}(\hat{y}_i - \overline{\hat{y}})^2}}, \tag{15}$$

$$RMSE = \sqrt{\frac{1}{N}\sum_{i=1}^{N}(y_i - \hat{y}_i)^2}, \tag{16}$$

where N is the number of paired data, y is the simulated forward slip, $\hat{y}$ is the predicted forward slip, $\overline{y}$ and $\overline{\hat{y}}$ are the mean values of y and $\hat{y}$, respectively. In accordance with Ref. [45], the relationship between two groups of data can be considered strong, provided that their r value is greater than 0.7; therefore, as is noted from Figure 13 that a good correlation with r of 0.9106 exists between the simulated and predicted data. Furthermore, the equation between simulated and predicted forward slip can be written based on the fitting of the data plotted in Figure 13, as follows:

$$y = 0.8205x + 0.8586. \tag{17}$$

In addition, the root mean square error was calculated to be 1.7383, which indicates that the developed regression model has reasonable capability and accuracy in predicting the forward slip during micro flexible rolling of aluminium alloy 5052 strips with regard to a range of reduction combinations, rolling speeds, friction coefficients and T/D ratios.

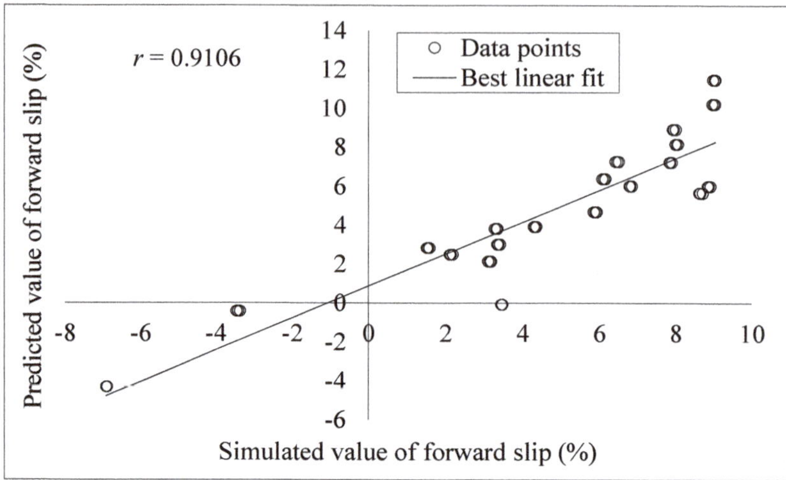

Figure 13. Correlation between the simulated and predicted values of forward slip under different process conditions.

Finally, in order to investigate the magnitudes of each process parameter affecting forward slip, variance analysis has been carried out in accordance with [38], and the results are listed in Table 2.

Table 2. Variance analysis for identification of significance of each variable.

Source	S	df	V	F	Contribution	Rank
x_1	279.737	5	55.947	16.974	25.82%	2
x_2	13.453	2	6.727	2.041	1.24%	4
x_3	248.439	3	82.813	25.125	22.93%	3
x_4	360.338	2	180.169	54.663	33.26%	1
e	181.298	55	3.296		16.74%	
Total	1083.265	67			100.00%	

S, df and V in Table 2 denote the sum of squared deviation, degree of freedom and variance, respectively, and F value is the ratio of variance of each variable to that of the error (e) between the simulated and predicted forward slip which is 3.296 in the present study. According to the size of F value, the ratio of strip thickness to grain size is believed to be the factor that has the most significant influence on forward slip.

Based on the value of *Contribution* in Table 2 which is defined as the ratio of sum of squared deviation for each factor to the total sum of squared deviation, the ratio of strip thickness to grain size is ranked first among the four process parameters, with most contribution of 33.26% to the forward slip, which is followed by the reduction with its contribution of 25.82%. The next influential parameter is the estimated friction coefficient with the contribution of 22.93%, while the rolling speed has little significant influence on the forward slip with the contribution of only 1.24%.

5. Conclusions

This study aims to identify the influence of process parameters on forward slip within thickness transition zone in micro flexible rolling of submillimetre thick strips based on the finite element method, through which numerical estimations of forward slip for a range of process conditions can be obtained in an intuitive and accurate manner as compared with theoretical calculations. A summary of the main points in the present study is presented as follows:

1. In view of the force equilibrium condition, as well as the material flow characteristics within the roll bite, the 2D analytical models are established to determine the neutral angle and consequently the forward slip of the thin workpiece during micro flexible rolling phases with thicker and thinner thicknesses, respectively; the theoretical results obtained from mathematical models indicate that an increase of reduction, rolling speed or estimated friction coefficient leads to an increase of the forward slip, while a larger initial strip thickness suggests a smaller value of it.

2. In order to provide a more intuitive and practical evaluation of forward slip, a novel 3D finite element model with bumped ridges on the roll is created based upon strip marking method to mimic the rolled strip with two line markings indented by the ridges within the thickness transition zone; meantime, the 3D Voronoi tessellation polyhedrons equipped with different mechanical properties are employed to represent the inhomogeneous grain structure in the material, through which the qualitative and quantitative estimation of the microstructural effect on the forward slip can be conducted in terms of the ratio of strip thickness to grain size, namely that the forward slip decreases as the initial strip thickness increases with regard to a constant average grain size within thickness transition zones; however, for the rolling phase with thinner thickness, the forward slip shows a negative value of -6.89% when the T/D ratio is 2, which suggests a relative sliding between the strip and the roll.

3. Numerical simulations are performed with regard to the mechanical properties of aluminium alloy 5052, whereof the results are found to have reasonable accuracy by comparison with the theoretical estimates, viz. that the difference between the numerical and theoretical results ranges from 5.95% to 10.92% among all cases, which reveals that the newly developed finite element model is suitable to analyse the forward slip and its influential factors respecting the thickness transition area in micro flexible rolling.

4. The variation of simulated forward slip in response to the variations in process parameters is in reasonable agreement with theoretical results, to wit, that for either rolling phase with thinner or thicker thickness, the forward slip generally exhibits an increasing trend along with increased reduction, rolling speed or estimated friction coefficient, whereas it shows a decrease in response to rising T/D ratio; nevertheless, negative values of forward slip are observed for cases with estimated friction coefficient of 0.08 and T/D ratio of 2, respectively, during the rolling phase with thinner thickness, namely that the neutral point no longer resides within the roll bite, and the relative sliding between the strip and roll is incurred under these conditions.

5. The scatter effect of forward slip is qualitatively assessed taking advantage of 3D Voronoi modelling which highlights the heterogeneous nature of the material, viz. that each individual grain contributes more to the macro deformation of a thinner strip consisting of fewer grains, leading to a larger scatter of material flow in the rolling direction and consequently a bigger difference during determination of forward slip in the thickness transition zone, and vice versa.

6. A multilinear regression prediction model is constructed based on simulation results obtained from both cases with thinner and thicker thicknesses, which is afterwards validated through both correlation and error analyses, since a good correlation with r of 0.9106 exists between the simulated and predicted data, and the root mean square error is estimated to be 1.7383; the developed statistical model displays reasonable capability and accuracy, and can, therefore, be utilised to reasonably predict the forward slip of thin aluminium alloy 5052 strips under various process conditions during micro flexible rolling phase with a variable strip thickness.

7. According to the analysis of variance, the ratio of strip thickness to grain size is regarded to be the most significant factor with its contribution of 33.26% to the forward slip, which is followed by the reduction with the contribution of 25.82% and the estimated friction coefficient with that of 22.93%, whilst the rolling speed hardly significantly influences the forward slip with the contribution of only 1.24%.

Author Contributions: F.Q.: Conceptualization, Formal analysis, Investigation, Methodology, Validation, Writing—original draft; J.X.: Supervision, Writing—review and editing; Z.J.: Funding acquisition, Project administration, Supervision, Writing—review and editing.

Funding: This study received grants from the Australian Research Council (FT120100432, DP190100738) and National Natural Science Foundation of China under Grant No. 51474127.

Acknowledgments: The authors would like to express their sincere appreciation to the University of Wollongong, Australian Research Council (FT120100432, DP190100738) and National Natural Science Foundation of China under Grant No. 51474127 for their support to this study.

Conflicts of Interest: The authors declare that there is no conflict of interests.

References

1. Hot and cold rolled strips in aluminium for industrial applications. Available online: https://aludium.com/industrial-strips/ (accessed on 13 August 2019).
2. Common uses for cold rolled strip steel. Available online: https://www.meadmetals.com/blog/common-uses-for-cold-rolled-strip-steel (accessed on 13 August 2019).
3. Jiang, Z.Y.; Wei, D.; Tieu, A.K. Analysis of cold rolling of ultra thin strip. *J. Mater. Process. Technol.* **2009**, *209*, 4584–4589. [CrossRef]
4. Jiang, Z.Y.; Zhu, H.T.; Tieu, A.K. Mechanics of roll edge contact in cold rolling of thin strip. *Int. J. Mech. Sci.* **2006**, *48*, 697–706. [CrossRef]
5. Xie, H.B.; Jiang, Z.Y.; Yuen, W.Y.D. Analysis of friction and surface roughness effects on edge crack evolution of thin strip during cold rolling. *Tribol. Int.* **2011**, *44*, 971–979. [CrossRef]
6. Jiang, Z.Y.; Xiong, S.W.; Tieu, A.K.; Wang, Q.J. Modelling of the effect of friction on cold strip rolling. *J. Mater. Process. Technol.* **2008**, *201*, 85–90. [CrossRef]
7. Jiang, Z.Y.; Tieu, A.K. Elastic-plastic finite element method simulation of thin strip with tension in cold rolling. *J. Mater. Process. Technol.* **2002**, *130–131*, 511–515. [CrossRef]
8. Li, E.B.; Tieu, A.K.; Yuen, W.Y.D. Forward slip measurements in cold rolling by laser Doppler velocimetry: Uncertainty analysis and accuracy improvement. *J. Mater. Process. Technol.* **2003**, *133*, 348–352. [CrossRef]
9. Wang, Q.Y.; Zhu, Y.; Zhao, Y. Friction and forward slip in high-speed cold rolling process of aluminum alloys. *Appl. Mech. Mater.* **2012**, *229–231*, 361–364. [CrossRef]
10. Liu, Y.J. Friction at Strip-Roll Interface in Cold Rolling. Ph.D. Thesis, University of Wollongong, Wollongong, Australia, 2002.
11. Vini, M.H.; Farhadipour, P. Analytical modified model of cold rolling process and investigation of the effect of work roll flattening on the rolling force. *J. Mod. Process. Manuf. Prod.* **2017**, *6*, 5–13.
12. Pawelski, H. An analytical model for dependence of force and forward slip on speed in cold rolling. *Steel Res.* **2003**, *74*, 293–299. [CrossRef]
13. Bayoumi, L.S. A kinematic analytical approach to predict roll force, rolling torque and forward slip in thin hot strip continuous rolling. *Ironmak. Steelmak.* **2007**, *34*, 444–448. [CrossRef]
14. Tieu, A.K.; Jiang, Z.Y.; Lu, C.; Kosasih, B. Friction and asperity contact in strip rolling. In Proceedings of the 3rd Symposium on Advanced Structural Steels and New Rolling Technologies, Shenyang, China, 3–6 November 2005; pp. 49–64.
15. Yuen, W.Y.D. Determination of friction from measured forward slip and its applications in hot strip rolling. In Proceedings of the First Australasian Congress on Applied Mechanics, Melbourne, Australia, 21–23 February 1996; pp. 927–932.
16. Fang, Z.; Jiang, Z.; Wang, X.; Zhou, C.; Wei, D.; Liu, X. Grain size effect of thickness/average grain size on mechanical behaviour, fracture mechanism and constitutive model for phosphor bronze foil. *Int. J. Adv. Manuf. Technol.* **2015**, *79*, 1905–1914. [CrossRef]
17. Lee, R.S.; Chen, C.H.; Gau, J.T. Effect of thickness to grain size ratio on drawability for micro deep drawing of AISI 304 stainless steel. In Proceedings of the Ninth International Conference on the Technology of Plasticity, Gyeongju, Korea, 7–11 September 2008; pp. 7–11.
18. Ananda, D.; Kumar, D.R. Effect of thickness and grain size on flow stress of very thin brass sheets. *Procedia Mater. Sci.* **2014**, *6*, 154–160. [CrossRef]

19. Raulea, L.V.; Goijaerts, A.M.; Govaert, L.E.; Baaijens, F.P.T. Size effects in the processing of thin metal sheets. *J. Mater. Process. Technol.* **2001**, *115*, 44–48. [CrossRef]

20. Regression analysis. Available online: https://en.wikipedia.org/wiki/Regression_analysis (accessed on 27 July 2018).

21. Vorkov, V.; Aerens, R.; Vandepitte, D.; Duflou, J.R. Two regression approaches for prediction of large radius air bending. *Int. J. Mater. Form.* **2019**, *12*, 379–390. [CrossRef]

22. Schmid, M.; Hackl, B.; Pauli, H.; Grünsteidl, A. A linear regression model to predict FLCs from tensile test data and sheet thickness. In Proceedings of the 5th Forming Technology Forum, Zurich, Switzerland, 5–6 June 2012; pp. 57–62.

23. Strano, M.; Colosimo, B.M. Logistic regression analysis for experimental determination of forming limit diagrams. *Int. J. Mach. Tools Manuf.* **2006**, *46*, 673–682. [CrossRef]

24. Hubbard, J.B.; Stoudt, M.R.; Possolo, A.M. Topographic analysis and Weibull regression for prediction of strain localisation in an automotive aluminium alloy. *Mater. Sci. Technol.* **2011**, *27*, 1206–1212. [CrossRef]

25. Kopp, R.; Böhlke, P.; Hohmeier, P.; Wiedner, C. Metal forming of lightweight structures. In Proceedings of the Fifth ESAFORM Conference, Kraków, Poland, 14–17 April 2002; pp. 13–22.

26. Kopp, R.; Wiedner, C.; Meyer, A. Flexibly rolled sheet metal and its use in sheet metal forming. *Adv. Mater. Res.* **2005**, *6–8*, 81–92. [CrossRef]

27. Liu, X.H. Prospects for variable gauge rolling: Technology, theory and application. *J. Iron Steel Res. Int.* **2011**, *18*, 1–7. [CrossRef]

28. Zhang, G.J.; Liu, X.H.; Hu, X.L.; Zhi, Y. Horizontal velocity of variable gauge rolling: Theory and finite elements simulation. *J. Iron Steel Res. Int.* **2013**, *20*, 10–16. [CrossRef]

29. Prakash, R.S.; Dixit, P.M.; Lal, G.K. Steady-state plane-strain cold rolling of a strain-hardening material. *J. Mater. Process. Technol.* **1995**, *52*, 338–358. [CrossRef]

30. Ginzburg, V.B.; Ballas, R. *Flat Rolling Fundamentals*; Marcel Dekker, Inc.: New York, NY, USA, 2000.

31. Tarnovskii, I.Y.; Pozdeyev, A.A.; Lyashkov, V.B. *Deformation of Metals during Rolling*; Elsevier: Amsterdam, The Netherlands, 2013.

32. Qu, F.J.; Jiang, Z.Y.; Wei, D.B.; Chen, Q.Q.; Lu, H.N. Study of micro flexible rolling based on grained inhomogeneity. *Int. J. Mech. Sci.* **2017**, *123*, 324–339. [CrossRef]

33. Modeling of Sheet Metal—The "Known Unknowns". Available online: http://formingworld.com/constitutive-modeling-of-sheet-metal-the-known-unknowns/ (accessed on 9 September 2019).

34. Bhattacharyya, D. *Composite Sheet Forming*, 1st ed.; Elsevier: Amsterdam, The Netherlands, 1997.

35. Qu, F.J.; Jiang, Z.Y.; Lu, H.N. Analysis of micro flexible rolling with consideration of material heterogeneity. *Int. J. Mech. Sci.* **2016**, *105*, 182–190. [CrossRef]

36. Qu, F.J.; Xie, H.B.; Jiang, Z.Y. Finite element method analysis of surface roughness transfer in micro flexible rolling. *MATEC Web Conf.* **2016**, *80*, 04002. [CrossRef]

37. Properties of aluminum alloy AA 5052. Available online: http://www.efunda.com/Materials/alloys/aluminum/show_aluminum.cfm?ID=AA_5052&show_prop=all&Page_Title=AA%205052 (accessed on 25 June 2018).

38. Qu, F.J.; Jiang, Z.Y.; Xia, W.Z. Evaluation and optimisation of micro flexible rolling process parameters by orthogonal trial design. *Int. J. Adv. Manuf. Technol.* **2018**, *95*, 143–156. [CrossRef]

39. Qu, F.J.; Jiang, Z.Y.; Wang, X.G.; Zhou, C.L. Analysis of springback behaviour in micro flexible rolling of crystalline materials. *Adv. Mater. Sci. Eng.* **2018**, *2018*, 5287945. [CrossRef]

40. Wang, T.T.; Qi, K.M. *Plastic Forming of Metals: Principles and Technology of Rolling*, 2nd ed.; Metallurgical Industry Press: Beijing, China, 2001.

41. The forward slip and backward slip in rolling process. Available online: https://wenku.baidu.com/view/a4324dbd65ce050876321350 (accessed on 9 July 2019).

42. Multiple linear regression. Available online: http://www.stat.yale.edu/Courses/1997-98/101/linmult.htm (accessed on 12 July 2019).

43. Pearson correlation coefficient. Available online: https://en.wikipedia.org/wiki/Pearson_correlation_coefficient (accessed on 29 July 2019).

44. Samui, P.; Roy, S.S.; Balas, V.E. *Handbook of Neural Computation*, 1st ed.; Academic Press: London, UK, 2017.
45. Moore, D.S.; Notz, W.I.; Fligner, M.A. *The Basic Practice of Statistics*, 6th ed.; W.H. Freeman: New York, NY, USA, 2013.

© 2019 by the authors. Licensee MDPI, Basel, Switzerland. This article is an open access article distributed under the terms and conditions of the Creative Commons Attribution (CC BY) license (http://creativecommons.org/licenses/by/4.0/).

Article

Examination of Formability Properties of 6063 Alloy Extruded Profiles for the Automotive Industry

Athanasios Vazdirvanidis [1,*], Ioannis Pressas [1], Sofia Papadopoulou [1], Anagnostis Toulfatzis [1], Andreas Rikos [1], Marianna Katsivarda [1], Grigoris Symeonidis [2] and George Pantazopoulos [1,*]

[1] ELKEME Hellenic Research Centre for Metals S.A., 61st km Athens—Lamia National Road, 32011 Oinofyta, Greece; ipressas@elkeme.vionet.gr (I.P.); spapadopoulou@elkeme.vionet.gr (S.P.); atoulfatzis@elkeme.vionet.gr (A.T.); arikos@elkeme.vionet.gr (A.R.); mkatsivarda@elkeme.vionet.gr (M.K.)

[2] ETEM Extrusion Company, bul. "Iliyantsi" 119, 1220 Sofia, Bulgaria; gsimeonidis@etem.com

* Correspondence: avazdirvanidis@elkeme.vionet.gr (A.V.); gpantaz@elkeme.vionet.gr (G.P.); Tel.: +30-22626-04436 (A.V.); +30-22626-04463 (G.P.)

Received: 19 August 2019; Accepted: 30 September 2019; Published: 7 October 2019

Abstract: Bendability is a crucial property of automotive parts, which describes the ability of extruded profiles to be formed to shape, without the appearance of discontinuities that will have an adverse effect on the mechanical properties and their energy absorption capacity (crashworthiness). Anisotropic behavior exhibited by Al alloy extrusions has been documented due to the nature of the hot working process. In the present article an attempt is made in order to determine the main factors influencing bending properties in longitudinal and transverse to extrusion direction, which are also related with anisotropic behavior. Mechanical testing, microscopic examination and finite element analysis were applied in the frame of the current investigation. The findings were indicative that grain morphology and orientation (texture), especially on the surface zone of the extruded profiles, as well as morphology and distribution of the constituent particles are strongly related with formability.

Keywords: automotive; 6063 Alloy; EBSD; bendability; fractography; modeling; texture

---

1. Introduction

Previous works in this field have highlighted the role of process parameters on microstructure and texture and it has been found that the fracture mode of tensile specimens and therefore the ductility is strongly related with the cooling rate that followed the soaking thermal treatment [1–4]. More specifically, slower cooling rates, i.e., air cooling, resulted in a higher amount of intergranular fracture and lower ductility, relatively to the amount of transgranular ductile fracture that was observed after water quenching.

In one of the studies, the effect of microstructure on formability is examined, where the anisotropic behavior of extruded samples is attributed to both grain morphology and distribution of primary particles [5]. The role of the constituent particles, as well as the dispersoid particles is also discussed by S. Das et al. [6] and A. Vazdirvanidis et al. [7]—while investigation of AlFeSi particles formation and their transformation during heat treatment is discussed in [8–10]. In reference [11], the role of directionality in mechanical properties is mentioned. Lower bending angles were reported with the bending axis parallel to extrusion direction, while a correlation between bendability and ductility was found. D.J. Lloyd noticed macroscopic cracking on the outer bent surface with the following sequence and by increasing bending strain: (i) slip band and shear bands formation within the grains, (ii) rotation and displacement of grain boundaries, (iii) increase of the number of grain boundaries exhibiting displacement, (iv) final rupture assisted by cracking of intermetallics, and development of discontinuities by shear bands and also on grain boundaries due to precipitates [12].

The role of texture is also crucial in the examination of automotive alloys properties [13]. It has been reported that a high amount of deformation texture compared to random texture on the surface zone of the profiles leads to an increase of shear bands formation during bending within each grain [14]. In the same work, the role of the size and distribution of constituent particles, especially on the free surface of bent specimens, is highlighted. Texture examination including Taylor factor maps are correlated with bendability in work presented by H. Inoue [15]. Extrusion temperature also plays a significant role in mechanical strength development of 6xxx Al-alloys. According to Reference [16], extrusions processing at a temperature below recrystallization (120 °C), offers a great advantage towards strength and ductility, stimulating dynamic precipitation and/or accelerated precipitation processes. Based on Reference [17], pre-aging treatment of extruded 6082 Al alloy at 175 °C for 30 min, after solution heat treatment, has a beneficial role, eliminating the negative delay effect. It results in a stable alloy microstructure when kept at room temperature for 24 h, improving the final mechanical properties.

In this paper, the factors that influence formability of 6xxx series alloy extruded profiles for the automotive industry are examined by means of (i) microstructure characterization by optical, electron microscopy and electron backscatter diffraction (EBSD), (ii) mechanical testing involving tensile and bending tests and (iii) finite element analysis of stresses and strains in extrusions during mechanical testing.

The effects of grain structure and texture, especially near the surface and size and distribution of AlFeSi intermetallic particles on bendability are examined. The aim of the project is to determine the microstructure properties that are related with the mechanical properties of the material and through the proper configurations to optimize the manufacturing process in order to achieve high formability. The outcome of this study has significant value in understanding the contribution of texture and second phase particle orientation to the formability and failure susceptibility of Al 6xxx profile for automotive applications. This is an original industrial R&D work, while there is only limited published research addressing these issues from a consolidated point of view.

2. Materials and Methods

2.1. Materials Production

6xxx series extruded profiles were manufactured by industrial trials. The alloy studied in this work is 6063 (0.20–0.60% Si, 0.35% max Fe, 0.10% max Cu, 0.10% max Mn, 0.45–0.90% Mg, 0.10% max Cr, Zn, Ti, remainder Al, all are expressed in wt.%). The samples have been provided by ETEM Bulgaria S.A. Extrusions with complicated cross-sections differing in the wall thickness (2.9 mm vs. 3.2 mm) were produced. The profiles had been artificially aged under the same, industrial, aging process to peak age, reaching a T6 metallurgical condition.

2.2. Mechanical Testing

Tensile and three-point bending tests were performed using an Instron 5567 30 kN ((Instron, Norwood, MA, USA) electromechanical testing machine at ambient temperature. Tensile tests were conducted at a constant speed of 10 mm/min according to ISO 6892-1 standard. Tensile specimens were extracted from the original aluminum profile for testing parallel to the extrusion direction. Three-point bending tests (transverse and longitudinal) were performed using 20 mm/min speed, according to the instructions of VDA 238–100 and ISO 7438 standards. Tests stop criteria was set at 10% load drop. Roller span distance was set at two times the specimen thickness plus 0.2 mm. Specimens were extracted from the middle of the external profile face. Due to the profile geometry sample sizes of 50×25 mm^2 were used. Measurement of bending angles was manually performed after the end of test. Higher bendability is associated with high bending angle values.

2.3. Microstructure Characterization

Macroscopic examination of the fracture surfaces from tensile tests and 3-point bend tests was performed by a Nikon SMZ 1500 (Nikon, Tokyo, Japan) stereo-microscope. Optical microscopy examination was conducted using a Nikon Epiphot 300 inverted metallographic microscope equipped with an image analysis software (Image Pro Plus, Rockville, MD, USA). The samples were first examined in as-polished condition for revealing AlFeSi intermetallics density and morphology. This information was provided as input to FEM software for modeling the 3D response of the material to tensile stresses. Polarized light illumination was applied after Barker's electrolytic etch for revealing grain structure and bright field illumination after HF immersion etching.

High magnification microscopic and fractographic observation was performed using a FEI XL40 SFEG SEM (FEI, Eindhoven, The Netherlands). EBSD analysis was performed with an EDAX Hikari XP camera (EDAX, Mahwah, NJ, USA) on longitudinal cross-sections (extrusion direction coincides with TD and normal direction with ND coordinates).

2.4. Finite Element Analysis

The morphology, size and distribution of the AlFeSi intermetallic particles together with their influence on the response of the different areas of the material to tensile stresses were simulated via finite element analysis (FEA). The present approach is to simulate the Al-matrix behaviour based primarily on the second phase particle morphology and the stress- fields, derived by the presence of harder particles distributed in the microstructure. Hence, it is considered that the density of the intermetallic phase is not a predominant factor of the presented results, as it is also supported by the non-coherency (absence of strong metallurgical bonding) of the second phase particles with the matrix. Further information concerning the influence of impurities on properties and processing of Al alloys are presented in Refs. [18,19]. The aim of this task was the detection of (i) areas where high stress concentration can occur, (ii) potential differences of stress and strain distribution between areas with dense or sparse intermetallic particles distribution and (iii) differences in the response of the material in the longitudinal and transverse to extrusion direction. For the simulation, the commercial FEA software LS-DYNA™ was used. The model consisted of a rectangular aluminum body with dimensions of 38.1 μm × 38.1 μm × 4.57 μm, while eleven particles were scattered inside the aluminum body. The distribution of the particles is presented in Figure 1.

The size of the particle ranges from approximately 0.4 to 3.2 μm in length and 0.4 to 1.5 μm in width. In order to be more realistic with the simulation, the size, shape and distribution of the particles were based on actual metallographic observations of the experimental samples, as presented in Figure 2. The particle shapes involved wedge type, needle type and circular type particles. The needle and circular type particles were modeled via icosahedra, in order to simplify the meshing process.

The numerical model was meshed using approximately 1.1 million tetrahedral elements. The generated mesh in both the aluminum body and the particles is presented in Figure 3. It is worth noting that the mesh in the particle cavities of the aluminum bodies matched that of the corresponding particles, in order to avoid any numerical flaws caused by mesh incompatibility.

The mechanical properties used for the current numerical analysis are presented in Table 1. Both for the aluminum body and the particles a bilinear elasto-plastic material model (MAT 24—PIECEWISE_LINEAR_PLASTICITY [20]) was used.

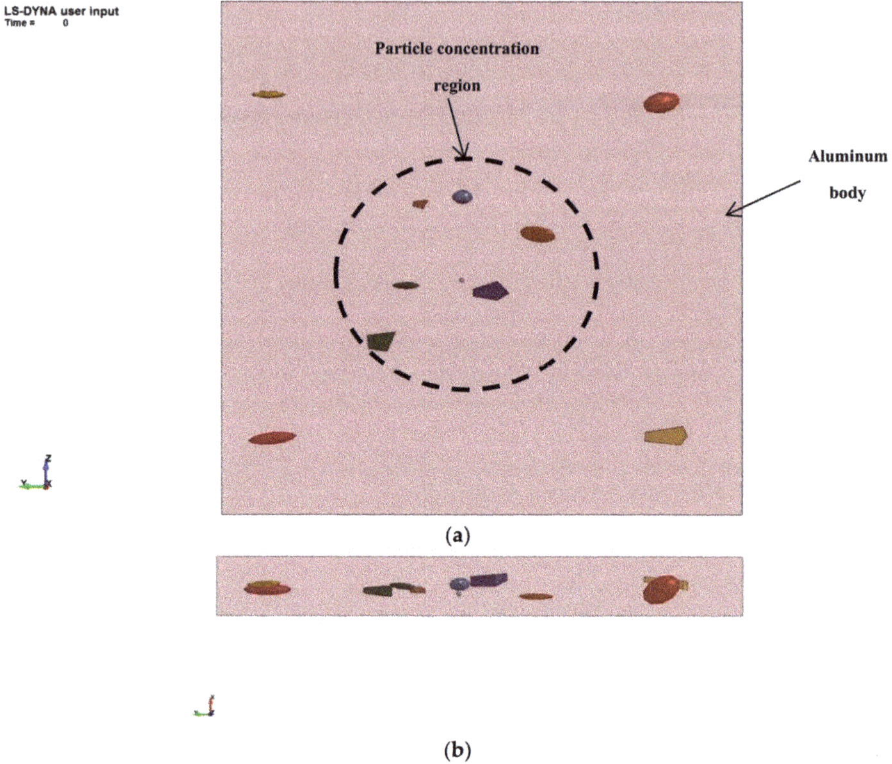

Figure 1. Particles distribution in aluminum body for FEA: (**a**) Top view and (**b**) side view imaging.

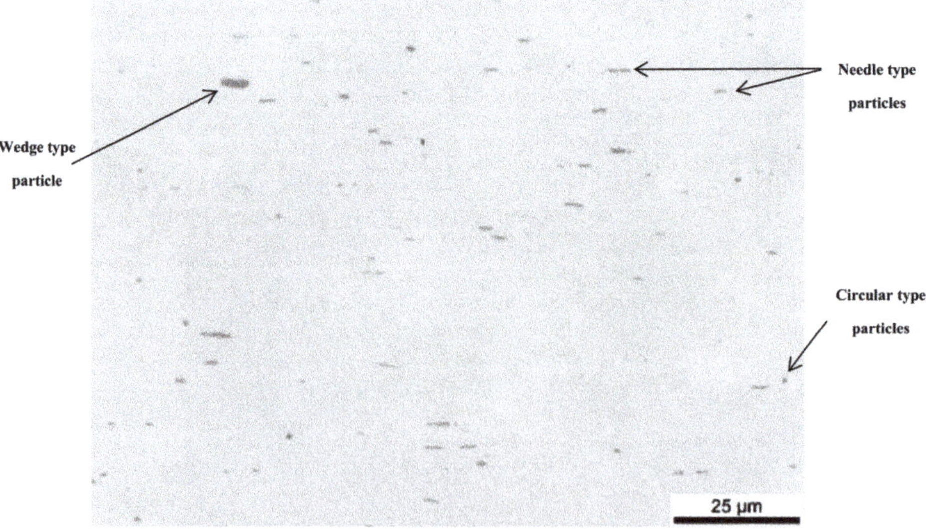

Figure 2. Optical micrograph showing intermetallic particles in Al 6063 alloy sample (observation parallel to extrusion). Wedge, needle and circular type particles can be seen.

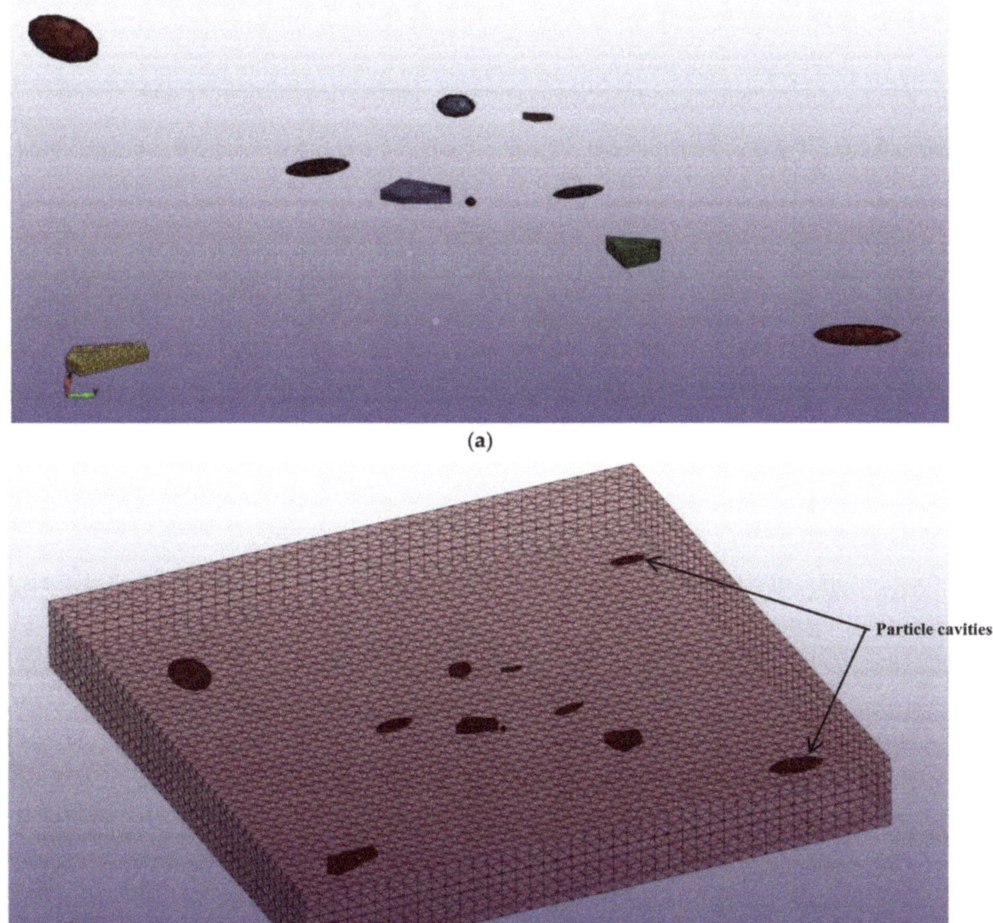

(a)

(b)

Figure 3. Generated mesh: (**a**) Particle mesh and (**b**) Aluminum body mesh.

Table 1. Mechanical properties used in the FE models.

Material	Density (kg/m^3)	Young's Modulus (GPa)	Poisson's Ratio	Yield Strength—YS (MPa)	Tensile Strength—TS (MPa)	Tangent Modulus (MPa)
Al 6063	2700	70	0.33	234	272	680.85
Particle material	3580 [18]	139.2 [21]	0.31496 [21]	2700 [18,19]	-	-

It is worth noting that due to the lack of material property references for the AlFeSi phase particles in literature and because this analysis is mainly aimed at the mechanical behavior of the Al body, the mechanical properties of Al$_3$Fe mentioned by Belov et al. [18] and Gaillac et al. [21] were used for the particles. For the yield strength of the Al$_3$Fe particles, the empirical rule deeming it equal to approximately HV/3 was used, as presented by Belov et al. [18] and Glazoff et al. [19]. The aluminum mechanical properties were exported from the tensile tests performed on the samples (see Section 3.1).

Via the finite element model, two different loading scenarios were investigated. In the first scenario, a uniform tensile load of 150 MPa was applied along two opposite sides of the aluminum body, aligned with the extrusion axis. In the second scenario, the same uniform tensile load of 150 MPa was applied in two opposite sides of the aluminum body, but transverse to the axis of extrusion. The two different loading scenarios are presented in Figure 4.

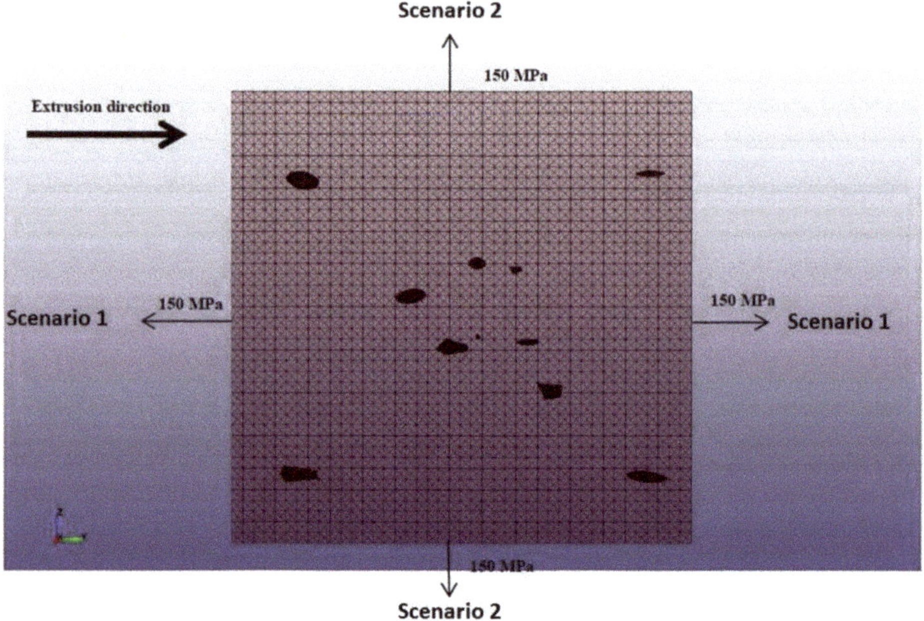

Figure 4. Loading scenarios: Load aligned with the extrusion direction (Scenario 1) or transverse to the extrusion direction (Scenario 2), (Top view). Each scenario was separately simulated.

3. Results

3.1. Tensile Testing and Fracture Modes

Ten (10) tensile specimens were retrieved from each profile section (10 from the profile with 2.9 mm wall thickness and 10 from the profile with 3.2 mm wall thickness). The results of the tensile tests are summarized in Table 2. One representative stress-strain curve from each profile is given in Figure 5a (see also Figure 5b for extrusion section). It is shown that the 2.9 mm thick samples attained higher yield and tensile stress values (239 and 271 MPa), but lower elongation at fracture (10.3% vs. 12%) signifying lower ductility as this can be determined by elongation at fracture. The respective values for the 3.2 mm thick sample were 229 MPa yield strength, 261 MPa tensile strength and 12.0% elongation at fracture. The two materials were in the same metallurgical condition and had similar tensile to yield strength ratio values (1.13), which leads to the conclusion that the differences in the mechanical properties should be related with grain morphology. Stereoscopic micrographs of the fracture surfaces are given in Figure 5c,d. It is readily observed that the 3.2 mm thick sample exhibited higher amount of area reduction at fracture.

Table 2. Tensile tests results (mean and standard deviation of ten measurements).

YS (MPa)		TS (MPa)		Percentage Elongation at Fracture (%)	
2.9 mm	3.2 mm	2.9 mm	3.2 mm	2.9 mm	3.2 mm
239 ± 14	229 ± 9	271 ± 2	261 ± 1	10.3 ± 0.6 (95% Confidence level = ±1.2%)	12.0 ± 0.7 (95% Confidence level = ±1.4%)

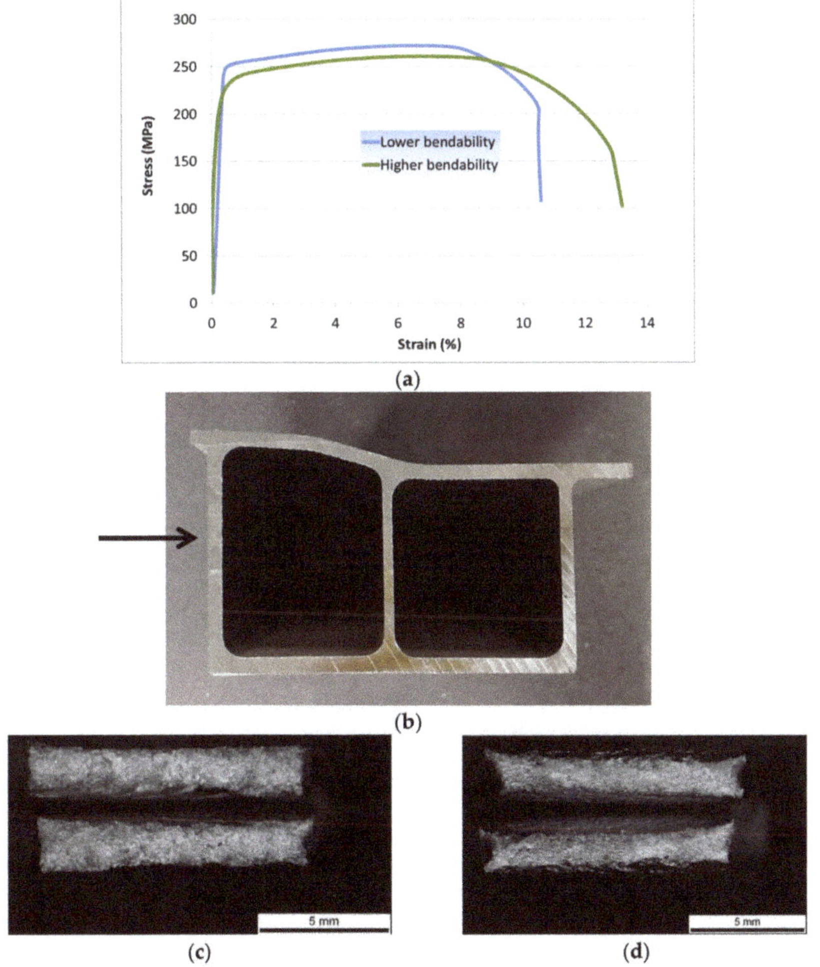

Figure 5. (**a**) Stress-strain curves showing lower strength and higher ductility for the thicker sample. (**b**) Macrograph of the aluminum profile cross-section, showing the wall area from which the mechanical testing specimens were retrieved, the wall thickness corresponds to the initial tensile specimen thickness. Stereoscopic micrographs of the fracture surfaces from tensile tests, (**c**) 2.9 mm thick sample and (**d**) 3.2 mm thick sample. Higher reduction of area is observed in (**d**).

The fracture surfaces were also examined with an electron microscope. In Figure 6, it can be observed that the less ductile sample exhibited a high amount of intergranular fracture, especially on the subsurface regions and a low amount of dimple fracture (mixed fracture mode). The prevailing mechanism is microvoid coalescence on the grain boundaries caused by Mg_2Si precipitation and nanoscale precipitates free zones formation [1,7]. Ductile dimpled fracture follows a significant extent of permanent plastic deformation, which is shown even macroscopically by cross section reduction (necking). At the microscopic level, voids are generated around second phase particles, and at the centre of the necked region, where the stress triaxiality is maximum. A recent review of the fracture mechanisms is presented in [22].

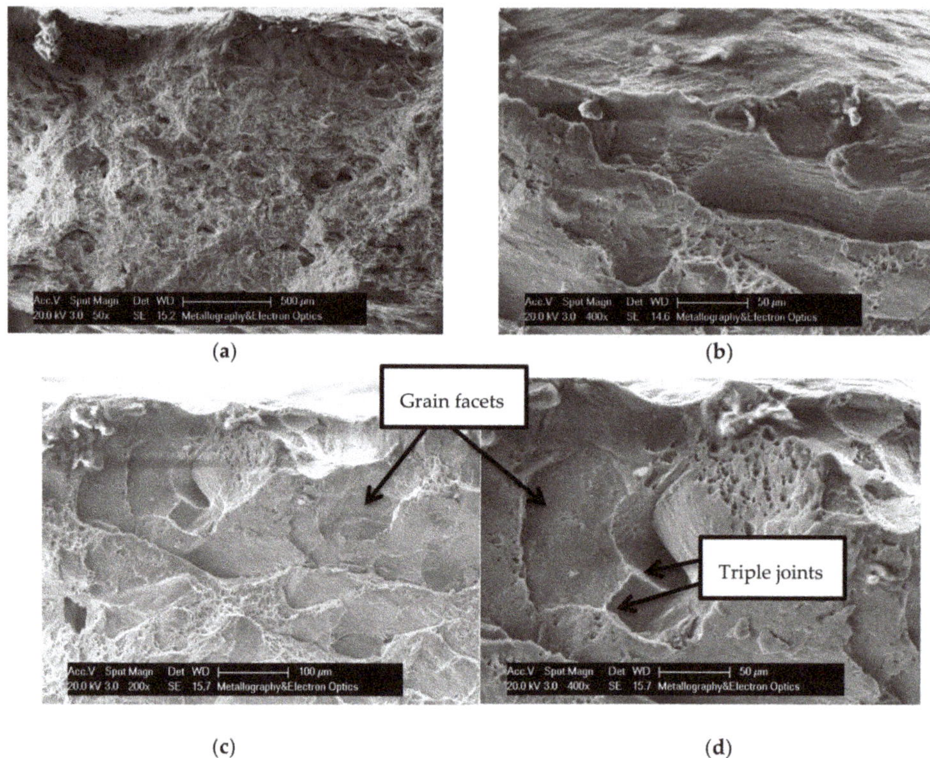

(a)

(b)

(c)

(d)

Figure 6. (**a**–**d**) SEM fractographs, fracture surface of tensile specimen with 2.9 mm thickness. Mixed fracture mode consisting mainly of intergranular fracture and lower percentage exhibiting dimpled morphology. Grain facets and triple joints are highlighted.

It is noteworthy that higher amount of intergranular fracture relatively to dimple fracture is observed on the surface region of the 2.9 mm sample.

On the other hand, the more ductile, 3.2 mm sample exhibited a uniform, equiaxed dimpled, fracture surface (Figure 7). No significant variation in the fracture mode between surface and inner regions was observed.

In both cases, no other microstructural defects such as inclusions or porosity, that could impair ductility, were identified with electron fractography in both samples.

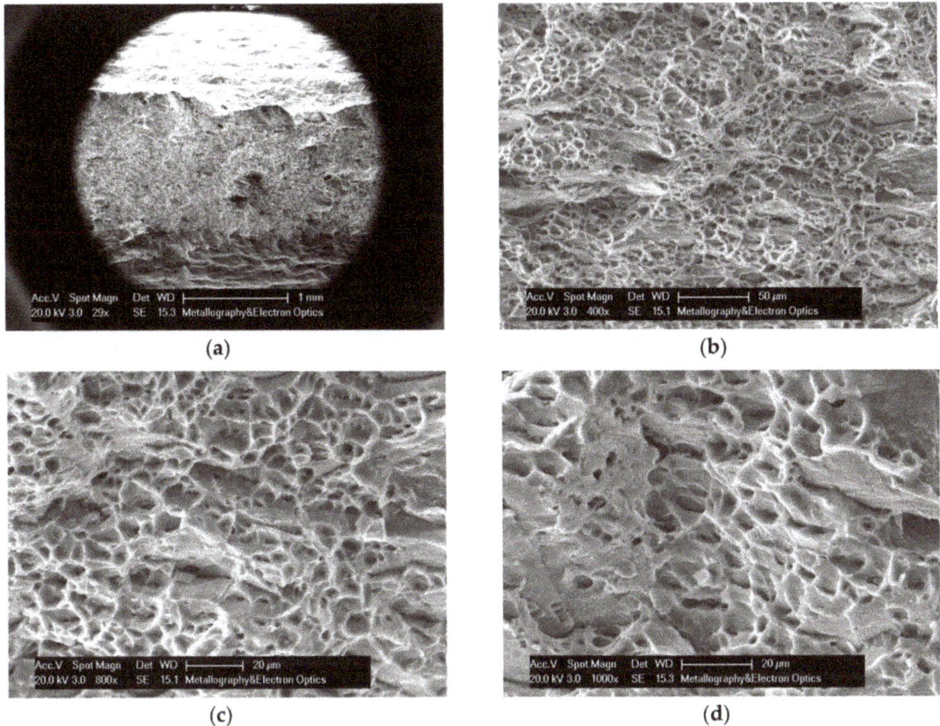

Figure 7. (a–d) SEM fractographs, fracture surface of tensile specimen with 3.2 mm thickness. Mostly dimpled ductile fracture is observed.

3.2. Three Point Bending Testing

Three (3) transverse and three (3) longitudinal bend tests were performed with tests being interrupted at 10 percent load drop. The results of three-point bend tests are given in Table 3. Similar bending angle values were reported in longitudinal tests for all specimens (approximately 102°). Regarding the transverse tests, lower bending angle values were exhibited in 2.9 mm thick specimens (96°) and higher in the 3.2 mm thick specimens (115°). This is an indication that transverse testing could be preferred in the specific alloy and metallurgical condition, when highlighting of bendability in different samples is sought.

One representative load-distance curve from each test is given in Figure 8. Samples with 3.2 mm thickness attained higher load values in both longitudinal and transverse directions, as it was anticipated by their higher thickness values—especially in longitudinal direction where the bending angle was almost identical.

Metallographic examination of bent specimens revealed "roughening" at the surface and shear bands formation, accompanied by formation of intergranular discontinuities, mainly on the surface, but also in the subsurface regions (Figure 9). It is also shown that in 2.9 mm thick sample excessive cracking occurred in transverse tests compared to longitudinal tests which accounts for the drop of the load and the lower bending angle values (Figure 9a,b). In 3.2 mm sample transverse testing led to more intense surface "roughening", but longer subsurface cracks were formed in longitudinal testing (Figure 9c,d).

In the same figure, it is also observed that even though both materials are fully recrystallized, with equiaxed microstructure, near the surface coarser grains that reduce mechanical properties can be found. The grain boundary interface of such grains is in many cases the stress concentration and cracking initiation region (see Figure 9d).

In Figures 10–13 representative (a) macrographs of the bent areas at 10% load drop, (b) high magnification optical micrographs (cross-sections) of the bent areas, (c) and (d) SEM micrographs of the microstructure in HF etched condition and (e) macrographs of the bent areas at 180° bend angle, are given from longitudinal and transverse tests. From the examination it could be deduced that:

- Formation of discontinuities can occur on grain boundaries as a result of different accommodation of strain in neighboring grains. Shear bands are observed traversing different grains on the surface and subsurface regions (see Figure 13b). In Figures 10b, 11b and 13b it is readily seen that the prevailing fracture mode is intergranular and cracking was preferentially formed in the subsurface regions.

- Formation of discontinuities also occurred between constituent intermetallic particles inter- and transgranularly. The size and morphology and general appearance of the intermetallic particles for this type of Al-alloy, as they have been identified by optical and electron microscopy constitute typical evidence of Al-Fe-Si secondary phase that have been subjected to deformation processing (extrusion). In Figures 10d and 12d representative SEM micrographs are given showing discontinuities formed between adjacent particles inside the grains. In Figures 11d and 13d it can be seen that during bending testing intermetallic particles also promoted cracking since strain could not be accommodated by such brittle phases.

- Mg_2Si precipitates contribute to voids nucleation on grain boundaries favoring the intergranular fracture mode. Precipitates are preferentially formed on double and triple grain boundary regions during artificial aging. Grain boundary debonding observed in Figure 10d, Figure 11c,d, Figures 12c and 13c is associated with this precipitation behavior as well as the potential occurrence of precipitate free zones (PFZs), which are very narrow (expected to be tenths of nanometers thick in T6 condition) and therefore they can be only discerned by transmission electron microscopy (TEM).

- The outer surfaces of the samples bent by 180 degrees were completely fractured for 2.9 mm sample in transverse testing (Figure 10e), but were only roughened in the other three cases (2.9 mm—longitudinal test, Figure 11e, 3.2 mm sample transverse and longitudinal tests, Figures 12e and 13e).

Table 3. Results of bending tests (three tests in every category). Tests were interrupted at 10% load drop.

Longitudinal Tests				Transverse Tests			
2.9 mm wall thickness		3.2 mm wall thickness		2.9 mm wall thickness		3.2 mm wall thickness	
Bending Angle (°)	Max Load (N)	Bending Angle (°)	Max Load (N)	Bending Angle (°)	Max Load (N)	Bending Angle (°)	Max Load (N)
102	3340	102	3725	96	3820	115	4830

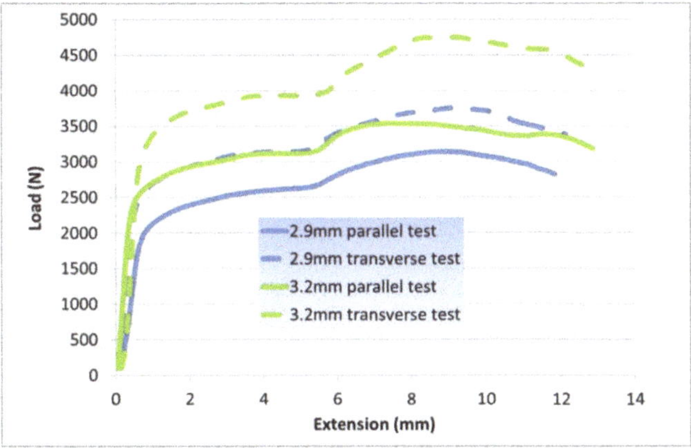

Figure 8. Indicative load—extension curves for longitudinal (solid lines) and transverse (dashed lines) bend tests for 2.9- and 3.2-mm thick samples.

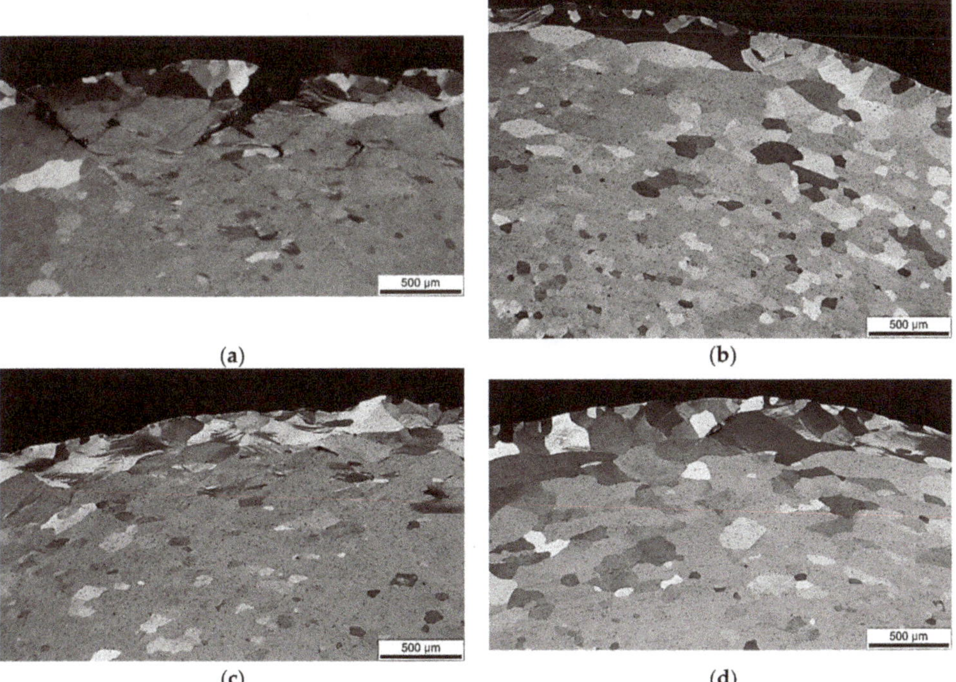

Figure 9. Bent specimens with tests being interrupted at 10% load drop, optical micrographs, polarized light illumination. Cross-section transverse to bending line showing roughening of the outer surface, shear bands formation and intergranular cracking initiation. (**a**) Sample 2.9 mm thickness (transverse test), (**b**) sample 2.9 mm thickness (longitudinal test), (**c**) sample 3.2 mm thickness (transverse test) and (**d**) sample 3.2 mm thickness (longitudinal test).

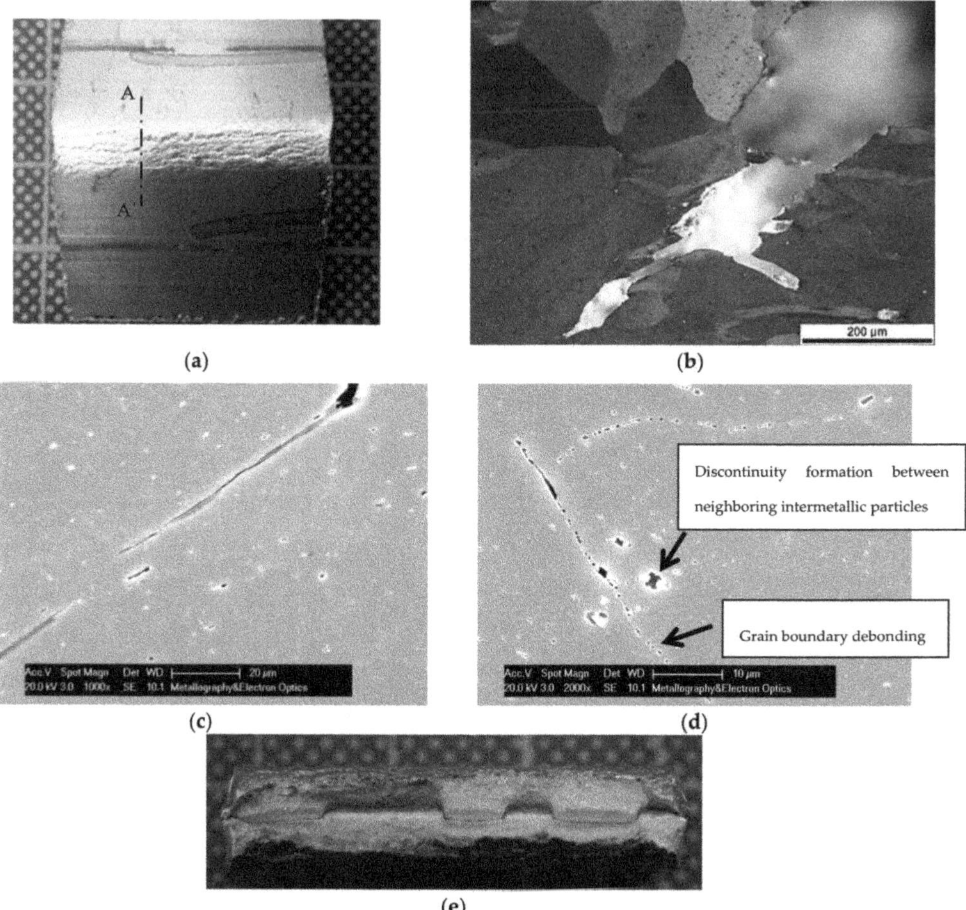

(a)

(b)

(c)

(d)

(e)

Figure 10. (**a**) Macroscopic image of bent specimen sample with 2.9 mm thickness (end of bending test at 10% load drop, transverse bend test with bend line parallel to extrusion direction). (**b**) Optical micrograph, cross-section A-A' showing heavy, partly intergranular cracking on the surface. (**c**,**d**) SEM micrographs, grain boundary debonding and rupture. (**e**) Macrograph, end of bending test at 180 degrees.

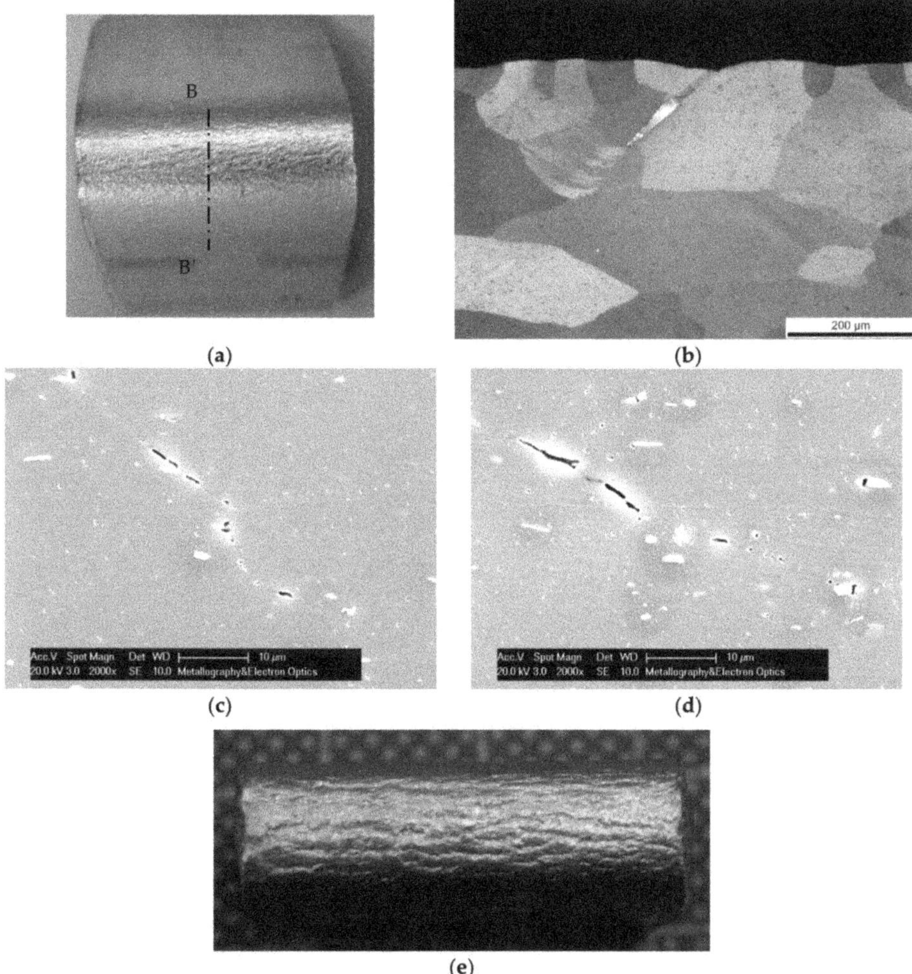

Figure 11. (**a**) Macroscopic image of bent specimen sample with 2.9 mm thickness (end of bending test at 10% load drop, longitudinal bend test with bend line transverse to extrusion direction). (**b**) Optical micrograph, cross-section B-B' showing mainly intergranular cracking on the surface. (**c**,**d**) SEM micrographs, grain boundary debonding and rupture. (**e**) Macrograph, end of bending test at 180 degrees.

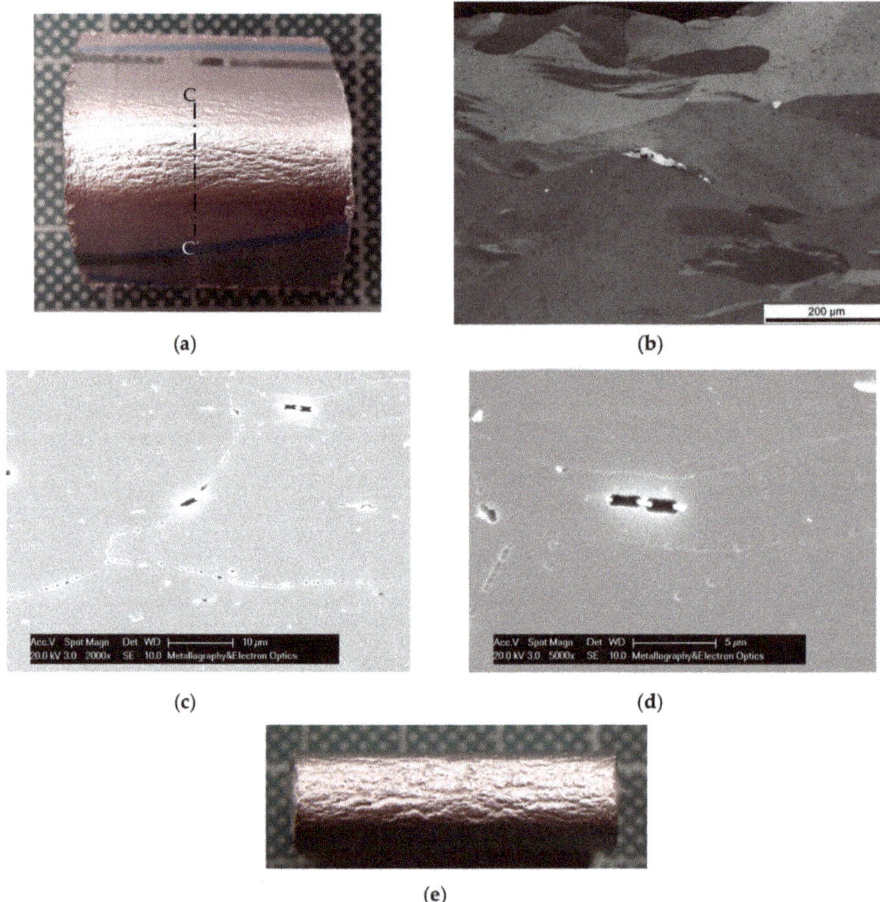

(a)

(b)

(c)

(d)

(e)

Figure 12. (**a**) Macroscopic image of bent specimen sample with 3.2 mm thickness (end of bending test at 10% load drop, transverse bend test with bend line parallel to extrusion direction). (**b**) Optical micrograph, cross-section C-C′ showing partly intergranular cracking on the surface. (**c**,**d**) SEM micrographs, grain boundary debonding and rupture. (**e**) Macrograph, end of bending test at 180 degrees.

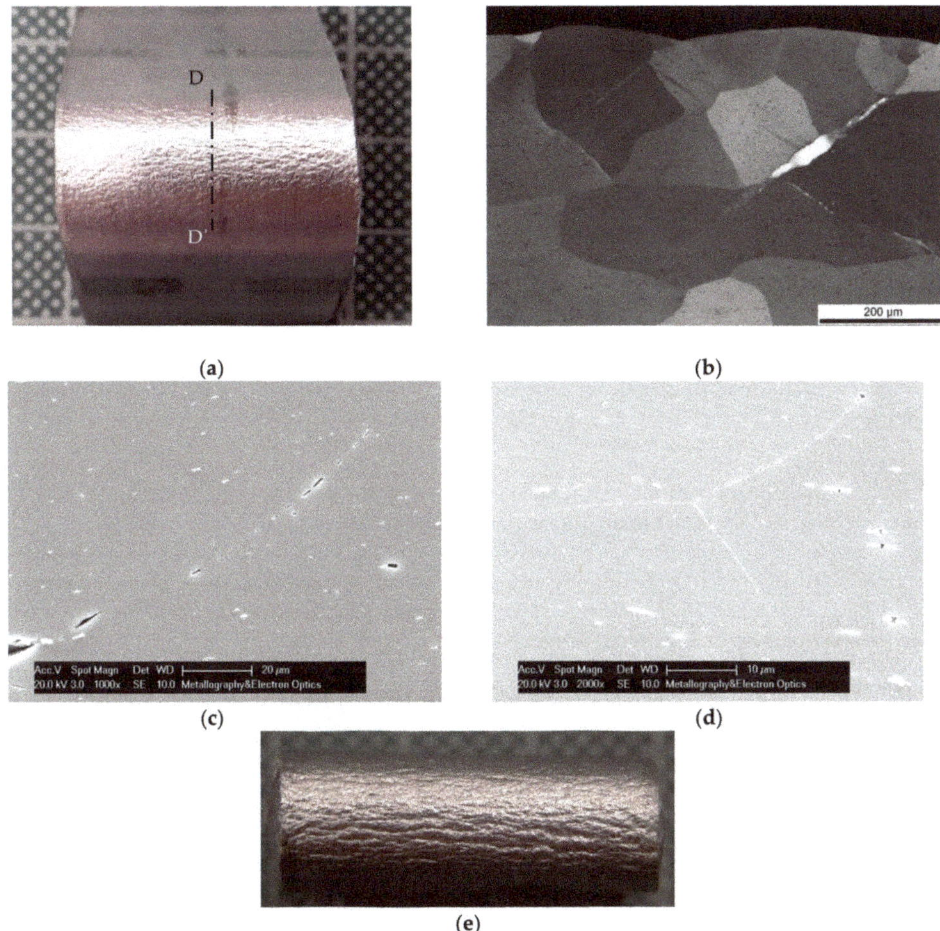

Figure 13. (**a**) Macroscopic image of bent specimen sample with 3.2 mm thickness (end of bending test at 10% load drop, longitudinal bend test with bend line transverse to extrusion direction). (**b**) Optical micrograph, cross-section D-D′ showing partly intergranular cracking on the surface. (**c,d**) SEM micrographs, grain boundary debonding and rupture. (**e**) Macrograph, end of bending test at 180 degrees.

3.3. Electron Backscatter Diffraction (EBSD)

The microstructure and the texture of the extrusions was examined by electron backscatter diffraction. In Figure 14 the grain structure maps from a mid-thickness location for both 2.9- and 3.2-mm thick samples are given. The latter exhibited a finer microstructure with 59 µm average grain size versus 85 µm for the thinner section (both with equiaxed morphology).

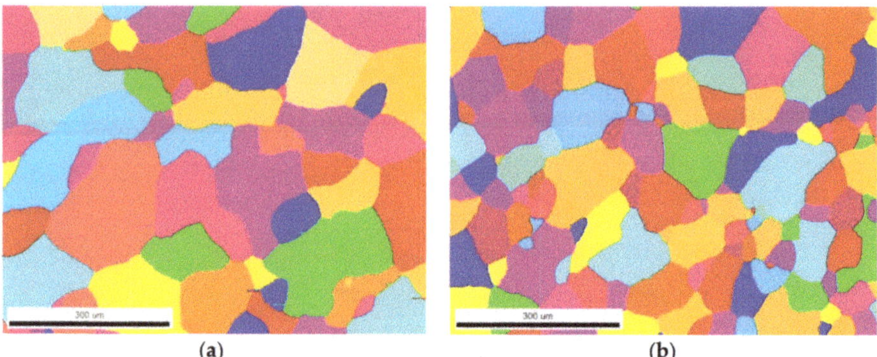

(a)   (b)

Figure 14. Grain structure derived by electron backscatter diffraction (EBSD): (**a**) 2.9 mm thickness sample and (**b**) 3.2 mm thickness sample. The grain structure was finer in 3.2 mm sample in the middle of the section, but also on the surface region.

Additionally, Taylor factor maps from the surface of the profiles were retrieved (Figure 15). Examination revealed secondary grain growth phenomena in 2.9 mm thick sample, which are expected to lead to premature failure in mechanical testing. Purple arrows indicate potential locations for stress concentration, attributed to the specific orientation of the neighboring grains. The higher the difference in Taylor factor values and the higher frequency of such interfaces the more possible is that the material will fail during bending or any other forming process.

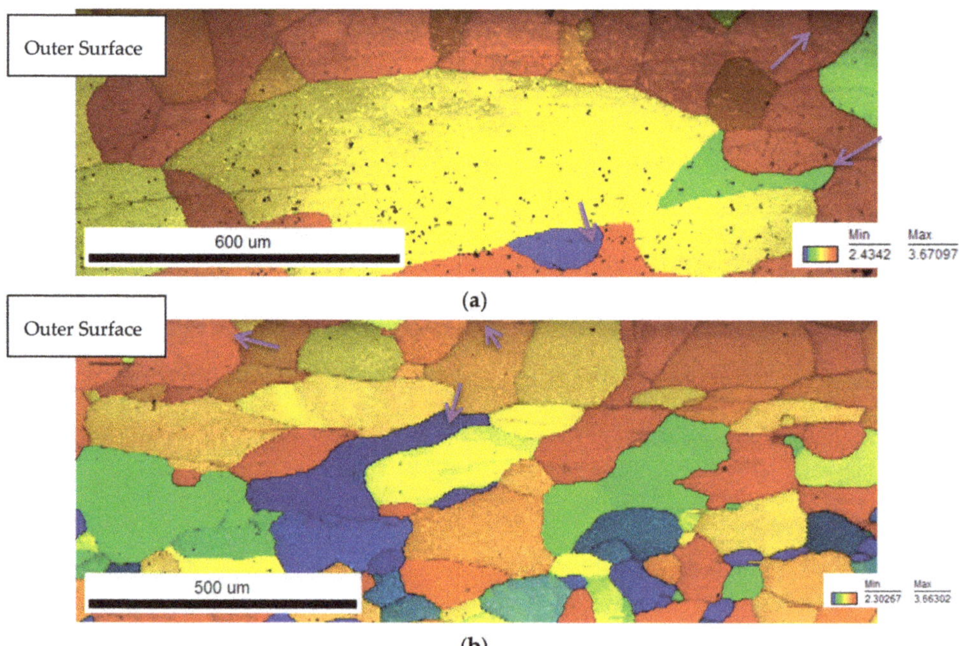

(a)

(b)

Figure 15. Taylor factor maps, secondary recrystallization was observed in 2.9 mm thick sample (**a**). Arrows indicate areas that exhibit anisotropic response to deformation and high stress concentration (**a**) sample with 2.9 mm thickness, (**b**) sample with 3.2 mm thickness. More prone to crack initiation are boundaries between blue and red colored grains.

3.4. Finite Element Analysis

3.4.1. Effect of Load Application Direction

From the finite element analysis, the Von Mises stress fields on the Al 6063 body were depicted and evaluated. Since the particles were located in different positions along the thickness (4.57 µm) of the aluminum body, various cross-sections of the sample were examined. In Figure 16, the Von Mises stress fields of a cross-section around the middle of the sample's thickness for both loading scenarios are presented. The cross-sections presented in Figure 16 are located in a height of approximately 1.4 µm towards the positive values of the X-axis (the bottom of the aluminum body is located at a height of X = 0 µm in the model).

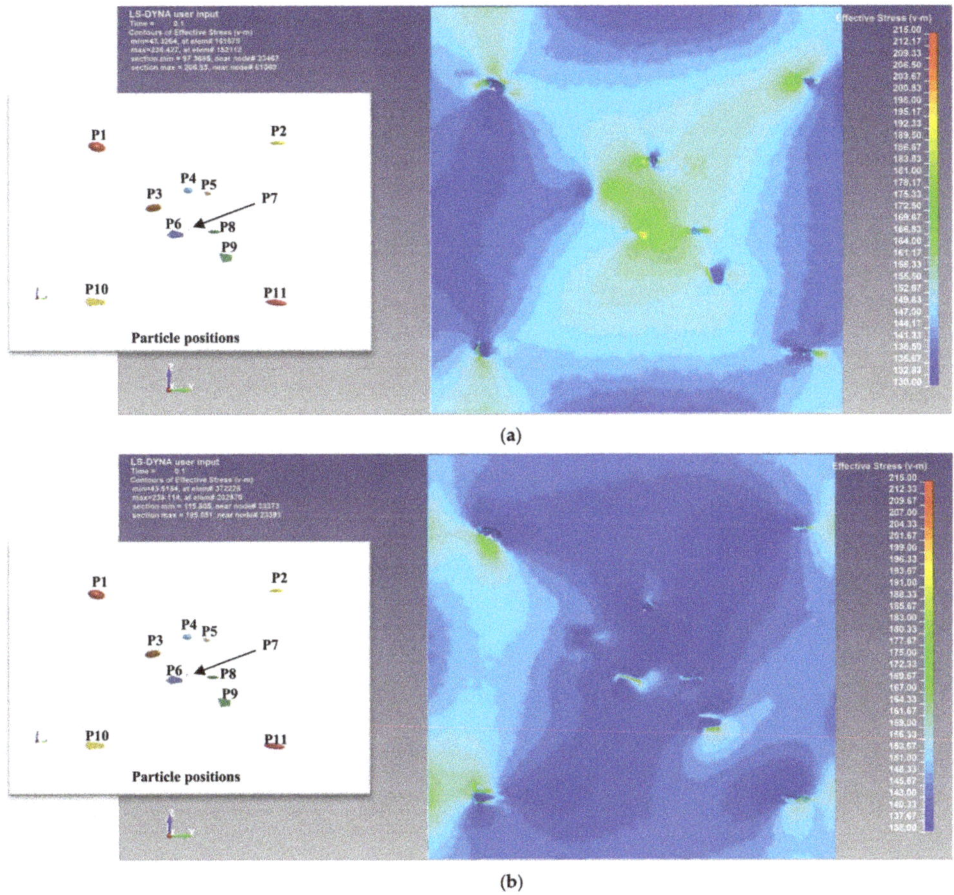

(a)

(b)

Figure 16. Mid-thickness cross-section Von Mises stress fields (units in MPa): (**a**) Load aligned with extrusion direction and (**b**) load transverse to extrusion direction. The particle positions are given as reference.

Observation over Figure 16 revealed a correlation between the loading direction and the induced stress fields on the sample. More specifically, in Figure 16a, where the load axis is aligned with the extrusion direction, and thus with the approximate alignment of the particles as well, some high stress values can be seen in the particles concentration region (also see Figure 1a). On the contrary, in Figure 16b, where the load axis is transverse to the extrusion direction, the lowest stress values are those observed in the particle concentration region, as the induced stress fields are "moving" around this area, thus indicating that higher energy is required to cross the aforementioned particle concentration area. This fact indicates that, given the direction of the loading, the presence of the particles can act either as a "path" or a "hindrance" for the induced stresses, as the particle elongation caused by the extrusion leads to high stress fields in the elongation direction over its transverse axis.

Furthermore, the aforementioned particles elongation can lead to stress fields merging in the case of collinear loading-to-extrusion direction scenario, given that the distance between two different particles in the direction of the loading is relatively small and the stress field intensity is high enough. This fact can be seen between the stress fields of "P3" and "P4", "P4" and "P5" and between "P6", "P7" and "P8", although the addition of the stress field of "P7" in this case gets overshadowed from the wider "P6" stress field. On the other hand, a lower intensity, stress fields merging is observed in the case of the transverse load axis-to-extrusion direction scenario, only between the "P1" and "P10" and between the "P2" and "P11" particles, which are outside the particles concentration region.

Further examination of Figure 16 reveals a rather expected behavior, where high stress loci appear in the edges of the particles that are relatively (given that the particles are rotated from the extrusion direction axis by a few degrees) aligned to the loading axis, while low stress loci appear in the corresponding transverse direction. By taking into account that the relative rotations of the particles are in all three directions of the workspace, the existence of particles in different depths along the thickness of the sample can affect the induced stresses in higher of lower cross-sections. This is clarified in Figure 16a, where the existence of particle "P3" causes a branching of the stress merged in "P6" and "P4" stress fields.

Regarding the stress values observed in Figure 16, in both loading scenarios the induced stresses have locally surpassed the applied 150 MPa. The stress fields surpassing the applied value are mainly a result of stress concentration near the ends of the particles (with respects to the loading direction in each loading scenario), with the majority of stress fields being approximately 160–185 MPa in the first loading case (Figure 16a) and 150–170 MPa in the second loading scenario (Figure 16b). The maximum stress values observed in these scenarios appear near some sharp edges of certain particles, with their values approximately 190–210 MPa in Figure 16a (observed in "P4", "P6""P7" and "P8") and 170–185 MPa in Figure 16b (observed in "P1", "P6", "P7" and "P10"). Given that all of the above-mentioned stress fields are present in the interfaces between the particles and the aluminum body, a debonding between the two is considered highly possible. Based on the definition of stress concentration factor K (Equation (1)), given by Carvill [23], the stress concentration factors for the above-mentioned stress fields are calculated in Table 4.

$$K = \frac{\sigma_{max}}{\sigma_{applied}} \tag{1}$$

Table 4. Stress concentration factors (K).

Stress Fields	Loading Scenario 1	Loading Scenario 2
Majority of observed stress fields	1.07–1.23	1.00–1.13
Maximum valued stress fields	1.27–1.40	1.13–1.23

Especially for the particles concentration region, a comparison between Figure 16a,b reveals that stress concentration in the first loading scenario is severely amplified, while the stress fields in the same region in the second loading scenario are rather relaxed. The respective stress concentration factors are approximately 1.07–1.23 for the first scenario, compared to 1.00–1.13 for the second scenario. Thus, it can be concluded that the particle elongation from the extrusion combined with their dense distribution can lead to high intensity stress field, which is not observed when the load is applied transversely to the particle elongation axis.

3.4.2. Effect of Particle Aspect Ratio

The importance of particles shape and aspect ratio is further clarified from the results presented in Figures 17 and 18, respectively.

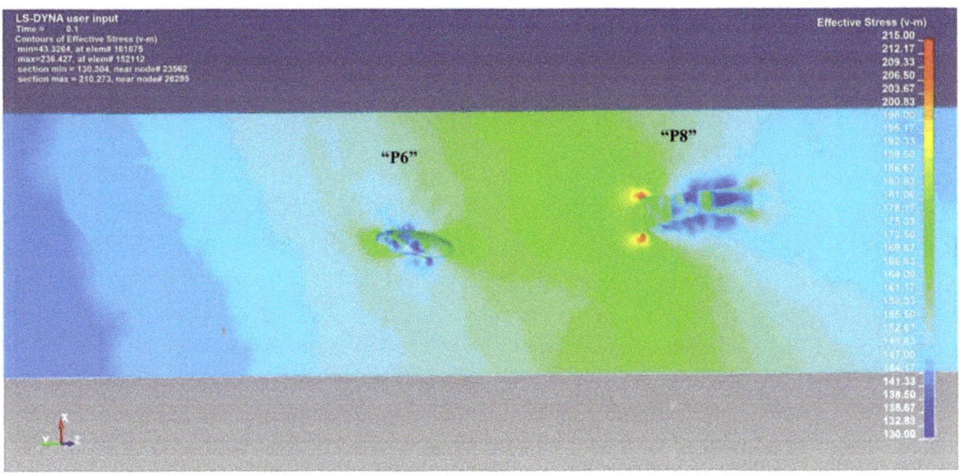

(a)

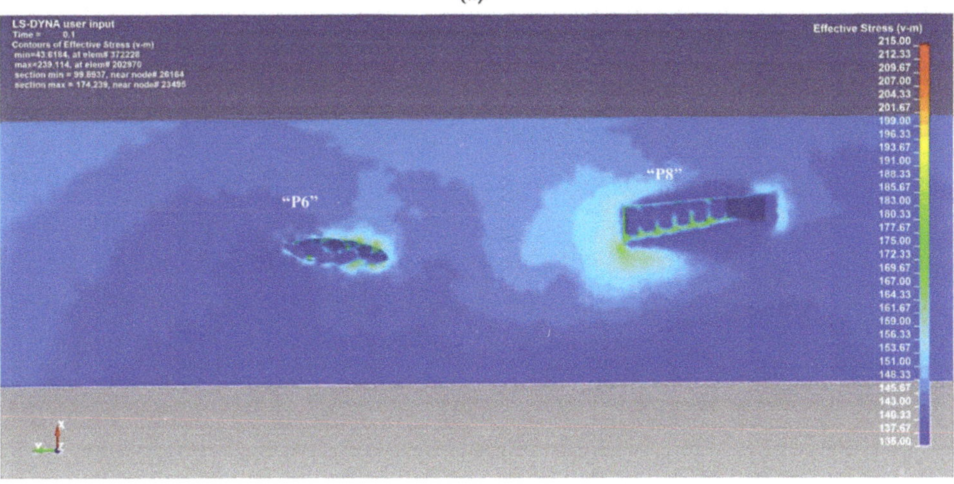

(b)

Figure 17. Comparison between the induced stress fields of "P6" and "P8" (units in MPa): (**a**) first loading scenario and (**b**) second loading scenario.

(a)

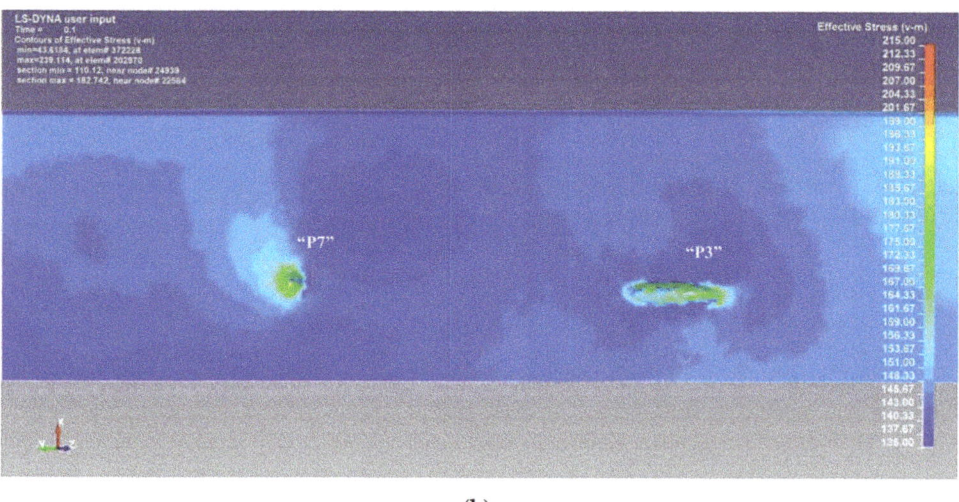

(b)

Figure 18. Comparison between the induced stress fields of "P3" and "P7" (units in MPa): (a) first loading scenario and (b) second loading scenario.

From the results presented in Figure 17, the importance of particle shape is examined. In this case, the two particles have similar alignment and comparable sizes, thus the induced stress fields are overall close in value. However, since "P8" is a wedge type particle with sharp edges, some significant stress fields are concentrated in the corners of the wedge. On the other hand, no such stress concentration exists in the case of the much smoother needle type particle "P6". This observation is similar to both loading scenarios, although the induced stress values between the two are different.

In Figure 18, a comparison between the induced stress field in particles "P3" and "P7" is presented. In this case, the two particles are a needle type and a circular type respectively, with different aspect ratios and in both cases no sharp corners exist. From the comparison between the two, the induced stresses around the two particles have similar maximum values. However, because of the equiaxed morphology of "P7", its induced stress field is continuous and has mostly the same intensity throughout its periphery, while the corresponding stress field of "P3" is varying across its periphery. Additionally, the stress fields of particle "P7" are very similar in both loading scenarios, while the same is not true for the stress fields of "P3". From the above analysis, it can be concluded that globular particles can act as constant stress raiser points regardless of the direction of loading, while the effectiveness of load direction becomes more important in non-equiaxed particles.

4. Discussion

The fracture surface of 2.9 mm thick sample from tensile testing was mainly intergranular but also exhibited dimples, with the higher ratio of intergranular to transgranular fracture occurring on the surface regions. This could be explained by the high stress concentration occurring due to presence of secondary recrystallized grains with approximately millimeter size which lower the ductility. A very coarse grain is very likely to have many neighboring grains with high difference in Taylor factor values. As a consequence, these grains will respond in a different manner to applied stresses during forming (in this case bending) leading to high stress concentration and ultimately fracture. This is the case of the grain interfaces highlighted in Figure 15.

From the obtained results, it is clear that the sample with the finer mean grain size in the matrix and the surface layer with no coarse grains had better formability. It is therefore necessary to attain a fine, equiaxed microstructure when high bendability is sought. It was observed that the strength of the sample with the coarser structure was higher. This is attributed to the higher amount of precipitation that occurred on the 2.9 mm thick sample. The examination of the size and distribution of Mg_2Si precipitates remains as future work (TEM) since their size could not be resolved with optical and scanning electron microscopy. The amount of natural aging that also affects seriously the resulting strength and that could lead to the observed differences in strength was not provided and should always be taken into account when mechanical properties of 6xxx series alloys is examined.

Both extrusion samples exhibited high bendability in longitudinal tests. However different behavior and generally bigger variation was observed on transverse tests. It is assumed that transverse bend testing is a more decisive mechanical test for determination of formability performance and for comparing different materials. Results also revealed that except from the 10 percent load drop tests it is suggested to perform 180 degrees bend tests, as the sample with the poor bendability was catastrophically fractured and the samples with the acceptable bendability were folded, exhibiting only orange peeling.

The formation of discontinuities in bending testing is strongly related with shear banding phenomena. Metallographic examination of bent samples showed that once stresses are concentrated on the grain boundaries, shearing can progress easily by "cutting" the neighboring grains in a mixed inter- and transgranular, fracture mode (low ductility fracture). The intergranular fracture mode is assisted by cracks formed between neighboring AlFeSi intermetallic particles producing microvoids and also by the fact that these particles are fractured during straining as a result of their brittle behavior. Consequently, the chemical composition of the alloy, which is directly related with the number of Fe-containing phases has a significant role in bendability, but should be regarded when comparing samples or alloys with different amount of Fe and Si.

Bendability is seriously affected by the metallurgical condition since this governs (i) the amount of Mg_2Si precipitation and the potential of grain boundary dissipation which occurs due to the cross linking of grain boundary precipitates and (ii) the presence of PFZs that constitute lower strength inter-crystalline regions. Higher bendability could be expected in T7 condition where the matrix is considerably softened or in underaged condition where no appreciable hardening has been performed. Si content is crucial in reducing the amount of Si grain boundary crystals precipitation that decrease ductility.

The results of FEM analysis showed that high stress concentration occurs at the edges/tips of the intermetallic particles, especially on the more elongated and/or sharply angular particles. A maximum concentration factor of 1.4 was calculated for the collinear load axis-to-extrusion direction loading scenario, while 1.23 was calculated for transverse load axis loading scenario.

Noticeable differences in the stress fields and the response of the metal in the longitudinal and transverse to extrusion direction were observed. More specifically, the direction of loading and its relative angle to the extrusion direction can completely change the pathing of stress fields inside the aluminum body, and thus any possible crack path formed. However, the direction of loading seems to have little effect on the stress induced fields of smaller particles, which operate as constant stress raiser points.

Comparing the areas with dense and sparser distribution of intermetallic particles it is readily observed that between closely spaced particles the stress fields are quite more intensive by a factor of 1.07–1.23 in the collinear load axis-to-extrusion direction loading scenario and 1.00–1.13 in the transverse load axis-to-extrusion direction loading scenario. Also, in the case of the former loading scenario, stress field merging occurs, thus the induced stress field paths completely change.

Based on the calculated stress values and the concentration of high stresses around the aluminum-particle interfaces, the possibility of debonding between the two is highly probable. In this case, any such point can be considered as a crack initiation point.

According to these results it could be said that:

- The resulting strength in tensile testing is strongly affected not only by grain structure but also by the amount of precipitation and the alloying addition as well as other factors affecting precipitation behavior such as natural aging, homogenization conditions, cooling rate after homogenization, solution heat treatment or extrusion (thermo-mechanical treatment) etc. This could provide an explanation for the higher strength of 2.9 mm thick sample, which had coarser microstructure than the 3.2 mm sample.

- Ductility in tensile testing and bendability are strongly related with grain size and morphology especially on the subsurface zones as well as in the bulk material. It is believed that cracks are easily nucleated in coarse grained material and once formed they can propagate through low ductility grain boundary regions or highly stressing shear bands cutting the grains transgranularly. It is easier to reveal any differences in bendability through transverse tests either by selecting 10% load drop as stop criterion or simply by bending by 180 degrees.

- Fe and Si should be considered when comparing the bendability of different alloy samples. Constituent particles have been proved by FEM to create high stress intensity fields during stressing, in both longitudinal and transverse directions. Especially when they have wedge-type morphology, significant stress fields were derived in their corners' area. This finding comes in agreement with metallographic examination findings where microvoids are formed between adjacent particles and this is the reason for the desired low Fe content in alloys with high crash performance and high ductility.

- FEM was successfully performed by applying tensile stresses parallel and transverse to extrusion direction in order to simulate the performance of the α-AlFeSi particles in the surface region of the profiles during 3-point bending tests in the longitudinal and transverse direction respectively. A possible explanation for the fact that the maximum load values in 3-point bending tests were observed in transverse direction could be given by finite element analysis. The model showed that among closely spaced elongated intermetallic particles the stress fields that develop are higher in intensity in the longitudinal tests and therefore the profiles are more prone to crack initiation and propagation in the respective tests.

- The role of globular particles is the same for transverse and longitudinal testing and this could apply not only for constituent particles but also for similar morphology precipitates.

- EBSD analysis and Taylor factor mapping can provide a valuable means in characterization of bendability not only by presenting an accurate grain structure but mainly by revealing crystallographic data related with the response of the material to stresses signifying potential locations for crack initiation.

5. Conclusions and Further Research

A thorough study including microstructure characterization, mechanical testing and finite element analysis of 6063 alloy profile samples was performed in order to determine the factors governing strength, ductility, bendability and also de-codify the role of the intermetallic particles in applied stresses. It is proposed that bendability of 6xxx series alloy extrusions can be easily evaluated by three-point bending tests, especially on the transverse direction but also with ductility as this is calculated by tensile tests. Precipitation behavior and metallurgical condition are the more crucial parameters for strength while grain morphology is more important in bendability, when comparing samples with the same metallurgical condition.

1. The load direction was proved to be important by both actual tests and FEM. Transverse testing proved to be more decisive in the determination of the more formable material as both samples had the same bendability in the longitudinal tests. Statistical analysis of bending results was not in the scope of the present research, but has been planned as a future work.
2. Texture analysis should be included in formability related projects in order to perform a more complete characterization of the role of the resulting texture from extrusion especially in the more sensitive surface regions where stress is concentrated and fracture initiation is expected. This type of analysis would include Taylor factor mapping for revealing the frequency and location of interfaces with high Taylor factor values, mean grain size analysis which can be precisely performed by EBSD in aluminum alloys (fine grains cannot be resolved by optical microscopy affecting calculation of the mean grain size), calculation of the amount of recrystallization by grain boundary angle misorientation maps, etc.
3. Finite element analysis has been proved very useful in the explanation of resulting loads in 3-point bend tests by revealing the respective stress fields formed between the intermetallic particles in the aluminum matrix, while the role of these particles as stress raisers especially according to their morphology was revealed. It can be inferred that globular particles can act as constant stress raiser points regardless of the direction of loading, while the effectiveness of load direction becomes more important in non-equiaxed type particles.
4. Further testing and analysis by EBSD will follow in order to reveal the role of various recrystallization texture components, especially cube or rotated cube texture, as well as the role of grain boundaries misorientation in crack assisting or arrestment. Additional FEA will be also performed by simulation of bending testing in order to verify the observed bending performance of the extrusions and optimize their performance by selecting the proper heat treatment after extrusion.

Author Contributions: Conceptualization, A.V.; Methodology, A.V.; Software, I.P.; Validation, A.V., Formal Analysis, I.P.; Investigation, S.P., A.T., A.R. and M.K.; Resources, G.S. and G.P.; Writing – Original Draft Preparation, A.V. and I.P.; Writing – Review & Editing, G.P.; Visualization, A.V. and I.P.; Supervision, A.V.; Project Administration, G.P.

Funding: This research received no external funding.

Acknowledgments: Special thanks are directed to ELKEME S.A. General Director Mr. Kimon Daniilidis for supporting the interdepartmental work on automotive extrusions.

Conflicts of Interest: The authors declare no conflict of interest.

References

1. Vazdirvanidis, A. Study on the Thermal Treatment of Aluminum 6xxx Series Extruded Products Aiming at the Optimization of Strength and Crashworthiness. Ph.D. Thesis, National Technical University of Athens (NTUA), Athens, Greece, 2016.

2. Ye, Y.; Sanders, R., Jr.; Yang, X. The Influence of Microstructure on the Fracture Resistance of 6xxx Alloy Sheet. In *Proceedings of the 16th International Aluminum Alloys Conference (ICAA16)*; Canadian Institute of Mining, Metallurgy & Petroleum: Westmount, QC, USA, 2018; ISBN 978-1-926872-41-4.

3. Vazdirvanidis, A.; Koumarioti, A.; Pantazopoulos, G.; Rikos, A.; Toulfatzis, A.; Kostazos, P.; Manolakos, D. Examination of Buckling Behavior of Thin-Walled Al-Mg-Si Alloy Extrusions. In Proceedings of the 12th International Conference on Aluminium Alloys, Carnegie Mellon University. Pittsburgh, PA, USA, 3–7 June 2012. [CrossRef]

4. Koumarioti, I.; Ping, S.; Vazdirvanidis, A.; Pantazopoulos, G.; Zormalia, S. Influence of Homogenizing and Ageing Practices on Microstructure and Dynamic Compression of Crash Relevant Al-Alloy Extrusions. In Proceedings of the 12th International Conference on Aluminium Alloys, The Japan Institute of Light Metals, Yokohama, Japan, 5–9 September 2010; pp. 1124–1129.

5. Westermann, I.; Snilsberg, K.; Holmedal, B.; Hopperstad, O. Bendability and Fracture Behaviour of Heat-Treatable Extruded Aluminium Alloys. In Proceedings of the 12th International Conference on Aluminium Alloys, ICAA12, The Japan Institute of Light Metals, Yokohama, Japan, 5–9 September 2010; pp. 595–600.

6. Das, S.; Heyen, M.; Kamat, R.; Hamerton, R. *Improving Bendability of Al-Mg-Si Alloy Sheet by Minor Alloying Element Addition*; Light Metals; Martin, O., Ed.; TMS, Springer: Berlin/Heidelberg, Germany, 2018; pp. 325–331.

7. Vazdirvanidis, A.; Pantazopoulos, G.; Toulfatzis, A.; Rikos, A.; Manolakos, D. Effect of Natural Aging of 6xxx Series Extrusions on the Energy Absorbance Capacity, Materials Science Forum. In Proceedings of the 15th International Conference on Aluminium, ICAA 2017, Tsinghua University, Beijing, China, 6–9 July 2017; Volume 87, pp. 315–321. [CrossRef]

8. Katsivarda, M. Homogenization Study of as Cast 6063 Aluminum Alloy Billets and Examination of the Effect of Cooling Rate on the Resulting. Ph.D. Thesis, National Technical University of Athens (NTUA), Athens, Greece, 2018.

9. Katsivarda, M.; Vazdirvanidis, A.; Pantazopoulos, G.; Kolioubas, N.; Papadopoulou, S.; Rikos, A.; Spiropoulou, E.; Papaefthymiou, S. Investigation of the effect of homogenization practice of 6063 alloy billets on beta to alpha transformation and of the effect of cooling rate on precipitation kinetics. *Mater. Sci. Forum* **2018**, *941*, 884–889. [CrossRef]

10. Vazdirvanidis, A.; Pantazopoulos, G.; Kolioubas, N.; Papadopoulou, S.; Katsivarda, M.; Rikos, A.; Spiropoulou, E. Investigation of the Effect of Homogenization Practice on the Microstructure of 6060 and 6082 Series Alloy Billets. In *Proceedings of the 16th International Aluminum Alloys Conference (ICAA16)*; Canadian Institute of Mining, Metallurgy & Petroleum: Westmount, QC, USA, 2018; ISBN 978-1-926872-41-4.

11. Snilberg, K.; Westermann, I.; Holmedal, B.; Hopperstad, O.S.; Langsrud, Y.; Marthinsen, K. Anisotropy of Bending Properties in Industrial Heat-Treatable Extruded Aluminium Alloys. In *Materials Science Forum*; Trans Tech Publications: Switzerland, 2010; Volume 638, pp. 487–492. [CrossRef]

12. Lloyd, D. Aspects of Plasticity and Fracture Under Bending in Al Alloys. In *Proceedings of the 16th International Aluminum Alloys Conference (ICAA16)*; Canadian Institute of Mining, Metallurgy & Petroleum: Westmount, QC, USA, 2018; ISBN 978-1-926872-41-4.

13. Vazdirvanidis, A.; Skordilis, I.; Katsivarda, M.; Stavroulakis, P.; Papaefthymiou, S. Study of Grain structure and crystallographic orientation of extruded 6xxx series alloy profiles. In Proceedings of the 14th International Scientific Congress Machines, Technologies, Materials (MTM 2017), Varna, Bulgaria, 13–16 September 2017; Volume 11, pp. 457–459.

14. Dao, M.; Li, M. A micromechanics study on strain-localization-induced fracture initiation in bending using crystal plasticity models. *Philos. Mag. A* **2001**, *81*. [CrossRef]

15. Inoue, H. Prediction of in-plane anisotropy of bendability based on orientation distribution function for polycrystalline face-centered cubic metal sheets with various textures. *Mater. Trans.* **2015**, *56*. [CrossRef]

16. Berndt, N.; Frint, P.; Wagner, M.F.-X. Influence of extrusion temperature on the aging behavior and mechanical properties of an AA6060 aluminum alloy. *Metals* **2018**, *8*, 51. [CrossRef]

17. He, X.; Pan, Q.; Li, H.; Huang, Z.; Liu, S.; Li, K.; Li, X. Effect of artificial aging, delayed aging, and pre-aging on microstructure and properties of 6082 aluminum alloy. *Metals* **2019**, *9*, 173. [CrossRef]

18. Belov, N.A.; Aksenov, A.A.; Dmitry, G. *Eskin Iron in Aluminium Alloys: Impurity and Alloying Element*; CRC Press: Boca Raton, FL, USA, 2002.

19. Glazoff, M.V.; Zolotorevsky, V.S.; Belov, N.A. *Casting Aluminum Alloys*; Elsevier: Amsterdam, The Netherlands, 2010.

20. LSTC. *LS-DYNA Keyword User's Manual*; Volume II, Material Models; Livermore Software Technology Corporation (LSTC): Troy, MI, USA, 2001.

21. Gaillac, R.; Pullumbi, P.; Coudert, F.-X. ELATE: An Open-Source Online Application for Analysis and Visualization of Elastic Tensors. *J. Phys. Condens. Matter* **2016**, *28*, 275201. Available online: http://progs.coudert.name/elate/mp?query=mp-984873 (accessed on 23 September 2019). [CrossRef] [PubMed]

22. Pantazopoulos, G.A. A short review on fracture mechanisms of mechanical components operated under industrial process conditions: Fractographic analysis and selected prevention strategies. *Metals* **2019**, *9*, 148. [CrossRef]

23. Carvill, J. 1—Strengths of Materials. In *Mechanical Engineer's Data Handbook*; Butterworth-Heinemann: Oxford, UK, 1993; pp. 1–55.

© 2019 by the authors. Licensee MDPI, Basel, Switzerland. This article is an open access article distributed under the terms and conditions of the Creative Commons Attribution (CC BY) license (http://creativecommons.org/licenses/by/4.0/).

Article

Research of Tribological Properties of 34CrNiMo6 Steel in the Production of a Newly Designed Self-Equalizing Thrust Bearing

Marek Urban [1] and Katarina Monkova [2,3,*

[1] Faculty of Mechanical Engineering, West Bohemia University in Pilsen, Univerzitni 22, 301 00 Pilsen, Czech Republic; maarc@kto.zcu.cz

[2] Faculty of Manufacturing Technologies with the Seat in Presov, Technical University of Kosice, Sturova 31, 080 01 Presov, Slovakia

[3] Faculty of Technology, UTB Tomas Bata University in Zlin, Vavreckova 275, 760 01 Zlin, Czech Republic

* Correspondence: katarina.monkova@tuke.sk; Tel.: +421-55-602-6370

Received: 30 November 2019; Accepted: 28 December 2019; Published: 3 January 2020

Abstract: There are many cases when in large power equipment (such as a turbine or compressor) asymmetrically loading on bearings due to thermal deformations, production inaccuracies, or simple deflection of the shaft occurs. This asymmetrically loading means misalignment of rotor against stator in angle more than several tenths of a degree and it has an influence on a journal and thrust bearings. Over the last few years, thanks to increasingly precise manufacturing, solutions that can eliminate this phenomenon have been revealed. In the case of the thrust bearing, it is a system of very precise manufactured levers, which are in close contact each to other, so they have to be not only properly designed from the geometrical point of view but the important role plays a quality of the functional surfaces of these levers. The article deals with the surface treatment effect on tribological properties of 34CrNiMo6 steel used for the production of bearing levers, which are the critical parts of a newly developing self-equalizing thrust bearing. The samples with cylindrical and plate shapes were produced from 34CrNiMo6 steel as representatives of the most suitable geometries for contact surfaces. All samples were heat-treated. The surfaces of some samples were treated by electroless nickel plating or nitriding, some of the samples were treated by tumbling. Gradually, the surface roughness, microhardness, metallographic analysis and the influence of selected types of surface treatments on the wear for individual samples were evaluated within the research presented in the article. As the testing methods for evaluation of tribological properties were selected Pin-on-disc test and frequency tribological test. The results showed that the best tribological properties achieved samples treated by electroless nickel plating compared with the nitrided or only heat-treated samples.

Keywords: tribological properties; wear; surface treatment; self-equalizing bearing

---

1. Introduction

Today, the energy industry must deal with two interrelated trends. The first trend is the ever-increasing demand for electricity worldwide and the second trend is the continuously the rising price of power. This situation can be solved in two ways, either to build new installations (or whole power plants) with greater efficiency or to improve (increase efficiency) existing installations [1]. One way or another, the worldwide annual need for rotating machines (e.g., turbines, compressors, generators, turbo-gearboxes, etc.), which is already now in several thousands, will grow [2].

In order to manufacturers of rotary machines flexibly respond to the growing demand for high-performance machines with high efficiency, designed for specific applications in specific operating conditions, and to maintain their market share, they must use top quality components. Undoubtedly, one of the key parts of the power machine are bearings that make a significant contribution to the overall power loss of the machine. It is a reason, why the manufacturers of such rotary machines place the highest demands on the bearings [3].

In machinery, to absorb the axial forces, the hydrodynamic thrust bearings are used, especially where the use of antifriction bearings is unsuitable in terms of dimensional limits, service life, high load, or difficult access during assembly [4]. When applied correctly, the plain bearing saves weight, space, bears more load, requires less maintenance and resists vibrations better than a roller bearing. On the other hand, an incorrect application can cause the relative surface interaction of the sliding bearing components that can lead to material loss at the points of contact, i.e., wear. Depending on the type of contact and movement of the bodies, wear is realized by various mechanisms, e.g., adhesive, abrasive, erosive or cavitation [5,6]. Wear resistance is a critical property directly affecting the life of bearings and surface properties. Therefore, knowledge of the mechanical and tribological properties of a material makes it possible to predict to a large extent its surface resistance to various forms of wear [7].

2. State of the Art

It often happens in real practice, especially in large rotary machines, that between stator and rotor necessary parallelism of active bearing and thrust collar is not guaranteed, what results in reduced bearing capacity. This misalignment can be caused by many factors (thermal dilatation, shaft misalignment, "inaccuracy" in production, etc.), the magnitude of which is dependent on the peripheral speed and the load [8]. The phenomenon of deflection of the shaft collar from the bearing axis during operation (up to several tenths of a degree) has been known for several decades, but at present, the only solution, that is capable to equalize misalignment between shaft and bearing axis, is the solution using the system of levers inside so-called a self-equalizing bearing, presented in Figure 1 [9,10].

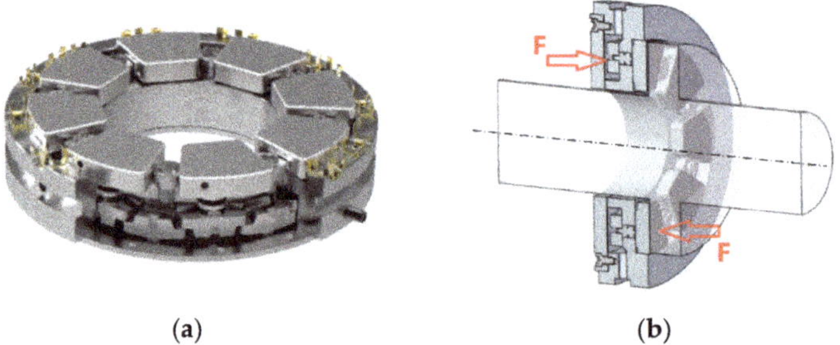

(a) (b)

Figure 1. Axial slide bearing with the self-equalizing segment, (**a**) real view; (**b**) illustration of force transfer by means of levers.

Due to the different mergers between global bearing manufacturers in recent years, there are currently only three major global suppliers that produce this type of bearings as standard—i.e., not as a special project. They are Waukesha (USA, Scotland), Kingsbury (USA, GB, Germany), Miba (John Crane) (USA, Germany). Each of these manufacturers uses, to a greater or lesser degree, a different design and manufacturing technology.

At least two elements are required for the slide bearing—the bearing itself and the rotor collar. The sliding bearing contains segments to which lubricating oil is supplied. Each segment is a separate carrier part of the bearing. The bearing surfaces (both bearings and shaft collars) are completely separated by an oil film with a thickness of approx. 20–40 μm. The oil film avoids the risk of contact and therefore abrasion of the bearing surfaces [11].

Self-equalizing thrust pad bearings consist of the following basic parts (Figure 2):

(1) bearing body/housing;

(2) thrust pad;
(3) self-equalizing element (lever);
(4) nozzle;
(5) floating pressure element.

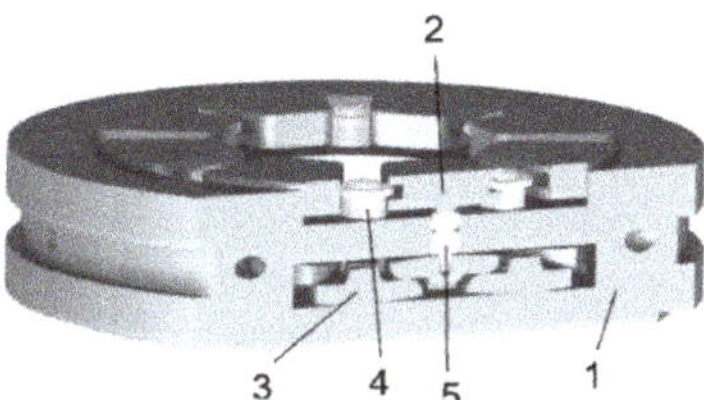

Figure 2. Self-equalizing bearing with levers (1—bearing body/housing, 2—thrust pad, 3—self-equalizing element (lever), 4—nozzle, 5—floating pressure element).

The most important (and critical) part of the axial self-aligning bearing assembly is the equalizing element—the lever. For every thrust, the pad is necessary to use 2 levers. It means that for example at 18 pads bearing is necessary to use 36 levers.

The aim of the set of levers is to distribute the load evenly over the entire circumference of the bearing. The evenly distributed load is then transferred to the bearing housing and subsequently to the machine frame. The lever system must be designed and manufactured so that it is capable of transferring the pressure/force exerted on the lower part of the bearing to the upper part of the bearing [12], i.e., that the thrust pads are always in contact with the shaft collar. (Figure 3).

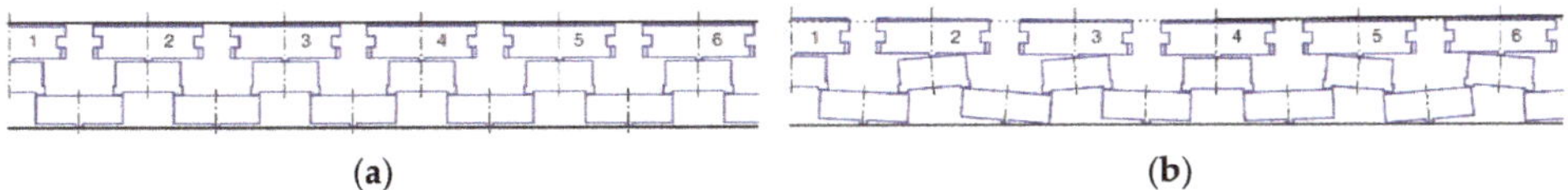

(a)   (b)

Figure 3. The principle of the lever tilting, (**a**) without the shaft deflection (misalignment); (**b**) at the shaft deflection (misalignment).

Ideally, the forces acting on the lever should be the same everywhere, but because of the passive resistance and the inappropriate geometry of the levers, these forces are different. In the case of an incorrectly designed bearing construction, it occurs to the undesirable wear of levers [13,14], as shown in Figure 4.

Figure 4. Comparison of the unworn surface of the lever (left side) with worn (right side highlighted in red).

The biggest problem is often incorrect machining of the geometrical/micro-geometrical shape of the functional surfaces [15]. For example, if the machining technology or strategy is incorrectly selected, the surfaces may be undulating and thus do not meet the theoretical requirements for an ideal wedge surface, respectively for an ideally designed shape. The functional surfaces can also be deformed by thermal loading of the components [16].

The roughness of the functional sliding bearing surfaces plays an important role. Too much roughness can cause rapid wear, damage or destruction of the functional surfaces if the lubrication is insufficient [17]. It is always better for the flow of medium if roughness on the sliding surface of slide bearings with white metal is low. (The white metal is a bearing metal, usually on a tin or lead basis, which is applied to the sliding surface of the bearing due to a certain "fuse" if the oil supply would fail or there would be some overload in the bearing and the oil film is broken. In this case, the bearing will be damaged and not the shaft that is many times more expensive). Regarding surface on levers (i.e., without white metal), low roughness in the range Ra 0.6–4.5 is secure (it doesn't occur there to micro-welds, and either adhesive part of the wear is not significant) [18,19]. At high sliding speeds, the surface roughness can change the flow pattern of the medium from laminar to turbulent. A great influence on the bearing function has the choice of a suitable oil, which under the operating conditions of the sliding bearing should have such a viscosity so that a hydrodynamic wedge is created [19].

The material of the levers must, therefore, possess sufficient mechanical and tribological properties, taking into account also aspects of surface integrity. According to Pantazopoulos [20], if are known the loading conditions and the maximum (undetected) crack or minimum (detectable) crack size, specified that can be accurately measured by quality control, then also a minimum fracture resistance (toughness) of the material can be ascertained and the information can be used for material selection or during the design stage. Based on this approach, the wear and tribological properties of the material of newly developed self-equalizing bearing have been investigated with the goal to evaluate the quality of a contact surface and, already in the design stage, to judge the suitability of the material, including the selected types of its surface treatments, for production of this bearing and its implementation into a large power equipment.

There are several materials that can be used for this purpose, but one of the most commonly used in the production of bearings and which meets the required characteristics of a highly stressed functional component is DIN 34CrNiMo6 steel [21]. Based on the long-term experience of the bearing producers with this material, and based on the current studies, this steel was also selected as a material for the production of bearing prototype and for the study within this research.

It is one of heat-treated low-alloy steel with a high hardenability and strength, which contains nickel, chromium, and molybdenum. Moreover, the 34CrNiMo6 steel exhibits very good toughness properties with a Charpy V-notch at a very low temperature. In the actual production process, the typical heat treatment of this steel involves two stages, quenching and tempering. The 34CrNiMo6 steel can obtain high strength after quenching to a fully martensitic microstructure, while the ductility and toughness can be improved with tempering [22].

Popescu et al. [23] studied the effects of bulk tempering on the hardenability and temper-ability of 34CrNiMo6 steel. The experiment investigated the correlation between the hardness achieved after high tempering of products and their equivalent diameter and the heat and time parameters of tempering. Cochet et al. [24] investigated the heat-treatment parameters of the 34CrNiMo6 steel used for shackles. The results firstly provided a validation of the input data and the prediction of the phase volume fractions and the resulting hardness, which showed that the proposed approach could yield a very good representation of the material properties. The results showed that the heat-treatment method could significantly improve the mechanical properties by changing the nucleation rate and growth rate of austenite. Researchers Ge & Wang [25] studied the effect of the tempering temperature on the microstructure and mechanical properties of the 34CrNiMo6 steel during a tempering process. Low-cycle fatigue behaviour of 34CrNiMo6 high strength steel was investigated systematically under fully reversed strain-controlled conditions at room temperature by Branco [26]. Li et al. [27] found that the mechanical properties of the 34CrMo4 steel gas cylinders were significantly improved after hot drawing and flow forming plus a designed heat treatment, compared with the base material. The observations of microstructure features such as grain size, sub-grain boundaries, and residual strain support the increase in mechanical properties due to the proposed manufacturing process.

Maniee [28] carried out a comparative study of tribological and corrosion behaviour of plasma nitrided 34CrNiMo6 steel under hot and cold wall conditions. The wear test was performed by pin-on-disc method. The results showed that nitriding under hot wall condition at the same temperature provided slightly better tribological and corrosion properties in comparison with cold wall condition.

Microstructure and Tribological Properties of Laser Forming Repaired 34CrNiMo6 Steel were studied by Huang [29]. He also observed that the wear mechanism of laser forming repaired samples is abrasive wear; whereas that of the substrate is abrasive wear and fatigue wear.

Despite the fact that 34CrNiMo6 steel is used quite often in bearing production [30], there are not many studies that deal with its tribological properties in detail, and therefore, it was necessary to investigate whether this material could be used to design and to manufacture a new type of the self-equalizing bearing that would be able to operate a longer time, more safely and with higher efficiency compared to the existing bearings implemented into a high-power rotary machines (such as turbines or compressors). The study was performed as a part of extensive and long-term research that was divided into several stages: from the development of new geometry of self-equalizing bearing elements, through the investigating their kinematic, dynamic and tribological behaviour, up to final testing of the newly designed bearing prototype including its implementation into real operation.

The specific goal of investigation that is elaborated in this article was to study the tribological properties of the self-equalizing bearing elements produced from EN/DIN 34CrNiMo6 steel and to find the most suitable type of their functional surface with respect of technology of design. To find the optimum variant of contact surfaces from both geometry and wear rate (bearing life) points of views, the influence of selected types of surface treatment on the functional surface properties was also analysed. The obtained results will make it possible to predict a large extent the surface resistance of the material to various forms of wear and it can give an answer on the question if it is reasonable to manufacture the functional surfaces of the levers with low roughness in a higher cost or to produce them with higher roughness in less expensive cost.

3. Experimental Samples Design

3.1. Geometry of Samples

In addition to the quality of functional surfaces and some aspects of technology, as the most important criteria affecting the wear and contact behaviour of functional surfaces, the geometry of the levers had to be taken into account when designing samples for experimental investigation of the tribological properties of 34CrNiMo6 steel.

The geometry of the samples for experimental testing was built upon the research carried out in a previous stage (not presented in this paper), in which several variants of the lever geometry were gradually designed to create a functional model of the newly developed bearing. The kinematic functionality of all variants was verified by a simple test to measure the maximum misalignment of the levers. Based on the test results, which confirmed that the stiffness and overall kinematics of the rocker arm were found to be the best, the following lever shapes (profiles) and their contact were considered for subsequent tribological tests (Figure 5):

(1) "Cylinder/Cylinder" + "Cylinder/Cylinder";
(2) "Cylinder/Cylinder" + "Cylinder/Plane";
(3) "Cylinder/Plane" + "Cylinder/Plane".

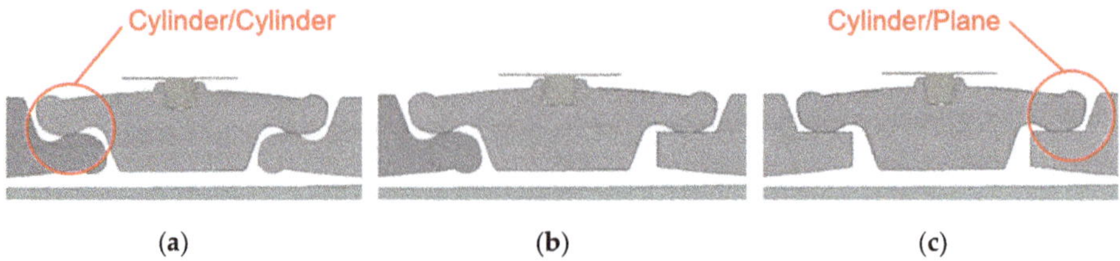

(a)   (b)   (c)

Figure 5. Variants of contact surfaces of lever, (**a**) Cylinder/Cylinder contacts on both sides of a lever, (**b**) Cylinder/Cylinder contact on the left side and Cylinder/Plane contact on the right side, (**c**) Cylinder/Plane contacts on both sides of a lever.

It would be unnecessary to produce complete levers for functional surface wear tests, and therefore, a simplified solution was adopted and the samples were designed in the form of a roller with the dimensions that corresponded to the rounded surface of the lever and in the form of a plate that represented the lever contact plane. Thus, in the experiment, the specimens had the shape of a 12 mm diameter cylinder and a 25 mm length, or a plate of 25×25 mm^2, which correlated with the size of a lever for a reference bearing with 18 pads. An example of the cylindrical sample is in Figure 6.

Figure 6. A tested cylindrical sample.

3.2. Material of Samples

In view of the above, EN/DIN 34CrNiMo6 chromium-nickel-molybdenum heat-treated steel was chosen for levers and therefore also for the production of experimental samples. The practical use of chromium-nickel-molybdenum steel is based on the long-term operational reliability of the supplied products. It is a steel with high hardenability for highly stressed machine parts. In the treated state, it has a very favourable strength-to-yield ratio and a high toughness which hinders the propagation of fatigue cracks. The steel is therefore characterized by high fatigue limit values for alternating and combined stresses. It has high strength, high toughness and good hardenability [31].

Chemical composition of the 34CrNiMo6 steel is in Table 1.

Table 1. Chemical composition of the 34CrNiMo6 steel.

Component	C	Mn	Si	P	S	Cr	Ni	Mo	V	Cu	Al
(wt.%)	0.34	0.793	0.282	0.0196	0.0052	1.72	1.55	0.221	0.0092	0.193	0.0194

3.3. Technological Conditions of Samples Production

Since the selected sample shape is in the form of a cylinder, turning technology has been selected as the sample production technology. The surface roughness of the sample, which plays an important role in tribological tests, can be influenced by suitably selected machining conditions.

The machining was carried out on a machine EMCO MAXXTURN 25 (EMCO GmbH, Hallein, Austria). A bar with a diameter of 25 mm was selected as a semi-finished product. In the first step, the bar was roughened to φ D = 13 mm using a changeable insert ISCAR CCMT 09T304-SM (ISCAR Ltd., Haiger, Germany); consequently it was finished to a diameter of φ D = 12 mm using a changeable insert ISCAR DCMT 11T304-F3P (ISCAR Ltd., Haiger, Germany) and finally it was cut to a length of L = 25 mm. The machining process of sample production is presented in Figure 7.

Figure 7. Production of cylinders for experimental testing.

The parameter that most influences the surface roughness during turning is a feed [32], therefore 5 different feeds per revolution f = 0.08; 0.12; 0.16; 0.2; 0.24 mm were chosen to monitor the wear of the cylinders. The next machining conditions were:

cutting speed: v_c = 120 m·min^{-1};

cutting depth: a_p = 0.5 mm;

Coolant: Blasocut BC35 Kombi −5% + 95% water;

feed per revolution: f = 0.08; 0.12; 0.16; 0.2; 0.24 mm.

All samples were subsequently treated after turning according to the recommendations for this material. The quenching was carried out in oil from a temperature of 830–860 °C and tempering during the temperatures ranged from 630 °C to 660 °C. The hardness of the base material was 330 HB.

At the same time, due to the load requirements of the levers, it was necessary to harden the functional surface of the levers. Based on existing research, scientific studies [33–40] and practical applications, the electroless nickel plating and nitriding were chosen as technologies for a surface reinforcement to enhance the tribological properties of the levers. These technologies can be used as a final surface treatment after finishing without additional grinding. Tumbling on an OTEC DF3 machine (OTEC GmbH, Straubenhardt, Germany) was also used as an intermediate between possible surface hardening for half of the samples (Samples No. 55-108). After several tests, the following tumbling parameters were selected.

Tumbling type	Towed;
Rotor speed	40 rpm;
Rotation bracket	90 rpm;
Immersion depth	420 mm;
Lift	Not used;
Medium	H4/400;
Clockwise tumbling time	3 min;
Counterclockwise tumbling time	3 min;
Total time	6 min.

3.4. Number of Experimental Samples

Each process is influenced by a number of specific factors that can be actively managed. At the output of these processes is then expected a certain result. Such a result is called a response. DoE (Design of Experiments) is a strategy in which the effects of several factors are studied at once by testing them at different levels. The task of DoE is then to find such a combination of factors that the process response is as accurate as possible. The response should then be monitored at several points in the experimental space. Tracking each point requires both time and cost, and this is important to realize. Therefore, it depends very much on how many points and how they are located in the experimental space [41].

With regard to the above-mentioned factors entering the process of production of experimental test samples (Section 3.3), it was not appropriate to choose a central composite plan, an application of which would require a very large number of samples, that would mean, besides the economic demands, a great time of experiment realization. Due to this reason, a "custom plan" was designed for the experiment. Orthogonality, rotation, etc., are also maintained in the design. According to this custom-designed plan, the production of testing cylinders was also planned, taking into account the following factors:

(a) Load contact pairs "Cylinder/Cylinder = C/C" and "Cylinder/Plane = C/P".
(b) Parameter x_1—feed per revolution: $f_1 = 0.08$ mm; $f_2 = 0.12$ mm; $f_3 = 0.16$ mm; $f_4 = 0.2$ mm; $f_5 = 0.24$ mm.
(c) Parameter x_2—surface hardening: with or without tumbling, nitriding or electroless nickel plating.

A total of 108 cylinders with an assigned number and associated characteristics according to Table 2 were produced. In order to reduce the number of experiments, the plates were only ground and possibly surface reinforced. The wear was then monitored on a roller.

Table 2. Samples with an assigned number and associated characteristics (C/C—Cylinder/Cylinder contact type; C/P—Cylinder/Plane contact type; R—refinement; N—nitriding, ENP—Electroless Nickel Plating).

Without Tumbling	With Tumbling	Feed	Type of Surface	Heat Treatment	Without Tumbling	With Tumbling	Feed	Type of Surface	Heat Treatment	Without Tumbling	With Tumbling	Feed	Type of Surface	Heat Treatment
Sample (Cylinder) Number					Sample (Cylinder) Number					Sample (Cylinder) Number				
1	55			R	19	73			R	37	91			R
2	56			R	20	74			R	38	92			R
3	57		C/C	R + N	21	75			R	39	93		C/C	R + N
4	58			R + N	22	76			R	40	94			R + N
5	59	f_1		R + ENP	23	77			R + N	41	95	f_4		R + ENP
6	60			R + ENP	24	78		C/C	R + N	42	96			R + ENP
7	61			R	25	79			R + N	43	97			R
8	62		C/P	R + N	26	80			R + N	44	98		C/P	R + N
9	63			R + ENP	27	81	f_3		R + ENP	45	99			R + ENP
10	64			R	28	82			R + ENP	46	100			R
11	65			R	29	83			R + ENP	47	101			R
12	66		C/C	R + N	30	84			R + ENP	48	102		C/C	R + N
13	67	f_2		R + N	31	85			R	49	103	f_5		R + N
14	68			R + ENP	32	86			R	50	104			R + ENP
15	69			R + ENP	33	87		C/P	R + N	51	105			R + ENP
16	70			R	34	88			R + N	52	106			R
17	71		C/P	R + N	35	89			R + EN	53	107		C/P	R + N
18	72			R + ENP	36	90			R + ENP	54	108			R + ENP

4. Preliminary Tests

4.1. Surface Roughness Investigation

After machining, it was measured the surface roughness of every cylindrical sample. The following parameters have been measured within the research of surface roughness: Ra—arithmetical average deviation from a mean line, Rz—ten-point height of irregularities, Rk—core roughness depth, Rpk—reduced peak height, Rvk—reduced valley depths, profile, Mr1—material portion 1 determined for the intersection line which separates the protruding peaks from the roughness core profile, Mr2—material portion 2 determined for the intersection line which separates the deep valleys from the roughness core profile, A1—Peak area and A2—Valley area.

In the beginning, on a selected set of samples, the surface roughness was measured by means of both Mahr MarSurf M300 (MAHR GmbH, Göttingen, Germany) and 3D scanning ALICONA measuring devices (Alicona Imaging GmbH, Graz, Austria). After comparison of all parameters recorded by both measuring devices, no significant differences between measures were found. So, due to a large number of samples and since the measuring by means of ALICONA was significantly longer, authors decided to use 2D analysis and measure surface on 3 spots and calculate an average value of every parameter. The representative parameters Ra (µm) (arithmetical average deviation from a mean line) and Rz (µm) (ten-point height of irregularities) have been chosen for presentation in the paper because they clearly express the behaviour of surface roughness achieved after machining with different cutting feeds. The results have been plotted by means of the graph shown in Figure 8. The parameters Rk, Rpk, Rvk, A1, A2, Mr1 and Mr2 were used for the Abbott curves construction (see Section 5.1.2, Figure 20).

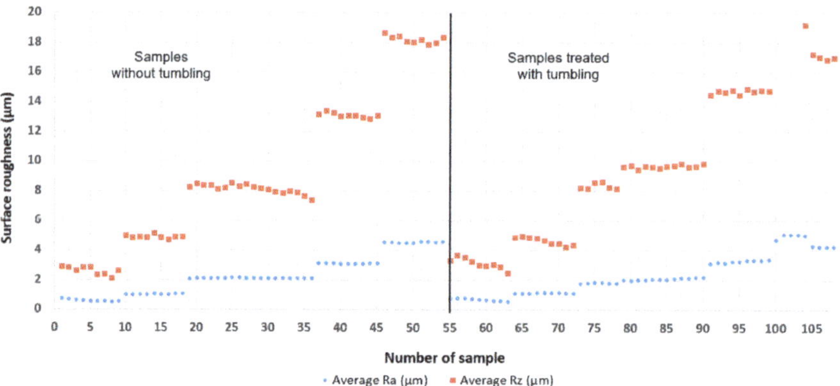

Figure 8. The surface roughness of tested samples.

From the graph bellow, it is evident that it managed to produce the cylinders, which surface roughness differed in a jumping way and which unambiguously corresponded to the used feed rate during machining as follows:

$$f_1 = 0.08 \text{ mm} \geq {\sim}\text{Ra } 0.6$$

$$f_2 = 0.12 \text{ mm} \geq {\sim}\text{Ra } 1$$

$$f_3 = 0.16 \text{ mm} \geq {\sim}\text{Ra } 2$$

$$f_4 = 0.2 \text{ mm} \geq {\sim}\text{Ra } 3$$

$$f_5 = 0.24 \text{ mm} \geq {\sim}\text{Ra } 4.5.$$

At the same time, it can be stated from the measurement results that the surface roughness of the cylinders decreased after tumbling. For the samples with a roughness that was initially low, this drop was not as striking as for cylinders that had initially greater roughness.

4.2. Microhardness and Metallographic Analysis

Although the selected steel EN/DIN 34CrNiMo6 is not considered to be nitriding, it is often nitrided in practice. The samples treated after machining and also samples with a hardened surface by nitriding and electroless nickel plating were analysed through the micro-hardness test HV 0.05 (or HV 0.025) to determine to which the depth could be expected a higher material resistance.

Within the preliminary tests, a microstructure of the tested samples was also evaluated. The samples were cut longitudinally and crosswise—perpendicular to the intended direction of damage by means of a metallographic saw. They were also embedded in the preparation mass, metallographically ground and polished. Metallographic sections were subsequently etched using Nital 3% and then documented using the metallographic microscope MULTICHECK PC 500 (AVYAC MACHINES SAS, Veauche, France). The evaluation only several types of samples with different surface treatment is presented in the article. Their microstructures are shown in Figures 9–12 and they are described below.

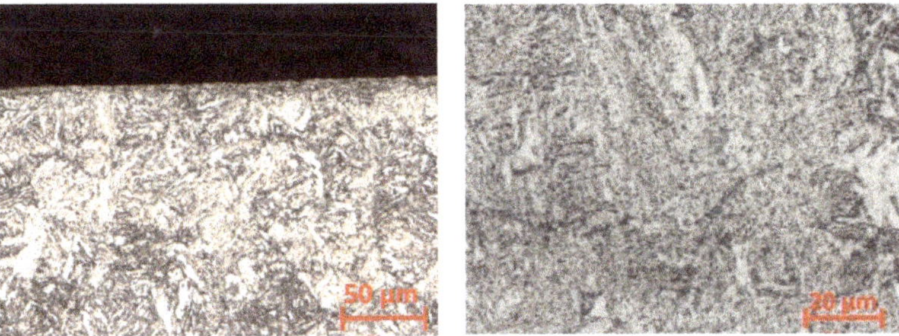

Figure 9. Sample without surface treatment, observation on the edge of the cylinder—magnified 100× (**left**); detail—magnified 500× (**right**).

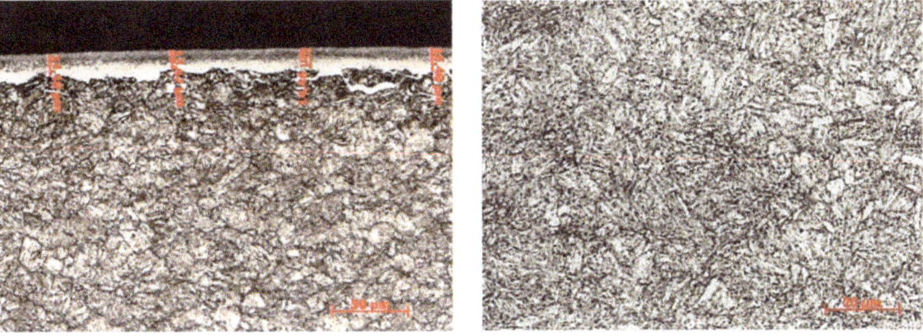

Figure 10. Nitrided sample—white layer thickness magnified 200× (**left**); sample centre magnified 500× (**right**).

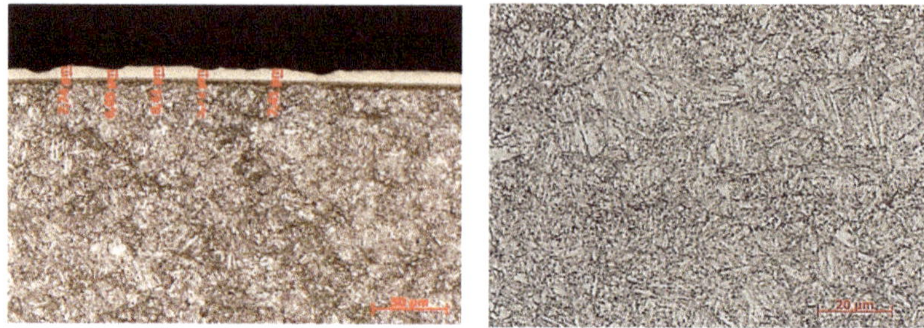

Figure 11. Sample electroless nickel-plated, turned with a feed of $f_1 = 0.08$ mm—surface with marked layer thickness—magnified 200× (**left**); sample center—magnified 500× (**right**).

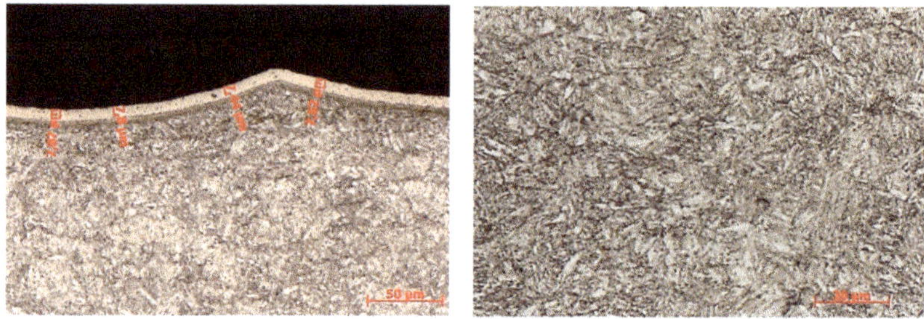

Figure 12. Sample electroless nickel-plated, turned with a feed of $f_1 = 0.24$ mm—surface with marked layer thickness—magnified 200× (**left**); sample center—magnified 500× (**right**).

The microstructure of a sample without surface treatment, machined with the feed per revolution $f_1 = 0.08$ mm is presented in Figure 9. The sample was treated by refinement without next surface hardening. According to Figure 9, the structure of the outer surface is the same as in the core of the material.

Nitriding sample, machined with the feed per revolution $f_1 = 0.08$ mm is presented in Figure 10. The sample exhibited a white layer thickness of about 15–20 μm, which was compact and didn't show an inconsistency. The average microhardness of the sample in the core was HV 0.05 = 395.

The microstructures of the samples hardened by electroless nickel plating machined with the feed per revolution $f_1 = 0.08$ mm (the surface roughness ~Ra 0.6 μm) is presented in Figure 11 and with the feed per revolution $f_5 = 0.24$ mm (the surface roughness ~Ra 4.5 μm) is presented in Figure 12.

Cylinders hardened by electroless nickel plating have the same layer thickness of approximately 7–8 μm. The layer follows the relief of the surface, it is consistent and without any visible defects. The base material corresponds to the hardness values required in the material tables, but the surface micro-hardness values were lower than expected. The measured microhardness (HV 0.025) of both fine and roughly turned samples with Ni coating deposition is presented in Table 3.

Table 3. Microhardness (HV 0.025) of fine and roughly turned samples with Ni coating deposition.

HV 0.025	$f_1 = 0.08$ mm				$f_1 = 0.24$ mm			
	Cross-Cut		Longitudinal Cut		Cross-Cut		Longitudinal Cut	
	Surface	Core	Surface	Core	Surface	Core	Surface	Core
Average Value	621	392	225	382	451	375	671	386
Deviation	18	19	23	13	108	17	11	26

It is clear from Table 3 that the surfaces of cylinders machined with the feed per revolution $f_1 = 0.08$ mm are harder compared to the samples machined with $f_5 = 0.24$ mm. The hardness of the roughly turned surface (and especially of the crosswise cut) is lower. However, this can be influenced by the fact that at a higher roughness at the measured point, the layer is not compact, but there is a "softer" base underneath.

5. Tribological Behaviour of 34CrNiMo6 Steel

5.1. Pin-on-Disk Test

At the first stage of tribological properties of 34CrNiMo6 steel investigation, a testing device—tribometer "PIN-ON-DISC" was designed and manufactured at the WBU in Pilsen. The principle of testing was as follows. Two cylinders were placed in the machine. One specimen was fixed firmly to the tribometer arm holder and the second to the movable holder. The operational load was carried out using weights (utilization of gravitational force). By a movement of the samples relative to each other, a contact load was simulated. After reaching the prescribed number of cycles of contact, the test was discontinued and both samples were turned to be tested in another contact area or to be the samples exchanged. The samples clamping in a holder and the complex view on tribometer used at the testing process are shown in Figure 13.

Figure 13. The samples clamping in the holder (**left**) and the complex view on tribometer (**right**).

In order to follow the development of the damage, a gradual increase of cycles in the series was chosen: 1.8×10^3; 5×10^3; 10×10^3 and 20×10^3 cycles. The sample movement speed was 3 rpm, and the load was selected at 5 N or 10 N. At the end of the test, the generated tribological traces were examined by the stereo magnifying glass. To compare a degree of the wear of all samples, the wear area was selected as a measured parameter. The area better interprets the relative wear than e.g., the track width. The example of wear is presented in Figure 14.

Figure 14. The example of a wear.

5.1.1. Wear Measuring and Observation

Measured were always the samples that were fixed to the bottom bracket because these samples showed a more measurable and more regular trace of wear. The measurements were done by making photographs where the length and width of the track were measured. If the track tapered, the track was divided into multiple rectangles, and their area was then added together. It was necessary to use 2–3 images per one track evaluation, depending on the created length. In the case of nickel-plated specimens, there was generally no continuous measurable trace. Mostly, only the peaks were worn out after machining. In this case, the tracks had to be measured, averaged and then summed.

Since there is a possibility that the oil layers between the levers may break due to very high Hertzian pressure and there may be limit lubrication between the contact surfaces [42] (see Figure 15), authors decided not to use the oil lubricant. The second reason for not using oil or other lubricant was to accelerate the wear process during the test. It can also be said at this point that the finally produced levers were tested under real conditions (i.e., with lubricant) and the measured wear after long-term tests were very low (or none), so the levers comply with the standard requirements for their safety and reliable operation.

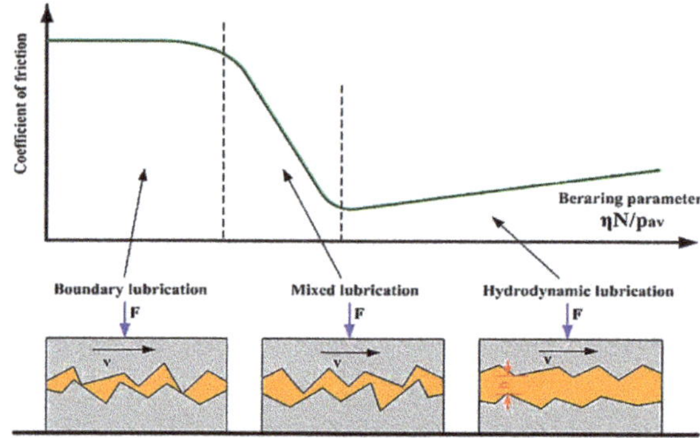

Figure 15. Lubrication regimes.

The wear results of all samples are shown in Figure 16 and examples of cylinders' wear for individual types of surface hardening are shown in Figures 17–19.

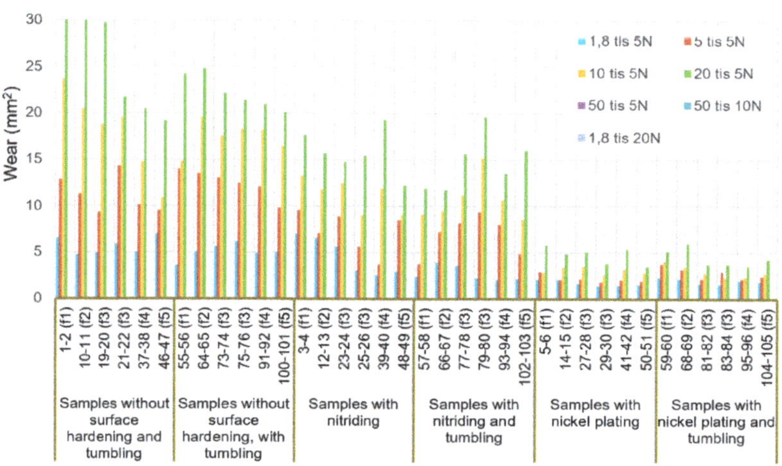

Figure 16. Graphs of all cylinders' wears—measured on the tribometer.

Figure 17. Sample No. 11—20× magnification, 1.8 × 103 cycles of 5 N load (**left**); Sample No. 55—10× magnification, 5 × 103 cycles, 5 N load (**right**).

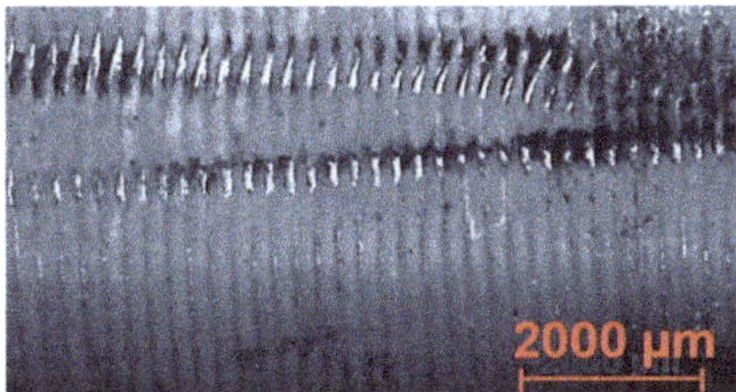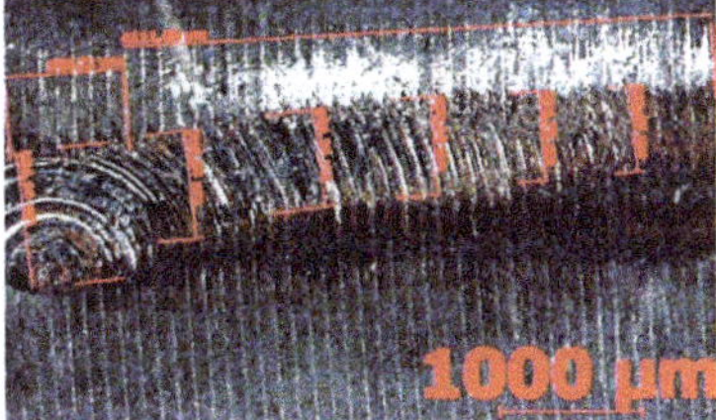

Figure 18. Sample No. 49—10× magnification—5 × 10^3 cycles, 5 N load (**left**); Sample No. 13—20× magnification—5 × 10^3 cycles 5 N (**right**).

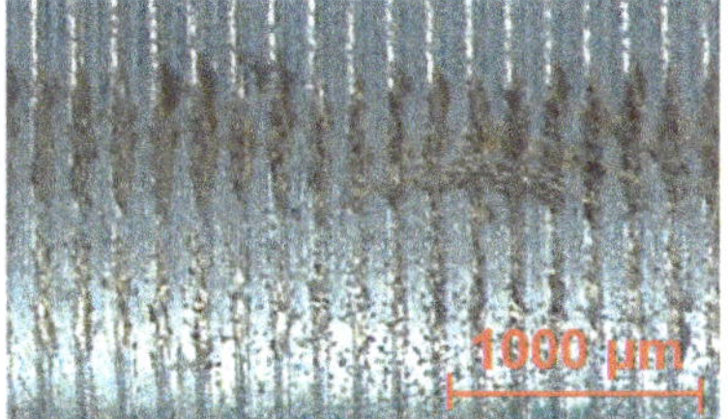

Figure 19. Sample No. 5—20× magnification, 1.8 × 10^3 cycles of 5 N load (**left**); Sample No. 27—20× magnification, 20 × 10^3 cycles, 5 N load (**right**).

The samples without surface treatment showed relatively similar abrasion resistance, despite differences in surface roughness. The following pictures in Figure 17 show the damage development of samples No. 11 and 55, which were evaluated as the worst in the tribological test.

For presentation in the article, the sample No. 49 was selected as a representative of nitriding samples. In Figure 18, the picture on the left side shows the beginning of the damage and on the right, there is a significant breakage (the layer has ceased to function). The effect of surface roughness was manifested only in the samples with higher roughness values.

The smallest wear was measured at the samples, which were hardened by electroless nickel plating. In Figure 19 is presented wear of the sample No. 5 reached after 1.8×10^3 cycles at a force of 5 N load. As another example can be selected sample num. 27, which didn't show any wear even after 50×10^3 cycles at 5 N load, because only a discontinuous trace was created that was difficult to measure. Therefore, the sample was loaded again at 100×10^3 cycles by a force of 10 N load. On this sample, in the most stressed area, it was already possible to find a continuous measurable trace with measured surface damage of approximately 920 μm. However, even in this case, it cannot be said that this sample would behave as worn out.

The measured values were statistically processed by ANOVA analysis, which shows the best results for surface and wears analysis. This statistical method consists of evaluating the relationships between the variance of the selected sets. The essence of the method is to perform the so-called decomposition of the total variance into two sub-variations, namely the variation caused by the influence of individual factors and on a part called "noise", which can be assumed to occur accidentally. The statistical significance of the ratio between these components is then tested in this method. Generally, this method makes it possible to verify whether the monitored factors have a statistically significant effect on the monitored quantities [43].

The monitored factors included: displacement, surface condition (surface hardening) and the number of cycles. In this analysis, some pairs of factors, at which were supposed they could be statistically significant were also studied. When evaluating all these parameters, it was found that the factors Feed, Surface Condition, Number of Cycles, and the factor pairs "Feed-Surface Condition" and "Surface Condition-Number of Cycles" have been statistically significant.

5.1.2. Discussions

Summary from the used surface treatments point of view

It is clear from the table and graphs above that the best wear performance was achieved on samples treated with electroless nickel plating. These coatings, whose hardness is not significantly higher than that of the other types of coatings tested, exhibit excellent tribological properties due to very good lubricating properties. However, due to the relatively thin deposited layers, it is to be assumed that in the event of the layer destruction, the next damage will be rapidly increasing.

The hardness of the nitrided layer was recorded in relatively small depths. However, samples with the nitrided surfaces showed higher wear resistance compared to samples without surface treatment. Due to the small thickness of the hard layer, rapid delamination and spreading of damage occurred. In the early stages of the test, nitrided specimens seemed to have an effect of surface roughness, where a concentration of damage has occurred primarily on the profile peaks that prevented a relatively long time of spreading of the next degree of layer degradation.

Samples only heat-treated showed no difference in a surface quality in dependence on tumbling. Already in the early stages of loading, a worn area was formed on the surface, which rapidly increased.

Summary from the surface roughness point of view

It is also evident from the above graphs that the samples machined with higher feeds per revolution (f_3–f_5) came out best from the surface roughness point of view. Subsequent surface tumbling had little or no effect on wear. Within the experimental study, the Abbott-Firestone's curves (AFC) were used for surface roughness evaluation. These curves graphically describe the distribution of material within the profile height and they can be also used for assessing the functional properties of surfaces and their possible exploitation. The principle of AFC construction (according to the standard ISO13565-2:1996) is in Figure 20a, while an example of an AF curve obtained by means of the software belonging to the Mahr MarSurf M300 device (MAHR GmbH, Göttingen, Germany) for sample number 55 is in Figure 20b.

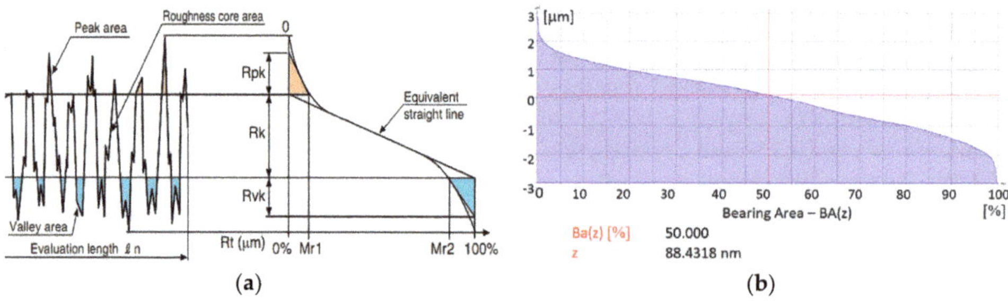

Figure 20. Abbott-Firestone's curves, (**a**) construction according to the standard ISO13565-2:1996; (**b**) AFC obtained by means of software belonging to the Mahr MarSurf M300 device for sample No. 55.

Problem of Mahr MarSurf M300 (MAHR GmbH, Göttingen, Germany) or Alicona (Alicona Imaging GmbH, Graz, Austria) software for data processing is that it can generate only one particular Abbott-Firestone's curve, and so it is impossible to compare different AF curves in the same ratio. So, the measured values Rk, Rpk, Rvk, A1, A2, Mr1 and Mr2 were used for the bearing area curves construction using MS Excel software (version 1911, Microsoft corporation Inc., Redmond, WA, USA) application. They were plotted based on the principle presented in Figure 20a, what provided a possibility to compare them in one chart and easy to evaluate. The bearing area curves of selected cylinders are presented in Figure 21.

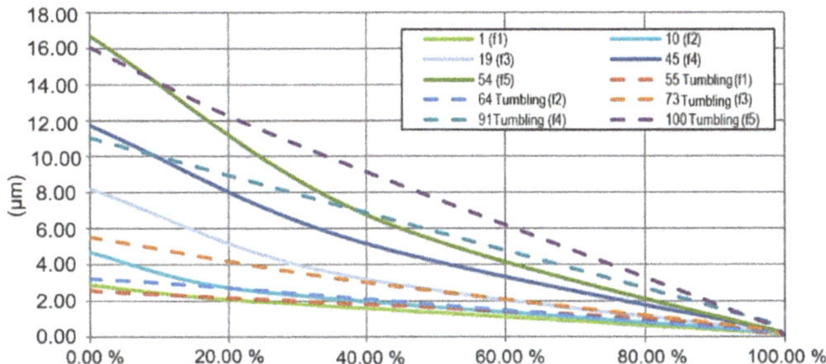

Figure 21. The bearing area curves of selected cylindrical samples (No. 1–54 without tumbling and No. 55–108 with tumbling).

Looking at the bearing area curves in Figure 21, it can be seen how at feeds f_4 and f_5 the curves are steep. The peaks at the feed f_5 reach 16–17 µm. During the initial run-in, the "peaks" usually reduced to about 0.07 µm. Since the levers are in contact with each other, this value needs to be doubled. Such a value (~0.015 µm) is already critical, as the production of the levers should be within an absolute accuracy of about 0.03 mm. The bearing at worse accuracy can no longer absorb deflections higher than 0.1°. If almost 50% of the manufacturing tolerance is "decreased" only due to surface roughness, it unnecessarily increases the cost and complicates production, as more levers would have to be thrown out due to inaccurate.

Another important aspect is that when using higher feeds (and hence higher roughness values), a higher coefficient of friction between the levers results. Given that the levers are flooded with oil, it is not exactly possible to specify how much the roughness of the functional surfaces affects the mutual rolling movement of the levers.

5.2. Frequency Tribological Test

Considering the need to specify the real tribological properties of the levers, a special test facility (Figure 22) was built at WBU in Pilsen, technical prerequisites of which are close to the real conditions of bearing operation.

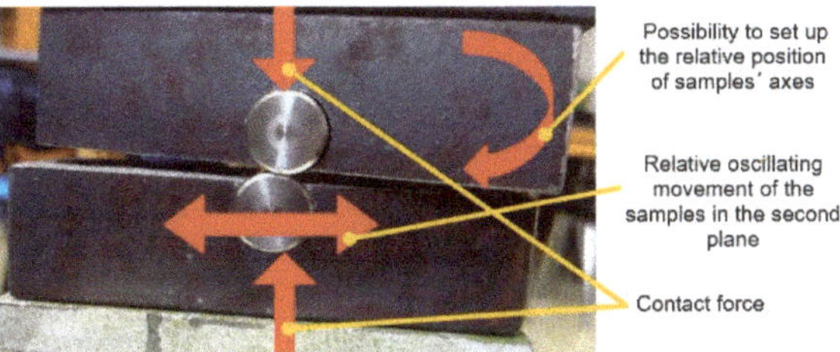

Figure 22. Frequency tribological testing device: Detail of adjustable holders and vibration exciter; (**left**), detail of samples positioning with the indication of movement possibilities (**right**).

The oscillating motion was selected on the principle of a frequency exciter with a frequency generator that can work in a fluent frequency change from 0 Hz to 10 MHz. Scanning of vibration intensity was continuously performed using an accelerometer set with a possibility to display measured results at an oscilloscope and to record measured data on PC.

The aim of the proposed frequency load experiments was to compare the behaviour of selected surface treatments and to monitor their response to the frequency load. The frequency test was performed for both the Cylinder/Cylinder assembly and the Cylinder/Plate assembly without lubrication and with the following loading conditions:

contact force: 15 N;
feed: ±0.2 mm;
oscillation speed: 50 Hz;
a number of oscillations: 1×10^6.

Two samples were relatively positioned to each other. The upper sample was firmly clamped in a horizontal bracket and the lower sample was mounted on a movable arm where an operating load was applied through the lever. The core of the electromagnetic coil was resiliently attached to the arm, causing the respective oscillations. Their frequencies were set at 50 Hz, which are the closest to the expected operating frequency of future machine parts.

Within the frequency tribological tests, the samples made from 34CrNiMo6 steel were used, which were either only heat-treated, either nitrided or nickel-plated. Wear documentation was made using a stereo magnifying glass. Due to the fact that very different roughness of samples was in the experiment, using the volume of wear for the evaluation would be burdened with a high error. Since all samples had the same diameter, the wear area and width seemed to be sufficient value for compare. The measurement was carried out by dividing the track according to pictures and then the obtained areas were summed up. The wear width was averaged based on the figures.

It can be stated that all deviations were less than 10%. A minimum of three repetitions of measurements was performed on each sample and if there were any doubts (e.g., the deviation was in the range of 8 to 10%), the samples were tested five times.

5.2.1. Configuration "Cylinder/Cylinder"

The purpose of this test was to achieve a point contact between the samples and thereby achieve a simulated probable load during the operation of the levers in the turbine when contacting the Cylinder/Cylinder types of surfaces.

The configuration of tested samples was as it is shown in Figure 23 below. Two 12 mm diameter cylinders with the same roughness, provided with the same surface treatment, were offset at 15° to each other and loaded against each other by the weights.

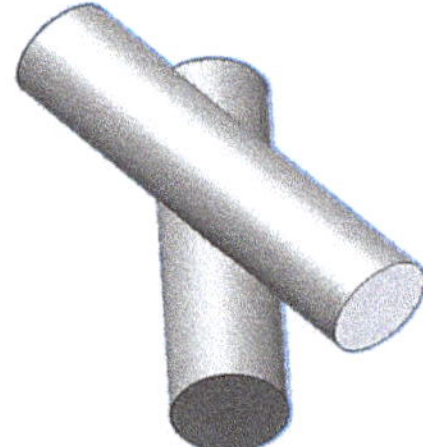

Figure 23. Test configuration–position of cylinders with respect to each other.

The graph in Figure 24 provides a wear comparison between individual samples. The best results achieved the samples treaded by electroless nickel plating. Their worn area was about half in comparison to nitrided samples and a quarter in comparison with samples, which surface was not treated. However, it cannot be unequivocally said that these surfaces will exhibit four times the resistance to untreated real parts, which will, in addition, be intensively lubricated.

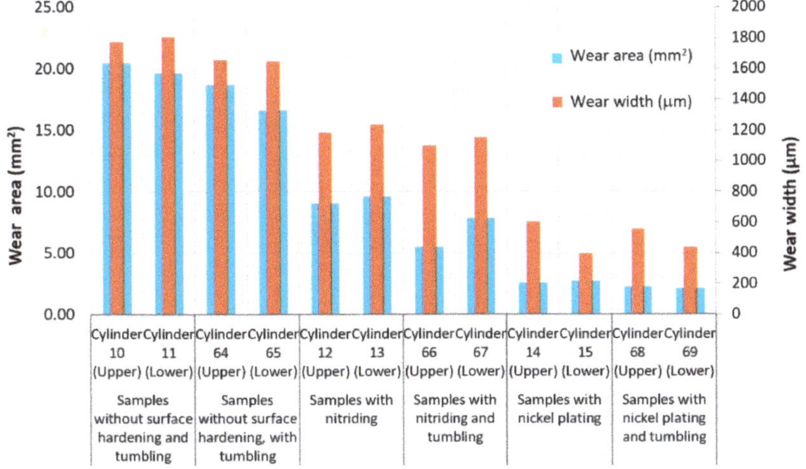

Figure 24. Frequency load (all samples of configuration Cylinder/Cylinder together).

Moreover, some samples showed some fragmentation of the nickel-plated coating at the edge of the trace. The size of these fragments was in the range of units up to tens of micrometre. Due to the oil film thickness of the bearing, which is in the range of 25–40 µm, this value is already critical.

The nitrided specimens showed about a half less wear than the untreated specimens, so the nitride layer partially retained its properties despite the revelation of the base material.

Significant amounts of oxides formed in the process of rubbing the samples against each other appeared on the untreated and nitrided samples. For nickel-plated samples, this amount was smaller. This can be attributed to the very good natural lubricating properties of the nickel surface. Documented wears on selected samples are shown in Figure 25.

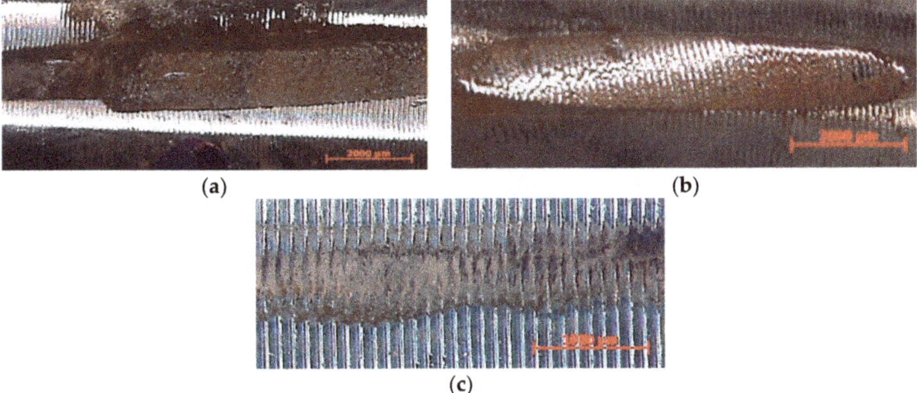

Figure 25. The example of wears of selected samples at tested configuration Cylinder/Cylinder. (**a**) Sample No. 10 without surface treatment (only heat-treated), 10× magnification; (**b**) sample No. 66 nitrided, 10× magnification; (**c**) sample No. 15 treated by electroless nickel plating, 10× magnification.

5.2.2. Configuration "Cylinder/Plate"

The configuration of tested samples for testing the Cylinder/Plate combination of surfaces is shown in Figure 26. A 12 mm diameter cylinder, provided with the same finish as the plate, was loaded against the plate with weights. Comparison of wear between individual samples is evaluated by means of the graph presented in Figure 27.

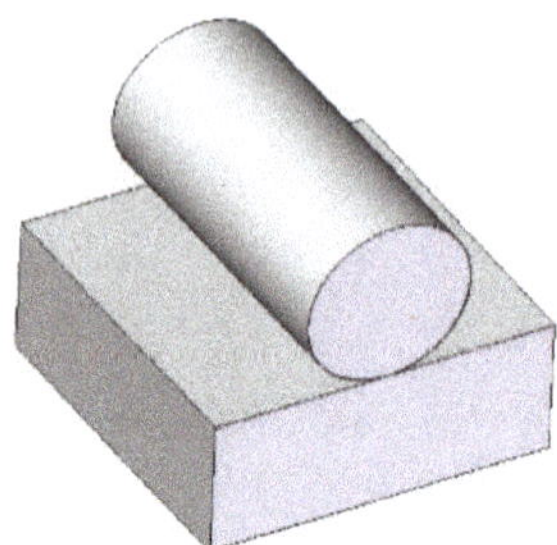

Figure 26. Test configuration–position of a cylinder with respect to the plate.

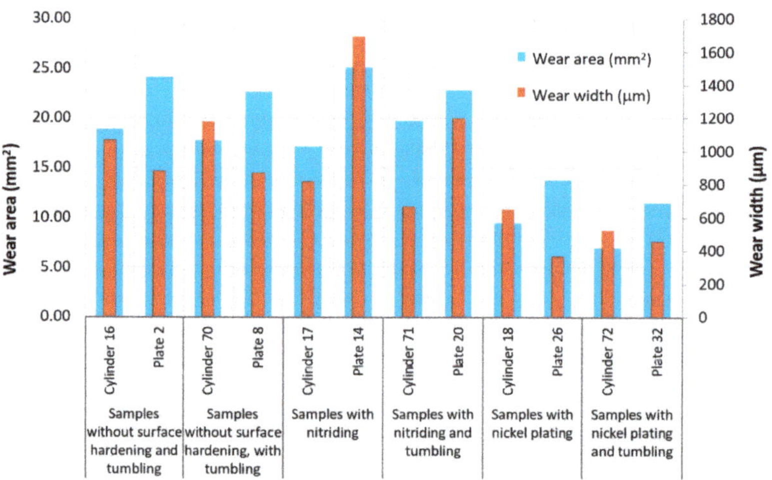

Figure 27. Frequency load (all samples of configuration Cylinder/Plate together).

From the point of view of worn-out surfaces, the nickel-plated samples were again the best. The wear area of these samples was about half, compared to uncoated and nitrided samples. In this configuration (i.e., Cylinder/Plate), the nickel-plated surface did not prove to be easily susceptible to delamination and dividing in larger portions and thus to the risk of an accident related to oil lubrication. The wear area of the nitrided and uncoated samples was not very different. Here, it depends probably on the thickness of the layer [44].

The worn area of the plate was in all cases about 20% larger than that of the cylinders. Documented wears on selected samples are shown in Figure 28a–c.

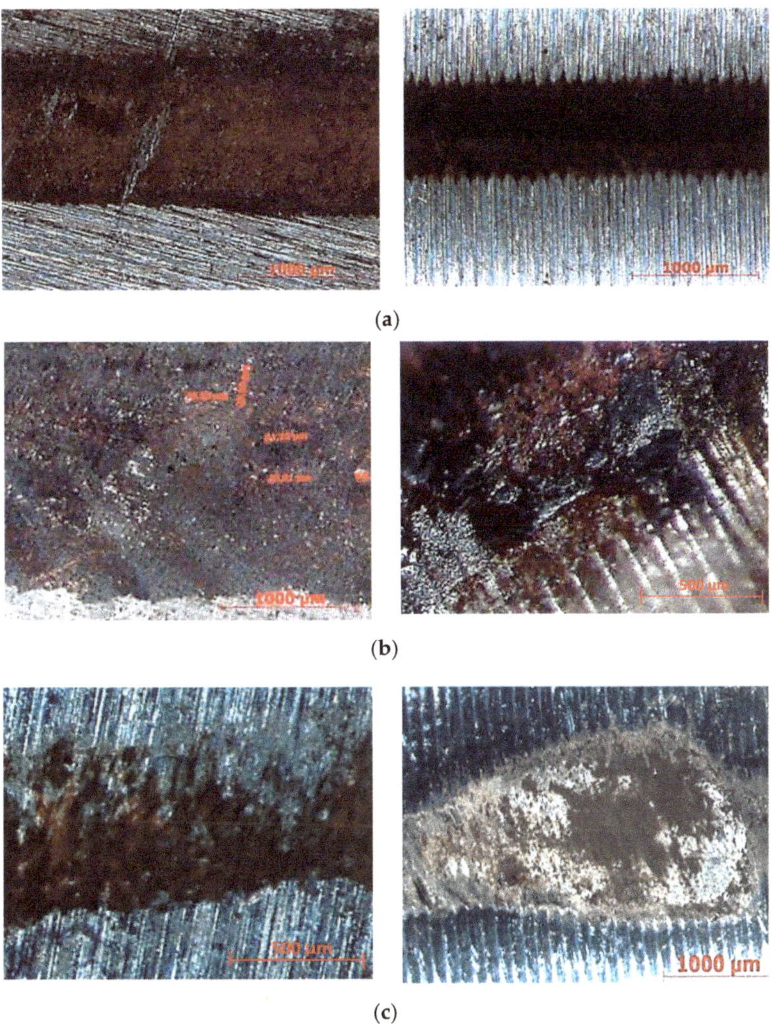

Figure 28. The example of wears of selected samples at tested configuration Cylinder/Plate. (**a**) Samples without surface treatment (only heat-treated): Plate No. 2—30× magnification (left); Cylinder No. 16, 30× magnification (right). (**b**) Nitrided samples: Plate No. 14—measurement of layer fragments, 30× magnification (left); Cylinder No. 17—detail of degradation of nitrided layer, 63× magnification (right). (**c**) Samples treated by electroless nickel plating: Plate No. 32, 40× magnification, (left); Cylinder No. 72, 20× magnification (right).

6. Conclusions

Two types of contact surfaces (Cylinder/Cylinder and Cylinder/Plate) were selected for investigation tribological properties of 34CrNiMo6 steel in the production of a newly designed self-equalizing thrust bearing. Except that all samples were heat-treated, and some of the samples were treated with tumbling, next two types of surface treatment were selected for the surface quality of bearing levers improvement, i.e., nitriding and electroless nickel plating.

Within preliminary tests, the surface roughness, microhardness HV and microstructure of the samples were analysed. The tests confirmed the high influence of the feed on surface roughness. At the same time, it could be stated from the measurement results that the surface roughness of the cylinders decreased after tumbling.

When studied the microhardness (HV 0.025) of fine and roughly turned samples with Ni coating deposition, the results show that surfaces of cylinders machined with the feed per revolution $f_1 = 0.08$ mm are harder compared to the samples machined with $f_5 = 0.24$ mm. The hardness of the roughly turned surface (and especially of the crosswise cut) is lower. However, this can be influenced by the fact that at a higher roughness at the measured point, the layer is not compact, but there is a "softer" base underneath.

Tribological properties of samples were investigated experimentally within Pin-on-Disc and Frequency tests focusing on the wear and surface roughness evaluation. It was observed that the most likely mechanism of wear was abrasive and partially adhesive. It can be also stated based on the results that in all cases the best tribological properties have achieved samples treated by electroless nickel plating compared with the nitrided or only heat-treated samples. The effect of tumbling was not significant.

Within the frequency test of contact surfaces pair Cylinder/Cylinder, the best results achieved the samples treaded by electroless nickel plating. Their worn area was about half in comparison to nitrided samples and a quarter in comparison with samples, which surface was not treated. However, the problem needs to be looked at comprehensively—whether in terms of the kinematics of the mechanism it seems to be the most suitable variant of the Cylinder/Cylinder assembly, and in wear tests, this assembly already showed defects in layer delamination—which is undesirable. This claim was confirmed in subsequent static tests that are not part of this article. Therefore, it was decided to use levers with kinematic Cylinder/Plate pairs when testing prototype self-aligning bearings.

Author Contributions: Conceptualization, M.U. and K.M.; methodology, K.M. and M.U.; validation, K.M.; investigation, M.U. and K.M.; resources, K.M. and M.U.; data curation, M.U.; writing—original draft preparation, K.M.; writing—review and editing, M.U.; supervision, K.M. All authors have read and agreed to the published version of the manuscript.

Funding: This article was funded by the Ministry of Education of the Slovak Republic by grants APVV-19-0550 and KEGA 007TUKE-4/2018.

Acknowledgments: The article was prepared thanks to the Project TE02000232 "Special rotary machine engineering center" with the financial support of TA CZ and thanks to supporting by the Ministry of Education of the Slovak Republic by grants APVV-19-0550 and KEGA 007TUKE-4/2018.

Conflicts of Interest: The authors declare no conflict of interest.

Nomenclature

a_p	depth of cut (mm)
v_c	cutting speed (mmin^{-1})
f	feed per revolution (mm)
rpm	revolution per minute (min^{-1})
Ra	arithmetical average deviation from a mean line (μm)
Rz	ten-point height of irregularities (μm)
C/C	Cylinder/Cylinder contact surfaces
C/P	Cylinder/Plate contact surfaces
R	refinement
N	nitriding
ENP	Electroless Nickel Plating
Rk	core roughness depth: Depth of the roughness core profile
Rpk	reduced peak height Average height of protruding peaks above the roughness core profile
Rvk	reduced valley depths: Average depth of valleys projecting through roughness core profile
Mr1	material portion 1: Level in (%): determined for the intersection line which separates the protruding peaks from the roughness core profile

Mr2	material portion 2: Level in (%), determined for the intersection line which separates the deep valleys from the roughness core profile.
A1	Peak area
A2	Valley area

References

1. DeCamillo, S. Current issues regarding unusual conditions in high-Speed turbomachinery. In *5th EDF/LMS Poitiers Workshop Proceedings*; Université de Poitiers: Paris, France, 2006; pp. A1–A10.
2. Kumar, D.A.; Manisha, D.; Shailendra, J. Study of shaft voltage & bearing currents in electrical machines. In Proceedings of the IEEE Students' Conference on Electrical, Electronics and Computer Science (SCEECS), Bhopal, India, 1–2 March 2012; pp. 1–4.
3. Tanaka, T. Approaches to the safer operation of thrust and journal bearings used in turbomachinery. In *10th EDF/Pprime Poitiers Workshop*; Université de Poitiers: Paris, France, 2011; pp. A1–A22.
4. Wodtke, M.; Schubert, A.; Fillon, M.; Wasilczuk, M.; Pajączkowski, P. Large hydrodynamic thrust bearing—Comparison of the theoretical prediction and measurements. In *9th EDF/LMS Poitiers Workshop Proceedings*; Université de Poitiers: Paris, France, 2010; pp. M1–M8.
5. Gregory, R.S. Factors Influencing Power Loss of Tilting-Pad Thrust Bearings. *J. Lubr. Technol.* **1979**, *101*, 154–160. [CrossRef]
6. Zhu, S.; Huang, P. Influence mechanism of morphological parameters on tribological behaviors based on bearing ratio curve. *Tribol. Int.* **2017**, *10*, 10–18. [CrossRef]
7. Malotová, Š.; Čep, R.; Kratochvíl, J.; Šajgalík, M.; Czán, A. Dependence of the Resistance of the Integrated Layers on the Wear of Ceramic Cutting Tool. *Manuf. Technol.* **2018**, *18*, 444–448. [CrossRef]
8. Ettles, C.M.; Knox, R.T.; Ferguson, J.H.; Horner, D. Test Results for PTFE-Faced Thrust Pads, With Direct Comparison Against Babbitt-Faced Pads and Correlation with Analysis. *J. Tribol.* **2003**, *125*, 814–823. [CrossRef]
9. Martsinkovsky, V.; Yurko, V.; Tarelnik, V.; Filonenko, Y. Designing Thrust Sliding Bearings of High Bearing Capacity. *Procedia Eng.* **2012**, *39*, 148–156. [CrossRef]
10. Mikula, A.M. The Leading-Edge-Groove Tilting-Pad Thrust Bearing: Recent Developments. *J. Tribol.* **1985**, *107*, 423–428. [CrossRef]
11. Rohatgi, P.K.; Tabandeh-Khorshid, M.; Omrani, E.; Lovell, M.R.; Menezes, P.L. *Tribology for Scientists and Engineers*; Springer: New York, NY, USA, 2013; pp. 233–268.
12. Straka, F. *Static Analysis of Self-Equalizing System in Tilting Pad Thrust Bearing*; Pilsen Doosan Skoda Power: Pilsen, Czech Republic, 2013.
13. Pitel, J.; Matiskova, D.; Marasova, D. A new approach to evaluation of the material cutting using the artificial neural networks. *TEM J.* **2019**, *8*, 325–332.
14. Branagan, L.A. Survey of Damage Investigation of Babbitted Industrial Bearings. *Lubricants* **2015**, *3*, 91–112. [CrossRef]
15. Ferroudji, F. Static Strength Analysis of a Full-scale 850 kW wind Turbine Steel Tower. *Int. J. Eng. Adv. Technol.* **2019**, *8*, 403–406.
16. Pantazopoulos, G.; Toulfatzis, A.; Vazdirvanidis, A.; Rikos, A. Analysis of the Degradation Process of Structural Steel Component Subjected to Prolonged Thermal Exposure. *Met. Microstruct. Anal.* **2016**, *5*, 149–156. [CrossRef]
17. Glavatskih, S.B. Tilting Pad Thrust Bearings. In *Tribological Research and Design for Engineering Systems*; Elsevier: Amsterdam, The Netherlands, 2003; pp. 379–390.
18. Noda, S.; Zenitani, S.; Yamada, Y.; Sasaki, T. Improved Technologies of Steam Turbine for Long Term Continuous Operation. *Mitsubishi JUKO GIHO* **2004**, *41*, 161–165.
19. Urban, M.; Skopeček, T.; Dolejš, J. Measurement of axial bearings with self-equalized elements and the fixture design for the measuring their maximal misalignement. *Proc. Mech. Eng. Technol.* **2015**, *1*, 263–270.
20. Pantazopoulos, G.A. A Short Review on Fracture Mechanisms of Mechanical Components Operated under Industrial Process Conditions: Fractographic Analysis and Selected Prevention Strategies. *Metals* **2019**, *9*, 148. [CrossRef]

21. Lehocká, D.; Simkulet, V.; Klich, J.; Štorkan, Z.; Krejčí, L.; Kepič, J.; Birčák, J. Evaluation of possibility of AISI 304 stainless steel mechanical surface treatment with ultrasonically enhanced pulsating water jet. In *Lecture Notes in Mechanical Engineering*; Springer Nature: Basel, Switzerland, 2019; pp. 163–172.

22. Battez, A.H.; González, R.; Felgueroso, D.; Fernández, J.; Fernández, M.D.R.; García, M.; Peñuelas, I.; Fernandez, R. Wear prevention behaviour of nanoparticle suspension under extreme pressure conditions. *Wear* **2007**, *263*, 1568–1574. [CrossRef]

23. Popescu, N.; Cojocaru, M.; Mihailov, V. Experimental studies on bulk tempering of 34CrNiMo6 steel. *Surf. Eng. Appl. Electrochem.* **2012**, *48*, 28–34. [CrossRef]

24. Cochet, J.; Thuillier, S.; Loulou, T.; Decultot, N.; Carré, P.; Manach, P.Y. Heat treatment of 34CrNiMo6 steel used for mooringshackles. *Int. J. Adv. Manuf. Technol.* **2017**, *91*, 2329–2346. [CrossRef]

25. Ge, Y.; Wang, K. Effect of tempering temperature on precipitate evolution and mechanical properties of 34CrNiMo6 steel. *Mater. Tehnol.* **2019**, *53*, 527–534. [CrossRef]

26. Branco, R.; Costa, J.; Antunes, F. Low-cycle fatigue behaviour of 34CrNiMo6 high strength steel. *Theor. Appl. Fract. Mech.* **2012**, *58*, 28–34. [CrossRef]

27. Li, Y.; Fang, W.; Lu, C.; Gao, Z.; Ma, X.; Jin, W.; Ye, Y.; Wang, F. Microstructure and Mechanical Properties of 34CrMo4 Steel for Gas Cylinders Formed by Hot Drawing and Flow Forming. *Materials* **2019**, *12*, 1351. [CrossRef]

28. Maniee, A.; Mahboubi, F.; Soleimani, R. The study of tribological and corrosion behaviour of plasma nitrided 34CrNiMo6 steel under hot and cold wall conditions. *Mater. Des.* **2014**, *60*, 599–604. [CrossRef]

29. Huang, C.; Lin, X.; Yang, H.; Liu, F.; Huang, W. Microstructure and Tribological Properties of Laser Forming Repaired 34CrNiMo6 Steel. *Materials* **2018**, *11*, 1722. [CrossRef] [PubMed]

30. Abd El-Azim, M.E.; Ghoneim, M.M.; Nasreldin, A.M.; Soliman, S. Effect of various heat treatments on microstructure and mechanical properties of 34CrNiMo6 steel. *Z. Metallkd* **1997**, *88*, 502–507.

31. Costa, J.; Ferreira, J.; Ramalho, A. Fatigue and fretting fatigue of ion-nitrided 34CrNiMo6 steel. *Theor. Appl. Fract. Mech.* **2001**, *35*, 69–79. [CrossRef]

32. Selmy, A.I.; El-Sonbaty, I.; Shehata, F.; Khashaba, U.A. Some factors affecting the accuracy of turned parts. *Sci. Bull. Fac. Eng.* **1989**, *24*, 356–368.

33. Baron, P.; Dobránsky, J.; Pollák, M.; Kočiško, M.; Cmorej, T. The Parameter Correlation of Acoustic Emission and High-Frequency Vibrations in the Assessment Process of the Operating State of the Technical System. *Acta Mech. Autom.* **2016**, *10*, 112–116. [CrossRef]

34. Xiao, L.; Rosen, B.G.; Amini, N.; Nilsson, P.H. A study on the effect of surface topography on rough friction in roller contact. *J. Wear* **2003**, *254*, 1162–1169. [CrossRef]

35. Borghi, A.; Gualtieri, E.; Marchetto, D.; Moretti, L.; Valeri, S. Tribological effects of surface texturing on nitriding steel for high-performance engine applications. *Wear* **2008**, *265*, 1046–1051. [CrossRef]

36. Liu, D.; Zhang, Q.; Qin, Z.; Luo, Q.; Wu, Z.; Liu, L. Tribological performance of surfaces enhanced by texturing and nitrogen implantation. *Appl. Surf. Sci.* **2016**, *363*, 161–167. [CrossRef]

37. Sedláček, M.; Podgornik, B.; Vizintin, J. Influence of surface preparation on roughness parameters, friction and wear. *Wear* **2009**, *266*, 482–487. [CrossRef]

38. Polcar, T.; Parreira, N.; Novak, R. Friction and wear behaviour of CrN coating at temperatures up to 500 °C. *Surf. Coat. Technol.* **2007**, *201*, 5228–5235. [CrossRef]

39. Kuduzović, A.; Poletti, M.C.; Sommitsch, C.; Dománková, M.; Mitsche, S.; Kienreich, R. Investigations into the delayed fracture susceptibility of 34CrNiMo6 steel, and the opportunities for its application in ultra-high-strength bolts and fasteners. *Mater. Sci. Eng. A* **2014**, *590*, 66–73. [CrossRef]

40. Bayes, D.M. The Physical Properties of Electroless Nickel Coatings. In Proceedings of the EN 95 Conference, Singapore, 21–23 November 1995.

41. Kiran, K.; Manohar, B.; Divakar, S. A central composite rotatable design analysis of lipase catalyzed synthesis of lauroyl lactic acid at bench-scale level. *Enzym. Microb. Technol.* **2001**, *29*, 122–128. [CrossRef]

42. Wang, Y.; Wang, Q.J.; Lin, C.; Shi, F. Development of a Set of Stribeck Curves for Conformal Contacts of Rough Surfaces. *Tribol. Trans.* **2006**, *49*, 526–535. [CrossRef]

43. Filippov, A.; Nikonov, A.; Rubtsov, V.; Dmitriev, A.; Tarasov, S. Vibration and acoustic emission monitoring the stability of peakless tool turning: Experiment and modeling. *J. Mater. Process. Technol.* **2017**, *246*, 224–234. [CrossRef]

44. Panda, A.; Dobransky, J.; Jančik, M.; Pandova, I.; Kačalova, M. Advantages and effectiveness of the powder metallurgy in manufacturing technologies. *Metalurgija* **2018**, *57*, 353–356.

© 2020 by the authors. Licensee MDPI, Basel, Switzerland. This article is an open access article distributed under the terms and conditions of the Creative Commons Attribution (CC BY) license (http://creativecommons.org/licenses/by/4.0/).

MDPI
St. Alban-Anlage 66
4052 Basel
Switzerland
Tel. +41 61 683 77 34
Fax +41 61 302 89 18
www.mdpi.com

Metals Editorial Office
E-mail: metals@mdpi.com
www.mdpi.com/journal/metals

www.ingramcontent.com/pod-product-compliance
Lightning Source LLC
LaVergne TN
LVHW071502180726
843512LV00014B/1003